Jasna Jankovic, Jürgen Stumper (Eds.)
PEM Fuel Cells

Also of Interest

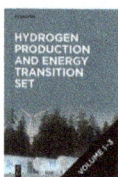

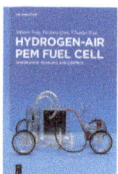

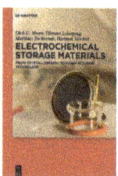

PEM Fuel Cells

Characterization and Modeling

Edited by
Jasna Jankovic and Jürgen Stumper

DE GRUYTER

Editors

Prof. Jasna Jankovic
University of Connecticut
Center for Clean Energy Engineering
44 Weaver Road unit 5233
Storrs, 06269 CT
USA
jasna.jankovic@uconn.edu

Dr. Jürgen Stumper
Juergen Stumper & Assoc.
Vancouver, BC V5M 1H5
Canada
juergen.stumper@gmail.com

ISBN 978-3-11-062262-1
e-ISBN (PDF) 978-3-11-062272-0
e-ISBN (EPUB) 978-3-11-062360-4

Library of Congress Control Number: 2023930654

Bibliographic information published by the Deutsche Nationalbibliothek
The Deutsche Nationalbibliothek lists this publication in the Deutsche Nationalbibliografie;
detailed bibliographic data are available on the Internet at http://dnb.dnb.de.

© 2023 Walter de Gruyter GmbH, Berlin/Boston
Cover image: luchschen / iStock / Getty Images Plus
Typesetting: VTeX UAB, Lithuania
Printing and binding: CPI books GmbH, Leck

www.degruyter.com

Preface

Fuel cells are the only devices capable of directly converting chemical energy into electrical energy. H_2-fueled Polymer Electrolyte Membrane Fuel Cells (PEMFCs) are currently in the early stages of commercialization, with close to 1000 megawatts in power shipped worldwide in 2020, roughly a doubling since 2016.[1] High volume fuel cell system costs have also decreased dramatically since 2008 to about $80/kW in 2020, reaching about twice the DOE 2025 target.[2] Applications in the transportation sector dominate and it is predicted that decreasing hydrogen supply costs will make most road transportation segments competitive with conventional options by 2030.

While battery technology is also advancing rapidly, fuel cell electric vehicles (FCEVs) are emerging as a complementary solution, in particular, for heavy-duty trucks and other long-range applications. Whereas batteries and fuel cells are both electrochemical devices, automotive applications of the latter represent additional challenges due to (i) a highly dynamic load profile and (ii) a strong dependence of fuel cell performance on operating conditions such as gas pressure, temperature, and humidity. Therefore, the fuel cell management system (FCMS) necessary for the control of reactant supply/H_2O removal, water balancing, cooling, etc., which is in addition to the monitoring of electrical parameters such as cell voltage or impedance, exhibits a significantly higher complexity compared to the management system of a battery electric vehicle. Furthermore, fuel cell stack and fuel cell management system need to be designed in concert so that stack and system complexity and requirements can be traded off against each other in order to optimize the whole fuel cell drivetrain for a given vehicle application. To enable such trade-off studies, appropriate fuel cell stack (and system) models are required providing stack performance as a function of (i) stack operation parameters and (ii) stack/system design parameters.

While there has been significant cost reduction for PEM fuel cell systems, significant gaps remain for system cost and durability. These gaps can be addressed through further reductions in Pt loading, increased stack power density, optimized system components and design and improved manufacturing processes to enable full commercialization. A key pre-condition for the achievement of these goals is the capability to rationally design the fuel cell and its components, i. e., the creation of new fuel cell component structures with desired properties based on physical models rather than empirical correlations. Toward this end, this book focuses on three key areas: (i) fuel cell performance modeling and characterization on a level ranging from single cell to stack, (ii) fundamental and advanced techniques for structural and compositional characterization and (iii) electrocatalyst design, which the editors hope will support progress toward this goal.

1 The Fuel Cell Industry Review (2020).

2 D. Papageorgopoulos, *Fuel Cell Technologies Overview*, AMR, US DOE, Washington, D. C., 2021.

https://doi.org/10.1515/9783110622720-201

The book consists of ten chapters, with cross references to related subjects in other chapters. In each chapter, emphasis has been placed on connections between theory and experiment. As the replacement of a precious metal catalyst is an important avenue of materials cost reduction, a special chapter (Chapter 3) discusses experimental techniques for the structural and compositional characterization of platinum group metal (PGM)-free catalyst layers. Chapters 9 and 10 are unique contributions summarizing the highly specialized methods for fuel cell performance diagnostics and fuel-to-air transfer leak development, respectively, developed by one of the fuel cell technology pioneers, Ballard Power Systems.

Overall, this book brings together contributions by leading experts from the industry and academia and is directed toward scientists and engineers in the rapidly growing field of PEMFCs and its applications.

Contents

Part I: Structural and compositional characterization

Part II: Performance modeling/characterization

Part III: **Catalysis/CL components**

List of Contributing Authors

Maryam Ahmadi
Center for Clean Energy Engineering, Institute of Materials Science, and Department of Materials Science and Engineering
University of Connecticut
44 Weaver Rd. Unit 5233
Storrs
CT 06269
USA

Mariah Batool
Center for Clean Energy Engineering, Institute of Materials Science, and Department of Materials Science and Engineering
University of Connecticut
44 Weaver Rd. Unit 5233
Storrs
CT 06269
USA

Aimy Bazylak
Thermofluids for Energy and Advanced Materials (TEAM) Laboratory, Department of Mechanical and Industrial Engineering, Institute for Sustainable Energy, Faculty of Applied Science and Engineering
University of Toronto
Toronto
ON M5S 3G8
Canada

Max Cimenti
Ballard Power Systems Inc.
Burnaby
BC V5J 5J9
Canada

David A. Cullen
Center for Nanophase Materials Sciences
Oak Ridge National Laboratory
Oak Ridge
TN 37831
USA

Kieran Doyle-Davis
Department of Mechanical and Materials Engineering
The University of Western Ontario
London
ON N6A 5B9
Canada

Qing Du
State Key Laboratory of Engines
Tianjin University
135 Yaguan Road
Tianjin 300350
China

Michael Eikerling
Theory and Computation of Energy Materials (IEK-13), Institute of Energy and Climate Research
Forschungszentrum Jülich GmbH
52428 Jülich
Germany
Chair of Theory and Computation of Energy Materials, Faculty of Georesources and Materials Engineering
RWTH Aachen University
52062 Aachen
Germany
JARA Energy
Jülich Aachen Research Alliance
52425 Jülich
Germany

Dietmar Gerteisen
Fraunhofer Institute for Solar Energy Systems ISE
Heidenhofstr. 2
Freiburg 79110
Germany

Ahmed M. Hasan
Bazylak Group, Department of Mechanical and Industrial Engineering, Faculty of Applied Science and Engineering
University of Toronto
Toronto
ON M5S 3G8
Canada

Adam P. Hitchcock
Dept. of Chemistry & Chemical Biology, Brockhouse Institute for Materials Research
McMaster University

Hamilton
ON L8S 4M1
Canada

Yuze Hou
State Key Laboratory of Engines
Tianjin University
135 Yaguan Road
Tianjin 300350
China

Jasna Jankovic
Center for Clean Energy Engineering, Institute of Ma-
terials Science, and Department of Materials Science
and Engineering
University of Connecticut
44 Weaver Rd. Unit 5233
Storrs
CT 06269
USA
E-mail: jasna.jankovic@uconn.edu

Kui Jiao
State Key Laboratory of Engines
Tianjin University
135 Yaguan Road
Tianjin 300350
China

Thomas Kadyk
Theory and Computation of Energy Materials (IEK-
13), Institute of Energy and Climate Research
Forschungszentrum Jülich GmbH
52428 Jülich
Germany
JARA Energy
Jülich Aachen Research Alliance
52425 Jülich
Germany

Aslan Kosakian
Department of Mechanical Engineering
University of Alberta
Edmonton, AB T6G 1H9
Canada

Jason K. Lee
Bazylak Group, Department of Mechanical and
Industrial Engineering, Faculty of Applied Science
and Engineering

University of Toronto
Toronto
ON M5S 3G8
Canada

Junjie Li
Department of Mechanical and Materials
Engineering
The University of Western Ontario
London
ON N6A 5B9
Canada

Deborah J. Myers
Chemical Sciences and Engineering Division
Argonne National Laboratory
Lemont
IL 60439
USA

Wolfgang Olbrich
Corporate Research
Robert Bosch GmbH
71272 Renningen
Germany

Richard Andres Ortiz Godoy
Center for Clean Energy Engineering
Institute of Materials Science, and Department of
Materials Science and Engineering, University of
Connecticut
44 Weaver Rd. Unit 5233
Storrs
CT 06269
USA

Sara Pedram
Center for Clean Energy Engineering
Institute of Materials Science, and Department of
Materials Science and Engineering, University of
Connecticut
44 Weaver Rd. Unit 5233
Storrs
CT 06269
USA

Amir Peyman Soleymani
Center for Clean Energy Engineering
Institute of Materials Science, and Department of
Materials Science and Engineering, University of

Connecticut
44 Weaver Rd. Unit 5233
Storrs
CT 06269
USA

Ulrich Sauter
Corporate Research
Robert Bosch GmbH
71272 Renningen
Germany

Marc Secanell
Department of Mechanical Engineering
University of Alberta
Edmonton, AB T6G 1H9
Canada

Pranay Shrestha
Bazylak Group, Department of Mechanical and
Industrial Engineering, Faculty of Applied Science
and Engineering
University of Toronto
Toronto
ON M5S 3G8
Canada

Jürgen Stumper
Juergen Stumper & Assoc.
University of British Columbia
Vancouver
BC V5M 1H5
Canada
E-mail: juergen.stumper@gmail.com

Xulei Sui
Department of Mechanical and Materials
Engineering
The University of Western Ontario

London
ON N6A 5B9
Canada

Xueliang Sun
Department of Mechanical and Materials
Engineering
The University of Western Ontario
London
ON N6A 5B9
Canada

Nada Zamel
Fraunhofer Institute for Solar Energy Systems ISE
Heidenhofstr. 2
Freiburg 79110
Germany

Piotr Zelenay
Materials Physics and Applications Division
Los Alamos National Laboratory
Los Alamos
NM 87545
USA

Lei Zhang
College of Chemistry and Environmental
Engineering
Shenzhen University
Shenzhen
Guangdong 518060
P.R. China

Jake deVaal
Ballard Power Systems Inc.
Burnaby
BC V5J 5J9
Canada

Jasna Jankovic and Jürgen Stumper

1 Introduction

In order for PEMFCs to become fully commercial, further reductions of cost and improvements in performance and durability of are still required. In particular, since catalyst cost is the largest single component of stack cost, the Pt loading of the cathode catalyst layer (CCL) needs to be reduced to 0.2–0.1 mg/cm^2 while increasing the efficiency of Pt-utilization at high power densities. Consequently, it becomes increasingly important to not only develop new catalyst materials, but also to optimize the 3D structural arrangement of CCL components so that all critical functionalities can be achieved simultaneously, and precious metal utilization can be maximized. In general terms, this entails provision of a maximum of (i) sites catalytically active for the oxygen reduction reaction (ORR), as well as (ii) optimum transport to/from these sites for the reactants O_2, protons, electrons and products H_2O and heat, respectively.

For the design of catalyst layer (CL) structures that meet the performance and durability requirements, it is necessary to obtain a better understanding of structure versus performance relationships, enabling the rational design of such structures.

Rational design of the fuel cell stack or one of its components is possible once the relationships between process, structure, properties and performance (PSPP) are known through a systematic approach linking component structure, performance and manufacturing process. These relationships can be either empirical correlations or based on models using first principles (see Figure 1). The latter have the advantage that, compared to purely empirical correlations, such analytical models also enable the creation of designs unconstrained by existing materials and structures. With the PSPP approach, specific performance targets can be cascaded back into property and structural requirements. Based on these requirements, appropriate changes can be implemented for existing processes or, ultimately, a new material or process can be found to create the desired new structure that fulfils the requirements.

Supported by model-based PSPP relationships, even advanced performance targets, which are often beyond the range of established empirical correlations, can be translated into structural requirements for the fuel cell components. Or, alternatively, in some cases fundamental capability limitations can be identified for the existing technologies and utilized to direct future R&D efforts. Thus, the PSPP approach constitutes one of the key pillars of the product development and design process.

However, the rational design of fuel cell electrodes (as well as many other PEMFC components) is still in its infancy due to

Jasna Jankovic, Center for Clean Energy Engineering, Institute of Materials Science, and Department of Materials Science and Engineering, University of Connecticut, 44 Weaver Rd. Unit 5233, Storrs, CT 06269, USA, e-mail: jasna.jankovic@uconn.edu
Jürgen Stumper, Juergen Stumper & Assoc., Vancouver, BC V5M 1H5, Canada, e-mail: juergen.stumper@gmail.com

https://doi.org/10.1515/9783110622720-001

Process $\quad$ *Model* $\quad$ Structure $\quad$ *Model* $\quad$ Properties $\quad$ *Model* $\quad$ Perfor-mance

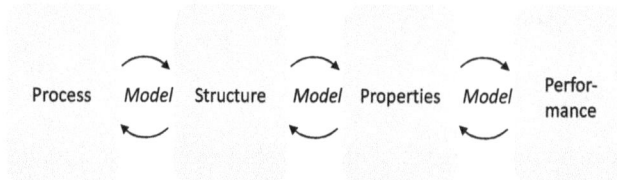

Figure 1: Schematic illustration of a Process-Structure-Property-Performance (PSPP) approach for fuel cell component design. Using models/correlations (i) performance can be predicted based on stack component physico-chemical properties or (ii) performance targets can be cascaded down into requirements for physico-chemical properties and further to parameters for the structure and manufacturing process.

- incomplete understanding of the effects of ink formulation, coating physics and drying dynamics on structure/properties during electrode manufacturing
- limitations in characterization capabilities of structure/composition over scales from nm to µm
- lack/limitations of suitable measurement tools for the measurement of relevant physico-chemical properties
- model complexity

and a complete translation of performance requirements to all relevant physico-chemical properties is not yet possible.

In order to achieve rational design capability for the catalyst layer (CL), for example, a systematic approach is needed comprising a combination of empirical and theoretical studies such as (i) fabrication of different CL structures, (ii) characterization of the spatial distribution of all constituent phases within the catalyst layer (carbon, Pt, ionomer and void), (iii) the measurement of physico-chemical properties (both ex situ and in situ) and finally (iv) the utilization of these experimental data as inputs for the development of a model-based understanding of the PSPP relationships between CL structure and CL performance/durability.

A comprehensive review of all four capabilities mentioned above for all components of the fuel cell is beyond the scope of this book, which is focusing instead on three key areas: structural characterization of the CL (Part I), fuel cell performance modeling and advanced performance characterization (Part II) and electrocatalyst design (Part III).

Since the structural design of the CL requires the capability to measure the 3D distribution and composition of its components over a wide range of length scales (from nm to tens of µm), Part I highlights the recent technological advances in the area of structural and compositional characterization of fuel cell components. For structural characterization, one can distinguish between microscopy, which provides 2D or 3D structural imaging, or spectroscopy, which provides information on chemical state/composition. The most powerful is the combination (spectroscopic imaging) of the two, providing spatial resolution together with materials contrast. Chapters 2, 3 and 5 describe such techniques in more detail.

In Chapter 2, electron-microscopic and spectroscopic methods are described that are useful for the characterization of PEMFCs, such as Scanning Electron Microscopy (SEM), Transmission Electron Microscopy (TEM), coupled with Scanning Transmission Electron Microscopy (STEM) , Energy Dispersive X-Day Spectroscopy (EDS) and Electron Energy Loss Spectroscopy (EELS), Identical Location TEM (IL-TEM), and in-situ TEM, as well as the associated image and data processing.

Chapter 3 reviews characterization approaches for Platinum Group Metal (PGM) free catalysts, which are attracting increasing attention as a replacement for the incumbent Pt-based materials. As the most promising of these catalysts are carbon based, the focus is on characterization techniques capable of providing insight into reactions on these catalysts, which are dramatically different from the thoroughly studied and well-understood platinum-based systems. Among the techniques discussed are X-ray Photoelectron Spectroscopy (XPS), X-ray Absorption Spectroscopy (XAS), Scanning Transmission Electron Microscopy (STEM) and Mössbauer Spectroscopy.

In Chapter 4, recent progress in the characterization of PEMFCs and its components using in-operando synchrotron X-ray imaging is reviewed. Synchrotron X-rays are ideal for visualizing and studying material distributions within typically opaque PEMFCs owing to the following features: (a) high photon flux, (b) collimated beam and (c) monochromatism. As a result, material distributions within fuel cells (including liquid water) can be quantified accurately using synchrotron X-rays with high spatial (several μm) and temporal resolution.

Chapter 5 describes a suite of X-ray based characterization techniques that also rely on a synchrotron source: Soft X-ray Spectromicroscopy (SXM). In contrast to the hard X-rays used for in-operando X-ray imaging (see Chapter 4), soft (low energy) X-rays are especially suited to study radiation-sensitive PEMFC components such as the perfluorosulfonic acid (PFSA) ionomer, which is an important component of the fuel cell electrodes. The spectroscopic capability provides materials contrast and enables separation of the ionomer, carbon and catalyst. The spatial resolution of several tens of nm makes SXM an ideal tool for example to 2D-map the materials distribution in fuel cell electrodes.

Part II focuses on the analysis and characterization of fuel cell performance, where Chapter 6 describes OpenFCST, an open-source fuel cell simulation software for the membrane electrode assemble (MEA) implementing a transient, two-phase, nonisothermal MEA model, capable of predicting experimentally measured polarization curves, cyclic voltammograms, linear sweeps at varying scan rates, electrochemical impedance spectra, liquid-water crossover and long-term physical, chemical and electrochemical degradation. To introduce the reader to the software, the simulation setup and results obtained using the single-phase, nonisothermal MEA model are discussed. In addition, parametric studies are performed to illustrate the effect of operating conditions (temperature and relative humidity) and microscale transport parameters.

Chapter 7 provides an analytical 1-D macro-homogeneous performance model for the cathode catalyst layer of a PEMFC, distinguishing three performance regimes: (i) kinetic, (ii) intermediate and (iii) oxygen depletion. The model was employed to analyze a wide range of fuel cell performance data from the literature. The analysis reveals the role of the "tipping water balance" that affects the interplay of transport and re-action in the catalyst layer and gas diffusion media and represents an alternative to the widespread ionomer-film hypothesis used to explain the observed power losses at low Pt loadings.

Chapter 8 reviews pore-scale simulations of PEMFC electrodes and compares popu-lar pore-scale numerical tools available through the literature. A tutorial on pore-scale simulations in the electrode is presented, including how to reconstruct the GDL, MPL and CL, and how to develop single- and two-phase pore-scale models.

In Chapter 9, methods for the analysis of fuel cell performance are presented, pro-viding essential tools for the analysis of polarization curves. Polarization curve analysis is a fundamental for the advancement of PEMFC technology because on one hand it al-lows for the attribution of cell voltage losses to the different physico-chemical processes operating in a fuel cell, and on the other, it enables the association of these losses with individual fuel cell components. Three different approaches for the breakdown of the voltage losses (VLB) are discussed in detail. An important application of polarization curve analysis is fuel cell diagnostics, for example, of low-performing or failing cells, providing important information for root-cause analysis, or to provide direction for fu-ture development efforts. Some of these methods may also be useful for on-board diag-nostics to reduce the complexity of the fuel cell management system (see Chapter 10).

Chapter 10 discusses methods for the diagnostics and quantification of fuel-to-air (transfer) leak formation, one of the leading failure modes for fuel cell stacks, as well as the suitability of the different diagnostic methods for on-board diagnostics. Transfer leaks can lead to H_2 emissions in the exhaust gas, posing potential safety risks. After a brief discussion of the fundamentals of gas flow through pinholes, the different meth-ods for the quantification of transfer leaks are introduced and research assessing the capability of Electrochemical Impedance Spectroscopy (EIS) to diagnose a stack perfor-mance decrease due to transfer leaks is presented. Finally, the H_2 concentration in the cathode exhaust gas of a fuel cell is calculated and compared with experimental results obtained during H_2 injection into the cathode inlet.

Chapter 11 focuses on recent developments in the area of Pt-based and nonprecious metal catalysts, as well as their applications in PEMFCs. It is shown how by tuning parti-cle size, surface structure and composition, the ORR activity of Pt-based catalysts can be significantly enhanced and compared with commercial Pt/C catalysts. For nonprecious metal catalysts, Metal-Organic Framework (MOF)-derived transition metal catalysts are discussed as a promising alternative to substitute Pt catalysts due to the remarkable ad-vantages of a high surface area, well-tuned coordination environment and high Atom Utilization Efficiency (AUE). Furthermore, metal/nitrogen-doped carbon (M-N-C) elec-trocatalysts, which provide remarkable electrocatalytic activity in alkaline media, are reviewed together with their advantages and disadvantages.

Part I: **Structural and compositional characterization**

Sara Pedram, Richard Andres Ortiz Godoy, Maryam Ahmadi,
Amir Peyman Soleymani, Mariah Batool, and Jasna Jankovic

2 Microstructural and spectroscopic characterization of PEMFCs

Understanding the microstructure and properties of PEMFCs components, especially the ones employed in the catalyst layers of the electrodes (cathode and anode), is necessary to enhance their performance and longevity. The catalyst layer microstructure, composition, morphology, spatial distribution and interactions play a crucial role in performance efficiency and degradation of a fuel cell, directly affecting the cost and durability. Inadequate electrode and catalyst layer design can lead to ineffective transport and reaction processes that hinder efficiency of the fuel cell, as well as cause degradation of the components, hence lowering performance and durability. These challenges could be addressed by rational selection of component materials and electrode microstructure tailored to specific fuel cell operational conditions, a process currently done mainly by costly and time-consuming trial and error. To enable rational and educated fuel cell design, we need fundamental understanding of the processes that happen in the electrodes from the nano to macro level, and the effects that the spatial distribution and properties of the components have on electrode performance and degradation. Due to the complexity of the system, involving multiple materials and physical/electrochemical processes that span a wide range of length scales (Å to mm), and time scales (fractions of a second to thousands of hours), a range of characterization techniques need to be employed to enable full understanding. It is essential to characterize the structure and distribution of components in the catalyst layers (e. g., characterize catalyst particles, carbon support, ionomer and pores), observe morphology and layer interfaces between gas diffusion layers, microporous layers and catalyst layers in MEAs and measure the properties of interest for catalyst inks and catalyst layers, to understand the relationships between dispersion ink, MEA structure, transport properties and FC performance and degradation. This chapter reviews the application of a range of microscopy and spectroscopy techniques used to understand PEMFCs.

2.1 Scanning electron microscopy characterization

In scanning electron microscopy (SEM) imaging, the electron beam is rastered over the surface of the sample, and due to the interaction with the sample, a signal is formed

Sara Pedram, Richard Andres Ortiz Godoy, Maryam Ahmadi, Amir Peyman Soleymani,
Mariah Batool, Jasna Jankovic, Center for Clean Energy Engineering, Institute of Materials Science, and
Department of Materials Science and Engineering, University of Connecticut, 44 Weaver Rd. Unit 5233,
Storrs, CT 06269, USA, e-mail: jasna.jankovic@uconn.edu

https://doi.org/10.1515/9783110622720-002

and detected, producing an image. Resolutions down to 0.5 nm can be achieved in more advanced systems. Two types of signal electrons are typically detected: secondary electrons (SEs) and backscattered electrons (BSEs). SEs originate from the surface sample atoms, produced due to inelastic interactions between the incident electron beam and the sample. SEs have lower energy than the BSEs and help study the sample's surface topography [1]. BSEs are the high-energy electrons caused by the elastic interactions between the beam and the sample. They are sensitive to the atomic weight of the elements in the sample, and hence provide information on the elemental distribution; the image is brighter for the material with a higher atomic number.

Catalyst powders

Sample preparation is the first essential step in order to obtain good-quality SEM images. In some cases, catalyst powders need to be images by SEM. One method to prepare powders for SEM imaging is to prepare a diluted alcohol solution of the powder and sonicate it for 30 minutes. Then drop coat the powder dispersion on the clean substrate or conductive tape and let it dry. Another alternative is to use a double-sided sticky carbon/copper tape attached to a flat sample holder. A small amount of powder is placed onto the sticky tape, and the sample holder is tilted and taped onto a surface to remove the excess powder. It is also suggested to remove all loose particles from the sample by blowing dry air or nitrogen onto the sample.

An extra step is required for specific types of samples to collect high-quality images. This additional step includes sample coating (typically by sputter-coating) with a thin layer of a few nanometers (1–3 nm) of a conductive material such as gold, silver, platinum, palladium or chromium. Nonconductive materials and beam-sensitive samples are the most prominent examples of samples that require sputter coating. The surface of a nonconductive specimen acts as an electron trap due to its nonconductive nature. In other words, a negative charge develops on the surface of a nonconductive material at normal electron accelerating voltages because more electrons enter the specimen than leave in the form of SEs or BSEs. The buildup of electrons on the surface is known as "charging," which generates strong bright areas on the sample and can influence the image quality. The thin conductive coating on the sample surface enables removal of the charging. The type of the metal sputtering target is determined based on the analysis, e. g., low-magnification or high-magnification imaging, to achieve the optimum performance [2]. However, additional time and effort are necessary to determine the optimal coating parameters to obtain accurate and high-quality images. Furthermore, this technique should be employed carefully because intense coating may alter the topography and morphology of the surface or result in false elemental information. Beam-sensitive samples are generally biological samples or materials made from plastic. The interaction between the high-energy electrons from the SEM beam and the surface of these samples releases energy in the form of heat, which can damage the sample structure. Sputter coating with not-sensitive materials is beneficial to prevent such damage.

However, PEMFC catalyst powders are typically conductive and not very beam sensitive, hence can be imaged without sputtering or additional powder sample preparation. PEMFC catalyst powder samples are often simply dispersed on conductive support on a sample holder, most typically an aluminum stub. These SEM stubs are available commercially in different sizes and geometry [3].

The sample must adhere completely to the surface of the stub before placing it on the sample holder or SEM stage. Different methods are applied to attach the sample to the pin stub, including:
– Double-sided carbon or copper tape
– Conductive paint
– Special clamps
– or a combination of the above [4]

Surface of the catalyst layer
Surface of the FC electrodes could be imaged similar to the powders. A small piece of the electrode should be attached to the stub using double-sided tape. The sample is sputter-coated to obtain high-quality images. However, since the electrode is conductive, often this step is not needed. SEM images of the catalyst surface can provide information about the surface morphology and homogeneity. Figure 1 demonstrates a set of SEM images of the Pt/C catalyst layer (CL). As seen in Figure 1a, a CL surface is homogenous with some cracks and debris. It is also demonstrated that CL has relatively rough surface characteristics, which could cause substantial interfacial gaps at the MPL-CL interface. SE images at low magnification (Figure 1a) can also be used to identify the orientation, size, shape, depth and density of cracks on CL surfaces. The high magnification SE image (Figure 1b) shows the pores' structure, size and topography. It reveals that the catalyst particles are actually agglomerates of smaller particles. On the other hand, the BSE image shows platinum agglomerates of approximately 100 nm in diameter on the catalyst layer. The Pt particles in the CL have an average diameter of 4 nm in a typical PEMFC, and SEM is usually not able to distinguish these nanoparticles. This would be accomplished using TEM or high resolution SEM.

Figure 1: SEM image of Pt/C catalyst layer (a) Low magnification SE mode, (b) High magnification SE mode and (c) High magnification BSE mode.

Membrane electrode assembly cross-section

SEM can also be utilized to measure different layer thicknesses (mean thickness and standard deviation) and to study the uniformity of the layers (cathode, membrane and anode, as well as GDL and MPL) in the membrane electrode assembly (MEA). In this regard, the cross-section of the MEA, which consists of the membrane and the catalyst layers, is typically imaged. The backscattered detector is usually used for better contrast, and MEA samples are prepared by embedding them in epoxy resin, surface polishing and gold coating. The samples should be cut to about 1 cm × 3 cm pieces and fixed inside a plastic cup using a metal spring mounting clip. Then the cup is filled with epoxy resin. When the epoxy is cured, the section is removed from the plastic cup and is polished with silicon carbide grinding papers starting from coarse (60-) to fine (1200-) grit. It is followed by polishing by 0.5 and 0.2 μm diamond and alumina polishing paste, respectively. The mirror-like surface of the sample is then gold-coated to provide a conductive surface. The details can be found in [5].

An example of SEM cross-section analysis is shown in Figure 2. In this study, the effect of the catalyst ink dispersion medium on the structure, properties and performance of the resulting MEA is investigated by Giner, Inc [6, 7]. Catalyst layers are, as usual, prepared from ink-based techniques, and the electrode structure and morphology can be optimized by controlling the ink formula and fabrication method. To study the effect

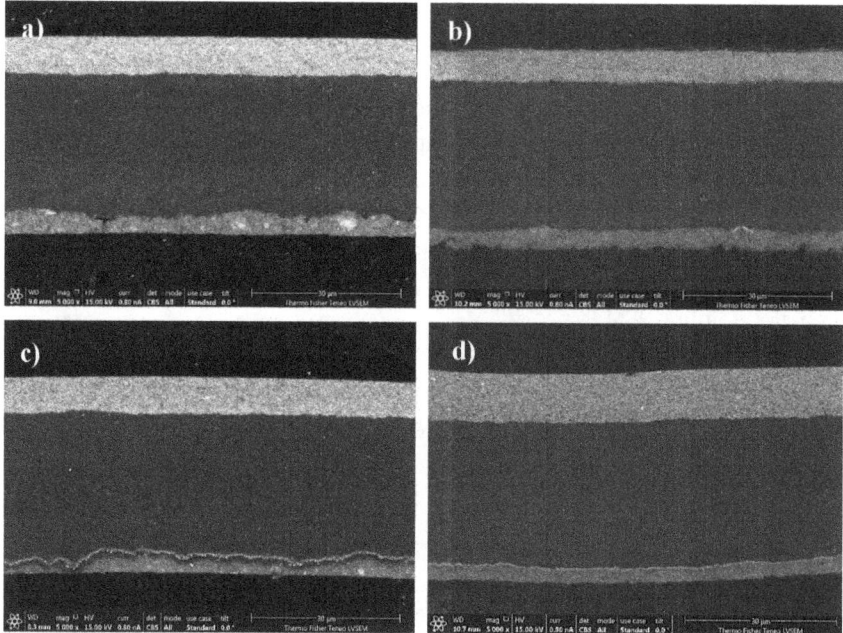

Figure 2: SEM cross-sectional images of FC MEAs with (a) NPA-based beginning of life (BOL) MEA and (b) EG-based BOL, (c) NPA-based end of life (EOL) MEA and (d) EG-based EOL MEA. (The anode, membrane and cathode are the top, middle and bottom layers, respectively.)

of solvent on uniformity and morphology of the cathode, the cathodes are prepared with two different solvents, i. e., 1-propanol (NPA) and ethylene glycol (EG). Decal transfer method is used to make two MEA samples from Pt/C anode, perfluorosulfonic acid (PFSA) membrane and Pt/C cathode. The anode and membrane of these two samples are similar. The anode, membrane and cathode are the top, middle and bottom layers in Figure 2. As can be seen, both samples have a uniform anode and membrane morphology and thickness. NPA MEA resulted in a thicker and uniform cathode with some inhomogeneity in composition compared to EG MEA (Figure 2b). Although both MEAs show Pt-band formation after a 30,000 square wave accelerated stress test (SWAST) cycles, the effect is more severe for the NPA MEA, suggesting a more heavily degraded sample. Cathode thinning is also observed largely in the NPA sample, suggesting more severe carbon corrosion. This sample also has a Pt band located farther into the membrane, whereas for the EG MEA (Figure 2d), it is closer to the CL/membrane interface. The layers' thicknesses could be measured with an image analysis software such as ImageJ.

2.2 Scanning electron microscopy with energy dispersive spectroscopy

As the electron interactions between incident electron beam and sample atoms are utilized for imaging, another physical phenomenon occurs in the electron and specimen interaction volume. When an incident electron beam strikes the sample surface, it knocks off an electron from the atom's inner shell, leaving a hole. A displaced electron attracts another electron from a neighboring shell to fill the gap. As the shell transition occurs, energy difference can be released in the form of an X-ray. A specific element and transition produce a unique X-ray spectrum. Elemental chemical information can be derived from these interactions. This, so-called, energy dispersive spectroscopy (EDS) technique can provide qualitative and semiquantitative materials analysis. This SEM instrument is equipped with an X-ray detector that converts the emitted X-ray's energy into the elemental composition of the area. This technique can identify carbon (C) to uranium (U) elements with concentrations as low as 0.1 wt.% [8]. A standard SEM-EDS spectrum of a typical FC cathode is shown in Figure 3. The spectrum shows expected elements such as Pt, C and Au. Au peaks correspond to the coating applied on the surface of the sample. Si peaks could result from contamination (polishing papers (SiC) or surrounding dust) or the Si-Li detector in the EDS system. Fluorine and oxygen peaks come from the ionomer (binder) used to prepare the catalyst layer.

The SEM-EDS technique can also produce elemental distribution maps. However, SEM-EDS mapping does not have high enough resolution to distinguish nanoscale metal particles such as Pt and Co in FC catalyst samples.

Focused ion beam/scanning electron microscopy (FIB/SEM) has recently been used as a microscopy technique to detect more microstructure information about the fuel

Area 27

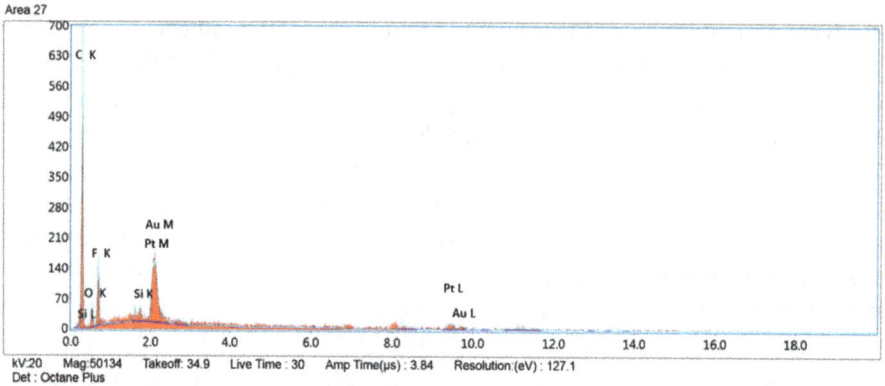

kV:20   Mag:50134   Takeoff: 34.9   Live Time : 30   Amp Time(µs) : 3.84   Resolution:(eV) : 127.1
Det : Octane Plus

Figure 3: SEM/EDS spectrum of a typical FC cathode.

cell CL, or in other words, 3D tomography characterization [9, 10]. In a FIB instrument, a beam of charged ions (mostly Ga or He) directly mills the specimen surface with a nanometer precision while an electron beam (SEM) is used to construct the image of the material surface. A dual-beam FIB/SEM enables a sequential removal of the surface and imaging, which can be employed to reconstruct the CL morphology. This technique [11] is beyond the scope of this chapter.

2.3 Transmission electron microscopy and scanning transmission electron microscopy with energy dispersive spectroscopy

Fuel cell catalyst powders and layers detailed morphology could be evaluated by transmission electron microscopy (TEM) and scanning transmission electron microscopy with energy dispersive spectroscopy (STEM-EDS). High resolutions can be achieved, with point resolution of e. g., 0.25 nm for TEM and 0.16 nm for STEM. Atomic level resolutions can be achieved with aberration corrected TEM microscopes. Catalysts in powder form and MEAs can be investigated using this technique. TEM technique can be applied in various aspects of the PEMFC research, including catalyst powder and ink evaluation, as well as catalyst layer microstructure, optimization, degradation mechanism and so on.

Powders

Commercial PEMFC catalysts and the ones synthesized in labs are generally in a powder form. TEM is a powerful technique to study the powder sample's uniformity, particle size and size distribution. Powder samples will be prepared for TEM by one of the following methods:

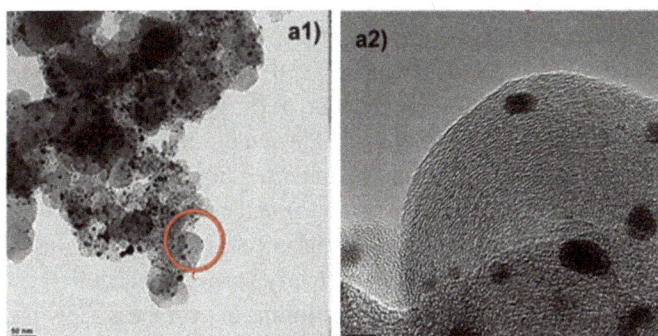

Figure 4: TEM Images of a Pt/C catalyst powder [14]. Reprinted from Int. J. Hydrogen Energy, Copyright 2019, with permission from Elsevier.

- Dipping a holey carbon grid in the dried powdered sample [12, 13].
- By adding a small amount of the sample to the 1:1 vol. percentage deionized water: Isopropanol (IPA) solution, sonicate for about 15 minutes and drop-cast the aqueous nanoparticle suspension onto the TEM grid.

The TEM image of the commercial Pt/C catalyst (20 wt.% Pt on Vulcan XC-72R, ETEK Inc., USA) is shown in Figure 4 [14]. The circled part is illustrated in Figure 4b with a higher magnified image. In bright-field TEM images, the semispherical black particle indicates the dispersed Pt particles on gray carbon spheres. Large Pt agglomerates detected in the image denote a non-uniform distribution of Pt particles through the catalyst.

Inks

The ink solution of a binder/ionomer and Pt/C catalyst is typically used for CL formation through different solution-coating methods. It is important to adjust the formulations and processes of catalyst inks to achieve the properties required to form an adherent, continuous layer. Catalyst ink properties significantly affect the interaction between catalyst and ionomer, particle size, catalyst utilization, surface morphology, thickness and ionomer distribution of the catalyst layer. Therefore, it is important to study the ink properties to understand its correlation to CL particle size and structure. Ink processing is primarily concerned with breaking up agglomerates of catalyst/support aggregates to the desired size for further electrode fabrication and obtaining uniform dispersion.

TEM sample preparation for ink samples is relatively straightforward. A small amount of the ink dispersion should be diluted using the appropriate solvent (the solvent used for ink preparation), sonicated for about 15 minutes and drop-casted onto the TEM grid. The grid may be dried in an ambient atmosphere or under mild UV light. In some cases, when ink dilution needs to be avoided, the TEM grid is gently dabbed on a surface of a small ink droplet to transfer a minute amount of the ink, but very dense and nontransparent areas of the ink on the TEM grid should be avoided when imaging.

M. Wang et al. [6] systematically investigated the effects of the ultrasonic dispersing procedure and time on catalyst agglomerate size in PEMFC catalyst ink dispersions and CL electrochemical properties. The direct correlation of catalyst agglomeration to fuel cell performance has been obtained by combined TEM and *in situ* electrochemical testing. An efficient method for dissolving carbon agglomerates into primary particles without detaching Pt nanoparticles (NPs) has been demonstrated in which ionomers are distributed more uniformly on aggregates, resulting in higher catalyst utilization. Insufficient sonication will result in highly agglomerated primary particles, which prevent interior Pt NPs from interacting with ionomers, hence resulting in poor mass activities. The detachment of Pt NPs was severe with excessive sonication.

The size and properties of the catalyst layer material are crucial factors in understanding how these materials work together. In other words, the performance properties of nanoparticles often depend on the products' size, shape and particle size distributions (PSD). The stability investigations that examine particle size effects cannot yet provide solid conclusions because uncertainty exists in analogy to the activity studies. In many studies, increases in electrochemical surface area (ECSA) and area-specific activity have been observed with decreasing particle size. However, small NPs showed less stability than the larger NPs, more rapid dissolution, more pronounced ECSA loss and reduced Pt utilization [15–18]. On the contrary, some studies have illustrated that particles of different sizes do not affect degradation when they have narrow particle size distributions within each material and suggest that larger particle size distributions increase degradation through Ostwald ripening [19–21].

Various analytical methods are employed to determine ink or catalyst particle size, including electron microscopy, dynamic light scattering, atomic force microscopy and small-angle X-ray scattering [22]. Particle size measurements on the nanoscale mainly focus on determining the sample's average particle size. One popular particle sizing method for determining the target particles' size distribution is through TEM analysis. In TEM, nanoparticles are visualized, and these images can be used to generate number-based size distributions. In this method, an electron beam passes through the sample, enabling direct observation and imaging of the particles. The compiled images can be processed subsequently, and particle size statistics such as mean, maximum, minimum and particle size distribution histograms can be determined. There are three types of image processing approaches manual, software-assisted and fully automatic. An analysis can be done manually using a calibrated micrograph, a ruler and magnifying glass. It also can be done with the aid of an image analysis software that offers a simple line-measurement method, such as Digital Micrograph or ImageJ. These two methods benefit from having control over analyzing single particles, having a less strict signal-to-noise ratio and being straightforward. These methods are more convenient for the fuel cell samples that have supported nanoparticles. A nonuniform background is expected for supported nanoparticles due to the variation in thickness of the support. However, it is essential to employ a consistent measurement methodology (e. g., horizontally or vertically or the shortest diameter measurement, etc.) and present it appropriately when reporting results. There are a number of considerable limitations with manual measurements, comprising user bias, the need to repeat measurements and the limited sampling statistics that can be obtained by measuring only a few hundred particles within a suitable time frame. It is especially challenging to perform manual measurements when comparing different samples is required. The application of a fully automated process is described later in this chapter.

An example of catalyst PSD measurement in inks is given by B. Li et al. [23], who investigated how the properties of Pt/GC (graphitized carbon) catalyst ink will affect CL morphology and permeability. Mechanical shear mixing or high-pressure homogenization were used to prepare the catalyst inks. Figure 5 indicates the TEM images, particle size distribution histograms and the average value of particle diameter of the ink prepared at different dispersion processes. As visualized from the TEM images, the catalyst particles of the ink prepared at lower sheer speed (10,000 and 13,000 RPM) develop considerable agglomeration. Likewise, increasing the homogenization pressure from 10 MPa to 50 MPa results in a more uniform ink and smaller agglomerated particles.

Membrane electrode assemblies

To image MEAs using TEM, a small piece of the sample is cut for TEM sample preparation, and then the GDLs are peeled off from the catalyst-coated membrane (CCM). The CCM is embedded in epoxy, and thin sections of approximately 100 nm are cut using an ultramicrotome sectioning setup with a diamond knife [24]. Several slices are placed onto a few 200 mesh or 300 mesh Cu/Pd grids to determine the most appropriate one for imaging. TEM or STEM-EDS are typically utilized for imaging and elemental mapping of the sample. Some of the typical applications are described in [25–28].

High-angle annular dark-field (HAADF)-STEM mode is applied by N. Macauley et al. [29] to study the catalyst layer thickness and porosity changes during accelerated stress. The results reveal a ~50 % reduction in the cathode CL thickness and a 22 % porosity reduction over the 1000 hr. wet drive cycle test. However, SEM measurements would provide more accurate overall thickness information than TEM cross-sectional imaging since the latter can only present local thickness.

In recent years, aberration-corrected STEMs equipped with secondary electron (SE) detectors have yielded some exciting outcomes [30–32]. A key advantage of SE images is that they illustrate the topography of the specimen, which is very valuable for particulate specimens like catalysts. A scintillator-photomultiplier (PM) detector situated in a STEM's upper objective polepiece detects SEs emitted at the surface of the specimen in the top few nanometers [33].

The effect of Pt and ionomer distributions on the surface of carbon black (CB) support with various nanopore volumes on PEMFC performance and durability is investigated by A. Kobayashi et al. [30]. The TEM and STEM-SE imaging are employed to study the size and distribution of Pt particles located on the CB support surface (exterior) and CB support pores (interior). To ensure that all of the Pt particles located on the exterior

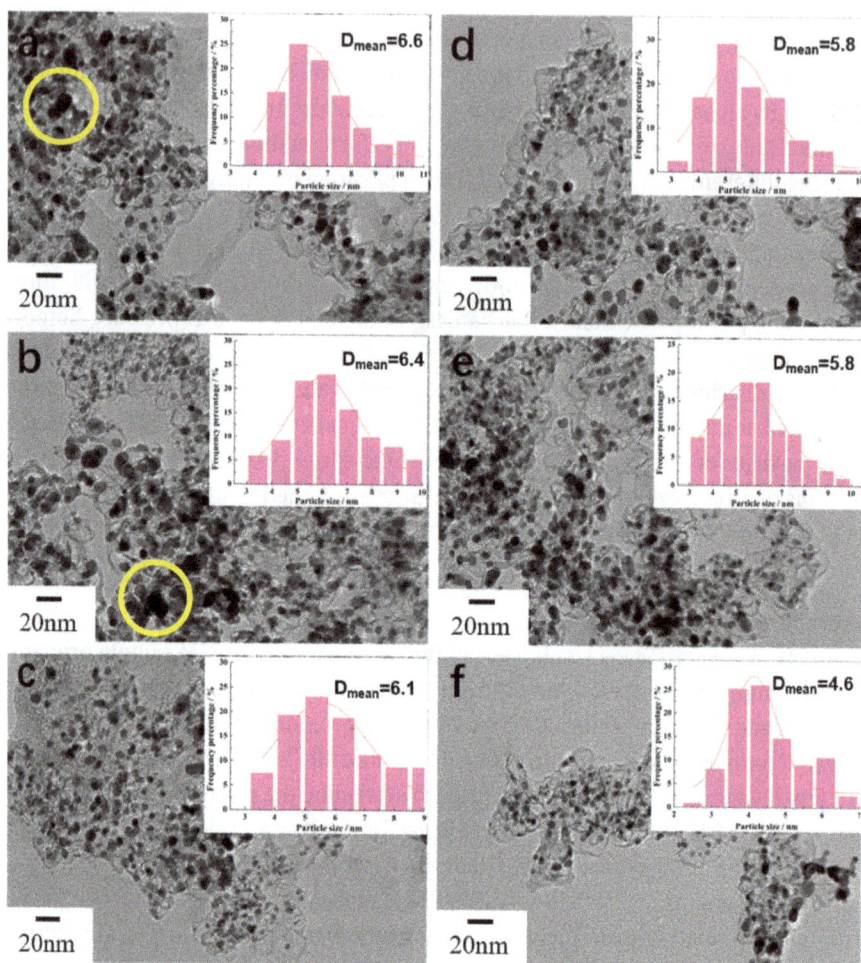

Figure 5: TEM images of catalyst ink for different dispersion processes, (a–c) mechanical shear mixing: 10,000 RPM, 13,000 RPM, 16,000 RPM (d–f) high-pressure homogenization: 10 MPa, 30 MPa, 50 MPa [23]. Reprinted from Int. J. Hydrogen Energy, Copyright 2021, with permission from Elsevier.

are counted, the Pt particles on the reverse side of the CB particles should be observed. Therefore, STEM-SE images are recorded at 0° and 180° holder rotation, enabling the separation of the exterior and interior Pt particles on the CB support. For illustrative purposes, the 0 and 180° SE image (p, q), 0 and 180° TE image (r, s) and (t) Pt distribution of Pt/ECO600JD obtained by STEM are shown in Figure 6. TEM images demonstrated dispersed small Pt particles as the Pt particle number decreased in STEM-SE images with wide interparticle distances. These results indicate that most of the Pt particles are supported on the surfaces of the interior nanopores ECO600JD support (interior 73 %, exterior 27 %). In addition, the measured Pt particles on the surfaces of the interior nanopores (2.2 ± 0.3 nm) are smaller than those on the exterior surfaces (2.8 ± 0.4 nm).

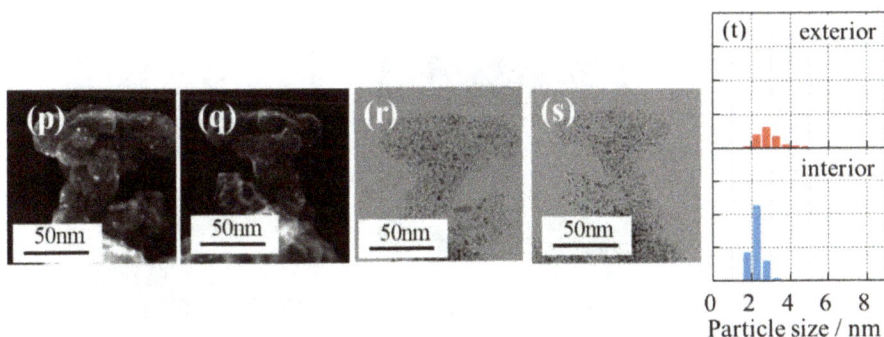

Figure 6: (p, q) 0 and 180 ° STEM-SE images, (r, s) 0 and 180 ° TEM images, and (t) Pt distribution of particles interior and exterior of the catalysts Pt/ECO600JD [30]. Reprinted from ACS Appl. Energy Mater., Copyright 2021. Published by American Chemical Society with CC-BY license.

The STEM-EDS elemental mapping technique can be employed to characterize the structure and elemental distribution of the catalyst and ionomer in catalyst layers, before and after testing. Catalyst layer component distribution (e. g., Pt, Co, Pd catalyst, carbon support and ionomer) can be visually observed at various magnifications, as well as quantified. Nafion, a fluoropolymer-copolymer based on sulfonated tetrafluoroethylene, is the most common ionomer used in PEMFC applications. So, the fluorine X-ray signal can be used to map the ionomer distribution within the different CLs.

Jankovic (reported in Kneer et al. [25]) has developed a method for the quantitative study of the CL's porosity and Pt loading using STEM-EDS net count data sets. This method determines the amount of Pt in the measured volume derived from an X-ray net count signal from the catalyst layer region. A characteristic X-ray intensity measured in thin-film specimens is predicted to be proportional to its mass-thickness and element concentration based on the ζ-factor approach. For more details, please refer to [25].

C. Wang et al. [5] investigated the origin of performance loss in accelerated startup and shutdown (SU/SD) cycling, performing elemental mapping under STEM mode of beginning of life (BOL) and end of life (EOL) samples. The STEM-EDS maps of the FC cathode before and after the SU/SD test at low magnification (5kx) are shown in Figure 7. Signals from EDS can be identified by color, where green, red and blue indicate fluorine, platinum and cobalt signals. Low magnification STEM-EDS images clearly reveal the presence of Pt and Co in the membrane for the EOL sample. In addition, the images show that at the cathode-membrane interface, a Pt depleted layer developed during degradation. A quantitative analysis of Pt loading and loss from the cathode is also conducted on STEM-EDS net count data sets. The results specify a 50 % decrease in Pt loading in the cathode for the tested samples, which agrees with the visual observations. Increased Pt/Co ratio in EOL samples is an interesting supplementary result that could be extracted from these images. The increased Pt/Co ratio in EOL samples could be attributed to the Co faster-leaching rate due to its thermodynamic instability in fuel cell conditions.

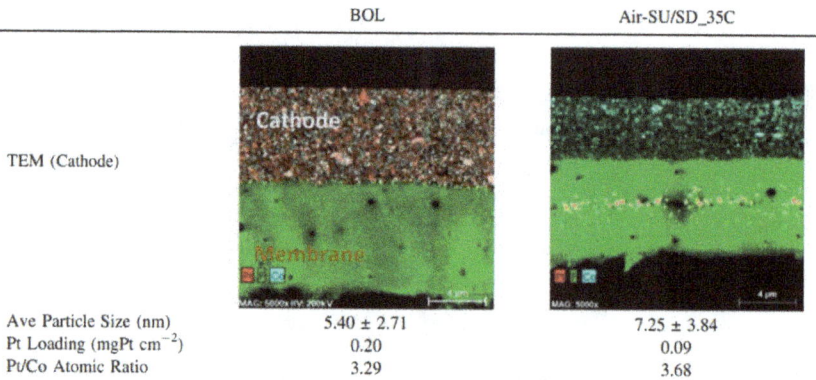

	BOL	Air-SU/SD_35C
TEM (Cathode)		
Ave Particle Size (nm)	5.40 ± 2.71	7.25 ± 3.84
Pt Loading (mgPt cm^{-2})	0.20	0.09
Pt/Co Atomic Ratio	3.29	3.68

Figure 7: STEM-EDS maps of the cathode catalyst layer during SU/SD processes at 35 °C [5]. Reprinted from J. Electrochem. Soc., Copyright 2021. Published by IOP Publishing with 4.0 CC-BY license.

In addition to Pt loading, low magnification STEM-EDS maps also can be used to determine the porosity and ionomer to carbon weight ratio (I/C) in the catalyst layer. S. Pedram et al. [7] investigated the effect of ink preparation conditions on the morphology and structures of the catalyst layer. Microscopy evaluation reflects ionomer network connection and higher porosity in the electrode made from a 5-day mixing ink compared to the one made from a 3-day. The MEA fabricated from a longer mixing time displays improved performance and durability, contributing to enhanced proton conductivity and mass transport due to higher porosity. The calculated I/C ratio agrees well with the ratio used in ink preparation for all the electrodes.

High magnification STEM-EDS mapping would allow one to visualize the distribution of different elements in the submicron system. C. Lei et al. [6] acquired STEM-EDS maps to investigate the impact of dispersing solvents in catalyst inks on the resulting electrode morphology. The Pt + C + F maps (red, blue and green, respectively) obtained from each of the MEAs with cathode CLs (CCLs) prepared from Pt/C inks and Nafion ionomer are displayed in Figure 8, comparing BOL and end of test (EOT) conditions. The n-propanol (NPA) and 1,2-butanediol (BUT) CCLs' ionomer distribution at BOL is qualitatively more uniform than in the iso-propanol (IPA) and ethylene-glycol (EG) CCLs. However, the BUT cathode with a uniform ionomer and Pt distribution yielded inferior performance to the EG cathode. These observations suggest that the uniformity of the ionomer distribution is not a key factor for initial fuel cell performance. After 30k voltage cycles, a large aggregation of the Pt-catalyst nanoparticles is observed in the CCLs of the NPA and IPA MEAs. Moreover, the presented data also implies that the NPA and IPA cathodes experience more ionomer distribution shifts between BOT and EOT relative to BUT and EG. Ionomer distribution changes could be related to the catalyst redistribution during cycling.

The influence of ionomer distribution and coverage in electrospun Pt/Vulcan nanofiber electrodes was examined by S. Kabir et al. [34]. Electrospun nanofiber elec-

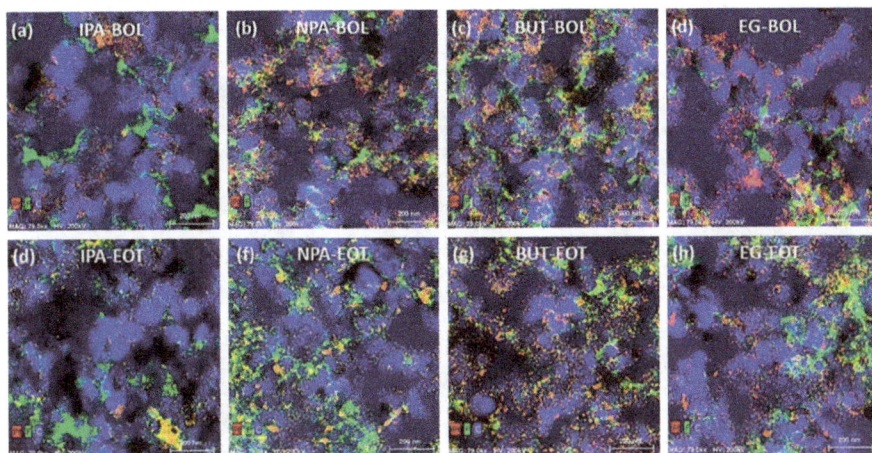

Figure 8: Elemental distribution of the different MEAs before and after voltage cycling from STEM-EDS mapping at high magnification IPA: 2-propanol, NPA: 1-propanol, EG: ethylene glycol, BUT: 1,2-butanediol (Red: Pt, Blue: Carbon, Green: Fluorine.) [6]. Reprinted from J. Electrochem. Soc., Copyright 2021. Published by IOP Publishing with 4.0 CC-BY-NC-ND license.

trodes' morphological properties and ionomer distribution across the fiber diameter and cross-section are studied using electron microscopy techniques. Poly(acrylic acid) (PAA) as the carrier polymer at varying concentrations (with respect to total ink solids) affects the morphology of nanofibers. A higher PAA concentration (15 wt.%) results in a conformally coated ionomer film on the exterior of the Pt/Vulcan nanofibers, and consequently, lower ionomer coverage on both Pt and carbon throughout the fiber diameter (Figure 9). With 10 wt.% PAA, the ionomer is uniformly distributed within each fiber, resulting in an improved overall ionomer coverage and proton accessibility.

The application of STEM-EDS mapping is not restricted to the PEMFC catalyst layer. W. Pan et al. [35] performed microstructure characterization of the MPL layer with different PTFE loadings and solvent compositions. It is found that CB/PTFE interaction can take two distinct forms, i. e., covering and bridging (Figure 10). As PTFE loading increases, bridging dominates.

Both TEM and STEM imaging can be performed in 3D through the electron tomography approach. In electron tomography, the sample holder with the sample is tilted from −70 degrees to +70 degrees, or for the full 360 degrees, and a micrograph is taken at every 1 or 2 degrees. The set of images is then used to reconstruct the 3D volume using a reconstruction algorithm or software such as Inspect 3D. Finally, a 3D model can be constructed after component (Pt, carbon and ionomer) segmentation. This approach faces a number of challenges related to beam damage, wedge effects, high noise presence, time-consuming manual segmentation, etc. but with advances in imaging and use of machine learning and automatic image processing, it holds a promise for advanced catalyst and electrode characterization. Details of this approach are beyond the scope of this chapter, but more details can be found in the literature [36–43].

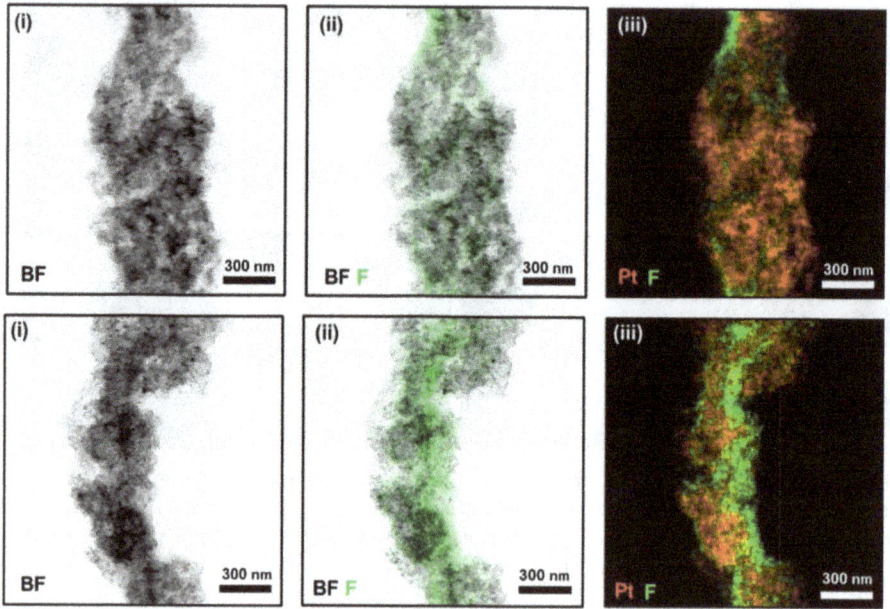

Figure 9: (i) STEM Bright Field (BF), (ii) Overlay of BF and EDS map of fluorine, (iii) EDS maps showing platinum and fluorine overlays (Pt-red, F-green) in Pt/Vu nanofiber electrospun with (top row) 10 wt.% PAA and (bottom row) 15 wt.% PAA [34]. Reprinted from ACS Appl. Energy Mater., Copyright 2021, with permission from the American Chemical Society.

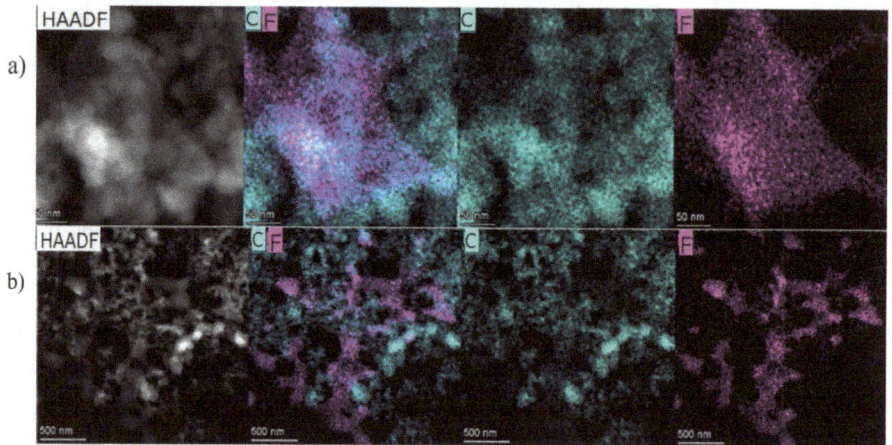

Figure 10: HAADF-STEM-EDS mapping images showing different CB/PTFE interaction modes: (top) Covering, (bottom) Bridging [35]. (Carbon black is light blue; PTFE is magenta color). Reprinted from Chem. Eng. Sci., Copyright 2021, with permission from Elsevier.

Pt-group metal-free catalysts

STEM-EDS characterization is also used for Pt-group metal-free (PGM-free) catalysts. Here, we touch on some basic approaches, with extended details given in the next chapter. With Pt's high cost and scarcity, a substantial reduction in Pt usage is desirable in PEMFCs. There are two approaches to reducing Pt utilization: ultralow Pt content catalysts and PGM-free catalysts [44–46]. Although both strategies demonstrate considerable progress in PEMFCs ORR activities, there comprise some intrinsic drawbacks. Ultralow Pt loading attempts are focused on dispersing Pt at the atomic level, such as single atoms, core-shell structures or tiny clusters. These structures present ultrahigh mass activity but deficient catalytic activity and stability due to fast Pt dissolution or agglomeration [46–48].

Various PGM-free designs have been investigated, comprising metal–organic framework, nitrogen-doped carbon, a couple of earth-abundant transition metals (Fe, Co, and Mn), nonprecious-metal chalcogenides, and metal oxides [45, 49, 50]. Despite the significant progress over the last decade, their inferior mass activity and long-term durability hinder use of PGM-free catalysts in practical PEMFC applications. Unlike PGM catalysts that lose activity typically over Pt atoms dissolution or agglomeration, PGM-free catalysts have a complicated deactivation mechanism that is not fully understood. In this regard, microscopy techniques have been used widely to characterize the catalyst layer [45, 48, 51] and investigate the catalyst deactivation mechanism of PGM-free or low-loading Pt/C catalysts in PEM fuel cells [52–54].

L. Zhang et al. [48] benefited from the atomic resolution HAADF-STEM technique to confirm the synthesis of Pt–Ru dimers on nitrogen-doped carbon nanotubes through the atomic layer deposition (ALD) method. The presence of two different atoms is reflected through distinct contrast of atoms in the dimer-like structure. Y. Cheng et al. [55] studied the stability of synthesized high-density iron single atoms (FeSA) supported carbon nanotubes (CNTs) as a PGM-free cathode. TEM and STEM-EDS images exhibit a uniform distribution of Fe elements and no Fe NPs or agglomerates, as shown in Figure 11.

Figure 11: TEM images and STEM-EDS mapping of FeSA (a) Before stability test, (b) after stability test and (c) STEM-EDS mapping after stability test [55]. Reprinted from Nano Energy, Copyright 2021, with permission from Elsevier.

A hierarchical porous Fe-N-C catalyst (cyanamide (CM)+polyaniline (PANI)-Fe-C) synthesized by H. Chung [56] has been reported to be one of the most active catalysts ($0.016\,A\,cm^{-2}$ at $0.9\,V_{iR\text{-free}}$) under the Department of Energy (DOE) testing protocol. STEM and EELS techniques were used to examine the atomic-level structure and chemistry of the catalyst (CM+PANI)-Fe-C to better understand the origin of high ORR activity. The general morphology of the (CM+PANI)-Fe-C catalyst was imaged using a bright field (BF)-STEM mode. The catalyst layer is constituted of two phases of fibrous carbon particles and a few-layer graphene sheets. Fibrous carbon particles contain randomly oriented entwined turbostratic graphitic domains a few nanometers in size. Due to high graphitic field density, graphite (002) basal planes are mostly exposed on the surfaces of fibrous carbon particles. The high ORR activity is attributed mainly to the exposure of such basal-plane edges in the (CM+PANI)-Fe-C catalyst. More discussion on PGM-free catalysts and their characterization is offered in the next chapter.

2.4 Electron energy loss spectroscopy

The electron energy loss spectroscopy (EELS) technique analyzes the content of the sample by analyzing the distribution of electron energy that passes through it and creating images with unique contrast benefits. EELS instruments are typically integrated into transmission electron microscopes (TEMs) or scanning electron microscopes (STEMs). This technique measures how the kinetic energy of an electron changes after interacting with a specimen. Electrons can interact with a sample elastically or inelastically, and EELS uses these interactions to develop information about the sample. As a result of this technique, the types and quantities of atoms present in a specimen, their chemical states and the electronic structure of the sample's atoms can be determined. Although traditionally a difficult technique, EELS can measure atom composition and chemical bonding, electronic properties of valence and conduction bands. Compared to EDS elemental mapping, EELS is a more sensitive technique for light elements and offers better spatial resolution and analytical sensitivity, making it a desirable method for atomic-level chemical mapping to study the microstructural changes of the monolayer catalyst. It is also possible to achieve an energy resolution of <1 eV through EELS detectors [33, 49].

In fuel cell studies, EELS is mainly used for elemental mapping and conformation of the catalyst structure, e. g., core-shell. For example, Kongkanand et al. [50] studied the fuel cell performances and microscopic properties of a Pt monolayer core–shell catalyst with low Pt loading. Figure 12 shows a typical analysis for determining the thickness of the Pt shell. Line profiles of the area indicated in Figure 12d are shown in Figure 12e. It was discovered that Pt shell thickness is not truly uniform, having thick, thin or infrequently missing Pt.

Y. Kang et al. [51] synthesized low-Pt-loading (PtNi$_3$ alloy) catalysts showing high ECSA Pt ($55–60\,m^2/g$) and high mass activities for high power density PEMFCs. Two dif-

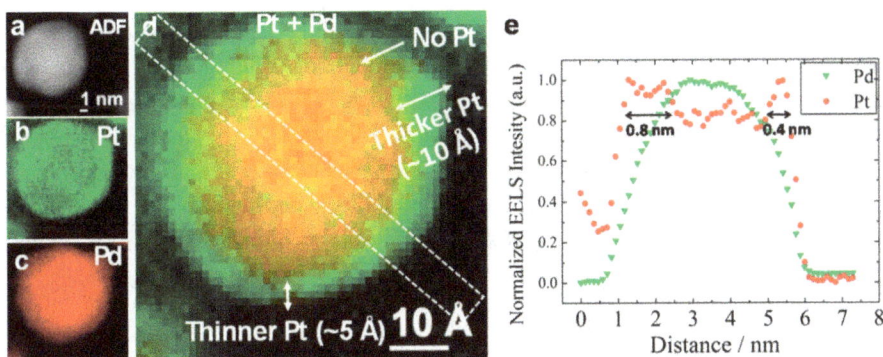

Figure 12: (a) HAADF-STEM of an as-received PtML/Pd/C nanoparticle along a zone axis where the atoms were not directly visible, along with EELS analysis of (b) Pd M4,5 edges, (c) Pt M4,5 edges and (d) a composite showing both Pd and Pt elemental distribution. (e) Line profile drawn to measure the Pt shell thickness across the direction that is shown in (d) [50]. Reprinted from Nat. Catal., Copyright 2016, with permission from the American Chemical Society.

ferent catalyst-design strategies were adopted. The first approach involves dealloying large PtNi$_3$ alloy NPs prepared by conventional high-temperature annealing that leads to nonporous structures (referred to as D-PtNi$_3$/C-600). The second approach was through fluidics-bed synthesis followed by a dealloying process leading to the solid core-shell structure (referred to as D-PtNi$_3$/C-600F). Elemental mapping via electron energy loss spectroscopy (EELS) reveals a mainly uniform Pt/Ni skeleton in the nonporous particles and a Pt-rich surface for solid core-shell structure. EELS analysis was also conducted on the mixed catalyst/Nafion ionomer powder. The results showed that the D-PtNi$_3$/C-600 catalyst has a semiimmersed nonporous structure that facilitates rapid O$_2$ diffusion onto the Pt surface and fast proton conductivity. The D-PtNi$_3$/C-600F catalyst layer, however, generates considerable resistance to O$_2$ transport due to its "fully-immersed core-shell-Pt/ionomer" structure.

Electron energy loss spectroscopy was also utilized to study the oxidation state of catalyst materials in PEMFC. For instance, Chinchilla et al. [52] have employed EELS to investigate the structural and chemical changes of electrochemically tested PEMFC cathode materials composed of platinum nanoparticles on a niobium oxide-carbon black support. The oxidation state evolution of Niobium particles, after an accelerated stress test (AST) potential cycles, was investigated considering Niobium L2 and L3 lines. Results illustrated a slight increase in the number of Nb particles with lower oxidation states and a narrowing distribution after electrochemical cycling. It is hypothesized that the Nb oxidation state evolution is related to the increased oxygen vacancy concentration on the surface of NbOx, which facilitates oxygen atom release from NbOx to Pt NPs and accelerates ORR.

Ultimately, it will take significant improvements in PEMFCs to make them a mainstream energy source. The structural and elemental composition of MEAs affects their structure, transport properties and FC performance, so better understanding of these factors is crucial. To accomplish this goal, a variety of dedicated techniques have been used to characterize the properties, structure and performance of each component. As this section reviewed, microscopy techniques such as SEM, TEM and STEM, combined with spectroscopy approaches like EDS and EELS are of immense importance to fully understand these systems. Further sections review some more advanced approaches such as identical location TEM, *in situ* TEM, as well as advanced image processing methodologies, applied to PEMFCs.

2.5 Identical location TEM

Identical location TEM (IL-TEM) is a technique in which an identical location of a sample (typically a dispersed catalyst powder on a TEM grid) is observed before and after the electrochemical test performed in a specially designed, ex-situ electrochemical cell and changes are visually tracked. Figure 13 shows an IL-TEM setup. Imaging by IL-TEM can be used to complement the half-cell electrochemical studies to provide a qualitative and quantitative insight into the different types of degradation mechanisms that are promoted after applying potential cycling to the catalyst powders supported on the TEM grid, which acts as the working electrode (WE). The electrochemical procedures are typically based on ex-situ AST protocols performed in a classical three-electrode electrochemical cell (or half-cell), using a liquid electrolyte. These ex-situ studies in an electrochemical half-cell can be performed by using Cyclic Voltammetry (CV), a methodology used to obtain information about convoluted reactions taking place in the conventional half-cell setup, which is usually comprised of a reference (RE), working (WE) and counter (CE) electrode [57]. These electrodes can be placed within the cell in the same or three different compartments to prevent the diffusion and exchange of unwanted ions in some components of the cell.

The choice of the right type of electrode will depend on the protocol that needs to be studied. In general, this three-electrode-setup CV is based on varying the applied potential at WE by a specific scan rate while the current is monitored (relative to RE) to guarantee an exact control of the potential at the WE. The RE has a stable and well-known potential and CE is the auxiliary electrode through which the current from and to the WE flow. In this way, it is possible to survey the current between the WE and RE. When choosing these electrodes it is essential to consider: (i) their stability (example.g., the Standard Calomel Electrode (SCE) and the Ag/AgCl electrode can last for more than 12 hr when working as RE); (ii) their ability to introduce impurities into the cell such as chlorides and (iii) the potentials and other conditions at which the electrode materials may dissolve (example.g., when severe Pt dissolution occurs in a Pt mesh that is used as CE, redeposition of Pt ions onto the catalyst in the WE can occur drastically altering the make-up of the catalyst film). In this particular last scenario, a glassy carbon rod with a high surface area can be used as CE in the degradation tests.

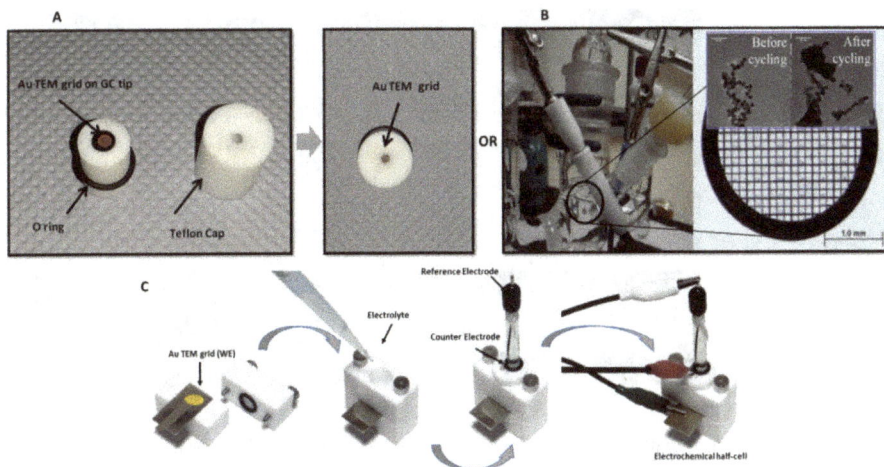

Figure 13: Two different set-ups to accommodate the gold (Au) TEM grid used as support for the NP catalyst material and as a working electrode in the electrochemical half-cell. (A) TEM gold grid mounted on the RDE tip using a Teflon cup [58], Reprinted from Electroanalytical Chemistry, Copyright (2011), with permission from Elsevier. And (B) punched Au TEM grid with a fine gold wire and attached to a platinum holder to get electrical contact [59]. Reprinted from J. Am. Chem. Soc., Copyright (2015) with permission from the American Chemical Society. (C) Plate material evaluating cell or electrochemical cell for identical location experiment [60]. Reprinted from Nano Lett., Copyright (2019), with permission from the American Chemical Society.

Figure 13a–c shows some set-ups that enable the use of the Au TEM grid as WE, where the catalyst NPs are deposited onto a very thin carbon film that coats the Au TEM grid. These NPs should be (i) free of capping agent or ligand and (ii) randomly distributed on the top of the carbon film with very little overlapping and agglomeration to allow for an unobstructed scrutiny of the degradation processes [62]. The TEM grid material has to be nobler than the catalyst materials because it has to be stable in the electrolyte at the voltage range (and other conditions) used during the AST process. Once the catalyst is ready for TEM imaging, to trace a NP in the same location before and after the AST, a series of TEM micrographs must be acquired of specific features in the TEM grid at different magnifications to provide the map of the areas and NPs of interest in the pristine catalyst. The next step is to perform the AST once the conditions of the pristine catalyst NPs on the Au TEM grid are established. The AST is performed in an ex-situ half-cell setup, such as the one shown in Figure 13. To use the TEM Au-grid as a WE, it must be attached to a stable and electrically conductive material, like an Au support, Au wire, or a designed Rotation Disk Electrode (RDE) holder made of Glassy Carbon (GC) with a Teflon cap, as shown in Figure 13a–c. When the AST is over, the TEM Au-grid must be first rinsed with deionized (DI) water, then dried under certain conditions, ultraviolet lamp or oven for 30 min or overnight and later transferred to the TEM for more imaging of the same location (Figure 14). A note that the effect of the electron beam by the TEM

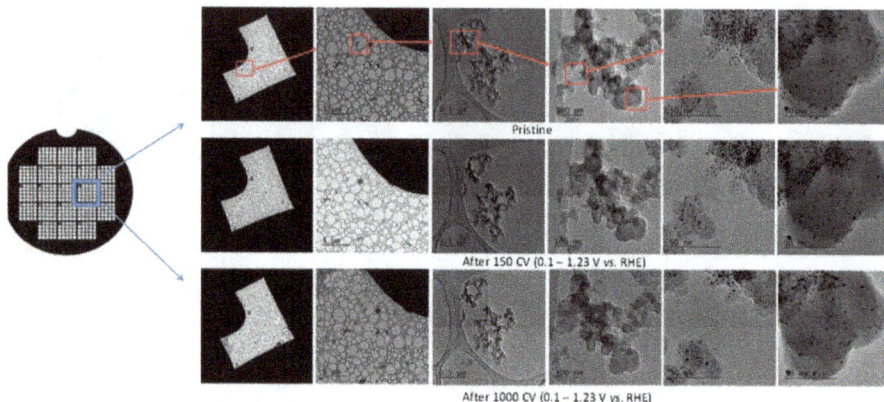

Figure 14: Identical location process in the TEM for pristine and degraded Pt/C catalysts after AST in an Au TEM grid [61]. Reprinted from ACS Catal., Copyright (2018), with permission from the American Chemical Society.

can induce some degradation because of the high dose of electrons applied. Such dose should be minimized as much as possible, and hence 120 kV TEM accelerating voltage is commonly used. Identical location studies can also be carried out using SEM, which is, of course, of lower resolution than TEM. However, in this method, a modified carbon SEM holder can be used that makes imaging of higher catalyst loadings possible. However, the application of IL-TEM is not limited to PEMFC and it is used in other systems like acidic and alkaline water electrolyzers, alkaline fuel cells, regenerative fuel cells and batteries [63].

The first study using IL-TEM was carried out in 2008 by Mayrhofer et al. [64], where they subjected a commercial Pt/C catalyst to a potential cycling range from 0.4–1.4 V_{RHE} with a scan rate of 1 V/s for 3600 CV cycles at room temperature having the Au TEM grid attached to an RDE as a WE. Their findings presented no sign of severe carbon corrosion, but for the first time, Pt NPs detachment from the carbon support was observed (Figure 15a–c). These results led them to conclude that those detached Pt NPs dissolved into the electrolyte without re-deposition, going in clear contrast to the widely assumed degradation processes in fuel cell catalysts, Ostwald ripening and particle agglomeration [65]. They considered these mechanisms to be minimal since the size distributions of the Pt NPs before and after treatment were comparable to each other. However, further IL-TEM studies, by Schlögl et al. [66], on different commercial catalysts and using the same test conditions as Mayrhofer [64] with 0.1 M $HClO_4$ as the electrolyte, showed evidence that these diverse catalysts materials can experience considerably different degradation processes. They also observed a strong influence on such degradation by the catalyst itself and its synthesis mechanisms, even though detailed information about the synthesis mechanisms is usually proprietary and undisclosed in the industry. In their paper, they confirmed that in a diverse set of catalysts, having Pt NPs with different average particle sizes supported on amorphous carbon, the detachment of Pt NPs was more prevalent in larger average particle sizes (~5 nm), whereas in lower average particle sizes (~3 nm) migration and coalescence was more favored. They also confirmed the positive effect of decreasing degradation when the carbon support is graphitized, which increases the durability of PEMFC catalysts. This can be seen in the IL-TEM micrographs in (Figure 15a–c), showing that Pt NPs detachment on graphitic substrate seems to be less significant than on the amorphous one. Since no obvious carbon corrosion was observed, they attributed this degradation resistance to better anchoring of the NPs onto the graphitic planes at the carbon particle edges, which are favorable sites for Pt NPs to bind to. Another important conclusion from this study was that a clear identification of Pt dissolution was not possible because either (i) the dissolved Pt^{z+} ions might diffuse into the vast electrolyte solution lowering the probability of re-deposited onto larger particles or (ii) the applied AST conditions could decrease the amount of dissolved Pt and those few generated Pt^{z+} ions might be more prone to re-deposit onto the particle they were dissolved from.

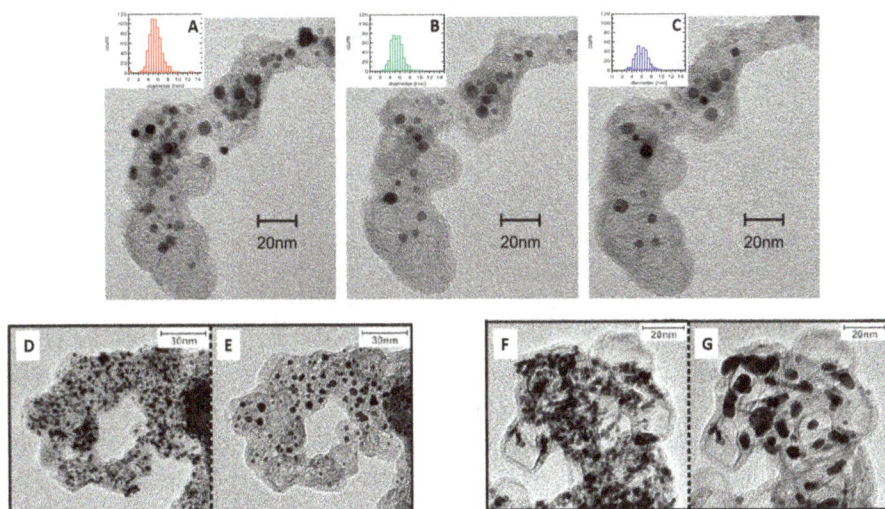

Figure 15: IL-TEM micrographs of a commercial, high surface area carbon-supported Pt catalyst: (A) before electrochemical treatment, (B) after 2 hr of treatment [64] and (C) after 4 hr of treatment; insets show histograms of particle size distribution. Below, IL-TEM (120 kV) micrographs of the Pt NPs deposited on amorphous (D, E) and graphitized (F, G) carbon, before (D, F) and after (E, G) the AST [67]. Reprinted from J Electrochem Soc., Copyright (2012), with permission from IOP Publication Ltd.

Later on, Perez-Alonso et al. [68] modified Mayrhofer et al.'s pivotal IL-TEM experiment, (in O_2 saturated 0.1 M $HClO_4$, for 3000 or 30,000 cycles between 0.6 to 1.1 or 1.2 V_{RHE} at different scan rates, 50 mV/s, and 200 mV/s, and at room temperature), by using the Au TEM grid as only a single working electrode in a RDE cell, in contrast to Mayrhofer study in which the TEM grid was connected in parallel to another working electrode. Alonso's group used a 10 wt% Pt/C catalyst with Pt mean particle size of ~ 2.3 ± 0.4 nm, which were of a lower loading and smaller NPs size than in the Mayrhofer study (~50 wt% Pt, and average particle diameter of 5 nm). The reason behind choosing such a small NPs size, apart from being so close to the optimal size (~3 nm) for a maximum ORR mass activity [69], was because they can be very unstable (the dissolution potential for some NPs has an almost linear increase with particle size) [70–72]. Hence, their degradation can be observable easier without much difficulty, compared to larger nanoparticles. From their results, after performing the AST, Pt dissolution was proven to be the most dominant type of degradation with some NP sintering through coalescence. They also demonstrated how the upper voltage limit value plays an important role during the accelerated corrosion test in which higher upper potential (1.2 V_{RHE} versus 1.1 V_{RHE}) accelerated the Pt NPs dissolution, although similar effects were also observed when the sample was cycled between 0.6 and 1.1 V_{RHE} but for longer periods. Additionally, using IL-TEM (200kV) they also explored the effect of three important but rarely studied conditions: the electrolyte effect, the electron beam and the scan rate. IL-TEM did not show changes in the catalyst nanoparticles after testing the effect of the electrolyte by submerging the Au TEM grid in 0.1 M $HClO_4$ and keeping it in the solution for one night. Variations in the rates (50 mV/s and 200 mV/s) for both potential windows (0.6–1.1 V_{RHE} and 0.6–1.2 V_{RHE}) seemed to have a minimal effect on the observed phenomena of degradation, suggesting that the number of oxidation–reduction cycles have a more essential role in the Pt degradation than the length of each cycle. Regarding the effect of the electron beam radiation, TEM analysis was performed at areas in the grid not previously exposed to the electron beam, identified as "different locations (DL)." When compared, there were slight differences between the NPs size distributions from IL and DL, but in general, they both had similar trends in the sizes. The slight difference was attributed to statistical artifacts since a significant number of NPs were lost after the AST, so the total count of NPs for IL can be much less than for DL micrographs.

A year later and using the same AST conditions as Mayrhofer et al., Meier et al. [73], not only provided more evidence of the same degradation mechanisms but also observed that some Pt NPs decreased in size after AST, demonstrating that some level of dissolution could take place. In their catalyst, particle growth appeared to be the more dominant degradation mechanism, even though evidence of several degradation mechanisms (dissolution, agglomeration, detachment and carbon corrosion) occurring simultaneously were also demonstrated (Figure 16a–b). They also noticed that this inhomogeneous degradation behavior could be found at a single location in the catalyst, especially where local differences of the carbon support are present, like morphology, functionalization and degree of graphitization, which are critical for the interaction between Pt NPs and the carbon support. This last discovery would have been almost impossible without IL-TEM characterization, showcasing the essential role of the carbon support in controlling the degradation pathways of the catalyst NPs.

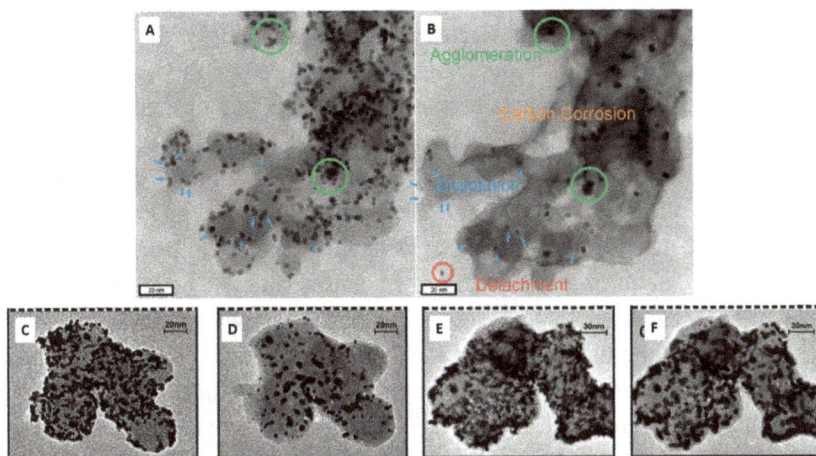

Figure 16: IL-TEM (200kV) micrographs of Pt NPs supported on Vulcan XC72R (A) at 0 and (B) after 3600 degradation cycles [73]. Copyright (2012) with permission from American Chemical Society. Below, are IL-TEM micrographs of catalyst I (C, D) and catalyst II (E, F), before (C, E) and after (D, F) the AST treatment. BET surface areas of 30 and 53 m^2/g, for catalysts I and II, respectively [74]. Reprinted from Energy Environ Sci., Copyright (2011), with permission from The Royal Society of Chemistry.

It is known that the NPs detachment is linked to the oxidation of the carbon support to carbon dioxide (CO_2), therefore it may be possible to improve the resistance to degradation of a catalyst system by using supports that are less susceptible to oxidation. For this reason, conventional high surface area carbon supports can be replaced by graphitized or low surface area carbons. On this line of thought, using IL-TEM, Hartl et al., [74] studied the degradation behavior of Pt-based catalysts supported on two different Low Surface Area (LSA) carbon materials, one conventional "or catalysts I" and another modified by a transition metal prior to Pt impregnation "or catalysts II," and then compared to a Pt catalyst supported on a "standard" High Surface Area (HSA) carbon. After using the same AST conditions as Schlögl et al., [66], when compared to the catalyst on "standard" support, there was an improvement in the performance of "catalysts I" and "II." From the IL-TEM study (Figure 16c–f), "catalyst I" displayed particle detachment as the main degradation mechanism, but coupled with migration, coalescence and/or electrochemical Ostwald ripening, even though the disappearance of small Pt NPs was not evident. On other hand, "catalyst II" showed a considerable reduction in the loss of Pt NPs and far superior improvements compared to "catalyst I." However, the authors did not offer a more integral evaluation of the effect that the transition metal-induced modifications on the LSA carbon offer a higher ability to anchor Pt NPs. The authors also reported that both catalysts on LSA carbons demonstrated an almost identical resistance toward complete oxidation and the IL-TEM measurements did not expose any significant change in the carbon support after cycling treatment, indicating (according to the authors) that oxidation of the carbon support did not play a decisive part in the particle detachment process of the catalyst material.

The catalyst aging conditions, or selected protocols, can induce different degradation mechanisms even on the same catalyst materials. Therefore, the aging protocols must be carefully chosen to fundamentally investigate the degradation mechanisms presented on the catalyst. Speder et al. [20] demonstrated that depending on the accelerated degradation treatment, the reduction in the amount of Pt NPs on the carbon support was induced in different ways. Their study, conducted on 30 wt.% Pt on Ketjen black at room temperature in Ar saturated 0.1 M HClO$_4$ using a home-built multielectrode setup, employed two degradation treatments. One to simulate load cycle conditions experienced in an in-service fuel cell for vehicles by applying square-wave potential steps of 0.6–1.0 V$_{RHE}$ with a resting time of 3 sec at each potential for 9000 cycles: The second treatment tried to simulate start-up/shutdown conditions at potential cycling between 1–1.5 V$_{RHE}$ with a sweep rate of 500 mV/s for 27,000 cycles. At Start-Stop conditions (Figure 17a–b), NPs detachment mechanism was observed, most likely as a result of carbon corrosion, but Pt NPs coalescence can also occur. For load cycle conditions (Figure 17c–f), no massive particle detachment occurs but instead NPs growth is observed mainly due to NPs migration and coalescence (characteristic of dendritic-shaped NPs) or due to electrochemical Oswald ripening (characteristic of rounded NPs).

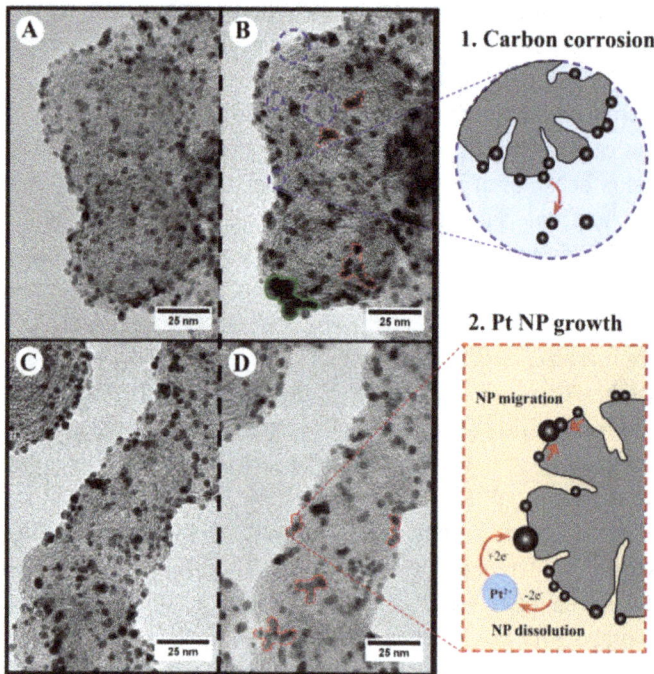

Figure 17: IL-TEM (120 kV) micrographs of 30 wt.% Pt on Ketjen black sample before and after AST protocols. (A, B) Effect of Start-Stop conditions, (C, D). Effect of load cycle conditions [20]. Reprinted from J. Power Sources, Copyright (2014), with permission from Elsevier.

Similarly, Lafforgue et al. [61] further confirmed that in a given electrocatalyst, the degradation processes would not only be governed by its nature but also by the values of the AST applied potentials and by the type of electrolyte (Figure 18). Their study was carried out on an alkaline electrochemical system (to simulate Anion Exchange Membrane Fuel Cells, AEMFCs), in which (i) the complex electrochemical reactions are faster than in an acidic medium and many electrocatalysts materials can easily promote them [75], also (ii) many metals/oxides used as catalysts are more stable at high pH [76]. However, degradation can still occur in alkaline fuel cells and electrolyzers. They [61] studied the durability of a 10 wt.% Pt/C electrocatalyst in two different media: liquid alkaline electrolyte (0.1 M NaOH) and a solid polymer electrolyte (using a dry cell), both at $T = 25\,°C$, 150 (or 1000 in the dry cell) cycles, and 100 mV/s scan rate in RDE. When the catalysts were tested in liquid media at a full potential range (0.1–1.23 V_{RHE}), the degradation occurred mostly by the detachment of the Pt NPs from the carbon support, similar to the already described acidic medium of PEMFC. However, they suggested a more specific degradation mechanism: first (i) Pt NPs can assist the corrosion process of the carbon surface by locally producing CO_2, which (ii) reacts with OH^{--} anions in the electrolyte, forming $CO_3{}^{2-}$ anions, that eventually (iii) generate solid alkali–metal carbonates Na_2CO_3 by precipitation with the Na^+ in $NaOH^-$ based electrolyte; (iv) These carbonates are preferentially localized at the interface between the Pt NPs and the carbon support; Finally, (v) the Na_2CO_3 crystals mechanically eject the Pt NPs from the carbon support (Figure 18b). The authors did not provide any chemical characterization, like Energy Dispersive X-Ray (EDS) spectroscopy, to support this claim but instead used as a frame of reference the conclusions by Zadick et al. [77, 78] that studied the harsh degradation phenomena that carbon-supported Pt or Pd NPs experience when liquid alkaline is used as electrolyte. They [61] also investigated effect of cycling potential range. They performed AST at reduced potential windows to prevent Pt NPs from completely oxidizing (stripping CO_{ad}-like species) or completely reducing (adsorbing CO_{ad}-like groups). Based on the IL-TEM micrographs, they observed that at the low potentials range (0.1–0.6 V_{RHE}) (Figure 18a), the degree of degradation was considerably lower compared to the full potential range, with Ostwald ripening (recrystallization) as the main degradation mechanism. The high potential range (0.6–1.23 V_{RHE}) (Figure 18c), also showed a reduced degradation degree than the full potential range. At these potentials, Pt recrystallization is not likely and mainly Pt NP agglomeration led to the degradation. Pt/Pt-oxides will enhance this process and corrosion processes, which leads to the mobility of the Pt NPs at the same time. It means that Pt nanoparticles on carbon substrate migrates to the functionalized areas on the carbon surface that ultimately forms the agglomerated NPs structures. They also performed the same study to simulate dry hydrogen-fed alkaline fuel cells using no liquid alkaline solution but a solid polymer Anion Exchange Membrane (AEM) as an electrolyte. After comparing the IL-TEM micrographs from (Figure 18d), it was evident that the extent of degradation was notably reduced when the AST took place in an AEM-based dry cell than in the liquid NaOH electrolyte, even after 1000 cycles in the dry cell. Pt NPs aged in dry-cell AEM showed

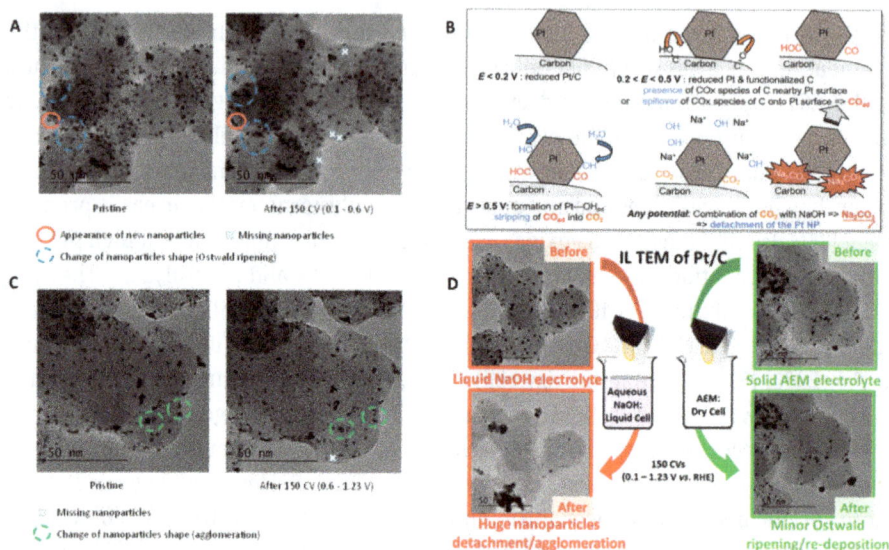

Figure 18: IL-TEM micrographs of Pt/C NPs before (pristine) and after 150 CV cycles performed at V = 100 mV/s, T = 25 °C in 0.1 M NaOH between (A) 0.1–0.6 V_{RHE} at reducing conditions, and (C) 0.6–1.23 V_{RHE} at oxidizing conditions. (D) Comparing IL-TEM micrographs of the same AST treated electrocatalyst before (pristine) and after 150 CV performed in liquid 0.1 M NaOH electrolyte and the interface with an anion-exchange membrane in the dry cell at T = 25 °C both [61]. Reprinted from ACS Catal., Copyright (2018), with permission from the American Chemical Society. (B) Schematic representation of Pt NPs detachment from their carbon support in an alkaline electrochemical system [79]. Reprinted from Energy Environ. Sci., Copyright (2020), with permission from The Royal Society of Chemistry.

"darker" and "rounder" NPs after AST compared to the liquid cell. This suggested that in the dry cell, Pt dissolution and redeposition occurred to a larger extent than in the liquid electrolyte because in the later, the formed Pt^{z+} ions (by oxidation of Pt NPs) had a lower chance to redeposit on the surface of any Pt NPs. Similarl to Schlögl et al. [66] previously stated conclusions, this is due to the large excess of liquid electrolyte and the fast diffusion mass-transport of those cations in a liquid that allows them to move away from catalytic material after their formation. Meanwhile, in the dry cell, the Pt^{z+} species are likely caught easier in the area of their formation, allowing their redeposition once Pt^{z+} reduction potentials are reached.

Along those lines, Dubau et al. [61] went as far as to address the influence of the upper potential limit and the gas atmosphere, used in the ASTs, on the degradation mechanism of Pt/Vulcan XC72 electrocatalysts, which were aged in an RDE at 800 CV from 0.05–0.50 V_{RHE} or to 1.23 V_{RHE} and in different gas atmospheres, including Ar, CO or O_2 but in an acidic electrolyte solution, 0.1 M $HClO_4$. Their IL-TEM micrographs (not shown here) after a 0.05–0.50 V_{RHE} potential sweep illustrated that every gas atmosphere had a predominant degradation mechanism. Minor morphological changes were detected under inert Ar, carbon corrosion and the Pt NPs detachment were more visible in O_2-containing electrolyte, and Pt nanocrystallites migration/aggregation was the major degradation mechanism when CO was presented in solution. Increasing the upper potential limit to 1.23 V_{RHE}, degradation trends were maintained, but Ostwald ripening processes began to interfere and overlapped with other degradation mechanisms, hindering an isolated identification of the more dominant mechanisms taking place. Their evidence for that claim was the observed increase in the mean diameter of isolated Pt NPs that could not be induced by carbon corrosion or NPs migration/aggregation.

Zadick et al. [77] compared, by IL-TEM imaging (Figure 19) and electrochemical characterizations, the stability of a state-of-the-art Pt/C electrocatalyst in alkaline and acidic media. They intended to study whether alkaline media are more aggressive or not than acidic ones. Their AST was performed using 150 cycles between 0.1–1.23 V_{RHE} at 100 mV/s in three different electrolyte solutions: 0.1 M H_2SO_4, $HClO_4$ or NaOH. They demonstrated that an alkaline medium is more aggressive for Pt/C catalysts than acidic ones at the conditions studied with a high loss of ECSA (~60 %) observed for AST in 0.1 M NaOH, whereas in acidic media the ECSA loss is limited to (~20 %). In the acid medium, Pt dissolution/redeposition and low NPs detachment were observed. In contrast, massive degradation was noticed in the alkaline medium even at a very small number of cycles, e. g., 150, with a large loss of number of Pt NPs (~63 %). The author explained that the detachment of NPs from the carbon surface was not due to carbon corrosion, as proven by their combined use of IL-TEM (no changes in the carbon shape after AST was observed), Raman spectroscopy (no significant changes in the D1-band and G-band vibrational modes was observed), and XPS (the C1s band was unchanged after AST). However, they did not rule out the possibility that the modification of the carbon surface chemistry in alkaline media occurs, which modifies the anchoring sites of the particles on the carbon support and could explain the severe degradation and the larger loss of nanoparticles in the alkaline medium than in acidic ones.

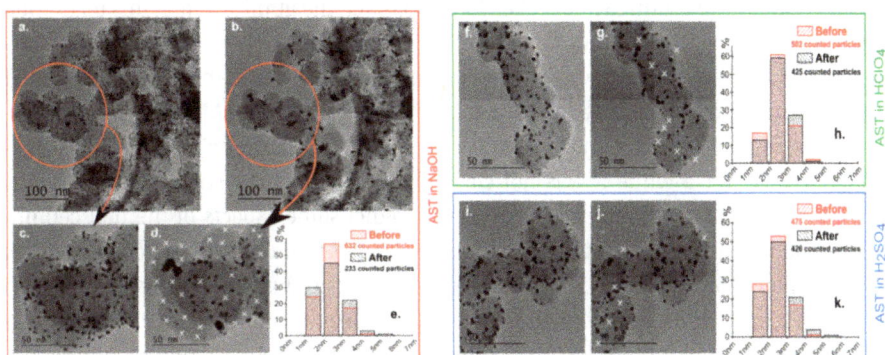

Figure 19: IL-TEM micrographs of pre- and post-AST were performed at 25 °C in NaOH (a–d), $HClO_4$ (f, g) and H_2SO_4 (i, j). The white ×'s represent locations of NPs loss. NP size distribution histograms are displayed in (e) NaOH, (h) $HClO_4$ and (k) H_2SO_4 [77]. Reprinted from ACS Catal., Copyright (2015), with permission from American Chemical Society.

In general, the efficient operation of any type of fuel cell will greatly depend on the chemical reactions that generate electricity and the nature of the electrodes involved in the process. These reactions are known for being structure sensitive, so by controlling the size, morphology, and/or composition of the catalysts material it is possible to enhance its activity [80–82]. Arán-Ais et al. [59], by using IL-TEM reported on the influence of the electrochemical activation method on the morphology of strategically shaped Pt NPs (initially hexagonally-shaped) by applying potential cycles between 0.05–1.3 V_{RHE} at 50 mV/s in a 0.5 M H_2SO_4 solution at room temperature in a conventional three-electrode electrochemical cell. This upper potential was used because it can "perturb the well-ordered surface domains of the NPs." In Figure 20a–c, it can be observed that apart from Pt NPs coalescence and migration, these potential cycling conditions can induce dramatic surface modification leading to the loss of their characteristic faceted structure, site-selective activity, and hence limit any practical application. Although, according to the Fourier transforms (FFT) of some selected nanocrystals, some crystallinity was maintained even after cycling. Figure 10 further confirmed that electrochemical activation on the shaped nanocatalysts can result in drastic changes to their structures even after a small number of cycles. These interesting breakthroughs that IL-TEM has unveiled in the catalyst morphology are indeed very important. However, changing the morphology of Pt NPs to reach more efficient catalytic activities is just one of the many efforts devoted to enhancing PEMFC technology, and the ultimate goal is the reduction or elimination of the expensive noble catalyst metal [83, 84]. Pt-based alloy electrodes can reduce the noble metal content while offering promising mass activities 2–5 times higher than the pure Pt [69, 84]. The structural ordering of these alloys seems to have a significant impact on the final electrochemical properties of the catalyst [85–89]. IL-TEM was used to study Pt alloy catalysts as well. The reader is referred to the original work reported in the literature [60, 90–92].

It is imperative to highlight that for a while IL-TEM was performed at samples tested at room temperature, until Schlögl et al. [58] expanded the application of this technique to elevated temperatures, comparable to PEMFCs operating temperatures., They performed IL-TEM at elevated electrolyte temperatures such as 60 °C and 75 °C, and applied fixed electrode potentials (1.0, 1.1, 1.2, 1.3, 1.4 V_{RHE}) for 16 h in 0.1 M $HClO_4$. This significant step now allows for a more reliable ex-situ characterization of the catalyst materials because for a while it was not clear if the results of the studies performed at room temperature can be used in practice to predict some aspects of the fuel cell under operation, such as its lifetime. They demonstrated that under these conditions' Pt NPs agglomeration was mainly induced by the oxidative shrinking of the HSA carbon support (Figure 21a–d), probably one of the main causes of catalyst layer collapse in an MEA. Additionally, as a direct result of the elevated temperatures, oxidation of the carbon particles at rather positive potentials of 1.3 V and above was observed to start from their core and progress toward their structural borders. This is because the shell region of carbon black particles contains a higher density of corrosion resistant graphitic layers; this can ultimately lead to the formation of hollow graphitic structure of the carbon

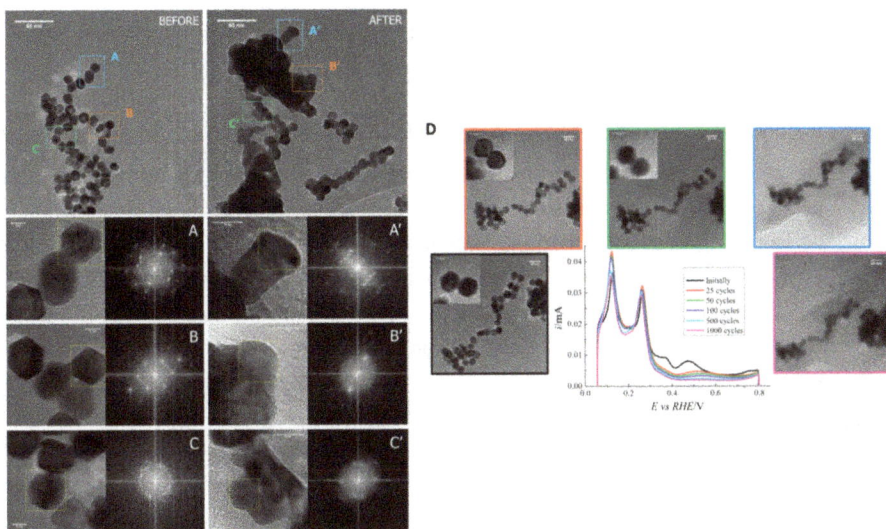

Figure 20: IL-TEM (200 kV) and IL-HRTEM micrographs (with FFT of the selected areas) of the shaped Pt NPs before (A, B, C) and after (A', B', C') 1000 potential cycles. (D) CVs and IL-TEM micrographs before and after 25, 50, 100, 500 and 1000 cycles up to 1.3 V, 50 mV/s [59]. Reprinted from J. Am. Chem. Soc., Copyright (2015), with permission from the American Chemical Society.

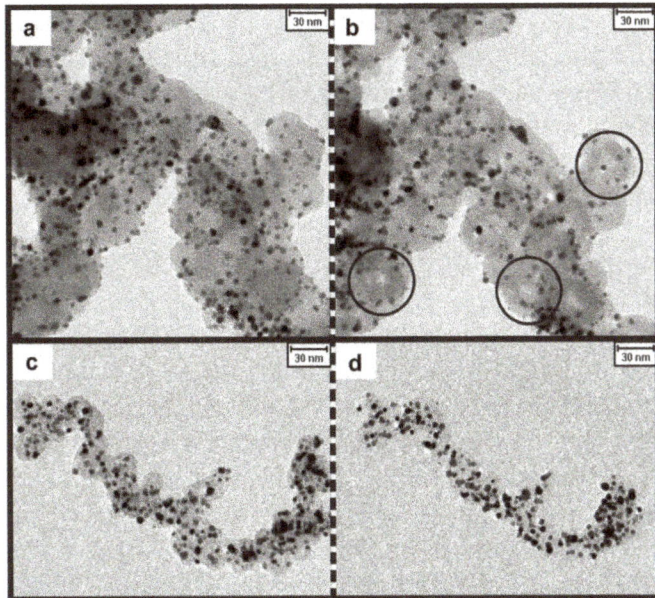

Figure 21: IL-TEM (120 kV) micrographs before (a), (c) and after (b, d) potentiostat degradation measurements at 1.3 V_{RHE} and 75 °C K for 16 h in 0.1 M HClO$_4$ [58]. Reprinted from J. Electroanal. Chem., Copyright (2011), with permission from Elsevier.

primary particles after corrosion and degradation. Meanwhile, the Pt NPs are not as severely affected as the HSA carbon support. These observations are in direct disagreement with degradation studies performed at room temperature where mainly catalyst NPs degradation is observed without obvious alterations in the carbon support. IL-TEM studies at high temperatures are quite challenging because the integrity of the TEM grid can be compromised. For this reason, choosing the right potential window is crucial to avoid the degradation of the Au and the carbon film in the TEM grid while at the same time allowing for the accelerated degradation of the carbon-supported Pt NPs system. Schlögl et al. [58], also documented that under their experimental conditions the maximum positive potential limit was $1.3\,V_{RHE}$, above which Au dissolved and redeposited on the Pt NPs. On the other hand, at or below $1.2\,V_{RHE}$ (60 °C) the catalyst materials experience no changes.

IL-TEM tomography

Visualization of structures at the nanoscale can be accomplished with TEM micrographs, although certain information can be lost in the 2D projections of a real 3D structure. For this purpose, the concept of IL-TEM can be extended to identical location tomography (IL-tomography). These techniques visualize the 3D structure of an identical catalyst location before and after degradation by recording a series of images at different angles and reconstructing them to render a stereoscopic representation of the catalyst [89]. Meier et al. [73] using IL- tomography, determine similar degradation in 2D (Figure 16a–b) as in 3D projections (Figure 22a–h) where severe particle growth and additionally, as an advanced effect of the corrosion, a change in the structure of the carbon support can be observed in the tomograms. The authors argued that with this new approach, even though imagining is in principle possible, significant signal-to-noise ratio can obstruct an appealing and unambiguous representation of the reconstructed images of the carbon support. For this reason, cross-sections from the reconstructed tomograms are taken before and after degradation to elucidate the changes in the support more precisely. Small changes in the carbon structure were visible with some thinning due to carbon corrosion. Agglomeration of Pt NPs was observed unobscured by overlapping features as in the case of 2D IL-TEM projections. IL-tomography also allowed them to confirm that Pt NPs were still located at the carbon support surface after degradation, so their accessibility to electrolyte and reaction gases would continue unaffected. The 3D reconstructions also shed light on the fact that Pt NPs did not appear to be cubooctahedrons, but instead exhibited an extended contact area with the carbon surface by forming "string-shaped" or "T-shaped" Pt NPs clusters located at contact points of two or more primary carbon particles, regions that can provide extra stability because after electrochemical treatment Pt NPs agglomerates can be found in these sites more frequently than in more exposed sites.

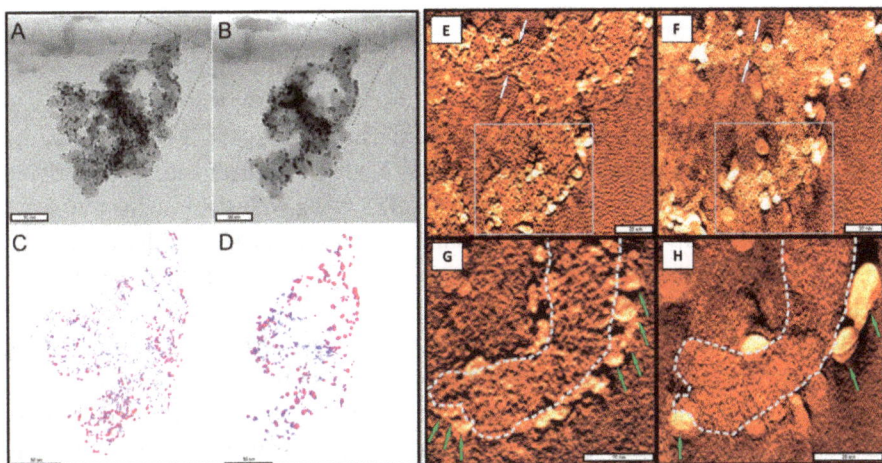

Figure 22: Standard IL-TEM micrographs (A, B) and IL-tomography reconstructed images (C, D) of the same catalyst location after 0 (A, C) and 3600 (B, D) degradation cycles between 0.4–1.4 V_{RHE} with a scan rate of 1 V/s. The dashed box in A and B highlights the area where cross-sections were taken through reconstructed tomograms before (E, G) and after (F, H) 3600 degradation cycles. The white arrows point out a location that indicates thinning of the support due to carbon corrosion. The gray boxes highlight areas magnified in (G, H), which illustrates the carbon support (dashed white lines) and platinum particle agglomeration within the plane (green arrows). The image acquisition covered the tilting angular range of −74° to +77° with a 1° increment [73]. Reprinted from ACS Catal., Copyright (2012), with permission from the American Chemical Society.

On the other hand, by way of offering more flexibility on the type of study and scale, this IL process has also been performed using Scanning Electron Microscopy (SEM) to analyze samples that have been tested for the activity or ECSA loss in RDE.

As the presented section reviews, the identical location transmission electron microscopy approach is being applied extensively in fuel cell degradation studies because of its simplicity and capability to study morphological and structural changes in the same location on the same nanoparticles in catalysts. However, the technique is still lacking observation in operando, while processes are occurring. This can be accomplished by the *in situ* TEM approach, which is reviewed in the next section.

2.6 *In situ* TEM characterization

The microscopy field is experiencing emerging developments of the *in situ* TEM holders made by the leading companies [93–95]. These holders enable implementation and control of external stimuli, e. g., heat, stress or electrochemical biasing while observing changes in the studied materials. In addition, progress made in manufacturing of the powerful direct electron sensing TEM cameras, with the rapid image acquisition has made it possible and feasible to study myriad nanoscale phenomena by *in situ* TEM techniques [96, 97]. An *in situ* sample holder coupled with the imaging modes—TEM and STEM—and spectroscopy techniques—EELS and EDS–can reveal the time-resolved alteration of the heterogeneous catalyst materials in PEMFCs [96]. By this means, the research community can have an opportunity to study and understand the underlying mechanisms of the nanoscale processes in PEMFCs along with the insights into ways to improve functional performance by modifications to the system. In this section, an overview on the available *in situ* TEM technologies, with the focus on PEMFC characterization, is presented.

Being an electrochemical system, most PEMFC studies have been performed using *in situ* electrochemical liquid cell TEM (LCTEM) [98]. Even though a fuel cell operation is not performed in a liquid environment, due to unavailability of gas-based electrochemical holders, the liquid *in situ* cell is currently the only viable approach to study dynamic processes in a TEM. The current status of techniques available to study materials in liquid environments inside the TEM are summarized in two main approaches: open-cell and closed-cell systems. Evaporation of the liquid involved in an open-cell TEM system, which is part of high vacuum column (10^{-5} Pa or 10^{-8} Torr inside the TEM stage chamber compared to atmospheric pressure of 10^5 Pa or 750 Torr), is considered as one of the major obstacles of the technique. To address this issue system, the experiments can be carried out with low-vapor-pressure ionic solutions, which can withstand the high vacuum inside the TEM column [99, 100]. However, most relevant PEMFC studies use high-vapor-pressure electrolytes (e. g., a catalyst powder dispersed in a mixture of water and hydrochloric acid, sulfuric acid) [101, 102], and are not compatible with the open-cell *in situ* TEM platform [103, 104]. The "environmental" approach with the aid of a differential pumping system has been introduced [105] and developed [106–108] as a solution to overcome the above-mentioned limitations. The differential pumping system inside an environmental transmission electron microscope (ETEM) can provide the system with the "environmental" conditions, which enable pressures up 100 Torr in the sample chamber. Although the pressure inside the ETEM sample chamber is far lower than ambient conditions, it can be sufficient for various fundamental studies [108]. However, even utilization of an ETEM system could cause a rapid evaporation of most aqueous solutions used for the PEMFCs catalysts studies.

"Closed liquid-cells" in liquid *in situ* TEM experiments relevant to PEMFCs have been developed as an approach to tackle the application limitations of the open-cell configurations [109]. In a closed liquid-cell configuration, two electron-transparent solid windows are used to confine the liquid solution and seal it from the high vacuum column. Graphene liquid-cell and SiN_x windowed liquid-cell TEM (LCTEM) have been two different closed-cell configurations widely used to study the liquid specimens. Graphene encapsulation of specimens and liquids studied under the TEM in graphene liquid-cells has emerged in the recent decades as the result of developments in thin and flexible graphene sheets production [110]. However, the issues like difficulties of handling nanometer thick graphene sheets [111], small volume of liquid and stagnant flow formation inside the graphene capsule [111, 112] and inability of performing electro-chemical testing due to lack of electrical circuit have hindered the mass adoption of this system for PEMFC materials studies. The windowed cell has been considered as a more common approach for closed-cell *in situ* liquid TEM experiments using aqueous elec-trolyte solutions [113, 114]. The advanced microfabrication and lithography technologies and procedures [94, 115, 116] have nowadays paved the way to pattern the electrical cir-cuits on small silicon chips that contain very thin (ca. 50 nm) electron transparent SiN_x windows. The LCTEM holder not only secures a sealed system but also can provide the necessary electrical connections for the biasing, heating and cooling capability [96]. An example of an electrochemical-LCTEM (EC-LCTEM), being used to study PEMFCs catalyst materials, is shown in Figure 23. As can be seen, a 3-electrode configuration of working electrode, counter electrode and reference electrode with a SiNx window and a confined space for electrolyte circulation are engraved on the Silicon wafer. The configuration of electrodes, specifically reference electrode, is discussed further in next section. Al-though LCTEM has proven to be a useful tool in studying the PEMFC catalyst materials, there are issues like the image resolution limitations, electrode configuration inside the confined volume and beam damage and radiolysis that are elaborated on further.

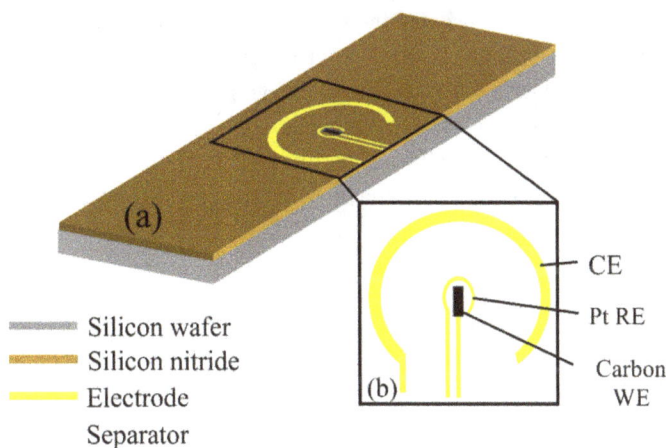

Figure 23: (a) Schematic view of an EC-LCTEM chip and (b) expanded view of the electrode's configuration on the chip [98]. Reprinted from Advanced Functional Materials, Copyright (2022), with permission from Wiley.

Imaging resolution

De Jong et al. [114, 117] have reviewed the fundamentals of TEM imaging resolution and signal-to-noise ratio of objects within liquid media in details. The optimum distance be-tween two resolvable points (d_{TEM}) is governed by the following equation:

$$d_{TEM} = 6 \times 10^{12} \frac{aC_cT}{E^2},$$ (1)

where d_{TEM} is the distance between two fully-resolved points in TEM mode, a is objective semi-angle, C_c is the chromatic aberration coefficient, E is the electron beam energy and T is the liquid thickness. The takeaway message of Equation (1) is that the resolution is inversely proportional to the thickness of the liquid media.

Calculating the resolution of an object being imaged in the STEM mode in the liquid environment is more complicated compared to that of TEM mode [114]. It should be noted that the resolution of an object in the STEM mode is theoretically defined as the smallest nanoparticle distinguishable from the background noise. Technically, d_{STEM}, the distance between two fully-resolved points in the STEM mode, depends on different parameters such as the mean free path length of the electron scattering, the total number of the electrons reaching the sample or electron dose [117] and the liquid thickness. Despite presenting different equations for the resolution of an object in the liquid media in STEM mode [114], calculating the d_{STEM} remains one of the major challenges of the technique. The consensual result of all available equations is that the d_{STEM} is directly proportional to square root of the liquid thickness. Therefore, in both TEM and STEM modes, thicker liquid is always less beneficial for imaging resolution. The LCTEM system in which an electron scattering solution is confined in an electrolyte circulation space between two windows can suffer from a low image resolution in both TEM and STEM modes. This condition can be exacerbated by the pressure difference inside the chips with the column and bowing of the windows. Considering Equation (1) for TEM and the relationship of STEM resolution with thickness, achieving atomic and/or lattice resolution of Pt particles in a PEMFC related system is challenging, if not impossible, in standard commercial LCTEM systems. There have been some efforts so as to overcome the liquid thickness issue and relatively improve the image resolution [118–120].

Reference electrode

As a PEMFC is an electrochemical system operating under the different pHs, voltages and reactants concentrations, its materials have been typically studied using electrochemical LCTEM (EC-LCTEM). A common EC-LCTEM uses a three-electrode configuration-working electrode (WE) made of Pt or glassy carbon, Pt counter electrode (CE) and Pt reference electrode (RE). Since PEMFCs widely use Pt-based catalyst materials, it is recommended to avoid the application of Pt working electrode, whose catalytic activity can interfere with that of an actual catalyst. Therefore, glassy carbon is a more common working electrode for the EC-LCTEM systems. However, the reference electrode faces more complexities that need more deliberation.

Commonly, the electrochemical experiments, including EC-LCTEM, report the potential values in reference to a conventional and standard reference electrode, such as the reversible hydrogen electrode (RHE) [121], standard hydrogen electrode (SHE) [122], Ag/AgCl and saturated calomel reference electrode. However, the size constraints of an EC-LCTEM chip interferes with employment of a complicated reference electrode system like RHE or SHE on chips. As a result, the community has settled for a Pt electrode as the reference electrode in the EC-LCTEM setup. The main issue of a Pt reference electrode is its instability during the experiment. H_2 gas is generated or consumed on Pt electrodes due to reduction of the protons from the electrolyte and reverse H_2 oxidation reactions, respectively, catalyzed by Pt during the potential cycling. Therefore, H_2 pressure around the reference electrode can be unstable resulting in potential variations of reference electrode. As a result of that phenomenon, comparison of the EC-LCTEM results with the ex-situ experiments that can provide one with an understanding about the relationship between applied electrochemical potential and changes in nano/microstructure [122] is burdensome. Some research groups have devised systems (such as a connecting external reference electrode to the chip [118]) that can address the reference electrode issues [118].

Beam damage and radiolysis

Beam damage is conceived as the damage of any component inside LCTEM chips caused by scattering collisions, i. e., transfer of portion of the incident electrons' energy to the sample, during bombardment of the electrolyte and nanomaterial [123]. As a result of the beam damage to the electrolyte, unwanted artefactual reactions can be induced by radiolysis of the liquid [124–126], causing alterations to the measured electrochemical signal [127]. The interaction of the incoming electrons with the water molecules results in decomposition of electrolyte and formation of components such as hydrated electrons (e_h^-), hydrogen and hydrogen radicals, hydronium ions, hydroxyl radicals, peroxide, etc., affecting the behavior of the nanoparticles under study [128–130]. It is critical to PEMFC studies, where the dissolution/re-deposition of the Pt nanoparticle is essential, to understand such beam damages and artefacts. Indeed, an additional reaction of reduction of Pt ions to Pt deposits, a spike in the cyclic voltammetry (CV) and the potential shift can occur by interference of primary product of electron irradiation of aqueous solutions (radiolysis), i. e., solvated electrons [127, 131] with the main process. Consequently, the interpretation of the electrochemical behavior of the nanoparticles can be altered.

The aqueous solution in the system is not the only component affected by radiolysis. The organic ionomer used in the catalyst inks and layers of PEMFCs is susceptible to the electron beam and undergoes the radiolysis too [123]. The reports show that the radiation by electron beam for longer exposure time during STEM imaging mode, specifically EDS mapping at high dwell time, can cause a severe F depletion in PFSA backbone [132], leaving behind a disordered structure. A clear and comprehensive chemical mechanism on the PFSA beam damage has not been reported yet [133]; however, an increase of the temperature of the ionomer due to the collision of the incoming electrons with valence electrons in the PFSA structure is believed to lead to major damages to PFSA. In addition to that, there is a possibility that the chemical degradation proceeds through the secondary reactions of PFSA with water radiolysis products, such as aqueous electrons and hydroxyl radicals. A few technical and fundamental issues in current *in situ* LCTEM technologies make them poorly suitable to study the PFSA structural change. First, surface poisoning of catalyst by the application of the ionomer in an aqueous experiment can negatively affect the catalyst performance [134] and, therefore, it is suggested to avoid addition of the ionomer to the ink. Second, aqueous experiments are not representative of the actual role of an ionomer in a porous network of the PEMFC catalyst layer [135]. Finally, tracking an organic material, sensitive to water and convection of the liquid in the system, is challenging. Therefore, while the community has settled for the current LCTEM technologies, new designs and further developments of *in situ* TEM platforms are necessary in order to obtain a deeper understanding about the mechanisms of reaction at atomic and molecular level in MEAs.

In situ LCTEM studies of PEMFC catalyst materials

Despite the challenges discussed above, the *in situ* LCTEM is one of the valuable tools currently available to the research community to understand the mechanisms of catalyst degradation and promote the development of more efficient and durable PEMFCs. The advances in the field of *in situ* LCTEM holders and chips microfabrication resulted in the emergence of promising reports on the PEMFC's PGM and PGM-free catalyst materials in the last decade. Here, we review the LCTEM research reported so far related to the PEMFCs catalyst materials (beam damage and degradation) and reactions (ORR, HER, corrosion, etc.) to depict the current status of the field and outline possible future research opportunities.

To perceive the effect of electrolyte radiolysis on the PEMFC related catalyst materials, Wu et al. [136] carried out a beam-induced solution radiolysis in an oxidative reaction environment (0.1 M $HAuCl_4$ and 0.1 M KCl solution) on presynthesized icosahedral and cubic nanoparticles of Pt, with no electrodes or applied potential used. A circulating solution was used by a dynamic liquid flow cell in order to remove the Pt ions produced in the system and suppress the beam-induced particle growth. As evidenced in the progress map of the Pt nanoparticles etching process in Figure 24, after 10 min of introduction of solution to the cell, noticeable changes to the Pt icosahedron (initially enclosed by {111} facets) and nancubes (initially enclosed by {100} facets) started to show up. However, the rate of the dissolution was different along the corner, edge and terrace atoms such that the overall dissolution rate was dominated and controlled by each of them at different time steps. The Pt nanocube was transformed to a pseudospherical icosahedron after 20 min due to the progress of dissolution along the {100} terraces. The authors utilized the results of the experiment to develop a kinetic model for the selective surface etching, demonstrating the capability of LCTEM in unravelling the mechanism and kinetics of the reactions.

The facet-dependent rate of etching of PGM materials under the beam was demonstrated in the work by Shan et al. [137] who studied the cubic Pd@Pt core-shell nanocubes corrosion in a 0.1 M NaBr solution by the LCTEM technique (with no voltage applied). The dissolution started from corners, continued to the center and left behind a Pt cage

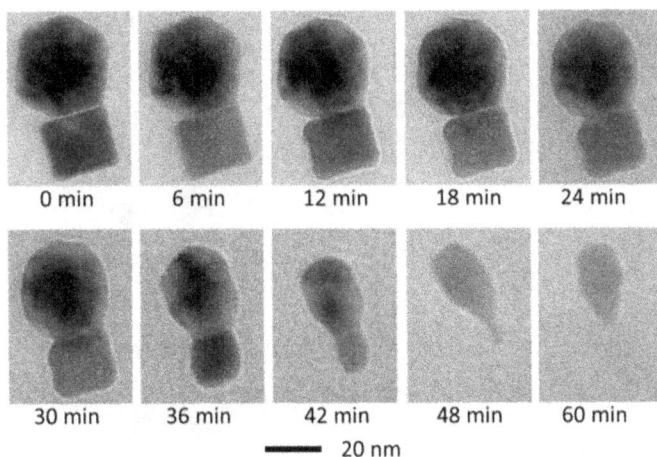

Figure 24: The progress of Pt icosahedral and cubic nanoparticles dissolution in the mixture of HAuCl$_4$ and KCl solution after 60 min [136]. Reprinted from ACS Nano, Copyright (2017), with permission from ACS Publications.

that finally created a hollow sphere. In addition to the face dependency of the etching rate, the surface energy and presence of defects in the structure of core-shell played important roles in the preferential and site-dependent corrosion of the structure.

Activity and degradation of PEMFC catalyst materials have been the focus of the majority of *in situ* EC-LCTEM studies in the recent years. Zhu et al. [138] studied the changes of Pt-Fe nanoparticles supported on Vulcan carbon, drop-casted on the electrochemical chips provided by Protochips [94], cycled in the ranges of −0.5 to 1.2 V vs. Pt in 0.1 M HClO$_4$ electrolyte. The authors reported the addition of 0.025 wt% Nafion drop-casted on the nanoparticles inside the chips and air-dried prior to the experiment. The results showed some evidence of the conventional behavior of PGM catalysts during potential cycling. First, the carbon corrosion was directly proportional to the amount of the Pt-contained catalyst materials deposited on the carbon particle. As shown in Figure 25a–f, looking at the region in the red circle (with low Pt loading) and with the red arrow (with high Pt loading) indicated the higher carbon corrosion rate in the regions with higher Pt loading, pointing that Pt can catalyze corrosion of the carbon particles in its vicinity. That finding is in-line with a long-believed effect of Pt on the carbon corrosion rate [139]. Second, the corresponding CV (Figure 25g) showed the decrease of the ECSA of the catalyst materials, presuming that the growth of the particles could be the main reason of the ECSA drop. The authors attributed the coarsening of the nanoparticels to acid leaching of the Fe from Pt-Fe nanocatalysts. Finally, the nucleation and growth of the Pt nanoparticles showed to be random in both space (any appropriate site) and time (any potential cycle) [140].

As previously mentioned, the Pt reference electrode can be one of the sources of error. As a solution to the issue, Nagashima et al. [118] designed and implemented an RHE

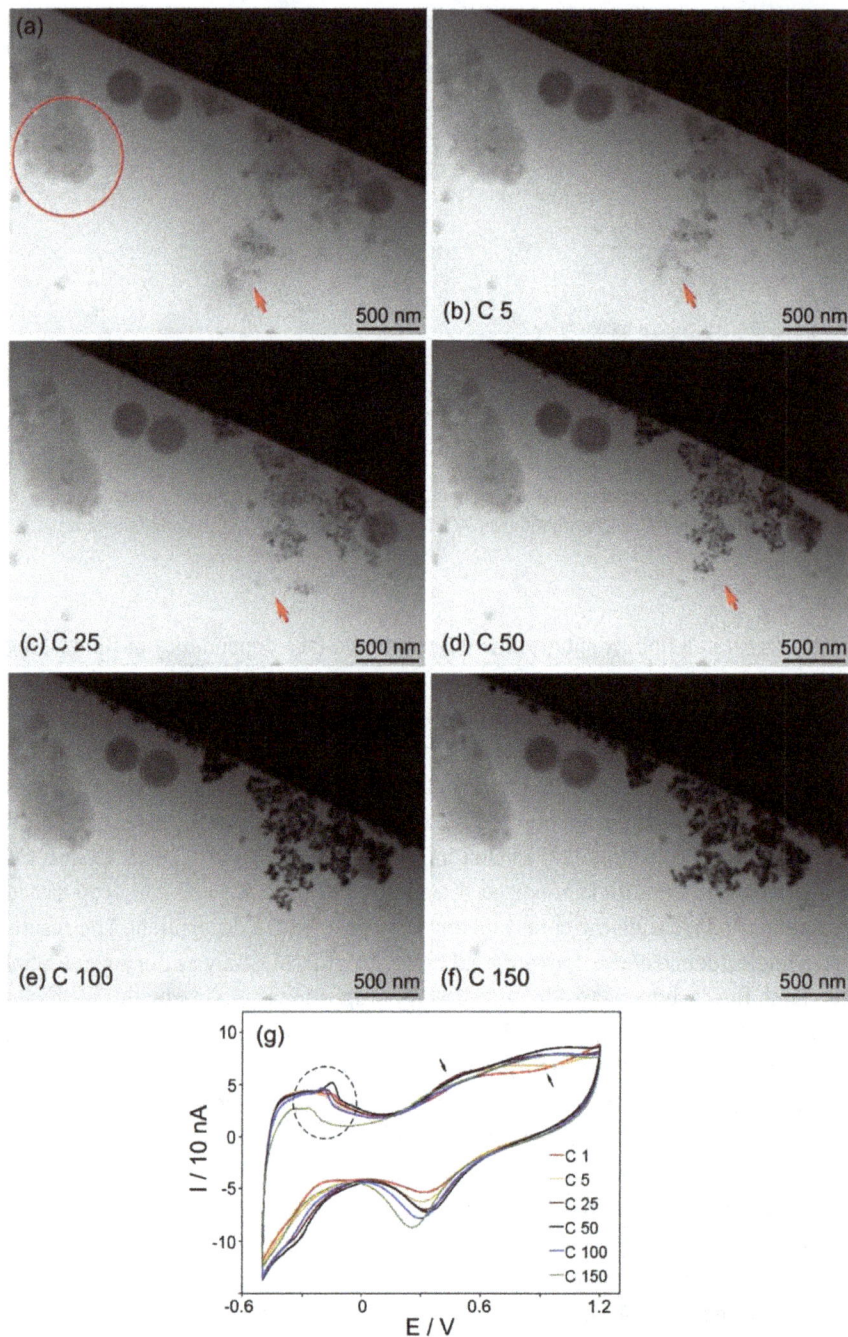

Figure 25: Changes in the Pt-Fe nanoparticles (a)–(f) at 0, 5th, 25th, 50th, 100th and 150th cycles after an initial 130 cycles of stabilization (g) the corresponding CV curves at 1st, 5th, 25th, 50th, 100th and 150th cycles [138]. Reprinted from The Journal of Physical Chemistry C, Copyright (2014), with permission from Elsevier.

reference electrode junction to avoid the instability of Pt reference electrode during EC-LCTEM. The liquid junction enabled an external RHE for the LCTEM holder-in which the electrolyte was fed into the system through a syringe connected to RHE and discharged through the LCTEM holder outlet line. By this means, the LCTEM holder with external RHE cell system stabilized the H_2 partial pressure around the Pt electrode inside the chip (Figure 26a) to minimize the error in the electrochemical measurements. In addition, Nagashima et al. [118] introduced a design to reduce the thickness of the liquid in LCTEM by minimizing the effect of pressure difference inside the chip with the TEM column, which causes the windows to bow. A 25 nm thin film of polycrystalline Pt in a hole array pattern (Figure 26b) was deposited on the SiN_x window and underwent a potential hold protocol presented in Figure 26c as the working electrode in a 0.1 M $HClO_4$ electrolyte solution. The thin film of polycrystalline Pt in a hole array pattern in addition to use of energy filtered TEM (EF-TEM) improved the resolution of the imaging drastically, up to the point of lattice resolution and observing the facets changes during the dissolution and redeposition processes (Figure 26d–h).

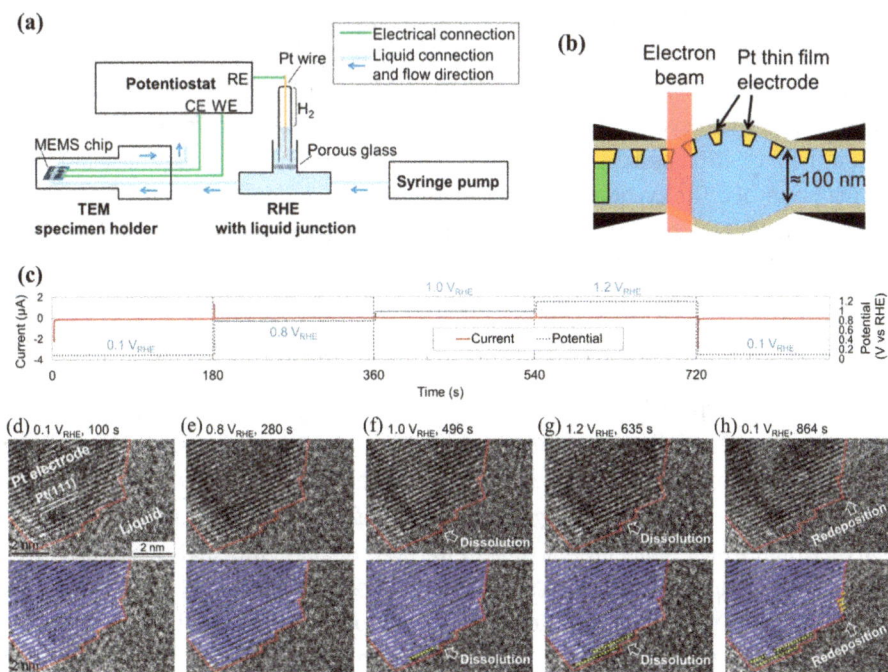

Figure 26: (a) RHE with liquid junction used by Nagashima et al. [118], (b) A schematic cross-sectional representation of the MEMS chips with Pt patterned in a hole array, (c) the potential protocol imposed on the working electrode and (d–h) a series of EFTEM images of Pt surface showing the dissolution and redeposition processes [118]. Reprinted from Nano Letters, Copyright (2019), with permission from ACS Publications.

The in-house developed chips by Nagashima et al. [118] highlighted the importance of the thickness of the liquid in an *in situ* EC-LCTEM experiment. Despite commercially available micro-electromechanical systems (MEMS) chips with at least 500 nm gap between two windows, Nagashima et al. had ~100 nm gap, which made the lattice resolution imaging possible. Also, the RHE with liquid junction improved the sensitivity of the electrochemical measurements and reliability of the interpretations made based of the EC-LCTEM results.

Kato [119] reported the *in situ* LCTEM to study Pt degradation within a carbon-supported Pt catalyst as a valuable analyzing tool. However, it was reported that the technique suffers from both low electrochemical signal-to-noise ratio and imaging resolution. The author used a "Toyota" MEMS modified chip, which had thinner SiN windows and electrolyte to elucidate Pt nanoparticles degradation, dissolution in the electrolyte. In addition to that, an-extended-surface working electrode was utilized in the MEMS chips to increase the amount of the Pt loading, and consequently, improve the electrochemical signal-to-noise ratio in the cyclic voltammetry.

The degradation of octahedrally shaped synthesized Pt-Ni catalyst supported by Vulcan carbon was studied in a 0.1M $HClO_4$ electrolyte using a commercial EC-LCTEM system by Beermann et al. [96, 141]. Different potential cycling protocols were implemented to understand the response of the nanoparticles to the harsh environment present in the PEMFCs. Potential cycling in the range of 0.0 to 1.0 V vs. RHE for 20 cycles to activate the catalyst, a potential cycling in the range of 0.0 to 1.2 V vs. RHE followed by the potential hold at 1.4 V vs. RHE, and a startup/shut down- mimicked cycling were three different testing protocols. The cyclic activation showed the dissolution of Ni in the electrolyte; however, further dissolution of the Ni was impeded by passivation of the surface of the catalyst by the Pt content. This phenomenon was a function of the uniformity of Ni distribution in a different particle (which is subject to change due to the synthesize process of nanoparticles) so that the spongy particles with higher Ni content were more prone to ultimate dissolution and disappearing from the structure (Figure 27).

Cycling the activated nanoparticles in the range of 0.0 to 1.2 V vs. RHE showed a minimal effect on the nanoparticles; some of the Pt-Ni catalysts experienced motions and in some cases, nanoparticles coalesced with close neighboring particles. The authors attributed the phenomenon to the high upper vortex of potential, which provided the carbon corrosion with a good kinetics, causing the movements of the catalyst nanoparticles due to the motion/removal of catalyst support. The hold at 1.4 V vs. RHE showed high mobility, collision of the particles, agglomeration, and consequently, a global catalyst structure changes due to the high rate of carbon corrosion. This study confirmed the effect of potential on the carbon corrosion [139, 142]. The startup/shut down-mimicked potential cycling started with a hold at 0.8 V vs. RHE where no significant structural changes were observed. As expected from the previous experiment, the cycling in the range of 0.0 to 1.2 V vs. RHE was accompanied by the corrosion and motion of the carbon support and particles agglomeration. During the potential cycling step, the redeposition of mobile

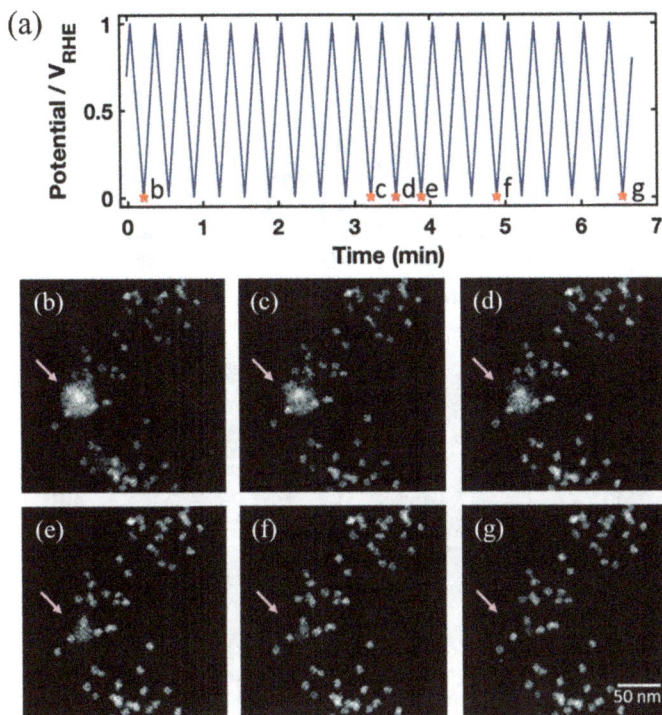

Figure 27: (a) Activation potential cycling of the catalyst, (b–g) corresponding images of the marked points on the potential profile. The marked Ni-rich particles by arrow dissolved over the time under the potential cycling [96]. Reprinted from Energy & Environmental Science, Copyright (2019), with permission from The Royal Society of Chemistry.

Pt-Ni clusters was observed in the system. The mobile clusters showed movements during the hold at 1.2 V vs. RHE. In addition, some localized growth of the particles at this stage was reported. More aggressive cycling up to 1.4 V vs. RHE increased the mobility of both carbon support and Pt-Ni materials. The final potential hold at 1.4 V vs. RHE caused severe carbon corrosion, the particles mobility and Pt-Ni particle growth (Figure 28).

According to the results, carbon showed a high level of sensitivity to the potential holds, although it had negligible corrosion during the potential cycling. As for the catalyst particles, the high potential of the experiment caused the detachment of the catalyst particles from the support, agglomeration and dissolution/redeposition to be the predominant degradation effects. The authors proposed two mechanisms at high potentials [96, 141]: (i) decrease of the particle attachment to the carbon support due to formation of Pt and Ni surface oxides and increase in the mobility, or more likely, (ii) detachment and migration of Pt and Ni particles after losing the physical support of the carbon due to corrosion. Although the report provided the community with valuable results on the degradation and carbon corrosion, it suffers from several drawbacks of the technique. The high potential of the startup/shot down-mimicking experiment was

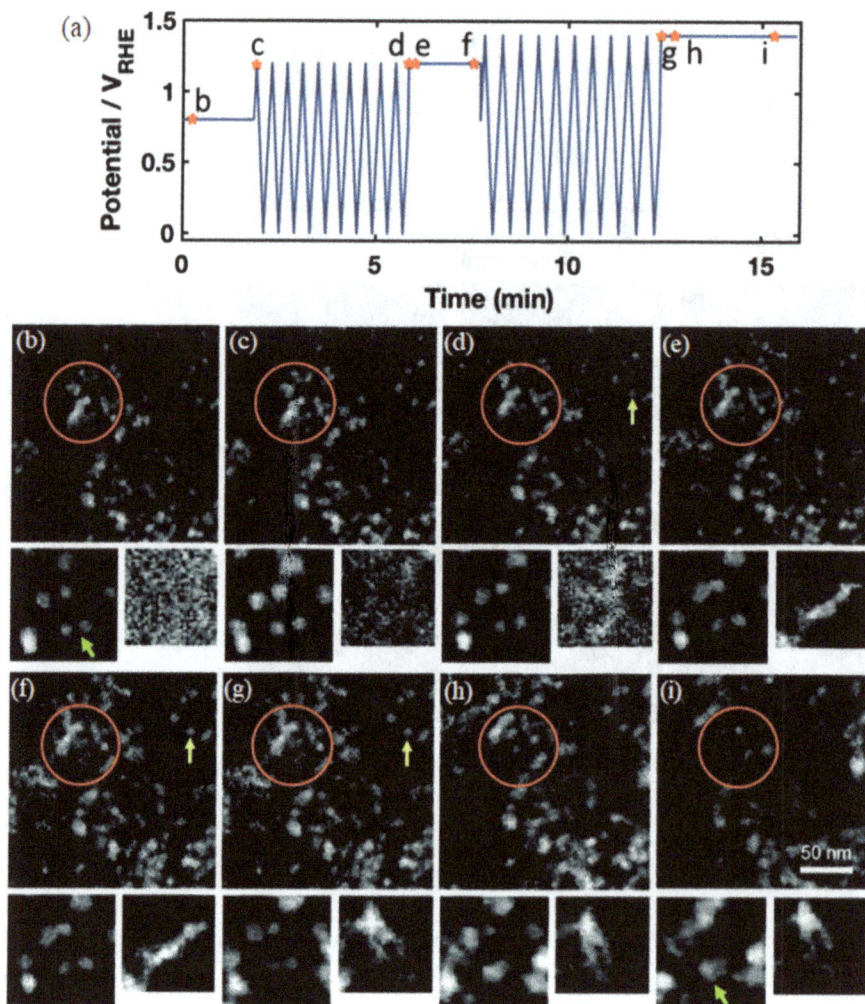

Figure 28: (a) Startup/shut down-mimicking potential cycling of the catalyst, (b–i) corresponding images of the marked points on the potential profile with cutouts on the left to track individual particles and the Pt growth on the right side. Green arrows indicate two Pt-Ni particles coalescing slowly, yellow arrows mark nucleation and growth of stringy Pt-rich deposits and red circles point to abrupt detachment of rede-posited Pt/Ni during hold at elevated potential, 1.2 V vs. RHE [96]. Reprinted from Energy & Environmental Science, Copyright (2019), with permission from Royal Society of Chemistry.

the only parameter comparable to that of SUSD conditions that an actual MEA may ex-perience. The limitation in the number of cycles in that protocol was the other issue of the experiment, limiting the viability of the results to mimic the actual MEA. The other issue stemmed from the limitation of the imaging resolution of the technique, which is too low to discuss the mechanism on the lattice and facet level, and even to perform a reliable particle size distribution (PSD) analysis. For instance, involvement of the dissolution/redeposition processes suggests the possibility of Ostwald ripening as the degradation mechanism; to confirm this, the analysis needs high resolution images to perform PSD analysis and discuss the changes in the shape of PSD histogram. Therefore, the design and development of a system, which can carry out close-to-reality experiments, with higher imaging resolution to understand the involvement of the Ostwald ripening in the real startup/shut down process, is crucial.

Moving forward, Impagnatiello et al. [120] studied the Tanaka catalyst consisting of Pt nanoparticles supported by the Vulcan XC72 carbon in the potential cycles of –1.0 to 0.7 V vs. Pt within an *in situ* LCTEM holder, with a dynamic flow of a 0.1 M HClO$_4$ through the liquid cell. The ink of catalyst containing Nafion perfluorinated ionomer was deposited on the electrode chip of an EC-LCTEM holder. The authors used two approaches to improve the imaging resolution of the experiment. First, the experiment was started by illuminating the liquid electrolyte in the STEM mode (converging beam) to maximize the beam-induced reactions inside the electrolyte and produce bubbles (H$_2$ and O$_2$) from water radiolysis. In fact, the electron scattering by the electrolyte was reduced under the mixed effect of liquid and gas. After the bubble saturation of the liquid, the beam was switched to the parallel mode, TEM, to record the changes in the catalyst materials during potential cycling. The approach can improve the resolution of imaging; however, it causes errors in the electrochemical measurements due to H$_2$ pressure instability and non-equilibrium conditions between the electrolyte and solvated components. Second, energy-EF-TEM, with a zero-loss peak, was used for the imaging to improve the resolutions, which has already proven to be effective [118].

Nanoparticles stabilization and surface cleaning by precycling was employed in these experiments to achieve a stable distribution of Pt particles on the working electrode of the EC-LCTEM system. The authors suggested degradation mechanisms of the Pt catalysts in two different regions. Ostwald ripening is suggested to be a dominant degradation mechanism in Pt/C agglomerates and clusters, caused by Pt concentration gradient between interior regions of the Pt/C agglomerate (lower Pt concentration) and the surface of the particles (higher Pt concentration). As a result of the concentration gradient, the small particles can dissolve during the electrochemical cycling and increase the population of larger particles in the system (Figure 29f–h), indicating Ostwald ripening as the dominant degradation mechanism by widening the PSD histogram. It was suggested that Pt ions can be reduced and aggregate on the bigger Pt particles after diffusing and settling inside the carbon particles.

The second region considered the area far from the initial Pt particles. The authors suggested that after dissolution of the Pt particles, Pt ions could be carried away by either the electrolyte or ionomer and then could precipitate in the electrolyte/ionomer phase. The visual evidence supported the claim about the dissolution/redeposition of Pt particles. (Figure 30e–h). The critical concentration of the Pt ions to overcome the thermodynamics conditions of nucleation and form a stable nucleus found to be around 150–200 cycles of CV (compare Figure 29f and g). After nucleation of a stable nuclei and growth of it in the range of 150 and 250 cycles, coalescence beyond 250 cycles resulted

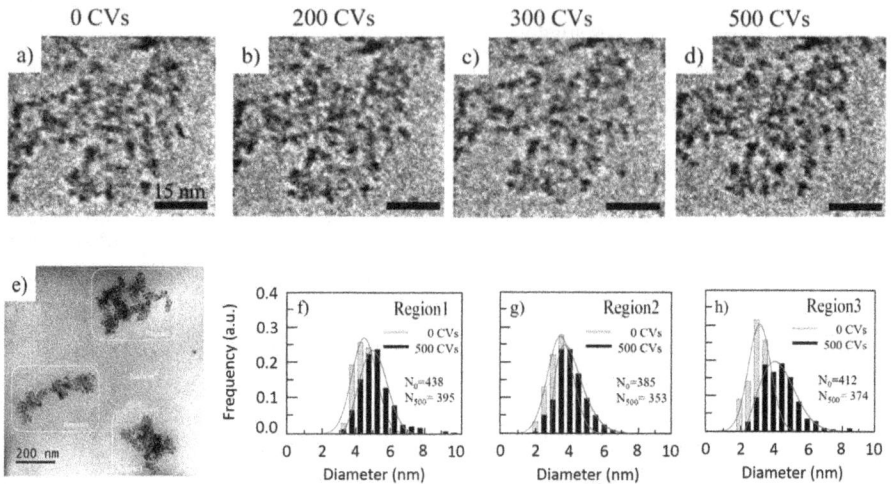

Figure 29: Pt/C catalyst degradation studied by Impagnatiello: (a–d) A Pt/C aggregate changes upon potential cycling in a selected region. (e) three different Pt/C regions in the electrolyte/ionomer phase. (g–h) PSD histogram of Pt particles for three different aggregates shown in (e) at 0 cycle (gray) and 500 cycles (black) of CV [120]. Reprinted from ACS Applied Energy Materials, Copyright (2020), with permission from ACS Publications.

in a slight increase of the average particles size (Figure 30i). Although the predominant mechanisms of degradation the Pt ions were identified through a PSD analysis, the image resolution was not enough to distinguish the boundary of the particles and confirm the reliability of the PSD analysis [120]. Therefore, improvements in imaging signal-to-noise ratio can result in a more reliable and conclusive analysis of the system.

Conclusion and opportunities for future work

The discussions on the presented studies have demonstrated that *in situ* LCTEM and *in situ* EC-LCTEM can be valuable analyzing tools to the PEMFC community to study the materials temporal evolution inside the electrode. However, the current technologies suffer from several drawbacks that need to be addressed prior to prevalent adoption of the technique. With all the advances in the electron transparent materials and fabrication procedures, the imaging resolution remains as one of the crucial issues. The matter needs to be addressed and improved by either reducing the thickness of the liquid or fabricating windows that are more effective, in order to better resolve the Pt particles and their facets. Although the role of the ionomer in carrying the Pt ions has been hypothesized [143], the community has not been presented with tangible evidence of the process. It is believed that some critical parameters including contact between Pt and support, Pt dissolution rate and carbon corrosion may be affected by type of the ionomer, dispersion of the ionomer in the structure of catalyst, ionomer to carbon weigh ratio and the thickness of the ionomer on the nanoparticles. However, it is challenging

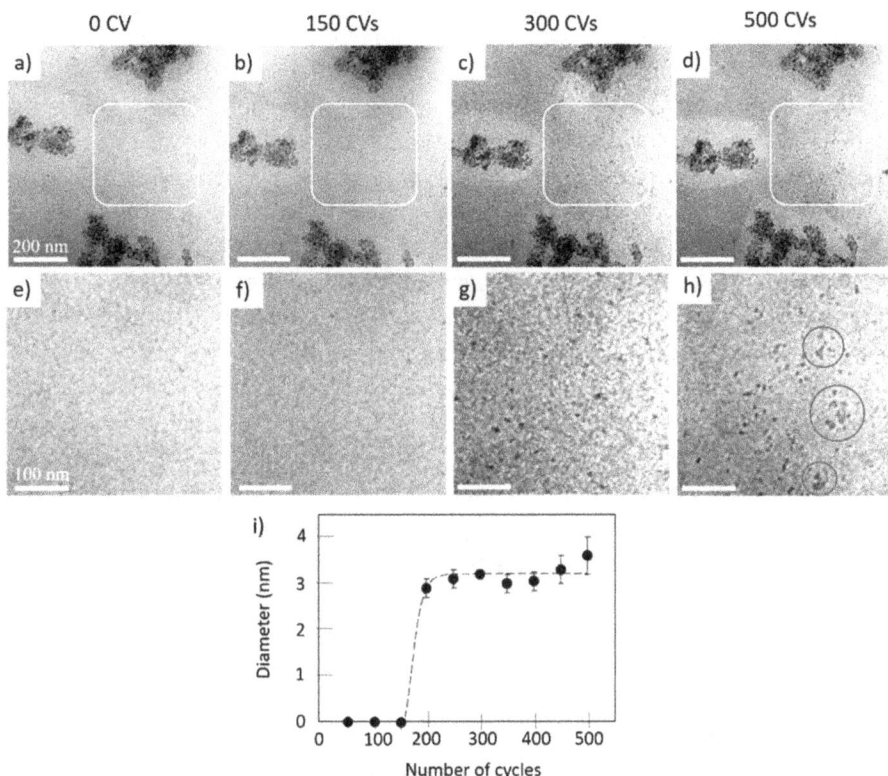

Figure 30: Pt/C catalyst degradation studied by Impagnatiello (a–d) Pt clusters formation in electrolyte/ionomer phase. (e–h) blown-up view of the Pt clusters in the electrolyte/ionomer phase. (i) Pt particles average size changes with the number of the cycles [120]. Reprinted from ACS Applied Energy Materials, Copyright (2020), with permission from ACS Publications.

to achieve the visual evidence of ionomer role, a low-density organic compound sensitive to the electron beam, on the mentioned critical parameters. Although the current *in situ* LCTEM and *in situ* EC-LCTEM have challenges in imaging the ionomer and illustrate the role it plays in *the* structure, perhaps utilization of ultrafast imaging approaches [144, 145] and/or low dose imaging could help researchers in this matter. The main degradation mechanisms of the Pt have been, for long, recognized as dissolution/precipitation, Ostwald ripening or species migration and agglomeration. However, understanding the share of each of those mechanisms in the overall degradation process and the conditions to trigger each of those mechanisms need a temporal observation to understand the pathways of the Pt redistribution/movement under the dominant mechanisms. The degradation mechanism and effect of enhanced activity and stability of the Pt alloy catalysts (e. g., using transition metals [146] specifically Co [147–149]) on the overall system performance needs to be perceived in detail, too. Carbon corrosion and the structural changes of the carbon material during the cycling is the other daunting issue of PEMFC systems. The effect of the crystallinity degree on the corrosion and stability of carbon [150] should be elaborated on by a temporal observation. However, the sensitivity of carbon to the working conditions and its imaging difficulties requires a fine-tuning of the experimental parameters of an *in situ* LCTEM experiment to achieve the best imaging resolution. Finally, understanding the structure-properties-performance relationship, as the ultimate goal of the characterization techniques, requires the perception about the effect of microstructural properties such as porosity, tortuosity, ionomer coverage, Pt distribution, agglomeration of catalysts and C in the actual electrodes (not only on the dispersed catalyst particles) on the overall performance. In addition, the liquid environment should be avoided and replaced by a gaseous one, still in the electrochemical cell, to provide a more realistic environment for PEMFC operation. This approach would, at the same time, improve the resolution. The current *in situ* LCTEM technologies cannot provide the community with the current demands on hand, and design of novel *in situ* holders are a must.

It can be concluded that the promise of the current *in situ* LCTEM technique, for a complex system such as PEMFCs, should be considered as a starting point of designing a novel, and more system-specific *in situ* LCTEM platforms, which can study degradation and corrosion processes in a real PEMFC devices during operation. To be more specific, the need for an *in situ* TEM system that makes operation under H_2/O_2 conditions possible is clearly perceptible. Finally, correlating the findings of *in situ* TEM experiments and those observed in the relevant ex-situ MEA tests is of a high importance, which can be achieved by the aid of a more close-to-reality system.

2.7 Image processing in PEMFC characterization

The last decade has seen a considerable increase in studies focusing on ways/methods to extract information in terms of structural and compositional descriptors from characterization images/data through additional processing of the raw characterization data. This series of operations/steps, collectively termed image processing, results in a set of parameters useful in the quantitative description of the system. Image processing in general can be defined as the procedure of processing the images/data by using various techniques, which can serve to improve the overall quality of the image and enable the extraction of data for easier interpretation of the image. Mostly, the input images used in fuel cell applications are two-dimensional (2D) or in some cases three-dimensional (3D) digital images acquired directly from advanced microscopy and characterization techniques frequently used by materials scientists and engineers. Every digital image, whether 2D or 3D, consists of an assortment of many smaller areas/squares termed pixels, which are either filled with a single gray level or digitally defined color [151, 152]. An image may contain thousands or even millions of these pixels so fittingly placed against one another that they give the impression of a smooth realistic image [153]. The greater the number of pixels, the clearer and higher the definition of the image would be [154]. The pixels make up the basics of digital image processing [155]. The gray images, as acquired by several microscopy techniques, commonly have 256 gray levels ascribed to their pixels [156]. According to this notation, the value 0 is assigned to the black color while the value 255 is assigned to the white color [157]. In general, each pixel of a digital image could be translated into different numbers from 0–255 according to the color they carry as demonstrated in Figure 31. Colored images such as gradient, thickness or elemental maps are usually stored in the RGB format, which simply refers to red, green and blue. In a digital RGB image, each image is subdivided into red, green and blue color planes, each having its own 0 to 255 pixel values range, and any specified color in the image is represented by the combination of the amount of each of these three primary colors and their corresponding pixel values [158, 159].

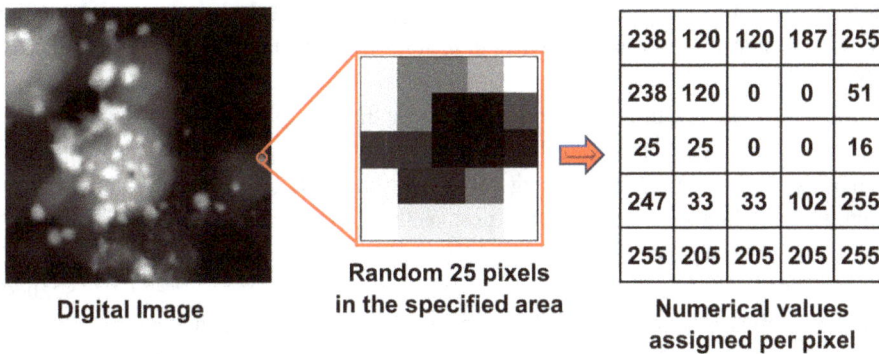

Digital Image **Random 25 pixels in the specified area** **Numerical values assigned per pixel**

238	120	120	187	255
238	120	0	0	51
25	25	0	0	16
247	33	33	102	255
255	205	205	205	255

Figure 31: A general example of how a digital image is interpreted by a computer.

Types of image processing

Image processing covers a broad range of techniques; it can be as simple as identifying particles/inclusions in an image, can include image filtering and sharpening operations, all the way to complex approaches of distinguishing between agglomerates and particles and effectively measuring their size based on morphological data alone. Image analysis and processing can be subdivided into two main types, i. e., manual and automatic. A third type, combining the two basic types, has also been used and can be termed the semiautomatic approach.

Manual approach

The manual approach is based on the concept that visual perception of certain features in an image, especially when carried out by an experienced operator, is still the most reliable and finest tool of image analysis. However, in most cases, it is also subject to bias in the operator's understanding and judgment regarding the presence of a microstructural feature [160, 161]. This approach is commonly used for size measurements carried out on images obtained from microscopy techniques, typically scanning electron microscopy (SEM) and transmission electron microscopy (TEM) [162, 163]. The TEM micrograph is preferred, e. g., catalyst nanoparticles because of its exceptional capability of aptly picturing high-resolution individual particles at nanometric scales [164]. The process involves manually drawing a linear intercept across the diameter of individual particles, one at a time preferably in a similar direction, and measuring the length of that intercept. The final average particle size is then acquired by averaging the total sum of the measurements by the total number of particles measured. This is a standard procedure of particle size measurement by the American Society for Testing and Materials (ASTM) and usually, only a subset of particles in the micrograph is considered using this method [165]. It has been widely used by researchers for particle size or pore size measurements for a wide range of materials, and hence also quite frequently used for the study of materials for different fuel cells, e. g., by Holby et al. to find particle size distribution of platinum (Pt) nanoparticle in proton exchange membrane (PEM) fuel cells [166].

Nevertheless, it is important to mention here that the manual approach is affected by certain other factors as well such as quality, magnification power and type of image as well as user's preferences in carrying out measurements [12, 167]. A high-quality image magnified at a higher resolution, with proper illumination can help easily identify distinct boundaries of particles making it easier to perform the measurements. On contrary, low-quality images not taken under favorable conditions, images of irregularly shaped particles or those with excessive agglomeration may be very difficult to analyze. A comparison among two different types of TEM micrographs is given in Figure 32. Besides, sample preparation and image acquisition through TEM is itself a complex procedure that requires time and experience [12]. Also, since micrographs may contain a wide variety of particles, and while the procedure involves the operator selectively measuring a small subset of them to find the average value, the preference of the operator to measure more small or large particles for the analysis can skew the measurements toward either ends. As a consequence, different operators might get different values. For the convenience, the particle size measurements are generally carried out by loading the image under consideration in the software, most commonly FIJI ImageJ, performing the above-described manual measurement, while the software automatically records all the measurements and offers a copy of data for quick calculation of average size [168, 169]. For this procedure, the operator needs to first set the scale for the measurements and although it is a straightforward procedure, a slight miscalculation might seriously impact the final measurements. As evident from the described manual procedure, though simple and cost-effective, single-handedly measuring individual particles can be a very long, tiresome and generally inefficient way that requires the practice of caution and experience on the part of the operator for ensuring precision and repeatability/reproducibility of the measurements [170–172]. Besides catalyst particle measurement, man-

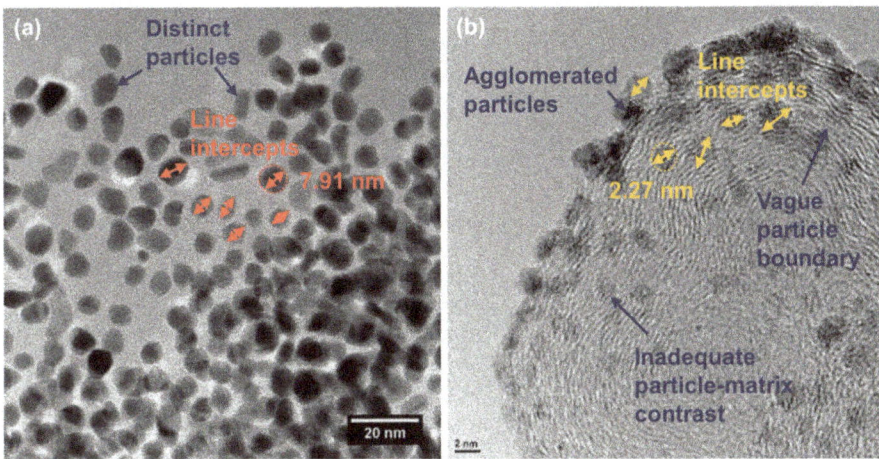

Figure 32: Comparison of two different types of TEM micrographs for manual particle size measurements of Pt nanoparticles (a) Clearer image, precise calculations possible. Reproduced with permission [48]. Reprinted from Nat. Commun., Copyright 2020, with permission from Elsevier. (b) Vague image, calculations subject to fluctuations [177]. Reproduced from Int. J. Hydrogen Energy, Copyright 2021, with permission from Elsevier.

ual measurements are typically used for determining the thickness of membrane electrode assembly (MEA) layers (cathode, membrane, anode, etc.), catalyst layer (CL), crack length, gas diffusion layer (GDL) fiber diameters, etc. using SEM images [173–176].

Automatic approach

Computers allow implementing a prespecified set of image analysis instructions at high processing speeds, significantly reducing the time taken for manual analysis. Since we acquire digital images from materials characterization techniques used for any type of fuel cells, the computer translates them not as separate images but rather as an array or matrix of numbers hence making it easier for applying mathematical functions to enhance such images. A general flowchart of steps that the computer follows for automatic image analysis is shown in Figure 33 [178].

In the first step, the image is acquired from one of the characterization techniques and saved on the computer. The automatic approach to image analysis allows for use of images from a wide variety of different sources including but not limited to SEM, TEM, STEM, micro-CT, nano-CT, magnetic resonance spectroscopy (MRS), atomic force microscopy (AFM) and EDS [179–186]. Computers also allow the application of numerous automatic functions or operations on the images such as normalization, thresholding, binarization, segmentation, and brightness and contrast adjustment to either improve the quality of the image or acquire useful information from it [187]. Usually, these functions are available nowadays as a part of software packages, most of which are available free or can be downloaded by purchasing a license, e. g., ImageJ, MATLAB, Avizo, AVS,

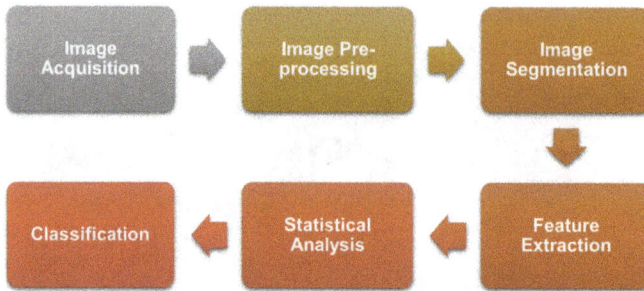

Figure 33: General steps for executing automatic image analysis method.

PAX-it, SPIDER, etc. [188–192]. Several libraries and toolkits in programming languages such as OpenCV, Sci-kit-image, Pillow in Python, Magick, EBImage, OpenImageR in R, CImg, Simd in C++ and AWT, BoofCV, AlgART in Java, also allow operators to perform basic image processing operations [193–196]. Once the acquired image is loaded into the computer, it is pre-processed. Preprocessing performs the preliminary operations for making the image more suitable for advanced processing. The steps may involve slight contrast and illumination adjustments and/or blurring or sharpening of the image, a few of which are explained here. Digital images are prone to noise, which can be defined shortly as a random signal that may falsify the information acquired from an image [197]. Increased brightness or contrast or other unanticipated conditions during image acquisition such as fluctuations in heat, illumination, sensors, etc. may lead to the inclusion of noise in the image resulting in unwanted alteration or damage to the image data and can be avoided if a suitable balance of brightness and contrast adjustments is achieved [198]. This step, in image preprocessing, is termed normalization [199]. Normalization, in definition, refers to the set of operations used to modify the range of the intensity values of pixels in the image under observation to a range easily understandable by the computer, i. e., the grayscale range for grayscale images or the RGB range for colored images [200]. The normalization of all images is carried out using the following formula:

$$\text{Output}_{\text{Channel}} = \frac{\text{Input}_{\text{channel}} - \text{Min}_{\text{value}}}{\text{Max}_{\text{value}} - \text{Min}_{\text{value}}} \times 255,$$

where $\text{Min}_{\text{value}}$ and $\text{Max}_{\text{value}}$ refer to the minimum and maximum intensity values of the input channel, respectively. For grayscale images, normalization is done for one channel only while normalization is performed for three separate RGB channels if the image is colored [201]. Other initial transformations, although less frequently used in image processing, include gamma modulation, histogram equalization, filtering, etc. [202]. Gamma modulation or gamma correction makes the image more sensitive toward the darker intensity values range in comparison with the brighter intensity values range making it more suitable for human's visual perception sensitivity region [203]. Histogram equalization, on the other hand, generates a high contrast image while retaining all the details of the initial image making it more suitable for advanced image processing or automatic thresholding [204]. Filtering is different from other transformations as it depends on the neighboring pixel values of all pixels and can be of many different types such as smoothing, sharpening, Gaussian, median, Fourier, etc. Smoothing, Gaussian and median filters are very effective in reducing the image noise and applying mathematical procedures, i. e., arithmetic mean, median calculation weighted average or Fourier transformations, etc. to achieve it [205]. A sharpening filter is rarely used in image processing as it adds unwanted noise to the image, but it is at times used to enhance the appearance of certain features in an image, e. g., as used by Y. Suzue. et al. for distinguishing the different phases in nickel-yttria stabilized zirconia (Ni-YSZ) based anode in SOFCs [206]. The next step in automatic image analysis is the binarization of the image. Binarization refers to the conversion of any image (grayscale or colored) into a two-toned black and white image, assigning a pixel value of either 0 or 255 to all the pixels in the image [207]. Usually, binarization is accompanied by thresholding where the division of the image into its binary mask is defined by a selected thresholding value [208]. The pixels above the specified threshold value are assigned 255 or "white color" while those below are assigned 0 or "black color." Inversion of the binary image, i. e., conversion of the black region into white and vice versa can also be done depending on the features of interest [209]. Binarization can be performed according to a user-defined thresholding value, or an automatic threshold value computed by clustering, spatial or entropy-based methods [210]. The most common method is the Otsu method, which divides the image into foreground and background pixels and then calculates the optimum threshold value differentiating the two classes of pixels minimizing the intraclass and/or maximizing the interclass intensity variance [211]. Binarization of the image is done to extract or segment the features of interest from an image and is part of the image segmentation step. There are many image segmentation techniques available that offer region-based segmentation such as watershed segmentation or boundary-based segmentation such as edge-based segmentation as per the requirements of the analysis [212, 213]. Segmentation is particularly important when we want to quantify a specific group of features on a microscopic image, e. g., the presence of electrocatalyst particles in the catalyst layer of a fuel cell whose size distribution can help reveal a lot about the processes and reactions occurring inside the fuel cell during operation as observed by a study carried out by Guilminot et al. [214]. Once the feature of interest is segmented from the image, statistical calculations of the size, area, number or distributions of the feature as per requirements is done simply using mathematical operations [215]. The statistical analysis is then followed by classification of the image or features based on either a well-established database or decisive algorithms giving the user a complete data set of information. Hence, automated image processing is often preferred over the manual approach due to its quick and time-saving image analysis.

Advanced approaches

Implementation of advanced artificial intelligence algorithms for image processing steps explained above has revolutionized the entire image analysis procedure lately, greatly reducing user dependency, and cutting back on analysis time, while providing more reliable and comprehensive result data sets [216, 217]. Various machine learning and deep learning approaches to artificial intelligence continue to gain the attention of researchers in fuel cells as they did for nearly every field of study [218, 219]. Deep learning is basically a subset of machine learning, which collectively falls under the domain of artificial intelligence. The main difference between artificial intelligence, machine learning and deep learning is shown in Figure 34 [220]. The current machine learning and deep learning models draw their core inspiration from the human perception model since it is believed to possess an excellent image analysis mechanism [221]. Machine learning by definition is the set of steps or models developed to perform a specific task by continuously making use of available data for optimization of its algorithms [222]. The classification of machine learning into its groups and subgroups is illustrated in Figure 35. As evident in Figure 35, machine learning algorithms are subdivided into three classes namely: supervised learning, unsupervised learning, reinforcement learning and deep learning [223].

Supervised learning refers to the use of labeled training sets for teaching the model to generate the required output [224]. The training sets include both labeled inputs as well as accurate outputs allowing the model to measure accuracy and learn over time until optimal error minimization is achieved [225]. The two subgroups of supervised learning, i. e., classification and regression, differ from each other based on the type of dependent attribute for result prediction [226]. For classification, the dependent attribute is categorical and refers to classifying the test data into certain categories and attempting to label or define them [227]. While, for regression, the dependent attribute is numerical and implies recognizing the interrelation between variables to put up projections [228]. Common classification and regression algorithms frequently employed in image processing include support vector machines (SVM), k-nearest neighbors (KNN) algorithm, decision tree, random forest, linear regression, polynomial regression, etc. [229]. SVM has been widely used for the classification of patterns and images by generating an optimal hyperplane by iterations for error minimization [230]. The KNN algorithm has also been used for categorizing images into respective classes by computing the distance between the feature vector of the input image to the training image [231]. Decision trees consist of a continuous framework of data splitting according to specified parameters and can be used to classify different features in an image such as the classification of different types of nanoparticle aggregates as attained by Martinez et al. [232, 233]. A random forest algorithm can be used for the same purpose and defined as a collection of unrelated decision trees merged for a more precise feature prediction [234]. Image classification on the base of image texture is possible via linear regression models predicting outcomes based on one dependent and one or more independent variables

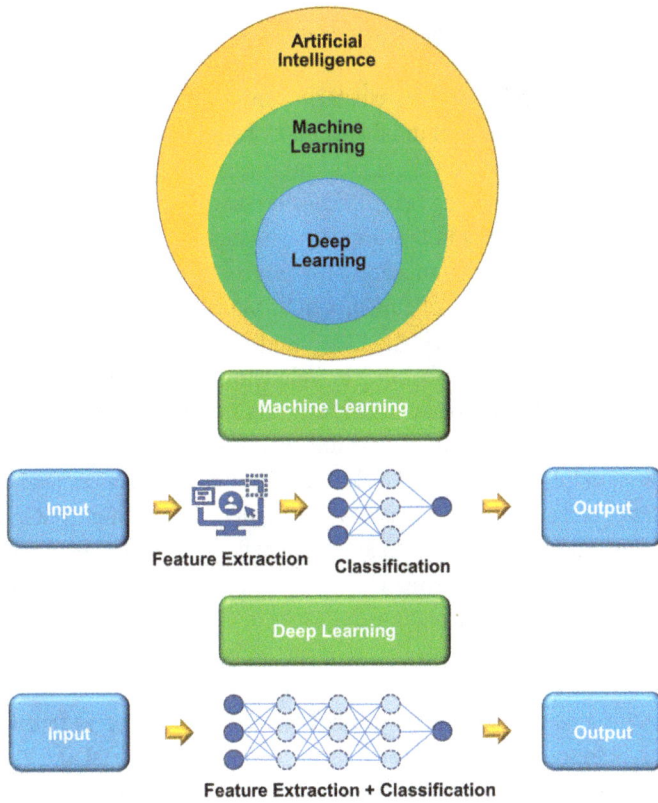

Figure 34: Relationship and classification of artificial intelligence, machine learning and deep learning.

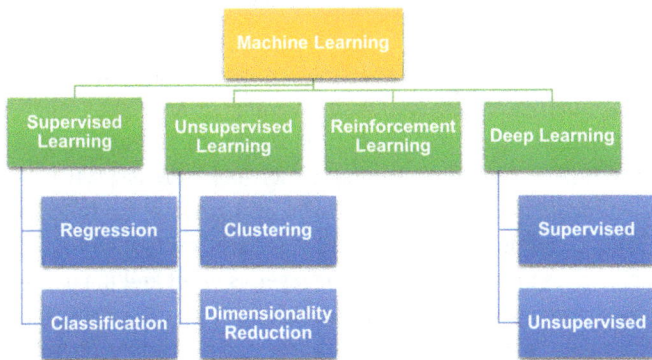

Figure 35: Classification of machine learning in general.

[235]. Polynomial regression models offer advancement to the linear regression models by modeling dependent and independent variables in nth-degree polynomials and proving very effective in solving image segmentation problems [236].

Contrary to supervised learning, unsupervised learning algorithms deal with unlabeled data and analyze it without the need for human intervention [237]. Unsupervised machine learning models can be subdivided into clustering and dimensionality reduction [238]. Clustering refers to the grouping of unlabeled data based on either similarity or dissimilarity [239]. A few important clustering algorithms include exclusive, overlapping, probabilistic and hierarchical clustering [240]. Exclusive clustering includes dividing data into exclusive groups, commonly in K groups by k-means clustering where K refers to the number of clusters based on distance from the centroid of each group while overlapping clustering refers to the division of data into multiple and overlapping clusters or groups [241, 242]. Probabilistic clustering, on the other hand, refers to the division or clustering of data based on their probability of belonging to a specific class or distribution [243]. Hierarchical clustering group data clusters either through agglomerative clustering involving merging data via iterations into a cluster based on similarity or divisive clustering involving dividing data into clusters based on differences [244]. All of these different types of clustering algorithms have been successfully employed by many researchers to cluster features of interest in an image and carry out the segmentation process [245–247]. Another form of unsupervised learning is the dimensionality reduction algorithm, which refers to the extraction and usage of data points or features of interest only to avoid complications, caused due to larger data sets and this type of algorithm is very effective in carrying out the preprocessing.

Another form of machine learning is reinforcement learning which refers to the attainment of the best possible solution by the model through an extensive trial and error method without the use of any labeled input or output dataset [248]. Reinforcement learning has been used in image processing to carry out image preprocessing, thresholding, and segmentation [249–251].

The most advanced subset of machine learning is deep learning, which consists of artificial neural networks (ANNs) with three or more layers, imitating the intricacy of the human brain. This approach possesses the ability to effectively recognize the process, predict data patterns and reach logical decisions without using structured or labeled data [252, 253]. There are different types of ANNs, but the most common ones include convolution neural networks (CNNs) and recurrent neural networks (RNNs) [254]. CNNs consist of multiple layers of artificial neurons in the form of mathematical functions which employ multiple inputs, compute their sum weightage and generate an activation value as output, which passes through the following layer as input and the process continues through all subsequent layers to generate a decisive output [255]. CNNs have been widely used in image analysis to perform a wide variety of operations including image detection, recognition, segmentation, etc. [256, 257]. RNNs on the other hand, also employ multiple artificial neuron layers but differ from CNNs in the way that they can reutilize the activation outputs from any neural network layer successively, and thus, can interpret sequential information to optimize the output [258]. RNNs in image processing have been utilized for solving more complex image analysis problems including image reconstruction, hyperspectral image classification, semantic segmentation, etc.

[259–261]. For instance, CNNs and RNNs-based image processing in fuel cells has been utilized for screening catalyst inks and investigating stack voltage degradation in PEM-FCs [262, 263]. Keeping in view such interesting prospects and the level of research being carried out involving machine and deep learning approaches, it can be aptly assessed that many more fascinating applications of artificial intelligence in fuel cells have yet to stem from these techniques in the near future.

Applications in fuel cells

Image processing using the aforementioned approaches has been implemented for wide-ranging applications in fuel cells research as demonstrated in Figure 36. It can be aptly judged that the manual approach of image processing is almost fully evolved, but there is a lot of room for improvement in the case of the automatic approach, with significant emphasis being placed on the advanced automation of image processing through machine learning and deep learning models and techniques. This section covers how different image processing techniques have managed to assist current researchers in investigating and resolving some key issues and challenges faced in engineering, manufacturing, operation and diagnostics of different types of fuel cells, mainly focusing on the two most common fuel cell technologies, i. e., proton exchange membrane fuel cells and solid oxide fuel cells, but applicable to other types of fuel cells.

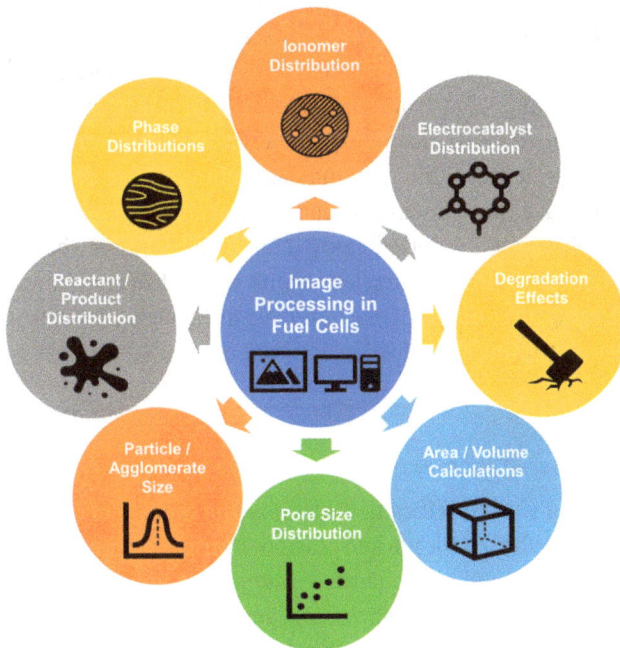

Figure 36: Applications of image processing in fuel cells.

Water distribution in channels or reactant/product distribution

Effective water management is vital to ensure optimal performance and long-term stability of a PEMFC. Inadequate water content is linked with insufficient proton conductivity while its surplus is related to flooding and impediment of reactant transport [264]. Sergio et al. reported using the image processing method for visualizing water distribution in the gas channels of the PEM fuel cell. The study introduced a water coverage ratio parameter that could differentiate between channel water and condensate in contrast with the previously applied algorithm offering a more accurate estimation of channel water distribution [265]. White et al. implemented different Matlab and Python-based 3D image processing techniques of thresholding, filtering and grayscale pixel-based segmentation for combined operando visualization of transient water distribution and catalyst layer degradation in GDL of PEMFC as acquired by micro-CT [266]. Satija et al. and Trabold et al. employed image masking and normalizing procedures on a collection of radiographs from neutron imaging to demonstrate the functioning of the water management system and to carefully identify, locate and calculate the water content at different locations in a PEMFC [267, 268]. A similar study by Chen et al. also used masking techniques for detecting and measuring water distribution from neutron radiographs at four different channel and rib combinations of anode and cathode of a PEM fuel cell [269]. Another study by Lee et al. reported a simpler method using a high spatial resolution X-ray CCD camera to investigate and quantify water distribution at different locations in PEMFC by applying a median filter and image segmentation based on grayscale values of image pixels [270]. Lister et al. made use of image processing of fluorescence micrographs through correlation of light intensity to the relative height of water surface to produce 3-D projections of water surface, surface velocities and interconnectivity of its flow paths inside GDL of a PEMFC [271]. Water agglomeration in a GDL sample of PEMFC was identified and quantified by 3D reconstruction of X-ray synchrotron radiographs after normalization, phase retrieval filtering and data subtraction of dry from wet GDL samples, further processed through binarization via global thresholding and a set of various filtering techniques in a different study [272]. A comparative characterization of two-phase flow pressure drop in correlation with water distribution was carried out by Chauhan et al. using three different machine learning models, i. e., logistic regression, SVM and ANN demonstrating classification accuracy of 63 %, 80 % and 95 %, respectively [273].

Ionomer distribution

Ionomer distribution is associated with the efficient dispersion and binding of catalyst particles, as well as with offering pathways for ion and reactant transport in the catalyst layer of PEMFC [274]. Suzuki et al. recently introduced an interesting approach of using image processing techniques of smoothing, binarization, and segmentation through p-tile thresholding, which determines threshold value based on the cumulative sum of pixel intensities closest to percentile to demonstrate the ionomer distribution and Pt/C in the catalyst layer in 3D image data from two-stage focused ion beam/scanning ion microscopy (FIB/SIM) of a PEMFC. The proposed method also made use of the BoneJ plugin as a part of ImageJ software to further quantify agglomerate sizes and secondary pores of Pt/C [275, 276]. In a different study, a series of sequential TEM images taken at tilt angles from −60 to 60 degrees were reconstructed in 3D space by template matching and modified simultaneous iterative reconstruction image processing techniques by Uchida et al. to investigate ionomer distribution over Pt nanoparticles dispersed on carbon black agglomerates [41]. A novel method of imaging ionomer distribution and segregating Pt particles in Cs-stained electrode of a PEMFC was also presented by Normile et al. using synchrotron transmission X-ray microscopy (STXM) whereby image processing steps of phase-contrast 3D reconstruction, median filtration, manual and default thresholding yielded the morphology, tortuosity factor and volume fraction of ionomer while the size distributions of ionomer and Pt were computed using BoneJ plugin of ImageJ [277].

Electrocatalyst distribution
Homogeneous catalyst distributions in fuel cells have been associated with improved mass activity during the ORR and an increase in the active catalytic area linked to an increase in overall fuel cell performance and catalyst stability [278]. Artyushkova et al. introduced an image processing program written in Matlab for investigating the distribution of Pt electrocatalyst particles from SEM micrographs. Using surface roughness based on image intensity histogram, the authors located pores, while texture analysis based on varying contrast levels was used to investigate morphology and uniformity of electrocatalyst distribution in PEMFCs [279]. A study by Serov et al. observed platinum group metal-free (PGM-free) Fe-N-C electrocatalyst distribution in a PEMFC by phase retrieval and tomographic reconstructions of micro-CT images of the catalyst layer by TomoPy. This was followed by image segmentation to visualize electrocatalyst particle distribution by manual and Otsu thresholding in ImageJ and Avizo. The study revealed an inhomogeneous distribution with smaller catalyst aggregates near the interface of the MPL and larger aggregates toward the opposite end of the catalyst layer, near the membrane caused by drying and solvent evaporation [280]. Khedekar et al. used synchrotron micro X-ray diffraction (Micro-XRD) images processed using XMAS software to examine the distribution of Pt catalyst particles at the inlet, middle and outlet of MEA in a PEMFC under dry as well as wet conditions. The study concluded a more homogeneous distribution of Pt under land and flow field channels at the inlet and middle MEA region, but an inhomogeneous distribution under the same at the outlet [281]. Pt dispersion and distribution upon reduced graphene oxide (RGO) support was studied by Pushkareva et al. using ImageJ on TEM micrographs revealing ethylene glycol (EG) as a good dispersing agent while RGO as useful carbon support for promoting Pt utilization [282]. Chao et al. used image processing tools from ImageJ and Matlab on TEM micrographs of electrode/electrolyte interface of SOFCs to acquire Pt coverage and boundary density for Pt fabricated from combined atomic-layer deposition (ALD) and sputtering to

reduce current-collecting resistance and improve triple-phase boundaries (TPB). The results indicated the uniform distribution of Pt particles on the substrate with an increase in coverage with the increase in ALD cycles along with a TPB density up to 8 times better than that achieved from sputtering only [283].

Pore size distributions

The study of pore size or porosity distribution is vital for a wide range of fuel cells, particularly PEMFCs and SOFCs, where an optimum porosity level in electrodes, MPL and GDL is required for ensuring good electronic and ionic conductivity, better catalyst particle distribution, good water and reactant management and transport, and preservation of mechanical integrity of the fuel cell [284]. Wejrzanowski et al. performed quantitative image analysis on micro-CT images of the cathode of an molten carbonate fuel cell (MCFC) to compute the pore size distributions using a commercial software from Zeiss for 3D reconstruction and Skyscan CTan for computation of porosity percentage, average pore size and its distribution, giving valuable insight to reactant and product transport as well as reaction area in the fuel cell [285]. Thiele et al. utilized different tools of image processing in ImageJ, Matlab and Avizo for geometric registration, intensity distribution-based thresholding and segmentation of pore and solid structure and 3D visualization of the segmented areas of the cathode catalyst layer of a PEMFC imaged by FIB/SEM. The study concluded that the pore diameters lie in the range of 5 to 205 nm with 95 % of the pore volume being contributed by pores with diameters between 15 to 120 nm and pointed out the flow mechanisms governing the gas and liquid transport in the fuel cell [286]. Santoro et al. analyzed 6 different air-breathing gas-diffusion cathodes of a microbial fuel cell (MFC) prepared under different pressures and temperatures using image processing techniques of low and high pass Gaussian filters, background intensity surface deviation and grey-level based correlations of neighboring pixels to acquire morphological parameters of porosity and surface roughness using a custom-coded program in Matlab [287]. James et al. measured the porosity profiles for channel, land and overall region of transport layers in a PEMFC under 0 %, 20 % and 40 % nonhomogeneous compression to investigate geometric and transport properties using image processing tools of contrast adjustments, binarization, Otsu thresholding, area morphology opening and 3D reconstruction by Xradia, Matlab, Fiji and MicroFOAM on the image data from micro-CT [288]. Bercero et al. utilized Gatan Digital Micrograph software to perform image analysis on SEM micrographs of Ni-YSZ-based anode and LSM-YSZ-based cathode of a solid oxide regenerative fuel cell (RFC) to compute average pore size and pore size distributions and corroborated the result achieved with Hg porosimetry [289].

Particle/agglomerates size analysis

Particle or agglomerate size is directly associated with the electrochemical properties of fuel cells [290–292]. For instance, in PEMFCs, particle/agglomerate size analysis of Pt nanoparticles gives a clearer understanding of the available electrochemical surface area (ECSA) and affects the long-term stability of the catalyst layer and electrochemical performance of the fuel cell. In SOFCs, the particle size of ceramic-based electrodes affects the final pore structure influencing the reactant distribution [293, 294]. Particle size analysis, especially that of the catalysts is also a critical parameter of study for other fuel cell technologies such as anion exchange membrane fuel cells (AEMFCs), MCFCs, direct methanol fuel cells (DMFCs), regenerative fuel cells (RFCs), etc., and is directly linked with the respective fuel cell's performance [295–298]. Lee et al. used the manual image processing method of line intercept included in Image-Pro for calculating particle sizes of Ni, YSZ and pores from SEM micrographs of the anode of a SOFC. The analysis revealed an increase in average Ni and YSZ particle size with heat treatment while also indicating that the microstructural evolution of the anode composite structure, in general, is greatly influenced by the Ni coarsening phenomena [299]. The same method of manual line intercept of over 300 particles was used by Zhou et al. for investigating the particle size distribution of Pt over carbon in cathode CL of a DMFC using TEM micrographs. The results were later confirmed by the CO chemisorption method also indicating that the particle sizes and distributions were greatly influenced by the amount of water in the reaction system [300]. Yu et al. used images from high angle annular dark-field (HAADF) imaging mode of STEM to extract the particle sizes of Pt in the catalyst layer of PEM fuel cell via operations of thresholding, Gaussian blurring, background subtraction, contrast adjustment in ImageJ. The study revealed that low Pt loading is linked with low electrochemical surface area and high Pt degradation, especially at the interface of the cathode/membrane. They concluded that gradient-based particle sizes of Pt in the catalyst layer can help resolve this issue to some extent [301]. Granero et al. used a deep learning U-Net model with StarDist formulation to carry out complex segmentation of particles in TEM images of a carbon-supported Pt catalyst in a PEMFC and generated precise automatic computation of their particle sizes, which when compared to another recently developed deep learning based ImageDataExtractor software, greatly surpassed in accurate segregation of overlapping particles [302].

Volume/area calculations

Quantification of the area or volume of a specific feature in fuel cells can sometimes predict how a certain material will perform under different operating conditions and such information is valuable for fuel cell research [303]. Cronin et al. studied the active regions of an anode-supported SOFC via 3D reconstruction by back-projection of nano-TXM data further processed by water-shedding, histogram seeding, region-growing and threshold-based segmentation to segregate Ni-YSZ anode and lanthanum strontium manganite (LSM)-YSZ cathode regions. The surface area, volume and TPB density of Ni, LSM and YSZ were quantified by algorithms written in Matlab and Interactive Data Language (IDL), showing good agreement to values computed experimentally from FIB-SEM of the same sample [304]. Singh et al. used image processing tools of color thresholding, image segmentation and binarization in ImageJ to calculate the interfacial contact area between cerium (Ce) and palladium (Pd) particles in STEM images of CeO_x-Pd/C based catalysts for AEMFC and found the results to correlate with the hydrogen oxidation reaction (HOR) result obtained experimentally [305]. The authors used the image and data processing techniques in Esprit and Python to study the same CeO_x-Pd/C-based catalyst concluding that the overlap area calculations are profoundly influenced by different image processing parameters such as resolution, filters and thresholding [306]. Hwang et al. used the semantic segmentation-based deep learning approach to compute the surface volume fractions of each phase in gadolinium-doped ceria (GDC)/lanthanum strontium cobaltite (LSC) based cathode of SOFCs. The algorithm named DeepLabV3+ was trained via 49 binarized images of the cathode from FIB-SEM for segmentation of phases, which were reconstructed in a 3D space by Avizo yielding pixel-based accuracy of 0.78 [307]. White et al. calculated the area fraction of the solid in the cathode layer of SOFC by applying a Gaussian histogram fitting using MagicPlot software on 3D reconstructed micro-CT data for segmentation [308].

Phase distributions

Phase distribution and segregation analysis is an important parameter in developing structure-performance correlations for different types of fuel cells and aid in the thorough study of any particular phase of interest [309]. Prass et al. characterized different phase distributions from micro-CT images of compressed GDL-catalyst coated membrane (CCM) assemblies of a PEM fuel cell by image processing tools of ImageJ. They segregated the Nafion membrane, catalyst layer (CL), carbon fibers, polytetrafluoroethylene (PTFE) phases and MPL based on density variations, followed by filtering and binarization using MaxEntropy threshold to segment solid from void space for analysis of gap area between CL and MPL and crack density of MPL [310]. A study by Tadbir et al. visualized the phase distribution of carbon, PTFE and pores within the MPL of a PEMFC in 3D space by using image processing on nano-CT results. Phase segregation was achieved by 3D reconstruction of tomographic data sets, filtration by histogram equalization and identification of phases based on grayscale intensity values of pixels thresholded by maximum entropy and Otsu methods [311]. Phase segregation understanding is vital for the performance of high-temperature fuel cells such as SOFCs. Yu et al. studied the relationship between an increase in strontium (Sr) content and degradation of strontium-doped lanthanum cobalt ferrite (LSCF) based cathode material in SOFCs by analyzing the surface segregation of Sr phases using image processing techniques on atomic force microscopy (AFM) images. The study concluded that the percentage surface area coverage of Sr secondary phase decreased from decreasing Sr content from 40 % to 30 % but increased when reduced below 20 %. This trend was found to agree with the data obtained from synchrotron-based total-reflection X-ray fluorescence (TXRF) [312]. Villanova et al. visualized different phases inside the Ni/YSZ-based anode of

SOFC by filtering, Otsu thresholding and segmentation of 3D images obtained from X-ray-holotomography of the electrode [313]. Harris et al. investigated silicon contamination suffered by the Nd-nickelate (NNO) $Nd_{1.95}NiO_{4+\delta}$ based cathode of SOFCs by identifying two poisoning phases, i. e., NiO and silicate using image processing techniques of phase correlation, water-shedding and filtration and representative volume elements (RVE) based segmentation on images obtained by X-ray nanotomography of cathode [314].

Degradation effects

Long-term operation or exposure to rigorous operating conditions can cause degradation of several components in different types of fuel cells, and hence the quest for better material fabrication and engineering necessitates an in-depth study of the different degradation mechanisms in a fuel cell [315]. Simwonis et al. used the image processing technique to study the degradation phenomena of Ni-YSZ-based anode cermets used in SOFCs using the KS400 Carl Zeiss software package. The study reportedly segmented the different phases and features in the cermet allowing phase fraction, porosity and Ni particle size calculations before and after being exposed to 1000 °C in a mixture of Ar–H_2–H_2O for 4000 cycles for a comprehensive degradation analysis [316]. White et al. imaged both beginning-of-life (BOL) and end-of-life (EOL) samples of the catalyst layer of a PEM fuel cell with the help of micro- and nano-CT. They visualized macro and nanoscale crack formation, as well as catalyst thickness by Zeiss and ImageJ, followed by quantification with a Fourier power spectrum analysis in Python [317]. Holzer et al. compared and studied the microstructures of separate phases of the Ni-ceria gadolinium oxide (CGO) based anode in SOFCs, obtained from SEM in back-scattered electron mode, via image processing-based segmentation and feature extraction in dry and humid conditions for correlation of grain growth leading to Ni coarsening and subsequent degradation of the anode [318]. Eberhardt et al. applied median filtering, grayscale threshold segmentation and area opening filtering techniques of image processing in ImageJ and Matlab on a synchrotron-based XTM image data set for imaging and quantification of phosphoric acid redistribution under dynamic load conditions in high-temperature phosphoric acid (PA) doped with polybenzimidazole (PBI) electrolyte membrane-based HT-PEFC intending to study fuel cell degradation. The study demonstrated a high degree of electrolyte migration from cathode to anode under dynamic load correlating to the degradation [319]. Pokhrel et al. performed the failure analysis of a PEM fuel cell via an image processing-based 3D segmentation of pore, Pt/C, and ionomer for both BOL and EOL samples from micro-CT data indicating agglomeration of Pt/C particles, acute carbon corrosion at the interface of CL and membrane in the EOL sample and an increase in volume fraction of ionomer [320].

Challenges and prospects

In essence, image processing has an important role to play in the development and improvement of basic material components of nearly every type, including fuel cells, and this type of advancement is vital for the research, development and commercialization of fuel cell technology. However, automatic image processing is susceptible to certain downsides and may misrepresent, mislead or generate indecisive or inaccurate results if not suitably programmed or executed. For instance, the image acquisition software or equipment must be compatible with the image processing software or algorithms being used and should be able to provide the required resolution to identify or investigate the region of interest to prevent undesirable results. A higher resolution microscope should be used for the quantification of nanoparticles in the catalyst layer of fuel cells. In the same way, excessive or unnecessary use of various filtering techniques can partially or completely distort or destroy image data and can complicate the process of automatic feature extraction, thus should be used with caution. For example, oversmoothing can hide edges and boundaries of pores or particles while over-sharpening can result in inclusion of the noise. The choice of image segmentation method (e. g., approach-based, region-based or boundary-based) is also important and should always be decided keeping in view the image input data and its relationship with the features or properties of interest. Different types of segmentation methods may be required when opting for phase segregating and reactant distribution quantification. Furthermore, for machine learning-based models, an appropriate amount of training data set should be chosen to facilitate fast, uncomplicated yet accurate classification of image features. Thus, with the right amount of knowledge and expertise in how image processing algorithms work, scientists and engineers can overcome these issues. Currently, several free and open-source image processing software based on automatic, semi-automatic or AI-based algorithms are available online with nearly thousands of new plugins being introduced every year pointing out the continuous evolving and perfecting of image processing algorithms. Keeping this in mind, we can undoubtedly perceive a future of partial or complete automation in image analysis with a more intelligent and precise approach. Nevertheless, advancements in image and data processing, and especially in AI-based approaches, are undoubtedly required for automated, time-effective and statistically relevant structure-property-performance correlations in the fuel cell field, enabling their continuous progress and commercialization.

2.8 Conclusions

This chapter offers an overview of the microscopy and spectroscopy techniques employed to characterize PEMFCs. With SEM and TEM approaches most commonly used, microstructural characterization can offer information on catalyst powders, inks and catalyst layers surface and cross-section morphology and defects, for both fresh and degraded samples. EDS and EELS offer elemental analysis and distribution of the components on the nanometric scale. 3D FIB-SEM and electron tomography, even though just shortly explained, provide useful information on spatial morphology and component distribution, as well as pore morphology. To obtain more detailed and time resolved understanding of reactions and degradation processes in the system, advanced approaches like IL-TEM and *in situ* TEM are used. These techniques offer a range of advantages, but also carry a number of challenges that need to be addressed in order to track the changes with high resolution and in more realistic operating conditions. Finally, image processing, aided by machine learning and artificial intelligence, offers an immense opportunity to automatically process the data and extract valuable parameters. Only with morphological and compositional descriptors defined from the microscopy and spectroscopy data, the applied techniques are fully utilized to provide fundamental understanding and aid in development of PEMFCs. Further development of the microscopy techniques and image processing, including operando imaging close to realistic conditions, improving resolutions, enabling fast imaging and reducing beam damage, as well as operator-independent image analysis, automatic sampling, imaging and large data processing, are the future prospects and needs in characterization approaches supporting PEMFC development.

Bibliography

[1] Y. Leng, *Materials Characterization Introduction to Microscopic and Spectroscopic Methods*, 2nd edition. Wiley-VCH Verlag GmbH & Co. kGaA, 2008.

[2] R. Heu, S. Shahbazmohamadi, J. Yorston, and P. Capeder, Target Material Selection for Sputter Coating of SEM Samples. *Micros. Today*, **27**(4), 32–36, 2019. https://doi.org/10.1017/s1551929519000610.

[3] SEM Sample Stubs.

[4] The extended guide to sample preparation to obtain. ThermoFisher Sci.

[5] C. Wang et al., Improved Carbon Corrosion and Platinum Dissolution Durability in Automotive Fuel Cell Startup and Shutdown Operation. *J. Electrochem. Soc.*, **168**(3), 034503, 2021. https://doi.org/10.1149/1945-7111/abe6ea.

[6] C. Lei et al., Impact of Catalyst Ink Dispersing Solvent on PEM Fuel Cell Performance and Durability. *J. Electrochem. Soc.*, **168**(4), 044517, 2021. https://doi.org/10.1149/1945-7111/abf2b0.

[7] S. Pedram, N. Macauley, S. Zhong, H. Xu, and J. Jankovic, Microscopy Studies of the Catalyst Inks for PEM Fuel Cells. *ECS Meet. Abstr.*, MA2020-02(33), 2111, 2020. https://doi.org/10.1149/ma2020-02332111mtgabs.

[8] S. Nasrazadani and S. Hassani, *Modern analytical techniques in failure analysis of aerospace, chemical, and oil and gas industries.* Elsevier Ltd., 2016. https://doi.org/10.1016/B978-0-08-100117-2.00010-8.

[9] N. Goswami, A. N. Mistry, J. B. Grunewald, T. F. Fuller, and P. P. Mukherjee, Corrosion-Induced Microstructural Variability Affects Transport-Kinetics Interaction in PEM Fuel Cell Catalyst Layers. *J. Electrochem. Soc.*, **167**(8), 084519, 2020. https://doi.org/10.1149/1945-7111/ab927c.

[10] M. Prokop, M. Vesely, P. Capek, M. Paidar, and K. Bouzek, High-temperature PEM fuel cell electrode catalyst layers part 1: Microstructure reconstructed using FIB-SEM tomography and its calculated effective transport properties. *Electrochim. Acta*, **413**, 140133 2022. https://doi.org/10.1016/j.electacta.2022.140133.

[11] K. J. Lange, H. Carlsson, I. Stewart, P. C. Sui, R. Herring, and N. Djilali, PEM fuel cell CL characterization using a standalone FIB and SEM: Experiments and simulation. *Electrochim. Acta*, **85**, 322–331, 2012. https://doi.org/10.1016/j.electacta.2012.08.082.

[12]   W. D. Pyrz and D. J. Buttrey, Particle size determination using TEM: A discussion of image acquisition and analysis for the novice microscopist. *Langmuir*, **24**(20), 11350–11360, 2008. https://doi.org/10.1021/la801367j.

[13]   E. N. Coker, W. A. Steen, J. T. Miller, A. J. Kropf, and J. E. Miller, The preparation and characterization of novel Pt/C electrocatalysts with controlled porosity and cluster size. *J. Mater. Chem.*, **17**(31), 3330–3340, 2007. https://doi.org/10.1039/b703916f.

[14]   G. S. Avcioglu, B. Ficicilar, A. Bayrakceken, and I. Eroglu, High performance PEM fuel cell catalyst layers with hydrophobic channels. *Int. J. Hydrog. Energy*, **40**(24), 7720–7731, 2015. https://doi.org/10.1016/j.ijhydene.2015.02.004.

[15]   D. Li et al., Environmental Science Functional links between Pt single crystal morphology and nanoparticles with different size and shape: the oxygen reduction reaction case †. *Energy Environ. Sci.*, **7**, 4061–4069, 2014. https://doi.org/10.1039/c4ee01564a.

[16]   D. J. S. Sandbeck et al., Particle Size Eﬀect on Platinum Dissolution: Considerations for Accelerated Stability Testing of Fuel Cell Catalysts. *ACS Catal.*, **10**(11), 6281–6290, 2020. https://doi.org/10.1021/acscatal.0c00779.

[17]   E. Padgett et al., Mitigation of PEM Fuel Cell Catalyst Degradation with Porous Carbon Supports Mitigation of PEM Fuel Cell Catalyst Degradation with Porous Carbon Supports. *J. Electrochem. Soc.*, **166**, F198, 2019. https://doi.org/10.1149/2.0371904jes.

[18]   E. Carcadea, M. Varlam, A. Marinoiu, M. Raceanu, and M. S. Ismail, Influence of catalyst structure on PEM fuel cell performance e A numerical investigation. *Int. J. Hydrog. Energy*, **44**(25), 12829–12841, 2018. https://doi.org/10.1016/j.ijhydene.2018.12.155.

[19]   H. Yano, M. Watanabe, A. Iiyama, and H. Uchida, Particle-size effect of Pt cathode catalysts on durability in fuel cells. *Nano Energy*, **29**, 323–333, 2016. https://doi.org/10.1016/j.nanoen.2016.02.016.

[20]   J. Speder et al., Comparative degradation study of carbon supported proton exchange membrane fuel cell electrocatalysts e The inﬂuence of the platinum to carbon ratio on the degradation rate. *J. Power Sources*, **261**, 14–22, 2014. https://doi.org/10.1016/j.jpowsour.2014.03.039.

[21]   M. Watanabe, H. Yano, H. Uchida, and D. A. Tryk, Achievement of distinctively high durability at nanosized Pt catalysts supported on carbon black for fuel cell cathodes. *J. Electroanal. Chem.*, **819**, 359–364, 2018. https://doi.org/10.1016/j.jelechem.2017.11.017.

[22]   S. B. Rice et al., Particle size distributions by transmission electron microscopy: An interlaboratory comparison case study. *Metrologia*, **50**(6), 663–678, 2013. https://doi.org/10.1088/0026-1394/50/6/663.

[23]   B. Li et al., Controlling the microscopic morphology and permeability of catalyst layers in proton exchange membrane fuel cells by adjusting catalyst ink agglomerates. *Int. J. Hydrog. Energy*, **46**(63), 32215–32225, 2021. https://doi.org/10.1016/j.ijhydene.2021.06.216.

[24]   D. A. Blom, J. R. Dunlap, T. A. Nolan, and L. F. Allard, Preparation of Cross-Sectional Samples of Proton Exchange Membrane Fuel Cells by Ultramicrotomy for TEM. *J. Electrochem. Soc.*, **150**(4), A414, 2003. https://doi.org/10.1149/1.1556593.

[25]   A. Kneer et al., Correlation of Changes in Electrochemical and Structural Parameters due to Voltage Cycling Induced Degradation in PEM Fuel Cells. *J. Electrochem. Soc.*, **165**(6), F3241–F3250, 2018. https://doi.org/10.1149/2.0271806jes.

[26]   C. Wang, M. Ricketts, A. P. Soleymani, J. Jankovic, J. Waldecker, and J. Chen, Effect of Carbon Support Characteristics on Fuel Cell Durability in Accelerated Stress Testing. *J. Electrochem. Soc.*, **168**(4), 044507, 2021. https://doi.org/10.1149/1945-7111/abf265.

[27]   A. O. Godoy et al., Synthesis and Characterization of Platinum on Carbon Nanoparticles Selectively Coated with Titanium Nitride (TiN), *ECS Meet. Abstr.*, MA2020-02(36), 1021, 2021. https://doi.org/10.1149/ma2021-02361021mtgabs.

[28]   M. J. Dzara et al., Physicochemical Properties of ECS Supports and Pt/ECS Catalysts. *ACS Appl. Energy Mater.*, **4**(9), 9111–9123, 2021. https://doi.org/10.1021/acsaem.1c01392.

[29] N. Macauley et al., Carbon Corrosion in PEM Fuel Cells and the Development of Accelerated Stress Tests. *J. Electrochem. Soc.*, **165**(6), F3148–F3160, 2018. https://doi.org/10.1149/2.0061806jes.

[30] A. Kobayashi et al., Effect of Pt and Ionomer Distribution on Polymer Electrolyte Fuel Cell Performance and Durability. *ACS Appl. Energy Mater.*, **4**(3), 2307–2317, 2021. https://doi.org/10.1021/acsaem.0c02841.

[31] A. Kobayashi, T. Fujii, K. Takeda, K. Tamoto, K. Kakinuma, and M. Uchida, Effect of Pt Loading Percentage on Carbon Blacks with Large Interior Nanopore Volume on the Performance and Durability of Polymer Electrolyte Fuel Cells. *ACS Appl. Energy Mater.*, **5**(1), 316–329, 2022. https://doi.org/10.1021/acsaem.1c02836.

[32] G. M. Leteba, D. R. G. Mitchell, P. B. J. Levecque, and C. I. Lang, Solution-grown dendritic pt-based ternary nanostructures for enhanced oxygen reduction reaction functionality. *Nanomaterials*, **8**(7), 462, 2018. https://doi.org/10.3390/nano8070462.

[33] D. B. Williams and C. B. Carter, *Transmission electron microscopy: A textbook for materials science*, 2009. https://doi.org/10.1007/978-0-387-76501-3.

[34] S. Kabir et al., Toward Optimizing Electrospun Nanofiber Fuel Cell Catalyst Layers: Microstructure and Pt Accessibility. *ACS Appl. Energy Mater.*, **4**(4), 3341–3351, 2021. https://doi.org/10.1021/acsaem.0c03073.

[35] W. Pan, Z. Chen, D. Yao, X. Chen, F. Wang, and G. Dai, Microstructure and macroscopic rheology of microporous layer nanoinks for PEM fuel cells. *Chem. Eng. Sci.*, **246**, 117001, 2021. https://doi.org/10.1016/j.ces.2021.117001.

[36] S. Takahashi et al., Electrochimica Acta Observation of ionomer in catalyst ink of polymer electrolyte fuel cell using cryogenic transmission electron microscopy. *Electrochim. Acta*, **224**, 178–185, 2017. https://doi.org/10.1016/j.electacta.2016.12.068.

[37] E. Padgett et al., Editors' Choice — Connecting Fuel Cell Catalyst Nanostructure and Accessibility Using Quantitative Cryo-STEM Tomography Connecting Fuel Cell Catalyst Nanostructure and Accessibility Using Quantitative Cryo-STEM Tomography. *J. Electrochem. Soc.*, **165**, F173, 2018. https://doi.org/10.1149/2.0541803jes.

[38] B. T. Sneed et al., 3D Analysis of Fuel Cell Electrocatalyst Degradation on Alternate Carbon Supports. *ACS Appl. Mater. Interfaces*, **9**(35), 29839–29848, 2017. https://doi.org/10.1021/acsami.7b09716.

[39] D. Rossouw, L. Chinchilla, N. Kremliakova, and G. A. Botton, The 3D Nanoscale Evolution of Platinum–Niobium Oxide Fuel Cell Catalysts via Identical Location Electron Tomography. *Part. Part. Syst. Charact.*, **34**(7), 1–7, 2017. https://doi.org/10.1002/ppsc.201700051.

[40] B. T. Sneed, D. A. Cullen, K. S. Reeves, and K. L. More, Electron Tomography of PEM Fuel Cell Catalyst Coarsening on Alternate Carbon Supports. *Microsc. Microanal.*, **23**(S1), 2090–2091, 2017. https://doi.org/10.1017/s1431927617011114.

[41] H. Uchida, J. M. Song, S. Suzuki, E. Nakazawa, N. Baba, and M. Watanabe, Electron tomography of nafion ionomer coated on Pt/carbon black in high utilization electrode for PEFCs. *J. Phys. Chem. B*, **110**(27), 13319–13321, 2006. https://doi.org/10.1021/jp062678s.

[42] E. C. S. Transactions, *Catalyst-Layer Ionomer Imaging of Fuel Cells*, 2015.

[43] J. Jankovic, S. Zhang, A. Putz, M. S. Saha, and D. Susac, Multiscale imaging and transport modeling for fuel cell electrodes. *J. Mater. Res.*, **34**(4), 579–591, 2019. https://doi.org/10.1557/JMR.2018.458.

[44] X. Yang et al., PGM-Free Fe/N/C and Ultralow Loading Pt/C Hybrid Cathode Catalysts with Enhanced Stability and Activity in PEM Fuel Cells. *ACS Appl. Mater. Interfaces*, **12**(12), 13739–13749, 2020. https://doi.org/10.1021/acsami.9b18085.

[45] X. Yang et al., SiO2-Fe/N/C catalyst with enhanced mass transport in PEM fuel cells. *Appl. Catal. B, Environ.*, **264**, 118523, 2020. https://doi.org/10.1016/j.apcatb.2019.118523.

[46] M. Li et al., Ultrafine jagged platinum nanowires enable ultralhigh mass activity for the oxygen reduction reaction. *Science*, **354**(6318), 1414–1419, 2016. https://doi.org/10.1126/science.aaf9050.

[47] T. K. Fatma and M. Yurtsever, Mesut Yurukcu Mahbuba Begum, Fumiya Watanabe, "Stacked and Core–Shell Pt Ni-WC Nanorod Array Electrocatalyst for Enhanced Oxygen Reduction Reaction in Polymer Electrolyte Membrane Fuel Cells. *ACS Appl. Energy Mater.*, **1**(11), 6115–6122, 2018.

[48] L. Zhang et al., Atomic layer deposited Pt-Ru dual-metal dimers and identifying their active sites for hydrogen evolution reaction. *Nat. Commun.*, **10**(1), 1–11, 2019. https://doi.org/10.1038/s41467-019-12887-y.

[49] X. Wan et al., Fe–N–C electrocatalyst with dense active sites and efficient mass transport for high-performance proton exchange membrane fuel cells. *Nat. Catal.*, **2**(3), 259–268, 2019. https://doi.org/10.1038/s41929-019-0237-3.

[50] J. Li et al., Atomically dispersed manganese catalysts for oxygen reduction in proton-exchange membrane fuel cells. *Nat. Catal.*, **1**(12), 935–945, 2018. https://doi.org/10.1038/s41929-018-0164-8.

[51] T. Reshetenko et al., Design of PGM-free cathodic catalyst layers for advanced PEM fuel cells. *Appl. Catal. B, Environ.*, **312**, 121424, 2022. https://doi.org/10.1016/j.apcatb.2022.121424.

[52] K. Kumar et al., On the Influence of Oxygen on the Degradation of Fe-N-C Catalysts. *Angew. Chem., Int. Ed. Engl.*, **59**(8), 3235–3243, 2020. https://doi.org/10.1002/anie.201912451.

[53] X. Wan, X. Liu, and J. Shui, Stability of PGM-free fuel cell catalysts: Degradation mechanisms and mitigation strategies. *Prog. Nat. Sci. Mater. Int.*, **30**(6), 721–731, 2020. https://doi.org/10.1016/j.pnsc.2020.08.010.

[54] E. Proietti et al., Iron-based cathode catalyst with enhanced power density in polymer electrolyte membrane fuel cells. *Nat. Commun.*, **2**(1), 416, 2011. https://doi.org/10.1038/ncomms1427.

[55] Y. Cheng et al., A template-free method to synthesis high density iron single atoms anchored on carbon nanotubes for high temperature polymer electrolyte membrane fuel cells. *Nano Energy*, **80**, 105534, 2021. https://doi.org/10.1016/j.nanoen.2020.105534.

[56] H. T. Chung et al., Direct atomic-level insight into the active sites of a high-performance PGM-free ORR catalyst. *Science*, **357**(6350), 479–484, 2017. https://doi.org/10.1126/science.aan2255.

[57] S. Maier. Plasmonics: fundamentals and applications, 2007. Accessed: Nov. 06, 2022. [Online]. Available: https://link.springer.com/978-0-387-37825-1.

[58] K. Schlögl, K. J. J. Mayrhofer, M. Hanzlik, and M. Arenz, Identical-location TEM investigations of Pt/C electrocatalyst degradation at elevated temperatures. *J. Electroanal. Chem.*, **662**(2), 355–360, 2011. https://doi.org/10.1016/J.JELECHEM.2011.09.003.

[59] R. M. Arán-Ais et al., Identical Location Transmission Electron Microscopy Imaging of Site-Selective Pt Nanocatalysts: Electrochemical Activation and Surface Disordering. *J. Am. Chem. Soc.*, **137**(47), 14992–14998, 2015. https://doi.org/10.1021/JACS.5B09553/ASSET/IMAGES/LARGE/JA-2015-095537_0006.JPEG.

[60] S. Rasouli, D. Myers, N. Kariuki, K. Higashida, N. Nakashima, and P. Ferreira, Electrochemical Degradation of Pt-Ni Nanocatalysts: An Identical Location Aberration-Corrected Scanning Transmission Electron Microscopy Study. *Nano Lett.*, **19**(1), 46–53, 2019. https://doi.org/10.1021/ACS.NANOLETT.8B03022/ASSET/IMAGES/MEDIUM/NL-2018-030222_M002.GIF.

[61] C. Lafforgue, M. Chatenet, L. Dubau, and D. R. Dekel, Accelerated Stress Test of Pt/C Nanoparticles in an Interface with an Anion-Exchange Membrane – An Identical-Location Transmission Electron Microscopy Study. *ACS Catal.*, **8**(2), 1278–1286, 2018. https://doi.org/10.1021/ACSCATAL.7B04055.

[62] M. Arenz and A. Zana, Fuel cell catalyst degradation: Identical location electron microscopy and related methods. *Nano Energy*, **29**, 299–313, 2016. https://doi.org/10.1016/J.NANOEN.2016.04.027.

[63] N. Hodnik and S. Cherevko, Spot the difference at the nanoscale: identical location electron microscopy in electrocatalysis. *Curr. Opin. Electrochem.*, **15**, 73–82, 2019. https://doi.org/10.1016/J.COELEC.2019.03.007.

[64] K. J. J. Mayrhofer et al., Fuel cell catalyst degradation on the nanoscale. *Electrochem. Commun.*, **10**(8), 1144–1147, 2008. https://doi.org/10.1016/J.ELECOM.2008.05.032.

[65] S. G. Rinaldo, J. Stumper, and M. Eikerling, Physical theory of platinum nanoparticle dissolution in polymer electrolyte fuel cells. *J. Phys. Chem. C*, **114**(13), 5773–5785, 2010. https://doi.org/10.1021/JP9101509/ASSET/IMAGES/MEDIUM/JP-2009-101509_0011.GIF.

[66] K. Schlögl, M. Hanzlik, and M. Arenz, Comparative IL-TEM study concerning the degradation of carbon supported Pt-based electrocatalysts. *J. Electrochem. Soc.*, **159**, B677, 2012. https://doi.org/10.1149/2.035206jes.

[67]   K. Schlögl, M. Hanzlik, and M. Arenz, Comparative IL-TEM Study Concerning the Degradation of Carbon Supported Pt-Based Electrocatalysts. *J. Electrochem. Soc.*, **159**(6), B677–B682, 2012. https://doi.org/10.1149/2.035206JES.

[68]   F. J. Perez-Alonso, C. F. Elkjær, S. S. Shim, B. L. Abrams, I. E. L. Stephens, and Ib Chorkendorff. Identical locations transmission electron microscopy study of Pt/C electrocatalyst degradation during oxygen reduction reaction. *J. Power Sources*, **196**(15), 6085–6091, 2011. https://doi.org/10.1016/j.jpowsour.2011.03.064.

[69]   H. A. Gasteiger, S. S. Kocha, B. Sompalli, and F. T. Wagner, Activity benchmarks and requirements for Pt, Pt-alloy, and non-Pt oxygen reduction catalysts for PEMFCs. *Appl. Catal. B, Environ.*, **56**(1–2), 9–35, 2005. https://doi.org/10.1016/J.APCATB.2004.06.021.

[70]   L. Tang et al., Electrochemical stability of nanometer-scale Pt particles in acidic environments. *J. Am. Chem. Soc.*, **132**(2), 596–600, 2010. https://doi.org/10.1021/JA9071496.

[71]   L. Tang, X. Li, R. C. Cammarata, C. Friesen, and K. Sieradzki, Electrochemical stability of elemental metal nanoparticles. *J. Am. Chem. Soc.*, **132**(33), 11722–11726, 2010. https://doi.org/10.1021/JA104421T/SUPPL_FILE/JA104421T_SI_001.PDF.

[72]   Y. Shao-Horn, W. C. Sheng, S. Chen, P. J. Ferreira, E. F. Holby, and D. Morgan, Instability of supported platinum nanoparticles in low-temperature fuel cells. *Top. Catal.*, 46, no. 3–4, 285–305, 2007. https://doi.org/10.1007/S11244-007-9000-0.

[73]   J. C. Meier et al., Degradation Mechanisms of Pt/C Fuel Cell Catalysts under Simulated Start–Stop Conditions. *ACS Catal.*, **2**(5), 832–843, 2012. https://doi.org/10.1021/cs300024h.

[74]   K. Hartl, M. Hanzlik, and M. Arenz, IL-TEM investigations on the degradation mechanism of Pt/C electrocatalysts with different carbon supports. *Energy Environ. Sci.*, **4**(1), 234–238, 2011. https://doi.org/10.1039/C0EE00248H.

[75]   J. S. Spendelow and A. Wieckowski, Electrocatalysis of oxygen reduction and small alcohol oxidation in alkaline media. *Phys. Chem. Chem. Phys.*, **9**(21), 2654–2675, 2007. https://doi.org/10.1039/B703315J.

[76]   M. Pourbaix, H. Zhang, and A. Pourbaix, Presentation of an Atlas of Chemical and Electrochemical Equilibria in the Precence of a Gaseous Phase. *Mater. Sci. Forum*, 251–254, 143–148, 1997. https://doi.org/10.4028/www.scientific.net/msf.251-254.143.

[77]   A. Zadick, L. Dubau, N. Sergent, G. Berthomé, and M. Chatenet, Huge Instability of Pt/C Catalysts in Alkaline Medium. *ACS Catal.*, **5**(8), 4819–4824, 2015. https://doi.org/10.1021/ACSCATAL.5B01037/ASSET/IMAGES/LARGE/CS-2015-01037X_0005.JPEG.

[78]   A. Zadick, L. Dubau, U. B. Demirci, and M. Chatenet, Effects of Pd Nanoparticle Size and Solution Reducer Strength on Pd/C Electrocatalyst Stability in Alkaline Electrolyte. *J. Electrochem. Soc.*, **163**(8), F781–F787, 2016. https://doi.org/10.1149/2.0141608JES/XML.

[79]   W. E. Mustain, M. Chatenet, M. Page, and Y. S. Kim, Durability challenges of anion exchange membrane fuel cells. *Energy Environ. Sci.*, **13**(9), 2805–2838, 2020. https://doi.org/10.1039/D0EE01133A.

[80]   J. Solla-Gullón, F. J. Vidal-Iglesias, and J. M. Feliu, Shape dependent electrocatalysis. *Annu. Rep. Prog. Chem., Sect. C, Phys. Chem.*, **107**(0), 263–297, 2011. https://doi.org/10.1039/C1PC90010B.

[81]   Y. Bing, H. Liu, L. Zhang, D. Ghosh, and J. Zhang, Nanostructured Pt-alloy electrocatalysts for PEM fuel cell oxygen reduction reaction. *Chem. Soc. Rev.*, **39**(6), 2184–2202, 2010. https://doi.org/10.1039/B912552C.

[82]   V. R. Stamenkovic et al., Improved oxygen reduction activity on Pt3Ni(111) via increased surface site availability. *Science*, **315**(5811), 493–497, 2007. https://doi.org/10.1126/SCIENCE.1135941.

[83]   A. Rabis, P. Rodriguez, and T. J. Schmidt, Electrocatalysis for polymer electrolyte fuel cells: Recent achievements and future challenges. *ACS Catal.*, **2**(5), 864–890, 2012. https://doi.org/10.1021/CS3000864/ASSET/IMAGES/LARGE/CS-2012-000864_0020.JPEG.

[84]   M. K. Debe, Electrocatalyst approaches and challenges for automotive fuel cells. *Nature*, **486**(7401), 43–51, 2012. https://doi.org/10.1038/nature11115.

[85] J. Kim, Y. Lee, and S. Sun, Structurally ordered FePt nanoparticles and their enhanced catalysis for oxygen reduction reaction. *J. Am. Chem. Soc.*, **132**(14), 4996–4997, 2010. https://doi.org/10.1021/JA1009629.

[86] J. Liang et al., Atomic Arrangement Engineering of Metallic Nanocrystals for Energy-Conversion Electrocatalysis. *Joule*, **3**(4), 956–991, 2019. https://doi.org/10.1016/J.JOULE.2019.03.014.

[87] S. Koh, M. F. Toney, and P. Strasser, Activity-stability relationships of ordered and disordered alloy phases of Pt3Co electrocatalysts for the oxygen reduction reaction (ORR). *Electrochim. Acta*, 52, no. 8 SPEC. ISS, 2765–2774, 2007. https://doi.org/10.1016/j.electacta.2006.08.039.

[88] M. Z. Martin, S. D. Wullschleger, C. T. Garten, and A. V. Palumbo, Material concepts in surface reactivity and catalysis. In: *Laser-Induced Break Spectrosc.*, pages 341–351, 1990. https://doi.org/10.1016/B978-044451734-0.50018-1.

[89] L. C. Gontard, R. E. Dunin-Borkowski, and D. Ozkaya, Three-dimensional shapes and spatial distributions of Pt and PtCr catalyst nanoparticles on carbon black. *J. Microsc.*, **232**(2), 248–259, 2008. https://doi.org/10.1111/J.1365-2818.2008.02096.X.

[90] N. Hodnik, M. Zorko, M. Bele, S. Hočevar, and M. Gaberšček, Identical location scanning electron microscopy: A case study of electrochemical degradation of PtNi nanoparticles using a new nondestructive method. *J. Phys. Chem. C*, **116**(40), 21326–21333, 2012. https://doi.org/10.1021/JP303831C/SUPPL_FILE/JP303831C_SI_001.PDF.

[91] C. Baldizzone et al., Confined-Space Alloying of Nanoparticles for the Synthesis of Efficient PtNi Fuel-Cell Catalysts. *Angew. Chem., Int. Ed. Engl.*, **53**(51), 14250–14254, 2014. https://doi.org/10.1002/ANIE.201406812.

[92] N. Hodnik et al., Effect of ordering of PtCu3 nanoparticle structure on the activity and stability for the oxygen reduction reaction. *Phys. Chem. Chem. Phys.*, **16**(27), 13610–13615, 2014. https://doi.org/10.1039/C4CP00585F.

[93] Hummingbird Scientific | In-Situ TEM Specimen Holders. http://hummingbirdscientific.com/ (accessed Apr. 10, 2020).

[94] Advanced TEM Solutions for In Situ Microscopy Characterization | Protochips. https://www.protochips.com/ (accessed Apr. 10, 2020).

[95] DENSsolutions – In situ Microscopy Innovative Solutions. https://denssolutions.com/ (accessed Apr. 10, 2020).

[96] V. Beermann, M. E. Holtz, E. Padgett, J. F. De Araujo, D. A. Muller, and P. Strasser, Real-time imaging of activation and degradation of carbon supported octahedral Pt-Ni alloy fuel cell catalysts at the nanoscale using: In situ electrochemical liquid cell STEM. *Energy Environ. Sci.*, **12**(8), 2476–2485, 2019. https://doi.org/10.1039/c9ee01185d.

[97] M. L. Taheri et al., Current status and future directions for in situ transmission electron microscopy. *Ultramicroscopy*, **170**, 86–95, 2016. https://doi.org/10.1016/j.ultramic.2016.08.007.

[98] A. P. Soleymani, L. R. Parent, and J. Jankovic, Challenges and Opportunities in Understanding Proton Exchange Membrane Fuel Cell Materials Degradation Using In-Situ Electrochemical Liquid Cell Transmission Electron Microscopy. *Adv. Funct. Mater.*, **32**(5), 2105188, 2022. https://doi.org/10.1002/ADFM.202105188.

[99] S. K. Eswaramoorthy, J. M. Howe, and G. Muralidharan, In situ determination of the nanoscale chemistry and behavior of solid-liquid systems. *Science*, **318**(5855), 1437–1440, 2007. https://doi.org/10.1126/science.1146511.

[100] T. Miyata and T. Mizoguchi, Fabrication of thin TEM sample of ionic liquid for high-resolution ELNES measurements. *Ultramicroscopy*, **178**, 81–87, 2017. https://doi.org/10.1016/j.ultramic.2016.10.009.

[101] S. Zhang, X. Z. Yuan, J. N. C. Hin, H. Wang, K. A. Friedrich, and M. Schulze, A review of platinum-based catalyst layer degradation in proton exchange membrane fuel cells. *J. Power Sources*, **194**(2), 588–600, 2009. https://doi.org/10.1016/J.JPOWSOUR.2009.06.073.

[102] R. L. Borup et al., Recent developments in catalyst-related PEM fuel cell durability. *Curr. Opin. Electrochem.*, **21**, 192–200, 2020. https://doi.org/10.1016/j.coelec.2020.02.007.

[103] J. Y. Huang et al., In situ observation of the electrochemical lithiation of a single SnO_2 nanowire electrode. *Science*, **330**(6010), 1515–1520, 2010. https://doi.org/10.1126/science.1195628.

[104] C. M. Wang et al., In situ transmission electron microscopy and spectroscopy studies of interfaces in Li ion batteries: Challenges and opportunities. *J. Mater. Res.*, **25**(8), 1541–1547, 2010. https://doi.org/10.1557/jmr.2010.0198.

[105] H. Hashimoto, T. Naiki, T. Eto, and K. Fujiwara, High Temperature Gas Reaction Specimen Chamber for an Electron Microscope. *Jpn. J. Appl. Phys.*, **7**(8), 946–952, 1968. https://doi.org/10.1143/jjap.7.946.

[106] R. T. K. Baker, F. S. Feates, and P. S. Harris, Continuous electron microscopic observation of carbonaceous deposits formed on graphite and silica surfaces. *Carbon*, **10**(1), 93–96, 1972. https://doi.org/10.1016/0008-6223(72)90014-0.

[107] R. T. K. Baker, In Situ Electron Microscopy Studies of Catalyst Particle Behavior. *Catal. Rev.*, **19**(2), 161–209, 1979. https://doi.org/10.1080/03602457908068055.

[108] R. Sharma, An environmental transmission electron microscope for in situ synthesis and characterization of nanomaterials. *J. Mater. Res.*, **20**(7), 1695–1707, 2005. https://doi.org/10.1557/JMR.2005.0241.

[109] I. M. Abrams and J. W. Mcbain, A closed cell for electron microscopy. *Science*, **100**(2595), 273–274, 1944. https://doi.org/10.1126/science.100.2595.273.

[110] K. Nishijima, J. Yamasaki, H. Orihara, and Tanaka. Development of microcapsules for electron microscopy and their application to dynamical observation of liquid crystals in transmission electron microscopy. *Nanotechnology*, **15**(6), S329, 2004. Accessed: Apr. 10, 2020. [Online]. Available: https://iopscience.iop.org/article/10.1088/0957-4484/15/6/001.

[111] M. Textor and N. De Jonge, Strategies for Preparing Graphene Liquid Cells for Transmission Electron Microscopy. *Nano Lett.*, **18**(6), 3313–3321, 2018. https://doi.org/10.1021/acs.nanolett.8b01366.

[112] J. Yang, S. B. Alam, L. Yu, E. Chan, and H. Zheng, Dynamic behavior of nanoscale liquids in graphene liquid cells revealed by in situ transmission electron microscopy. *Micron*, **116**, 22–29, 2019. https://doi.org/10.1016/j.micron.2018.09.009.

[113] M. J. Williamson, R. M. Tromp, P. M. Vereecken, R. Hull, and F. M. Ross, Dynamic microscopy of nanoscale cluster growth at the solid-liquid interface. *Nat. Mater.*, **2**(8), 532–536, 2003. https://doi.org/10.1038/nmat944.

[114] N. De Jonge and F. M. Ross, Electron microscopy of specimens in liquid. *Nat. Nanotechnol.*, **6**(11), 695–704, 2011. https://doi.org/10.1038/nnano.2011.161.

[115] J. M. Grogan and H. H. Bau, The nanoaquarium: A platform for in situ transmission electron microscopy in liquid media. *J. Microelectromech. Syst.*, **19**(4), 885–894, 2010. https://doi.org/10.1109/JMEMS.2010.2051321.

[116] S. Azim et al., Environmental Liquid Cell Technique for Improved Electron Microscopic Imaging of Soft Matter in Solution. *Microsc. Microanal.*, **27**(1), 44–53, 2020. https://doi.org/10.1017/S1431927620024654.

[117] N. de Jonge, L. Houben, R. E. Dunin-Borkowski, and F. M. Ross, Resolution and aberration correction in liquid cell transmission electron microscopy. *Nat. Rev. Mater.*, **4**(1), 61–78, 2019. https://doi.org/10.1038/s41578-018-0071-2.

[118] S. Nagashima et al., Atomic-Level Observation of Electrochemical Platinum Dissolution and Redeposition. *Nano Lett.*, **19**(10), 7000–7005, 2019. https://doi.org/10.1021/acs.nanolett.9b02382.

[119] H. Kato, In-Situ Liquid TEM Study on the Degradation Mechanism of Fuel Cell Catalysts. *SAE Int. J. Altern. Power.*, **5**(1), 189–194, 2016. https://doi.org/10.4271/2016-01-1192.

[120] A. Impagnatiello et al., Degradation Mechanisms of Supported Pt Nanocatalysts in Proton Exchange Membrane Fuel Cells: An Operando Study through Liquid Cell Transmission Electron Microscopy. *ACS Appl. Energy Mater.*, **3**(3), 2360–2371, 2020. https://doi.org/10.1021/acsaem.9b02000.

[121] G. Jerkiewicz, Standard and Reversible Hydrogen Electrodes: Theory, Design, Operation, and Applications. *ACS Catal.*, **10**(15), 8409–8417, 2020. https://doi.org/10.1021/ACSCATAL.0C02046.

[122] C. H. Hamann, A. Hamnet, and Q. Vielstich, *Electrochemistry*. WileyVCH, Weinheim, Germany, 2007.

[123] R. F. Egerton, P. Li, and M. Malac, Radiation damage in the TEM and SEM. *Micron*, **35**(6), 399–409, 2004. https://doi.org/10.1016/j.micron.2004.02.003.

[124] N. Bhattarai, D. L. Woodall, J. E. Boercker, J. G. Tischler, and T. H. Brintlinger, Controlling dissolution of PbTe nanoparticles in organic solvents during liquid cell transmission electron microscopy. *Nanoscale*, **11**(31), 14573–14580, 2019. https://doi.org/10.1039/c9nr04646a.

[125] Y. Jiang et al., Probing the oxidative etching induced dissolution of palladium nanocrystals in solution by liquid cell transmission electron microscopy. *Micron*, **97**, 22–28, 2017. https://doi.org/10.1016/j.micron.2017.03.003.

[126] T. Su, Z. L. Wang, and Z. Wang, In Situ Observations of Shell Growth and Oxidative Etching Behaviors of Pd Nanoparticles in Solutions by Liquid Cell Transmission Electron Microscopy. *Small*, **15**(14), 1900050, 2019. https://doi.org/10.1002/smll.201900050.

[127] R. R. Unocic et al., Quantitative Electrochemical Measurements Using In Situ ec-S/TEM Devices. *Microsc. Microanal.*, **20**(2), 452–461, 2014. https://doi.org/10.1017/S1431927614000166.

[128] J. M. Grogan, N. M. Schneider, F. M. Ross, and H. H. Bau, Bubble and pattern formation in liquid induced by an electron beam. *Nano Lett.*, **14**(1), 359–364, 2014. https://doi.org/10.1021/nl404169a.

[129] E. R. White, M. Mecklenburg, S. B. Singer, S. Aloni, and B. C. Regan, Imaging nanobubbles in water with scanning transmission electron microscopy. *Appl. Phys. Express*, **4**(5), 055201, 2011. https://doi.org/10.1143/APEX.4.055201.

[130] N. M. Schneider, M. M. Norton, B. J. Mendel, J. M. Grogan, F. M. Ross, and H. H. Bau, Electron-Water interactions and implications for liquid cell electron microscopy. *J. Phys. Chem. C*, **118**(38), 22373–22382, 2014. https://doi.org/10.1021/jp507400n.

[131] Y. Liu and S. J. Dillon, In situ observation of electrolytic H_2 evolution adjacent to gold cathodes. *Chem. Commun.*, **50**(14), 1761–1763, 2014. https://doi.org/10.1039/c3cc46737f.

[132] L. G. A. Melo, A. P. Hitchcock, J. Jankovic, J. Stumper, D. Susac, and V. Berejnov, Quantitative Mapping of Ionomer in Catalyst Layers by Electron and X-ray Spectromicroscopy. *ECS Trans.*, **80**(8), 275–282, 2017. https://doi.org/10.1149/08008.0275ecst.

[133] I. Martens, L. G. A. Melo, D. P. Wilkinson, D. Bizzotto, and A. P. Hitchcock, Characterization of X-ray Damage to Perfluorosulfonic Acid Using Correlative Microscopy. *J. Phys. Chem. C*, **123**(26), 16023–16033, 2019. https://doi.org/10.1021/acs.jpcc.9b03924.

[134] K. Shinozaki, Y. Morimoto, B. S. Pivovar, and S. S. Kocha, Suppression of oxygen reduction reaction activity on Pt-based electrocatalysts from ionomer incorporation. *J. Power Sources*, **325**, 745–751, 2016. https://doi.org/10.1016/J.JPOWSOUR.2016.06.062.

[135] A. Hartig-Weiss, M. F. Tovini, H. A. Gasteiger, and H. A. El-Sayed, OER Catalyst Durability Tests Using the Rotating Disk Electrode Technique: The Reason Why This Leads to Erroneous Conclusions. *ACS Appl. Energy Mater.*, **3**(11), 10323–10327, 2020. https://doi.org/10.1021/ACSAEM.0C01944.

[136] J. Wu, W. Gao, H. Yang, and J. M. Zuo, Dissolution Kinetics of Oxidative Etching of Cubic and Icosahedral Platinum Nanoparticles Revealed by in Situ Liquid Transmission Electron Microscopy. *ACS Nano*, **11**(2), 1696–1703, 2017. https://doi.org/10.1021/acsnano.6b07541.

[137] H. Shan et al., Nanoscale kinetics of asymmetrical corrosion in core-shell nanoparticles. *Nat. Commun.*, **9**(1), 1–9, 2018. https://doi.org/10.1038/s41467-018-03372-z.

[138] G. Z. Zhu, S. Prabhudev, J. Yang, C. M. Gabardo, G. A. Botton, and L. Soleymani, In situ liquid cell TEM study of morphological evolution and degradation of Pt-Fe nanocatalysts during potential cycling. *J. Phys. Chem. C*, **118**(38), 22111–22119, 2014. https://doi.org/10.1021/jp506857b.

[139] L. M. Roen, C. H. Paik, and T. D. Jarvi, Electrocatalytic Corrosion of Carbon Support in PEMFC. Cathodes," *Electrochem. Solid-State Lett.*, **7**(1), 19, 2004. https://doi.org/10.1149/1.1630412.

[140] K. L. More, D. Cullen, R. Unocic, and S. Reeves, *Characterization of Fuel Cell Materials Project ID FC020*, 2015.

[141] V. Beermann, Studies on Shape Defined Pt-alloy Based Nanoparticles for the Electrochemical Reduction of Oxygen vorgelegt von Master of Science, Technischen Universität Berlin, 2018. Accessed: May 18, 2020. [Online]. Available: https://pdfs.semanticscholar.org/362d/b0168e1b0831867ed992b8b3cec50d3f9857.pdf.

[142] S. C. Ball, S. L. Hudson, D. Thompsett, and B. Theobald, An investigation into factors affecting the stability of carbons and carbon supported platinum and platinum/cobalt alloy catalysts during 1.2 V potentiostatic hold regimes at a range of temperatures. *J. Power Sources*, **171**(1), 18–25, 2007. https://doi.org/10.1016/j.jpowsour.2006.11.004.

[143] P. J. Ferreira, G. J. la O', Y. Shao-Horn, D. Morgan, R. Makharia, S. Kocha, and Gasteiger. Instability of Pt/C Electrocatalysts in Proton Exchange Membrane Fuel Cells: A Mechanistic Investigation. *J. Electrochem. Soc.*, **152**(11), A2256–A2271, 2005. https://doi.org/10.1149/1.2050347.

[144] V. Ortalan, Catching them in Action: Ultrafast Transmission Electron Microscopy. *Microsc. Microanal.*, **27**(S1), 3124–3126, 2021. https://doi.org/10.1017/S1431927621010813.

[145] V. Ortalan and A. H. Zewail, 4D Scanning Transmission Ultrafast Electron Microscopy: Single-Particle Imaging and Spectroscopy. *J. Am. Chem. Soc.*, **133**(28), 10732–10735, 2011. https://doi.org/10.1021/JA203821Y.

[146] C. Domínguez et al., Effect of transition metal (M: Fe, Co or Mn) for the oxygen reduction reaction with non-precious metal catalysts in acid medium. *Int. J. Hydrog. Energy*, **39**(10), 5309–5318, 2014. https://doi.org/10.1016/j.ijhydene.2013.12.078.

[147] H. Du et al., High-Quality and Deeply Excavated Pt3Co Nanocubes as Efficient Catalysts for Liquid Fuel Electrooxidation. *Chem. Mater.*, **29**(22), 9613–9617, 2017. https://doi.org/10.1021/acs.chemmater.7b03406.

[148] E. Flores-Rojas et al., Electrocatalysis of oxygen reduction on CoNi-decorated-Pt nanoparticles: A theoretical and experimental study. *Int. J. Hydrog. Energy*, **41**(48), 23301–23311, 2016. https://doi.org/10.1016/j.ijhydene.2016.10.009.

[149] Z. Liu et al., Efficient synthesis of Pt-Co nanowires as cathode catalysts for proton exchange membrane fuel cells. *RSC Adv.*, **10**(11), 6287–6296, 2020. https://doi.org/10.1039/d0ra00264j.

[150] Y. Hashimasa, Y. Matsuda, and T. Shimizu, Comparison of Carbon Corrosion Test Methods for Polymer Electrolyte Fuel Cell. *Electrochim. Acta*, **179**, 119–125, 2015. https://doi.org/10.1016/j.electacta.2015.03.147.

[151] S. L. Tanimoto, *An Interdisciplinary Introduction to Image Processing: Pixels, Numbers, and Programs*. MIT Press, 2012.

[152] I. Young, J. Gerbrands, and L. van Vliet, In *Fundamentals of Image Processing*, pages 1–85, 2009. https://doi.org/10.1201/9781420046090-c13.

[153] J. Nakamura, *Image Sensors and Signal Processing for Digital Still Cameras*. CRC Press, 2017.

[154] A. J. Patti and Y. Altunbasak, Artifact reduction for set theoretic super resolution image reconstruction with edge adaptive constraints and higher-order interpolants. *IEEE Trans. Image Process.*, **10**(1), 179–186, 2001. https://doi.org/10.1109/83.892456.

[155] T. Acharya and A. K. Ray, *Image Processing: Principles and Applications*, Wiley, 2005.

[156] C. Solomon and T. Breckon, *Fundamentals of Digital Image Processing: A Practical Approach with Examples in Matlab*. Wiley, 2011. [Online]. Available: https://books.google.com.pk/books?id=NoJ15jLdy7YC.

[157] R. H. Alwan, F. J. Kadhim, and A. T. Al-Taani, Data Embedding based on Better Use of Bits in Image Pixels. *Int. J. Signal Process.*, **2**(2), 104–107, 2005.

[158] H. J. Trussell, E. Saber, and M. Vrhel, Color image processing [basics and special issue overview. *IEEE Signal Process. Mag.*, **22**(1), 14–22, 2005. https://doi.org/10.1109/msp.2005.1407711.

[159] S. Rege, R. Memane, M. Phatak, and P. Agarwal, 2D Geometric Shape and Color Recognition Using Digital Image Processing. *Int. J. Adv. Res. Electr. Electron. Instrum. Eng.*, **2**, 2278–8875, 2013.

[160] T. Searles et al., Rapid characterization of titanium microstructural features for specific modelling of mechanical properties. *Meas. Sci. Technol.*, **16**(1), 60–69, 2005. https://doi.org/10.1088/0957-0233/16/1/009.

[161] A. Campbell, P. Murray, E. Yakushina, S. Marshall, and W. Ion, New methods for automatic quantification of microstructural features using digital image processing. *Mater. Des.*, **141**, 395–406, 2018. https://doi.org/10.1016/j.matdes.2017.12.049.

[162] D. Basu and S. Basu, Performance studies of Pd-Pt and Pt-Pd-Au catalyst for electro-oxidation of glucose in direct glucose fuel cell. *Int. J. Hydrog. Energy*, **37**(5), 4678–4684, 2012. https://doi.org/10.1016/j.ijhydene.2011.04.158.

[163] P. M. Anderson, H. Guo, and P. B. Sunderland, Repeatability and reproducibility of TEM soot primary particle size measurements and comparison of automated methods. *J. Aerosol Sci.*, **114**, 317–326, 2017. https://doi.org/10.1016/j.jaerosci.2017.10.002.

[164] J. E. Bonevich and W. K. Haller, *Measuring the Size of Nanoparticles Using Transmission Electron Microscopy (TEM)*, 2010.

[165] ASTM E112, Standard Test Methods for Determining Average Grain Size E112-10, *Astm E112-10*, 96(2004), 1–27, 2010. https://doi.org/10.1520/E0112-10.Copyright.

[166] E. F. Holby, W. Sheng, Y. Shao-Horn, and D. Morgan, Pt nanoparticle stability in PEM fuel cells: Influence of particle size distribution and crossover hydrogen. *Energy Environ. Sci.*, **2**(8), 865–871, 2009. https://doi.org/10.1039/b821622n.

[167] D. B. Williams and C. B. Carter, *Transmission Electron Microscopy Electron Materials Science*, page 729, 1996.

[168] C. A. Schneider, W. S. Rasband, and K. W. Eliceiri, NIH Image to ImageJ: 25 years of image analysis. *Nat. Methods*, **9**(7), 671–675, 2012. https://doi.org/10.1038/nmeth.2089.

[169] C. Igathinathane, L. O. Pordesimo, E. P. Columbus, W. D. Batchelor, and S. R. Methuku, Shape identification and particles size distribution from basic shape parameters using ImageJ. *Comput. Electron. Agric.*, **63**(2), 168–182, 2008. https://doi.org/10.1016/j.compag.2008.02.007.

[170] S. Kook et al., Automated Detection of Primary Particles from Transmission Electron Microscope (TEM) Images of Soot Aggregates in Diesel Engine Environments. *Int. J. Engines*, **9**(1), 279–296, 2016. https://doi.org/10.2307/26284812.

[171] B. Foster and B. Fookes, Image analysis for materials science. *Adv. Mater. Process.*, **149**(2), 23–25, 1996.

[172] K. M. Kolltveit, T. Solheim, and S. I. Kvaal, *Methods of measuring morphological parameters in dental radiographs Comparison between image analysis and manual measurements*, 1998.

[173] R. W. Atkinson, Y. Garsany, B. D. Gould, K. E. Swider-Lyons, and I. V. Zenyuk, The Role of Compressive Stress on Gas Diffusion Media Morphology and Fuel Cell Performance. *ACS Appl. Energy Mater.*, **1**(1), 191–201, 2018. https://doi.org/10.1021/acsaem.7b00077.

[174] S. Wang, S. Guan, L. Zhang, F. Zhou, J. Tan, and M. Pan, Enhancing the properties of water and gas management for proton exchange membrane fuel cells via tailored intersected cracks in a microporous layer. *J. Power Sources*, **533**, 231402, 2022. https://doi.org/10.1016/j.jpowsour.2022.231402.

[175] I. V. Zenyuk, D. Y. Parkinson, L. G. Connolly, and A. Z. Weber, Gas-diffusion-layer structural properties under compression via X-ray tomography. *J. Power Sources*, **328**, 364–376, 2016. https://doi.org/10.1016/j.jpowsour.2016.08.020.

[176] M. P. Arcot, K. Zheng, J. McGrory, M. W. Fowler, and M. D. Pritzker, Investigation of catalyst layer defects in catalyst-coated membrane for PEMFC application: Non-destructive method. *Int. J. Energy Res.*, **42**(11), 3615–3632, 2018. https://doi.org/10.1002/er.4107.

[177] N. A. Ivanova et al., Comparison of the performance and durability of PEM fuel cells with different Pt-activated microporous layers. *Int. J. Hydrog. Energy*, **46**(34), 18093–18106, 2021. https://doi.org/10.1016/j.ijhydene.2020.08.234.

[178] M. Roumi, *Implementing Texture Feature Extraction Algorithms on FPGA*, 2009. [Online]. Available: https://www.researchgate.net/publication/46140875.

[179] D. Saladra and M. Kopernik, Qualitative and quantitative interpretation of SEM image using digital image processing. *J. Microsc.*, **264**(1), 102–124, 2016. https://doi.org/10.1111/jmi.12431.

[180] P. Karin, Y. Songsaengchan, S. Laosuwan, C. Charoenphonphanich, N. Chollacoop, and H. Katsunori, Nanostructure investigation of particle emission by using TEM image processing method. *Energy Proc.*, **34**, 757–766, 2013. https://doi.org/10.1016/j.egypro.2013.06.811.

[181] J. Ohyama, A. Esaki, T. Koketsu, Y. Yamamoto, S. Arai, and A. Satsuma, Atomic-scale insight into the structural effect of a supported Au catalyst based on a size-distribution analysis using Cs-STEM and morphological image-processing. *J. Catal.*, **335**, 24–35, 2016. https://doi.org/10.1016/j.jcat.2015.11.021.

[182] K. Schladitz, Quantitative micro-CT. *J. Microsc.*, **243**(2), 111–117, 2011. https://doi.org/10.1111/j.1365-2818.2011.03513.x.

[183] S. H. Wang et al., A user-friendly nano-CT image alignment and 3D reconstruction platform based on LabVIEW. *Chin. Phys. C*, **39**(1), 018001, 2015. https://doi.org/10.1088/1674-1137/39/1/018001.

[184] L. Landini, V. Positano, and M. F. Santarelli, *Advanced Image Processing in Magnetic Resonance Imaging*, 2005.

[185] Q. Liu, H. Wang, J. Liu, and H. Huang, AFM image processing for estimating the number and volume of nanoparticles on a rough surface. *Surf. Interface Anal.*, **43**(10), 1354–1359, 2011. https://doi.org/10.1002/sia.3722.

[186] M. Batool et al., Evaluation of Automatic Microstructural Analysis of Energy Dispersive Spectroscopy (EDS) Maps via a Python-Based Data Processing Framework. *Meet. Abstr.*, MA2021-02, 1041, 2021.

[187] R. C. Gonzalez and R. E. Woods, *Digital Image Processing*. Pearson, 2018. [Online]. Available: https://books.google.com/books?id=0F05vgAACAAJ.

[188] M. D. Abràmoff, P. J. Magalhães, and S. J. Ram, Image processing with imageJ. *Biophoton. Int.*, **11**(7), 36–41, 2004. https://doi.org/10.1201/9781420005615.ax4.

[189] Q. X. Meng, W. Y. Xu, H. L. Wang, X. Y. Zhuang, W. C. Xie, and T. Rabczuk, DigiSim — An Open Source Software Package for Heterogeneous Material Modeling Based on Digital Image Processing. *Adv. Eng. Softw.*, **148**, 102836, 2020. https://doi.org/10.1016/j.advengsoft.2020.102836.

[190] S. McCaslin and A. Kesireddy, Metallographic Image Processing Tools Using Mathematica Manipulate. In *Innovations and Advances in Computing, Informatics, Systems Sciences, Networking and Engineering*, pages 357–363, 2015.

[191] K. Radlak, M. Frackiewicz, M. Szczepanski, M. Kawulok, and M. Czardybon, Adaptive Vision Studio – Educational tool for image processing learning. In *Proc. – Front. Educ. Conf. FIE*, vol. 2015, pages 1–8, 2015. https://doi.org/10.1109/FIE.2015.7344309.

[192] T. R. Shaikh et al., Spider image processing for single-particle reconstruction of biological macromolecules from electron micrographs. *Nat. Protoc.*, **3**(12), 1941–1974, 2008. https://doi.org/10.1038/nprot.2008.156.

[193] S. Dey, *Hands-On Image Processing with Python: Expert techniques for advanced image analysis and effective interpretation of image data*. Packt Publishing, 2018. [Online]. Available: https://books.google.com/books?id=gC59DwAAQBAJ.

[194] C. Thölen, *Optimization of image processing operators for the AI-based discrimination of planktonic morphotypes*, 2020. https://doi.org/10.1002/9783527693009.ch60.

[195] M. Wang and C. H. Lai, *A Concise Introduction to Image Processing using C++*. CRC Press, 2016. [Online]. Available: https://books.google.com/books?id=fp7SBQAAQBAJ.

[196] D. A. Lyon, *Image Processing in Java*. Prentice Hall PTR, USA, 1999.

[197] A. K. Boyat and B. K. Joshi, A Review Paper: Noise Models in Digital Image Processing. *Signal Image Process., Int. J.*, **6**(2), 63–75, 2015. https://doi.org/10.5121/sipij.2015.6206.

[198] O. Joshua, T. Ibiyemi, and B. Adu, A Comprehensive Review On Various Types of Noise in Image Processing. *Int. J. Sci. Eng. Res.*, **10**, 388–393, 2019.

[199] A. Rai, E. Shivhare, and D. Mandloi, *Segmentation Based Denoising of Color Images using Morphological Operation*, 2018. https://doi.org/10.9790/9622-0807018084.

[200] P. L. Delisle, B. Anctil-Robitaille, C. Desrosiers, and H. Lombaert, Realistic image normalization for multi-Domain segmentation. *Med. Image Anal.*, **74**, 102191, 2021. https://doi.org/10.1016/j.media.2021.102191.

[201] G. D. Finlayson, B. Schiele, and J. L. Crowley, Comprehensive colour image normalization. In *Computer Vision – ECCV'98*, pages 475–490, 1998.

[202] M. Veluchamy and B. Subramani, Image contrast and color enhancement using adaptive gamma correction and histogram equalization. *Optik*, **183**, 329–337, 2019. https://doi.org/10.1016/j.ijleo.2019.02.054.

[203] R. Jenkin, R. E. Jacobson, N. R. Axford, and E. Allen, Effect of Gamma Correction in Discrete Imaging Systems upon Determination of MTF. *Soc. Imaging Sci. Technol. Image Process. Image Qual. Image Capture, Syst. Conf.*, **4**, 142–147, 2002.

[204] S. S. Bagade, Use of Histogram Equalization in Image Processing for Image Enhancement. *Int. J. Softw. Eng. Res. Pract.*, **1**(2), 6–10, 2011.

[205] G. Gupta, Image Filtering Algorithms and Techniques: A Review, *Int. J. Adv. Res. Comput. Sci. Softw. Eng.*, 2018.

[206] Y. Suzue, N. Shikazono, and N. Kasagi, Micro modeling of solid oxide fuel cell anode based on stochastic reconstruction. *J. Power Sources*, **184**(1), 52–59, 2008. https://doi.org/10.1016/j.jpowsour.2008.06.029.

[207] C. Tensmeyer and T. Martinez, Historical Document Image Binarization: A Review. *SN Comput. Sci.*, **1**(3), 2020. https://doi.org/10.1007/s42979-020-00176-1.

[208] T. R. Singh, S. Roy, O. I. Singh, T. Sinam, and K. M. Singh, A New Local Adaptive Thresholding Technique in Binarization. *IJCSI Int. J. Comput. Sci. Issues*, **8**(6), 2011.

[209] M. M. P. Petrou and C. Petrou, *Image Processing: The Fundamentals*. Wiley, 2010. [Online]. Available: https://books.google.com/books?id=w3BpSIxN9ZYC.

[210] P. Guruprasad, Overview Of Different Thresholding Methods in Image Processing. In *TEQIP Spons. 3rd Natl. Conf. ETACC*, 2020.

[211] T. Y. Goh, S. N. Basah, H. Yazid, M. J. Aziz Safar, and F. S. Ahmad Saad, Performance analysis of image thresholding: Otsu technique. *Meas. J. Int. Meas. Confed.*, **114**, 298–307, 2018. https://doi.org/10.1016/j.measurement.2017.09.052.

[212] M. A. Gaheen, E. Ibrahim, and A. A. Ewees, Chapter 8 – Edge detection-based segmentation for detecting skin lesions. In: P. Kumar, Y. Kumar and M. A. Tawhid, Eds., *Machine Learning, Big Data, and IoT for Medical Informatics*, pages 127–142. Academic Press, 2021. https://doi.org/10.1016/B978-0-12-821777-1.00008-2.

[213] S. Lou, L. Pagani, W. Zeng, X. Jiang, and P. J. Scott, Watershed segmentation of topographical features on freeform surfaces and its application to additively manufactured surfaces. *Precis. Eng.*, **63**, 177–186, 2020. https://doi.org/10.1016/j.precisioneng.2020.02.005.

[214] E. Guilminot, A. Corcella, F. Charlot, F. Maillard, and M. Chatenet, Detection of Pt[sup z+] Ions and Pt Nanoparticles Inside the Membrane of a Used PEMFC. *J. Electrochem. Soc.*, **154**(1), B96, 2007. https://doi.org/10.1149/1.2388863.

[215] V. Kumar and P. Gupta, Importance of Statistical Measures in Digital Image Processing. *Int. J. Emerg. Technol. Adv. Eng.*, **2**(8), 2012. [Online]. Available: www.ijetae.com.

[216] X. Zhang and W. Dahu, Application of artificial intelligence algorithms in image processing. *J. Vis. Commun. Image Represent.*, **61**, 42–49, 2019. https://doi.org/10.1016/j.jvcir.2019.03.004.

[217] N. Bonnet, Artificial intelligence and pattern recognition techniques in microscope image processing and analysis. In: P. W. Hawkes, Ed., *Advances in Imaging and Electron Physics*, vol. 114, pages 1–77. Elsevier, 2000. https://doi.org/10.1016/S1076-5670(00)80020-8.

[218] W. Cai, K. L. Lesnik, M. J. Wade, E. S. Heidrich, Y. Wang, and H. Liu, Incorporating microbial community data with machine learning techniques to predict feed substrates in microbial fuel cells. *Biosens. Bioelectron.*, **133**, 64–71, 2019. https://doi.org/10.1016/j.bios.2019.03.021.

[219] X. Wang, J. Chen, S. Quan, Y. X. Wang, and H. He, Hierarchical model predictive control via deep learning vehicle speed predictions for oxygen stoichiometry regulation of fuel cells. *Appl. Energy*, **276**, 115460, 2020. https://doi.org/10.1016/j.apenergy.2020.115460.

[220] M. Q. Huang, J. Ninić, and Q. B. Zhang, BIM, machine learning and computer vision techniques in underground construction: Current status and future perspectives. *Tunn. Undergr. Space Technol.*, **108**, 103677, 2021. https://doi.org/10.1016/j.tust.2020.103677.

[221] I. Arel, D. C. Rose, and T. P. Karnowski, Deep Machine Learning – A New Frontier in Artificial Intelligence Research [Research Frontier]. *IEEE Comput. Intell. Mag.*, **5**(4), 13–18, 2010. https://doi.org/10.1109/MCI.2010.938364.

[222] M. I. Jordon and T. M. Mitchell, Machine learning: Trends,perspectives, and prospects. *Science*, **349**(6245), 253–255, 2015. https://doi.org/10.1126/science.aac4520.

[223] Y. Zhang, *New Advances in Machine Learning*. IntechOpen, 2010. [Online]. Available: https://books.google.com/books?id=XAqhDwAAQBAJ.

[224] A. Singh, N. Thakur, and A. Sharma, A review of supervised machine learning algorithms. In *Proc. 10th INDIACom., 2016 3rd Int. Conf. Comput. Sustain. Glob. Dev.*, pages 1310–1315, 2016.

[225] T. Jiang, J. L. Gradus, and A. J. Rosellini, Supervised Machine Learning: A Brief Primer. *Behav. Ther.*, **51**(5), 675–687, 2020.

[226] V. Nasteski, An overview of the supervised machine learning methods. *HORIZONS B*, **4**, 51–62, 2017. https://doi.org/10.20544/horizons.b.04.1.17.p05.

[227] F. Y. Osisanwo, J. E. T. Akinsola, O. Awodele, J. O. Hinmikaiye, O. Olakanmi, and J. Akinjobi, Supervised Machine Learning Algorithms: Classification and Comparison. *Int. J. Comput. Trends Technol.*, **48**(3), 128–138, 2017. https://doi.org/10.14445/22312803/IJCTT-V48P126.

[228] S. Rong and Z. Bao-Wen, The research of regression model in machine learning field. In *MATEC Web of Conferences*, vol. 176, 2018. https://doi.org/10.1051/matecconf/201817601033.

[229] G. Bonaccorso, *Machine Learning Algorithms*. Packt Publishing, 2017. [Online]. Available: https://books.google.com/books?id=%5C_-ZDDwAAQBAJ.

[230] W. S. Noble, What is a support vector machine?. *Nat. Biotechnol.*, **24**, 2006. [Online]. Available: http://www.nature.com/naturebiotechnology.

[231] Z. Zhang, Introduction to machine learning: K-nearest neighbors. *Ann. Transl. Med.*, **4**(11), 2016. https://doi.org/10.21037/atm.2016.03.37.

[232] R. Fernandez Martinez, A. Okariz, J. Ibarretxe, M. Iturrondobeitia, and T. Guraya, Use of decision tree models based on evolutionary algorithms for the morphological classification of reinforcing nano-particle aggregates. *Comput. Mater. Sci.*, **92**, 102–113, 2014. https://doi.org/10.1016/j.commatsci.2014.05.038.

[233] L. Rokach and O. Maimon, Decision Trees. In: O. Maimon and L. Rokach, Eds., *Data Mining and Knowledge Discovery Handbook*, pages 165–192. Springer, Boston, 2005. https://doi.org/10.1007/0-387-25465-X_9.

[234] N. Horning, Random Forests: An algorithm for image classification and generation of continuous fields data sets. In *Int. Conf. Geoinformatics Spat. Infrastruct. Dev. Earth Allied Sci.*, pages 1–6, 2010.

[235] Z. Z. Wang and J. H. Yong, Texture analysis and classification with linear regression model based on wavelet transform. *IEEE Trans. Image Process.*, **17**(8), 1421–1430, 2008. https://doi.org/10.1109/TIP.2008.926150.

[236] Z. Li, N. Xiong, J. Liu, W. Gao, and R. Shamey, Determining the colorimetric attributes of multicolored materials based on a global correction and unsupervised image segmentation method. *Appl. Opt.*, **57**(26), 7482, 2018. https://doi.org/10.1364/ao.57.007482.

[237] R. Gentleman and V. J. Carey, Unsupervised Machine Learning. In *Bioconductor Case Studies*, pages 137–157. Springer, New York, 2008. https://doi.org/10.1007/978-0-387-77240-0_10.

[238] T. Jo, *Machine Learning Foundations: Supervised, Unsupervised, and Advanced Learning*. Springer, 2021. [Online]. Available: https://books.google.com/books?id=0egdEAAAQBAJ.

[239] A. Kassambara, *Practical Guide to Cluster Analysis in R: Unsupervised Machine Learning*. STHDA, 2017. [Online]. Available: https://books.google.com/books?id=pIEyDwAAQBAJ.

[240] S. Kaushik and S. Tiwari, *Soft Computing: Fundamentals, Techniques and Applications*. McGraw-Hill Education, 2018. [Online]. Available: https://books.google.com/books?id=BFOkDwAAQBAJ.

[241] V. Bhatnagar and S. Kaur, Exclusive and Complete Clustering of Streams. In *Database and Expert Systems Applications*, pages 629–638, 2007.

[242] A. Banerjee, C. Krumpelman, J. Ghosh, S. Basu, and R. J. Mooney, Model-based overlapping clustering. In *Proc. ACM SIGKDD Int. Conf. Knowl. Discov. Data Min.*, pages 532–537, 2005. https://doi.org/10.1145/1081870.1081932.

[243] R. Zass and A. Shashua, *A Unifying Approach to Hard and Probabilistic Clustering*, 2005.

[244] R. C. Camargos, P. R. Nietto, and M. do Carmo Nicoletti, Agglomerative and Divisive Approaches to Unsupervised Learning in Gestalt Clusters. In *Intelligent Systems Design and Applications*, pages 35–44, 2017.

[245] G. Nasierding, G. Tsoumakas, and A. Z. Kouzani, Clustering based multi-label classification for image annotation and retrieval. In *Conf. Proc. – IEEE Int. Conf. Syst. Man Cybern.*, pages 4514–4519, 2009. https://doi.org/10.1109/ICSMC.2009.5346902.

[246] V. K. Dehariya, S. K. Shrivastava, and R. C. Jain, Clustering of image data set using k-means and fuzzy k-means algorithms. In *Proceedings – 2010 International Conference on Computational Intelligence and Communication Networks, CICN 2010*, pages 386–391, 2010. https://doi.org/10.1109/CICN.2010.80.

[247] S. Noraini Sulaiman and N. Ashidi Mat Isa, *Adaptive Fuzzy-K-means Clustering Algorithm for Image Segmentation*, 2010.

[248] R. S. Sutton and A. G. Barto, *Reinforcement Learning, second edition: An Introduction*. MIT Press, 2018. [Online]. Available: https://books.google.com/books?id=uWV0DwAAQBAJ.

[249] A. V. Bernstein and E. V. Burnaev, Reinforcement learning in computer vision. In *Tenth International Conference on Machine Vision (ICMV 2017)*, vol. 10696, pages 458–464, 2018. https://doi.org/10.1117/12.2309945.

[250] M. Shokri and H. R. Tizhoosh, Using reinforcement learning for image thresholding. *Can. Conf. Electr. Comput. Eng.*, **2**, 1231–1234, 2003. https://doi.org/10.1109/ccece.2003.1226121.

[251] F. Sahba, H. R. Tizhoosh, and M. M. M. A. Salama, *Application of Opposition-Based Reinforcement Learning in Image Segmentation*, 2007.

[252] I. Goodfellow, Y. Bengio, and A. Courville, *Deep Learning*. MIT Press, 2016. [Online]. Available: https://books.google.com/books?id=omivDQAAQBAJ.

[253] Y. Lecun, Y. Bengio, and G. Hinton, Deep learning. *Nature*, **521**(7553), 436–444, 2015. https://doi.org/10.1038/nature14539.

[254] Y. Guo, Y. Liu, A. Oerlemans, S. Lao, S. Wu, and M. S. Lew, Deep learning for visual understanding: A review. *Neurocomputing*, **187**, 27–48, 2016. https://doi.org/10.1016/j.neucom.2015.09.116.

[255] S. Albawi, T. A. Mohammed, and S. Al-Zawi, Understanding of a convolutional neural network. In *Proc. 2017 Int. Conf. Eng. Technol. ICET 2017*, pages 1–6, 2018. https://doi.org/10.1109/ICEngTechnol.2017.8308186.

[256] R. Chauhan, K. K. Ghanshala, and R. C. Joshi, Convolutional Neural Network (CNN) for Image Detection and Recognition. In *ICSCCC 2018 – 1st Int. Conf. Secur. Cyber Comput. Commun.*, pages 278–282, 2018. https://doi.org/10.1109/ICSCCC.2018.8703316.

[257] T. M. Quan, D. G. C. Hildebrand, and W.-K. Jeong, FusionNet: A Deep Fully Residual Convolutional Neural Network for Image Segmentation in Connectomics. *Front. Comput. Sci.*, **3**, 2021. https://doi.org/10.3389/fcomp.2021.613981.

[258] W. Zaremba, I. Sutskever, and O. Vinyals. Recurrent Neural Network Regularization, 2015. [Online]. Available: http://arxiv.org/abs/1409.2329.

[259] R. Cierniak, A new approach to image reconstruction from projections using a recurrent neural network. *Int. J. Appl. Math. Comput. Sci.*, **18**(2), 147–157, 2008. https://doi.org/10.2478/v10006-008-0014-y.

[260] X. Zhang, Y. Sun, K. Jiang, C. Li, L. Jiao, and H. Zhou, Spatial Sequential Recurrent Neural Network for Hyperspectral Image Classification. *IEEE J. Sel. Top. Appl. Earth Obs. Remote Sens.*, **11**(11), 4141–4155, 2018. https://doi.org/10.1109/JSTARS.2018.2844873.

[261] T. H. N. Le, K. G. Quach, K. Luu, C. N. Duong, and M. Savvides, Reformulating Level Sets as Deep Recurrent Neural Network Approach to Semantic Segmentation. *IEEE Trans. Image Process.*, **27**(5), 2393–2407, 2018. https://doi.org/10.1109/TIP.2018.2794205.

[262] M. J. Eslamibidgoli, F. P. Tipp, J. Jitsev, J. Jankovic, M. H. Eikerling, and K. Malek, Convolutional neural networks for high throughput screening of catalyst layer inks for polymer electrolyte fuel cells. *RSC Adv.*, **11**(51), 32126–32134, 2021. https://doi.org/10.1039/d1ra05324h.

[263] F. K. Wang, T. Mamo, and X. B. Cheng, Bi-directional long short-term memory recurrent neural network with attention for stack voltage degradation from proton exchange membrane fuel cells. *J. Power Sources*, **461**, 228170, 2020. https://doi.org/10.1016/j.jpowsour.2020.228170.

[264] M. M. Daino and S. G. Kandlikar, Evaluation of Imaging Techniques Applied to Water Management Research in PEMFCs. In *ASME 2009 7th International Conference on Nanochannels, Microchannels, and Minichannels*, pages 467–479. ASME, 2009. https://doi.org/10.1115/ICNMM2009-82031.

[265] J. M. Sergi and S. G. Kandlikar, Quantification and characterization of water coverage in PEMFC gas channels using simultaneous anode and cathode visualization and image processing. *Int. J. Hydrog. Energy*, **36**(19), 12381–12392, 2011. https://doi.org/10.1016/j.ijhydene.2011.06.092.

[266] R. T. White et al., Four-dimensional joint visualization of electrode degradation and liquid water distribution inside operating polymer electrolyte fuel cells. *Sci. Rep.*, **9**(1), 1843, 2019. https://doi.org/10.1038/s41598-018-38464-9.

[267] R. Satija, D. L. Jacobson, M. Arif, and S. A. Werner, In situ neutron imaging technique for evaluation of water management systems in operating PEM fuel cells. *J. Power Sources*, **129**(2), 238–245, 2004. https://doi.org/10.1016/j.jpowsour.2003.11.068.

[268] T. A. Trabold, J. P. Owejan, D. L. Jacobson, M. Arif, and P. R. Huffman, In situ investigation of water transport in an operating PEM fuel cell using neutron radiography: Part 1 - Experimental method and serpentine flow field results. *Int. J. Heat Mass Transf.*, **49**(25–26), 4712–4720, 2006. https://doi.org/10.1016/j.ijheatmasstransfer.2006.07.003.

[269] Y. S. Chen et al., Water distribution measurement for a PEMFC through neutron radiography. *J. Power Sources*, **170**(2), 376–386, 2007. https://doi.org/10.1016/j.jpowsour.2007.03.076.

[270] S. J. Lee, N. Y. Lim, S. Kim, G. G. Park, and C. S. Kim, X-ray imaging of water distribution in a polymer electrolyte fuel cell. *J. Power Sources*, **185**(2), 867–870, 2008. https://doi.org/10.1016/j.jpowsour.2008.08.101.

[271] S. Litster, D. Sinton, and N. Djilali, Ex situ visualization of liquid water transport in PEM fuel cell gas diffusion layers. *J. Power Sources*, **154**(1), 95–105, 2006. https://doi.org/10.1016/j.jpowsour.2005.03.199.

[272] M. G. George et al., Effects of compression on water distribution in gas diffusion layer materials of PEMFC in a point injection device by means of synchrotron X-ray imaging. *Int. J. Hydrog. Energy*, **43**(1), 391–406, 2018. https://doi.org/10.1016/j.ijhydene.2017.11.047.

[273] V. Chauhan, M. Mortazavi, J. Z. Benner, and A. D. Santamaria, Two-phase flow characterization in PEM fuel cells using machine learning. *Energy Rep.*, **6**, 2713–2719, 2020. https://doi.org/10.1016/j.egyr.2020.09.037.

[274] D. Lee and S. Hwang, Effect of loading and distributions of Nafion ionomer in the catalyst layer for PEMFCs. *Int. J. Hydrog. Energy*, **33**(11), 2790–2794, 2008. https://doi.org/10.1016/j.ijhydene.2008.03.046.

[275] T. Suzuki, S. Okada, and S. Tsushima, Analysis of Ionomer Distribution and Pt/C Agglomerate Size in Catalyst Layers by Two-Stage Ion-Beam Processing. *J. Electrochem. Soc.*, **167**(12), 124513, 2020. https://doi.org/10.1149/1945-7111/abad6a.

[276] T. Suzuki, T. Koyama, and S. Tsushima, Analysis of Ionomer Distribution in Catalyst Layers by Two-Stage Ion Beam Processing. *ECS Trans.*, **80**, 419, 2017.

[277] S. J. Normile and I. V. Zenyuk, Imaging ionomer in fuel cell catalyst layers with synchrotron nano transmission x-ray microscopy. *Solid State Ion.*, **335**, 38–46, 2019. https://doi.org/10.1016/j.ssi.2019.02.017.

[278] M. Uchida et al., Effect of the state of distribution of supported Pt nanoparticles on effective Pt utilization in polymer electrolyte fuel cells. *Phys. Chem. Chem. Phys.*, **27**, 11236–11247, 2013. https://doi.org/10.1039/c3cp51801a.

[279] K. Artyushkova, S. Pylypenko, M. Dowlapalli, and P. Atanassov, Use of digital image processing of microscopic images and multivariate analysis for quantitative correlation of morphology, activity and durability of electrocatalysts. *RSC Adv.*, **2**(10), 4304–4310, 2012. https://doi.org/10.1039/c2ra00574c.

[280] A. Serov et al., Nano-structured platinum group metal-free catalysts and their integration in fuel cell electrode architectures. *Appl. Catal. B, Environ.*, **237**, 1139–1147, 2018. https://doi.org/10.1016/j.apcatb.2017.08.067.

[281] K. Khedekar et al., Probing Heterogeneous Degradation of Catalyst in PEM Fuel Cells under Realistic Automotive Conditions with Multi-Modal Techniques. *Adv. Energy Mater.*, **11**(35), 202101794, 2021. https://doi.org/10.1002/aenm.202101794.

[282] I. V. Pushkareva et al., Reduced graphene oxide-supported pt-based catalysts for pem fuel cells with enhanced activity and stability. *Catalysts*, **11**(2), 1–14, 2021. https://doi.org/10.3390/catal11020256.

[283] C. C. Chao, M. Motoyama, and F. B. Prinz, Nanostructured platinum catalysts by atomic-layer deposition for solid-oxide fuel cells. *Adv. Energy Mater.*, **2**(6), 651–654, 2012. https://doi.org/10.1002/aenm.201200002.

[284] N. P. Brandon and D. J. Brett, Engineering porous materials for fuel cell applications. *Philos. Trans. R. Soc., Math. Phys. Eng. Sci.*, **364**(1838), 147–159, 2006. https://doi.org/10.1098/rsta.2005.1684.

[285] T. Wejrzanowski et al., Multi-modal porous microstructure for high temperature fuel cell application. *J. Power Sources*, **373**, 85–94, 2018. https://doi.org/10.1016/j.jpowsour.2017.11.009.

[286] S. Thiele, R. Zengerle, and C. Ziegler, Nano-morphology of a polymer electrolyte fuel cell catalyst layer-imaging, reconstruction and analysis. *Nano Res.*, **4**(9), 849–860, 2011. https://doi.org/10.1007/s12274-011-0141-x.

[287] C. Santoro et al., Parameters characterization and optimization of activated carbon (AC) cathodes for microbial fuel cell application. *Bioresour. Technol.*, **163**, 54–63, 2014. https://doi.org/10.1016/j.biortech.2014.03.091.

[288] J. P. James, H. W. Choi, and J. G. Pharoah, X-ray computed tomography reconstruction and analysis of polymer electrolyte membrane fuel cell porous transport layers. *Int. J. Hydrog. Energy*, **37**(23), 18216–18230, 2012. https://doi.org/10.1016/j.ijhydene.2012.08.077.

[289] M. A. Laguna-Bercero, R. Campana, A. Larrea, J. A. Kilner, and V. M. Orera, Performance and aging of microtubular YSZ-based solid oxide regenerative fuel cells. *Fuel Cells*, **11**(1), 116–123, 2011. https://doi.org/10.1002/fuce.201000069.

[290] D. J. S. Sandbeck et al., Particle Size Effect on Platinum Dissolution: Practical Considerations for Fuel Cells. *ACS Appl. Mater. Interfaces*, **12**(23), 25718–25727, 2020. https://doi.org/10.1021/acsami.0c02801.

[291] K. Sugihara et al., A quantitative analysis of influence of Ni particle size of SDC-supported anode on SOFC performance: Effect of particle size of SDC support. *Solid State Ion.*, **262**, 433–437, 2014. https://doi.org/10.1016/j.ssi.2014.02.012.

[292] N. Giordano et al., Effect of platinum particle size on the performance of PAFC O2 reduction electrocatalysts. *Int. J. Hydrog. Energy*, **19**(2), 165–168, 1994. doi. https://doi.org/10.1016/0360-3199(94)90122-8.

[293] Z. Yang, S. Ball, D. Condit, and M. Gummalla, Systematic Study on the Impact of Pt Particle Size and Operating Conditions on PEMFC Cathode Catalyst Durability. *J. Electrochem. Soc.*, **158**(11), B1439, 2011. https://doi.org/10.1149/2.081111jes.

[294] J. H. Lee et al., The impact of anode microstructure on the power generating characteristics of SOFC. *Solid State Ion.*, **158**(3–4), 225–232, 2003. https://doi.org/10.1016/S0167-2738(02)00915-3.

[295] X. Peng et al., High-performing pgm-free aemfc cathodes from carbon-supported cobalt ferrite nanoparticles. *Catalysts*, **9**(3), 1–12, 2019. https://doi.org/10.3390/catal9030264.

[296] S. Shawuti and M. A. Gulgun, Solid oxide-molten carbonate nano-composite fuel cells: Particle size effect. *J. Power Sources*, **267**, 128–135, 2014. https://doi.org/10.1016/j.jpowsour.2014.05.010.

[297] A. A. Serov, M. Min, G. Chai, S. Han, S. Kang, and C. Kwak, Preparation, characterization, and high performance of RuSe/C for direct methanol fuel cells. *J. Power Sources*, **175**(1), 175–182, 2008. https://doi.org/10.1016/j.jpowsour.2007.08.089.

[298] K. J. Yoon et al., Nano-tailoring of infiltrated catalysts for high-temperature solid oxide regenerative fuel cells. *Nano Energy*, **36**, 9–20, 2017. https://doi.org/10.1016/j.nanoen.2017.04.024.

[299] J.-H. Lee, H. Moon, H.-W. Lee, J. Kim, J.-D. Kim, and K.-H. Yoon, Quantitative analysis of microstructure and its related electrical property of SOFC anode, Ni-YSZ cermet. *Solid State Ion.*, **148**, 15–26, 2002.

[300] Z. Zhou et al., Novel synthesis of highly active Pt/C cathode electrocatalyst for direct methanol fuel cell. *Chem. Commun.*, **3**(3), 394–395, 2003. https://doi.org/10.1039/b211075j.

[301] H. Yu et al., Strategies to mitigate Pt dissolution in low Pt loading proton exchange membrane fuel cell: I. A gradient Pt particle size design. *Electrochim. Acta*, **247**, 1155–1168, 2017. https://doi.org/10.1016/j.electacta.2017.07.093.

[302] A. Colliard-Granero et al., Deep Learning for the Automation of Particle Analysis in Catalyst Layers for Polymer Electrolyte Fuel Cells. *Nanoscale*, **14**(1), 10–18, 2022. https://doi.org/10.1039/d1nr06435e.

[303] D. Yang and Z. Liu, Quantification of Microstructural Features and Prediction of Mechanical Properties of a Dual-Phase Ti-6Al-4V Alloy. *J. Mater.*, **9**(8), 628, 2016. https://doi.org/10.3390/ma9080628.

[304] J. S. Cronin, Y. C. K. Chen-Wiegart, J. Wang, and S. A. Barnett, Three-dimensional reconstruction and analysis of an entire solid oxide fuel cell by full-field transmission X-ray microscopy. *J. Power Sources*, **233**, 174–179, 2013. https://doi.org/10.1016/j.jpowsour.2013.01.060.

[305] R. K. Singh et al., Synthesis of CeOx-Decorated Pd/C Catalysts by Controlled Surface Reactions for Hydrogen Oxidation in Anion Exchange Membrane Fuel Cells. *Adv. Funct. Mater.*, **30**(38), 2002087, 2020. https://doi.org/10.1002/adfm.202002087.

[306] M. Batool, A. O. Godoy, M. Birnbach, D. R. Dekel, and J. Jankovic, Evaluation of Automatic Microstructural Analysis of Energy Dispersive Spectroscopy (EDS) Maps via a Python-Based Data Processing Framework. *ECS Trans.*, **104**(8), 137–153, 2021. https://doi.org/10.1149/10408.0137ecst.

[307] H. Hwang et al., Integrated application of semantic segmentation-assisted deep learning to quantitative multi-phased microstructural analysis in composite materials: Case study of cathode composite materials of solid oxide fuel cells. *J. Power Sources*, **471**, 228458, 2020. https://doi.org/10.1016/j.jpowsour.2020.228458.

[308] R. T. White, A. Wu, M. Najm, F. P. Orfino, M. Dutta, and E. Kjeang, 4D in situ visualization of electrode morphology changes during accelerated degradation in fuel cells by X-ray computed tomography. *J. Power Sources*, **350**, 94–102, 2017. https://doi.org/10.1016/j.jpowsour.2017.03.058.

[309] K. Epping Martin, J. P. Kopasz, and K. W. McMurphy, Status of Fuel Cells and the Challenges Facing Fuel Cell Technology Today. In *Fuel Cell Chemistry and Operation*, 2010. [Online]. Available: https://pubs.acs.org/sharingguidelines.

[310] S. Prass, S. Hasanpour, P. K. Sow, A. B. Phillion, and W. Mérida, Microscale X-ray tomographic investigation of the interfacial morphology between the catalyst and micro porous layers in proton exchange membrane fuel cells. *J. Power Sources*, **319**, 82–89, 2016. https://doi.org/10.1016/j.jpowsour.2016.04.031.

[311] M. Andisheh-Tadbir, F. P. Orfino, and E. Kjeang, Three-dimensional phase segregation of micro-porous layers for fuel cells by nano-scale X-ray computed tomography. *J. Power Sources*, **310**, 61–69, 2016. https://doi.org/10.1016/j.jpowsour.2016.02.001.

[312] Y. Yu et al., Effect of Sr Content and Strain on Sr Surface Segregation of La1-xSrxCo0.2Fe0.8O3-δ as Cathode Material for Solid Oxide Fuel Cells. *ACS Appl. Mater. Interfaces*, **8**(40), 26704–26711, 2016. https://doi.org/10.1021/acsami.6b07118.

[313] V. Julie et al., 3D phase mapping of solid oxide fuel cell YSZ/Ni cermet at the nanoscale by holographic X-ray nanotomography. *J. Power Sources*, **243**, 841–849, 2013. https://doi.org/10.1016/j.jpowsour.2013.06.069.

[314] W. M. Harris et al., Three-dimensional microstructural mapping of poisoning phases in the Neodymium Nickelate solid oxide fuel cell cathode. *Solid State Ion.*, **237**, 16–21, 2013. https://doi.org/10.1016/j.ssi.2013.01.020.

[315] J. Thangavelautham, Degradation in PEM Fuel Cells and Mitigation Strategies Using System Design and Control. In: T. Taner, Ed., *Proton Exchange Membrane Fuel Cell*. IntechOpen, Rijeka, 2018. https://doi.org/10.5772/intechopen.72208.

[316] D. Simwonis, F. Tietz, and D. Stöver, Nickel coarsening in annealed Ni/8YSZ anode substrates for solid oxide fuel cells. *Solid State Ion.*, **132**(3), 241–251, 2000. https://doi.org/10.1016/s0167-2738(00)00650-0.

[317] R. T. White et al., Correlative X-ray Tomographic Imaging of Catalyst Layer Degradation in Fuel Cells. *J. Electrochem. Soc.*, **166**(13), F914–F925, 2019. https://doi.org/10.1149/2.0121913jes.

[318] A. Zekri, M. Knipper, J. Parisi, and T. Plaggenborg, Microstructure degradation of Ni/CGO anodes for solid oxide fuel cells after long operation time using 3D reconstructions by FIB tomography. *Phys. Chem. Chem. Phys.*, **19**(21), 13767–13777, 2017. https://doi.org/10.1039/c7cp02186k.

[319] S. H. Eberhardt, M. Toulec, F. Marone, M. Stampanoni, F. N. Büchi, and T. J. Schmidt, Dynamic Operation of HT-PEFC: In-Operando Imaging of Phosphoric Acid Profiles and (Re)distribution. *J. Electrochem. Soc.*, **162**(3), F310–F316, 2015. https://doi.org/10.1149/2.0751503jes.

[320] A. Pokhrel, M. El Hannach, F. P. Orfino, M. Dutta, and E. Kjeang, Failure analysis of fuel cell electrodes using three-dimensional multi-length scale X-ray computed tomography. *J. Power Sources*, **329**, 330–338, 2016. https://doi.org/10.1016/j.jpowsour.2016.08.092.

David A. Cullen, Deborah J. Myers, and Piotr Zelenay

3 Characterization approaches for atomically dispersed platinum group metal-free catalysts

3.1 Introduction

The long-term success of polymer electrolyte fuel cells (PEFCs) will require significant cost reductions for the affordable production, storage, distribution and use of hydrogen [1]. With respect to the fuel cell stack, much of the cost is associated with the scarce precious metal catalysts used to drive the oxygen reduction reaction (ORR) in the cathode of the membrane electrode assembly (MEA) [2]. Replacing the costly platinum-based ORR catalysts with platinum group metal-free (PGM-free) alternatives is an important long-term goal to ensure the sustainability of the hydrogen economy [3]. The first PGM-free ORR catalysts, discovered in 1964, were metal phthalocyanines [4]. This was followed by improvements in activity through the heat treatment of well-defined organic macrocycles [5–7]. The pioneering effort by Yeager et al. in 1981 of mixing carbon and polyacrylonitrile with transition metals salts gave birth to most current PGM-free ORR catalyst synthesis approaches [8], with much of the field currently focused on metal organic frameworks as precursor materials. For additional background on the nature of the active sites, synthesis approaches, reaction pathways for the ORR and degradation mechanisms, we direct the reader to one of the many excellent reviews in the literature [9–15]. While much progress has been made over the last 50 years, significant advances in activity and durability are still required before this class of catalysts can begin to replace platinum in commercial applications. In this book chapter, we review the unique techniques developed to characterize electrocatalysts, explore the impact of synthesis parameters on catalyst structure and understand degradation mechanisms. We review established techniques along with promising emerging methods that aim to accelerate PGM-free catalyst development not only for the sustainable hydrogen economy, but a number of other emerging applications including anion-exchange membrane fuel cells, electrochemical CO_2 reduction and ammonia synthesis [16–18].

Acknowledgement: These authors would like to recognize support from the U. S. Department of Energy (DOE) Office of Energy Efficiency and Renewable Energy (EERE), Hydrogen and Fuel Cell Technologies Office, through the Electrocatalysis Consortium (ElectroCat).

David A. Cullen, Center for Nanophase Materials Sciences, Oak Ridge National Laboratory, Oak Ridge, TN 37831, USA
Deborah J. Myers, Chemical Sciences and Engineering Division, Argonne National Laboratory, Lemont, IL 60439, USA
Piotr Zelenay, Materials Physics and Applications Division, Los Alamos National Laboratory, Los Alamos, NM 87545, USA

https://doi.org/10.1515/9783110622720-003

3.2 X-ray photoelectron spectroscopy

One of the earliest and most common methods employed for correlating the composition and structure of PGM-free catalysts with electrochemical performance is X-ray photoelectron spectroscopy (XPS) [19]. This surface-sensitive technique relies on the excitation and emission of electrons using an X-ray beam with a spot size on the order of hundreds of microns. Most XPS studies use lab-scale instruments, although the technique is also available at synchrotron facilities, which provide higher X-ray brightness, improved signal-to-noise ratios and tunable X-ray energy to adjust probing depth. It should be noted that although XPS is a surface sensitive technique, due to the high micro and mesoporosity of the carbon-based catalysts, the signal contributing to the spectra is not strictly limited to the outermost surface of the catalyst particles.

XPS can be used to determine both the average near-surface elemental composition of the catalysts and the chemical states of the individual elements. Compositional information is extracted from survey spectra, and typical PGM-free ORR electrocatalysts contain a few tenths of an atomic percent of transition metal (Fe, Co, Ni, etc.), with 2–4 at.% N and O, and the remainder C. Other common residual elements include Cl, S, K, Si and Zn, depending on the synthesis route. Details on the chemical or electronic state of individual elements can be obtained from deliberate, informed curve fitting of core level spectra. Such fitting has been guided by density functional theory (DFT) calculations, as performed in early XPS studies of PGM-free catalysts to elucidate the chemical bonding of C, N, O and the transition metal, typically Fe [20, 21].

In order to make sense of the complex, three-dimensional structures comprising PGM-free catalysts, the material is often simplified to a depiction of graphene sheets containing a range of possible C, N and Fe defects, as shown in Figure 1a. Fitting of the high-resolution XPS core level spectra can provide critical insights into the degree to which these various moieties may be present. Due to the low transition metal loading, core level fitting is typically focused on the C 1s and N 1s edges for lab-scale instruments, although increasing the acquisition time or using synchrotron-based XPS can yield an Fe 2p spectrum with sufficient signal-to-noise for reliable peak fitting. The C 1s spectrum is often fit with five peaks representing sp^2, sp^3, C–N, C–O and O=C–O bonds, as shown in Figure 1b, with attention given to the relative degree of sp^2 and sp^3 carbon as a measure of the degree of graphitization, which has been shown to impact activity and durability alike [22]. The N 1s spectra is most commonly fit with four peaks indicative of nitrogen in the pyridinic, pyrrolic, quaternary and graphitic states, as well as a fifth peak relating to a N_x-Fe bond that many attribute to the active species (Figure 1c). The Fe 2p can be fit with metallic Fe, Fe oxides and FeN_x peaks (Figure 1d), but due to the low concentration (typically < 0.3 at.%), such peak fitting is not typically performed or published.

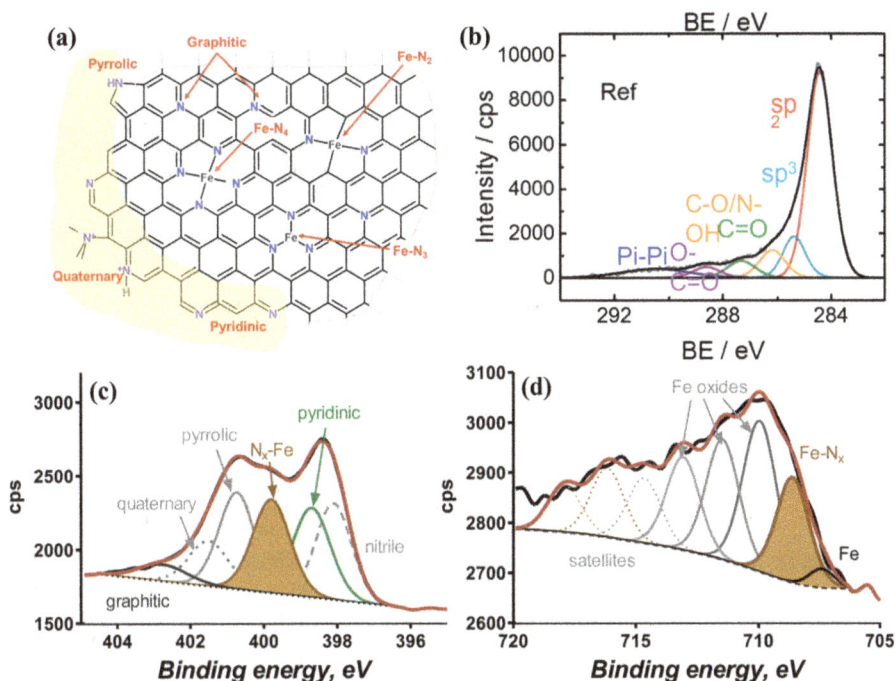

Figure 1: (a). Diagram of some of the potential functionalized N and Fe species present in heterogeneous PGM-free catalysts. Typical XPS peak fitting of (b) C 1s, (c) N 1s and (d) Fe 2p core level spectra for pyrolyzed PGM-free catalysts. (a, c, d) Adapted with permission from [23]. Copyright 2015, American Chemical Society. (b) Adapted from [24] with permission from The Royal Society of Chemistry.

A key finding from early XPS studies is the presence of "multitudinous" active sites, as coined by Artyushkova and coauthors, meaning a variety of FeN and CN moieties exist that contribute to the activity of a given PGM-free catalyst [23]. Figure 2 shows the range of possible active sites along with their DFT-determined binding energy [21]. While many efforts have aimed to limit the diverse atomic arrangements of the N and Fe to those predicted to have the highest contributions to activity, such control has been elusive given the complex, three-dimensional structures resulting from the high temperature pyrolysis of these materials. The understanding of the complexity of these materials, obtained from characterization efforts using XPS and other techniques, highlights the importance of not oversimplifying the nature of the active site to a single type of graphene-hosted FeN_x moiety.

Thanks to many pioneering efforts, XPS is now routinely used to understand the impact of different synthesis strategies on the catalyst composition, degree of carbon graphitization and nitrogen speciation, which can then be correlated with changes in electrochemical performance and durability to help formulate new synthesis strategies.

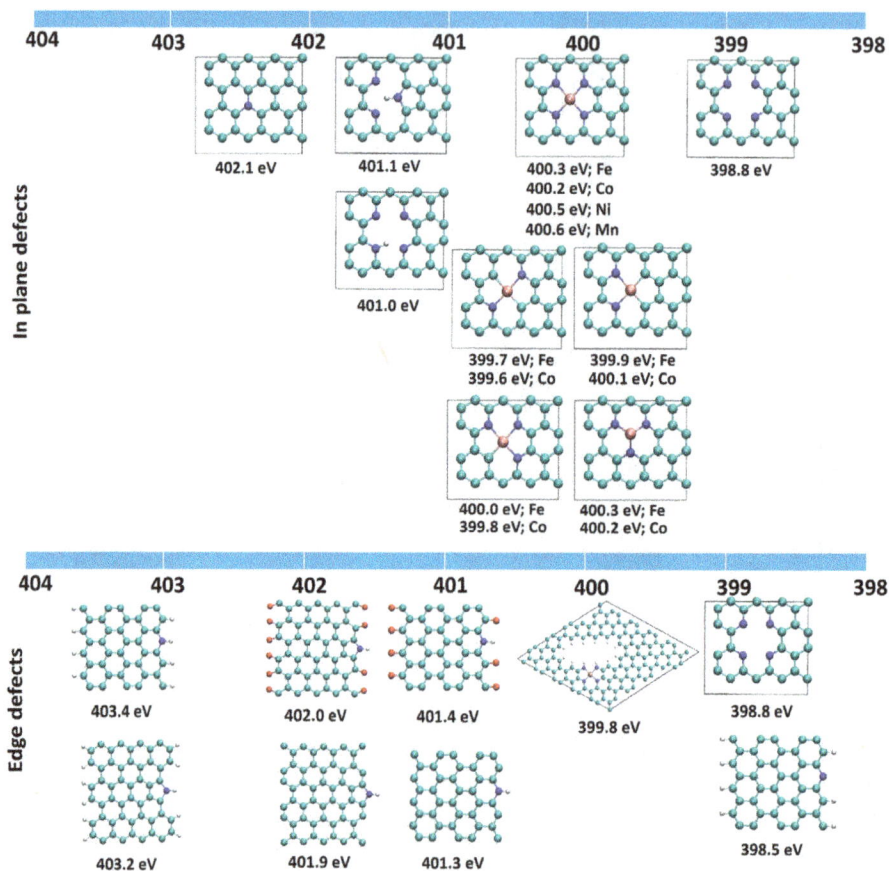

Figure 2: DFT-calculated N 1s binding energies for a range of possible N defects within graphene. Reprinted from [21]. Copyright 2018, with permission from Elsevier.

The pioneering efforts continue as more advanced techniques, such as *operando* and near-ambient pressure XPS, come online to further elucidate the nature of active sites and degradation mechanisms in this class of catalysts [25, 26]. Early efforts along this direction are shown in Figure 3 where changes to the N 1s spectra in the presence of oxygen and water, coupled with *in situ* soft X-ray absorption spectroscopy, confirmed the presence of multiple Fe and N moieties in electrocatalysts prepared under different synthetic and processing conditions and indicated the most active species readily adsorb oxygen when exposed to air [27, 28].

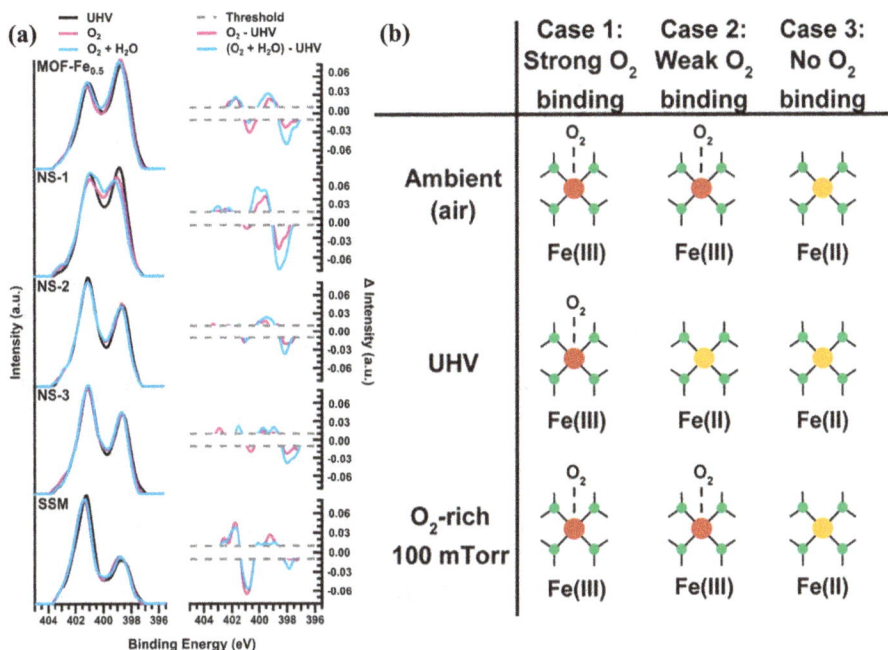

Figure 3: (a) Comparison of ultrahigh vacuum and near-ambient pressure XPS N-1s spectra in the presence of O_2 and O_2+H_2O. (b) Comparison of effect of oxygen absorption on oxidation state of Fe in FeN_xC_y sites under different environments. The most active catalysts had been pyrolyzed at higher temperatures (>1000 °C) and exhibited strong oxygen bonding on the FeN_x sites when exposed to air. Reprinted with permission from [28]. Copyright 2020, American Chemical Society.

3.3 X-ray absorption spectroscopy

X-ray absorption spectroscopy (XAS) has proven to be a highly valuable technique to characterize the nature of the active species in M–N–C catalysts [29]. Using a synchrotron source, the energy of X-rays can be tuned to interact only with the element of interest in catalytic materials. The transition metals typically employed in PGM-free electrocatalysts, e. g., Fe, Co, Mn, have K absorption edges in the 6.5 to 8 keV energy range. In this energy range, other materials in the PEFC cathode environment, such as water and carbon, have low absorption cross-sections. This property allows X-rays to penetrate through the flow field, cathode gas (air, N_2 or Ar), and gas diffusion layer in the fuel cell environment or through aqueous electrolyte and carbon-based substrates (e. g., glassy carbon, graphene, carbon paper) to probe just the transition metal in the cathode electrocatalyst. This makes XAS a powerful and common tool for characterizing the transition metals in PGM-free electrocatalysts during fuel cell operation (*operando*) or in aqueous electrolyte (*in situ*), which is often used to mimic the fuel cell environment.

An XAS spectrum consists of two main regions: the energy region near the absorption edge, termed the X-ray absorption near-edge structure (XANES) region, typically defined as the region within 30 eV of the absorption edge, and the extended X-ray absorption fine structure (EXAFS) region, which encompasses the region from approximately 30 to 1,000 eV above the absorption edge. The XANES region is particularly sensitive to the coordination environment and oxidation state of the absorbing atom and to its electronic structure [30]. For catalysts with transition metals in multiple oxidation states and coordination environments, it is often not possible to determine the structure of the transition metal species through fitting of the EXAFS region of the spectrum, as described below [18, 31, 32]. In these cases, linear combination fitting (LCF) of the XANES spectrum to the spectra of relevant standards can be used to deconvolute the catalyst's composition. Analysis of the EXAFS region of the XAS spectrum through Fourier transform and atomic-level fitting provides the number, identity and length of the bonds between the absorbing element and its nearest neighbors. EXAFS can be limited in the information it can provide regarding the bonding geometries, however, the structure of the XANES region, in particular the absorption intensity pattern in the pre-edge region, is sensitive to the bonding geometry and can be modeled and compared to those of standards to provide this essential information [33, 34]. This is illustrated in Figure 4, which summarizes a study involving a PGM-free catalyst obtained by mixing ZIF-8, Fe-acetate and phenanthroline, followed by pyrolysis in argon or ammonia, resulting in a PGM-free catalyst, which is free of crystalline iron. Fitting of the EXAFS edge indicated ligation of a FeN_4 center by oxygen. Theoretical XANES spectra calculated for various Fe-N-C structures were compared to the catalyst's spectrum, showing porphyrin-like FeN_4C_{12} moieties with axial oxygen atoms best fit the experimental result [35].

A main advantage of XAS is the ability to perform experiments *in situ* or *operando* [36, 37], which has been used extensively for PGM-free catalyst characterization [32, 38–43]. Such experiments have shown that the ORR activity of Fe-N-C catalysts prepared using different precursor materials have different behavior and undergo different structural changes with changing potential. However, despite their different structures, the activity of these catalysts can be correlated with the potential dependence of the Fe^{3+} to Fe^{2+} redox transition. It was found that the $Fe(III)N_4$ center has a site-blocking axial O(H) ligand, formed from water, which is removed during this redox transition, freeing the site for adsorption of O_2. Osmieri et al. expanded this model to the fuel cell environment and correlated the cathode potential dependence of the active site availability, as determined from hydrogen-air polarization curves acquired using a range of pressures and temperatures, with the Fe oxidation state determined using *operando* XAS (Figure 5) [32]. Active site availability was shown to be a function of both the change in oxidation state (i. e., the number of electrons transferred in the redox reaction) and the potential at which this transition occurs, which was found to be ~120 mV higher in the fuel cell environment as compared to the aqueous acidic (0.5 M H_2SO_4) environment.

In another *in situ* experiment, a range of PGM-free catalysts, which varied in precursor material, synthesis approach and pyrolysis temperature, were studied by XAS, among other methods [42, 44]. As summarized in Figure 6, it was shown that despite

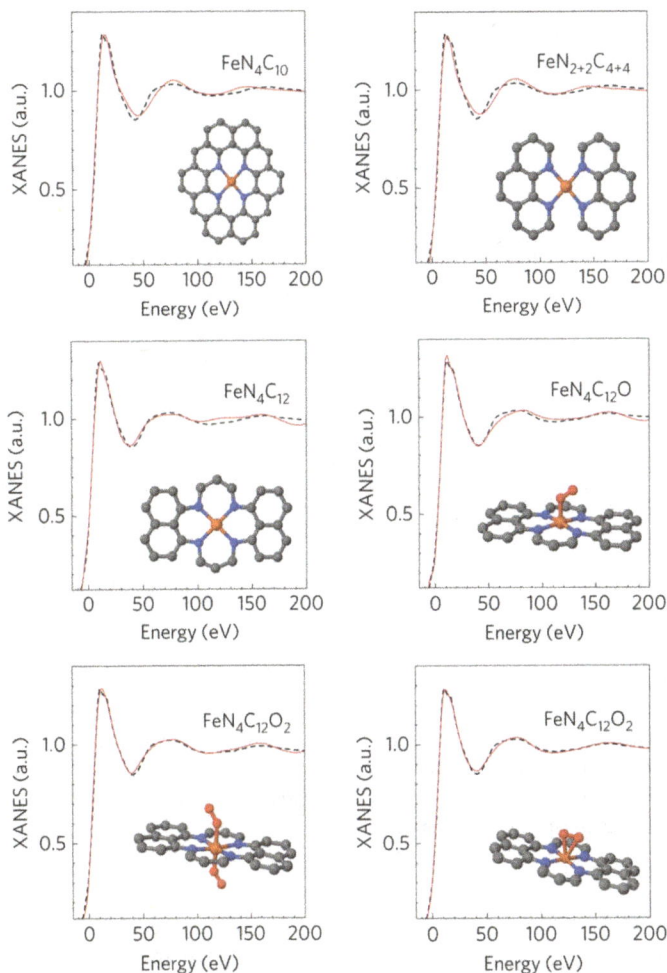

Figure 4: Graphical depictions of various FeN_xC_y moieties with and without adsorbed oxygen molecules along with the experimental Fe K-edge XANES spectrum for each structure fitted against theoretical spectra. Reprinted by permission from [35]. Copyright 2015, Springer Nature.

the differences in synthesis conditions and starting materials, each shared a common high potential for the Fe^{2+}/Fe^{3+} redox transition in acidic electrolyte, which the authors identified as a nonplanar FeN_4 moiety embedded in distorted carbon matrix.

Unlike near-surface sensitive XPS, XAS probes all the selected metal species in the beam path, and thus can be categorized as a bulk technique. However, surface sensitivity can be imparted to XAS by utilizing a technique termed delta-μ ($\Delta\mu$), pioneered by Ramaker and Koningsberger for fuel cell-relevant electrochemical systems [45]. For these systems, the $\Delta\mu$ technique involves taking the difference of XAS spectra acquired at two potentials: one where the transition metal is expected to be adsorbate-free and the other

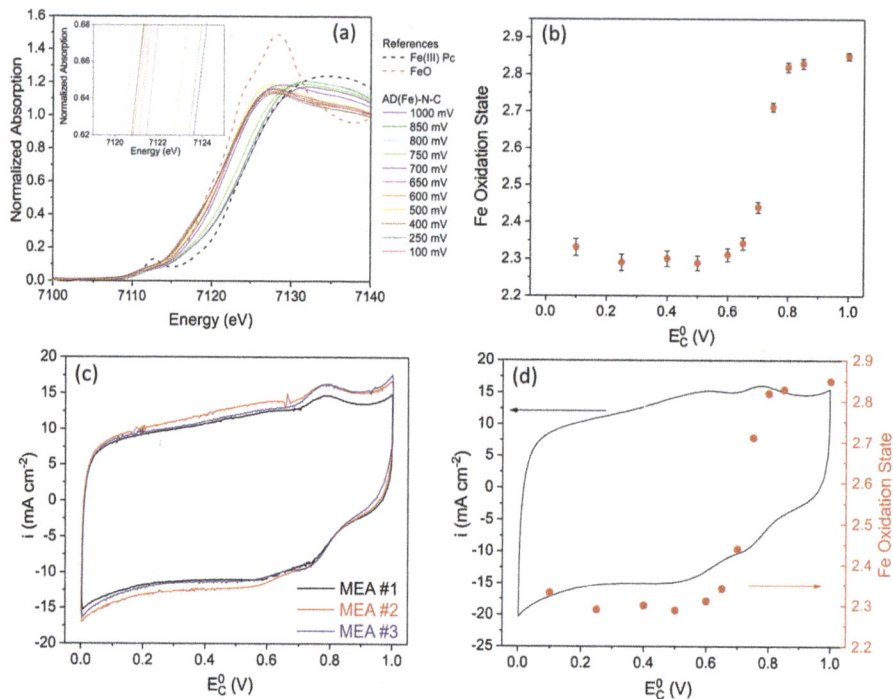

Figure 5: (a) *In situ* XANES spectra of a PGM-free cathode in a membrane electrode assembly (MEA) acquired with helium at the cathode and hydrogen at the anode. (b) Change in Fe oxidation state vs. cathode potential. (c) Cyclic voltammetry (CV) of the cathode. (d) Cathode CV superimposed on Fe oxidation state. Reprinted from [32]. Copyright 2019, with permission from Elsevier.

where it is coordinated with a reaction-relevant adsorbate. This difference spectrum can then be analyzed to determine atomic structures of the surface site(s) [45–50].

The $\Delta\mu$ technique was applied to Fe-N-C catalysts prepared from Fe tetraphenyl porphyrin and Fe polyaniline by Jia et al. [38]. Difference spectra were obtained by subtracting the spectrum taken at 0.1 V vs. RHE from the spectra obtained from PEFC cathode-relevant potentials of 1.0 to 0.7 V and down to 0.3 V. The potential of 0.1 V in both acidic and alkaline electrolyte was chosen because it was determined by EXAFS analysis that the Fe sites do not contain electrochemical adsorbates at this potential. The $\Delta\mu$ spectra of Figure 7 show a sinusoidal shape and a negative peak at 7126 eV, attributed to the charge transfer from the Fe site to the adsorbed oxygen species (i. e., associated with the conversion of Fe^{2+} to Fe^{3+}). Through calculation of the XANES spectra for a number of FeN_4 clusters, it was found that the experimental $\Delta\mu$ features match those calculated for FeN_4C_8 or FeN_4C_{10} clusters in divacancies in the graphitic carbon basal plane or armchair edge site. Comparison of the experimental and calculated spectra indicated that at least four coordinating nitrogen atoms were needed to reproduce the spectra and O(H) was adsorbed in the axial position, consistent with prior findings shown in Figure 6e.

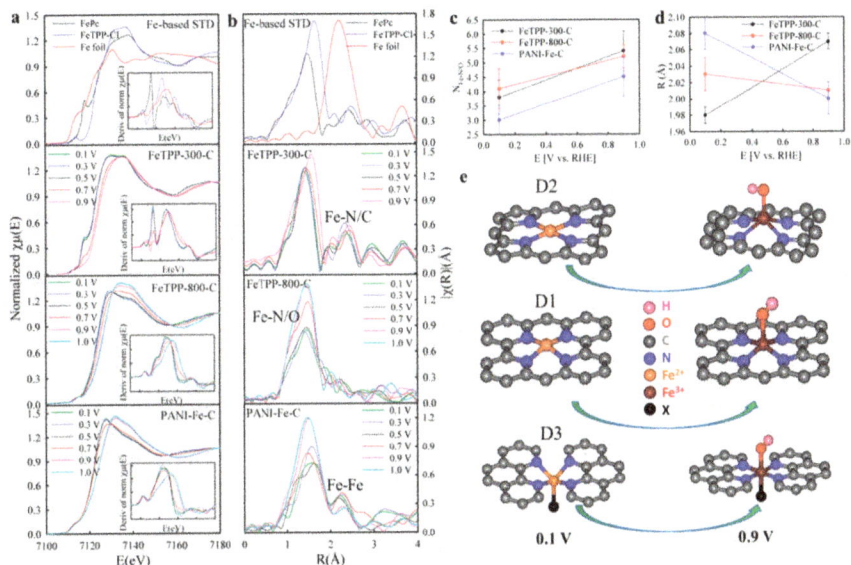

Figure 6: (a, b) *In situ* XANES performed at various potential in N_2-saturated 0.1 M $HClO_4$ (FeTPP-800-C and PANI-Fe–C) or 0.1 M KOH electrolyte (FeTPP-300-C) for three PGM-free ORR electrocatalysts. Changes in (c) bond distance and (d) coordination with potential were correlated with the adsorption of O–H ligand and change in Fe position to out of plane for (e) three potential FeN_xC_y moieties. Reprinted with permission from [42]. Copyright 2015, American Chemical Society.

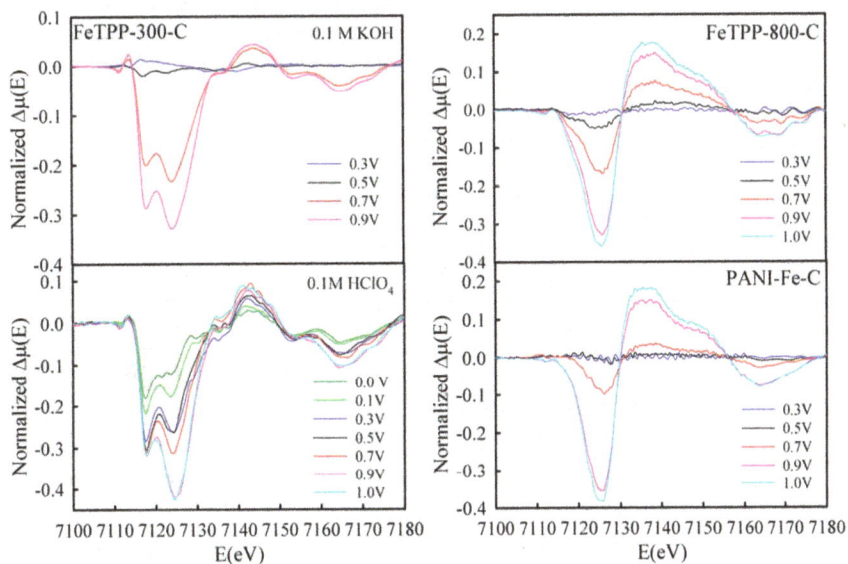

Figure 7: *In situ* $\Delta\mu$ spectra for three catalysts obtained by acquiring Fe K-edge in N_2- and O_2-saturated electrolyte at various potentials. The sinusoidal shape that emerges at various voltages is attributed to charge transfer between absorbed oxygen and the Fe sites. Reprinted with permission from [42]. Copyright 2015, American Chemical Society.

X-ray emission spectroscopy (XES) is an emerging X-ray scattering method, which can be used to investigate changes in the spin state of the transition metal atoms in PGM-free catalysts [51]. Unlike XAS, in which the attenuation of the incident X-ray beam is recorded, XES measures the X-rays, which are emitted from the sample following the decay of an excited electron state. The process involves using reference compounds to calibrate the relationship between the width of the $K\beta'$ line and transition metal spin state. *In situ* experiments can then be performed showing the influence of applied potential on spin state, as shown in Figure 8 [52]. A reversible change in spin state was observed, further validating the hypothesis of a potential-induced change in the geometry of the FeN_x site during fuel cell operation.

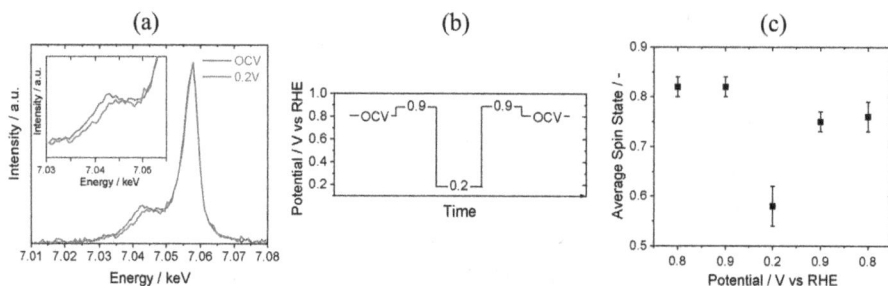

Figure 8: (a) *In situ* $K\beta$ main-line X-ray emission spectra aquired from a PGM-free catalyst in N_2-saturated 0.5 M H_2SO_4 at open circuit voltage (OCV) and at 0.2 V along with (b) the voltage profile vs. time and (c) average spin state of Fe calculated from changes in the $K\beta'$ partition of the spectra, confirming reversible changes in spin state with potential cycling. Reprinted with permission from [52]. Copyright 2021, John Wiley and Sons.

3.4 Scanning transmission electron microscopy

The transmission electron microscope (TEM) has long been employed to study materials at the nanoscale. For imaging heavier elements, such as transition metals, on lighter supports, such as carbon, the instrument is typically operated in the scanning mode (STEM) to enable Z-contrast sensitive imaging using annular dark field (ADF) detectors. The addition of aberration-correctors to the STEM at the turn of the century improved the instrument resolution to the sub-Ångstrom level and has led to new opportunities for visualizing single-atom and atomically dispersed catalysts [53–56]. These aberration-corrected electron probes can readily resolve dopants in graphitic structures, and when coupled with electron energy loss spectroscopy (EELS), can identify the metal site and surrounding species at the single atom level [57, 58]. This was demonstrated in a study aimed at measuring single atom spin states by EELS for graphene-hosted Fe atoms with and without nitrogen neighbors [59]. As shown in Figure 9a, the charge transfer from multiple pyridinic N atoms was observable in the near-edge fine structure of the Fe L

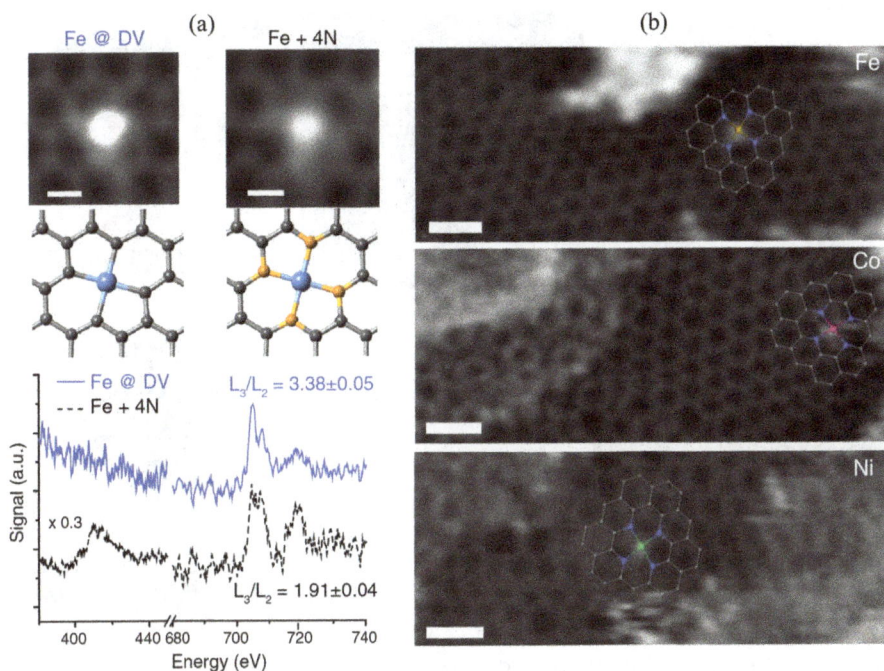

Figure 9: (a) FeC$_4$ and FeN$_4$ moieties observed in single layer graphene showing change in spin state determined by ratio of Fe L$_3$/L$_2$ EELS edges (scale bars 0.2 nm). Reprinted with permission from [59]. Copyright 2015 by the American Physical Society. (b) Direct visualization of the local atomic structure of single Ni, Fe and Co atoms in nitrogen-doped graphitic carbon (scale bars 0.5 nm). Reprinted by permission from [60]. Copyright 2018, Springer Nature.

edge. Another example involved the characterization of single transition metal atoms (Fe, Co or Ni) embedded in nitrogen-doped holey graphene frameworks as a model electrocatalyst for a broad range of applications, including oxygen evolution/reduction, hydrogen oxidation/evolution and CO$_2$/N$_2$ reduction. As shown in Figure 9b, the local coordination of the single metal sites with the graphene lattice could be directly determined from atomic resolution images.

The combination of atomic-scale imaging and EELS was also used on an atomically-dispersed Fe-N-C catalyst derived from a mix of cyanamide and polyaniline precursors [61]. Unlike the model graphene systems, this study revealed two distinct morphologies in this scalable material, as shown in Figure 10. The first structure, which represented the majority of the catalyst and is common to most PGM-free catalysts, was a complex nitrogen-doped mesoporous carbon with small, curved graphitic domains decorated with transition metal atoms. The secondary structure consisted of few-layer graphene sheets with transition metal atoms dispersed both on and within the sheets. The metal atoms hosted within the layers of graphene could be probed by the electron beam without significant changes to their structure and were used to show the colocation of Fe

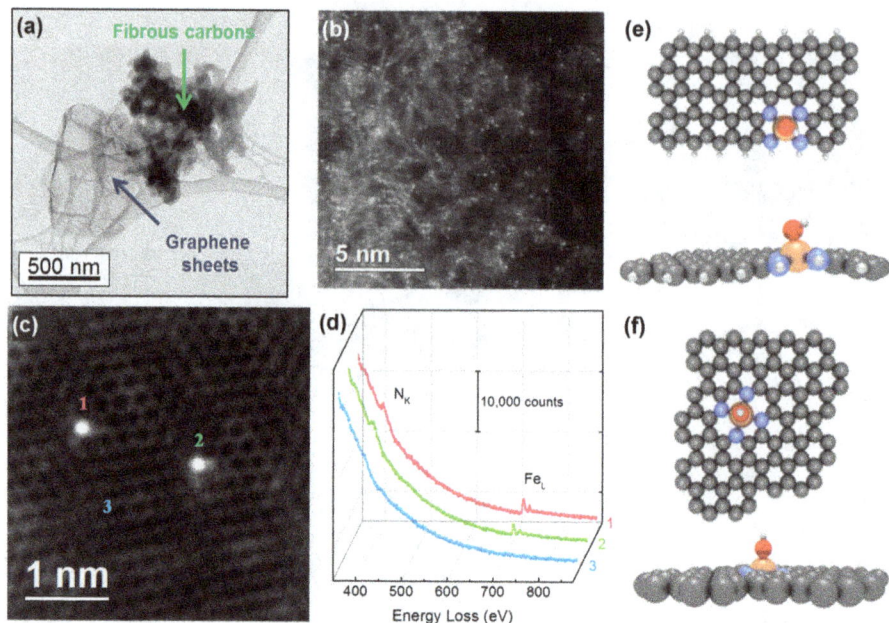

Figure 10: STEM images showing (a) two distinct morphologies consisting of (b) complex N-doped meso-porous carbon decorated with Fe predominantly at carbon edge sites and (c) Fe moieties embedded in multilayer graphene. (d) The stable bulk-hosted atoms could be readily probed by STEM-EELS to show co-location of Fe and N. The higher frequency of (e) edge-hosted FeN$_x$ sites versus (f) "bulk-hosted" sites led to DFT-modeling of the zig-zag edges, which when ligated with OH-, lead to highly ORR active structures. In (e, f), gray spheres represent C, bronze spheres represent Fe, white spheres represent H, blue spheres represent N and red spheres represent O. Adapted with permission from [61]. Copyright 2017, American Association for the Advancement of Science.

and N for the first time in a high-performance ORR catalyst. However, the majority of Fe atoms were present in the mesoporous phase and were found to reside at edge sites, which could be easily perturbed by the electron beam. This led to a DFT investigation of the role of the bulk- vs. edge-hosted Fe sites. The latter sites were found to spontaneously ligate with -OH (Figure 10e, f) and exhibited higher ORR activity than bulk-hosted sites. However, knock-on displacement threshold energy (KODTE) calculations showed edge sites had lower stability under the electron beam, giving birth to the use of KODTE as an early DFT-descriptor for electrochemical durability [11].

Efforts have since turned to finding new STEM methods capable of probing these metal sites without disturbing them. An obvious approach guided by KODTE calculations was to lower the accelerating voltage of the electron microscope below 60 kV. However, the reduced knock-on damage at lower voltages comes at the cost of spatial resolution, a limitation recently overcome by the use of the four-dimensional STEM (4D-STEM) methodology [62, 63]. This technique is achieved by recording the full 2D diffraction pattern, or ronchigram, at each pixel, then performing post-analysis to improve the image single-to-noise and deconvolute residual aberrations to maximize the resolution of the data set. This approach is shown in Figure 11, in which a noisy high-angle annular dark-field (HAADF) image can be improved by calculating the divergence of center-of-mass (dCoM), with further resolution improvement achieved by performing residual aberration-corrected Wigner distribution deconvolution ptychography to reveal the anticipated four-fold symmetry around the bulk-hosted Fe site [64].

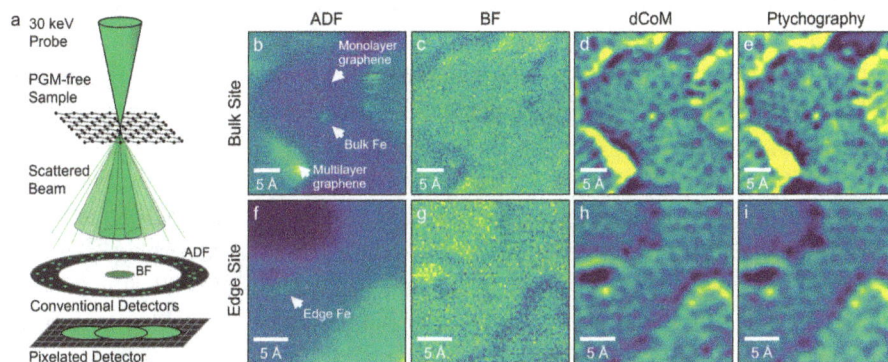

Figure 11: (a) Schematic of conventional bright-field (BF) and annular dark-field (ADF) versus 4D-STEM imaging. (b, f) ADF and (c, g) BF images of a model graphene-based PMG-free catalyst recorded with 30 keV electron probe showing limited spatial information around bulk and edg-hosted FeN moieties. (d, h) 4D-STEM divergence of center-of-mass (dCoM) and (e, i) residual aberration-corrected Wigner distribution deconvolution ptychography used to improve light-element contrast, reduce noise and improve resolution through residual aberrations minimization. Reprinted by permission from [64]. Copyright 2021, Cambridge University Press.

In situ and identical location STEM experiments also have important implications for PGM-free catalyst characterization. MEMS-based devices can be used to perform high-temperature pyrolysis either in the vacuum of the microscope or in the presence of a gas in special environmental cells [65, 66]. Although the SiN windows used in the latter case limit the ability to resolve the graphene lattice and track changes in N content, important morphological changes can be determined as a function of temperature during the pyrolysis process [67]. Identical-location experiments, in which the same region of the catalyst is characterized before and after a stimulus is applied *ex situ*, have also provided insights into how catalyst composition and morphology change with different processing methods [68]. This approach is even more impactful for aqueous electrochemical experiments used to elucidate degradation mechanisms [69]. Continued efforts into improving resolution and mitigating beam artifacts will further the applicability of these techniques to elucidate the structures governing catalyst activity and durability.

3.5 Nuclear resonance spectroscopy

Mössbauer spectroscopy and other nuclear resonance techniques are uniquely suited for the study of Fe-containing PGM-free catalysts. Mössbauer spectroscopy (MS) relies on the recoil-free emission and absorption of gamma rays in solids where the isotope of the element providing the gamma rays is the same as that absorbing the gamma rays. For example, ^{57}Co is used as a source for Fe MS as it decays to an excited state of ^{57}Fe that emits photons during its decay to the ground state. Nuclear inelastic spectroscopy (NIS) relies on the same gamma ray absorption process, but energy is lost or gained in the solid as phonons (quantized lattice vibrations). To shorten data acquisition time for both MS and NIS, ^{57}Fe-enriched compounds are typically used to prepare the Fe-N-C catalysts [70].

Mössbauer spectroscopy, also referred to as recoil-free nuclear resonance absorption, has provided valuable insights into the structure and prevalence of FeN_x moieties in PGM-free catalysts, as reviewed by Kramm [71]. Each gamma ray absorbing Fe species is characterized by unique Mössbauer parameters with values defined by the interactions of the s-electrons with the nucleus (isomer shift), the effect of the Fe species electric field gradient on the nuclear energy levels (quadrupole splitting) and the interaction of an internal or external magnetic field on the nucleus (magnetic hyperfine splitting). Spectral components in the MS are defined by their isomer shift (δ_{Iso}), which is affected by spin and oxidation states, providing information about interactions between the Fe center and adsorbates, such as oxygen, and quadrupole splitting (ΔE_Q), which provides information about the Fe coordination environment and geometry, and the full-width at half maximum (fwhm), which is related to the uniformity or order of the given structure. The table in Figure 12 demonstrates the range and complexity of possible sites present in PGM-free catalysts, including singlets and sextets related to the presence of metal nanoparticles. A table detailing δ_{Iso} and ΔE_Q for an extended range of molecular FeN_4 and inorganic species can be found in an additional publication [71]. Generally speaking, three doublets (D1, D2 and D3) observed in room temperature Mössbauer spectra have been linked to ORR activity and are attributed to FeN_4 moieties of varying spin and oxidation states [72–74]. The fitting and interpretation of Mössbauer spectra continue to be refined with contributions from DFT calculations and improved catalysts with controlled morphologies. In particular, recent low temperature studies (<5 K) have played an important role in the convergence of agreement on the fitting of spectra. Figure 13 highlights the importance of low-temperature Mössbauer spectroscopy in separating out the contribution of iron oxides to the D1 doublet due to an increase in magnetic splitting at these lower temperatures [74, 75].

Complementary to the XAS experiments discussed previously, *in situ* and *operando* Mössbauer experiments have also been performed to elucidate changes in structure under relevant potentials and loss of different sites under accelerated stress tests [43, 74, 76]. Using a combination of DFT and MS, Mineva et al. assigned the D1 doublet to a high

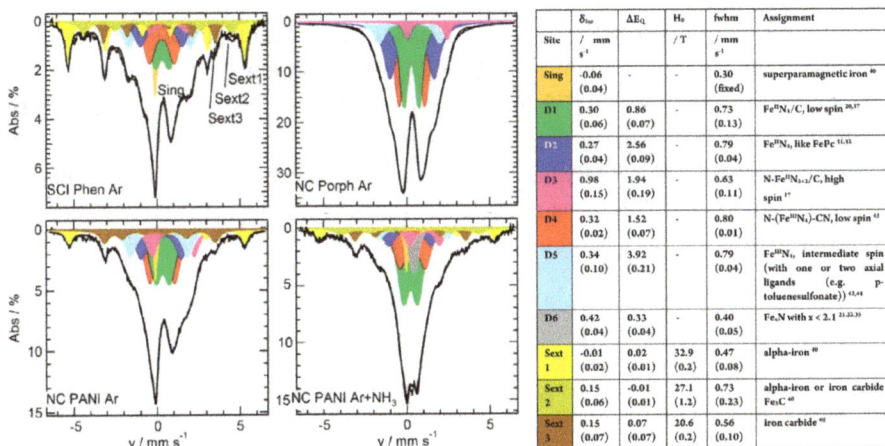

Site	δ_{Iso} / mm s^{-1}	ΔE_Q	H_0 / T	fwhm / mm s^{-1}	Assignment
Sing	-0.06 (0.04)	-	-	0.30 (fixed)	superparamagnetic iron [40]
D1	0.30 (0.06)	0.86 (0.07)	-	0.73 (0.13)	Fe^{III}N$_4$/C, low spin [20,47]
D2	0.27 (0.04)	2.56 (0.09)	-	0.79 (0.04)	Fe^{III}N$_4$, like FePc [11,42]
D3	0.98 (0.15)	1.94 (0.19)	-	0.63 (0.11)	N-Fe^{II}N$_{2+2}$/C, high spin [17]
D4	0.32 (0.02)	1.52 (0.07)	-	0.80 (0.01)	N-(Fe^{II}N$_4$)-CN, low spin [42]
D5	0.34 (0.10)	3.92 (0.21)	-	0.79 (0.04)	Fe^{II}N$_4$, intermediate spin (with one or two axial ligands (e.g. p-toluenesulfonate)) [42,44]
D6	0.42 (0.04)	0.33 (0.04)	-	0.40 (0.05)	Fe$_x$N with x < 2.1 [21,22,23]
Sext 1	-0.01 (0.02)	0.02 (0.01)	32.9 (0.2)	0.47 (0.08)	alpha-iron [40]
Sext 2	0.15 (0.06)	-0.01 (0.01)	27.1 (1.1)	0.73 (0.23)	alpha-iron or iron carbide Fe$_x$C [40]
Sext 3	0.15 (0.07)	0.07 (0.07)	20.6 (0.2)	0.56 (0.10)	iron carbide [40]

Figure 12: Fitting and interpretation of doublets and sextets for Mössbauer spectra of heated-treated catalysts derived from phenanthroline (Phen), porphyrin (Porp), and polyaniline (PANI). Reprinted with permission from [73]. Copyright 2014, American Chemical Society.

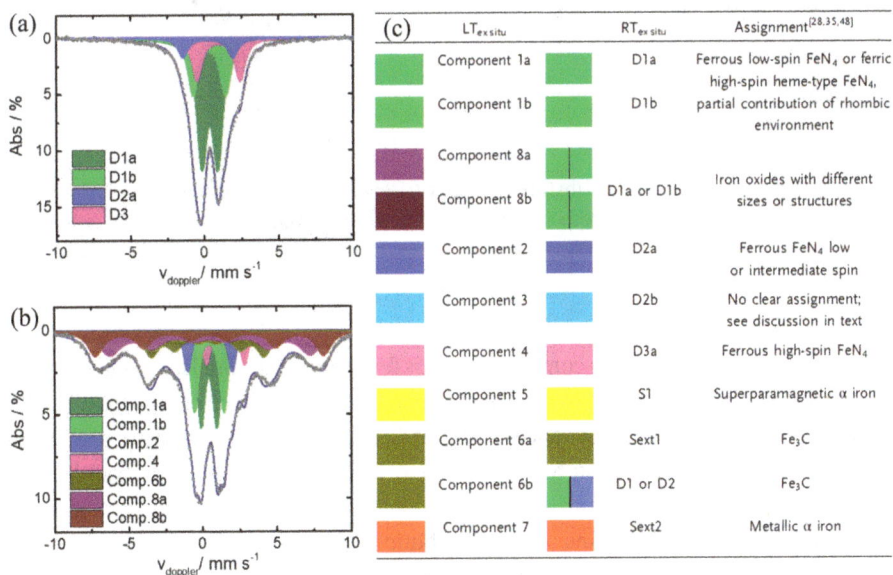

Figure 13: Comparison of Mössbauer spectra obtained at (a) room temperature and (b) 1.6K for a Fe-N-C catalyst using polypyrrole (ppy) as a nitrogen precursor. Low-temperature Mössbauer spectroscopy allows for (c) the identification of Fe oxides and carbides, which otherwise would be fitted as FeN$_x$C$_y$ species. Adapted with permission from [74]. Copyright 2021, John Wiley and Sons.

spin state Fe(III)N$_4$C$_{12}$ site sitting at exposed surface sites with high ORR turnover frequency and the D2 doublet to buried Fe(II)N$_4$C$_{10}$ sites with low O$_2$ binding energy for a ZIF-8-derived Fe-N-C catalyst [77]. In a follow-on study, the relative durability of these two different sites was studied using *in situ* and *operando* MS [43]. As shown in Figure 14, the D1 site to which high activity has been attributed, was found to quickly decompose to Fe-oxide, with much of the remaining performance attributed to the more stable, but less active, D2 site.

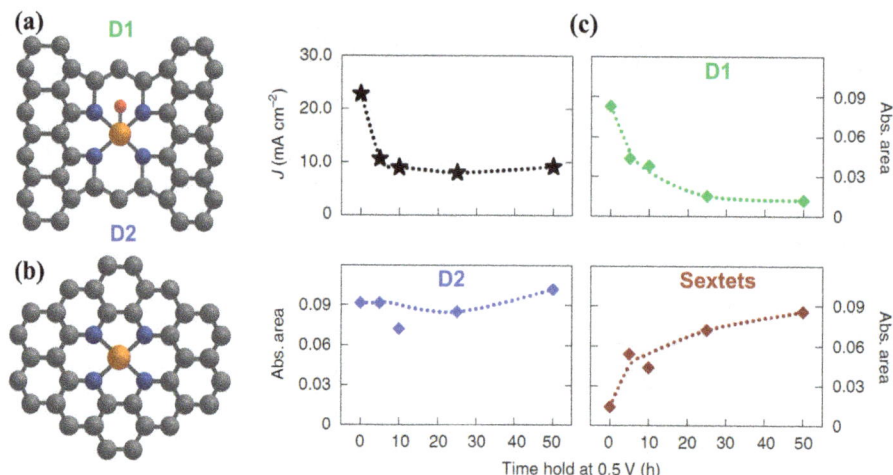

Figure 14: Depiction of the (a) D1 and (b) D2 Fe-N-C sites. (c) Relationship between fuel cell performance and density of D1, D2 and sextet sites, showing correlation between conversion of D1 sites to Fe carbides, concomitant with drop in current density. Adapted by permission from [43]. Copyright 2020, Springer Nature.

Resonance techniques which complement Mössbauer include nuclear inelastic scattering (NIS) [75, 78], also called nuclear resonance vibrational spectroscopy (NRVS), [79] and electron paramagnetic resonance (EPR) [75]. Solid state nuclear magnetic resonance (NMR) has also been employed for the characterization of PGM-free catalyst derived from aerogels, [80] but is primarily used for understanding the nanostructure of membranes and ionomers or imaging water distribution [81, 82].

Among these methods, NIS is seeing increased application as a means to elucidate bonding characteristics of Fe species in Fe-N-C catalysts [78, 83]. Like infrared and Raman spectroscopies, NIS probes the vibrational dynamics for Mössbauer-active nuclei, such as those in ^{57}Fe, but is not constrained by optical selection rules, which limit the application of the former methods to PGM-free catalysts. The NIS intensity is directly related to the magnitude and direction of the vibrational motion, providing a quantitative aspect to the spectral lines. In a pioneering study, NIS spectra of an Fe-PANI-based catalyst were acquired before and after treatment with a probe molecule (NO(g)) [78].

Differences in the experimental vibrational spectra were linked to edge-hosted FeN_4 or Fe_2N_5-dissociated-NO sites using DFT calculations. More recent NIS experiments on a catalyst having the majority of Fe in the form of single atoms of Fe coordinated with N in a graphene-like carbon, rather than the multiple types of Fe sites in the Fe-PANI-based catalyst, showed that the NO probe molecule does indeed bind to Fe surface sites at room temperature and that NO can bind to either Fe^{2+} or Fe^{3+} sites. These same experiments showed non-Debye-like behavior at low frequencies indicative of the presence of diverse FeN_x moieties. Wagner et al. also observed a broad NIS spectrum for a catalyst synthesized by heat treating a mixture of 5,10,15,20-Tetrakis(4-methoxyphenyl)-21H,23H-porphine iron(III) chloride (FeTMPPCl) and carbon black powder [75]. MS showed the Wagner et al. catalyst to contain alpha iron and iron oxide and thermogravimetry showed that only 50–70 % of the chloride ligand was removed during heat treatment; all of these species can contribute to the NIS spectrum in the low energy region, illustrating the need to use synthetic techniques or post-synthesis catalyst treatment to remove "spectator" Fe species prior to utilizing advanced characterization methods.

3.6 Complementary methods and synergistic studies

There are a host of additional methods that complement the characterization techniques discussed in the previous sections. This section highlights some of the more frequently used methods and the synergy among them. For instance, Raman spectroscopy, in which a laser is used to excite molecular vibrational frequencies, is often used to ascertain the graphitic content in PGM-free catalysts [84]. Recently, *in situ* and *operando* cells have been developed, which allow for improved catalyst degradation experiments [85]. For example, electrochemical tip-enhanced Raman spectroscopy (EC-TERS), shown in Figure 15a, has been used to study the degradation of Fe(II) phthalocyanine (FePc) as a model PGM-free ORR catalyst [86]. Along with Fourier transform infrared spectroscopy (FTIR), Raman spectroscopy can provide insight into the speciation of nitrogen dopants, although XPS is typically the more suitable technique for obtaining this information [87].

The low active site density and turnover frequency of PGM-free catalysts requires much thicker cathodes to achieve power densities required for commercial applications. This makes the control and optimization of pore size distributions critical, with the Brunauer–Emmett–Teller (BET) method playing a key role in determining porosity and surface area by nitrogen sorption (Figure 15b) [88, 89]. Time-of-flight secondary ion mass spectrometry (ToF-SIMS) has also been used as a marker for surface area differences [84], in addition to providing insight into the density and nature of the active species (Figure 15c) [90, 91]. Scanning electron microscopy (SEM) and X-ray computed tomography (CT) provide important information on electrode morphology toward understanding and improving mass transport in these thick electrode layers [92, 93]. As shown in Figure 15d, nano-CT was used to map and quantify the distribution of Cs-stained Nafion™

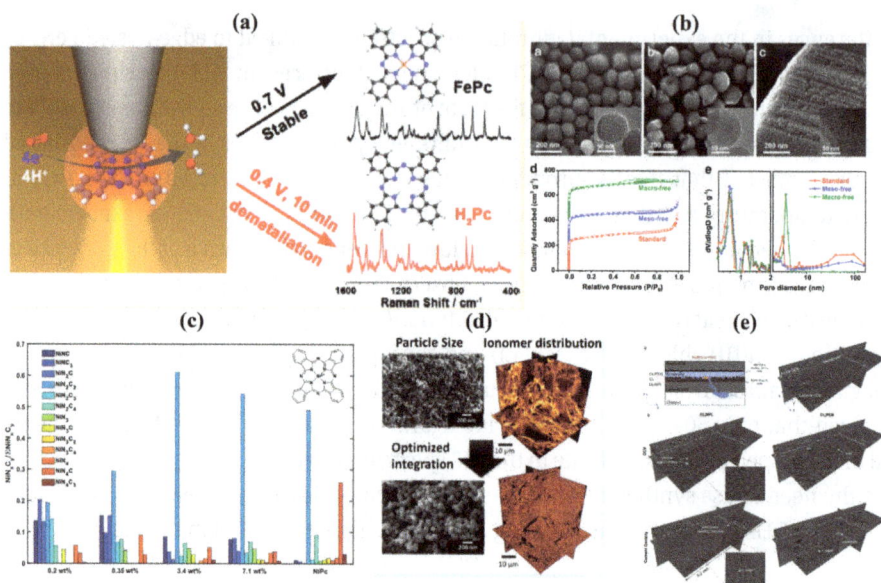

Figure 15: A sampling of the many additional methods used to characterize PGM-free catalysts: (a) Degradation studies by EC-TERS [86], (b) SEM and BET method for surface-are determination [89], (c) NiN$_x$C$_y$ fragments measured by TOF-SIMS [91], (d) nano-CT [94] and (e) *operando* CT [95] for visualizing electrode structure and interfaces. Reprinted with permission. (a, b, d) Copyright 2019, American Chemical Society. (c) Copyright 2020, John Wiley and Sons. (e) Copyright 2018, with permission from Elsevier.

in electrodes to show how power density could be increased by improving the distribution of the proton-conducting ionomer [94]. A special cell for *operando* CT has also been developed to track liquid water in operating cells and provides guidance for how interfaces can be improved to limit mass transport losses (Figure 15e) [95].

A powerful synergy emerges when the advanced methods detailed in this chapter come together. An early example of this was a study toward identifying the active site(s) in a polyaniline-derived Fe-based PGM-free catalyst [31]. The catalyst preparation procedure was systematically modified to correlate changes in rotating disk electrode (RDE)-measured ORR activity with changes in morphology and chemical speciation determined using a diverse set of characterization techniques, including XAFS, XPS, XRD, BET, inductively coupled plasma-optical emission spectrometry (ICP-OES), and ^{57}Fe Mössbauer spectroscopy. Principal component analysis (PCA) was also employed to draw out the correlations in speciation between various techniques. This multitechnique approach has since been implemented in many studies of PGM-free catalysts. An elegant example of this is included in Figure 16 from a study on the use of a chemical vapor deposition (CVD) method to increase the utilization of active sites. Multipanel figures, such as this combining electron microscopy, XPS, Mössbauer spectroscopy, XANES and EXAFS, attest to the range in expertise and advanced instrumentation required to solve this important materials development challenge.

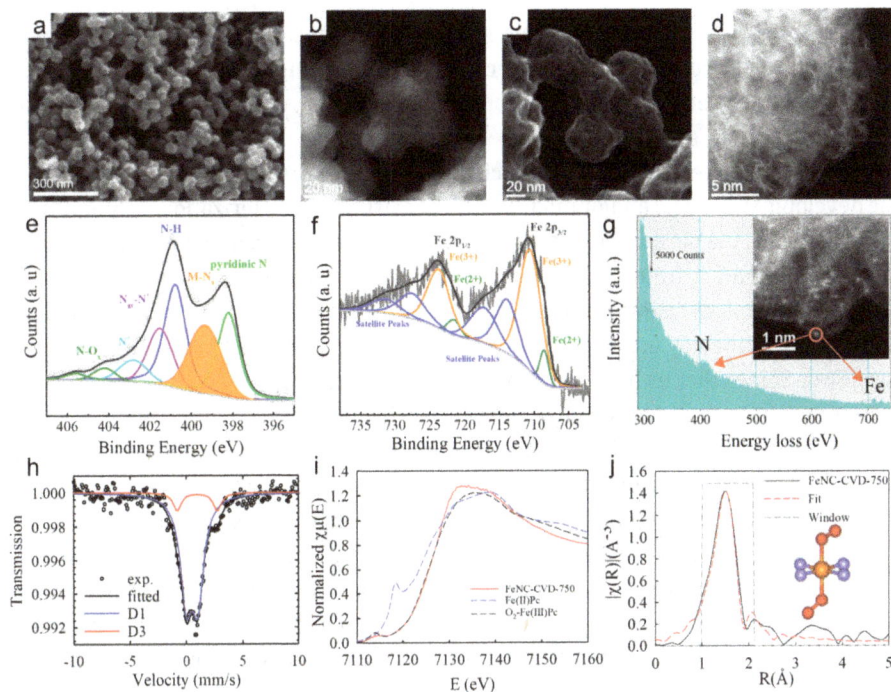

Figure 16: Example of multitechnique characterization combining (a) SEM, (b–d) STEM, (e–f) XPS, (g) EELS, (h) Mössbauer spectroscopy, (i) XANES and (j) EXAFS that has become common in studies of PGM-free ORR catalysts. Reprinted with permission from [96]. Copyright 2021, Spring Nature.

3.7 Conclusion and future directions

The need for lowering the overall cost of PEFCs has resulted in extraordinary progress in the development of PGM-free catalysts as replacements for the incumbent Pt-based materials for the oxygen reduction reaction, which currently account for 40 % to 50 % of the fuel cell stack cost in passenger cars. Initially, the PGM-free research revolved around increasing catalytic activity via an engineering approach to systematically, or randomly, varying catalyst precursors and synthesis conditions. As the field matured, it became obvious that further progress in the ORR activity, and especially durability, would not be possible without understanding of the reaction mechanism and factors governing performance of PGM-free catalysts. This called for the development of characterization techniques capable of providing much needed insight into reactions on the PGM-free catalyst surface that is dramatically different from the thoroughly studied and well-understood platinum-based systems. Compared to Pt and its alloys, which dominate practical fuel cell systems today, the best performing PGM-free cathode catalysts utilizing transition metals are carbon-based, which renders unusable some well-established spectroscopic techniques developed for precious metal catalysts. The biggest challenge facing mechanistic studies at PGM-free catalysts originates, however, from their chemical complexity, resulting in many potential active sites (nonexistent in the case of precious metal catalysts), and consequently, reaction mechanisms.

Challenges posed by the complex chemistry of PGM-free catalysts have called for the development of new and/or modification of existing characterization techniques for these catalysts, the most important of which have been reviewed in detail in this chapter. The information from these techniques is usually correlated with electrochemical responses, ORR activity in particular. At times, the techniques rely on theoretical modeling, and are limited by uncertainties associated with the assumed models and computing power. Some other techniques such as Mössbauer spectroscopy and nuclear inelastic scattering (NIS) are also limited to a single metal (Fe).

Consequently, in spite of all the immense progress in the development of the characterization techniques adapted to PGM-free catalysts, none of the available techniques is truly surface specific. Attempts have been made to extract surface-relevant information via the subtraction of spectroscopic information obtained at different potentials, various stages of catalyst degradation, or in the presence and absence of molecular probes (such as NO or NO_3^-) [97–99]. However, to date, no characterization technique has yielded unambiguous information on the origins of the ORR activity of metal-nitrogen-carbon PGM-free catalysts. Based on correlations between the metal content (usually the total content in the catalysts as that on the surface is virtually impossible to determine) and ORR performance, the widely accepted form of the active site in metal-nitrogen-carbon PGM-free catalysts is currently a nitrogen-coordinated transition-metal atom, capable of mediating the ORR via a redox process. While appealing per analogy with biological systems, this concept of the ORR active site lacks a direct experimental confirmation and fails short of explaining some of the behavior of PGM-free catalysts.

It is widely agreed that full understanding of the ORR mechanism is a prerequisite for further development of PGM-free catalysts, and ultimately, their practical viability. Research to date leaves no doubt that the much-needed progress cannot be achieved without developing additional, and likely more advanced, characterization techniques that would allow for a direct insight into the catalyst surface. Possible paths include, but are not limited to (i) development of truly active site-specific molecular or ionic probes amenable to spectroscopic and/or microscopic detection; (ii) elimination of the beam damage in STEM experiments; (iii) *operando* and *in situ* microscopic and spectroscopic characterization at atomic resolution while measuring the electrochemical response of the system in real time and (iv) high-fidelity modeling of possible active sites resulting in experimentally-verifiable outcomes. The remarkable technique development of the last few decades must continue to produce the next level of insights necessary to break the activity and durability glass ceiling and enable PGM-free ORR catalysts to contribute to the expansion of clean-energy technologies.

Bibliography

[1] E. L. Miller et al., US Department of Energy hydrogen and fuel cell technologies perspectives. *Mater. Res. Soc. Bull.*, **45**, 57–64, 2020. https://doi.org/10.1557/mrs.2019.312.

[2] S. T. Thompson and D. Papageorgopoulos, Platinum group metal-free catalysts boost cost competitiveness of fuel cell vehicles. *Nat. Catal.*, **2**, 558–561, 2019. https://doi.org/10.1038/s41929-019-0291-x.

[3] L. Osmieri et al., Status and challenges for the application of platinum group metal-free catalysts in proton-exchange membrane fuel cells. *Curr. Opin. Electrochem.*, **25**, 100627, 2021. https://doi.org/10.1016/j.coelec.2020.08.009.

[4] R. Jasinski, A New Fuel Cell Cathode Catalyst. *Nature*, **201**, 212, 1964. https://doi.org/10.1038/2011212a0.

[5] J. A. R. van Veen, J. F. van Baar, C. J. Kroese, J. G. F. Coolegem, N. De Wit, and H. A. Colijn, Oxygen Reduction on Transition-Metal Porphyrins in Acid Electrolyte I. Activity. *Ber. Bunsenges. Phys. Chem.*, **85**, 693–700, 1981. https://doi.org/10.1002/bbpc.19810850917.

[6] T. Marshall-Roth et al., A pyridinic Fe-N4 macrocycle models the active sites in Fe/N-doped carbon electrocatalysts. *Nat. Commun.*, **11**, 5283, 2020. https://doi.org/10.1038/s41467-020-18969-6.

[7] S. Dey et al., Molecular electrocatalysts for the oxygen reduction reaction. *Nat. Rev. Chem.*, **1**, 0098, 2017. https://doi.org/10.1038/s41570-017-0098.

[8] S. Gupta, D. Tryk, I. Bae, W. Aldred, and E. Yeager, Heat-treated polyacrylonitrile-based catalysts for oxygen electroreduction. *J. Appl. Electrochem.*, **19**, 19–27, 1989. https://doi.org/10.1007/BF01039385.

[9] W. Wang, Q. Jia, S. Mukerjee, and S. Chen, Recent Insights into the Oxygen-Reduction Electrocatalysis of Fe/N/C Materials. *ACS Catal.*, **9**, 10126–10141, 2019. https://doi.org/10.1021/acscatal.9b02583.

[10] X. X. Wang, M. T. Swihart, and G. Wu, Achievements, challenges and perspectives on cathode catalysts in proton exchange membrane fuel cells for transportation. *Nat. Catal.*, **2**, 578–589, 2019. https://doi.org/10.1038/s41929-019-0304-9.

[11] U. Martinez et al., Progress in the Development of Fe-Based PGM-Free Electrocatalysts for the Oxygen Reduction Reaction. *Adv. Mater.*, **31**, e1806545, 2019. https://doi.org/10.1002/adma.201806545.

[12] Y. Shao, J. P. Dodelet, G. Wu, and P. Zelenay, PGM-Free Cathode Catalysts for PEM Fuel Cells: A Mini-Review on Stability Challenges. *Adv. Mater.*, **31**, e1807615, 2019. https://doi.org/10.1002/adma.201807615.

[13] M. Liu et al., Atomically dispersed metal catalysts for the oxygen reduction reaction: synthesis, characterization, reaction mechanisms and electrochemical energy applications. *Energy Environ. Sci.*, **12**, 2890–2923, 2019. https://doi.org/10.1039/c9ee01722d.

[14] A. A. Gewirth, J. A. Varnell, and A. M. DiAscro, Nonprecious Metal Catalysts for Oxygen Reduction in Heterogeneous Aqueous Systems. *Chem. Rev.*, **118**, 2313–2339, 2018. https://doi.org/10.1021/acs.chemrev.7b00335.

[15] S. Zhuang, X. Shi, and E. S. Lee, A Review on Non-Pgm Cathode Catalysts for Polymer Electrolyte Membrane (PEM) Fuel Cell. In: *Proceedings of the ASME 2015 13th International Conference on Fuel Cell Science*, pages 1–11, 2015.

[16] N. Zion et al., Porphyrin Aerogel Catalysts for Oxygen Reduction Reaction in Anion-Exchange Membrane Fuel Cells. *Adv. Funct. Mater.*, **31**, 2100963, 2021. https://doi.org/10.1002/adfm.202100963.

[17] C. Zhang et al., Electrochemical CO2 Reduction with Atomic Iron-Dispersed on Nitrogen-Doped Graphene. *Adv. Energy Mater.*, **8**, 1703487, 2018. https://doi.org/10.1002/aenm.201703487.

[18] Z. Y. Wu et al., Electrochemical ammonia synthesis via nitrate reduction on Fe single atom catalyst. *Nat. Commun.*, **12**, 2870, 2021. https://doi.org/10.1038/s41467-021-23115-x.

[19] K. Artyushkova, S. Pylypenko, T. S. Olson, J. E. Fulghum, and P. Atanassov, Predictive Modeling of Electrocatalyst Structure Based on Structure-to-Property Correlations of X-ray Photoelectron Spectroscopic and Electrochemical Measurements. *Langmuir*, **24**, 9082–9088, 2008. https://doi.org/10.1021/la801089m.

[20] K. Artyushkova et al., Density functional theory calculations of XPS binding energy shift for nitrogen-containing graphene-like structures. *Chem. Commun.*, **49**, 2539–2541, 2013. https://doi.org/10.1039/c3cc40324f.

[21] I. Matanovic, K. Artyushkova, and P. Atanassov, Understanding PGM-free catalysts by linking density functional theory calculations and structural analysis: Perspectives and challenges. *Curr. Opin. Electrochem.*, **9**, 137–144, 2018. https://doi.org/10.1016/j.coelec.2018.03.009.

[22] M. Chen et al., Single Atomic Iron Site Catalysts via Benign Aqueous Synthesis for Durability Improvement in Proton Exchange Membrane Fuel Cells. *J. Electrochem. Soc.*, **168**, 044501, 2021. https://doi.org/10.1149/1945-7111/abf014.

[23] K. Artyushkova, A. Serov, S. Rojas-Carbonell, and P. Atanassov, Chemistry of Multitudinous Active Sites for Oxygen Reduction Reaction in Transition Metal–Nitrogen–Carbon Electrocatalysts. *J. Phys. Chem. C*, **119**, 25917–25928, 2015. https://doi.org/10.1021/acs.jpcc.5b07653.

[24] C. Prössl et al., Impact of Ir modification on the durability of Fe-N-C catalysts under start-up and shutdown cycle conditions. *J. Mater. Chem. A*, 2021. https://doi.org/10.1039/d1ta04668c.

[25] Y. Han, H. Zhang, Y. Yu, and Z. Liu, In Situ Characterization of Catalysis and Electrocatalysis Using APXPS. *ACS Catal.*, **11**, 1464–1484, 2021. https://doi.org/10.1021/acscatal.0c04251.

[26] T. Nakamura et al., Quick Operando Ambient Pressure Hard X-ray Photoelectron Spectroscopy for Reaction Kinetic Measurements of Polymer Electrolyte Fuel Cells. *J. Phys. Chem. C*, **124**, 17520–17527, 2020. https://doi.org/10.1021/acs.jpcc.0c04871.

[27] M. J. Dzara et al., Characterization of Complex Interactions at the Gas–Solid Interface with in Situ Spectroscopy: The Case of Nitrogen-Functionalized Carbon. *J. Phys. Chem. C*, **123**, 9074–9086, 2019. https://doi.org/10.1021/acs.jpcc.9b00487.

[28] M. J. Dzara et al., Characterizing Complex Gas–Solid Interfaces with in Situ Spectroscopy: Oxygen Adsorption Behavior on Fe–N–C Catalysts. *J. Phys. Chem. C*, **124**, 16529–16543, 2020. https://doi.org/10.1021/acs.jpcc.0c05244.

[29] J. E. Penner-Hahn, X-ray Absorption Spectroscopy. *Compr. Coord. Chem. II*, **2**, 159–186, 2004. https://doi.org/10.1016/B0-08-043748-6/01063-X.

[30] J. A. Kirby, D. B. Goodin, T. Wydrzynski, A. S. Robertson, and M. P. Klein, State of Manganese in the Photosynthetic Apparatus. 2. X-ray Absorption Edge Studies on Manganese in Photosynthetic Membranes. *J. Am. Chem. Soc.*, **103**, 5537–5542, 1981. https://doi.org/10.1021/ja00408a043.

[31] M. Ferrandon et al., Multitechnique Characterization of a Polyaniline–Iron–Carbon Oxygen Reduction Catalyst. *J. Phys. Chem. C*, **116**, 16001–16013, 2012. https://doi.org/10.1021/jp302396g.

[32] L. Osmieri et al., Elucidation of Fe–N–C electrocatalyst active site functionality via in-situ X-ray absorption and operando determination of oxygen reduction reaction kinetics in a PEFC. *Appl. Catal. B, Environ.*, **257**, 117929, 2019. https://doi.org/10.1016/j.apcatb.2019.117929.

[33] A. Zitolo et al., Identification of catalytic sites in cobalt-nitrogen-carbon materials for the oxygen reduction reaction. *Nat. Commun.*, **8**, 957, 2017. https://doi.org/10.1038/s41467-017-01100-7.

[34] T. E. Westre et al., A Multiplet Analysis of Fe K-Edge 1s to 3d Pre-Edge Features of Iron Complexes. *J. Am. Chem. Soc.*, **119**, 6297–6314, 1997. https://doi.org/10.1021/ja964352a.

[35] A. Zitolo et al., Identification of catalytic sites for oxygen reduction in iron- and nitrogen-doped graphene materials. *Nat. Mater.*, **14**, 937–942, 2015. https://doi.org/10.1038/nmat4367.

[36] J. Timoshenko and B. Roldan Cuenya, In Situ/Operando Electrocatalyst Characterization by X-ray Absorption Spectroscopy. *Chem. Rev.*, **121**, 882–961, 2021. https://doi.org/10.1021/acs.chemrev.0c00396.

[37] J. M. Yoo, H. Shin, S. Park, and Y.-E. Sung, Recent progress in in situ/operando analysis tools for oxygen electrocatalysis. *J. Phys. D, Appl. Phys.*, **54**, 173001, 2021. https://doi.org/10.1088/1361-6463/abd9a4.

[38] Q. Jia, E. Liu, L. Jiao, S. Pann, and S. Mukerjee, X-Ray Absorption Spectroscopy Characterizations on PGM-Free Electrocatalysts: Justification, Advantages, and Limitations. *Adv. Mater.*, **31**, e1805157, 2019. https://doi.org/10.1002/adma.201805157.

[39] J. Li, A. Alsudairi, Z. F. Ma, S. Mukerjee, and Q. Jia, Asymmetric Volcano Trend in Oxygen Reduction Activity of Pt and Non-Pt Catalysts: In Situ Identification of the Site-Blocking Effect. *J. Am. Chem. Soc.*, **139**, 1384–1387, 2017. https://doi.org/10.1021/jacs.6b11072.

[40] U. Tylus et al., Elucidating Oxygen Reduction Active Sites in Pyrolyzed Metal-Nitrogen Coordinated Non-Precious-Metal Electrocatalyst Systems. *J. Phys. Chem. C Nanomater. Interfaces*, **118**, 8999–9008, 2014. https://doi.org/10.1021/jp500781v.

[41] J. Li et al., Structural and mechanistic basis for the high activity of Fe–N–C catalysts toward oxygen reduction. *Energy Environ. Sci.*, **9**, 2418–2432, 2016. https://doi.org/10.1039/c6ee01160h.

[42] Q. Jia et al., Experimental Observation of Redox-Induced Fe-N Switching Behavior as a Determinant Role for Oxygen Reduction Activity. *ACS Nano*, **9**, 12496–12505, 2015. https://doi.org/10.1021/acsnano.5b05984.

[43] J. Li et al., Identification of durable and non-durable FeNx sites in Fe–N–C materials for proton exchange membrane fuel cells. *Nat. Catal.*, **4**, 10–19, 2020. https://doi.org/10.1038/s41929-020-00545-2.

[44] Q. Jia et al., Spectroscopic insights into the nature of active sites in iron–nitrogen–carbon electrocatalysts for oxygen reduction in acid. *Nano Energy*, **29**, 65–82, 2016. https://doi.org/10.1016/j.nanoen.2016.03.025.

[45] J. van Bokhoven, Recent developments in X-ray absorption spectroscopy. *Phys. Chem. Chem. Phys.*, **12**, 5502, 2010. https://doi.org/10.1039/c0cp90010a.

[46] J. T. Miller, C. L. Marshall, and A. J. Kropf, (Co)MoS2/Alumina Hydrotreating Catalysts: An EXAFS Study of the Chemisorption and Partial Oxidation with O2. *J. Catal.*, **202**, 89–99, 2001. https://doi.org/10.1006/jcat.2001.3273.

[47] M. Tada et al., In situ time-resolved dynamic surface events on the Pt/C cathode in a fuel cell under operando conditions. *Angew. Chem., Int. Ed. Engl.*, **46**, 4310–4315, 2007. https://doi.org/10.1002/anie.200604732.

[48] H. Imai et al., In Situ and Real-Time Monitoring of Oxide Growth in a Few Monolayers at Surfaces of Platinum Nanoparticles in Aqueous Media. *J. Am. Chem. Soc.*, **131**, 6293–6300, 2009. https://doi.org/10.1021/ja810036h.

[49] M. Teliska, W. E. O'Grady, and D. E. Ramaker, Determination of O and OH Adsorption Sites and Coverage in Situ on Pt Electrodes from Pt L23 X-ray Absorption Spectroscopy. *J. Phys. Chem. B*, **109**, 8076–8084, 2005. https://doi.org/10.1021/jp0502003.

[50] P. G. Allen et al., In Situ Structural Characterization of a Platinum Electrocatalyst By Dispersive X-Ray Absorption Spectroscopy. *Electrochim. Acta*, **30**, 2415–2418, 1994. https://doi.org/10.1016/0013-4686(94)E0196-U.

[51] F. de Groot, High-Resolution X-ray Emission and X-ray Absorption Spectroscopy. *Chem. Rev.*, **101**, 1779–1808, 2001. https://doi.org/10.1021/cr9900681.

[52] V. A. Saveleva et al., Potential-Induced Spin Changes in Fe/N/C Electrocatalysts Assessed by In Situ X-ray Emission Spectroscopy. *Angew. Chem., Int. Ed. Engl.*, **60**, 11707–11712, 2021. https://doi.org/10.1002/anie.202016951.

[53] D. J. Smith, Development of aberration-corrected electron microscopy. *Microsc. Microanal.*, **14**, 2–15, 2008. https://doi.org/10.1017/S1431927608080124.

[54] O. L. Krivanek, T. C. Lovejoy, and N. Dellby, Aberration-corrected STEM for atomic-resolution imaging and analysis. *J. Microsc.*, **259**, 165–172, 2015. https://doi.org/10.1111/jmi.12254.

[55] P. Tieu, X. Yan, M. Xu, P. Christopher, and X. Pan, Directly Probing the Local Coordination, Charge State, and Stability of Single Atom Catalysts by Advanced Electron Microscopy: A Review. *Small*, **17**, e2006482, 2021. https://doi.org/10.1002/smll.202006482.

[56] Y. Li et al., An oxygen reduction electrocatalyst based on carbon nanotube-graphene complexes. *Nat. Nanotechnol.*, **7**, 394–400, 2012. https://doi.org/10.1038/nnano.2012.72.

[57] O. L. Krivanek et al., Atom-by-atom structural and chemical analysis by annular dark-field electron microscopy. *Nature*, **464**, 571–574, 2010. https://doi.org/10.1038/nature08879.

[58] W. Zhou et al., Direct determination of the chemical bonding of individual impurities in graphene. *Phys. Rev. Lett.*, **109**, 206803, 2012. https://doi.org/10.1103/PhysRevLett.109.206803.

[59] Y. C. Lin, P. Y. Teng, P. W. Chiu, and K. Suenaga, Exploring the Single Atom Spin State by Electron Spectroscopy. *Phys. Rev. Lett.*, **115**, 206803, 2015. https://doi.org/10.1103/PhysRevLett.115.206803.

[60] H. Fei et al., General synthesis and definitive structural identification of MN4C4 single-atom catalysts with tunable electrocatalytic activities. *Nat. Catal.*, **1**, 63–72, 2018. https://doi.org/10.1038/s41929-017-0008-y.

[61] H. T. Chung, D. A. Cullen, D. Higgins, B. T. Sneed, E. F. Holby, K. L. More, and P. Zelenay, Direct atomic-level insight into the active sites of a high-performance PGM-free ORR catalyst. *Science*, **357**, 479–484, 2017. https://doi.org/10.1126/science.aan2255.

[62] Y. Jiang et al., Electron ptychography of 2D materials to deep sub-angstrom resolution. *Nature*, **559**, 343–349, 2018. https://doi.org/10.1038/s41586-018-0298-5.

[63] J. Song et al., Atomic Resolution Defocused Electron Ptychography at Low Dose with a Fast, Direct Electron Detector. *Sci. Rep.*, **9**, 3919, 2019. https://doi.org/10.1038/s41598-019-40413-z.

[64] M. Zachman et al., Atomic-scale Imaging of PGM-free Catalyst Active Sites by 30 keV 4D-STEM. *Microsc. Microanal.*, **27**, 2976–2977, 2021. https://doi.org/10.1017/s1431927621010357.

[65] S. Hwang, X. Chen, G. Zhou, and D. Su, In Situ Transmission Electron Microscopy on Energy-Related Catalysis. *Adv. Energy Mater.*, **10**, 1902105, 2019. https://doi.org/10.1002/aenm.201902105.

[66] R. R. Unocic and K. L. More, In: F. M. Ross, Ed., *Liquid Cell Electron Microscopy*, pages 237–257. Cambridge University Press, 2016.

[67] Y. Huang et al., Catalysts by pyrolysis: Direct observation of chemical and morphological transformations leading to transition metal-nitrogen-carbon materials. *Mater. Today*, **47**, 53–68, 2021. https://doi.org/10.1016/j.mattod.2021.02.006.

[68] D. Menga et al., Active-Site Imprinting: Preparation of Fe–N–C Catalysts from Zinc Ion–Templated Ionothermal Nitrogen-Doped Carbons. *Adv. Energy Mater.*, **9**, 1902412, 2019. https://doi.org/10.1002/aenm.201902412.

[69] K. Kumar et al., Fe–N–C Electrocatalysts' Durability: Effects of Single Atoms' Mobility and Clustering. *ACS Catal.*, **11**, 484–494, 2020. https://doi.org/10.1021/acscatal.0c04625.

[70] K. Ebner et al., (57)Fe-Enrichment effect on the composition and performance of Fe-based O2-reduction electrocatalysts. *Phys. Chem. Chem. Phys.*, **23**, 9147–9157, 2021. https://doi.org/10.1039/d1cp00707f.

[71] U. I. Kramm, L. Ni, and S. Wagner, (57) Fe Mossbauer Spectroscopy Characterization of Electrocatalysts. *Adv. Mater.*, **31**, e1805623, 2019. https://doi.org/10.1002/adma.201805623.

[72] U. I. Koslowski, I. Abs-Wurmbach, S. Fiechter, and P. Bogdanoff, Nature of the Catalytic Centers of Porphyrin-Based Electrocatalysts for the ORR: A Correlation of Kinetic Current Density with the Site Density of Fe-N4 Centers. *J. Phys. Chem. C*, **112**, 15356–15366, 2008. https://doi.org/10.1021/jp802456e.

[73] U. I. Kramm, M. Lefevre, N. Larouche, D. Schmeisser, and J. P. Dodelet, Correlations between mass activity and physicochemical properties of Fe/N/C catalysts for the ORR in PEM fuel cell via 57Fe Mossbauer spectroscopy and other techniques. *J. Am. Chem. Soc.*, **136**, 978–985, 2014. https://doi.org/10.1021/ja410076f.

[74] L. Ni et al., Active Site Identification in Fe-N-C Catalysts and Their Assignment to the Oxygen Reduction Reaction Pathway by In Situ 57 Fe Mössbauer Spectroscopy. *Adv. Energy Sustain. Res.*, **2**, 2000064, 2021. https://doi.org/10.1002/aesr.202000064.

[75] S. Wagner et al., Elucidating the Structural Composition of an Fe–N–C Catalyst by Nuclear- and Electron-Resonance Techniques. *Angew. Chem., Int. Ed. Engl.*, **58**, 10486–10492, 2019. https://doi.org/10.1002/anie.201903753.

[76] A. L. Bouwkamp-Wijnoltz et al., On Active-Site Heterogeneity in Pyrolyzed Carbon-Supported Iron Porphyrin Catalysts for the Electrochemical Reduction of Oxygen: An In Situ Mossbauer Study. *J. Phys. Chem. B*, **106**, 12993–13001, 2002. https://doi.org/10.1021/jp0266087.

[77] T. Mineva et al., Understanding Active Sites in Pyrolyzed Fe–N–C Catalysts for Fuel Cell Cathodes by Bridging Density Functional Theory Calculations and 57Fe Mössbauer Spectroscopy. *ACS Catal.*, **9**, 9359–9371, 2019. https://doi.org/10.1021/acscatal.9b02586.

[78] J. L. Kneebone et al., A Combined Probe-Molecule, Mössbauer, Nuclear Resonance Vibrational Spectroscopy, and Density Functional Theory Approach for Evaluation of Potential Iron Active Sites in an Oxygen Reduction Reaction Catalyst. *J. Phys. Chem. C*, **121**, 16283–16290, 2017. https://doi.org/10.1021/acs.jpcc.7b03779.

[79] W. R. Scheidt, J. Li, and J. T. Sage, What Can Be Learned from Nuclear Resonance Vibrational Spectroscopy: Vibrational Dynamics and Hemes. *Chem. Rev.*, **117**, 12532–12563, 2017. https://doi.org/10.1021/acs.chemrev.7b00295.

[80] N. Zion, D. A. Cullen, P. Zelenay, and L. Elbaz, Heat-Treated Aerogel as a Catalyst for the Oxygen Reduction Reaction. *Angew. Chem., Int. Ed. Engl.*, **59**, 2483–2489, 2020. https://doi.org/10.1002/anie.201913521.

[81] L. Ghassemzadeh, K.-D. Kreuer, J. Maier, and K. Muller, Chemical Degradation of Nafion Membranes under Mimic Fuel Cell Conditions as Investigated by Solid-State NMR Spectroscopy. *J. Phys. Chem. C*, **114**, 14635–14645, 2010. https://doi.org/10.1021/jp102533v.

[82] K. W. Feindel, S. H. Bergens, and R. E. Wasylishen, The use of 1H NMR microscopy to study proton-exchange membrane fuel cells. *ChemPhysChem*, **7**, 67–75, 2006. https://doi.org/10.1002/cphc.200500504.

[83] C. Gallenkamp, U. I. Kramm, and V. Krewald, Spectroscopic discernibility of dopants and axial ligands in pyridinic FeN4 environments relevant to single-atom catalysts. *Chem. Commun.*, **57**, 859–862, 2021. https://doi.org/10.1039/d0cc06237e.

[84] F. Jaouen, A. M. Serventi, M. Lefevre, J.-P. Dodelet, and P. Bertrand, Non-Noble Electrocatalysts for O2 Reduction: How Does Heat Treatment Affect Their Activity and Structure? Part II. Structural Changes Observed by Electron Microscopy, Raman, and Mass Spectroscopy. *J. Phys. Chem. C*, **111**, 5971–5976, 2007. https://doi.org/10.1021/jp068274h.

[85] C. Hess, New advances in using Raman spectroscopy for the characterization of catalysts and catalytic reactions. *Chem. Soc. Rev.*, **50**, 3519–3564, 2021. https://doi.org/10.1039/d0cs01059f.

[86] Z. Chen et al., Operando Characterization of Iron Phthalocyanine Deactivation during Oxygen Reduction Reaction Using Electrochemical Tip-Enhanced Raman Spectroscopy. *J. Am. Chem. Soc.*, **141**, 15684–15692, 2019. https://doi.org/10.1021/jacs.9b07979.

[87] P. Lazar, R. Mach, and M. Otyepka, Spectroscopic Fingerprints of Graphitic, Pyrrolic, Pyridinic, and Chemisorbed Nitrogen in N-Doped Graphene. *J. Phys. Chem. C*, **123**, 10695–10702, 2019. https://doi.org/10.1021/acs.jpcc.9b02163.

[88] M. Mazzucato et al., Effects of the induced micro- and meso-porosity on the single site density and turn over frequency of Fe–N–C carbon electrodes for the oxygen reduction reaction. *Appl. Catal. B, Environ.*, **291**, 120068, 2021. https://doi.org/10.1016/j.apcatb.2021.120068.

[89] S. H. Lee et al., Design Principle of Fe–N–C Electrocatalysts: How to Optimize Multimodal Porous Structures? *J. Am. Chem. Soc.*, **141**, 2035–2045, 2019. https://doi.org/10.1021/jacs.8b11129.

[90] M. Lefevre, J. P. Dodelet, and P. Bertrand, Molecular Oxygen Reduction in PEM Fuel Cell Conditions: ToF-SIMS Analysis of Co-Based Electrocatalysts. *Phys. Chem. B*, **109**, 16718–16724, 2005.

[91] D. M. Koshy et al., Direct Characterization of Atomically Dispersed Catalysts: Nitrogen-Coordinated Ni Sites in Carbon-Based Materials for CO_2 Electroreduction. *Adv. Energy Mater.*, **10**, 2001836, 2020. https://doi.org/10.1002/aenm.202001836.

[92] Y. He et al., Single Cobalt Sites Dispersed in Hierarchically Porous Nanofiber Networks for Durable and High-Power PGM-Free Cathodes in Fuel Cells. *Adv. Mater.*, **32**, e2003577, 2020. https://doi.org/10.1002/adma.202003577.

[93] J. Li et al., Designing the 3D Architecture of PGM-Free Cathodes for H2/Air Proton Exchange Membrane Fuel Cells. *ACS Appl. Energy Mater.*, **2**, 7211–7222, 2019. https://doi.org/10.1021/acsaem.9b01181.

[94] A. Uddin et al., High Power Density Platinum Group Metal-free Cathodes for Polymer Electrolyte Fuel Cells. *ACS Appl. Mater. Interfaces*, **12**, 2216–2224, 2020. https://doi.org/10.1021/acsami.9b13945.

[95] S. J. Normile et al., Direct observations of liquid water formation at nano- and micro-scale in platinum group metal-free electrodes by operando X-ray computed tomography. *Mater. Today Energy*, **9**, 187–197, 2018. https://doi.org/10.1016/j.mtener.2018.05.011.

[96] L. Jiao et al., Chemical vapour deposition of Fe–N–C oxygen reduction catalysts with full utilization of dense Fe-N4 sites. *Nat. Mater.*, 2021. https://doi.org/10.1038/s41563-021-01030-2.

[97] N. R. Sahraie et al., Quantifying the density and utilization of active sites in non-precious metal oxygen electroreduction catalysts. *Nat. Commun.*, **6**, 8618, 2015. https://doi.org/10.1038/ncomms9618.

[98] F. Luo et al., Surface site density and utilization of platinum group metal (PGM)-free Fe–NC and FeNi–NC electrocatalysts for the oxygen reduction reaction. *Chem. Sci.*, **12**, 384–396, 2021. https://doi.org/10.1039/d0sc03280h.

[99] P. Boldrin et al., Deactivation, reactivation and super-activation of Fe-N/C oxygen reduction electrocatalysts: Gas sorption, physical and electrochemical investigation using NO and O2. *Appl. Catal. B, Environ.*, **292**, 120169, 2021. https://doi.org/10.1016/j.apcatb.2021.120169.

Jason K. Lee, Pranay Shrestha, Ahmed M. Hasan, and Aimy Bazylak

4 Synchrotron X-ray imaging of PEMFCs

The synchrotron is a circular particle accelerator that provides a powerful source of X-rays. The X-rays are produced from high energy electrons that are accelerated in a closed loop. The electron beams are accelerated to a velocity near the speed of light (2.9×10^8 m/s) in a closed loop. As the electrons change direction using an array of magnets, electromagnetic radiation is produced and ranges from infrared radiation to hard X-rays. The high energy X-rays produced at synchrotron facilities are called synchrotron X-rays [1]. One of the numerous applications of synchrotron X-rays, and the focus of this chapter, is the use of synchrotron X-rays to image polymer electrolyte membrane fuel cells (PEMFCs), particularly during fuel cell operation (imaging conducted during cell operation is termed *operando* imaging). *Operando* imaging of fuel cells enables a thorough investigation of complex physiochemical phenomena occurring within fuel cells in realistic operating conditions and may be used to advance the understanding, accurate prediction, and optimization of fuel cells. Synchrotron X-rays are ideal for visualizing and studying material distributions within traditionally opaque PEMFCs owing to the following features: (a) high photon flux, (b) collimated (or parallel) beam and (c) availability of monochromatic beam (X-ray beam of a particular energy level). As a result, material distribution within fuel cells can be quantified accurately using synchrotron X-rays with high spatial (on the order of micrometers and nanometers) and temporal resolutions (on the order of seconds and subseconds).

In general, absorption imaging of fuel cells consists of 2-dimensional (2-D) radiography and 3-dimensional (3-D) computed tomography (CT). In this chapter, the fundamentals of synchrotron imaging are first introduced in Section 4.1, where the basics of image acquisition and image processing are discussed (for radiography). Next, the applications of synchrotron X-rays for imaging fuel cells are presented in Section 4.2 with examples of fuel cell research that use radiography and CT.

4.1 Synchrotron X-ray imaging: fundamentals

In this section, basic procedures for operando image acquisition and image processing of fuel cell radiographs are presented.

Jason K. Lee, Pranay Shrestha, Ahmed M. Hasan, Aimy Bazylak, Bazylak Group, Department of Mechanical and Industrial Engineering, Faculty of Applied Science and Engineering, University of Toronto, Toronto, ON M5S 3G8, Canada

https://doi.org/10.1515/9783110622720-004

4.1.1 Acquisition of synchrotron X-ray radiographs: through-plane and in-plane imaging

Insights gained from operando X-ray radiography depend on the imaging direction. Imaging orientation with respect to the membrane plane is used to categorize imaging modes: during *through-plane imaging*, the incident beam is oriented perpendicular to the membrane plane (Figure 1a) and during *in-plane imaging* the incident beam is parallel to the membrane plane (Figure 1b). It is important to note the difference between imaging direction and the directions of information obtained in the resulting radiographic images. For instance, to obtain material distribution in the through-plane direction (i. e., y-direction, perpendicular to the membrane plane), in-plane imaging is required since the incident beam is perpendicular to the axes in the obtained radiographs (Figure 1b). During in-plane imaging, we obtain information in the through-plane direction. It is also important to note that while the radiographic images are in 2-D, these images provide spatially distributed information beyond 2-D as each pixel carries information integrated over all the materials in the beam path. The differences in information captured during the two imaging modes are described in the following subsections.

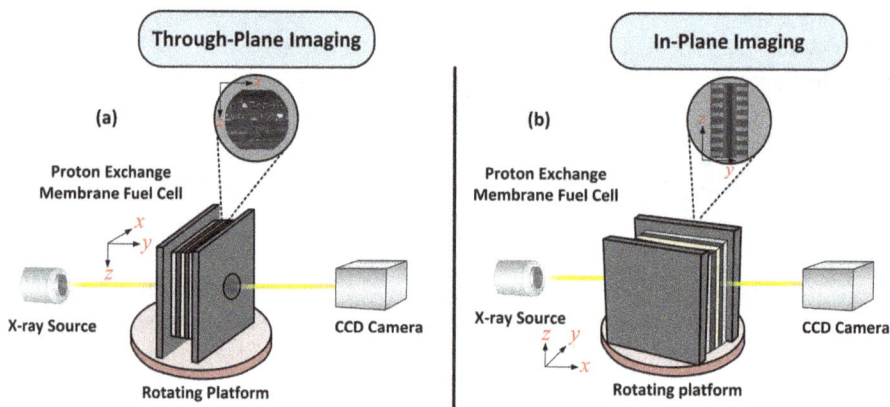

Figure 1: Representation of the two main imaging alignments (a) through-plane and (b) in-plane imaging. The y-direction is perpendicular to the membrane plane and represents the through-plane direction.

4.1.1.1 Through-plane imaging

Through-plane imaging is performed by aligning the fuel cell with the membrane perpendicular to the incident beam. Through-plane imaging enables direct visualization of mechanisms taking place in the in-plane directions, such as the evolution of water droplets in the fuel cell or the build-up of water along the length of serpentine channels. For instance, Hinebaugh et al. [2] used a customized fuel cell to visualize liquid water breakthrough during operation (Figure 2). Liquid water droplet formation throughout the channel and its transport from under the ribs were evident during operation. Visualizing these breakthrough points are crucial in investigating water transport properties because they are used to predict the liquid water pathways in the GDL. Through-plane imaging may also be utilized to evaluate transport properties of novel GDLs, such as perforated GDLs and GDLs with patterned wettability.

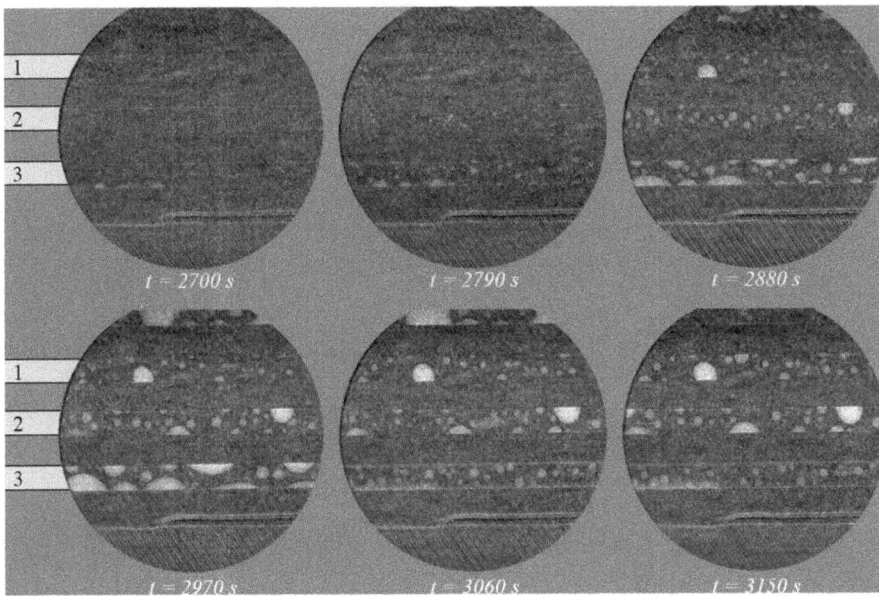

Figure 2: Six frames from the final stage ($\lambda_A = 2.8, \lambda_C = 1.4$) of an example experiment (GDL: Toray TGP-H 090 with 10 wt.% PTFE and proprietary MPL. Cell Temperature: 75 °C). Grey scale values correspond to the thickness of liquid water scaled between −0.2 mm and 0.6 mm. The positions of three cathode channels are highlighted on the left. For scale, each channel width is 1 mm [2]. Reprinted from *Electrochim. Acta*, 184, J. Hinebaugh, L. Jee, C. Mascarenhas, A. Bazylak, Quantifying Percolation Events in PEM Fuel Cell Using Synchrotron Radiography, 417–426, Copyright (2015), with permission from Elsevier.

An important design consideration for a through-plane imaging cell is to minimize highly attenuating materials in the X-ray beam path. Specifically, endplates and current collectors are main concerns for fuel cells, as they are typically fabricated from metals, which are highly absorbing for X-rays (corresponding to low signal in a 2-D radiograph). Hinebaugh et al. [2] successfully imaged a fuel cell using through-plane imaging by designing viewing holes in the endplates and current collectors as shown in Figure 3. It is important to strategically select the location of viewing holes, to fully capture the local effects near the inlet and outlet region. An alternative option to viewing holes would be to select appropriate materials for endplates and current collectors exhibiting low attenuation to X-rays, such as graphite. Another key design consideration is the orientation

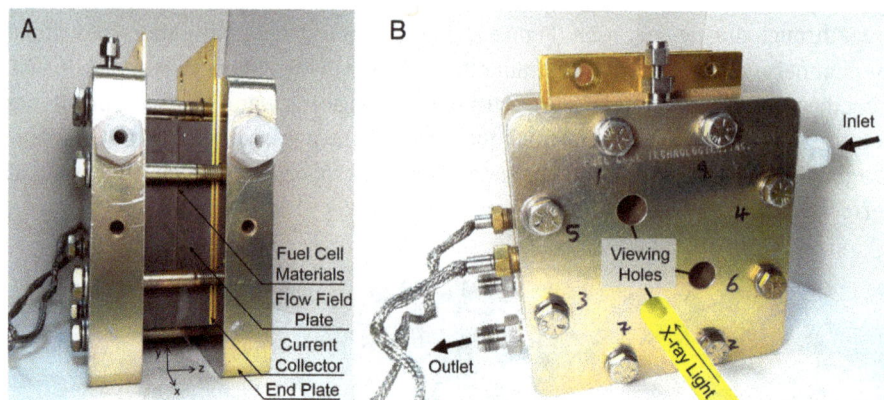

Figure 3: Images of modified 25 cm² Fuel Cell Technologies PEM fuel cell showing the fuel cell view per-pendicular to X-rays during: (a) in-plane imaging, and (b) through-plane imaging [2]. Reprinted from *Electrochim. Acta*, 184, J. Hinebaugh, L. Jee, C. Mascarenhas, A. Bazylak, Quantifying Percolation Events in PEM Fuel Cell Using Synchrotron Radiography, 417–426, Copyright (2015), with permission from Elsevier.

of fuel cell components when viewed in the through-plane direction. In a through-plane imaging radiograph, an integration is performed in the through-plane direction such that the anode and cathode compartments cannot be distinguished from each other. As such, if water in the anode and cathode channels need to be distinguished, the channels may be offset from each other, as was done by Hinebaugh et al. [2].

4.1.1.2 In-plane imaging

In-plane imaging is performed by aligning the fuel cell so that the membrane plane is parallel to the incident beam (Figure 4b). Radiographs obtained from in-plane imaging (Figure 4a) enables visualization and material quantification within multiple layers of the fuel cell, including flow channels/lands, gas diffusion layers (GDLs), microporous layers (MPLs) and catalyst coated membranes (CCMs) [3]. Material distribution such as operando liquid water distribution within these distinct layers of the fuel cell may be computed using image processing.

Operando in-plane synchrotron radiography is highly insightful when conducted at a high spatial resolution (on the order of a few μm) since individual layers within the fuel cell such as the GDL, MPL, and CCM may be resolved. However, a higher spatial resolution typically corresponds to a smaller field of view (on the order of mm). Due to a small field of view, custom miniaturized fuel cells are typically used in operando experiments instead of larger fuel cells or fuel cell stacks. These visualization cells are designed so the individual components including flow fields, GDLs, MPLs, CLs and membranes are positioned within the incident X-ray beam, as shown in Figure 5 [3]. It is important to minimize the attenuation by avoiding attenuative materials in the beam path when

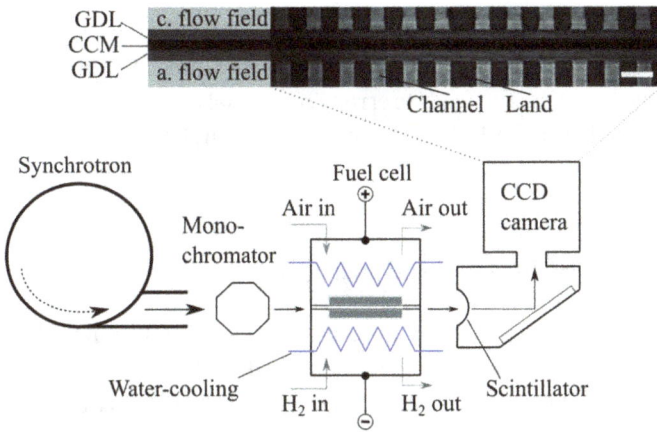

Figure 4: Schematic showing the orientation of the fuel cell apparatus and the incident beam during synchrotron X-ray radiographic imaging. The bottom inset illustrates fuel cell components in the obtained radiograph and the length bar is 0.4 mm [3]. The figure was reproduced with permission of the International Union of Crystallography, according to the terms and conditions of use of material published by the International Union of Crystallography.

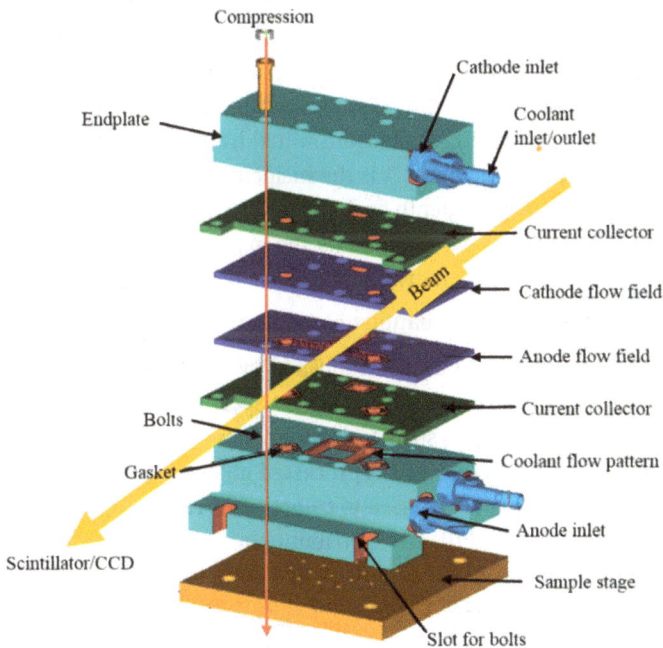

Figure 5: Schematic illustrating the components of the miniature PEM fuel cell. The yellow arrow represents the path of the beam [4].

designing the cell hardware. Moreover, alignment of the cell is crucial to resolve individual layers of the fuel cell and to avoid errors in the water thickness measurement. Movement during fuel cell operation also leads to errors in material quantification (correction procedures discussed in Section 4.1.2), and hence the fuel cell should be designed to mitigate any movements during operation.

4.1.2 Image processing of fuel cell radiographs

Raw radiographs from fuel cell experiments are processed to obtain useful material information, such as liquid water distributions, typically using the following procedure: (a) *Dark-field correction*: background camera noise is corrected for by subtracting dark-field radiographs from all experimental radiographs (sample corrected image in Figure 6a, Dark-field radiographs typically consists of an average of dark-field images, or radiographs without the application of incident X-ray beam. (b) *Beam intensity correction*: fluctuations or decay in incident beam is quantified and used to normalize the intensity of all radiographs. (c) *Movement correction*: unwanted movement such as translation or rotation during a fuel cell experiment cause artifacts in material quantification and are corrected for using procedures similar to [3]. (d) *Water quantification using Bouguer–Beer–Lambert law*: a dry reference radiograph (radiograph of fuel cell setup taken typically at open circuit voltage without current and water generation) and experimental radiographs are used to quantify the change in liquid water during fuel cell experiments. According to the Bouguer–Beer–Lambert law, the transmitted irradiance of a sample, I_T, is expressed as $I_T = I_0 \exp(- \sum_i \mu_i d_i)$, where I_0 is the incident irradiance, μ [m^{-1}] is the attenuation coefficient and d [m] is the depth or thickness of material in beam path of a material i [5]. If the reference (average frame) and the experimental irradiances are $I_{Ref}(x,y)$ and $I_{Exp}(x,y,t)$, respectively, the water thickness for each pixel location in the experimental radiographs, d_w [m] is quantified as $d_w(x,y,t) = -\frac{1}{\mu_w} \ln(\frac{I_{Exp}(x,y,t)}{I_{Ref}(x,y)})$.
The attenuation coefficient for water, μ_w [m^{-1}] is published by the National Institute of Standards and Technology [6]. However, calibrating μ_w for specific radiographic systems is highly recommended to avoid errors such as those due to scattering and higher harmonics (procedure detailed by N. Ge et al. [3]). A processed image with liquid water thickness (sample shown in Figure 6b) may be used to compute average through-plane liquid water distributions (Figure 6c and d). The distribution of water thickness may be further processed within porous materials (e. g., GDL and MPL) to obtain the liquid water saturation, $s_w(x,y,t) = \frac{d_w(x,y,t)}{L_z \varepsilon(y)}$ where L_z [m] is the thickness of porous material in the direction of the beam, and $\varepsilon(y)$ is the porosity profile in the through-plane direction. (e) *Uncertainty calculation*: the uncertainty in the measurement of liquid water from radiographic images is quantified and described in detailed by S. Chevalier et al. [7].

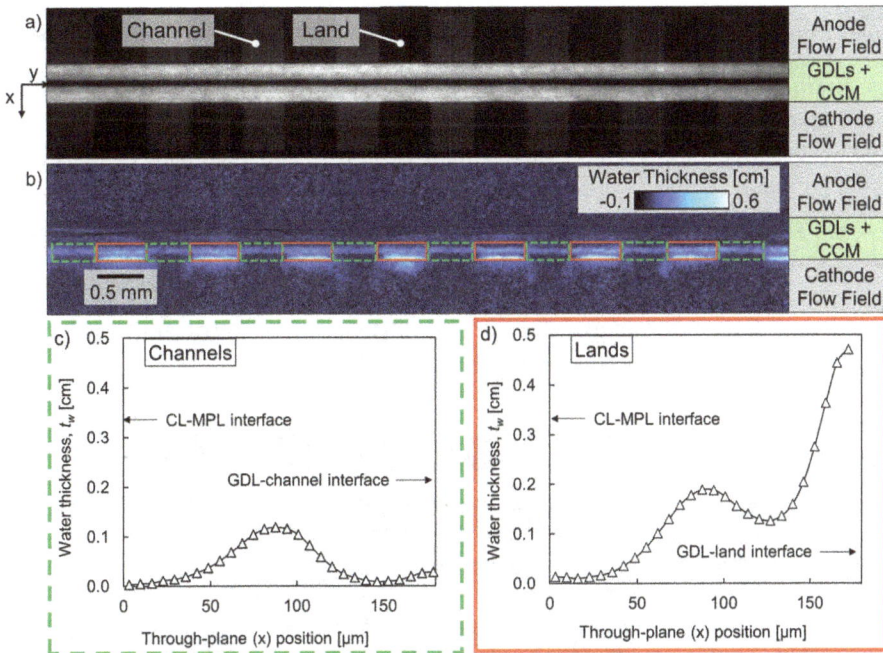

Figure 6: Sample quantification of liquid water from synchrotron X-ray radiography obtained at limiting current conditions (SGL 29 BC, 60 °C, 80 % relative humidity). (a) A raw absorption radiograph. (b) A processed image averaged over 10 frames (30 s) with water thickness (along beam direction) depicted on a color scale. Average water thickness distribution within the cathode GDL along through-plane direction, x, in (c) channel regions and (d) land regions, the locations of which are indicated by the regions with the dashed green outlines and red outlines, respectively, in (b) [8]. Reprinted from *Electrochim. Acta*, 274, D. Muirhead, R. Banerjee, M. G. George, N. Ge, P. Shrestha, H. Liu, J. Lee, A. Bazylak, Liquid water saturation and oxygen transport resistance in polymer electrolyte membrane fuel cell gas diffusion layers, 250–256, Copyright (2018), with permission from Elsevier.

4.2 Applications of synchrotron X-ray imaging on fuel cells

In this section, the application of synchrotron X-ray imaging to investigate and characterize mass transport phenomena in PEM fuel cells is presented. Section 4.2.1 presents the application of synchrotron X-ray radiography to water transport in fuel cell porous layers and flow channels. Section 4.2.2 describes the usage of synchrotron X-ray radiography and electrochemical impedance spectroscopy to quantify mass transport losses. Section 4.2.3 introduces methods to characterize degradation in fuel cells. Section 4.2.4 presents characterization of novel fuel cell materials using synchrotron X-ray radiography. Finally, in Section 4.2.5, the application of synchrotron X-rays for operando 3-D computed tomography is presented.

4.2.1 Water transport in fuel cell porous layer and flow channels

Operando synchrotron X-ray imaging is ideally suited for studying water transport mechanisms and dynamics within fuel cell porous layers (such as GDL and MPL) and flow field channels. Water distributions in the GDL, MPL, and flow field channels vary based on the operating conditions. For example, Chevalier et al. [9] illustrated drastic differences in the water thickness in the GDL and MPL at varying relative humidity conditions, as shown in Figure 7, where larger quantities of water thicknesses were measured at higher relative humidity conditions. Similarly, water distributions depend on the type of GDLs and MPLs due to the distinct wettability characteristics and structural properties, such as pore size distributions and porosity.

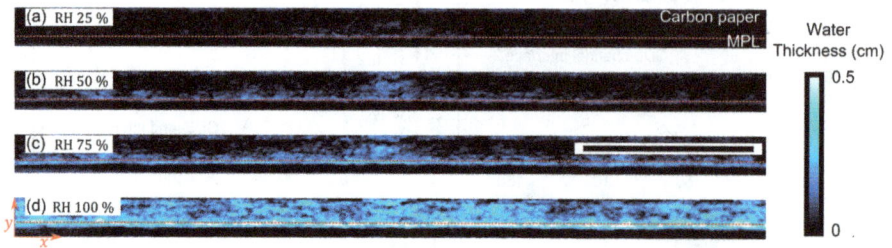

Figure 7: Water thickness cartographies at four inlet relative humidity (RH). Scale bar length in (c) is 1 mm. The red dashed line indicates the boundary between the MPL and the carbon paper substrate [9]. Reprinted from *Electrochim. Acta*, 210, S. Chevalier, J. Lee, N. Ge, R. Yip, P. Antonacci, Y. Tabuchi, T. Kotaka, A. Bazylak, "In operando measurements of liquid water saturation distributions and effective diffusivities of polymer electrolyte membrane fuel cell gas diffusion layers," 792–803, Copyright (2016), with permission from Elsevier.

Operando synchrotron X-ray imaging has enabled researchers to elucidate transport mechanisms in various components of the fuel cell. A salient example is the study of water transport mechanisms arising due to the addition of a microporous layer (MPL). The MPL is a backing layer between the catalyst layer (CL) and GDL and provides improved contact and enhanced liquid water management in the fuel cell. The MPL is often fabricated from carbon black and a hydrophobic agent, which leads to a pore structure exhibiting small pores and high hydrophobicity compared to the GDL. Operando imaging was crucial in uncovering transport mechanisms within the MPL. For instance, Lee et al. [10] investigated the impact of the MPL on liquid water transport by directly comparing the liquid water formation from a GDL with and without an MPL (as shown in Figure 8). Operando performance measurements showed that fuel cells consisting of GDLs with MPLs also exhibited more stable performance compared to cases without the MPL, indicating better mass transport behavior with MPLs. Counterintuitively, more water droplets were observed at the GDL-channel interface for GDLs with MPLs than for GDLs

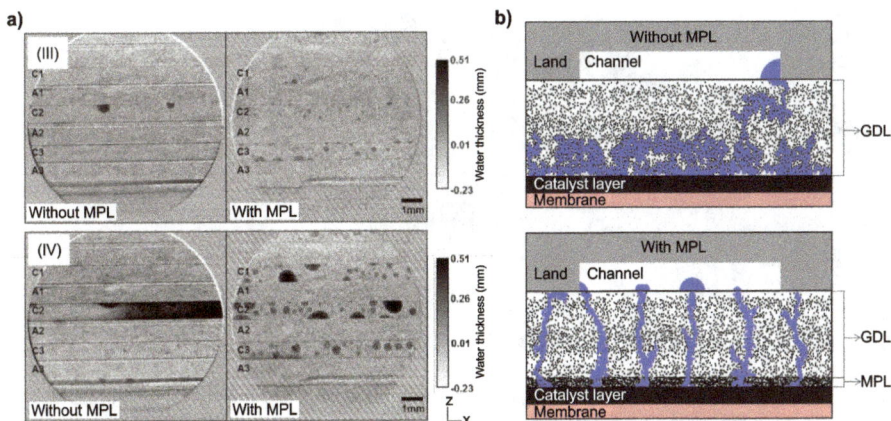

Figure 8: (a) Synchrotron X-ray radiographic images highlighting the appearance of liquid water at various times with Toray TGP-11-090 (left) and TGP-1-090 with MPL (right). (b) Comparison of liquid water evolution and transport in a GDL with and without MPL [10]. Reprinted from *J. Power Sources*, 227, J. Lee, J. Hinebaugh, A. Bazylak, Synchrotron X-ray radiographic investigations of liquid water transport behavior in a PEMFC with MPL-coated GDLs, 123–130, Copyright (2013), with permission from Elsevier.

without MPLs (Figure 8a). To explain the observation, Lee et al. [10] hypothesized that evenly distributed small pores in the MPL lead to a uniform distribution of liquid water at the CL-MPL interface before breakthrough, and a secondary breakthrough occurs when liquid water is expelled to the flow field channel. On the other hand, larger pores in the GDL-CL interface (for GDLs without an MPL) are prone to flooding due to the low threshold capillary pressure for water invasion (Figure 8b).

Through-plane operando imaging has also been used to investigate the role of cracks in the MPL on the liquid water transport by Markötter et al. [11]. The MPL typically consists of nanometer-sized pores and micrometer-sized cracks. Instead of invading smaller nanometer-sized pores with high capillary pressures, liquid water invades pores in micrometer-scale cracks, which are substantially larger in size. Markötter et al. [11] observed the continuous generation and snap-off of liquid water droplets at locations of MPL cracks (as indicated with arrows in Figure 9).

In-plane radiography is useful when quantifying liquid water separately in the various layers of the fuel cell, such as GDL, MPL, and flow-field channels. Hartnig et al. [12] quantified the liquid water distribution within the fuel cell at various operating current densities. They showed that liquid water accumulation corresponded to two distinct peaks, near the MPL-GDL interface and near the GDL-flow field interface (Figure 10a). They hypothesized important mechanisms of water transport, including condensation at MPL-GDL interface and the eruptive release of liquid water at GDL-channel interface (Figure 10b) based on their direct visualizations.

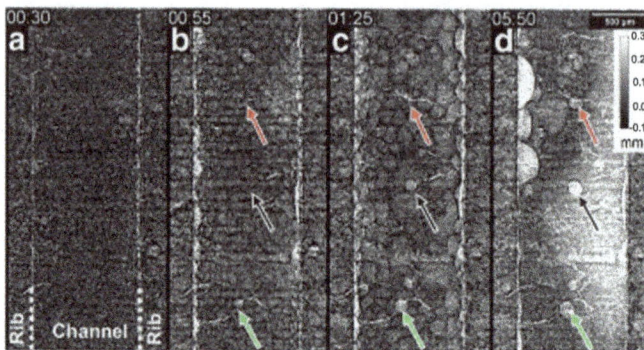

Figure 9: Radiographic image series showing the water distribution in one channel. Water emerges at MPL cracks and eventually builds droplets in the channel [11]. Reprinted from *Electrochem. Commun.*, 34, H. Markötter, J. Haußmann, R. Alink, C. Tötzke, T. Arlt, M. Klages, H. Riesemeier, J. Scholta, D. Gerteisen, J. Banhart, I. Manke, Influence of cracks in the microporous layer on the water distribution in a PEM fuel cell investigated by synchrotron radiography, 22–24, Copyright (2013), with permission from Elsevier.

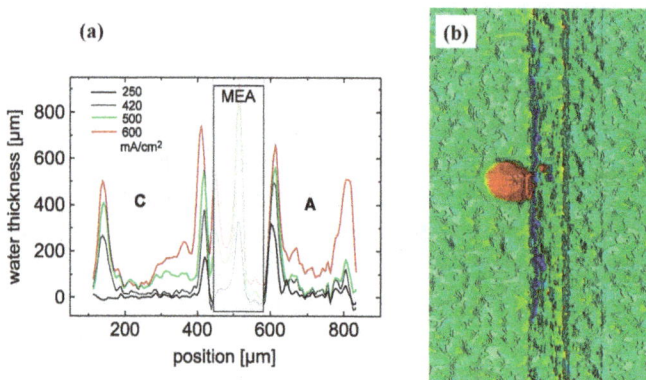

Figure 10: (a) Distribution of liquid water in cathodic (C) and anodic (A) gas diffusion layers. Two diffusion barriers are formed by liquid water depending upon operating temperature. (b) Eruptive water transport from the GDL to flow-field channel [12]. Reprinted from Appl. Phys. Lett. 92, C. Hartnig, I. Manke, R. Kuhn, N. Kardjilov, J. Banhart, W. Lehnert, Cross-sectional insight in the water evolution and transport in polymer electrolyte fuel cells, 134106, (2008) with the permission of AIP Publishing.

4.2.2 Combined synchrotron X-ray radiography and electrochemical impedance spectroscopy to study mass transport in fuel cells

Operando synchrotron X-ray radiography is a complementary technique to the electrochemical impedance spectroscopy (EIS) when investigating mechanisms of performance loss in PEM fuel cells. EIS is a noninvasive diagnostic tool that measures electrical energy dissipation within an electrochemical system [13]. EIS is a powerful tool for decomposing and measuring impedances attributed to various physiochemical processes in a PEM fuel cell. EIS measurements are often interpreted using equivalent circuit models composed of a network elements, such as resistors and capacitors, to represent the impedances related to performance losses in PEM fuel cells. For instance, EIS may be used to decompose and quantify performance losses, such as mass transport and ohmic resistances. Liquid water distributions obtained from operando synchrotron radiography pair insightfully with EIS measurements to reveal underlying mechanisms for individual losses.

Antonacci et al. [14] evaluated the effect of MPL thickness on fuel cell performance using a combination of EIS and operando synchrotron imaging. To capture the phenomena of proton transport, charge transfer, and mass transport within the cell, a current density of $1.0 \, \text{A/cm}^2$ (an intermediate current density for the cell setup, where none of the loss mechanisms was noticeably dominant) was chosen for the EIS calculation. A Randles equivalent circuit with a Warburg element was used to analyze the impedance response of the PEM fuel cell. Water volume within the GDL measured via synchrotron imaging was linked to ohmic and mass transport losses to explain the mechanisms that led to losses associated with oxygen mass transport and membrane dehydration, as shown in Figure 11. As such, the reason for the existence of an optimal MPL thickness was elucidated.

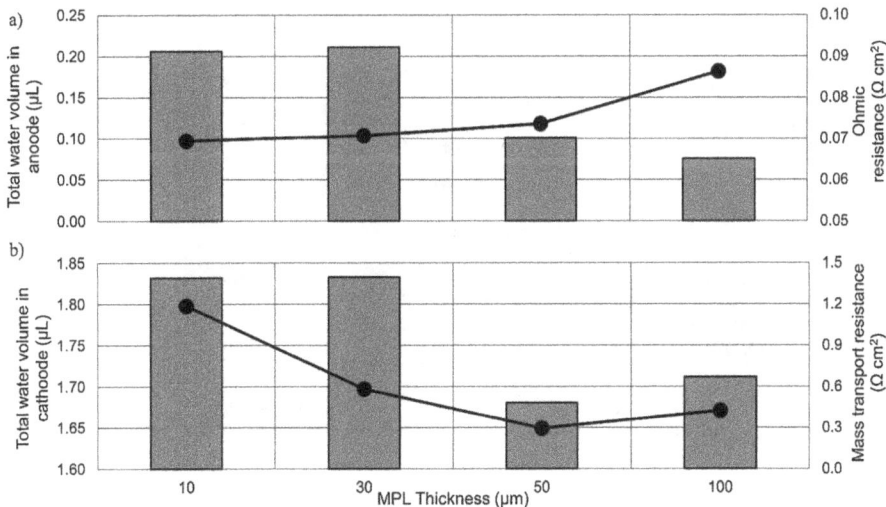

Figure 11: Total water volume content (represented by the bar graph) of (a) the anode substrate and (b) the cathode substrate for each cell build of 10, 30, 50 and 100 µm MPL thicknesses at 2.0 A/cm², compared with the (a) ohmic resistance and (b) mass transport resistance (represented by the points) [14]. Reprinted from *Electrochim. Acta*, 188, P. Antonacci, S. Chevalier, J. Lee, N. Ge, J. Hinebaugh, R. Yip, Y. Tabuchi, T. Kotaka, A. Bazylak, Balancing mass transport resistance and membrane resistance when tailoring microporous layer thickness for polymer electrolyte membrane fuel cells operating at high current densities, 888–897, Copyright (2016), with permission from Elsevier.

4.2.3 Synchrotron X-ray imaging to investigate degradation

Long term durability of PEM fuel cells is one of the most critical considerations in commercial adoption, especially in the heavy-duty automotive sector [15]. Synchrotron X-ray imaging aids in the analysis of degradation by providing operando characterization of changes in liquid water distribution and the associated mass transport as a result of degradation. In addition to in situ fuel cell degradation using hundreds to thousands of hours of fuel cell testing, several authors have utilized accelerated degradation procedures to analyze the degradation of specific components, such as the gas diffusion layer. For instance, Arlt et al. [16] investigated the effect of GDL degradation on liquid water distribution using high-resolution synchrotron radiography. The authors artificially aged commercial GDLs, Sigracet SGL 25 for 0 h, 16 h and 24 h in a hydrogen peroxide solution, then washed and dried the GDLs before testing them in a PEM fuel cell. Water distributions in GDLs were significantly altered by degradation, where liquid films were observed earlier in the GDLs and at the GDL-membrane and GDL-channel interfaces as a result of degradation, as shown in Figure 12. The water distributions were directly correlated to electrochemical measurements that were recorded during imaging. The measurements were repeated to verify reproducibility of findings. Following a similar procedure, Liu et al. [17] showed that the MPL was more prone to degradation than the GDL substrate. GDLs with degraded MPLs were more prone to flooding due to the loss

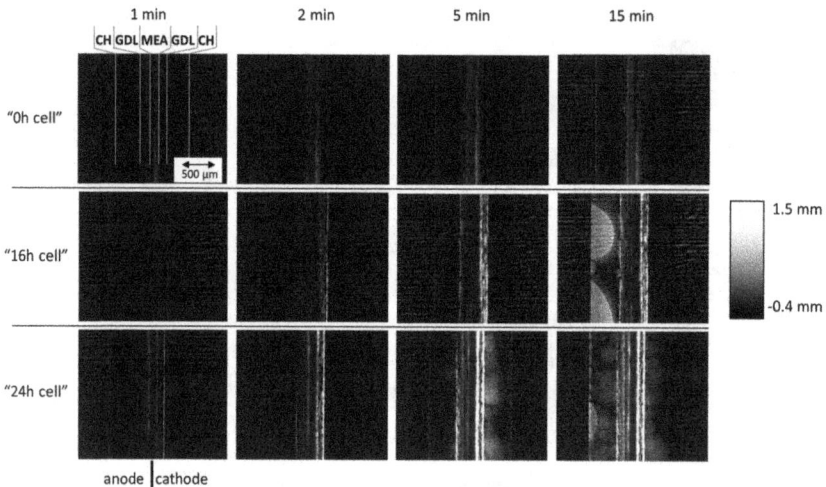

Figure 12: Water distribution in fuel cells for pristine ("0h cell") and degraded cells ("16h cell" and "24h cell"). Time (from the start of the fuel cell test) at which the radiographs are taken are shown on the top. The locations for fuel cell components are indicated: channels (CH), gas diffusion layer (GDL) and membrane electrode assembly (MEA), with anode and cathode compartments shown. Gray values represent water thickness in the range of −0.4 mm to 1.5 mm [16]. Reprinted from *Energy*, 118, T. Arlt, M. Klages, M. Messerschmidt, J. Scholta, I. Manke, Influence of artificially aged gas diffusion layers on the water management of polymer electrolyte membrane fuel cells analyzed with in-operando synchrotron imaging, 502–511, Copyright (2017), with permission from Elsevier.

of hydrophobicity of the MPLs; therefore, their use resulted in higher mass transport losses compared to pristine GDLs.

4.2.4 Synchrotron X-ray imaging to characterize novel fuel cell materials

Optimizing water management in fuel cells is key for enhancing fuel cell performance. Improved porous structure and wettability of the GDL are crucial in achieving superior

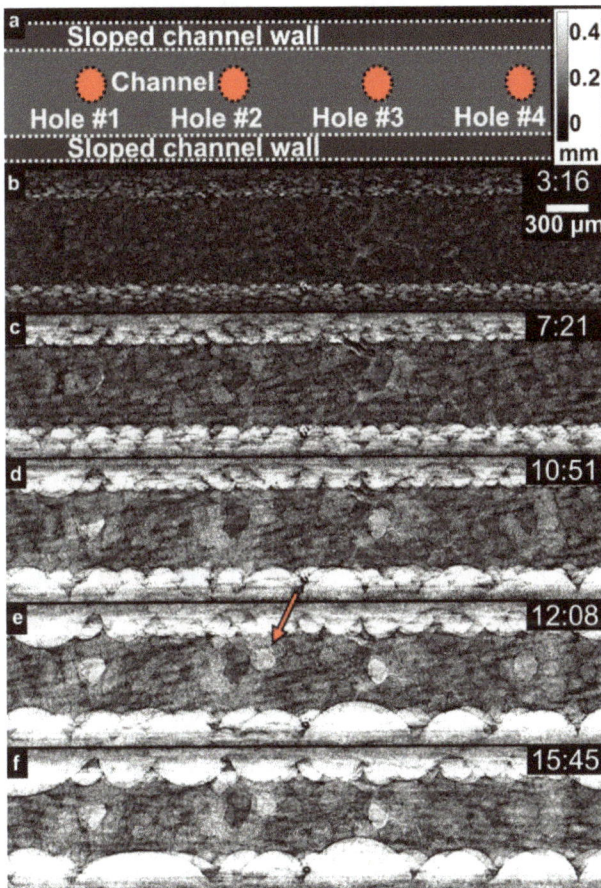

Figure 13: (a) Drawn position of the cell's channel and rib as well as the position of the imaged holes. Cathode gas flow points to the right. (b)–(f) Water distribution at holes of a perforated GDL subsequent to start up from 0 to 0.5 A/cm^2. The areas around the holes fill up with water as well as most of the holes [11]. Reprinted from *Int. J. Hydrogen Energy*, 37, H. Markötter, R. Alink, J. Haußmann, K. Dittmann, T. Arlt, F. Wieder, C. Tötzke, M. Klages, C. Reiter, H. Riesemeier, J. Scholta, D. Gerteisen, J. Banhart, I. Manke, Visualization of the water distribution in perforated gas diffusion layers by means of synchrotron X-ray radiography, 7757–7761, Copyright (2012), with permission from Elsevier.

water management properties. Synchrotron X-ray imaging provides a unique platform to evaluate and characterize water transport properties within rationally designed or novel GDL materials.

H. Markötter et al. [11] used synchrotron radiography to show that perforated GDLs provided high-porosity hydrophilic channels for liquid water accumulation (Figure 13). They suggested that these perforations may serve as drainage channels for effective water transport. M. Balakrishnan et al. [18] created electrospun GDLs with pore-size gradients. The novel GDLs showed enhanced fuel cell performance over a wide range of operating humidities. At dry (50 % RH) conditions, the graded GDL (using synchrotron radiography) retained liquid water at the CL-GDL interface and thereby enhanced membrane hydration (Figure 14). At a high inlet humidity (100 % RH), the graded GDL led to enhanced mass transport due to the directed removal of liquid water from the GDL.

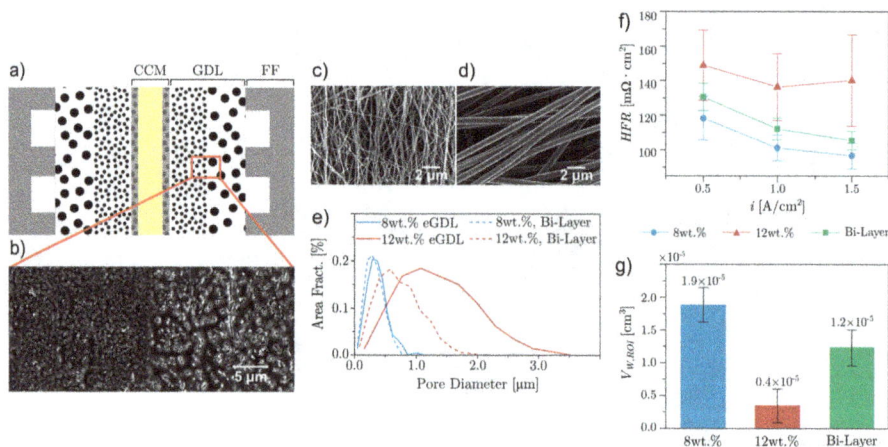

Figure 14: Schematic, scanning electron microscopy (SEM) images, and pore size distribution of tailored CGDLs. (a) Cross-sectional schematic of a PEM fuel cell with the graded Bi-layer eGDL with the 8 wt.% layer facing the catalyst-coated membrane (CCM) and the 12 wt.% layer facing the flow field (FF) forming a pore size gradient with increasing pore size from CL to FF. (b) SEM cross-section image of the Bi-layer eGDL. The left and right layers of the Bi-layer eGDL were electrospun from 8 and 12 wt.% PAN solution, respectively. Surface SEM image of (c) 12 wt.% eGDL and (d) 8 wt.% eGDL. (e) Area fraction weighted pore size distribution from SEM cross-section images. High-frequency resistance (HFR) and water content within the region of interest adjacent to the catalyst layer interface, $V_{W,ROI}$, at 50 % RH. (f) Average HFR at 50 % RH. Fuel cells with Bi-layer and 8 wt.% eGDLs exhibited lower HFRs compared to fuel cells with the 12 wt.% eGDL Error bars represent ±1 standard deviation. (g) Total liquid water content, $V_{W,ROI}$, in the region of interest at 1.0 A/cm^2 and 50 % RH. Fuel cells with Bi-layer and 8 wt.% eGDLs retained more liquid water adjacent to the CL interface compared to fuel cells with the 12 wt.% eGDL [18]. Reprinted with permission from *ACS Appl. Energy Mater.*, 3(3), M. Balakrishnan, P. Shrestha, N. Ge, C. H. Lee, K. F. Fahy, R. Zeis, V. P. Schulz, B. Hatton, B., A. Bazylak, Designing Tailored Gas Diffusion Layers with Pore Size Gradients via Electrospinning for Polymer Electrolyte Membrane Fuel Cells, 2695–2707, Copyright (2020) American Chemical Society.

A. Forner-Cuenca et al. [19] created dedicated liquid water removal pathways in GDLs using radiation grafting of hydrophilic compounds on hydrophobic-coated GDLs. The study showed that the wettability treatments significantly affected the liquid water distribution in these novel GDL designs. The distinct separation of water in the hydrophilic and hydrophobic domains was attributed to the narrow pore size distribution. In addition, they used synchrotron X-ray microtomography to evaluate the spatial distribution of water within the GDL at different inlet pressures (Figure 15). Shrestha et al. [20]

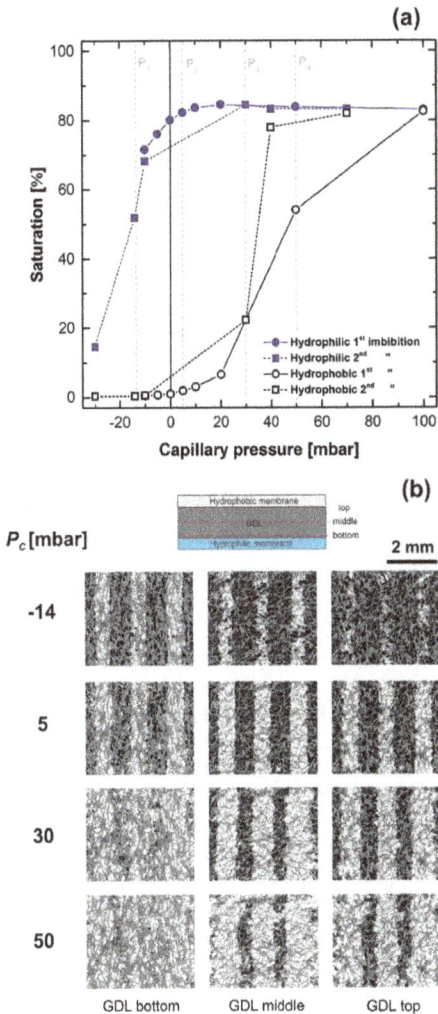

Figure 15: Capillary pressure characterization of Toray 70 % FEP-g-PNVF 500–930 km with homogeneous injection characterized with XTM. (a) Saturation versus capillary pressure curves, (b) XTM radiographs of three different slices (bottom, middle and top of the GDL sample) at four Pc (−14, 5, 30 and 50 mbar). The irradiated side of the GDL was placed toward the injection interface (bottom) [19].

also explored improved water management techniques by coating a thin hydrophilic MPL layer on a commercial hydrophobic GDL. The authors showed improved water retention at the CL-MPL interface (Figure 16a and b) through operando synchrotron X-ray radiography. Improved water retention led to decreased membrane resistance and enhanced fuel cell performance (Figure 16c and d) when the fuel cell was operated without anode humidification.

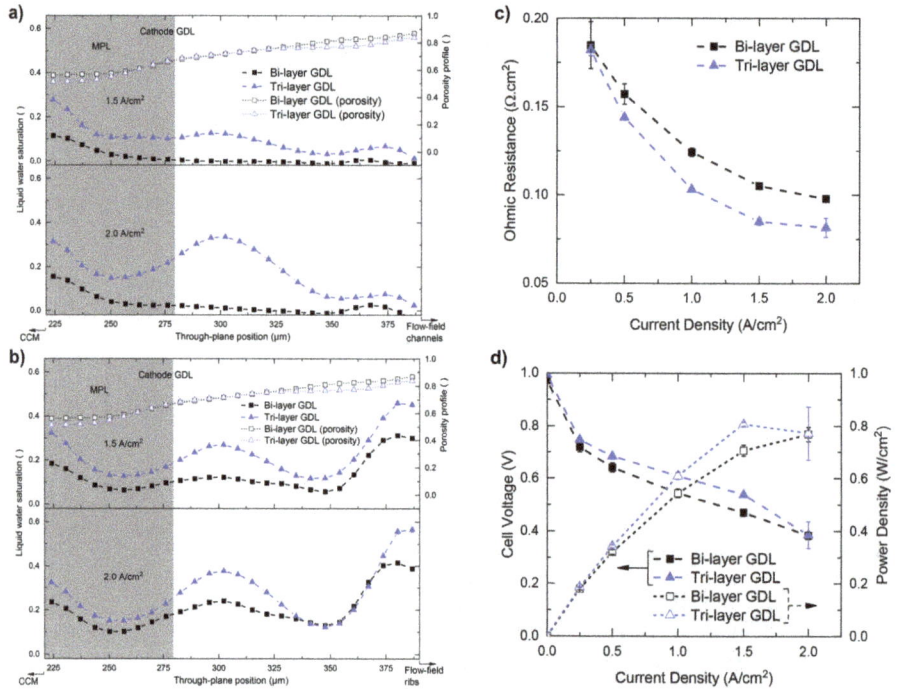

Figure 16: Liquid water saturation profiles along the through-plane position (y-direction or along the thickness of the GDL) under (a) channels and (b) ribs for cathode bi-layer and tri-layer GDLs. The inlet relative humidity was maintained at 0 % for the anode and 100 % for the cathode. The porosity profiles of the GDLs are shown in the secondary vertical axes using open symbols. Liquid water saturation at the cathode GDL increased at high current densities for the tri-layer GDL. (c) Ohmic resistance and (d) cell voltage (solid markers) and power density (hollow markers) for fuel cells with the bi-layer and tri-layer GDLs. The application of the hydrophilic MPL coating led to a decrease in the membrane resistance. The inlet relative humidity was maintained at 0 % for the anode and 100 % for the cathode [20]. Reprinted from *J. Power Sources*, 402, P. Shrestha, R. Banerjee, J. Lee, N. Ge, D. Muirhead, H. Liu, A. K. C. Wong, D. Ouellette, B. Zhao, A. Bazylak, Hydrophilic microporous layer coatings for polymer electrolyte membrane fuel cells operating without anode humidification, 468–482, Copyright (2018), with permission from Elsevier.

4.2.5 Operando synchrotron X-ray computed tomography

X-ray computed tomography (CT), also known as X-ray tomographic microscopy (XTM), is a nondestructive imaging method used to obtain 3-dimensional (3-D) material distributions. Two-dimensional (2-D) radiographs or projections of a sample are captured at multiple angles to obtain the required number of projection images. In the case of synchrotron tomography, the sample is rotated with respect to the X-ray beam and detector to obtain images at varying angles. Then the 2-D projections are combined (i. e., reconstructed) to provide a 3-D image. Visualizing the fuel cell in 3-D allows us to quantify material distributions in precise detail. For example, liquid water can be visualized and quantified within the 3-D pores of fuel cell materials, such as the GDL, MPL, catalyst layer, and their interfaces.

Furthermore, the high intensity of a synchrotron source allows for tomography with high temporal resolution (with subsecond tomography demonstrated at synchrotron facilities). Fast tomography allows researchers to incorporate an added dimension of time to yield 4-dimensional (4-D) images—3 spatial dimensions and 1 temporal dimension. 4-D tomography is particularly suited for operando studies since the time-resolved evolution of the fuel cell components can be visualized with simultaneous electrochemical operation of the fuel cell. Such operando studies typically require a custom fuel cell, which allows for rotation about the synchrotron stage while maintaining electrical and fluid connections required to operate the fuel cell. Two main configurations for fuel cell tomography are horizontal and vertical, where the fuel cell MEA is oriented either parallel or perpendicular to the X-ray beam (Figure 17).

Upon reconstruction, tomographic images may be further processed to quantify material distribution, such as water within GDL pores. For instance, a 3-D image of a dry GDL may be subtracted from an image of a GDL containing water (e. g., during fuel cell operation) to obtain a high-contrast image of the water distribution (Section 4.2.5.1). In addition, materials may be segmented based on pixel intensity to characterize the distribution of various materials (such as GDL substrate, MPL, void space, and water) within the fuel cell. Furthermore, material quantification during operando experiments may be further enhanced by morphological image correction techniques that account for local shifts in material distributions. For instance, H. Markötter et al. [22] detected local shifts in the morphology of fuel cell components due to membrane swelling/shrinkage using image correlation techniques. A correction technique was applied to account for these local shifts in morphology to yield accurate quantification of liquid water within GDLs (Figure 18).

4.2.5.1 Investigation of water in GDL pores

Synchrotron X-ray tomography provides a unique tool to investigate pore-scale water transport within the GDL with high spatial and temporal resolutions to reveal insights

a)

b)

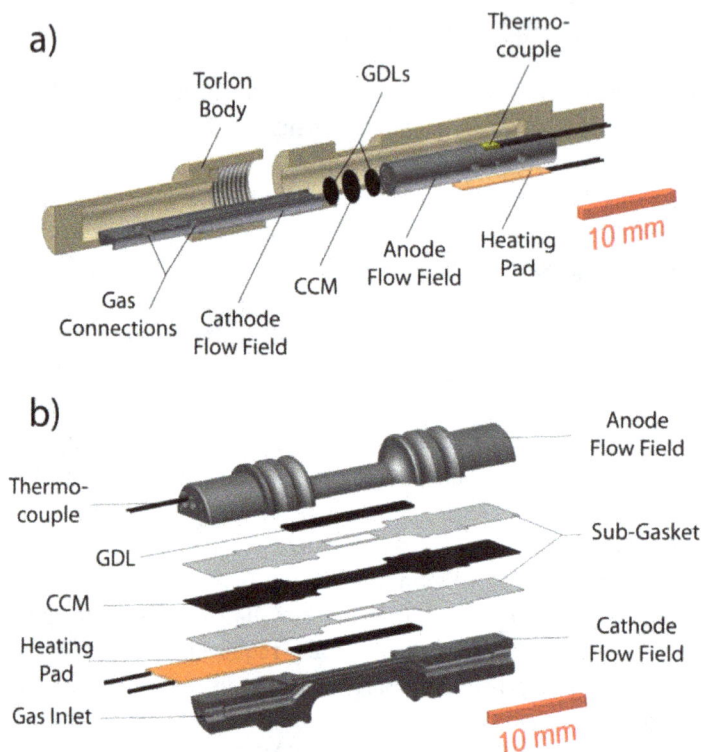

Figure 17: Schematics of XTM PEFC setups; (a) horizontal cell with an active area of 4.9 mm²; (b) vertical cell with an active area of 30.5 mm² without and of 11.0 mm² with sub-gaskets (as shown), respectively; flow fields are made of graphite composite; MEA components are described in the text; the flow fields of the vertical cell are clamped by small o-rings placed at the recesses of the thick flow field section, while the flow fields of the horizontal cell are clamped by an external Torlon body [21]. Reprinted from *J. Electrochem. Soc.,* 158, J. Eller, T. Rosén, F. Marone, M. Stampanoni, A. Wokaun, F. N. Büchi, Progress in In Situ X-Ray Tomographic Microscopy of Liquid Water in Gas Diffusion Layers of PEFC, B963, Copyright (2011), with permission from Elsevier.

into the mechanisms of water transport. For instance, H. Xu et al. [23] used submicron and subsecond X-ray tomography to investigate modes of water transport within GDLs at varying temperatures in an operating fuel cell. The subsecond resolution allowed the authors to reveal the dynamic evolution of water during start-up. In addition, the pore-scale distribution of water was utilized to reveal the connectivity of water clusters within the GDL. Using combined information on the connectivity and evolution of water clusters (as illustrated in Figure 19), the authors revealed the dominating mechanisms of water transport at each temperature condition. At 40 °C, capillary-fingering driven transport dominated water transport from start-up to stagnation, while at 60 °C, phase-change-induced water transport dominated the initial formation of water pathways, as shown in Figure 19.

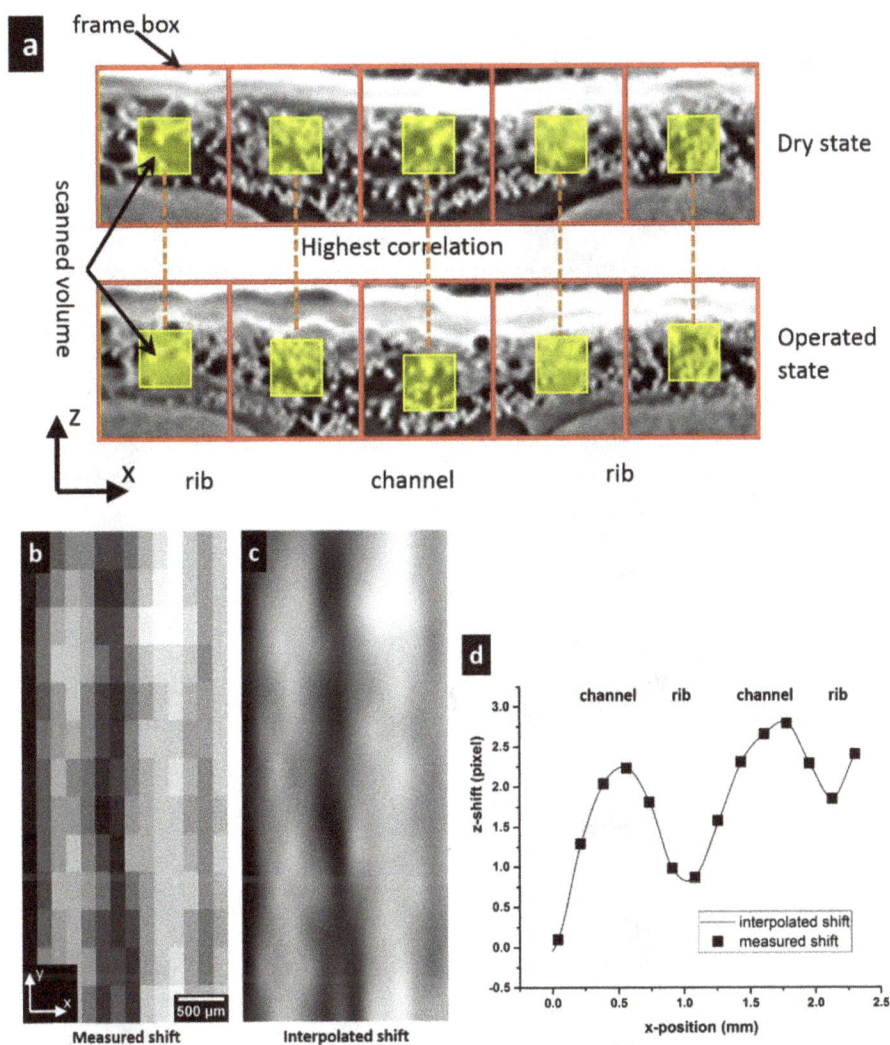

Figure 18: (a) The algorithm compares many small subvolumes, each within bigger frames and determines the shift via correlation [Shift of the yellow boxes has been amplified for visibility; Calculations are done in 3D, for simplicity the figure is sketched in 2D]. (b) Through plane component (z-axis) of the detected shift obtained from each box between the tomographic data sets. (c) The interpolated values calculated for each voxel and (d) a plot of the average z-shift against the x-position covering an area of two channels. For interpretation of the references to color in this figure legend, the reader is referred to the Web version of this article [22]. Reprinted from *J. Power Sources*, 414, H. Markötter, I. Manke, J. Böll, S. Alrwashdeh, A. Hilger, M. Klages, J. Haussmann, J. Scholta, Morphology correction technique for tomographic in-situ and operando studies in energy research, 8–12, Copyright (2019), with permission from Elsevier.

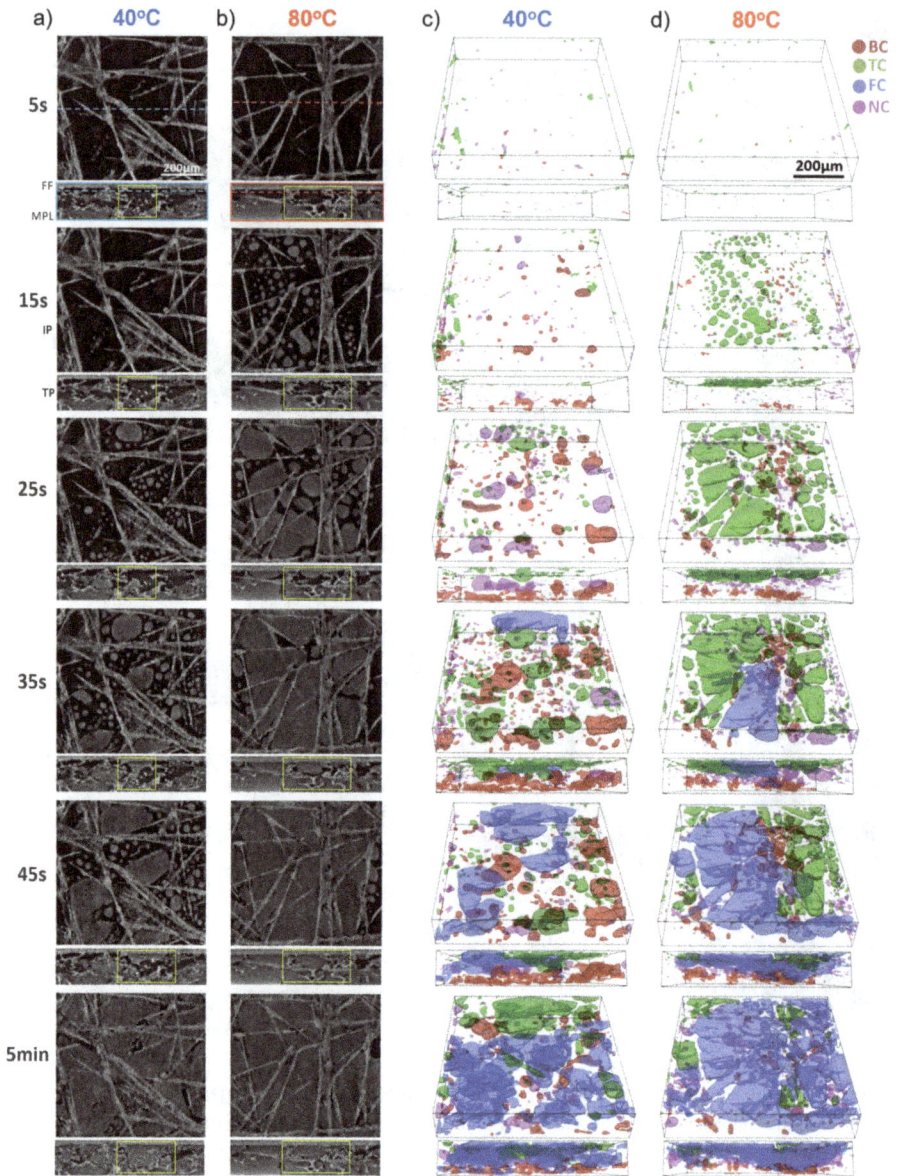

Figure 19: Through-plane (TP) and in-plane (IP) tomographic slices for 2 cells operated at (a) 40 °C and (b) 80 °C for selected early time-steps (5 s, 15 s, 25 s, 35 s, 45 s, 5 min) during the current ramp-up, and their corresponding 3D rendering presentations of cluster connectivity labeled water as (c) and (d). Blue and red dashed lines in (a) and (b) TP slices indicate the slicing positions for corresponding TP slices. All IP slices are taken from positions very close to the flow field. Yellow rectangles highlight the observed condensation and liquid water transport phenomena (for interpretation of the references to color in this figure legend, the reader is referred to the Web version of this article) [23].

4.2.5.2 Investigation of water within the MPL

Alrwashdeh et al. [24] used phase contrast synchrotron tomography to quantify water in the nanopores of the MPL during fuel cell operation. The collimated nature of synchrotron X-ray beams facilitates the enhancement of contrast between materials (in this case, water and MPL) using phase contrast tomography. The average water distribution over multiple pores was resolved with a spatial resolution of 1 μm, which provided a 3-D distribution of water within the MPLs. The MPL was shown to have an inhomogeneous distribution of water, where a sharp boundary existed between the MPL under the flow-field lands and channel. In addition, the MPL under the flow-field land exhibited a higher liquid water saturation (Figure 20) [24].

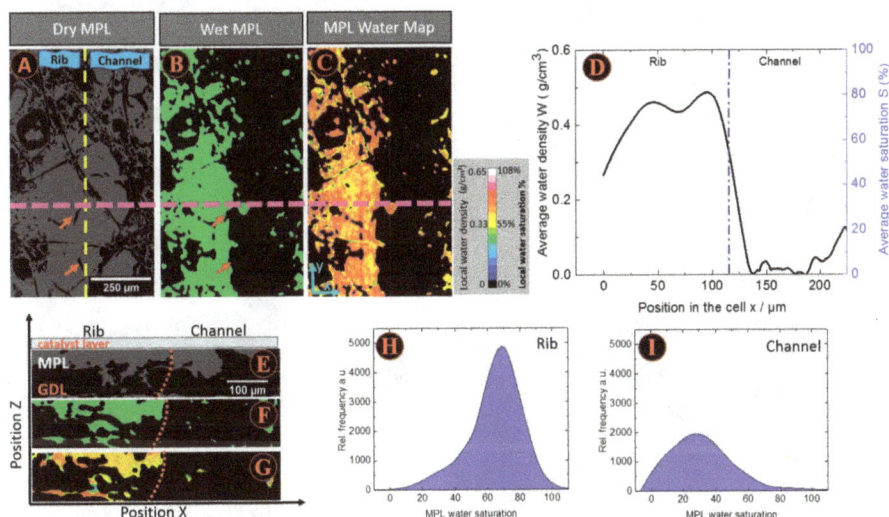

Figure 20: Selected area (same as in Figure 1A–C) showing the completely dry MPL (A), areas with wet MPL (B) and the local water density distribution in the MPL (C). Local MPL water density and local MPL water saturation profile along the x-axis for the whole volume (D). Three vertical cross-sections through the whole tomogram of the fuel cell along the line given in magenta in (B): completely dry MPL (E) corresponding to (A), wet MPL (F) corresponding to (B) and local MPL water density/saturation (G) corresponding to (C). Histograms of the local MPL water saturation in the rib (H) and channel region (I) (saturations below 0 % are caused by imaging artifacts and noise). Reprinted with permission from *ACS Nano*, 11, 6, S. S. Alrwashdeh, I. Manke, H. Markötter, M. Klages, M. Göbel, J. Haußmann, J. Scholta, J. Banhart, In Operando Quantification of Three-Dimensional Water Distribution in Nanoporous Carbon-Based Layers in Polymer Electrolyte Membrane Fuel Cells. Copyright (2017) American Chemical Society.

4.2.5.3 Investigation of catalyst layer and interfaces

J. Liu et al. [25] investigated the role of interfaces on performance and water management in fuel cells with platinum group metal (PGM)-free electrodes. The authors compared the three following catalyst layer configurations (CL): (a) CL coated on the membrane, i. e., catalyst coated membrane (CCM) configuration, (b) CL coated on the GDL, i. e., gas diffusion electrode (GDE) configuration, and (c) a hybrid CCM-GDE configuration. Using operando X-ray tomography, the authors found that interfacial porosity played a major role in water accumulation and flooding. Larger pores created due to surface roughness at interfaces led to higher liquid water accumulation. Specifically, water accumulation was observed at the CL-membrane interface and CL-GDL interface for GDE and CCM configurations, respectively (Figure 21). The study highlights that synchrotron X-ray tomography can be used to effectively characterize operando water management in fuel cells.

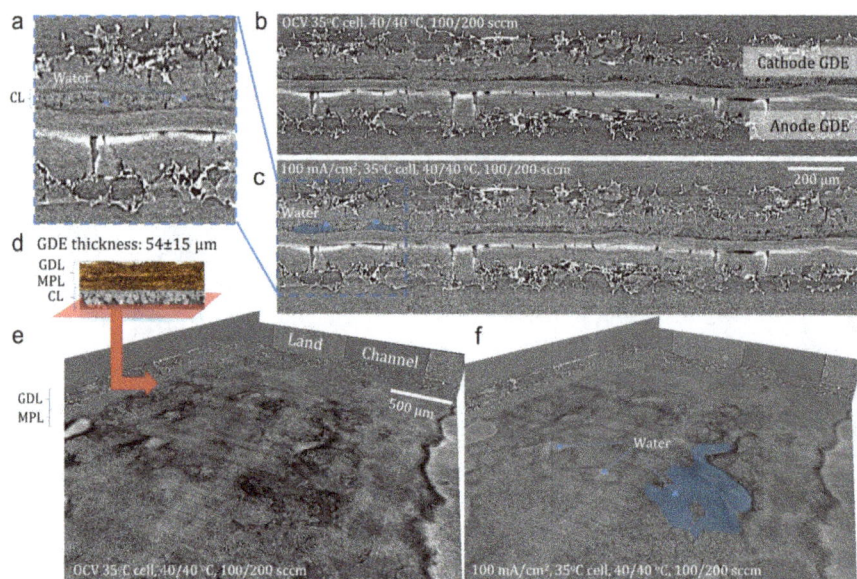

Figure 21: Operando X-ray CT cross sections for the cell with the GDE. The comparison of the cross sections under the channel (with part (a) the zoom-in) between the cell operating at (b) OCV and (c) 100 mA/cm². In-plane view (with part (d) the location) of the CL|PEM interface at (e) OCV and (f) 100 mA/cm². Selected locations of interfacial water accumulation are shown with false coloring [25]. Reprinted with permission from ACS Appl. Energy Mater., 2, 5, J. Liu, M. R. Talarposhti, T. Asset, D. C. Sabarirajan, D. Y. Parkinson, P. Atanassov, I. V. Zenyuk, Understanding the Role of Interfaces for Water Management in Platinum Group Metal-Free Electrodes in Polymer Electrolyte Fuel Cells, 3542–3553. Copyright (2019) American Chemical Society.

4.3 Summary/Conclusions

In this chapter, the fundamentals and applications of synchrotron X-ray imaging for PEMFCs were discussed. The main imaging configurations include 2-D radiography, consisting of in-plane and through-plane imaging and 3-D computed tomography (CT). Preferred imaging configurations depend upon the desired information, and special considerations must be made for the cell design to facilitate effective imaging. Image processing is required for radiography and CT to accurately quantify material distributions, such as that of liquid water within fuel cell components. Synchrotron X-rays have been successfully used for fuel cell research to accurately characterize liquid water transport within porous layers, interfaces, flow channels, and novel fuel cell materials. Additionally, imaging techniques lend unique insights into mechanisms, such as degradation within fuel cells, and may be combined and augmented with electrochemical techniques, such as electrochemical impedance spectroscopy. Overall, synchrotron X-ray imaging serves as a powerful technique to help characterize, diagnose, evaluate, and optimize PEMFC performance and durability, and there are tremendous opportunities to leverage the successes discussed in this chapter to further advance technologies in other areas of clean energy, such as electrolyzers and batteries.

Bibliography

[1]   N. Yagi, Synchrotron Radiation. In: *Comprehensive Biomedical Physics*, vol. 8, pages 17–33. Elsevier, 2014.
[2]   J. Hinebaugh, J. Lee, C. Mascarenhas, and A. Bazylak, Quantifying Percolation Events in PEM Fuel Cell Using Synchrotron Radiography. *Electrochim. Acta*, **184**, 417–426, 2015.
[3]   N. Ge et al., Calibrating the X-ray attenuation of liquid water and correcting sample movement artefacts during in operando synchrotron X-ray radiographic imaging of polymer electrolyte membrane fuel cells. *J. Synchrotron Radiat.*, **23**(2), 590–599, 2016.
[4]   J. Lee et al., Synchrotron Investigation of Microporous Layer Thickness on Liquid Water Distribution in a PEM Fuel Cell. *J. Electrochem. Soc.*, **162**(7), F669–F676, 2015.
[5]   T. G. Mayerhöfer, S. Pahlow, and J. Popp. The Bouguer–Beer–Lambert Law: Shining Light on the Obscure. *ChemPhysChem*, **21**(18), 2029–2046, 2020.
[6]   J. H. Hubbell and S. M. Seltzer, *X-Ray Mass Attenuation Coefficients*. NIST, 2004.
[7]   S. Chevalier et al., Synchrotron X-ray Radiography as a Highly Precise and Accurate Method for Measuring the Spatial Distribution of Liquid Water in Operating Polymer Electrolyte Membrane Fuel Cells. *J. Electrochem. Soc.*, **164**(2), F107–F114, 2017.
[8]   D. Muirhead et al., Liquid water saturation and oxygen transport resistance in polymer electrolyte membrane fuel cell gas diffusion layers. *Electrochim. Acta*, **274**, 250–265, 2018.
[9]   S. Chevalier et al., In operando measurements of liquid water saturation distributions and effective diffusivities of polymer electrolyte membrane fuel cell gas diffusion layers. *Electrochim. Acta*, **210**, 792–803, 2016.
[10] J. Lee, J. Hinebaugh, and A. Bazylak, Synchrotron X-ray radiographic investigations of liquid water transport behavior in a PEMFC with MPL-coated GDLs. *J. Power Sources*, **227**, 123–130, 2013.
[11] H. Markötter et al., Visualization of the water distribution in perforated gas diffusion layers by means of synchrotron X-ray radiography. *Int. J. Hydrog. Energy*, **37**(9), 7757–7761, 2012.
[12] C. Hartnig, I. Manke, R. Kuhn, N. Kardjilov, J. Banhart, and W. Lehnert, *Cross-sectional insight in the water evolution and transport in polymer electrolyte fuel cells*, 2008.
[13] X. Yuan, H. Wang, J. C. Sun, and J. Zhang, AC impedance technique in PEM fuel cell diagnosis—A review. *Int. J. Hydrog. Energy*, **32**(17), 4365–4380, 2007.

[14] P. Antonacci et al., Balancing mass transport resistance and membrane resistance when tailoring microporous layer thickness for polymer electrolyte membrane fuel cells operating at high current densities. *Electrochim. Acta*, **188**, 888–897, 2016.

[15] D. A. Cullen and et al., New roads and challenges for fuel cells in heavy-duty transportation. *Nat. Energy*, **6**(5), 462–474, 2021.

[16] T. Arlt, M. Klages, M. Messerschmidt, J. Scholta, and I. Manke, Influence of artificially aged gas diffusion layers on the water management of polymer electrolyte membrane fuel cells analyzed with in-operando synchrotron imaging. *Energy*, **118**, 502–511, 2017.

[17] H. Liu et al., Microporous Layer Degradation in Polymer Electrolyte Membrane Fuel Cells. *J. Electrochem. Soc.*, **165**(6), F3271–F3280, 2018.

[18] M. Balakrishnan et al., Designing Tailored Gas Diffusion Layers with Pore Size Gradients via Electrospinning for Polymer Electrolyte Membrane Fuel Cells. *ACS Appl. Energy Mater.*, **3**(3), 2695–2707, 2020.

[19] A. Forner-Cuenca et al., Advanced Water Management in PEFCs: Diffusion Layers with Patterned Wettability II. Measurement of Capillary Pressure Characteristic with Neutron and Synchrotron Imaging. *J. Electrochem. Soc.*, **163**(9), F1038, 2016.

[20] P. Shrestha et al., Hydrophilic microporous layer coatings for polymer electrolyte membrane fuel cells operating without anode humidification. *J. Power Sources*, **402**, 468–482, 2018.

[21] J. Eller, T. Rosén, F. Marone, M. Stampanoni, A. Wokaun, and F. N. Bü Chi, Progress in In Situ X-Ray Tomographic Microscopy of Liquid Water in Gas Diffusion Layers of PEFC. *J. Electrochem. Soc.*, **158**(8), 963–970, 2011.

[22] H. Markötter et al., Morphology correction technique for tomographic in-situ and operando studies in energy research. *J. Power Sources*, **414**, 8–12, 2019.

[23] H. Xu et al., Temperature dependent water transport mechanism in gas diffusion layers revealed by subsecond operando X-ray tomographic microscopy. *J. Power Sources*, **490**, 229492, 2021.

[24] S. S. Alrwashdeh et al., In Operando Quantification of Three-Dimensional Water Distribution in Nanoporous Carbon-Based Layers in Polymer Electrolyte Membrane Fuel Cells. *ACS Nano*, **11**(6), 5944–5949, 2017.

[25] J. Liu et al., Understanding the Role of Interfaces for Water Management in Platinum Group Metal-Free Electrodes in Polymer Electrolyte Fuel Cells. *ACS Appl. Energy Mater.*, **2**(5), 3542–3553, 2019.

Adam P. Hitchcock

5 PEMFC analysis using soft X-ray spectromicroscopy: methods and applications

5.1 Introduction

Polymer electrolyte membrane fuel cells (PEMFC) [1–3] offer many advantages over competing electric power systems for mobile applications. However, cost, durability and refueling issues need to be solved before widespread adoption will occur. In order to improve performance and reduce cost, a detailed understanding of the limiting factors is needed. In many cases quantitative, high spatial resolution, species-specific 2D and 3D imaging is required in order to understand the links between chemistry, structure/morphology and performance and thereby enable improvements through rational design. While analytical electron microscopes are superb at imaging and detailed analysis of radiation resistant components such as atomic-level studies of core-shell catalyst particles [4, 5], their capabilities are limited when applied to the very radiation sensitive perfluorosulfonic acid (PFSA)-ionomer, which is essential for proton conductivity in the electrodes. Reliable and comprehensive 2D and 3D structures at many different length scales are needed to improve modeling of PEMFC operation under normal and stressed conditions [6]. Areas where soft X-ray scanning X-ray microscopy (SXM) have played a role in achieving a better understanding include: (i) optimal use of Pt by optimization of the ionomer distribution in PEMFC electrodes; (ii) understanding degradation phenomena, which limit lifetime and efficiency; (iii) investigation of novel catalytic systems such as the 3M nanostructured thin film (NSTF) species and electroless deposition of Pt

Acknowledgement: I thank my students (Jian Wang, Adam Leontowich, Lis Melo, Juan Wu), postdocs (Jian Wang, Vincent Lee, Xiaohui Zhu) and collaborators (Viatcheslav Berezhnov, Darija Susac, Juergen Stumper, Shanna Knights, Sylvia Wessel, Monica Dutta) who were involved in the research described in this chapter. This research was supported financially by the Automotive Fuel Cell Co-operation Corporation (AFCC), the Natural Sciences and Engineering Research Council of Canada and the Catalysis Research for Polymer Electrolyte Fuel Cells (CaRPE-FC) Network. This research used resources of the Advanced Light Source (ALS), which is a DOE Office of Science User Facility under contract no. DE-AC02-05CH11231, and the Canadian Light Source (CLS), which is supported by the Canadian Foundation for Innovation. The SHARP code used for ptychographic data analysis was developed by the Center for Applied Mathematics for Energy Research Applications (CAM-ERA), led by Jamie Sethian in collaboration with Felipe Maia, Uppsala University. I particularly wish to thank Marcia West (McMaster) for excellent ultramicrotomy and other sample preparation innovations. I thank staff scientists, Jian Wang and Adam Leontowich at CLS, and David Kilcoyne, Tolek Tyliszczak and David Shapiro at ALS, for their scientific assistance and technical support of the beamlines and STXMs.

Adam P. Hitchcock, Dept. of Chemistry & Chemical Biology, Brockhouse Institute for Materials Research, McMaster University, Hamilton, ON, L8S 4M1, Canada

https://doi.org/10.1515/9783110622720-005

on membranes. This chapter describes SXM and its applications to these types of PEMFC studies.

My group has collaborated with industrial partners (Ballard Power Systems; the Automotive Fuel Cell Co-operation Corporation, AFCC; and 3M) and the Canadian Catalyst Research for Polymer Electrolyte Fuel Cells (CaRPE-FC) network over the past 12 years to develop SXM methods and apply them to PEMFC materials, systems and devices. Early work focused on developing and validating a method to use scanning transmission X-ray microscopy (STXM) [7–9] for quantitative 2D projection mapping of PFSA in PEMFC cathodes [10–12]. Ionomer mapping by STXM became a standardized analytical method for AFCC who used it extensively from 2013 to 2016 to help guide their research into cathode optimization. In parallel to that activity, and in collaboration with industry partners, we studied: Pt migration into the membrane (PTIM) [13, 14]; effect of the type of carbon support on carbon corrosion [15]; the fate of the polymer support in NSTF catalysts through their fabrication and implementation in an membrane electrode assembly (MEA) [16, 17]; and the Pt-membrane interface in electroless deposited Pt on membranes [18, 19]. In recent years, my group has concentrated on further improvements in SXM methodology as applied to PEMFC, including: evolving PFSA-in-cathode mapping from 2D to 3D through use of angle-scan tomography [20–23]; porosity mapping [24]; better understanding of radiation damage of PFSA by electron beams [25, 26], ion beams [27] and X-rays [28, 29]; improving spatial resolution through the emerging SXM method of ptychography [30, 31]; development of a STXM *in situ* environmental cell for water mapping in PEMFC membrane electrode assembly (MEA) under controlled (T, RH) [32, 33]; and design, construction and commissioning of a cryo-STXM [34, 35] with the significant financial support of AFCC. Cryo-STXM is a promising platform to improve 3D mapping accuracy by significantly reducing radiation damage rates [36]. Other groups who have used soft X-ray SXM methods to study PEMFC materials and devices include Bozzini et al. [37–41], Ohigashi et al. [42] and George et al. [43]. While this chapter focuses specifically on applications of SXM to PEMFC, SXM is a broad range of instruments and methods, which can be applied to many areas of science and technology. Jacobsen [44] has recently published a comprehensive book on X-ray microscopy, which provides an in-depth coverage of principles, instrumentation, analysis methods and applications of both soft and hard X-ray microscopy techniques. In the last decade, reviews of SXM have been reported by de Groot et al. [45], Hitchcock [8, 9] and Braun et al. [46].

Of course, many other powerful analytical methods are available and provide competitive and/or complementary capabilities. Some of these methods are presented in detail in other chapters, such as Jankovic et al. (Microstructural and Spectroscopic Characterization of PEMFCs, Chapter 2), Cullen et al. (Characterization methods for Atomically-Dispersed Platinum Group Metal-Free Catalysts, Chapter 3) and Bazylak et al. (Synchrotron X-ray imaging of PEMFCs, Chapter 4). In order to place the SXM studies described in this chapter in a broader context of advanced analysis of PEMFC, a number of these methods are listed here, with references to recent studies, which relate to the main subject of this chapter. Where appropriate, more details of alternate techniques are given in later sections describing SXM studies.

The closest equivalent to SXM is electron energy loss spectromicroscopy (EELS) in a transmission electron microscope (TEM). The energy resolution of modern TEM-EELS is quite similar to that achieved in synchrotron-based X-ray absorption spectroscopy and SXM. TEM has far better spatial resolution than SXM, and it has been shown that for very thin samples, molecular imaging can be done at lower radiation dose in electron microscopy than X-ray microscopy [47]. However, for analytical measurements involving TEM-EELS or X-ray fluorescence by energy dispersive spectroscopy (TEM-EDS), there is a significant advantage for SXM. TEM-EELS has considerable challenges for analytical studies of highly radiation sensitive PEMFC components such as PFSA or other ionomers. Several direct comparisons of SXM and TEM-EELS in terms of radiation damage per unit of analytical, core level excitation signal ("G-factor") have concluded there is a 100–500 fold advantage of SXM over TEM-EELS [48, 49], which is mostly related to the fact that the absorption of each X-ray photon at an analytically useful photon energy provides useful information while the vast majority of inelastic electron scattering events involve valence electron ionization, which cause bond breaking and damage, without providing analytically-useful core loss signal. Similarly, direct comparisons of SXM spectromicroscopy and TEM-EDS of ionomer in PEMFC cathodes have shown an enormous advantage for SXM [50]. Despite the analytical challenges, there are many excellent examples of electron microscopy—both secondary electron microscopy (SEM) and TEM—applied to PEMFC materials and systems [51–53]. Electron tomography (ET) has also been applied to PEMFC [54]. Discussion of the relative ability of TEM versus SXM to map ionomer in PEMFC electrodes is given elsewhere in this chapter. Another method very closely related to soft X-ray SXM is hard X-ray spectromicroscopy. The many variants of synchrotron based hard X-ray microscopy applied to materials research have been presented by Holt et al. [55]. Several groups have studied PEMFC devices under normal operating conditions [56–58], including the use of tomography [59] and laminography [60–62] for 3D mapping of water or Pt Lab-based hard X-ray tomography has also been used to study the 3D structure of operating fuel cells [63, 64].

This chapter is organized as follows. Section 5.2 describes SXM methods, including standard 2D-projection STXM; the application of STXM to mapping ionomer in PEMFC electrodes and MEAs; radiation damage monitoring and mitigation; 3D chemical mapping by tilt-series tomography and ptychography. The application of these methods to PEMFC issues is discussed as the methods are presented. Section 5.3 presents several specific applications of SXM to PEMFC, including optimization of distributions of ionomer in the cathode layer and STXM studies related to PEMFC degradation mechanisms, specifically Pt migration and carbon corrosion. Section 5.4 identifies future directions for SXM studies of PEMFC. Use of cryogenic STXM to reduce the radiation damage rate promises to allow more detailed studies of the 2D and 3D structure of ionomer in MEA electrodes. First steps toward *in situ* and *operando* studies of MEA under conditions close to operational are also described.

5.2 Methods

5.2.1 STXM instrumentation and methods

Figure 1a is a not-to-scale cartoon of the spectromicroscopy (SM) beamline at the Canadian Light Source (CLS) [65, 66] and interferometrically controlled STXM [67] including (from upper left to lower right): the elliptically polarizing undulator (EPU), which allows full control of the photon polarization (left/right circular and linear in any spatial orientation) [68]; some of the beamline optics (plane grating monochromator (PGM) to disperse the X-ray energies, and the exit slit, which is used to control coherence, energy resolution and incident flux; the M1 and M3 mirrors before and after the PGM are not shown) and the key STXM microscope elements—the Fresnel zone plate (ZP), order sorting aperture (OSA), sample and detector. To give a sense of the spatial distortion in Figure 1a, the distance from the source point (center of the EPU) to the sample in the ambient STXM is 33 meter [65] while the typical focal length of the STXM (ZP to sample) at the C 1s edge (300 eV) is 1 millimeter. Figure 1b is a photograph of the sample region showing (left to right): the aluminum snout of the UHV beamline, the brass carrier for the ZP, the OSA (laser drilled aperture in a piece of Mo foil) and the sample (Figure 1c, d), which, in this photograph, is a tomography system [21] consisting of a stepping motor for tilt angle rotation mounted on a standard trapezoidal aluminum STXM sample support plate [67]. The actual sample is a microtomed PEMFC-MEA supported on a formvar film, on a strip cut out of a 3 mm TEM Cu grid, as shown in Figure 1d. Figure 1e is a cartoon of how a STXM can be adapted to perform ptychography by adding a high-performance X-ray camera downstream of the sample to record the bright field and coherent scattered signal. More details on ptychography are given in a later section.

STXM is a complex, high precision, multiaxis mechanical device, which uses differential interferometry to monitor and control the (x, y) position of the sample with respect to the zone plate (ZP), and thus the X-ray beam. A full description of the microscope technology is given elsewhere [67]. Using typical high performance ZPs for focusing, a spot size of 20–30 nm is routinely achieved at most of the ~20 soft X-ray STXMs around the world (a complete list of all soft X-ray microscopes is available at http://unicorn.mcmaster.ca/xrm-biblio/xrm_bib.html). In 2020, the Swiss Light Source facility reported development of zone plates and associated improvements to the mechanical stability of the STXM, which allow them to routinely achieve 7 nm spatial resolution [69]. The photon energy range and ability to control the X-ray polarization depend on the source (bend magnet (BM), linear insertion device (ID) or EPU) and monochromator design (plane grating monochromator (PGM) or spherical grating monochromator (SGM)), and thus vary considerably among different soft X-ray STXMs. At the CLS-SM beamline, the EPU has four adjustable quadrants which provide: (i) fully linear polarized light with the linear E-vector orientation adjustable over 180 ° over the full photon energy range (130–3000 eV); (ii) fully circular polarized light over the EPU first har-

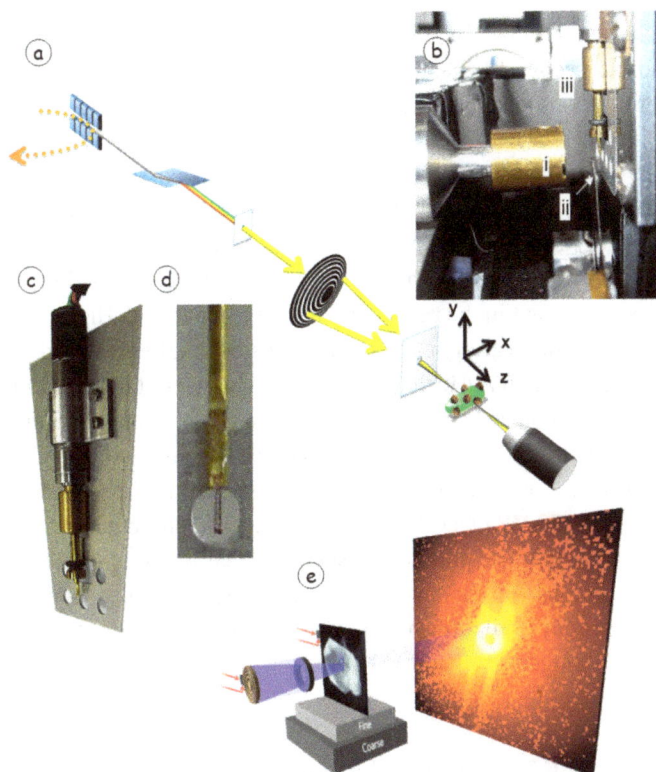

Figure 1: Schematic of a soft X-ray synchrotron beam line, and instrumentation for scanning transmission X-ray microscopy (STXM) and ptychography. (a) Cartoon of a typical soft X-ray synchrotron beamline and scanning transmission X-ray microscope (STXM). See text for details. (b) photograph of the sample region showing (i) the zone plate holder, (ii) order-sorting aperture, (iii) tomography sample. (c) photograph of a sample plate with stepper motor rotation stage [21] for tilt-angle tomography. (d) close-up of a grid strip as used for tomography. The formvar coating on the Cu grid supports an ultramicrotomed thin section of a membrane electrode assembly (MEA) of a polymer electrode membrane fuel cell (PEMFC). (e) Schematic of ptychography adaptation of STXM in which the phosphor-photo-multiplier integrating single channel detector is replaced by a high sensitivity X-ray camera in order to record coherent scattering patterns at each point in an (x, y) raster scan.

monic (130–1000 eV) and (iii) elliptical polarized light with variable circular/linear content above 1000 eV.

Detailed instrument description and operating procedures are published elsewhere [8, 67]. Briefly, after beamline tuning, instrument alignment, and sample focusing, the STXM is used to record transmission images at a single energy, to measure linescan spectra (intensity along a line at many photon energies), X-ray absorption spectra (XAS) at a single or multiple points and image sequences, also called stacks [70]. Imaging is typically performed by raster scanning the (x, y) positions of the sample when its z position is at the focal point of the X-rays. There are now several systems in which the ZP rather than the sample is scanned, which allows for faster scanning and heavier, more complex sample environments. Spectroscopy and chemical mapping are best done using stacks, since this allows the user to accurately align images post-acquisition. After alignment, typical stack processing involves converting the as-recorded transmission signal to optical density, OD $= -\ln(I/I_0)$, using an I_0 signal either measured as part of the stack, or recorded separately before or after the stack. The OD stack, which could include one or more core level XAS signals, can be further analyzed using multivariate statistical analysis methods [71, 72], or forward fitting with suitable reference spectra [8] to derive maps for the chemical species present in the region of the sample measured. High quality X-ray absorption spectra [73] of a variety of PFSA materials from different vendors have been measured in STXM and analyzed in detail to provide a base for understanding the links between ionomer morphology (specifically, short side chain, versus Nafion™ species) and developing efficient ionomer mapping techniques. Figure 2, which is discussed in detail in the next section, gives an example of STXM images and spectra relevant to mapping ionomer in PEMFC-MEA electrodes. These basic STXM functions can be combined with systematic scans of other parameters such as linear or circular photon polarization, sample polar tilt angle (for tomography or dichroism mapping), sample azimuthal angle (for laminography or dichroism mapping), voltage (*operando* studies), time (kinetics and dynamics), temperature, humidity, etc.

5.2.2 Quantitative mapping of ionomer in catalyst layers

Reliable, quantitative high spatial resolution 2D and 3D distributions of PFSA ionomer in the catalyst layers (CL) of PEMFC membrane electrode assemblies (MEA) are critical for advancing understanding of the role of ionomer in PEMFC, and how it can be optimized [74]. This is an area where SXM methods excel, and thus it is a central theme of this chapter. Recording full range (40–60 eV wide), detailed (100–150 energy points per edge) stacks at all core levels of the system within the photon energy range of the beamline (e. g., S 2p, C 1s, N 1s, O 1s, F 1s, Pt 3d for the PEMFC system at CLS) does provide the most analytical information. However, this is often not practical, either due to beamtime limitations, or more typically, radiation damage limitations. At 50 nm point spacing, a stack measuring at 100 photon energies, using a sampling time per pixel of 1 ms takes ~40 min/μm^2). Under normal flux (10–20 MHz) and full zone plate focusing (30 nm spot), PFSA suffers 30 % damage (F mass loss and chemical structure modifications) in about 100 ms [28]. Ways to maximize useful analytical information with a given radiation dose are explored in several sections of this chapter.

After performing enough detailed studies to identify the chemical species present, and their key spectral features, it is usually advantageous, from the perspectives of both time efficiency and to minimize radiation damage artifacts, to use a reduced set of photon energies which can provide the analytical information essential for a given project. In the case of analysis of PFSA ionomer in MEA cathodes, Figure 2 outlines a strategy of

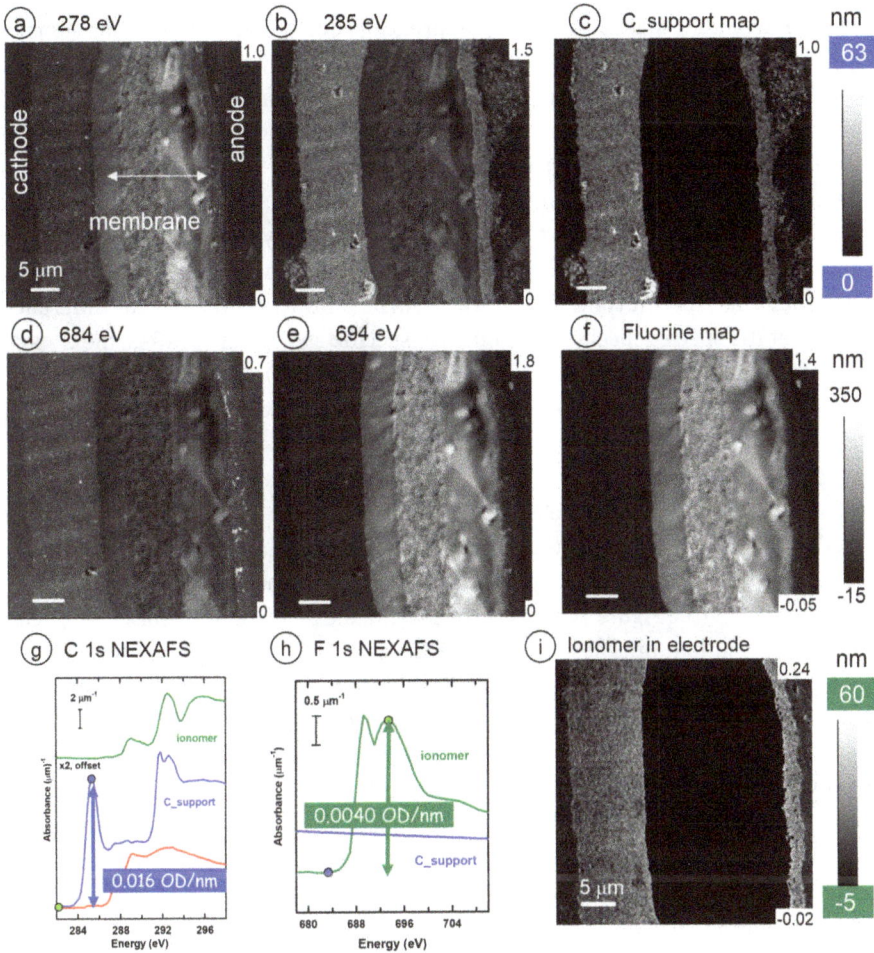

Figure 2: Quantitative, low-dose 2D mapping of ionomer in catalyst layers (CL) of a membrane electrode assembly (MEA) of a polymer-electrolyte membrane fuel cell (PEMFC) [10, 11]. Optical density (OD) images of an ultra-microtomed MEA extracted from a C 1s stack at (a) 278 eV and (b) 285 eV. (c) is the difference $(OD_{285} - OD_{278})$, which maps the carbon support (C_support) component. OD images from the F 1s stack at (d) 684 eV and (e) 694 eV. The numbers at the lower and upper right of each image are the limits of the OD grayscale. (f) is the difference $(OD_{694} - OD_{684})$ which maps the fluorinated regions, dominated by the membrane. (i) is the F-map derived by multiplying image (f) by a 0/1 mask with the threshold set to an OD intermediate between that of the membrane and that of the cathode/anode. (g) is the C 1s OD1 (optical density per nm) spectra of the PFSA, C_support and epoxy components. (h) is the F 1s OD1 spectra of the PFSA and C_support/ epoxy components. The difference in OD/nm indicated in (g) and (h) was used to convert the ΔOD scales in (c), (f) and (i) to an absolute thickness scale.

using only 4 specific photon energies to quantitatively map the PFSA ionomer relative to the carbon support. Figure 2 presents the images (Figures 2a, 2b, 2d, 2e), derived component maps (Figures 2c, 2f, 2i), and spectral basis (Figure 2g, 2h) for quantitative mapping of ionomer in PEMFC- MEA. The 4-energy approach to qualitatively map carbon support (C_support) and ionomer was first reported by Susac et al. [10], while a later publication [15] presented the method for using the same four energy images to quantify C support, ionomer and Pt. In principle, depending on the chemistry, and thus spectroscopy of the material used for embedding microtomed samples, adding a 5th energy would provide unambiguous, quantitative mapping of the embedding media, which is a reasonable surrogate for the porosity [24]. The key to successful few-energy stack map approaches is identifying photon energies where the absorption signal for each component dominates. Generic, non edge-specific contributions can then be subtracted by taking the difference of images at the species-specific and non edge-specific energies. For MEAs, the C_ support component is mapped at the C 1s edge by [OD(285.1 eV) – OD(278 eV)] (Figure 2c, 2g) (or [OD(284.4 eV) – OD(278 eV)] if polystyrene (PS) is used to hold the MEA during microtoming). The membrane and the PFSA ionomer are mapped at the F 1s edge by [OD(694 eV) – OD(684 eV)] (Figure 2f, 2i). Since the membrane is 100 % PFSA, in order to visualize the ionomer in the electrodes it is necessary to use a threshold masking method to identify the pixels in the membrane, which are then set to zero (Figure 2i). Since the absolute OD1 response (optical density per nm) of pure PFSA at the C 1s and F 1s edges is known (see Figure 2g and Figure 2h) [73], the ΔOD grayscale for each species can be converted to a thickness scale (see Figure 2c, 2f, 2h). In addition, by subtracting a scaled PFSA map from the pre-C1s image at 278 eV, an estimate of the quantitative distribution of Pt catalyst can be generated, as discussed in [16]. Finally, if there is negligible penetration of the embedding material into the cathode (which can be achieved with polystyrene sandwiching techniques), pores above ~50 nm can be seen directly. In contrast, if a fully penetrating resin such as that made by reacting trimethylolpropane triglycidyl ether and 4,4′-methylenebis(2-methyl-cyclo-hexylamine) (referred to as TTE resin) [75] is used, the resin can be mapped (e. g., at the N 1s edge) and the resin map used as a quantitative map of the porosity [24].

The ability of this method to map ionomer in CL is an important advantage of STXM in the context of analytical methods for PEMFC. There are many publications that report use of analytical electron microscopy methods for qualitative and/or quantitative mapping of ionomer in CL using high resolution imaging [76, 77], X-ray fluorescence (TEM-EDS) [77], STEM-EELS [52, 77], electron tomography [78, 79] or staining techniques [80]. Melo et al. [50] directly compared F-loss rates as a function of X-ray exposure in STXM versus electron beam-based TEM-EDS mapping of ionomer in MEA cathodes, clearly showing the significant advantage of STXM relative to TEM-EDS. Yakovlev et al. [81, 82] have carefully reviewed the case for use of electron beam methods for quantitative high resolution mapping of ionomer in CL and concluded that reliable, damage-free results can only be obtained if the spatial resolution is drastically reduced. Due to the rapid damage of the ionomer [50] by both electron [26] and ion [27] beams, results from focused ion beam, secondary electron microscopy (FIB-SEM) methods to map catalyst layer microstructure in 3D [83–85] should be treated with caution. Based on our comparisons of relative damage rates, models of transport and other properties of PEMFC cathodes, which are based on porosity or ionomer distributions derived from electron and ion beam methods [86, 87], may be flawed.

5.2.3 Radiation damage

In order to use STXM in a quantitative manner without artifacts associated with radiation damage, it is useful to estimate the radiation dose (in MGy, where $1\,\text{Gy} = 1\,\text{J/g}$) delivered in a given STXM analytical measurement and compare that to quantitative measurements of radiation damage as a function of dose. A systematic method to measuring both radiation damage and dose in STXM has been developed [88–90] and used to investigate X-ray damage of several polymer thin films, including polyethylene tereph-thalate (PET) [91], polymethyl-methacrylate (PMMA) [92] and PFSA ionomer [28]. Figure 3 presents OD images of a PFSA thin film before (Figure 3a) and after (Figure 3b) using the STXM beam to generate a 9-pad pattern by irradiating an initially uniform area of PFSA with controlled, systematically increasing doses. Figure 3c–3f compare C 1s, F 1s, O 1s and S 2p spectra of undamaged PFSA to those of the 3rd, 7th and 9th damage pad. At each edge, there is a systematic decrease in the intensity of the spectral features associated with intact PFSA (pre-C1s intensity; peaks at 173, 182, 292, 296, 690, 694 eV, the O 1s and F 1s continua) and a corresponding increase in spectral features associated with radiation damage products (C=C and C=O related peaks at 285, 287, 289 and 532 eV). Detailed analysis of the intensities of these features is used to derive critical doses for specific processes [28], which can be used to guide low-damage acquisition strategies.

In general, radiation damage by soft X-rays is found to obey first order kinetics [93]. Ideally, the dose used should be less than 20 % of the critical dose for any damage process (mass loss, electronic structure change, etc.) that would interfere with a quantitative analysis. The dose delivered by soft X-ray absorption can be calculated from the experimental conditions using

$$D = \frac{6.4 \times 10^5 \cdot E \cdot \text{OD1}(E) \cdot I_0 \left(1 - e^{-\text{OD}(E)}\right) \cdot t}{K \cdot \pi \cdot s^2 \cdot \rho} \tag{1}$$

where E is the photon energy (eV), OD1(E) is the optical density for absorption by 1 nm thickness at standard density at energy E, I_0 is the incident flux (photons/s), OD(E) is the OD for the sample at E, t is the length of time the absorption occurs (dwell time), K is the detector efficiency, s is the area of the X-ray beam at the sample and ρ is the sample density [90–92]. In fact, in cases where radiation damage causes mass loss and/or modification of the sample chemistry, such changes need to be taken into account in determining the dose, as described by Berejnov et al. [94, 95]. Within the approximation of first-order kinetics, a critical dose for a given radiation damage process can be derived by measuring changes as a function of increasing dose in the optical density at a photon energy sensitive to the type of damage of interest. For example, the main

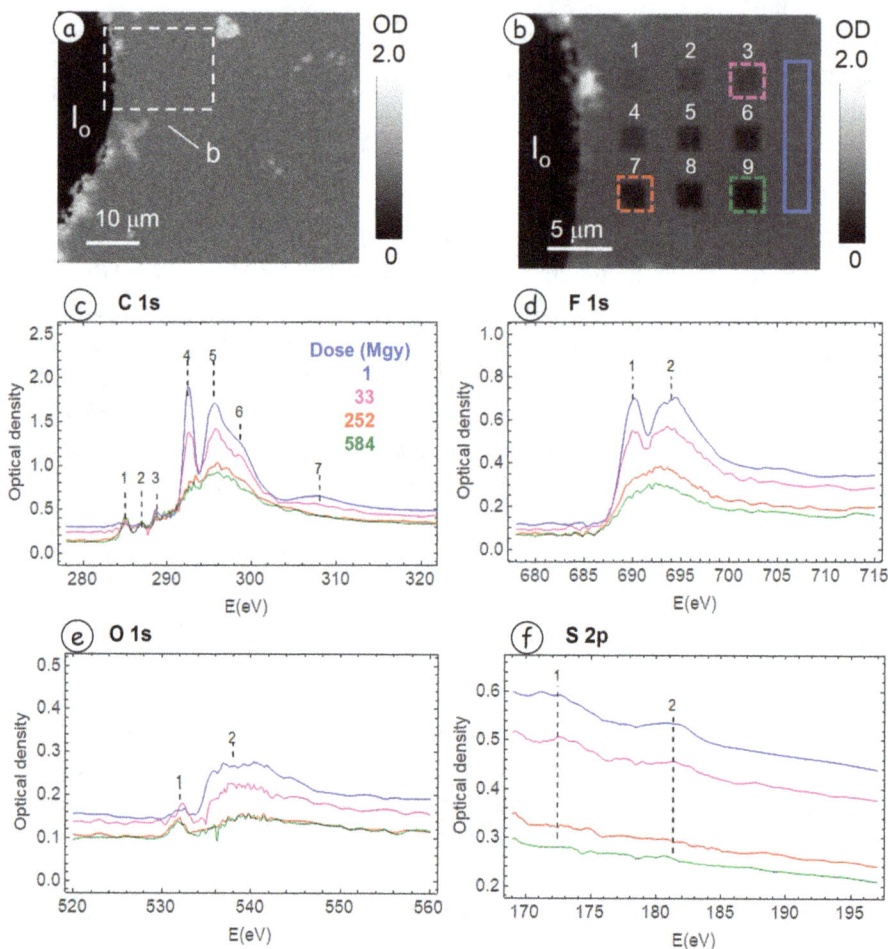

Figure 3: 9-pad pattern generation of radiation damage for quantitative understanding of radiation damage to perfluorosulfonic acid (PFSA) [28]. (a) STXM OD image of a spun cast thin film of PFSA on a silicon nitride support window, measured at 292.4 eV, before generation of a 9-pad radiation damage pattern. (b) STXM OD image at 690 eV of the 9-pad pattern generated at 320 eV in a He environment in the dashed rectangular area in (a). Io indicates the part of the image used for incident intensity measurement. Numbers denote the pad index. The near edge X-ray absorption (NEXAFS) spectra of the nondamaged area, blue (dose < 1 MGy); pad 3, pink (33 MGy dose); pad 7, red (252 MGy); and pad 9, green (584 MGy); are presented at the (c) C 1s, (d) F 1s, (e) O 1s and (f) S 2p edges. The color coding in the spectral plots identifying the pads corresponds to that in (b).

types of soft X-ray damage to PFSA are fluorine mass loss, ether-loss, sulfonate loss and degradation of the $-CF_2 -CF_2 -$ chain segments characteristic of the teflon™ backbone. The F-loss can be monitored at 280 eV (below C 1s) or from the difference in OD(710 eV) (F 1s continuum) and OD(680 eV) (pre-F1s). The sulfonate loss can be monitored at the S 2p edge (170–190 eV). The chain degradation can be monitored at characteristic C–F bonding peaks at the C 1s edge [OD(292 eV), and OD(296 eV)] or the F 1s edge [OD(690 eV), OD(694 eV)]. We have found it very useful to make systematic 9-pad patterns (as displayed in Figure 3) using the Pattern Generation function of the STXM control software, which then allows to track the changes in full spectra, or the OD at specific photon energies as a function of increasing dose [28, 90]. Fitting that data to a first-order kinetic model can then be used to derive a critical dose [28]. Starting with undamaged PFSA, the critical dose for F-loss and chain degradation is 80 ± 20 MGy while that for sulfonate loss is somewhat lower (50 ± 20 MGy).

Another aspect of STXM-based studies of X-ray damage is that the chemical changes induced by X-rays are similar to those resulting from electrochemical degradation of PFSA in PEMFC [29, 96]. Therefore, comparisons of materials degradation by chemical attack, such as through Fenton mechanism (Fe^{2+} and hydrogen peroxide) [97, 98] caused by operation of a PEMFC and X-ray damage [19, 29] can give insights into PEMFC degradation chemistry.

5.2.4 4D imaging—spectrotomography of PEMFC cathodes

The 3D distribution of the components in CLs, in particular in the cathode where the rate limiting oxygen reduction reaction (ORR) occurs, determines porosity, and thus permeability and transport kinetics for fuels and products; electrochemical effectiveness, and thermomechanical properties. Section 5.2.2 described how detailed quantitative maps of the C_support, ionomer, Pt catalyst and porosity could be determined by either full-range stacks at the C 1s and F 1s edges, or by 4-energy stack maps, which minimize radiation damage. In the soft X-ray regime, tilt angle STXM tomography at multiple photon energies was first demonstrated in 2006 by Johansson et al. [99], and first applied to PEMFC cathodes in 2013 [20]. However, when typical tilt angle series sampling 50 or more tilt angles is used with a fully focused spot size, there is a large radiation dose (>300 MGy), and thus extensive sample damage. At that large a dose, there are serious questions as to whether or not the measured 3D spatial distributions of ionomer have any relationship to that in the undamaged cathode. Working with Paul Midgley and his group (Cambridge) and Mirna Lerotic (Second Look Consulting, Hong Kong), we developed and tested a new procedure for STXM tomography reconstruction [22], based on the total variation principle [100] applied to cone beam tilt-angle tomography, which is a type of compressive sensing (CS). CS can provide high quality 3D spatial distributions from quite small numbers of tilt angles (as few as 10) and limited tilt angle ranges ($\pm 50°$). The CS results were demonstrated to be as good as results from treating data sets from the same sample, which had a much larger number of tilt angles and larger angle range, but reconstructed using the serial iterative reconstruction technique (SIRT) or Fourier back projection [22]. This greatly improved reconstruction method allowed us to reduce the dose significantly. When we combined reduced tilt angles (14–16 angles are now typically used), with a reduced incident intensity (2 MHz rather than 20 MHz) and defocusing the spot size from 30 to 50 nm, the overall dose for a 4-energy 2-edge stack map tomography tilt series was reduced from >300 MGy, to less than 15 MGy.

In order to evaluate the impact of radiation damage on 3D spatial distributions, we developed a multiscan procedure in which the same volume was sampled 3 times in succession with 14 tilt angles in each tilt series. The angles were interleaved so all 42 tilt angles could be used if desired. Figure 4 compares results from two, 3-pass spectro-tomography procedures, with different radiation dose and dose rates [23]. For the first measurement (Figure 4a–4d), full focus, large flux and a larger number of tilt angles (48) were used. The average F 1s spectrum of the whole cathode was measured before the start and after the finish of the 3-set measurement. The spectroscopy showed that about 40 % of the fluorine was lost, and all the remaining "PFSA" was transformed, as indicated by the disappearance of the characteristic double peaked structure (Figure 4d). The surface rendered face-on display of the reconstructed F-maps from the 1st, 2nd and 3rd set shows considerable modification of the spatial distribution.

In the second set (Figure 4e–4h), a lower dose rate, achieved by reducing the flux and using a slightly defocused X-ray beam, and fewer tilt angles (42) were used. The F 1s spectrum was measured 5 times, before the start, $\frac{1}{2}$ way through the first set, after the second set and finally after the third set of 14 tilt-angle, 2-energy acquisition. A defocused beam was used to get an averaged F 1s spectrum of the cathode, so negligible additional dose was imparted by these diagnostic measurements. In this case, due to the significant decrease in dose, there was relatively little change in the spectral shape and less than 10 % F loss (Figure 4h), even after all 3 sets of tilt-angle measurements. Figure 4e, 4f and 4g are slices through the center of the reconstructed volume. In this case, there is very little change in the spatial distribution with successive sets. These results are consistent with the estimated doses (a factor of ~10 less for the results shown in Figure 4(e–h) than those in Figure 4(a–d)). As summarized in Figure 4, the volume fraction of PFSA changes very little, reducing to 28 % from 31 vol% over the three tomography data sets. A movie of the 3D renderings of the second tomography data set shown in Figure 4 is available as supplemental material to ref. [23].

In addition to showing a rendering of the 3D structure of the PFSA from each of the 3 sets, the differences (set 2 – set 1) and (set 3 – set 2) are also displayed in the movie (see supplemental information of ref. [23]), which provides information about the location of the fluorine mass loss. Even with only 11 MGy total dose it is clear that thinner regions of PFSA are more susceptible to fluorine loss than thicker regions.

5.2.5 Pushing spatial resolution—ptychography

The improved acquisition methods that respect the radiation damage limits described in the last section have shown the feasibility of STXM tomography. However, the need for partial X-ray transmission, combined with high absorption cross-sections at the chemically specific energies, means that, for samples with a density of 0.5–2 g cm^{-3}, the thick-

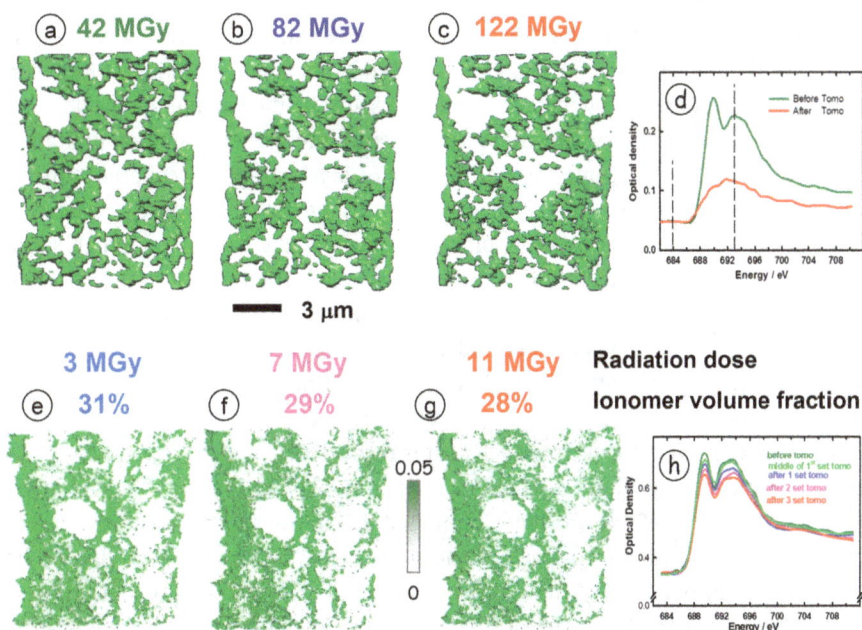

Figure 4: Use of multiset measurements to achieve ambient STXM three-dimensional (3D) mapping of ionomer in cathode layers with minimal radiation damage [23]. (a), (b), (c) Surface rendered 3D maps of PFSA in a PEMFC cathode, derived from a compressed sensing (CS) reconstruction of two energy F 1s spectrotomographic measurements. Three successive 16 tilt angle data sets (each consisting of STXM images at 684 eV and 705 eV) were recorded from the same volume, with integrated radiation doses of 42, 82 and 122 MGy for sets 1, 2 and 3. (d) Average F 1s spectra of the cathode prior to the first tomography measurement and after the end of all 3 measurements. (e), (f), (g) slices from the 3D distribution of the PFSA in a PEMFC cathode derived from CS reconstruction of two-energy F 1s spectrotomographic measurements of three, 14 tilt angle data sets on the same volume. By using fewer tilt angles, much lower incident flux, and a small defocus (increasing the spot size from 30 to 50 nm), the integrated radiation doses after each set were only 3, 7 and 11 MGy. The change in the average volume fraction of ionomer after each tilt series is indicated, amounting to about 10 % F-loss over the full data set. The intensity scale bar is voxel OD in a thickness of 10 nm. (h) Average F 1s spectra of the cathode, before measurements, at the middle of first tomography scan, and after the 1st, 2nd and 3rd tomography data sets.

est slices that can be measured by soft X-ray tilt series tomography are about 300 nm. With a 3D resolution of ~50 nm the information in the third dimension is limited to at most 6 discrete resolution elements. In addition, at high tilt angles the actual path length through a planar sample is 2–3 times larger than the sample thickness, which leads to artifacts associated with absorption saturation. The latter problem can be significantly reduced by use of laminography rather than tilt angle tomography. An apparatus optimized for soft X-ray STXM laminography measurements of planar samples has recently been commissioned at the PolLux beamline at the Swiss Light Source [101].

An approach to enhance 3D information is to improve the spatial resolution. The spatial resolution in zone plate-based STXMs is limited by the properties of the ZP. With highly coherent illumination, the diffraction limited resolution is $1.22 \cdot \delta_r$ where δ_r is the width of the outermost zone of the ZP [44]. While a spatial resolution of less than 8 nm has been demonstrated [69], the zone plate used is inefficient and the focal length very short, such that it is very difficult to use. Recently, the spatial resolution limitation of ZP optics has been overcome through the use of ptychography, a coherent diffraction imaging (CDI) technique [102]. Ptychography uses the same type of raster scanning as conventional STXM. At each (x, y) point in the scan, a low background, high sensitivity X-ray camera positioned 4–10 cm after the sample is used to record the coherent diffraction pattern, including both the bright field annulus, which contains diffraction signal as well as undiffracted photons, and all of the X-rays coherently scattered by the sample, out to as large a scattering angle where the diffraction signal can be differentiated from background. Using appropriate software such as PyNX [103], SHARP [104] or Ptypy [105], the CDI pattern from each spot is reconstructed into a real space image. By using a point spacing where the points overlap (by 30–50 % if using a zone plate focused X-ray spot, or 80–90 % if using a defocused 0.5–1.0 μm spot), the constraint that the reconstruction of the CDI pattern at one spot gives the same result as that in the areas which are also measured in adjacent spots, leads to a rapid and reliable convergence of the iterative method used for reconstruction. Added benefits of ptychography include provision of (a) the phase and spatial distribution of the X-ray probe, which can help optimize instrumentation performance; (b) the phase as well as amplitude of scattering, which can be analyzed together simultaneously to enhance the reliability of chemical analysis, resulting in refraction-based spectromicroscopy [106]. Although a zone plate is used to produce a cone-beam X-ray source with spot size at the focal point of 40–100 nm diameter, ptychography is not limited by the ZP properties and thus has the potential to reach the Rayleigh limit dictated by the wavelength of X-rays used (λ(nm) = 1239.8/E(eV), thus ~4 nm at the C 1s edge). Recently, the COSMIC beamline and two microscopes dedicated to soft X-ray ptychography have been developed at the Advanced Light Source. The record spatial resolution, which was achieved using 1500 eV photons, is 3 nm [107]. In addition to providing the highest spatial resolution in the world in the soft X-ray range, COSMIC is extensively equipped for *in situ, operando* [108] and tomographic experiments [108, 109].

Results from the first application of soft X-ray ptychography to a PEMFC cathode [31] are presented in Figure 5. Figures 5a and 5b show absorption images (OD scale) from ptychography measurements at 684 eV and 694 eV. Figure 5c, the difference between Figure 5b and Figure 5a, maps the PFSA in both the catalyst layer and the membrane. Note that the dark regions in the fluorine map (Figure 5c) correspond to the bright regions in Figure 5a, which are carbon-rich. The PFSA ionomer in the catalyst layer (Figure 5c) has a "feather" morphology superimposed on a fine network structure. We believe the rather uniform "feathery" areas correspond to a film of ionomer smeared over the cathode during the room temperature microtomy. Figure 5d is a color-coded composite of the F-map and the carbon support, with the latter derived from the pre-F 1s image and the F-map (carbon support = OD_{684} − 0.21 * F-map). Figure 5e, an expansion of the PFSA

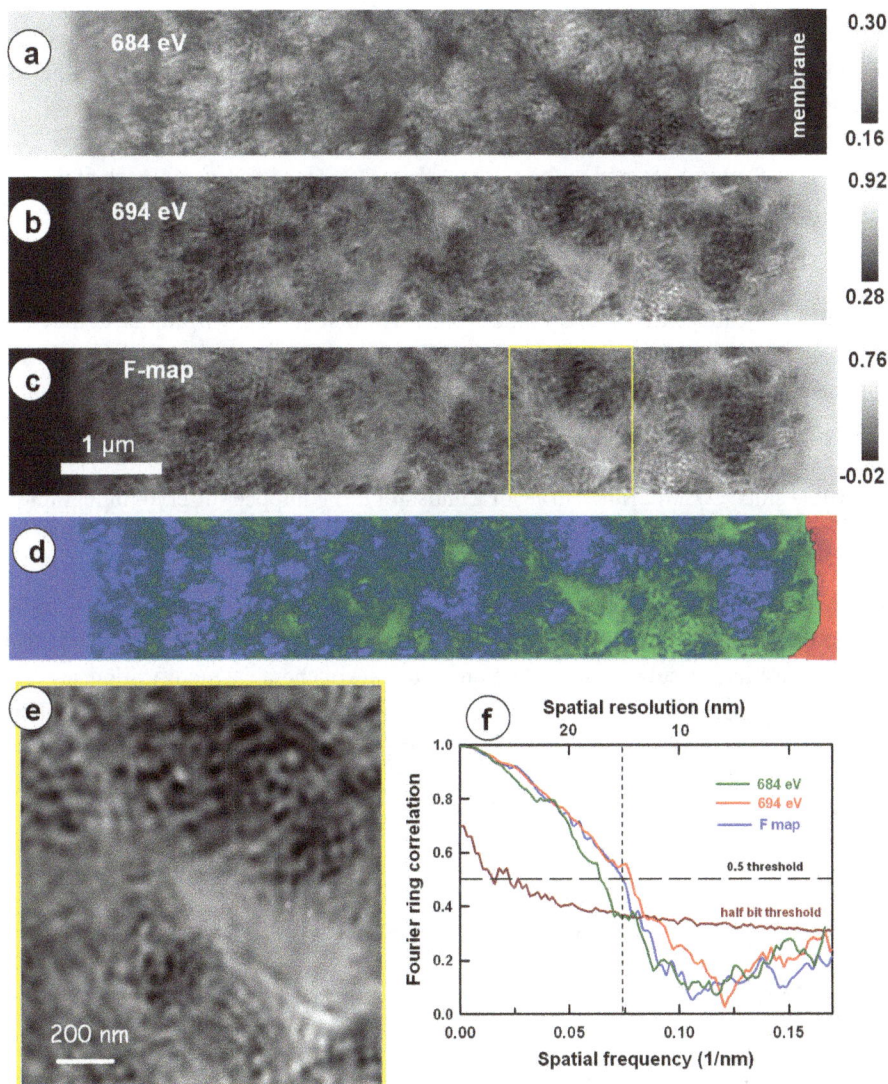

Figure 5: Improving spatial resolution of ionomer mapping in PEMFC MEAs by spectro-ptychography [31]. Ptychography-derived OD images of an MEA cathode at (a) 684 eV. (b) 694 eV. (c) difference (PFSA ionomer/membrane map). The optical density grayscale is indicated for each image. (d) Color-coded composite of the membrane (red), ionomer (green) and carbon (support & epoxy) (blue). The membrane and ionomer signals are derived from the total fluorine signal by threshold masking, and thus provide a quantitative map of the PFSA. The carbon signal is derived from the image at 684 eV [31]. (e) expansion of the PFSA map in the region of the yellow box in (c). (f) Fourier ring correlation (FRC) analysis of (c) indicating a spatial resolution of 15 nm. (adapted from Figures 5 and 6 of [31]; reproduced with permission from American Chemical Society, © 2018).

map in the region of the yellow box in Figure 5c, shows both the fine network structure and the "feathery" smear in greater detail. Fourier ring correlation (FRC) analysis (Figure 5f) of the PFSA map indicates a spatial resolution of ~15 nm. It is noteworthy that these results, recorded in 2015, in the early days of the development of ptychography at the ALS, were obtained with only 2 ptychographic measurements, each of which used only 20 ms to acquire the diffraction image at each spot. Including navigation set-up imaging and spot overlap, each area of the sample was exposed to only ~60 msec of beam with an estimated dose of 8 MGy, which is ~10 % of the critical dose for PFSA. COSMIC, a dedicated and highly optimized soft X-ray ptychography beamline at the ALS started operation on 2017. With its greater stability and higher coherent flux, it is possible to achieve <10 nm spatial resolution using 10 msec dwell per diffraction image [109]. Thus, 2D mapping of PFSA in PEMFC electrodes at sub-10 nm spatial resolution with negligible radiation damage is possible. Meaningful 3D ionomer distributions are also likely achievable since COSMIC Nanosurveyor 2 endstation is equipped for cryo-tomography. Using cryo conditions (sample T below −200 °C) is known to greatly reduce the rate of mass loss radiation damage [110]. Recently, the cryo-STXM at the CLS [35] has been used to compare rates for chemical modification of PFSA by X-rays with the sample at −175 °C versus at 25 °C [36]. The preliminary results show that cryo-cooling to near liquid N_2 temperatures results in a substantial reduction in the rate of chemical modification, as well as preventing mass loss. This is in considerable contrast with polymethylmethacrylate (PMMA). STXM spectromicroscopy of X-ray damaged PMMA with the sample at −160 °C showed identical chemical damage rates as at 25 °C [110], an observation reproduced in the recent cryo-damage measurements [36].

5.3 Examples of STXM applications to PEMFC optimization

5.3.1 Optimization of ionomer distributions in CL

While the balance of Pt coated carbon particles, Pt loading and the amount of PFSA ionomer is known to be optimized with a weight percent of ~30–40 % of ionomer [111, 112], the reason for this optimum was not well understood prior to use of STXM as a probe of the spatial distributions of ionomer in CL. The ionomer imaging methods described in Section 5.2.2 were applied to several different CL preparations with two different catalyst supports and systematic variation in the ionomer loading. Figure 6 presents color-coded composites of the ionomer and carbon support spatial distributions for low surface area carbon (LSAC), and high surface area carbon (HSAC) supports with ionomer loadings of 10–13, 33 and 45 weight % [113]. At the lowest ionomer loading the distributions are relatively uniform, with an increasing amount of ionomer on the membrane

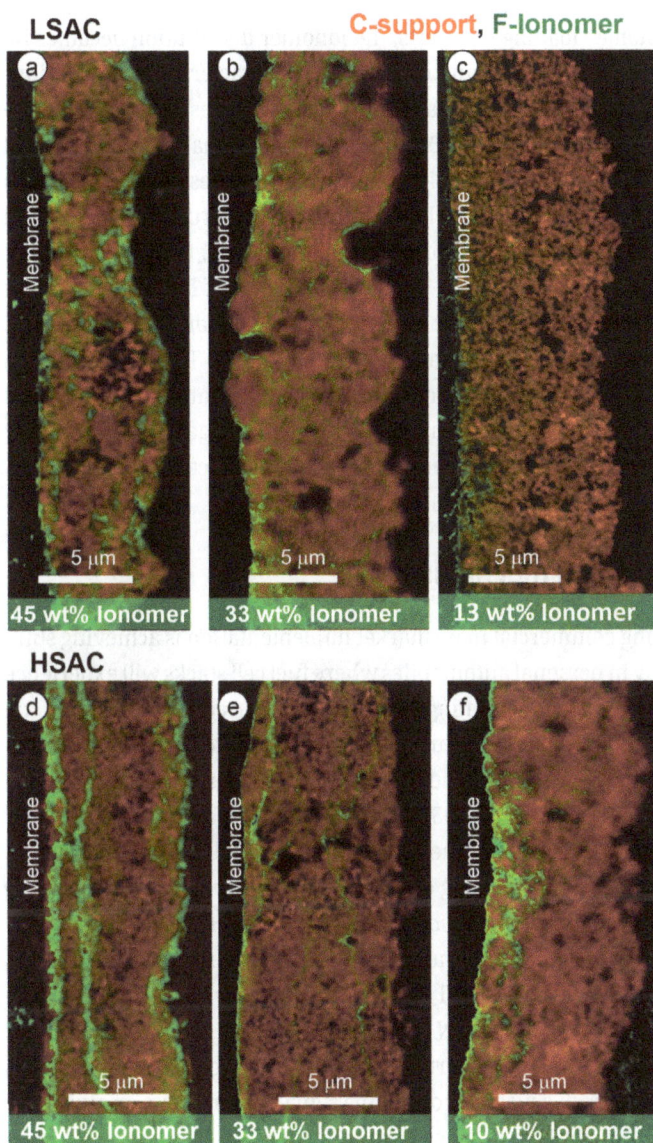

Figure 6: Use of STXM-based ionomer imaging to help understand optimization of catalyst ink composition by visualizing nonuniform ionomer distributions. (a)(b)(c) Carbon support (red) and ionomer (green) color composite maps derived from 4-energy (C 1s, F 1s) STXM measurements of PEMFC cathodes based on low surface area carbon (LSAC) carbon support, for formulations with (a) 45, (b) 33 and (c) 13 weight-% of ionomer in the catalyst ink. (d)(e)(f) Carbon support (red) and ionomer (green) color composite maps derived from 4-energy STXM measurements of PEMFC cathodes based on high surface area carbon (HSAC) carbon support, for formulations with (a) 45, (b) 33 and (c) 10 weight% of ionomer.

side. However, as the ionomer loading increases, the ionomer distribution becomes increasingly heterogeneous. For the LSAC support, there are characteristic ovoid regions of depleted ionomer content (Figure 6a, 6b) (similar ovoid structures are evident in the 3D mapping, see Figure 4 [23]). For the HSAC support, there are characteristic layered zones of high ionomer content, separated by ionomer depleted zones (Figure 6d, 6e). As the spatial distribution of ionomer gets less homogenous large portions of the CL will most likely underperform, due to reduced proton transport. Thus, the observation of an optimum ionomer loading of ~30 weight % is a consequence of having insufficient amounts of ionomer for adequate proton transport at low loadings, and too many "dead zones" with reduced gas access when the loading gets above 30 %. The heterogeneous distributions are most likely a consequence of partial phase separation in the catalyst inks during or after deposition, suggesting that improving the homogeneity and stability of catalyst inks would be an avenue to further optimize the CL structure.

5.3.2 PEMFC degradation mechanisms

A significant issue delaying commercial mass market implementation is achieving sufficient reliability, especially in personal automobiles where fuel cell stacks will experience a wider range of environments and operating conditions compared to for example stationary power applications. STXM has been used in several areas to investigate mechanisms of electrochemical performance degradation. Pt dissolution and migration, which can occur when there is cell reversal [114, 115], involves oxidation of Pt, migration of Pt ions from the CL into the membrane and then formation of metallic Pt deposits at the zone of zero charge [116, 117]. An interesting question is *"what aspects of the chemistry of the Pt in the CL might make the system more (or less) susceptible to Pt dissolution"*? When studying the Pt particles in the membrane in one system where the Pt catalyst was introduced by reducing a nitro-amine based Pt organometallic compound, STXM measurements showed there was a significant N 1s signal localized at the Pt particles [14]. Further, the N 1s NEXAFS spectrum of the particles was similar to that of the Pt particles in the cathode, but different from a weak N-containing signal (of unknown origin) in the membrane. Figure 7a–7c are component maps derived from N 1s stacks measured in an area of the membrane of an end-of-test sample with significant Pt migration. The analysis is based on fitting the N 1s stack to the N 1s spectra presented in Figure 7e and 7f. Figure 7d is a color-coded composite of the three component maps. When compared to the N 1s spectrum of the cathode (black curve in Figure 7d, 7e) the spectrum of the core of the Pt particles is very similar, suggesting that residual nitrogenous material from the nitro-amine based Pt precursor may have participated in, and perhaps facilitated migration of the Pt from the cathode into the membrane. N 1s STXM studies of Pt migration regions in an end-of-test CL where the Pt was derived from $PtCl_4$ did not show this N 1s signal, supporting the idea that the residual N from the amine ligand might be involved

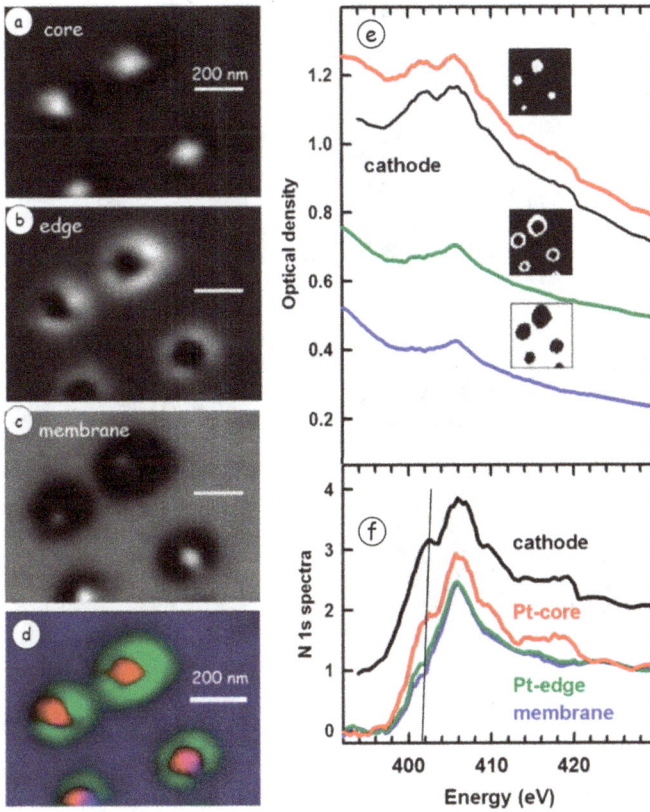

Figure 7: Example of a N 1s STXM study of degradation by Pt migration. (a–d) STXM spectromicroscopy: mapping an end-of-life PEMFC in the region of a Pt – in – membrane band at the N 1s edge [14]. (a) Component map of the Pt core, (b) component map of the signal at the edge of the particles, (c) component map of the membrane, (d) color coded composite of the three component maps. (e) Comparison of the N 1s spectra of the Pt core (red), compared to that of the cathode (black), the edge of Pt particles (green) and the membrane away from the particles (blue). The inset masks indicate the regions from which the spectra were obtained. (f) N 1s spectra of the membrane, Pt-edge, Pt-core and cathode after subtraction of a curved background extrapolated from the pre-N 1s region and unit normalization in the continuum. The spectrum of the cathode is offset by 1 unit (Reproduced from [14] with permission from the PCCP Owner Societies).

in Pt dissolution and migration. Unfortunately, both types of CL (with and without N) were found to be similarly susceptible to Pt dissolution [14].

A second important area of PEMFC degradation is that of carbon corrosion [118]. When PEMFCs are operated at high potentials as in start-up/shutdown conditions, the carbon black catalyst support is oxidized to CO_2. This can lead to performance decrease due to reduced catalyst surface area and/or alteration of pore morphology. A better understanding of the mechanism of carbon corrosion is needed to support rational design of mitigation strategies. One of the questions in this context is the extent to which the

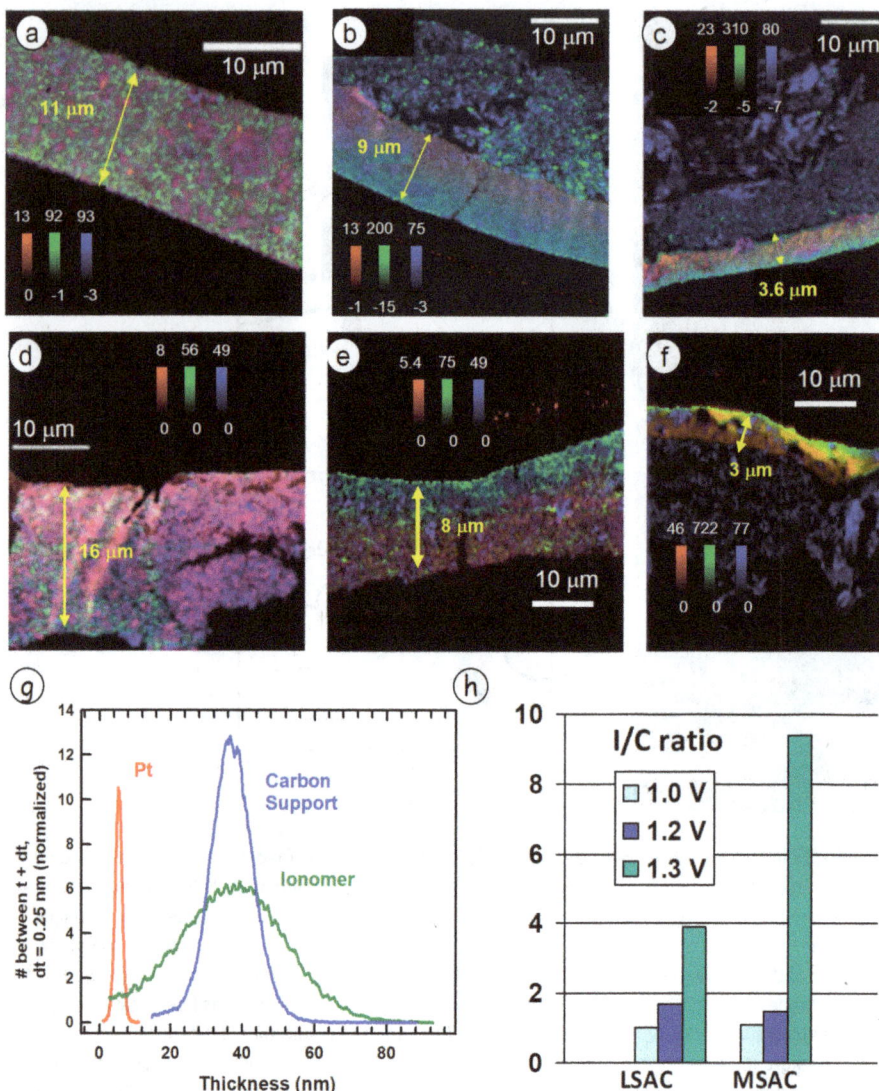

Figure 8: Carbon corrosion studied by STXM [15]. (a–f) are color coded composite maps of the thickness distributions (in nm) of Pt (red), ionomer (green) and carbon support (blue) in the cathode region of a PEMFC- MEA subjected to accelerated stress testing (end-of-test) with an upper voltage of 1.0 V (a, d), 1.2 V (b,e) and 1.3 V (c, f). The upper row (a–c) is from samples using a low surface area carbon (LSAC) support while the lower row (d–f) is from samples using a medium surface area carbon (MSAC) support. The membrane, which is on the smooth side of the cathode, and the embedding support, which shows particles in the microporous layer in some cases, have been removed by threshold masking (see Figure 2(f, i)). The dramatic shrinkage of the width of the cathode is due to carbon corrosion. (g) is an example of the histograms of the Pt, ionomer and carbon support thicknesses from (a). Similar quantitative results for all six samples were used to derive the ionomer to carbon support (I/C) ratios displayed as a bar graph in (h).

PFSA ionomer is lost or modified in extreme carbon corrosion events. In collaboration with scientists at Ballard Power Systems, STXM was used to study the chemical changes in the cathode CL in samples subjected to accelerated stress tests, consisting of many voltage cycles between 0 and V_{max} where V_{max} is between 1.0 and 1.3 V [15]. Figure 8 presents results from that study in the form of color-coded composites of the distributions of Pt, ionomer and carbon support in the cathodes of PEMFC samples, using LSAC (Figure 8a, b, c) and medium surface area carbon (MSAC) support (Figure 8d, e, f), which had been subjected to 1000s of electrochemical degradation cycles where the peak anodic potential was 1.0 V (Figure 8a, d), 1.2 V (Figure 8b, e) or 1.3 V (Figure 8c, f). Figure 8g is an example of the distribution of thicknesses of these components, extracted from the component maps of the cathode composition, in that case for the beginning of test (BOT) LSAC sample. Similar histograms were evaluated for the six samples and used to generate the ionomer to carbon (I/C) ratio, expressed as the mean of the ionomer thickness distribution divided by the mean of the carbon support thickness distribution. Figure 8h is a bar chart of the I/C ratio for the 6 end of test (EOT) samples. This study showed that the highly oxidative conditions caused extensive loss the carbon support, and the loss of carbon occurred to a similar extent with the LSAC and MSAC support [15]. However, there was relatively little loss of the ionomer, particularly in the MSAC samples, which then resulted in a large increase in the I/C ratio as the peak voltage increased. Further, measurements at the S 2p edge indicated that the sulfonate groups were largely resistant to the oxidative conditions causing the extensive carbon corrosion, implying that even after carbon loss, the ionomer could still be functional as a proton conductor [15].

5.4 Future directions for SXM studies of PEMFC

5.4.1 Low damage STXM of PEMFC MEAs

A lot of progress has been made over the past decade in evolving STXM techniques, which can measure 2D ionomer distributions with relatively little radiation damage induced changes in the fluorine levels, electronic structure or modification of the nanostructure [26, 90]. However, it is the 3D structure of the CLs, and how that may evolve under different operating conditions, that determines performance and stability. With few-angle acquisition methods and advanced tomographic reconstruction methods (see Section 5.2.4), it has been possible to measure 3D ionomer distributions at ambient temperature with a dose similar to that typically used in 2D measurements, although with somewhat reduced spatial resolution (50–80 nm in 3D versus 25–30 in 2D) [23]. Despite that technical achievement, only limited extents of 3D structure can be measured with conventional tilt-angle tomography due the strong absorption in the soft X-ray region, which limits samples to ~300 nm thickness and degrades 3D spatial resolution due to out-of-focus conditions at high tilt angles. Laminography, in which the 3rd dimension is accessed by rotating the sample azimuthally [119], offers one route to providing more extensive 3D results since (i) much larger volumes can be accessed, (ii) the absorption is uniform at all azimuthal angles, and is smaller than at high tilt angles in tilt-angle tomography; (iii) the focus does not change with angle. Laminography using tunable hard X-rays in the region of the Pt L_3 edge has been applied to PEMFC [60, 61]. Recently, a practical system for soft X-ray laminography has been implemented in the STXM at the Swiss Light Source [101].

Another approach to measuring the 3D structure of radiation sensitive systems with reduced damage is measurements under cryogenic conditions. This approach is very common in studies of biological samples but has been used relatively infrequently in materials science. An open question with regard to cryogenic studies of ionomer in CL is the extent to which maintaining a sample near liquid N_2 temperature will reduce the rate of radiation damage. As noted earlier in this chapter, the damage mechanism, which occurs at low doses and causes extensive reorganization of the material, is C–F bond breaking combined with fluorine loss. Reducing mass loss by "freezing in," and thus retarding loss of low molecular weight damage fragments is an accepted benefit of measuring dense materials under cryo conditions. This mechanism may be less effective in preventing loss of radiation damaged ionomer due to the open pore structure of CL. Recently, we have used the cryo-STXM at the CLS [35] to measure rates of fluorine loss and electronic structure changes in PFSA at 25 °C and −170 °C [36]. We also measured the radiation damage and mass loss at the O 1s edge of polymethylmethacrylate (PMMA) under cryo conditions. The temperature dependence of radiation damage to PMMA was studied much earlier by Beetz and Jacobsen [110]. They showed that cryo conditions prevented mass loss but did not affect the rate of electronic structure modification, as monitored by the intensity of the O 1s → $\pi^*_{C=O}$ transition at 532 eV. Our study [36] verified the earlier results on PMMA. Interestingly, both mass loss and electronic structure changes in PFSA, monitored at the F 1s → σ^* transition at 694 eV, proceeded at a much-reduced rate at −170 C as compared to 25 °C [36]. Although this observation holds promise for being able to improve the spatial resolution and signal quality of ionomer in CL measurements by using cryo-STXM, further study is needed, since there was significant ice build-up in those measurements, which could have played a role in reducing mass loss.

5.4.2 Progress toward *in situ* and operando studies of PEMFC by soft X-ray STXM

It is widely believed that significant improvements to PEMFC technologies are most likely to arise from operando studies [120–124] in which all or a subset of the PEMFC system is studied with spectromicroscopy methods under realistic operating conditions.

Issues which could be better understood with appropriate operando studies include op-timization of water distributions [125, 126], low temperature modifications to CL struc-ture ("cold start"), start-up/shut-down reversal induced Pt migration and carbon corro-sion [61].

Several *operando* or *in situ* soft X-ray STXM studies of electrocatalyst systems have been carried out using soft X-ray STXM and ptychography. Lithium batteries [127–129], Zn-air batteries [130], supercapacitors [40] and half-cell ionic liquid based systems [41] are among the systems studied. However, so far a full-working PEMFC system has not yet been probed by soft X-ray microscopy, which is perhaps not surprising given: (i) the extremely tight spatial constraints (see Figure 1); (ii) the requirement for introducing/ex-hausting reactant gases, and product water, as well as controlled heating to the typical 80 °C operating temperature and (iii) the large X-ray absorption coefficients in the soft X-ray region. Custom STXM devices for variable T [131]; variable humidity (RH [132, 133]; simultaneous (T, RH) control [32, 33]; liquid flow [132, 134]; electrochemistry [135–137]; and combined fluid flow and electrochemistry [138] have been built and tested under near-operating conditions for a variety of scientific studies.

As an example of *in situ* STXM, Figure 9 presents results from a study of hydration of a microtomed slice of PEMFC MEA in a controlled humidity environmental cell. Fig-ure 9a is a photo of an *in situ* cell with both (T,RH) feedback control [15] mounted in the CLS ambient STXM. A heating/cooling Peltier thermoelectric device and a thermocou-ple in good thermal contact with the sample are used to establish sample temperatures from −20 to +80 °C with feedback control. A humidity sensor in a flow of humidified He close to the sample is used to operate proportional valves in wet and dry He streams to maintain the user-selected humidity at the chosen operating temperature. One of the at-tractive aspects of the use of STXM for mapping water distributions in MEA is the ability of O 1s NEXAFS spectroscopy to easily differentiate the O 1s signal of PFSA and the gas, liquid and solid phases of water; see Figure 9b. Results from a study of liquid and gas dis-tributions in a microtomed slice of an MEA at T = 22 C and RH = 85 % [32] are presented in Figure 9c–f. Figure 9c is an OD difference map (OD_{694} − OD_{684}) of the fluorine, and thus the ionomer, in both the electrodes and the membrane. The latter is a Gore SELECT® ePTFE micro-reinforced membrane, where the central third of the membrane is a highly porous reinforcement region. The ultramicrotomy preparation used a novel polystyrene (PS) sandwich embedding process where beads of PS softened by exposure to toluene vapor, press against the MEA to hold it in place for ultra-microtoming without the PS penetrating into the electrodes. Unfortunately, the cathode fragmented during the mi-crotoming. Despite this, representative regions, free of embedding resin, were present and used for the humidification study. Figure 9d is an image at 540 eV, taken from an O 1s stack (524–564 eV, 60 energies) measured under high humidity conditions, The stack was fit to the O 1s spectra of PFSA, liquid and gaseous water, and a constant (for carbon and catalyst in the cathode). A color-coded composite of cathode, PFSA, and liquid water maps is presented in Figure 9e while the composite of the gaseous H_2O, PFSA and liquid H_2O maps is presented in Figure 9f. As expected, gaseous water was found in the highly

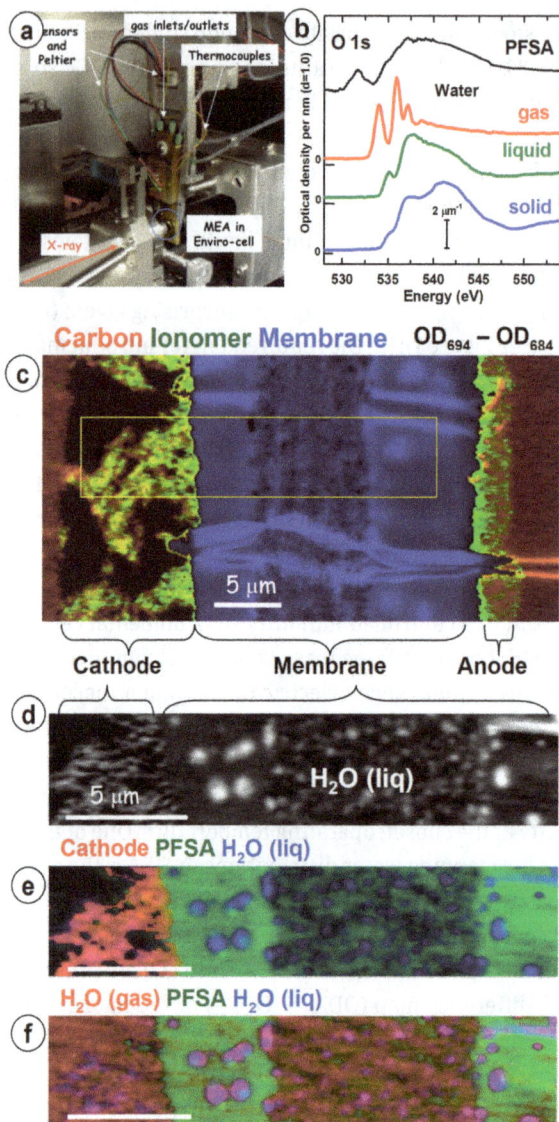

Figure 9: Example of an *in situ* STXM study of PEMFC MEA. (a) Photograph of the temperature and humidity (T,RH) controlled environment *in situ* STXM cell mounted in the polymer STXM on BL 5.3.2.2 at the Advanced Light Source. (b) O 1s spectra of gas, liquid and solid water. (c) Color-coded composite of component maps of the full width of an ultramicrotomed section of an MEA of a PEMFC derived from STXM OD images at 684 eV (pre-F1s) and 694 eV (F 1s maximum). The layers of the structure are labeled. The yellow rectangle indicates the area mapped under high RH conditions. (d) Component map of liquid water from fit of an O 1s stack recorded from the MEA in the environmental cell under saturated humidity conditions (*T* = 25 C, RH = 85 %). (e) Color-coded composite of component maps for cathode (no O 1s, red), PFSA (green) and liquid water (blue), derived by stack fit to the O 1s stack. (f) Color-coded composite of component maps for water vapor (red), PFSA (green) and liquid water (blue), derived by a fit of the O 1s stack to the O 1s spectra of PFSA, gas and liquid water, and a constant to map the carbon support in the cathode.

porous regions of the cathode and the e-PTFE micro-reinforced layer of the membrane. In contrast, rather than diffusing into the dense part of the membrane, the liquid water was observed as droplets on the surface of the dense regions of the membrane, filling the pores in the e-PTFE micro-reinforced layer, and the cathode (Figure 9d, 9f). The lack of penetration of liquid water into the membrane is most likely because the microtomed section was bone dry at the start of the experiment and not subjected to the peroxide conditioning typically used to activate membranes by making continuous hydrophilic pathways [139].

In general, operando STXM lags behind operando TEM [140] and operando hard X-ray [141, 142] capabilities, in both instrumentation and number of published studies. Hard X-ray synchrotron radiography [143] and lab-based hard X-ray tomography [63, 64] have been applied to study water distributions in full PEMFC using specialized thin cells with hard X-ray transparent windows. Such studies have provided useful insights into how liquid and vapor water distributions change under different operating conditions. However, it has been noted that radiation damage by hard X-rays, especially at the Pt L-edge, affects the operational electrochemical characteristics of the devices at a much faster rate than material damage [144, 145] probably due to hard X-rays being preferentially absorbed at the Pt catalyst resulting in modifications of the local structure and chemistry at the critical catalyst site.

5.5 Summary

Synchrotron radiation (SR) based scanning X-ray microscopy in the soft X-ray regime is a versatile tool for advanced chemical analysis and imaging of materials, devices and processes central to polymer electrolyte membrane fuel cells. In this chapter, I have presented the methods and some representative applications. Industrial, government and academic scientists can access STXM and ptychography at up to 20 different synchrotron facilities (see list at http://unicorn.mcmaster.ca/xrm-biblio/xrm_bib.html). More generally, for those interested in exploring SXM methods in their research, it might be useful to explore the http://lightsources.org web site, which is a clearing house for information about all 60+ synchrotron facilities worldwide, their capabilities and how to access them. Many SR facilities are upgrading to fourth generation technology, which provides a much brighter and more coherent beam, both factors which greatly enhance STXM and ptychography [9]. Over the next decade, there will be significant improvements in spatial resolution, acquisition speed and more capable instrumentation for *in situ* and *operando* studies by STXM and ptychography, all of which will significantly improve chemical analysis and imaging of PEMFC components and devices.

Bibliography

[1] Y. Wang, K. S. Chen, J. Mishler, S. C. Cho, and X. C. Adroher, A review of polymer electrolyte membrane fuel cells, Technology, applications, and needs on fundamental research. *Appl. Energy*, **88**, 981–1007, 2011.

[2] M. K. Debe, Electrocatalyst approaches and challenges for automotive fuel cells. *Nature*, **486**, 43–51, 2012.

[3] Y. Wang, D. F. Ruiz Diaz, K. S. Chen, Z. Wang, and X. C. Adroher, Materials, Technological Status, and Fundamentals of PEM Fuel Cells – A Review. *Mater. Today*, **32**, 178–203, 2020.

[4] N. Tian, B.-A. Lu, X.-D. Yang, R. Huang, Y.-X. Jiang, Z.-Y. Zhou, and S.-G. Sun, Rational Design and Synthesis of Low-Temperature Fuel Cell Electrocatalysts. *Electrochem. Energy Rev.*, **1**, 54–83, 2018.

[5] J. S. Walker, N. V. Rees, and P. M. Mendes, Progress towards the Ideal Core@shell Nanoparticle for Fuel Cell Electrocatalysis. *J. Exp. Nanosci.*, **13**, 258–271, 2018.

[6] A. Z. Weber, R. L. Borup, R. M. Darling, P. K. Das, T. J. Dursch, W. Gu, D. Harvey, A. Kusoglu, S. Litster, M. M. Mench, R. Mukundan, J. P. Owejan, J. G. Pharoah, M. Secanell, and I. V. Zenyuk, Critical Review of Modeling Transport Phenomena in Polymer-Electrolyte Fuel Cells. *J. Electrochem. Soc.*, **161**, F1254, 2014.

[7] H. Ade and A. P. Hitchcock, NEXAFS microscopy and resonant scattering: Composition and orientation probed in real and reciprocal space. *Polymer*, **49**, 643–675, 2008.

[8] A. P. Hitchcock, Soft X-Ray Imaging and Spectromicroscopy. In: *Handbook of Nanoscopy*, pages 745–791. John Wiley & Sons, Ltd, 2012.

[9] A. P. Hitchcock, Soft X-ray spectromicroscopy and ptychography. *J. Electron Spectrosc. Relat. Phenom.*, **200**, 49–63, 2015.

[10] D. Susac, V. Berejnov, A. P. Hitchcock, and J. Stumper, STXM Study of the Ionomer Distribution in the PEM Fuel Cell Catalyst Layers. *ECS Trans.*, **41**, 629–635, 2011.

[11] D. Susac, V. Berejnov, A. P. Hitchcock, and J. Stumper, STXM Characterization of PEM Fuel Cell Catalyst Layers. *ECS Trans.*, **50**, 405–413, 2013.

[12] A. M. V. Putz, D. Susac, V. Berejnov, J. Wu, A. P. Hitchcock, and J. Stumper, Doing More with Less, Challenges for Catalyst Layer Design. *ECS Trans.*, **75**, 3–23, 2016.

[13] D. Susac, J. Wang, Z. Martin, A. P. Hitchcock, J. Stumper, and D. Bessarabov, Chemical Fingerprint Associated with the Formation of Pt in the Membrane in PEM Fuel Cells. *ECS Trans.*, **33**, 391–398, 2010.

[14] V. Berejnov, Z. Martin, M. West, S. Kundu, D. Bessarabov, J. Stumper, D. Susac, and A. P. Hitchcock, Probing platinum degradation in polymer electrolyte membrane fuel cells by synchrotron X-ray microscopy. *Phys. Chem. Chem. Phys.*, **14**, 4835–4843, 2012.

[15] A. P. Hitchcock, V. Berejnov, V. Lee, M. M. West, V. Dutta, V. Colbow, and S. Wessel, Carbon corrosion of proton exchange membrane fuel cell catalyst layers studied by scanning transmission X-ray microscopy. *J. Power Sources*, **266**, 66–78, 2014.

[16] V. Lee, D. Susac, S. Kundu, V. Berejnov, R. T. Atanasoski, A. P. Hitchcock, and J. Stumper, STXM Characterization of Nanostructured Thin Film Anode Before and After Start-Up Shutdown and Reversal Tests. *ECS Trans.*, **58**, 473–479, 2013.

[17] V. Lee, V. Berejnov, M. West, S. Kundu, V. Berejnov, R. T. Atanasoski, A. P. Hitchcock, and J. Stumper, Scanning transmission X-ray microscopy of nano structured thin film catalysts for proton-exchange-membrane fuel cells. *J. Power Sources*, **263**, 163–174, 2014.

[18] L. Daniel, A. Bonakdarpour, and D. P. Wilkinson, Benefits of Platinum Deposited in the Polymer Membrane Subsurface on the Operational Flexibility of Hydrogen Fuel Cells. *J. Power Sources*, **471**, 228418, 2020.

[19] I. Martens, L. G. A. Melo, M. M. West, D. P. Wilkinson, D. Bizzotto, and A. P. Hitchcock, Imaging Reactivity of the Pt–Ionomer Interface in Fuel Cell Catalyst Layers. *ACS Appl. Energy Mater.*, **11**, 7772–7780, 2020.

[20] V. Berejnov, D. Susac, J. Stumper, and A. P. Hitchcock, 3D Chemical Mapping of PEM Fuel Cell Cathodes by Scanning Transmission Soft X-ray SpectroTomography. *ECS Trans.*, **50**, 361–368, 2013.

[21] G. Schmid, M. Obst, J. Wu, and A. P. Hitchcock, 3D Chemical Imaging of Nanoscale Biological, Environmental, and Synthetic Materials by Soft X-Ray STXM Spectrotomography. In: C. S. S. R. Kumar, Ed., *X-Ray and Neutron Techniques for Nanomaterials Characterization*, pages 43–94. Springer, Berlin, Heidelberg, 2016.

[22] J. Wu, M. Lerotic, S. Collins, R. Leary, Z. Saghi, P. Midgley, S. Berejnov, D. Susac, J. Stumper, G. Singh, and A. P. Hitchcock, Optimization of Three-Dimensional (3D) Chemical Imaging by Soft X-Ray Spectro-Tomography Using a Compressed Sensing Algorithm. *Microsc. Microanal.*, **23**, 951–966, 2017.

[23] J. Wu, L. G. A. Melo, X. Zhu, M. M. West, V. Berejnov, D. Susac, J. Stumper, and A. P. Hitchcock, 4D imaging of polymer electrolyte membrane fuel cell catalyst layers by soft X-ray spectro-tomography. *J. Power Sources*, **381**, 72–83, 2018.

[24] V. Berejnov, M. Saha, D. Susac, J. Stumper, M. West, and A. P. Hitchcock, Advances in Structural Characterization Using Soft X-ray Scanning Transmission Microscopy (STXM): Mapping and Measuring Porosity in PEMFC Catalyst Layers. *ECS Trans.*, **80**, 241–252, 2017.

[25] L. G. A. Melo, A. P. Hitchcock, J. Jankovic, J. Stumper, D. Susac, and V. Berejnov, Quantitative Mapping of Ionomer in Catalyst Layers by Electron and X-ray Spectromicroscopy. *ECS Trans.*, **80**, 275–282, 2017.

[26] L. G. A. Melo and A. P. Hitchcock, Electron Beam Damage of Perfluorosulfonic Acid Studied by Soft X-Ray Spectromicroscopy. *Micron*, **121**, 8–20, 2019.

[27] L. G. A. Melo, A. P. Hitchcock, V. Berejnov, D. Susac, J. Stumper, and G. A. Botton, Evaluating focused ion beam and ultramicrotome sample preparation for analytical microscopies of the cathode layer of a polymer electrolyte membrane fuel cell. *J. Power Sources*, **312**, 23–35, 2016.

[28] L. G. A. Melo and A. P. Hitchcock, Optimizing Soft X-ray Spectromicroscopy for Fuel Cell Studies: X-ray Damage of Ionomer. *Microsc. Microanal.*, **24**(S2), 460–461, 2018.

[29] I. Martens, L. G. A. Melo, D. P. Wilkinson, D. Bizzotto, and A. P. Hitchcock, Characterization of X-ray Damage to Perfluorosulfonic Acid Using Correlative Microscopy. *J. Phys. Chem. C*, **123**, 16023–16033, 2019.

[30] A. P. Hitchcock, Soft X-ray spectromicroscopy and ptychography. *J. Electron Spectrosc. Relat. Phenom.*, **200**, 49–63, 2015.

[31] J. Wu, X. Zhu, M. M. West, T. Tyliszczak, H.-W. Shiu, D. Shapiro, V. Berejnov, D. Susac, J. Stumper, and A. P. Hitchcock, High-Resolution Imaging of Polymer Electrolyte Membrane Fuel Cell Cathode Layers by Soft X-ray Spectro-Ptychography. *J. Phys. Chem. C*, **122**, 11709–11719, 2018.

[32] V. Berejnov, D. Susac, J. Stumper, and A. P. Hitchcock, Nano to Micro Scale Characterization of Water Uptake in The Catalyst Coated Membrane Measured by Soft X-ray Scanning Transmission X-ray Microscopy. *ECS Trans.*, **41**, 395, 2011.

[33] A. P. Hitchcock, V. Berejnov, V. Lee, D. Susac, and J. Stumper, In situ Methods for Analysis of Polymer Electrolyte Membrane Fuel Cell Materials by Soft X-ray Scanning Transmission X-ray Microscopy. *Microsc. Microanal.*, **20**, 1532–1533, 2014.

[34] J. Geilhufe, A. F. G. Leontowich, J. Wang, R. Berg, C. N. Regier, D. M. Taylor, D. Beauregard, J. Swirsky, C. Karunakaran, A. P. Hitchcock, and S. G. Urquhart, Soft X-ray Spectrotomographic Microscopy at Cryogenic Temperatures. *Microsc. Microanal.*, **24**(S2), 258–259, 2018.

[35] A. F. G. Leontowich, R. Berg, C. N. Regier, D. M. Taylor, J. Wang, D. Beauregard, J. Geilhufe, J. Swirsky, J. Wu, C. Karunakaran, A. P. Hitchcock, and S. G. Urquhart, Cryo scanning transmission x-ray microscope optimized for spectrotomography. *Rev. Sci. Instrum.*, **89**, 093704, 2018.

[36] A. P. Hitchcock, H. Yuan, L. G. A. Melo, and N. Bassim, Soft X-ray scanning transmission microscopy studies of radiation damage by electron, ion and X-ray beams. *Microsc. Microanal.*, **26**(S2), 2072–2074, 2002.

[37] B. Bozzini, A. Gianoncelli, B. Kaulich, M. Kiskinova, M. Prasciolu, and I. Sgura, Metallic Plate Corrosion and Uptake of Corrosion Products by Nafion in Polymer Electrolyte Membrane Fuel Cells. *ChemSusChem*, **3**, 846–850, 2010.

[38] B. Bozzini, C. Mele, A. Gianoncelli, B. Kaulich, M. Kiskinova, and M. Prasciolu, In situ X-ray spectromicroscopy study of bipolar plate material stability for nano-fuel-cells with ionic-liquid electrolyte. *Microelectron. Eng.*, **88**, 2456–2458, 2011.

[39] B. Bozzini, M. K. Abyaneh, M. Amati, A. Gianoncelli, L. Gregoratti, B. Kaulich, and M. Kiskinova, Soft X-ray Imaging and Spectromicroscopy: New Insights in Chemical State and Morphology of the Key Components in Operating Fuel-Cells. *Chemistry*, **18**, 10196–10210, 2012.

[40] B. Bozzini, M. Amati, A. Gianoncelli, L. Gregoratti, B. Kaulich, and M. Kiskinova, New Energy Sources: in-situ Characterisation of Fuel Cell and Supercapacitor Components. Complementary Studies using Transmission, Fluorescence and Photoelectron Microscopy and Imaging. *J. Phys. Conf. Ser.*, **463**, 012018, 2013.

[41] B. Bozzini, A. Gianoncelli, B. Kaulich, C. Mele, M. Prasciolu, and M. Kiskinova, In Situ Soft X-ray Microscopy Study of Fe Interconnect Corrosion in Ionic Liquid-Based Nano-PEMFC Half-Cells. *Fuel Cells*, **13**, 196–202, 2013.

[42] T. Ohigashi, Y. Inagaki, T. Horigome, and N. Kosugi, Observation of Morphology of a Fuel Cell by using a humidity control sample cell for STXM. *UVSOR Annual Rep.*, **43**, 69–70, 2015.

[43] M. G. George, J. Wang, R. Banerjee, and A. Bazylak, Composition analysis of a polymer electrolyte membrane fuel cell microporous layer using scanning transmission X-ray microscopy and near edge X-ray absorption fine structure analysis. *J. Power Sources*, **309**, 254–259, 2016.

[44] C. J. Jacobsen, *X-ray Microscopy*. Cambridge University Press, 2019.

[45] F. M. F. de Groot, E. de Smit, M. M. van Schooneveld, L. R. Aramburo, and B. M. Weckhuysen, In Situ Scanning Transmission X-ray Microscopy of Catalytic solds and Related Nanomaterials. *ChemPhysChem*, **11**, 951–962, 2010.

[46] A. Braun, *X-ray Studies on Electrochemical Systems: Synchrotron Methods for Energy Materials*. Walter de Gruyter GmbH & Co KG, 2017.

[47] M. Du and C. Jacobse, Relative Merits and Limiting Factors for X-Ray and Electron Microscopy of Thick, Hydrated Organic Materials. *Ultramicroscopy*, **184**(A), 293–309, 2018.

[48] E. G. Rightor, A. P. Hitchcock, H. Ade, R. D. Leapman, S. G. Urquhart, A. P. Smith, G. Mitchell, D. Fischer, H. J. Shin, and T. Warwick, Spectromicroscopy of Poly(Ethylene Terephthalate): Comparison of Spectra and Radiation Damage Rates in X-Ray Absorption and Electron Energy Loss. *J. Phys. Chem. B*, **101**, 1950–1960, 1997.

[49] J. Wang, G. A. Botton, M. M. West, and A. P. Hitchcock, Quantitative Evaluation of Radiation Damage to Polyethylene Terephthalate by Soft X-Rays and High-Energy Electrons. *J. Phys. Chem. B*, **113**, 1869–1876, 2009.

[50] L. G. A. Melo, A. P. Hitchcock, J. Jankovic, J. Stumper, D. Susac, and V. Berejnov, Quantitative Mapping of Ionomer in Catalyst Layers by Electron and X-ray Spectromicroscopy. *ECS Trans.*, **80**, 275–282, 2017.

[51] F. Scheiba, N. Benker, U. Kunz, C. Roth, and H. Fuess, Electron microscopy techniques for the analysis of the polymer electrolyte distribution in proton exchange membrane fuel cells. *J. Power Sources*, **177**, 273–280, 2008.

[52] C. Wang, V. Krishnan, D. Wu, R. Bledsoe, S. J. Paddison, and G. Duscher, Evaluation of the microstructure of dry and hydrated perfluorosulfonic acid ionomers, microscopy and simulations. *J. Mater. Chem. A*, **1**, 938–944, 2012.

[53] S. Yakovlev and K. H. Downing, Visualization of clusters in polymer electrolyte membranes by electron microscopy. *Phys. Chem. Chem. Phys.*, **15**, 1052–1064, 2012.

[54] J. Jankovic, D. Susac, T. Soboleva, and J. Stumper, Electron Tomography Based 3D Reconstruction of Fuel Cell Catalysts. *ECS Trans.*, **50**, 353–359, 2013.

[55] M. Holt, R. Harder, R. Winarski, and V. Rose, Nanoscale Hard X-Ray Microscopy Methods for Materials Studies. *Annu. Rev. Mater. Res.*, **43**, 183–211, 2013.

[56] R. T. White, A. Wu, M. Najm, F. P. Orfino, M. Dutta, and E. Kjeang, 4D in situ visualization of electrode morphology changes during accelerated degradation in fuel cells by X-ray computed tomography. *J. Power Sources*, **350**, 94–102, 2017.

[57] M. Povia, J. Herranz, T. Binninger, M. Nachtegaal, A. Diaz, J. Kohlbrecher, D. F. Abbott, B.-J. Kim, and T. J. Schmidt, Combining SAXS and XAS To Study the Operando Degradation of Carbon-Supported Pt-Nanoparticle Fuel Cell Catalysts. *ACS Catal.*, **8**, 7000–7015, 2018.

[58] S. J. Normile and I. V. Zenyuk, Imaging ionomer in fuel cell catalyst layers with synchrotron nano transmission x-ray microscopy. *Solid State Ion.*, **335**, 38–46, 2019.

[59] I. V. Zenyuk, D. Y. Parkinson, G. Hwang, and A. Z. Weber, Probing water distribution in compressed fuel-cell gas-diffusion layers using X-ray computed tomography. *Electrochem. Commun.*, **53**, 24–28, 2015.

[60] T. Saida, O. Sekizawa, N. Ishiguro, M. Hoshino, K. Uesugi, T. Uruga, S. Ohkoshi, T. Yokoyama, and M. Tada, 4D Visualization of a Cathode Catalyst Layer in a Polymer Electrolyte Fuel Cell by 3D Laminography–XAFS. *Angew. Chem., Int. Ed. Engl.*, **51**, 10311–10314, 2012.

[61] H. Matsui, N. Ishiguro, T. Uruga, O. Sekizawa, K. Higashi, N. Maejima, and T. M. Operando, 3D Visualization of Migration and Degradation of a Platinum Cathode Catalyst in a Polymer Electrolyte Fuel Cell. *Angew. Chem., Int. Ed. Engl.*, **56**, 9371–9375, 2017.

[62] O. Sekizawa, T. Uruga, N. Ishiguro, H. Matsui, K. Higashi, T. Sakata, Y. Iwasawa, and M. Tada, In-situ X-ray nano-CT System for Polymer Electrolyte Fuel Cells under Operating Conditions. *J. Phys. Conf. Ser.*, **849**, 012022, 2017.

[63] W. K. Epting, J. Gelb, and S. Litster, Resolving the Three-Dimensional Microstructure of Polymer Electrolyte Fuel Cell Electrodes using Nanometer-Scale X-ray Computed Tomography. *Adv. Funct. Mater.*, **22**, 555–560, 2012.

[64] R. T. White, S. H. Eberhardt, Y. Singh, T. Haddow, M. Dutta, F. P. Orfino, and E. Kjeang, Four-Dimensional Joint Visualization of Electrode Degradation and Liquid Water Distribution inside Operating Polymer Electrolyte Fuel Cells. *Sci. Rep.*, **9**, 1843, 2019.

[65] K. Kaznacheyev, I. Blomqvist, E. Hallin, S. Urquhart, D. Loken, T. Tyliszczak, T. Warwick, and A. P. Hitchcock, Principles of Optical Design of the SM Beamline at the CLS. In: *AIP Conference Proceedings*, vol. 705, pages 1303–1307, 2004.

[66] K. V. Kaznatcheev, C. Karunakaran, U. D. Lanke, S. G. Urquhart, M. Obst, and A. P. Hitchcock, Soft X-ray spectromicroscopy beamline at the CLS: commissioning results. *Nucl. Instrum. Methods Phys. Res., Sect. A*, **582**, 96–99, 2007.

[67] D. Kilcoyne, T. Tyliszczak, W. Steele, S. Fakra, P. Hitchcock, K. Franck, E. Anderson, B. Harteneck, E. Rightor, G. Mitchell, A. P. Hitchcock, L. Yang, T. Warwick, and H. Ade, Interferometer-Controlled Scanning Transmission X-Ray Microscopes at the Advanced Light Source. *J. Synchrotron Radiat.*, **10**, 125–136, 2003.

[68] K. V. Kaznacheyev, C. Karunakaran, F. He, M. Sigrist, M. Summers, M. Obst, and A. P. Hitchcock, CLS ID-10 Chicane Configuration: From "Simple Sharing" to Extended Performance with High Speed Polarization Switching, US Synchrotron Radiation Instrumentation, April 2007, Baton Rouge, LA. *Nucl. Instrum. Methods A*, **582**, 103–106, 2007.

[69] B. Rösner, S. Finizio, F. Koch, F. Döring, V. A. Guzenko, M. Langer, E. Kirk, B. Watts, M. Meyer, J. Loroña Ornelas, A. Späth, S. Stanescu, S. Swaraj, R. Belkhou, T. Ishikawa, T. F. Keller, B. Gross, M. Poggio, R. H. Fink, J. Raabe, A. Kleibert, and C. David, Soft X-Ray Microscopy with 7 nm Resolution. *Optica*, **7**, 1602–1608, 2020.

[70] C. Jacobsen, S. Wirick, G. Flynn, and C. Zimba, Soft X-ray spectroscopy from image sequences with sub-100 nm spatial resolution. *J. Microsc.*, **197**, 173–184, 2000.

[71] M. Lerotic, C. Jacobsen, T. Schäfer, and S. Vogt, Cluster analysis of soft x-ray spectromicroscopy data. *Ultramicroscopy*, **100**, 35–57, 2004.

[72] M. Lerotic, R. Mak, S. Wirick, F. Meirer, and C. Jacobsen, MANTiS: a program for the analysis of X-ray spectromicroscopy data. *J. Synchrotron Radiat.*, **21**, 1206–1212, 2014.

[73] Z. B. Yan, R. Hayes, L. G. A. Melo, G. R. Goward, and A. P. Hitchcock, X-ray Absorption and Solid-State NMR Spectroscopy of Fluorinated Proton Conducting Polymers. *J. Phys. Chem. C*, **122**, 3233–3244, 2018.

[74] A. Kusoglu and A. Z. Weber, New Insights into Perfluorinated Sulfonic-Acid Ionomers. *Chem. Rev.*, **117**, 987–1104, 2017.

[75] J. Li, A. P. Hitchcock, H. D. H. Stöver, and I. Shirley, A new approach to experimentally studying microcapsule wall growth mechanisms. *Macromolecules*, **42**, 2428–2432, 2009.

[76] K. L. More and K. S. Reeves, *Microsc. Microanal.*, **11**, 2104, 2005.

[77] D. A. Cullen, R. Koestner, R. S. Kukreja, Z. Y. Liu, S. Minko, O. Trotsenko, A. Tokarev, L. Guetaz, H. M. Meyer, C. M. Parish, and K. L. More, Imaging and Microanalysis of Thin Ionomer Layers by Scanning Transmission Electron Microscopy. *J. Electrochem. Soc.*, **161**, F1111–F1117, 2014.

[78] H. Uchida, J. M. Song, S. Suzuki, E. Nakazawa, N. Baba, and M. Watanabe, Electron Tomography of Nafion Ionomer Coated on Pt/Carbon Black in High Utilization Electrode for PEFCs. *J. Phys. Chem. B*, **110**, 13319–13321, 2006.

[79] F. I. Allen, L. R. Comolli, A. Kusoglu, M. A. Modestino, A. M. Minor, and A. Z. Weber, Morphology of hydrated as-cast Nafion revealed through cryo electron tomography. *ACS Macro Lett.*, **4**, 1–5, 2015.

[80] M. Lopez-Haro, G. L, T. Printemps, A. Morin, S. Escribano, P.-H. Jouneau, P. Bayle-Guillemaud, F. Chandezon, and G. Gebel, Three-Dimensional Analysis of Nafion Layers in Fuel Cell Electrodes. *Nat. Commun.*, **5**, 5229, 2014.

[81] S. Yakovlev, N. P. Balsara, and K. H. Downing, Insights on the Study of Nafion Nanoscale Morphology by Transmission Electron Microscopy. *Membranes*, **3**, 424–439, 2013.

[82] S. Yakovlev and K. H. Downing, Visualization of Clusters in Polymer Electrolyte Membranes by Electron Microscopy. *Phys. Chem. Chem. Phys.*, **15**, 1052–1064, 2013.

[83] L. Zielke, S. Vierrath, R. Moroni, A. Mondon, R. Zengerle, and S. Thiele, Three-Dimensional Morphology of the Interface between Micro Porous Layer and Catalyst Layer in a Polymer Electrolyte Membrane Fuel Cell. *RSC Adv.*, **6**, 80700–80705, 2016.

[84] H. Schulenburg, B. Schwanitz, N. Linse, G. G. Scherer, A. Wokaun, J. Krbanjevic, R. Grothausmann, and I. Manke, 3D Imaging of Catalyst Support Corrosion in Polymer Electrolyte Fuel Cells. *J. Phys. Chem. C*, **115**, 14236–14243, 2011.

[85] S. Zils, M. Timpel, T. Arlt, A. Wolz, I. Manke, and C. Roth, 3D Visualisation of PEMFC Electrode Structures Using FIB Nanotomography. *Fuel Cells*, **10**, 966–972, 2010.

[86] K. Karan, PEFC catalyst layer: Recent advances in materials, microstructural characterization, and modeling. *Curr. Opin. Electrochem.*, **5**, 27–35, 2017.

[87] E. H. Majlan, D. Rohendi, W. R. W. Daud, T. Husaini, and M. A. Haque, Electrode for proton exchange membrane fuel cells: A review. *Renew. Sustain. Energy Rev.*, **89**, 117–134, 2018.

[88] J. Wang, Radiation Chemistry Studied by Soft X-ray Microscopy. Ph. D. thesis, McMaster University, 2008.

[89] A. F. G. Leontowich, Tunable soft X-rays for patterning and lithography. Ph. D. thesis, McMaster University, 2012.

[90] L. G. A. Melo, Soft X-Ray Spectromicroscopy of Radiation Damaged Perfluorosulfonic Acid. Ph. D. thesis, McMaster University, 2018.

[91] J. Wang, G. A. Botton, M. M. West, and A. P. Hitchcock, Quantitative Evaluation of Radiation Damage to Polyethylene Terephthalate by Soft X-Rays and High-Energy Electrons. *J. Phys. Chem. B*, **113**, 1869–1876, 2009.

[92] J. Wang, C. Morin, L. Li, A. P. Hitchcock, X. Zhang, T. Araki, A. Doran, and A. Scholl, Radiation Damage in Soft X-ray Microscopy. *J. Electron Spectrosc. Relat. Phenom.*, **170**, 25–36, 2009.

[93] T. Coffey, S. Urquhart, and H. Ade, Characterization of the effects of soft X-ray irradiation on polymers. *J. Electron Spectrosc. Relat. Phenom.*, **122**, 65–78, 2002.

[94] V. Berejnov, B. Rubinstein, L. G. A. Melo, and A. P. Hitchcock, First principles X-ray absorption dose calculation for time dependent mass and optical density. *J. Synchrotron Radiat.*, **25**, 833–847, 2018.

[95] V. Berejnov, B. Rubinstein, L. G. A. Melo, and A. P. Hitchcock, Calculating Absorption Dose When X-ray Irradiation Modifies Material Quantity and Chemistry. *J. Synchrotron Radiat.*, **28**, 834–838, 2021.

[96]   T. Xie and C. A. Hayden, A kinetic model for the chemical degradation of perfluorinated sulfonic acid ionomers: Weak end groups versus side chain cleavage. *Polymer*, **48**, 5497–5506, 2007.

[97]   L. Ghassemzadeh, K. D. Kreuer, J. Maier, and K. Müller, Evaluating Chemical Degradation of Proton Conducting Perfluorosulfonic Acid Ionomer in a Fenton Test by Solid-state ^{19}F NMR Spectroscopy. *J. Power Sources*, **196**, 2490–2497, 2011.

[98]   K. Hongsirikarn, X. Mo, J. Goodwin, and S. Creager, Effect of H_2O_2 on Nafion (R) Properties and Conductivity at Fuel Cell Conditions. *J. Power Sources*, **196**, 3060–3072, 2011.

[99]   G. A. Johansson, T. Tyliszczak, G. E. Mitchell, M. H. Keefe, and A. P. Hitchcock, Three-dimensional chemical mapping by scanning transmission X-ray spectromicroscopy. *J. Synchrotron Radiat.*, **14**, 395–402, 2007.

[100] E. Y. Sidky and X. Pan, Image Reconstruction in Circular Cone-Beam Computed Tomography by Constrained, Total-Variation Minimization. *Phys. Med. Biol.*, **53**, 4777–4807, 2008.

[101] K. Witte, A. Späth, S. Finizio, C. Donnelly, B. Watts, B. Sarafimov, M. Odstrcil, M. Guizar-Sicairos, M. Holler, R. H. Fink, and J. Raabe, From 2D STXM to 3D Imaging: Soft X-ray Laminography of Thin Specimens. *Nano Lett.*, **20**, 1305–1314, 2020.

[102] F. Pfeiffer and X. Ptychography, *Nat. Commun.*, **12**, 9–17, 2018.

[103] O. Mandula, M. Elzo Aizarna, J. Eymery, M. Burghammer, and V. Favre-Nicolin, PyNX.Ptycho: A Computing Library for X-Ray Coherent Diffraction Imaging of Nanostructures. *J. Appl. Crystallogr.*, **49**, 1842–1848, 2016.

[104] S. Marchesini, H. Krishnan, B. J. Daurer, D. A. Shapiro, T. Perciano, J. A. Sethian, and F. R. N. C. Maia, SHARP: a distributed, GPU-based ptychographic solver. *J. Appl. Crystallogr.*, **49**, 1245–1252, 2016.

[105] B. Enders and P. Thibault, A Computational Framework for Ptychographic Reconstructions. *Proc. R. Soc. A*, **472**, 20160640, 2016.

[106] M. Farmand, R. Celestre, P. Denes, A. L. D. Kilcoyne, S. Marchesini, H. Padmore, T. Tyliszczak, T. Warwick, X. Shi, J. Lee, Y.-S. Yu, J. Cabana, J. Joseph, H. Krishnan, T. Perciano, F. R. N. C. Maia, and D. A. Shapiro, Near-Edge X-Ray Refraction Fine Structure Microscopy. *Appl. Phys. Lett.*, **110**, 063101, 2017.

[107] D. A. Shapiro, Y.-S. Yu, T. Tyliszczak, J. Cabana, R. Celestre, W. Chao, K. Kaznatcheev, A. L. D. Kilcoyne, F. Maia, S. Marchesini, Y. S. Meng, T. Warwick, L. L. Yang, and H. A. Padmore, Chemical composition mapping with nanometre resolution by soft X-ray microscopy. *Nat. Photonics*, **8**, 765–769, 2014.

[108] D. A. Shapiro, S. Babin, R. S. Celestre, W. Chao, R. P. Conley, P. Denes, B. Enders, P. Enfedaque, S. James, J. M. Joseph, H. Krishnan, S. Marchesini, K. Muriki, K. Nowrouzi, S. R. Oh, H. A. Padmore, T. Warwick, L. Yang, V. V. Yashchuk, Y.-S. Yu, and J. Zhao, An Ultrahigh-Resolution Soft x-Ray Microscope for Quantitative Analysis of Chemically Heterogeneous Nanomaterials. *Sci. Adv.*, **6**, eabc490, 2020.

[109] H. Yuan, H. Yuan, T. Casagrande, D. A. Shapiro, Y.-S. Yu, B. Enders, J. R. I. Lee, T. Van Buuren, M. M. Biener, S. A. Gammon, T. T. Li, T. F. Baumann, and A. P. Hitchcock, 4D Imaging of ZnO-Coated Nanoporous Al_2O_3 Aerogels by Chemically-Sensitive Ptychographic Tomography: Implications for Designer Catalysts. *ACS Appl. Nanomater.*, **4**, 621–632, 2021.

[110] T. Beetz and C. Jacobsen, Soft X-ray radiation-damage studies in PMMA using a cryo-STXM. *J. Synchrotron Radiat.*, **10**, 280–283, 2002.

[111] K.-H. Kim, K.-Y. Lee, K.-J. Kim, E. Cho, S.-Y. Lee, T.-H. Lim, S. P. Yoon, I. C. Hwang, and J. H. Jang, The Effects of Nafion® Ionomer Content in PEMFC MEAs Prepared by a Catalyst-Coated Membrane (CCM) Spraying Method. *Int. J. Hydrog. Energy*, **35**, 2119–2126, 2010.

[112] S. Mu and M. Tian, Optimization of Perfluorosulfonic Acid Ionomer Loadings in Catalyst Layers of Proton Exchange Membrane Fuel Cells. *Electrochim. Acta*, **60**, 437–442, 2012.

[113] A. P. Hitchcock, V. Lee, J. Wu, M. M. West, G. Cooper, V. Berejnov, T. Soboleva, D. Susac, and J. Stumper, Characterizing automotive fuel cell materials by soft x-ray scanning transmission x-ray microscopy. *AIP Conf. Proc.*, **1696**, 020012, 2016.

[114] Z. Chen and H. R. Catalyst, Support Degradation. In: *PEM Fuel Cell Failure Mode Analysis*. CRC Press, FL, USA, 2011.

[115] C. Qin, J. Wang, D. Yang, B. Li, and C. Zhang, Proton Exchange Membrane Fuel Cell Reversal: A Review. *Catalysts*, **6**, 197, 2016.

[116] F. Ettingshausen, J. Kleemann, A. Marcu, G. Toth, H. Fuess, and C. Roth, Dissolution and Migration of Platinum in PEMFCs Investigated for Start/Stop Cycling and High Potential Degradation. *Fuel Cells*, **11**, 238–245, 2011.

[117] M. Prokop, T. Bystron, P. Belsky, O. Tucek, R. Kodym, M. Paidar, and K. Bouzek, Degradation Kinetics of Pt during High-Temperature PEM Fuel Cell Operation Part III: Voltage-Dependent Pt Degradation Rate in Single-Cell Experiments. *Electrochim. Acta*, **363**, 137165, 2020.

[118] N. Macauley, D. D. Papadias, J. Fairweather, D. Spernjak, D. Langlois, R. Ahluwalia, K. L. More, R. Mukundan, and R. L. Borup, Carbon Corrosion in PEM Fuel Cells and the Development of Accelerated Stress Tests. *J. Electrochem. Soc.*, **165**, F3148–F3160, 2018.

[119] L. Helfen, A. Myagotin, P. Mikulík, P. Pernot, A. Voropaev, M. Elyyan, M. Di Michiel, J. Baruchel, and T. Baumbach, On the Implementation of Computed Laminography Using Synchrotron Radiation. *Rev. Sci. Instrum.*, **82**, 063702, 2011.

[120] S. Deabate, G. Gebel, P. Huguet, A. Morin, and G. Pourcelly, In Situ and Operando Determination of the Water Content Distribution in Proton Conducting Membranes for Fuel Cells: A Critical Review. *Energy Environ. Sci.*, **5**, 8824–8847, 2012.

[121] Q. Meyer, Y. Zeng, and C. Zhao, In Situ and Operando Characterization of Proton Exchange Membrane Fuel Cells. *Adv. Mater.*, **31**, 1901900, 2019.

[122] L. Wang, J. Wang, and P. Zuo, Probing Battery Electrochemistry with In Operando Synchrotron X-Ray Imaging Techniques. *Small Methods*, **2**, 1700293, 2018.

[123] D. Liu, Z. Shadike, R. Lin, K. Qian, H. Li, K. Li, S. Wang, Q. Yu, M. Liu, S. Ganapathy, X. Qin, Q.-H. Yang, M. Wagemaker, F. Kang, X.-Q. Yang, and B. Li, Review of Recent Development of In Situ/Operando Characterization Techniques for Lithium Battery Research. *Adv. Mater.*, **31**, 1806620, 2019.

[124] J. Li and J. Gong, Operando Characterization Techniques for Electrocatalysis. *Energy Environ. Sci.*, **13**, 3748–3779, 2020.

[125] M. Ji and Z. Wei, A Review of Water Management in Polymer Electrolyte Membrane Fuel Cells. *Energies*, **2**, 1057–1106, 2009.

[126] M. Sahraoui, Y. Bichiou, and K. Halouani, Three-Dimensional Modeling of Water Transport in PEMFC. *Int. J. Hydrog. Energy*, **38**, 8524–8531, 2012.

[127] Y.-S. Yu, C. Kim, D. A. Shapiro, M. Farmand, D. Qian, T. Tyliszczak, A. L. D. Kilcoyne, R. Celestre, S. Marchesini, J. Joseph, P. Denes, T. Warwick, F. C. Strobridge, C. P. Grey, H. A. Padmore, Y. S. Meng, R. Kostecki, and J. Cabana, Dependence on Crystal Size of the Nanoscale Chemical Phase Distribution and Fracture in Li_xFePO_4. *Nano Lett.*, **15**, 4282–4288, 2015.

[128] Y. Li, J. N. Weker, W. E. Gent, D. N. Mueller, J. Lim, D. A. Cogswell, T. Tyliszczak, and W. C. Chueh, Dichotomy in the Lithiation Pathway of Ellipsoidal and Platelet LiFePO4 Particles Revealed through Nanoscale Operando State-of-Charge Imaging. *Adv. Funct. Mater.*, **25**, 3677–3687, 2015.

[129] M. Wolf, B. M. May, and J. Cabana, Visualization of Electrochemical Reactions in Battery Materials with X-Ray Microscopy and Mapping. *Chem. Mater.*, **29**, 3347–3362, 2017.

[130] B. Bozzini, M. Kazemian, M. Kiskinova, G. Kourousias, C. Mele, and A. Gianoncelli, Operando Soft X-Ray Microscope Study of Rechargeable Zn–Air Battery Anodes in Deep Eutectic Solvent Electrolyte. *X-Ray Spectrom.*, **48**, 527–535, 2019.

[131] A. F. G. Leontowich and A. P. Hitchcock, Experimental investigation of beam heating in a soft X-ray scanning transmission X-ray microscope. *Analyst*, **137**, 370–375, 2012.

[132] T. Lefevre, M. Pézolet, D. Hernández Cruz, M. M. West, M. Obst, A. P. Hitchcock, C. Karunakaran, and K. V. Kaznatcheev, Mapping molecular orientation in dry and wet dragline spider silk. *J. Phys. Conf. Ser., Proc. 9th Int. Conf. on X-ray Microscopy*, **186**, 012089, 2009.

[133] S. T. Kelly, P. Nigge, S. Prakash, A. Laskin, B. Wang, T. Tyliszczak, S. R. Leone, and M. K. Gilles, An environmental sample chamber for reliable scanning transmission x-ray microscopy measurements under water vapor. *Rev. Sci. Instrum.*, **84**, 073708, 2013.

[134] C. Gosse, S. Stanescu, J. Frederick, S. Lefrançois, A. Vecchiola, M. Moskura, S. Swaraj, R. Belkhou, B. Watts, P. Haltebourg, C. Blot, J. Daillant, P. Guenoun, and C. A. Chevallard, Pressure-Actuated Flow Cell for Soft X-Ray Spectromicroscopy in Liquid Media. *Lab Chip*, **20**, 3213–3229, 2020.

[135] D. Guay, J. Stewart-Ornstein, X. Zhang, and A. P. Hitchcock, In situ spatial and time resolved studies of electrochemical reactions by scanning transmission X-ray microscopy. *Anal. Chem.*, **77**, 3479–3487, 2005.

[136] B. Bozzini, L. D'Urzo, A. Gianoncelli, B. Kaulich, M. Kiskinova, M. Prasciolu, and A. Tadjeddine, In Situ Soft X-Ray Dynamic Microscopy of Electrochemical Processes. *Electrochem. Commun.*, **10**, 1680–1683, 2008.

[137] Y. A. Wu, Z. Yin, M. Farmand, Y.-S. Yu, D. A. Shapiro, H.-G. Liao, W.-I. Liang, Y.-H. Chu, and H. Zheng, In-Situ Multimodal Imaging and Spectroscopy of Mg Electrodeposition at Electrode-Electrolyte Interfaces. *Sci. Rep.*, **7**, 42527, 2017.

[138] V. Prabu, M. Obst, H. Hosseinkhannazer, M. Reynolds, S. Rosendahl, J. Wang, and A. P. Hitchcock, Instrumentation for in situ flow electrochemical Scanning Transmission X-ray Microscopy (STXM). *Rev. Sci. Instrum.*, **89**, 063702, 2018.

[139] J. Parrondo, M. Ortueta, and F. Swelling, Swelling Behavior of PEMFC during Conditioning. *Braz. J. Chem. Eng.*, **24**, 411–419, 2007.

[140] N. Hodnik, G. Dehm, and K. J. J. Mayrhofer, Importance and Challenges of Electrochemical in Situ Liquid Cell Electron Microscopy for Energy Conversion Research. *Acc. Chem. Res.*, **49**, 2015–2022, 2016.

[141] A. Siebel, Y. Gorlin, J. Durst, O. Proux, F. Hasché, M. Tromp, and H. A. Gasteiger, Identification of Catalyst Structure during the Hydrogen Oxidation Reaction in an Operating PEM Fuel Cell. *ACS Catal.*, **6**, 7326–7334, 2016.

[142] Y. Wu and N. Liu, Visualizing Battery Reactions and Processes by Using In Situ and In Operando Microscopies. *Chem*, **4**, 438–465, 2018.

[143] R. Banerjee, N. Ge, C. Han, J. Lee, M. G. George, H. Liu, D. Muirhead, P. Shrestha, and A. Bazylak, Identifying in Operando Changes in Membrane Hydration in Polymer Electrolyte Membrane Fuel Cells Using Synchrotron X-Ray Radiography. *Int. J. Hydrog. Energy*, **43**, 9757–9769, 2018.

[144] J. Eller, F. Marone, and F. N. Büchi, Operando sub-second tomographic imaging of water in pefc gas diffusion layers. *ECS Trans.*, **69**, 523–531, 2015.

[145] A. Schneider, C. Wieser, J. Roth, and L. Helfen, Impact of synchrotron radiation on fuel cell operation in imaging experiments. *J. Power Sources*, **195**, 6349–6355, 2010.

Part II: **Performance modeling/characterization**

Aslan Kosakian and Marc Secanell

6 Membrane-electrode-assembly modeling using OpenFCST

6.1 Introduction

Mathematical models of membrane-electrode assemblies (MEAs) of proton-exchange-membrane fuel cells (PEMFCs) started to appear in the early 1990s. The works of Springer et al. [173] and Bernardi and Verbrugge [25] contain examples of such models, the governing equations of which describe the most critical transport and electrochemical processes occurring inside a PEMFC: gas transport, charge transport, water transport, and electrochemical reactions.

Early PEMFC MEA models were one-dimensional and, in many cases, did not take into account the finite thickness of the catalyst layer. Therefore, the early 2000s witnessed a proliferation of multidimensional PEMFC models. Two-dimensional models, such as those in [73, 107, 169] to name but a few, focused on better understanding of the land-channel effects and, in most cases, the reaction-rate distribution inside the catalyst layer with the aim of improving catalyst utilization. For example, Kulikovsky [107] showed the appearance of dead zones under the current collector indicating that the catalyst was not fully utilized in that area. Three-dimensional models, such as those in [26–28], and two-dimensional along-the-channel models, such as the one from [192], focused on understanding the impact of reactant depletion along the channel on current-density distribution.

Over the past decade, the majority of publications have used commercial PEMFC model implementations [36, 69, 70, 89, 99, 153, 168, 184, 185, 190, 191, 213]. The consistent use of well-established, commercial software hints at the idea that mathematical modeling of PEMFCs is a mature subject and that little change should be expected in the upcoming decades. However, a careful analysis of recent experimental insights and electrochemical techniques shows that the lack of development in mathematical modeling of MEAs is not due to the maturity of the models but the absence of the critical analysis needed to identify their limitations. Some of the limitations are due to the improved understanding of the transport and electrochemical processes inside the fuel cell. For example, in the early 2010s, researchers at several automotive companies [72, 88, 105, 106, 137, 140, 177, 200] showed that oxygen transport limitations were not only due to oxygen

Acknowledgement: The authors acknowledge the Natural Sciences and Engineering Council of Canada (NSERC) Discovery Grant for financial assistance.

Aslan Kosakian, Marc Secanell, Department of Mechanical Engineering, University of Alberta, Alberta, Canada

https://doi.org/10.1515/9783110622720-006

diffusivity in the gas-diffusion electrode, but also due to Knudsen diffusivity in the catalyst layer and, even more importantly, localized transport resistances at the ionomer-gas and ionomer-platinum interfaces, resulting in poor performance for low-loading electrodes. Similarly, academic research groups, such as those of Karan [94, 109, 145] and Weber [23, 109], showed that ion transport in ionomer thin films in the catalyst layer could be significantly different from that in the membrane. Electrochemical reactions, such as the oxygen reduction reaction, have also been shown not to follow a Tafel model by many researchers, including Markiewicz et al. [123] and Parthasarthy et al. [142–144]. Over the past decade, the knowledge of mechanical, chemical, and electrochemical degradation of MEAs has also improved, but it has seldom been transferred to mathematical models, especially of higher dimensionality (2D or 3D). Notable exceptions are, for instance, the works of T. Jahnke and coworkers [58, 85]. Finally, and perhaps most importantly, commercial, state-of-the-art models lack rigorous validation and the ability to simulate and reproduce the results from multiple electrochemical techniques. Therefore, despite the perceived maturity of PEMFC MEA models, a mathematical model capable of reproducing, with the same parameter set, experimentally measured polarization curves, cyclic voltammograms, linear sweeps at varying scan rates, electrochemical impedance spectra, liquid-water crossover and long-term physical, chemical and electrochemical degradation does not exist.

The lack of a consistent set of input parameters in PEMFC MEA models is becoming inexcusable considering recent progress in porous-media and electrode characterization. Advances in X-ray and electron microscopy now allow users to generate three-dimensional reconstructions of the porous media with micro- and nanometer resolution, respectively, making it possible to estimate transport properties in dry [21, 32, 55, 59–61, 83, 91, 92, 154, 156, 159] and partially saturated media [21, 32, 59, 60, 91, 92, 119, 154, 156, 159, 215]. Furthermore, electrochemical cyclic voltammetry using hydrogen and carbon monoxide can provide accurate estimations of active area [62, 186], and rotating-disk and floating electrodes have provided new insights regarding the oxygen reduction and hydrogen oxidation reactions (ORR and HOR) [117, 123, 210].

The development of multidimensional PEMFC MEA models requires not only the knowledge of mass transport and electrochemistry, but also numerical analysis, high-performance computing and programming languages. The solution of a variety of physical equations coupled with highly nonlinear source terms in multiple dimensions and complex geometries, e. g., a catalyst-layer microstructure, requires a careful understanding of mesh generation, discretization techniques, such as the finite-element method, and careful implementation into a general framework. The development of such a framework is time-consuming and, generally, beyond the scope of a single individual's contribution. It is for this reason that the open-source fuel cell simulation toolbox (OpenFCST) project was launched in 2013 with the aim of developing a multidimensional open-source framework to study PEMFCs. Over the past decade, OpenFCST has been extended to analyze effective transport properties and electrochemical performance in porous MEA components [91, 92, 156–159], to investigate liquid-water accumulation in MEAs [101, 221, 222], and to improve the understanding of the electrochemical impedance [101–103] and open-cell voltage [130] of PEMFCs. The aim of the developers is to make these extensions available in future releases. Access to the development version of the software is available upon request. To encourage collaboration between the fuel cell research communities around the world, OpenFCST is released under the MIT license [1] granting everyone the right to download, use, and modify the product to their needs.

This chapter provides a description of the governing equations implemented in the future (v1.0) and currently available releases of OpenFCST (v0.3). The objective of this chapter is to introduce the software to the community so that the readers interested in developing PEMFC MEA models can build upon and extend the basic governing equations and the readers interested in experiments can use OpenFCST to better interpret their results.

6.2 Governing equations and boundary conditions

Membrane-electrode-assembly models are required to fully relate electrochemical performance to mass, charge, heat, and water transport in fuel cells. Several MEA models have been proposed in the literature, and a thorough literature review on the different models was provided in [163, 197]. The aim of this section is to discuss the assumptions and governing equations of the two-phase and single-phase MEA models in OpenFCST. The current release, v0.3, contains a steady-state version of the model; however, a transient version of the code [101–103], as well as a two-phase MEA model [29, 101, 220–222], have already been implemented and will be a part of the next OpenFCST release. Therefore, the governing equations of the single-phase model are given in transient form, and the governing equations of the implemented two-phase flow model are also provided. Further, a generalized fluid-flow framework for channel-porous media interactions has also recently been implemented in OpenFCST [13, 87]. Therefore, these equations are also provided, even though a full coupling of the fluid-flow equations to species mass transport and electrochemical reactions, necessary to develop an MEA model, is still pending.

6.2.1 Transport in channels and porous media

Reactants and products are transported in fuel cells between the bipolar plates and the reaction sites via a collection of porous media. These media are designed to effectively carry mass, charge, and heat. The governing equations for the corresponding transport processes are discussed next.

6.2.1.1 Characterization

There are three main porous media in fuel cells: (a) gas-diffusion layers (GDLs; also known as porous transport layers), (b) microporous layers (MPLs), and (c) catalyst layers (CLs). Each one of these layers has a different composition, microstructure, and, as a result, different effective transport properties. In OpenFCST, transport properties can be input directly from experimental results or estimated based on the layer composition and, in the next release, pore-size distribution. Image-to-simulation tools are also available to estimate transport properties directly from tomography images [91, 92, 156–159].

GDLs are highly porous layers of 100–300 μm in thickness made of 5–15-μm carbon fibers. These layers are usually impregnated with a hydrophobic polymer, such as PTFE, to increase their hydrophobicity. MPLs are 10–50-μm porous layers usually fabricated by direct application of a slurry of 20–50-nm carbon-black nanoparticles, usually in aggregated state, and a polymer binder, usually PTFE, to the GDL. For both layers, the volume fractions of each phase are related by

$$\epsilon_V + \epsilon_S + \epsilon_{PTFE} = 1 \tag{1}$$

where ϵ_V is porosity, ϵ_S is the volume fraction occupied by the electrically conductive material (carbon fibers in GDLs and carbon particles in MPLs), and ϵ_{PTFE} is the volume fraction occupied by the binder.

Catalyst layers are 5–25-μm porous layers fabricated by applying a slurry/ink of supported catalyst, e. g., platinum on carbon nanoparticles, and ionomer, e. g., Nafion®, onto either the membrane or the MPL. In order to compute the effective properties of the CL, the volume fraction of each material in the catalyst layer, i. e., solid (support and catalyst), electrolyte and void space, need to be obtained. The CL ink composition and layer thickness must dictate the volume fraction of each one of these materials in the CL; therefore, OpenFCST has the option of providing the values directly or computing them based on the ink composition. The solid-phase volume fraction can be obtained from the platinum mass loading, m_{Pt} (g/cm²), and the mass ratio of platinum to the carbon/platinum mixture, Pt|C, using [171]

$$\epsilon_S^{CL} = \left(\frac{1}{\rho_{Pt}} + \frac{1}{\rho_c} \left[\frac{1 - Pt|C}{Pt|C} \right] \right) \frac{m_{Pt}}{L},$$

where ρ_{Pt} and ρ_c are the platinum and carbon densities (g/cm³) and L is the catalyst-layer thickness (cm). This last parameter is not known a priori, but it can be controlled during the electrode preparation [68] or obtained a posteriori.

The electrolyte volume fraction can also be easily obtained based on the weight percentage of Nafion® in the catalyst ink, w_N (Nafion® mass divided by the mass of Pt, carbon and Nafion®) [171]:

$$\epsilon_N^{CL} = \frac{1}{\rho_N} \frac{w_N}{(1-w_N)} \frac{1}{Pt|C} \frac{m_{Pt}}{L},$$

where the ionomer can be assumed to be covering the carbon particles or to be distributed inside and partly covering the carbon aggregates. Even though the latter was initially favored [164, 175] and might be a useful representation for non-PGM catalysts [12], microscopy images [120, 205] suggest that the idea of a film partially covering the supported catalyst particles may be more appropriate.

Once the electrolyte volume fraction, ϵ_N^{CL}, is estimated, porosity can be readily obtained using

$$\epsilon_V^{CL} = 1 - \epsilon_N^{CL} - \epsilon_S^{CL}.$$

In OpenFCST, the user can specify the platinum loading per unit volume, m_{Pt}/L, the platinum loading on support, $Pt|C$, and the electrolyte loading, w_N; the volume fractions are computed automatically. The amount of ionomer inside the agglomerates can also be specified if needed. Furthermore, for experimentalists' convenience, the ionomer-to-carbon ratio can be used instead of the Nafion®/electrolyte loading.

6.2.1.2 Gas transport

Gas transport in porous media is driven by both convection and diffusion. Convection can usually be neglected for parallel-channel flow fields; however, it is important in case of serpentine and interdigitated channels as shown by Pharoah [147] and, more recently, using OpenFCST, by Jarauta et al. [87]. When convection is considered, density, pressure, and velocity of the flow need to be obtained by solving the Navier–Stokes equations in the regular form in the channels and in the volume-averaged form in the porous media.

The following unified set of governing equations for the flow of a compressible fluid, which can be used in both channel and porous media, was recently implemented, in steady-state form, in OpenFCST and validated by Jarauta et al. [87]:

$$\frac{\partial \rho}{\partial t} + \nabla \cdot (\rho \mathbf{v}) = 0 \quad \text{in } \Omega, \tag{2}$$

$$\frac{\partial (\rho \mathbf{v})}{\partial t} + \nabla \cdot (\rho \mathbf{v} \otimes \mathbf{v}) = \nabla \cdot (-p\hat{\mathbf{I}} + \hat{\sigma}) + \mathbf{F} + \rho \mathbf{g} \quad \text{in } \Omega, \tag{3}$$

where ρ is density (kg/m³), $\mathbf{v}$ is velocity (m/s), p is pressure (Pa), $\hat{\mathbf{I}}$ is the unit tensor (–), $\hat{\sigma}$ is the shear-stress tensor (Pa), $\mathbf{F}$ is the drag force (Pa/m), and $\mathbf{g}$ is the gravitational acceleration (m/s²). Density and velocity fields are volume-averaged in the porous media. In order to achieve a formulation without discontinuities in either density or velocity, the superficial average density and the intrinsic average velocity are used in the derivation of the volume-averaged transport equations such that

$$\rho := \begin{cases} \rho & \text{in } \Omega_c; \\ \langle\rho\rangle = \frac{1}{V_{REV}} \int_{V_f} \rho \, dV & \text{in } \Omega_p, \end{cases}$$

$$v := \begin{cases} v & \text{in } \Omega_c; \\ \langle v\rangle^f = \frac{1}{V_f} \int_{V_f} v \, dV & \text{in } \Omega_p, \end{cases}$$

where V_{REV} and V_f are the representative elementary volume in the porous media and the volume of the fluid, and Ω_c and Ω_p represent the channel and porous-medium domains, respectively.

The stress tensor for compressible Newtonian fluids, $\hat{\sigma}$, is given by

$$\hat{\sigma} = 2\mu\nabla_s v - \frac{2}{3}\mu(\nabla \cdot v)\hat{I}, \tag{4}$$

where μ is the dynamic viscosity (kg/(m · s)) of the fluid (a function of the mixture composition), and $\nabla_s = (\nabla + \nabla^T)/2$ is the symmetric gradient.

The Darcy–Forchheimer drag force, F, is used to account for the resistance of the porous medium to the flow and is given by

$$F = \begin{cases} 0 & \text{in } \Omega_c; \\ -\mu\hat{K}^{-1}\epsilon_V v - \hat{\beta}\rho|\epsilon_V v|\epsilon_V v & \text{in } \Omega_p, \end{cases} \tag{5}$$

where $\hat{K}$ is the porous-matrix permeability (m^2) and $\hat{\beta}$ is the nonlinear Forchheimer correction tensor (m^{-1}).

Density and pressure are related by an equation of state. In this case, the ideal-gas law is used:

$$p = \rho RT/M,$$

where R is the gas constant (J/(mol · K)), M is the molar mass of the gas mixture (kg/mol) and T is temperature (K).

Using the equations above, mixture density, pressure and velocity can be solved for. Due to the dependence of density and viscosity on the composition of the mixture, the equations need to be solved iteratively with mass transport, even though in infinitely dilute mixtures it is commonly assumed that density and viscosity are constant. Note that the equations above are the most general governing equations and that, in some cases, simplifications might be possible. For example, if the Reynolds number in the porous medium is less than unity, the convective term, $\rho v \otimes v$, and the Forchheimer correction term could be neglected, resulting in the Brinkman equation.

Once mixture density, pressure, and velocity are known, the mass/molar fraction of the individual species in the mixture can be obtained using mass-/molar- and momentum-balance equations for all but one of the considered species. In the current version of the PEMFC MEA model, the mixture pressure, p, is assumed constant such that the velocity of the mixture, $\mathbf{v}$, can be considered negligible. Then the molar-balance equation of species i is given by

$$\frac{\partial}{\partial t}(\epsilon c_i) + \nabla \cdot \mathbf{N}_i = S_i, \tag{6}$$

where ϵ is the void fraction in the porous medium (–), $c_i = px_i/(RT)$ is the concentration of species i (mol/cm^3) with molar fraction x_i (–), $\mathbf{N}_i$ is the effective molar flux of species i (mol/(cm$^2 \cdot$ s)), and S_i represents source and sink terms (mol/(cm$^3 \cdot$ s)). The void fraction in the porous medium is a function of porosity, ϵ_V (–), and liquid saturation, s (–): $\epsilon = \epsilon_V(1 - s)$.

The flux $\mathbf{N}_i$ can be obtained using the Maxwell–Stefan equations for all $N - 1$ species [178]:

$$\nabla x_i + (x_i - \omega_i)\frac{\nabla p}{p} = -\sum_{j=1; i \neq j}^{N} \frac{x_j N_i - x_i N_j}{c_{tot} \mathcal{D}_{ij}},$$

where the external body forces have been neglected, ω_i is the mass fraction of species i (–), c_{tot} is the total mixture concentration (mol/cm^3), and $\mathcal{D}_{ij}$ is the binary diffusion coefficient between species i and j (cm^2/s).

In the current OpenFCST MEA model, in order to simplify the equations, infinite dilution is assumed so that the molar flux is given by [31]

$$\mathbf{N}_i = c_i \mathbf{v}_i = c_i \mathbf{v}^* - D_i^{\text{eff}} \nabla c_i, \tag{7}$$

where $c_i = c_{tot} x_i$ is the concentration of species i, $\mathbf{v}^*$ is the molar-averaged velocity (assumed negligible in the OpenFCST model), and D_i^{eff} is the effective diffusion coefficient (cm^2/s).

Substituting the molar-flux expression (7) into the molar-balance equation (6), the governing equation used for transport in the porous media is obtained:

$$\boxed{\frac{\partial}{\partial t}(\epsilon c_i) - \nabla \cdot (D_i^{\text{eff}} \nabla c_i) = S_i.} \tag{8}$$

This equation represents Fick's second law and is solved for $N - 1$ species, where the Nth species is considered to be the solvent. In the OpenFCST MEA model, hydrogen and nitrogen are considered solvents in the anode and cathode, respectively.

A more detailed discussion of the different mass-transport formulations in the context of fuel cells is given in Secanell et al. [162]. For a more thorough and general treatment of mass transport, see Bird et al. [31] and Taylor and Krishna [178].

To compute the effective diffusivity D_i^{eff}, its bulk value, D_i, must be known. The latter is computed in OpenFCST using the Bosanquet approximation [181, 197, 220]:

$$\frac{1}{D_i} = \frac{1}{D_i^m} + \frac{1}{D_i^K},\tag{9}$$

where D_i^m is the molecular diffusivity and D_i^K is the Knudsen diffusivity. The former accounts for the molecular interaction in the gas mixture, while the latter accounts for the collisions of the gas molecules with the pore walls.

The molecular diffusion coefficient of species i in the mixture, D_i^m, can be estimated using the experimental relationship [52]

$$D_i^m = \frac{1 - x_i}{\sum_{j \neq i}^{N} \frac{x_j}{D_{ij}}},$$

where D_{ij} are the binary diffusion coefficients computed using the Chapman–Enskog theory [31] for the given pressure and temperature. In OpenFCST, however, since an infinitely dilute approximation is used, it is assumed that D_i^m is equal to the molecular diffusion coefficient of species i in the solvent.

The Knudsen diffusion coefficient, D_i^K, is given by [47, 76, 118]

$$D_i^K = \frac{2r_p}{3} \sqrt{\frac{8RT}{\pi M_i}},\tag{10}$$

where r_p is the pore radius (cm). Since Knudsen diffusivity is proportional to the pore radius, its contribution is only important in CLs and MPLs.

The effective diffusion coefficient of GDLs, MPLs and CLs is estimated using appropriate semiempirical equations that relate effective diffusivity, D^{eff}, to bulk diffusivity, D_i, void volume fraction, and tortuosity of the porous material. This is done by using [31]

$$D_i^{\text{eff}} = \frac{\epsilon}{\tau} D_i,\tag{11}$$

where tortuosity, τ, is given by the generalized Archie's law [180, 181]:

$$\tau = \left(\frac{1 - \epsilon_{\text{p,th}}}{\epsilon - \epsilon_{\text{p,th}}} \right)^{\alpha}.\tag{12}$$

Here, $\epsilon_{\text{p,th}}$ is the percolation threshold and the exponent α depends on the material structure. No transport is assumed to take place if the void fraction of the medium is below $\epsilon_{\text{p,th}}$. This is enforced by multiplying equation (11) by a Heaviside function defined as

$$\Theta(\epsilon - \epsilon_{\text{p,th}}) = \begin{cases} 0, & \epsilon < \epsilon_{\text{p,th}}; \\ 1, & \epsilon \geq \epsilon_{\text{p,th}}. \end{cases}\tag{13}$$

Equations (11) and (12) can be applied to describe transport in various porous structures [179, 180]. Setting $\alpha = 1$ and $\alpha = 0.5$ with $\epsilon_{\text{p,th}} = 0$ results in the Bruggeman model for random porous media with cylindrical and spherical particles, respectively [179]. In the case of random in-plane orientation of cylindrical fibers (typical for nonwoven fuel cell GDLs), the in-plane and through-plane effective diffusivities can be obtained by taking $\alpha = 0.521$ and $\alpha = 0.785$, respectively, and $\epsilon_{p,th} = 0.11$ [181]. Other expressions with different values of the empirical constants in equation (11) for fuel cell materials are available in the literature [35, 53, 78, 112, 150]. For instance, the following relationship was proposed by Sabharwal et al. [156, 159] based on the analysis of stochastically reconstructed catalyst layers:

$$D_i^{\mathrm{eff}} = D_i \left(\frac{\epsilon - 0.05}{0.95} \right)^{1.9} \Theta(\epsilon - 0.05). \tag{14}$$

Other relationships are available in OpenFCST for computing effective transport properties of GDLs, MPLs, and CLs that are particular cases of equation (11).

In addition to liquid saturation, ionomer swelling in catalyst layers can also change the pore volume available for gas transport. However, in the current version of Open-FCST, changes in ϵ_V and ϵ_N due to swelling are not considered.

The source term of species i, S_i, is included in order to account for electrochemical reactions. Considering the current density per unit volume of catalyst layer is known, the reaction source term is governed by Faraday's law,

$$S_i = \pm \frac{j}{n_i F}, \tag{15}$$

where the positive and negative signs denote production and consumption of species i, respectively, j is the current produced per unit volume of the electrode (A/cm^3), n_i is the number of electrons produced/consumed to produce/consume a mole of species i, and $F = 96{,}485$ C/mol is Faraday's constant. For water vapor, term (15) is multiplied by a factor χ that controls whether water is produced in the vapor form ($\chi = 1$) or liquid form ($\chi = 0$, in which case the source term is moved to the liquid-water-transport equation):

$$S_{\mathrm{wv}} = \begin{cases} \chi \frac{j}{2F} & \text{in CCL;} \\ 0 & \text{everywhere else,} \end{cases} \tag{16}$$

where "CCL" stands for "cathode catalyst layer" (similarly, "ACL" refers to the anode catalyst layer). In this context, production of water in vapor form means immediate vaporization of the produced liquid water.

6.2.1.3 Electron transport

Electron transport in the porous electrodes of the MEA occurs through the carbon fiber in GDLs, carbon particles in MPLs, and through the carbon/catalyst phase in CLs. The volume-averaged charge-conservation equation for electrons in those domains is given by

$$\epsilon_S \frac{\partial \hat{\rho}_{e^-}}{\partial t} + \nabla \cdot \boldsymbol{i}_{e^-} = S_{e^-}, \tag{17}$$

where $\hat{\rho}_{e^-}$ is the electron (free-charge) density (C/cm^3), $\boldsymbol{i}_{e^-}$ is the electronic current density (A/cm^2), and

$$S_{e^-} = \begin{cases} j & \text{in CCL;} \\ -j & \text{in ACL;} \\ 0 & \text{everywhere else} \end{cases}$$

is an electron source/sink term due to electrochemical reactions (A/cm^3).

The transient term in equation (17) represents charging/discharging current. No charge is accumulated in the gas-diffusion layers, microporous layers, and in the membrane; therefore, this term is zero in those parts of the MEA. However, such current exists in catalyst layers, where the interface between the electronically conductive carbon/catalyst phase and the protonically conductive ionomer phase, usually referred to as the double layer, acts as a capacitor [18, 124, 139, 148]. Volume-averaged double-layer capacitance (F/cm^3) is defined as [124, 148]

$$C_{dl} = \epsilon_S \frac{\partial \hat{\rho}_{e^-}}{\partial (\phi_{e^-} - \phi_{H^+})}, \tag{18}$$

where ϕ_{e^-} and ϕ_{H^+} (V) are the potentials of the electronically and protonically conductive phases, respectively.

Therefore, the temporal derivative in equation (17) can be rewritten in the following form:

$$\epsilon_S \frac{\partial \hat{\rho}_{e^-}}{\partial t} = \epsilon_S \frac{\partial \hat{\rho}_{e^-}}{\partial (\phi_{e^-} - \phi_{H^+})} \frac{\partial (\phi_{e^-} - \phi_{H^+})}{\partial t} = C_{dl} \frac{\partial (\phi_{e^-} - \phi_{H^+})}{\partial t}. \tag{19}$$

This result represents the capacitive volumetric current density in the electronically conductive phase, $j_{c,e^-} = C_{dl}\partial(\phi_{e^-} - \phi_{H^+})/\partial t$.

The electronic current density, $\boldsymbol{i}_{e^-}$, is given by Ohm's law, i. e.,

$$\boldsymbol{i}_{e^-} = -\sigma_{e^-}^{eff} \nabla \phi_{e^-}, \tag{20}$$

where the effective electronic conductivity, $\sigma_{e^-}^{eff}$ (S/cm), depends on the structure and composition of the medium.

Combining equations (17), (19) and (20) results in the second-order governing equation for electron transport implemented in OpenFCST:

$$C_{dl} \frac{\partial(\phi_{e^-} - \phi_{H^+})}{\partial t} - \nabla \cdot (\sigma_{e^-}^{eff} \nabla \phi_{e^-}) = S_{e^-}. \tag{21}$$

A similar equation can be derived for proton transport in the electrolyte (see Section 6.2.2).

The effective electronic conductivity that appears in equation (21) can either be assigned a constant value in OpenFCST or be estimated based on the solid volume fraction using one of the available semiempirical formulas. The most common method for obtaining effective conductivities in porous media is to use equations (11) and (12) or the equivalent percolation model [48–50, 84, 174, 180, 212]:

$$\sigma_{e^-}^{eff} = \sigma_{e^-} \left(\frac{\epsilon_S - \epsilon_{S,th}}{1 - \epsilon_{S,th}} \right)^{\mu} \Theta(\epsilon_S - \epsilon_{S,th}), \tag{22}$$

where σ_{e^-} is the electronic conductivity of the bulk conducting material (usually carbon fibers or black supporting the catalyst) with volume fraction ϵ_S, and $\epsilon_{S,th}$ and μ are the percolation threshold and the universal exponent determined based on the percolating network structure.

In order to achieve realistic effective parameters in the GDL, either experiments or direct numerical simulations should be used that take into account the anisotropic nature of the fibrous media. OpenFCST contains a microscale simulation module to generate a mesh from a three-dimensional porous-medium reconstruction and to estimate electronic conductivity of the material. The percolation threshold and the universal exponent in equation (22) can then be adjusted so that the percolation model reproduces the experimentally measured effective conductivity or the conductivity that was extracted from pore-scale simulations.

6.2.1.4 Heat transport

Even if the cell is operated at a constant temperature, local thermal variations occur within the MEA due to a number of phenomena, such as reaction heat and phase change of water. Elevated temperature improves reaction rates and helps evaporate liquid water but reduces electrolyte conductivity due to dehydration. Therefore, thermal management in fuel cells is crucial, which motivates the inclusion of heat transport in mathematical models of PEMFCs.

The energy-conservation equation for a Newtonian fluid with constant heat capacity is given by [31]

$$\rho C_p \frac{DT}{Dt} = -\nabla \cdot \mathbf{q} - \boldsymbol{\tau} : \nabla \mathbf{v} - \left(\frac{\partial \ln \rho}{\partial \ln T} \right)_p \frac{Dp}{Dt}. \tag{23}$$

The heat flux, $\boldsymbol{q}$ (W/cm^2), combines Fourier conduction, enthalpy transport due to molecular interdiffusion, and a heat flux due to thermal transport with diffusion (Dufour effect) [29]:

$$q = -\kappa\nabla T + \sum_{i=\text{species}} \overline{H}_i N_i + q_D, \tag{24}$$

where κ is thermal conductivity (W/(cm $\cdot$ K)), $\overline{H}_i$ is molar enthalpy of species i (J/mol), and N_i is molar flux. Flux $\boldsymbol{q}_D$ can be neglected, as the effect of diffusion on heat flux is weak in PEMFCs [118, 201].

The term $\boldsymbol{\tau} : \nabla\boldsymbol{v}$ in equation (23), with $\boldsymbol{\tau}$ being the shear stress in the fluid, describes heat generation due to viscous dissipation [31]. This term is significant only for fluids with large viscosity or large velocity gradients [31], and, based on dimensional analysis, it is likely negligible in fuel cells.

Considering the assumptions above, and since gas pressure is assumed constant in OpenFCST, equation (23) for the gas phase becomes

$$\rho C_p \left(\frac{\partial T}{\partial t} + \boldsymbol{v} \cdot \nabla T \right) = -\nabla \cdot \boldsymbol{q}, \tag{25}$$

where the material derivative was expanded.

Rigorous treatment of heat transport calls for the use of multiple energy-conservation equations, one for each material phase in the MEA. Due to the large difference in the thermal conductivity of the solid, liquid, and gas phases in a fuel cell, a two- or three-equation approach can be taken, where the temperature of the solid and fluid phases is considered different [79–81, 93]. However, it is far more common to assume local thermal equilibrium between all phases in a porous medium due to its large interfacial area [4, 14, 15, 29, 30, 36, 37, 57, 64–66, 69, 70, 77, 89, 97, 99, 100, 114, 115, 141, 146, 155, 166, 172, 203, 204, 213, 216–222]. In that case, a single volume-averaged energy-conservation equation is used:

$$\sum_{i=\text{phase}} \rho_i C_{p,i} \frac{\partial(\epsilon_i T)}{\partial t} = -\nabla \cdot \boldsymbol{q}^{\text{eff}}, \tag{26}$$

where ϵ_i is the volume fraction of phase i and $\boldsymbol{q}^{\text{eff}}$ is the effective heat flux in a porous medium. Finally, noting that Bhaiya [29] determined, based on dimensional analysis, that convective heat transfer in the gas phase can be neglected compared to that due to conduction and interdiffusion, one can substitute equation (24) into the equation above and add a source/sink term to arrive at

$$\sum_{i=\text{phase}} \rho_i C_{p,i} \frac{\partial(\epsilon_i T)}{\partial t} - \nabla \cdot (\kappa^{\text{eff}}\nabla T) + \sum_{i=\text{gas}, \lambda} \nabla \cdot (N_i \overline{H}_i) = \hat{S}_T,$$

or, rearranging,

$$\sum_{i=\text{phase}} \rho_i C_{p,i} \frac{\partial(\epsilon_i T)}{\partial t} - \nabla \cdot (\kappa^{\text{eff}} \nabla T) + \sum_{i=\text{gas},\lambda} (N_i \cdot \nabla \overline{H}_i) = \hat{S}_T - \sum_{i=\text{gas},\lambda} (\overline{H}_i \nabla \cdot N_i)$$

$$= S_T, \tag{27}$$

where κ^{eff} is the effective thermal conductivity and the molar flux N_i is given by equations (7) and (50). The summation in the first term of equation (27) is performed over the solid, liquid, and gas phases; the summation in the third term and in the right-hand side is performed over the gas phase and the absorbed water. Equation (27) is used in OpenFCST to describe heat transport in fuel cells.

The right-hand side of equation (27) contains a number of terms that describe volumetric heat release and absorption in a fuel cell. These terms are discussed next.

The overall PEMFC reaction (72) is exothermic [51, 108], and most of the reversible heat is believed to be generated in the ORR [29, 202]. While the exact distribution of the reversible heat among the HOR (70) and ORR (71) is not known, the entropy change per mole of H_2 due to the overall reaction (72) can be found via [29]

$$\Delta \overline{S}_{\text{overall}} = 4.184(8(1 + \ln(T)) - 92.84) \text{ J/(mol} \cdot \text{K)}.$$

A coefficient f_{ORR} is used in OpenFCST to relate the entropy change in the two half-cell reactions to $\Delta \overline{S}_{\text{overall}}$: $\Delta \overline{S}_{\text{ORR}} = f_{\text{ORR}} \Delta \overline{S}_{\text{overall}}$ and $\Delta \overline{S}_{\text{HOR}} = (1 - f_{\text{ORR}}) \Delta \overline{S}_{\text{overall}}$. In that case, source terms due to reversible heat generation are given by [29, 30]

$$S_{T,\text{rev,ORR}} = \begin{cases} \frac{j}{2F}(-T f_{\text{ORR}} \Delta \overline{S}_{\text{overall}}) & \text{in CLs;} \\ 0 & \text{everywhere else,} \end{cases} \tag{28}$$

$$S_{T,\text{rev,HOR}} = \begin{cases} \frac{j}{2F}(-T(1 - f_{\text{ORR}}) \Delta \overline{S}_{\text{overall}}) & \text{in CLs;} \\ 0 & \text{everywhere else.} \end{cases} \tag{29}$$

Since electrochemical reactions take place only in catalyst layers, source terms (28) and (29) are zero in other MEA components.

Production of electrochemical current is accompanied with voltage loss (overpotential η) due to kinetic, ohmic, and mass-transport inefficiencies of a fuel cell. This voltage loss is transformed into irreversible heat, the amount of which is found by taking a product of the current density and the overpotential. The corresponding source terms in the cathode and anode catalyst layers are [29, 30]

$$S_{T,\text{irrev,ORR}} = \begin{cases} -j\eta = -j(\phi_{e^-} - \phi_{H^+} - E_{\text{ORR}}) & \text{in CCL;} \\ 0 & \text{everywhere else,} \end{cases}$$

$$S_{T,\text{irrev,HOR}} = \begin{cases} j\eta = j(\phi_{e^-} - \phi_{H^+} - E_{\text{HOR}}) & \text{in ACL;} \\ 0 & \text{everywhere else,} \end{cases}$$

respectively, where E_{ORR} and E_{HOR} are the reference half-cell potentials (V) for the ORR and HOR.

Joule (ohmic) heating takes place in all electrically conductive materials as current flows through them. The amount of the generated heat is equal to the product of the squared current and the resistance of the material. In differential form, this is described by the following source term [29, 30]:

$$
S_{T,\text{ohmic}} = \begin{cases} \sigma_{H^+}^{\text{eff}}(\nabla\phi_{H^+} \cdot \nabla\phi_{H^+}) & \text{in PEM;} \\ \sigma_{H^+}^{\text{eff}}(\nabla\phi_{H^+} \cdot \nabla\phi_{H^+}) + \sigma_{e^-}^{\text{eff}}(\nabla\phi_{e^-} \cdot \nabla\phi_{e^-}) & \text{in CLs;} \\ \sigma_{e^-}^{\text{eff}}(\nabla\phi_{e^-} \cdot \nabla\phi_{e^-}) & \text{in MPLs and GDL.} \end{cases}
\tag{30}
$$

Equation (30) contains one or two terms depending on whether the medium conducts protons, electrons, or both.

Calculation of Joule heating allows for in-operando estimation of ohmic resistance of MEA components in OpenFCST using the following equations [101–103, 165, 223] (in $\Omega \cdot cm^2$):

$$
R_{H^+}^{\text{eff}} = \frac{1}{i^2 A} \int \sigma_{H^+}^{\text{eff}} \nabla\phi_{H^+} \cdot \nabla\phi_{H^+} \, dV,
\tag{31}
$$

$$
R_{e^-}^{\text{eff}} = \frac{1}{i^2 A} \int \sigma_{e^-}^{\text{eff}} \nabla\phi_{e^-} \cdot \nabla\phi_{e^-} \, dV,
\tag{32}
$$

where i is the current density normalized per in-plane area A of the MEA (A/cm^2). This enables several unique analysis capabilities, such as thorough model calibration and validation with both experimental polarization and resistance data [101, 102, 165, 223], ohmic-resistance breakdown of an MEA [101, 102], and relating effective conductivity, ohmic resistance, and impedance of catalyst layers [101, 103].

The last two heat sources/sinks considered in the heat-transfer equation in Open-FCST are associated with the changes in the physical state of water. The source and sink terms describing the heat produced/released during water absorption/desorption from the electrolyte (heat of absorption/desorption) are given by [29, 30]:

$$
S_{T,\text{sorption}} = \begin{cases} \epsilon_N k_\lambda \frac{\rho_{N,\text{dry}}}{EW}(\lambda_{eq} - \lambda)\overline{H}_{\text{sorption}} & \text{in CLs;} \\ 0 & \text{everywhere else,} \end{cases}
\tag{33}
$$

where $\overline{H}_{\text{sorption}}$ is the enthalpy change in the absorption/desorption process (J/mol).

The source/sink term describing condensation/evaporation of water is obtained by multiplying the phase-change term, equation (39), by the molar latent heat of water vaporization, $\overline{H}_{lv}$ (J/mol) [29, 221]:

$$
S_{T,\text{evap/cond}} = \begin{cases} S_{H_2O}\overline{H}_{lv} & \text{in CLs, MPLs, GDLs;} \\ 0 & \text{in PEM.} \end{cases}
$$

The overall source/sink term in the heat-transfer equation (27) is

$$S_T = S_{T,\text{ohmic}} + S_{T,\text{sorption}} + S_{T,\text{evap/cond}}$$
$$+ S_{T,\text{rev,ORR}} + S_{T,\text{irrev,ORR}} + S_{T,\text{rev,HOR}} + S_{T,\text{irrev,HOR}}. \tag{34}$$

Further details on the thermal-transport model in OpenFCST can be found in [29, 30].

6.2.1.5 Liquid-water transport

Water in the porous components of the MEA can be in one of three states: gas, liquid, and ice. The current version of OpenFCST is aimed at applications at temperature above $0\,°C$, and thus the description of the formation and melting/sublimation of ice is not supported.

Under the assumption of a creeping flow of an incompressible Newtonian fluid, flow of liquid water through the porous media of fuel cells is commonly described with the Darcy equation [29, 43, 65, 66, 70, 134, 141, 166, 213, 214, 216, 220–222]:

$$\epsilon_V \frac{\partial(s\rho_1)}{\partial t} - \nabla \cdot \left(\frac{\rho_1 \kappa_1}{\mu_1} \nabla p_1 \right) = S_1, \tag{35}$$

where ρ_1, κ_1, p_1 are density, permeability and intrinsic average pressure of liquid water. Source/sink term S_1 $(g/(\text{cm}^3 \cdot s))$ describes liquid-water production, evaporation and condensation.

In fuel cell literature, capillary pressure is defined as the difference between the liquid and gas pressure, $p_c = p_1 - p_g$ [29, 65, 70, 141, 213, 214, 216, 220–222]. Under the assumption of constant gas pressure, it follows from this definition that $\nabla p_c = \nabla p_1$, and equation (35) can be rewritten in terms of capillary pressure instead of liquid pressure:

$$\boxed{\epsilon_V \frac{\partial(s\rho_1)}{\partial t} - \nabla \cdot \left(\frac{\rho_1 \kappa_1}{\mu_1} \nabla p_c \right) = S_1.} \tag{36}$$

It is common to solve equation (36) with respect to liquid saturation, s [29, 43, 66, 134, 166]. In that case, equation (36) is replaced with

$$\epsilon_V \frac{\partial(s\rho_1)}{\partial t} - \nabla \cdot \left(\frac{\rho_1 \kappa_1}{\mu_1} \frac{\partial p_c}{\partial s} \nabla s \right) = S_1, \tag{37}$$

where derivative $\partial p_c/\partial s$ is computed with one of the empirical or experimentally measured relationships between capillary pressure and saturation that are available in the literature [11, 29, 65, 66, 134, 141, 166, 193, 213, 214, 216]. Equation (37), implemented by Bhaiya et al. [29, 30] at steady state, is available in OpenFCST v0.3. Solution of equation (37) results in a continuous distribution of saturation in the porous layers of the

MEA, unless special treatment is taken, such as the use of the discontinuous Galerkin method as opposed to the more common continuous method used in OpenFCST. However, saturation is likely to be discontinuous between the layers due to the significant variation in the average pore size. Additionally, it is the liquid-pressure gradient, not the saturation gradient, that is the driving force for the liquid-water transport. Therefore, an alternative approach, where equation (36) is solved directly, is preferred.

Since the OpenFCST v0.3 release, a mixed-wettability pore-size-distribution-based model was developed by Zhou et al. [220, 221] based on the earlier works by Weber et al. [195, 196, 199] and Mateo Villanueva [125]. Similar models were also proposed by Eikerling [48], Balliet and Newman [14–16], Mulone and Karan [133], and Goshtasbi et al. [70]. The model implemented in OpenFCST enables the computation of saturation and of the effective transport properties, such as permeability and liquid-gas area, through integration of the experimentally measured pore-size distributions (PSD) of fuel cell materials. This calculation also takes the wettability information, i. e., the contact angles of hydrophilic and hydrophobic PSD modes, into account. This allows for the direct relation of the transport properties to the microstructural parameters of the porous MEA layers without the need for empirical relationships between capillary pressure and saturation, and equation (36) is used instead of equation (37). This makes it possible to analyze the effects of the electrode microstructure on fuel cell performance using OpenFCST [220–223]. The steady-state model by Zhou et al. [220, 221] was recently extended to the transient form by Kosakian [101].

Liquid-water production in the CCL and evaporation/condensation in the porous layers of the MEA are modeled through the source/sink term of equation (35), (36) or (37):

$$
S_1 = \begin{cases} (1-\chi)\frac{j}{2F}M_{H_2O} + M_{H_2O}S_{H_2O} & \text{in CCL;} \\ M_{H_2O}S_{H_2O} & \text{in ACL, GDLs, MPLs;} \\ 0 & \text{in PEM,} \end{cases} \tag{38}
$$

where

$$
S_{H_2O} = \begin{cases} k_e a_{l\text{-}g}(\frac{p_v - p_K^{sat}}{p_K^{sat}}), & p_v \le p_K^{sat}; \\ k_c a_{l\text{-}g}(\frac{p_v - p_K^{sat}}{p_K^{sat}}), & p_v > p_K^{sat}, \end{cases} \tag{39}
$$

k_e and k_c are the evaporation and condensation rate constants (mol/(cm^2 s)), $a_{l\text{-}g}$ is the interfacial area between liquid and gas phases (cm$^2_{l\text{-}g}$/cm$^3_{layer}$), and p_v and p_K^{sat} are vapor pressure and saturated-vapor pressure (Pa). The latter quantity is computed through the Kelvin equation to account for the curvature of the liquid-gas interface [220, 221]:

$$
p_K^{sat} = p^{sat} \exp\left(\frac{p_c M_{H_2O}}{\rho_l RT} \right),
$$

where [173]

$$\log_{10}(p^{\text{sat}}) = -2.1794 + 0.02953(T - 273.15)$$
$$- 9.1837 \cdot 10^{-5}(T - 273.15)^2 + 1.4454 \cdot 10^{-7}(T - 273.15)^3. \tag{40}$$

The units of pressure in equation (40) are atm. Source term (16) of the vapor-transport equation needs to be modified as follows to account for the phase change of water:

$$S_{\text{wv}} = \begin{cases} \chi \frac{j}{2F} - S_{H_2O} & \text{in CCL;} \\ -S_{H_2O} & \text{in ACL, GDLs, MPLs;} \\ 0 & \text{in PEM.} \end{cases} \tag{41}$$

6.2.2 Transport in the electrolyte

Transport of charged species in an electrolyte is described with a charge-balance equation and a flux expression derived from either concentrated-solution or diluted-solution theory. The former is more general, but it necessitates more knowledge of the interactions among the various species in the solution [135, 198]. Using the diluted-solution approximation, the flux of charged ions (mol/(cm$^2 \cdot$ s)) is given by the Nernst–Planck equation [18, 24, 38, 135, 183]:

$$N_i = -z_i \frac{F}{RT} c_i D_i \nabla \phi - D_i \nabla c_i + c_i v, \tag{42}$$

where z_i is the charge number for species i (–), c_i is the concentration of species i (mol/cm^3), D_i is the diffusion coefficient of species i in the solvent (cm^2/s), and v is the molar-averaged mixture velocity (cm/s). Multiplication of equation (42) by $z_i F$ and summation over species i gives current density (A/cm^2)

$$i = F \sum_i z_i N_i. \tag{43}$$

Using the definition of conductivity (S/cm),

$$\sigma = F^2 \sum_i z_i^2 c_i u_i = F^2 \sum_i z_i^2 c_i \frac{D_i}{RT},$$

and assuming electroneutrality, i. e.,

$$\sum_i z_i c_i = 0, \tag{44}$$

equation (43) is transformed into

$$i = -\sigma \nabla \phi - F \sum_i z_i D_i \nabla c_i. \tag{45}$$

In most polymer electrolytes used in PEMFCs, such as Nafion®, only two charged species are present: protons and immovable polymer functional groups (in the case of Nafion®, those are perfluorovinyl ether groups terminated with sulfonate groups). Assuming uniform concentration of the functional groups, equation (45) simplifies to Ohm's law,

$$i_{H^+} = -\sigma_{H^+}\nabla\phi_{H^+},\tag{46}$$

where all the current is due to the mobility of protons, ϕ_{H^+} (V) is the potential of the electrolyte, and σ_{H^+} (S/cm) is its protonic conductivity. In catalyst layers, protonic current density is a volume-averaged quantity given by

$$i_{H^+} = -\sigma_{H^+}^{eff}\nabla\phi_{H^+},\tag{47}$$

where $\sigma_{H^+}^{eff}$ is the effective protonic conductivity of the layer.

The transient term of the volume-averaged charge-conservation equation (17), written for the protonically conductive phase, can be rewritten in terms of the double-layer capacitance:

$$\epsilon_N\frac{\partial\hat{\rho}_{H^+}}{\partial t} = \epsilon_N\frac{\partial\hat{\rho}_{H^+}}{\partial(\phi_{H^+}-\phi_{e^-})}\frac{\partial(\phi_{H^+}-\phi_{e^-})}{\partial t} = -C_{dl}\frac{\partial(\phi_{e^-}-\phi_{H^+})}{\partial t}.\tag{48}$$

Substituting results (47) and (48) into the charge-conservation equation for a porous medium, one obtains

$$-C_{dl}\frac{\partial(\phi_{e^-}-\phi_{H^+})}{\partial t} - \nabla\cdot(\sigma_{H^+}^{eff}\nabla\phi_{H^+}) = S_{H^+}.\tag{49}$$

Term $S_{H^+} = -S_{e^-}$ (A/cm³) represents production and consumption of protons in HOR and ORR, respectively. Equation (49) is used in OpenFCST to describe transport of protons in the membrane and in the ionomer phase of catalyst layers.

The protonic conductivity of the electrolyte, $\sigma_{H^+}^{eff}$, depends on its water content. In order to estimate the amount of water in the electrolyte, a molar-balance equation and an expression for the water flux are required. A generalized expression for the water flux in the electrolyte is still not available, and several models have been proposed, e. g., those in [24, 25, 28, 160, 170, 173, 176]. In OpenFCST, the approach proposed by Springer et al. [173] is used. In this model, it is assumed that the electrolyte is homogeneous and nonporous and that transport of water in sorbed form is driven by electroosmosis and back-diffusion. In addition, thermoosmotic transport is also included [29, 30]. Then the total sorbed-water flux in the electrolyte is given by

$$N_\lambda = N_{\lambda,\text{electroosmosis}} + N_{\lambda,\text{diffusion}} + N_{\lambda,\text{thermoosmosis}}$$

$$= -n_d \frac{\sigma_{H^+}^{\text{eff}}}{F} \nabla\phi_{H^+} - \frac{\rho_{N,\text{dry}}}{EW} D_\lambda^{\text{eff}} \nabla\lambda - \frac{D_T^{\text{eff}}}{M_{H_2O}} \nabla T, \tag{50}$$

where $\rho_{m,\text{dry}}$ (g/cm^3) and EW (g/mol$_{SO_3^-}$) are the dry density and the equivalent weight (mass of ionomer per mole of ionic groups) of the electrolyte; n_d (–) is the electroosmotic drag coefficient; D_λ^{eff} (cm^2/s) and D_T^{eff} (g/(cm $\cdot$ s $\cdot$ K)) are the coefficients of back-diffusion and thermoosmosis of water in the electrolyte. Equation (50) is solved with respect to λ (mol$_{H_2O}$/mol$_{SO_3^-}$) that represents the ratio of moles of water per moles of ionic groups in the electrolyte.

Introducing flux (50) into the molar-balance equation for sorbed water, one obtains

$$\epsilon_N \frac{\rho_{N,\text{dry}}}{EW} \frac{\partial\lambda}{\partial t} - \nabla \cdot \left(n_d \frac{\sigma_{H^+}^{\text{eff}}}{F} \nabla\phi_{H^+} + \frac{\rho_{N,\text{dry}}}{EW} D_\lambda^{\text{eff}} \nabla\lambda + \frac{D_T^{\text{eff}}}{M_{H_2O}} \nabla T \right) = S_\lambda, \tag{51}$$

where ϵ_N is the volume fraction of the electrolyte (unity in PEM). Term S_λ (g/(cm^3 $\cdot$ s)) describes the finite-rate exchange of water between the vapor (or liquid) phase and the electrolyte, as well as production of a portion of ORR water directly in the electrolyte phase:

$$S_\lambda = \begin{cases} \epsilon_N k_\lambda \frac{\rho_{N,\text{dry}}}{EW} (\lambda_{eq} - \lambda) + \xi \frac{j}{2F} & \text{in CCL;} \\ \epsilon_N k_\lambda \frac{\rho_{N,\text{dry}}}{EW} (\lambda_{eq} - \lambda) & \text{in ACL;} \\ 0 & \text{everywhere else,} \end{cases} \tag{52}$$

where k_λ is the exchange rate (1/s), λ_{eq} is the water content in a vapor-equilibrated or liquid-equilibrated electrolyte, and $\xi \in [0, 1]$ controls the amount of water produced in the CCL pores or in its ionomer. It was argued by Kosakian et al. [101, 102] that some of the water produced in the ORR enters the ionomer directly. A small ξ of 0.03 was sufficient for the PEMFC model in OpenFCST to predict changes in experimentally measured ohmic resistance of the cell with current density and relative humidity (RH) [101, 102].

Water transport in the vapor, liquid, and electrolyte phases is coupled. Therefore, the source/sink terms in the transport equations for vapor (41) and liquid water (38) must be modified to account for the water uptake by the ionomer and for the water production in the ionomer phase in the catalyst layers. In general, water uptake from the liquid form should be described by an interfacial-exchange term similar to equation (52). Because the corresponding exchange rate is unknown, coefficient ξ can be increased instead to approximate a higher uptake of the produced liquid water in the CCL. Thus, the modified terms (41) and (38) are given by

$$S_{wv} = \begin{cases} (1-\xi)\chi\frac{j}{2F} - S_{H_2O} - \epsilon_N k_\lambda \frac{\rho_{N,dry}}{EW}(\lambda_{eq} - \lambda) & \text{in CCL;} \\ -S_{H_2O} - \epsilon_N k_\lambda \frac{\rho_{N,dry}}{EW}(\lambda_{eq} - \lambda) & \text{in ACL;} \\ -S_{H_2O} & \text{in GDLs, MPLs;} \\ 0 & \text{in PEM} \end{cases} \tag{53}$$

and

$$S_l = \begin{cases} (1-\xi)(1-\chi)\frac{j}{2F}M_{H_2O} + M_{H_2O}S_{H_2O} & \text{in CCL;} \\ M_{H_2O}S_{H_2O} & \text{in ACL, GDLs, MPLs;} \\ 0 & \text{in PEM,} \end{cases} \tag{54}$$

where the evaporation/condensation term S_{H_2O} is defined in equation (39).

A number of different relationships for the transport properties appearing in equations (51) and (52) are available in OpenFCST. They are discussed next.

6.2.2.1 Water uptake

In order to describe the exchange of water between the electrolyte and the pore space with the source/sink term (53), the rate of absorption/desorption, k_λ, and the equilibrium water content in the electrolyte, λ_{eq}, must be known. The water-exchange rate is commonly treated as a calibration parameter in PEMFC models. The equilibrium water content is computed through a sorption isotherm that has been fitted to experimental water-uptake data for a given type of polymer electrolyte.

Two sorption isotherms are available in OpenFCST v0.3 (in $mol_{H_2O}/mol_{SO_3^-}$):

$$\lambda_{eq} = \begin{cases} 0.3 + 10.8a_w - 16.0a_w^2 + 14.1a_w^3, & 0 < a_w < 0.975; \\ 9.2, & a_w \geq 0.975; \end{cases} \tag{55}$$

$$\lambda_{eq} = \begin{cases} (13.41a_w - 18.92a_w^2 + 14.22a_w^3) \\ \quad \cdot [1.0 + 0.2352a_w^2 \frac{T-303.15}{30.0}], & 0 < a_w < 0.975; \\ 8.71 + 0.0682864(T - 303.15), & a_w \geq 0.975. \end{cases} \tag{56}$$

Here, $a_w = p_{wv}/p_{sat}$ is water activity (equilibrium RH) and saturation pressure is given by equation (40). The former isotherm was measured by Hinatsu et al. [75] for Nafion® 117 and 125 membranes. As water activity approaches unity, maximum water content in the CL ionomer is capped at 9.2 (maximum vapor-equilibrated water content in the original isotherm [75]) in single-phase models. Isotherm (56) was fitted by Mittelsteadt and Liu [126] to the uptake data of various PFSA-based membranes, including Nafion® 112. It is capped at a temperature-dependent λ_{eq} that matches the original isotherm [126] at $a_w \approx 1$.

Recently, Kosakian et al. [101, 102] proposed the following sorption isotherm:

$$\lambda_{eq} = (18.37a_w - 37.46a_w^2 + 31.70a_w^3) \cdot \exp\left[-66.28\left(\frac{1}{T} - \frac{1}{303.15}\right)\right]. \tag{57}$$

Equation (57) was fitted to water-uptake data for $0 < a_w < 1$ and a variety of Nafion®
membranes at different temperature values. The isotherm is capped at 22 $mol_{H_2O}/mol_{SO_3^-}$,
a water content in liquid-equilibrated PFSA-based polymer electrolytes [111]. Temperature dependency of equation (57) is weak, and the predicted isotherms at 25–80 °C differ by less than 7 % from the uptake curve measured by Zawodzinski et al. [211] for
Nafion® 117 at 30 °C.

Isotherms (55), (56) and (57) were all obtained for the water uptake by polymer-
electrolyte membranes and may not be suitable for simulating water uptake by the
ionomer phase of catalyst layers. Kosakian et al. [101, 102] fitted water-uptake curves
of catalyst layers and pseudo catalyst layers at 25–80 °C and $0 < a_w < 1$ [82, 90, 110] and
illustrated their significantly stronger temperature dependence:

$$\lambda_{eq} = (6.932a_w - 14.53a_w^2 + 11.82a_w^3) \cdot \exp\left[-2509\left(\frac{1}{T} - \frac{1}{303.15}\right)\right]. \tag{58}$$

Relationship (58) predicts an uptake similar to equation (57) at 80 °C and a significantly
lower uptake at 25 °C. Sorption isotherms (57) and (58) will be available in the next release of OpenFCST.

6.2.2.2 Effective absorbed-water diffusivity

Apart from a constant diffusion coefficient for water in the polymer electrolyte that can
be provided by the user, OpenFCST contains multiple relationships that can be used to
compute it for different water content and temperature. The first of these expressions
is the one proposed by Springer et al. [173] for Nafion® 117 membranes (in cm^2/s):

$$D_\lambda = 10^{-6}(2.563 - 0.33\lambda + 0.0264\lambda^2 - 0.000671\lambda^3) \tag{59}$$

$$\cdot \exp\left[2416\left(\frac{1}{303} - \frac{1}{T}\right)\right]. \tag{60}$$

This relationship is valid for $\lambda > 4\ mol_{H_2O}/mol_{SO_3^-}$ [173]. Motupally et al. [132] presented
a different expression for Nafion® 115 membranes that is also available in OpenFCST:

$$D_\lambda = \begin{cases} 3.10 \cdot 10^{-3}\lambda(\exp(0.28\lambda) - 1)\exp[-\frac{2436}{T}], & 0 < \lambda \le 3; \\ 4.17 \cdot 10^{-4}\lambda(161\exp(-\lambda) + 1)\exp[-\frac{2436}{T}], & 3 < \lambda < 17. \end{cases} \tag{61}$$

The user can also choose from the following two expressions:

$$D_\lambda = 2.5 \cdot 10^{-3} \lambda \exp\left[-\frac{2436}{T}\right], \tag{62}$$

$$D_\lambda = (1.76 \cdot 10^{-5} + 1.94 \cdot 10^{-4}\lambda) \cdot \exp\left[-\frac{2436}{T}\right]. \tag{63}$$

Equation (62) was proposed by Fuller [56] for an unspecified Nafion® membrane. Relationship (63) was originally obtained by Nguyen and White [136] in terms of water activity and later rewritten in terms of water content by Motupally et al. [132] using a Nafion® 117 isotherm measured by Zawodzinski et al. [211] and fitted by Springer et al. [173].

All of the implemented diffusion coefficients support effective-transport correction for the ionomer phase in catalyst layers. This can be done through the general random-walk method (11) and its particular cases, such as the percolation method (22).

6.2.2.3 Effective protonic conductivity

Two protonic-conductivity relationships are available for polymer-electrolyte membranes in OpenFCST:

$$\sigma_{H^+} = (0.020634 + 0.01052\lambda - 1.0125 \cdot 10^{-4}\lambda^2) \exp\left[\frac{6248}{R}\left(\frac{1}{303} - \frac{1}{T}\right)\right] \tag{64}$$

and

$$\sigma_{H^+} = (0.005139\lambda - 0.00326) \exp\left[1268\left(\frac{1}{303} - \frac{1}{T}\right)\right]. \tag{65}$$

Equation (64) was fitted by Dobson et al. [44, 45] to experimental data for Nafion® NR-211 membranes. Relationship (65), valid for $\lambda > 1$, was reported by Springer et al. [173] for Nafion® 117 membranes. The numerical coefficients in equation (65) can be adjusted by the user for the resulting protonic conductivity to reflect the desired membrane type.

Equation (11) can be used to compute effective protonic conductivity of catalyst layers. In that case, porosity is replaced with the ionomer volume fraction, ϵ_N. Alternatively, the following expression is available in OpenFCST that was fitted by Domican [46] to the effective protonic conductivity of catalyst layers measured by Iden et al. [82] (in S/cm):

$$\sigma_{H^+}^{eff} = \epsilon_N^{1.6}\sigma_{H^+}, \tag{66}$$

where

$$\sigma_{H^+} = (-8 \cdot 10^{-3} + 7.5 \cdot 10^{-4}\omega - 6.375 \cdot 10^{-6}\omega^2 + 1.93 \cdot 10^{-7}\omega^3) \exp\left[\frac{6248}{R}\left(\frac{1}{353} - \frac{1}{T}\right)\right] \tag{67}$$

and

$$\omega = \begin{cases} 100(-0.1254 + 0.1832\lambda - 0.00865\lambda^2 + 0.000094\lambda^3), & 0 < \lambda < 13; \\ 100, & \lambda \geq 13. \end{cases}$$

Multiplier $e_N^{1.6}$ in equation (66) accounts for the volume fraction and tortuosity of the ionomer phase in the CL. Since absorbed water is transported through the same phase, relationship (66) can also be used to compute the effective water diffusivity in the electrolyte phase of catalyst layers.

6.2.2.4 Electroosmotic drag coefficient

While some authors reported a constant coefficient of electroosmotic drag of water in the electrolyte, $n_d \approx 1 \, mol_{H_2O}/mol_{H^+}$ [208, 209, 211], experimental data from a number of publications [33, 63, 111, 207] (and references therein) suggest a quasi-linear relationship between n_d and water content λ. This dependency can be captured reasonably well with the following expression originally proposed by Springer et al. [173] for Nafion® 117 membranes (in mol_{H_2O}/mol_{H^+}):

$$n_d = \frac{2.5\lambda}{22}. \tag{68}$$

Users can choose between a given constant value of n_d or equation (68) in OpenFCST.

6.2.2.5 Effective thermoosmotic diffusivity

The coefficient of thermoosmotic transport of water in the electrolyte is computed in OpenFCST through a relationship proposed by Kim and Mench [98] for Nafion® 112 membranes (in $g/(cm \cdot s \cdot K)$):

$$D_T = -1.04 \cdot 10^{-4} \exp\left(-\frac{2362.0}{T}\right). \tag{69}$$

A constant value can also be provided to OpenFCST.

The effective coefficient of thermoosmotic diffusion of water in the ionomer phase of catalyst layers can be computed from equation (69) with the same methods as those used for absorbed-water diffusion due to concentration gradient (Section 6.2.2.2) and for proton transport (Section 6.2.2.3), as all three phenomena occur in the same medium.

6.2.3 Electrochemical reactions and catalyst-level transport limitations

Electrochemical reactions in the fuel cell turn the chemical energy of the reactants, such as hydrogen, into electrical energy, resulting in the production of electric current. In addition, side reactions are responsible for the dissolution of platinum and platinum alloys, carbon decay, hydrogen peroxide formation, and the formation of other radicals that can bind to the membrane. Those side reactions are currently not considered in OpenFCST.

For a hydrogen PEMFC, the main half-cell reactions are the hydrogen oxidation reaction in the anode,

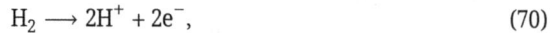

$$H_2 \longrightarrow 2H^+ + 2e^-, \tag{70}$$

and the oxygen reduction reaction in the cathode,

$$\frac{1}{2}O_2 + 2H^+ + 2e^- \longrightarrow H_2O. \tag{71}$$

The overall reaction is

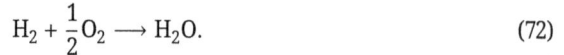

$$H_2 + \frac{1}{2}O_2 \longrightarrow H_2O. \tag{72}$$

Faradaic current density from the half-cell reactions appears in the source terms of the mass-, heat- and charge-transport equations in anode and cathode catalyst layers. This current density depends on the reactant/product concentration and the overpotential

$$\eta = \phi_{e^-} - \phi_{H^+} - E_{th},$$

where E_{th} is the theoretical half-cell potential (V).

Butler–Volmer and Tafel kinetic models are most commonly used to describe electrochemical reaction kinetics. Therefore, both models have been implemented in Open-FCST.

According to the Butler–Volmer equation, the volumetric current density j is given by [18, 138, 162]

$$
\begin{aligned}
j = A_v i_0^{\text{ref}} &\left\{ \left[\prod_{i=\text{reactants}} \left(\frac{c_i}{c_i^*} \right)^{\gamma_i} \right] \exp\left(-\frac{\alpha_R F \eta}{RT} \right) \right. \\
&\left. - \left[\prod_{i=\text{products}} \left(\frac{c_i}{c_i^*} \right)^{\gamma_i} \right] \exp\left(\frac{\alpha_P F \eta}{RT} \right) \right\},
\end{aligned}
\tag{73}
$$

where A_v is the volumetric active area (cm_{Pt}^2/cm_{layer}^3), c_i is the concentration of species i at the catalyst surface (mol/cm^3), c_i^* is the reference concentration of species i (mol/cm^3)

corresponding to the exchange current density, i_0^{ref} (A/cm^2), α_R and α_P are the transfer coefficients (–), and γ_i is the reaction order with respect to species i (–).

Since the ORR is a sluggish reaction that requires a large negative overpotential to proceed, the Butler–Volmer model (73) can be simplified to the Tafel equation [18, 130, 138]:

$$j = A_v i_0^{ref} \left\{ \left[\prod_{i=\text{reactants}} \left(\frac{c_i}{c_i^*} \right)^{\gamma_i} \right] \exp\left(-\frac{\alpha_R F \eta}{RT} \right) \right\}. \tag{74}$$

Butler–Volmer and Tafel kinetic models are limited to single-electron-transfer reactions and multistep reactions with a single rate-determining step [18, 138]. However, it has been experimentally observed [39, 104, 123, 144, 187] that the rate-determining step in HOR and ORR changes with overpotential. Therefore, more advanced kinetic models for these reactions are required that can accurately predict the reaction rates in the respective multistep mechanisms. OpenFCST offers two such models, namely the dual-path kinetic model for HOR [162, 187] and the double-path kinetic model for ORR [127–129, 188].

The dual-path HOR mechanism, proposed by Wang et al. [187], is based on the Tafel–Heyrovsky–Volmer mechanism:

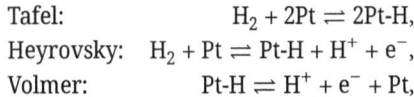

Tafel: $\quad\quad\quad$ $H_2 + 2Pt \rightleftharpoons 2Pt\text{-H}$,
Heyrovsky: $\quad$ $H_2 + Pt \rightleftharpoons Pt\text{-H} + H^+ + e^-$,
Volmer: $\quad\quad\quad$ $Pt\text{-H} \rightleftharpoons H^+ + e^- + Pt$,

where Pt-H denotes adsorbed species on the active Pt sites. Using this reaction mechanism and assuming steady state, small coverages, constant proton concentration due to the presence of ionomer, a significantly higher reaction rate for the Volmer step compared to the Tafel and Heyrovsky steps, and symmetric transfer coefficients for Heyrovsky and Volmer steps, the HOR current density is then computed as [187]

$$j_{HOR} = A_v i_{0T} \frac{c_{H_2}}{c_{H_2}^{ref}} \left[1 - \exp\left(\frac{-2F\eta}{\gamma RT} \right) \right]$$
$$+ A_v i_{0H} \frac{c_{H_2}}{c_{H_2}^{ref}} \left[\exp\left(\frac{F\eta}{2RT} \right) - \exp\left(\frac{-F\eta}{\gamma RT} \right) \exp\left(\frac{-F\eta}{2RT} \right) \right], \tag{75}$$

where $c_{H_2}^{ref}$ is the saturated hydrogen concentration at 1 atm (mol/cm^3), $c_{H_2}^0$ is hydrogen concentration at the catalyst surface at equilibrium ($\eta = 0$), $i_{0T} = 0.47$ A/cm^2 and $i_{0H} = 0.01$ A/cm^2 are the exchange-current densities of the Tafel and Heyrovsky steps, and $\gamma = 1.2$ (–) is the potential-range constant. The detailed derivation of the dual-path HOR model can be found in [187].

In the double-trap ORR mechanism by Wang et al. [188], two intermediate species, $OH_{(ads)}$ and $O_{(ads)}$, are assumed:

Dissociative Adsorption (DA): $\qquad \frac{1}{2}O_2 + Pt \rightleftharpoons Pt\text{-}O,$

Reductive Adsorption (RA): $\qquad \frac{1}{2}O_2 + Pt + H^+ + e^- \rightleftharpoons Pt\text{-}OH,$

Reductive Transition (RT): $\qquad Pt\text{-}O + H^+ + e^- \rightleftharpoons Pt\text{-}OH,$

Reductive Desorption (RD): $\qquad Pt\text{-}OH + H^+ + e^- \rightleftharpoons H_2O + Pt.$

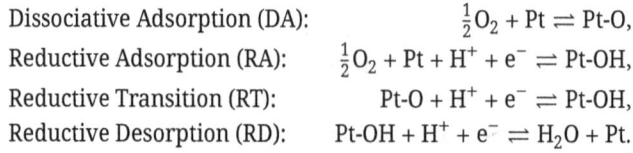

Then, assuming steady state, a constant concentration of protons, and Langmuir iso-therms for the concentration of the intermediate species, the ORR current density can be expressed as [127]

$$j_{ORR} = i^* A_v \left[\theta_{OH} \exp\left(-\frac{\Delta G^*_{RD}}{kT}\right) - \theta_{Pt} \exp\left(-\frac{\Delta G^*_{-RD}}{kT}\right) \right], \tag{76}$$

where i^* is a reference prefactor that should be adjusted based on the activity of the catalyst (it was set to 1,000 A/cm$^2_{Pt}$ in [129, 188]). The catalyst-surface coverages are given by [127, 129]

$$\theta_{Pt\text{-}OH} = \frac{Cg_{DA}B_1 - (Cg_{RA} + g_{\text{-}RD})B_2}{(Cg_{DA} - g_{\text{-}RT})B_1 - B_3B_2}, \tag{77}$$

$$\theta_{Pt\text{-}O} = \frac{Cg_{DA}B_3 - (Cg_{RA} + g_{\text{-}RD})(Cg_{DA} - g_{\text{-}RT})}{B_2B_3 - B_1(Cg_{DA} - g_{\text{-}RT})}, \tag{78}$$

and

$$\theta_{Pt} = 1 - \theta_{Pt\text{-}OH} - \theta_{Pt\text{-}O}, \tag{79}$$

where

$$B_1 = (Cg_{RA} + g_{\text{-}RD} - g_{RT}), \tag{80}$$

$$B_2 = (Cg_{DA} + g_{\text{-}DA} + g_{RT}), \tag{81}$$

$$B_3 = (Cg_{RA} + g_{\text{-}RA} + g_{\text{-}RT} + g_{RD} + g_{\text{-}RD}), \tag{82}$$

and

$$C = \left(\frac{c_{O_2}}{c^{ref}_{O_2}}\right)^{1/2} \tag{83}$$

is the oxygen concentration ratio. Quantities g_i in equations (77), (78), (80)–(82) are given by

$$g_i = e^{-\Delta G^*_i / kT}, \tag{84}$$

where ΔG^*_i is the potential-dependent free energy of activation of the ith step computed as discussed in [127, 129]. For more details and derivations, the reader is referred to publications [127, 129, 130, 188].

The equations above, even though they represent an improvement over the use of Tafel and Butler–Volmer equations in predicting steady-state ORR currents, still need further advancement. First, the equations were implemented in steady state and, therefore, they cannot predict the effects of platinum-oxide growth in cyclic voltammograms, dynamic polarization curves, and electrochemical impedance spectra. Secondly, it is well known that side reactions in anode and cathode affect the cell performance and durability. In the cathode, crossover currents are responsible for lowering the open-circuit voltage, and the production of peroxides at low cell voltage is responsible for membrane degradation. In both electrodes, platinum dissolution, Ostwald particle ripening, and carbon corrosion lead to long-term damage of the MEA. All of these phenomena need to be included in order to achieve a truly predictive model of the beginning- and end-of-life operation of fuel cells.

6.2.3.1 Catalyst-level transport limitations

The electrochemical models above require knowledge of the reactant concentration at the local catalyst site. However, the gas-transport models describe macroscale transport in the MEA. As reactants approach the catalyst sites, an additional transport resistance must be considered based on the experimental evidence that the reactant flux from the gas pores in the CL to the catalyst surface can be limited by transport through the ionomer and liquid-water films that may cover the platinum particles [105, 106, 137, 161].

In order to account for this nanometer-scale transport resistance, several submodels have been used in the literature that assume the existence of ionomer-filled [45, 86, 131, 164, 166, 169, 175, 189, 194, 206, 213] and water-filled agglomerates [45, 48, 189, 194] in catalyst layers. These agglomerates can be covered with a film of ionomer [45, 48, 57, 70, 86, 116, 131, 164, 166, 175, 189, 194, 206, 213] and/or water [57, 70, 116, 166, 206]. Reactants in these submodels need to first dissolve from the pore space into the ionomer/water films covering the agglomerates and then diffuse through the agglomerate filling to reach the catalyst surface. The user can specify if they want to use a model similar to the ones discussed above in OpenFCST or to assume no local transport limitations.

Experimental [105, 106, 137] and molecular-dynamics [88] studies suggest that interfacial transport resistance associated with the presence of ionomer films around catalyst particles dominates the overall local reactant-transport loss in CLs. OpenFCST offers a 1D nanometer-scale submodel that accounts for this interfacial resistance [165, 194]. In this model, catalyst layers are assumed to be collections of spherical ionomer-covered carbon particles (ICCPs). Before reaching the catalyst surface in each particle, reactants must first dissolve, with a finite rate, through the ionomer film and then diffuse through it. The current version of the ICCP model does not account for the liquid-water presence but can be improved to include it in the future. The ICCP model is described next for the example of oxygen transport in the cathode CL. A similar model is available for the anode CL.

The rate of oxygen dissolution into the ICCP ionomer (in mol/s) is given by [165, 194]

$$R_{O_2}^{int} = -4\pi(r_{core} + \delta)^2 k_{O_2}(c_{O_2,g|f} - c_{O_2,g|f}^{eq}),$$ (85)

where r_{core} is the radius of the carbon-platinum core (nm), δ is the ionomer-film thickness (nm) calculated from the catalyst and electrolyte loading of the CL [165, 194], $c_{O_2,g|f}$ is the oxygen concentration at the gas-film interface (mol/cm^3), and $c_{O_2,g|f}^{eq}$ is the equilibrium oxygen concentration from Henry's law (mol/cm^3).

At steady state, $R_{O_2}^{int}$ is equal to the rate of oxygen diffusion [165, 194]:

$$R_{O_2}^{diff} = 4\pi r^2 D_{O_2,m}\frac{dc_{O_2}}{dr},$$ (86)

where $4\pi r^2$ is the area of the diffusion front and $D_{O_2,m}$ is oxygen diffusivity in the ionomer. The rate of oxygen diffusion must be equal to the consumption rate of oxygen at the catalyst surface [165, 194]:

$$R_{O_2}^{ORR} = \frac{i(c_{O_2,c|f}, \eta)A_s}{4F},$$ (87)

where $i(c_{O_2,c|f}, \eta)$ is current density (A/cm$^2_{Pt}$), $c_{O_2,c|f}$ is oxygen concentration at the catalyst surface (mol/cm^3), and A_s is the active area of the ICCP (cm$^2_{Pt}$):

$$A_s = \frac{A_v}{1-\varepsilon_p}\frac{4}{3}\pi(r_{core} + \delta)^3.$$

The balance of rates (85)–(87) leads to an estimate of the oxygen concentration at the surface of the catalyst, which can be used in the kinetic models [165, 194]:

$$c_{O_2,c|f} = c_{O_2,g|f}^{eq} - \frac{i(c_{O_2,c|f}, \eta)A_s}{16F\pi}\left(\frac{1}{(r_{core} + \delta)^2 k_{O_2}} + \frac{\delta}{r_{core}(r_{core} + \delta)D_{O_2,m}}\right).$$ (88)

This equation is solved iteratively with respect to $c_{O_2,c|f}$ using Newton's method. The resulting current $i(c_{O_2,c|f}, \eta)A_s$ (A) includes the ICCP correction for the local transport resistance. The steady-state assumption of the ICCP model makes it quasi-stationary when it is incorporated in the transient PEMFC model (gas concentration and overpotential still vary with time). Further details on the ICCP model can be found in [165, 194].

6.2.4 Overall system of equations for a two-phase flow system

Equations (8), (21), (27), (36), (49), and (51) can be summarized in the following set of governing equations of a transient two-phase PEMFC model in OpenFCST:

$$\varepsilon_v\frac{\partial}{\partial t}((1-s)c_{O_2}) - \nabla \cdot (D_{O_2}^{eff}\nabla c_{O_2}) = S_{O_2},$$ (89)

$$\epsilon_V \frac{\partial}{\partial t}((1-s)c_{wv}) - \nabla \cdot (D_{wv}^{eff} \nabla c_{wv}) = S_{wv}, \tag{90}$$

$$-C_{dl} \frac{\partial(\phi_{e^-} - \phi_{H^+})}{\partial t} - \nabla \cdot (\sigma_{H^+}^{eff} \nabla \phi_{H^+}) = S_{H^+}, \tag{91}$$

$$C_{dl} \frac{\partial(\phi_{e^-} - \phi_{H^+})}{\partial t} - \nabla \cdot (\sigma_{e^-}^{eff} \nabla \phi_{e^-}) = S_{e^-}, \tag{92}$$

$$\epsilon_N \frac{\rho_{N,dry}}{EW} \frac{\partial \lambda}{\partial t} - \nabla \cdot \left(n_d \frac{\sigma_{H^+}^{eff}}{F} \nabla \phi_{H^+} + \frac{\rho_{N,dry}}{EW} D_\lambda^{eff} \nabla \lambda + \frac{D_T^{eff}}{M_{H_2O}} \nabla T \right) = S_\lambda, \tag{93}$$

$$\sum_{i=phase} \rho_i C_{p,i} \frac{\partial(\epsilon_i T)}{\partial t} - \nabla \cdot (\kappa^{eff} \nabla T) + \sum_{i=gas, \lambda} (N_i \cdot \nabla \overline{H}_i) = S_T, \tag{94}$$

$$\epsilon_V \frac{\partial(s\rho_l)}{\partial t} - \nabla \cdot \left(\frac{\rho_l \kappa_l}{\mu_l} \nabla p_c \right) = S_l. \tag{95}$$

Source/sink terms in the model are given by equations (15), (34), (39), (52), (53), and (54). Faradaic current density in those terms is computed through equation (73), (74), (75), or (76) depending on the compartment (anode or cathode) and the user's preference. Transport of nitrogen and hydrogen is not modeled, and their molar fractions are computed as $x_{N_2} = 1 - x_{O_2} - x_w$ in the cathode and $x_{H_2} = 1 - x_w$ in the anode.

6.2.4.1 Initial and boundary conditions

Boundary conditions for the model are listed in Table 1, where the zero superscript indicates quantities computed based on the operating conditions as discussed in [162]. If the problem is solved in transient form, initial conditions can be obtained by solving the system of equations (89)–(95) at steady state (without the temporal derivatives) and then importing the resulting solution into the transient simulation.

Table 1: Boundary conditions in the transient two-phase PEMFC model in OpenFCST.

Eq.	AGDL-channel	AGDL-land	CGDL-channel	CGDL-land	Symmetry
(89)	no flux	no flux	$x_{O_2} = x_{O_2}^0(t)$	no flux	no flux
(90)	$x_w = x_{w,a}^0(t)$	no flux	$x_w = x_{w,c}^0(t)$	no flux	no flux
(91)	no flux	no flux	no flux	no flux	no flux
(92)	no flux	$\phi_{e^-} = 0$	no flux	$\phi_{e^-} = V_{cell}(t)$	no flux
(93)	no flux	no flux	no flux	no flux	no flux
(94)	$\kappa^{eff}\nabla T \cdot n = 0$	$T = T_a^0(t)$	$\kappa^{eff}\nabla T \cdot n = 0$	$T = T_c^0(t)$	no flux
(95)	no-flux or (96)	no flux	no-flux or (96)	no flux	no flux

A dynamic boundary condition is used at the GDL-channel interfaces that describes liquid-water accumulation and drainage cycles observed in ex situ liquid-intrusion experiments with GDLs [71, 121, 224]:

$$-\frac{\rho_1 \kappa_1}{\mu_1} \nabla p_1 \cdot \boldsymbol{n} = k_{1,\,\mathrm{out}} \frac{p_c - p_{c,\,\mathrm{eq}}}{p_0}, \tag{96}$$

where $k_{1,\,\mathrm{out}}$ is the eruption rate constant (g/(cm$^2 \cdot$ s)), p_0 is a reference pressure chosen to be 1 Pa, and $p_{c,\,\mathrm{eq}}$ is the equilibrium capillary pressure at the GDL-channel interfaces that corresponds to $k_{1,\,\mathrm{out}} \rightarrow \infty$. At the start of the transient simulations, a no-flux condition is used instead of condition (96) until the breakthrough pressure, $p_{c,\,\mathrm{bt}}$, is reached. Liquid water then escapes from the parts of the GDL-channel boundaries where the breakthrough takes place. Once the capillary pressure reaches the given minimum value, $p_{c,\,\mathrm{min}}$, a no-flux condition is imposed again, and liquid-water accumulation continues.

6.3 Solution of governing equations and code structure

OpenFCST is developed as a toolbox with governing equations, constitutive models for each layer, and nonlinear and linear solvers. The user can adopt the equations provided to build their own applications. Each application already available in OpenFCST solves an appropriate subset of the governing equations described in Section 6.2.4. Further, the applications are accompanied by example folders that contain tutorials and the necessary input files to reproduce the example results.

In the next subsections, the transient and nonlinear solvers available to solve MEA models in OpenFCST are described. Then the software structure is discussed, including the location of the example folders that will be used in Section 6.4.

6.3.1 Solution approach

OpenFCST offers a variety of solvers for linearizing nonlinear problems, obtaining solutions to the discretized linear systems, and marching in time. The latter class of solvers will become available in the next release of the software (v1.0).

The governing equations of the PEMFC model in Section 6.2.4 can be summarized as a system of ODEs

$$\frac{d\boldsymbol{u}}{dt} = \boldsymbol{f}(t, \boldsymbol{u}(t)), \quad \boldsymbol{u}(t_0) = \boldsymbol{u}_0, \tag{97}$$

where $f(t, u(t))$ contains the spatial operators and the source/sink terms. Two classes of ODE solvers will be available in OpenFCST v1.0, namely the θ-scheme and the backward differentiation formulae (BDF methods). Since these solvers are designed to solve transient problems in OpenFCST, they are referred to as transient solvers.

The θ-scheme describes a family of single-step methods for solving ODE (97). The transient term is replaced with a first-order finite difference, and the rest of the equation is treated in a combined explicit/implicit fashion by introducing a variable $\theta \in [0, 1]$ [9]:

$$\frac{u^{(n+1)} - u^{(n)}}{\tau} = \theta f^{(n+1)} + (1 - \theta)f^{(n)}, \tag{98}$$

where $n + 1$ denotes the new transient solution, $\tau = t_{n+1} - t_n$ is the time-step size, and $f^{(n)} = f(t_n, u^{(n)})$. Three commonly used particular cases of this method are:

$\theta = 0$: the explicit Euler method (order 1, conditionally stable [9, 10]);
$\theta = 1/2$: the Crank–Nicolson (trapezoidal) method (order 2, absolutely stable [9, 10]);
$\theta = 1$: the implicit Euler method (order 1, absolutely stable [9, 10]).

The current implementation of the transient framework in OpenFCST limits the use of the θ-scheme to linear equations and nonlinear equations linearized with the Picard's method. For the solution of stiff nonlinear problems, such as the full-MEA models, it is recommended that the BDF transient solver is used instead.

A general k-step BDF method (BDFk) for solving problem (97) is given by [9, 10, 74]

$$\alpha_{n+1}u^{(n+1)} + \alpha_n u^{(n)} + \cdots + \alpha_{n-k+1}u^{(n-k+1)} = \tau f^{(n+1)}, \tag{99}$$

where $\alpha_i, i = \overline{n - k + 1, n + 1}$, are some real coefficients that can be found in [9, 10, 74]. The single-step BDF method ($k = 1$) is equivalent to the implicit Euler scheme. The order of a BDFk method is k [9, 10, 74]. These methods are absolutely stable for $k = 1$ and $k = 2$, conditionally stable for $3 \le k \le 6$, and unstable for $k > 6$ [9, 10, 74].

Both classes of transient solvers available in OpenFCST support adaptive time-stepping. In case of the θ-scheme and BDF1 solvers, absolute and relative solution errors are controlled by automatically adjusting the time-step size with the Richardson-extrapolation algorithm [9, 22, 74, 167]. A detailed description of the transient framework in OpenFCST can be found in [101].

Equations (98) and (99) are discretized in time but still highly nonlinear in both the effective transport properties and the source/sink terms. Two classes of nonlinear solvers are available in OpenFCST: Picard's method [6, 113, 182] and Newton's method [6, 54, 113, 122, 149, 182].

Assuming the implicit-Euler discretization in time of equation (97), Picard's method can be described as an iterative process

$$\frac{u^{(n+1,m+1)} - u^{(n)}}{\tau} = f(t_{n+1}, u^{(n+1,m)}), \tag{100}$$

$$f(t_{n+1}, u^{(n+1,m)}) = \nabla \cdot (A(u^{(n+1,m)})\nabla u_a^{(n+1,m+1)}) + S(u^{(n+1,m)}), \tag{101}$$

where $m+1$ denotes the new iteration of the nonlinear solver, $A(u)$ represents the transport coefficients in the model that may generally depend on the solution variables, and S contains the source/sink terms. The iterations start with the initial guess that is the transient solution from the previous time layer: $u^{(n+1,0)} = u^{(n)}$. It is clear from equations (100) and (101) that nonlinearities are treated explicitly in Picard's method. Because this method has linear convergence [6, 113, 182], it is not very effective in MEA simulations but is preferable, due to its relatively simple implementation, in simpler cases (such as single-equation models).

In Newton's method, the governing equations are represented as a set of residuals

$$R(u) = \frac{u^{(n+1,m+1)} - u^{(n)}}{\tau} - f(t_{n+1}, u^{(n+1,m+1)}) = 0 \tag{102}$$

and the solution iterate is defined as $u^{(n+1,m+1)} = u^{(n+1,m)} + \delta u$, $u^{(n+1,0)} = u^{(n)}$. Then equation (102) is solved with respect to the solution update, δu [122]:

$$J(u^{(n+1,m)})\delta u = -R(u^{(n+1,m)}), \tag{103}$$

where

$$J(u^{(n+1,m)}) = \left[\frac{\partial R_\alpha}{\partial u_\beta}\right]\bigg|_{u^{(n+1,m)}}$$

is the Jacobian matrix and the Greek subscripts denote the individual equations and solution variables. This method has a quadratic rate of convergence if the initial guess is sufficiently close to the solution; therefore, it provides an efficient way to solve fuel cell models.

The iterative process in Picard's and Newton's methods continues until the solution satisfies the nonlinear governing equations. This is controlled with three convergence criteria: the Euclidean norm of the system residual, $\|R(u^{(n+1,m+1)})\|_2$, satisfies a given tolerance (typically 10^{-8}), and the absolute and relative solution changes in the discrete l^2 norm defined as

$$\|u\|_{l^2} = \sqrt{\frac{\sum_{i=1}^{N} u_i^2}{N-1}} = \frac{\|u\|_2}{\sqrt{N-1}}$$

are below some given thresholds that depend on the problem at hand. Both Picard's and Newton's methods implemented in OpenFCST offer adaptive algorithms that control the solution update and aid the convergence. Details on these algorithms can be found in [101] and references therein. Should the nonlinear solver fail to converge, the timestep size is automatically adjusted by gradual halving until convergence is achieved, and then the original step size is recovered.

Equations (100) and (103) are discretized in time and are linearized; however, they still need to be discretized in space. This is done in OpenFCST with the Galerkin (Bubnov–Galerkin) finite-element method [6, 20, 151] using Lagrange shape functions of any order [20, 151] (second-order is recommended). The finite-element functionality is provided by the deal.II library [8, 17]. Spatial discretization yields a linear system of equations that can be solved with one of the linear solvers available in OpenFCST through deal.II and PETSc [2, 34]. Linear problems and nonlinear problems linearized with Picard's method can be solved, for instance, with the conjugate-gradient (CG) method [6, 95, 113, 182] if the resulting matrix is symmetric and positive-definite. However, the highly nonlinear electrode and MEA models typically result in nonsymmetric systems that can only be solved either by advanced iterative solvers with a precondi-tioner, such as the generalized method of minimum residuals (GMRES) in conjunction with an incomplete-LU-decomposition (ILU) preconditioner [17, 19] or by direct solvers, such as the unsymmetric-pattern multifrontal method (UMFPACK) [40–42] or the mul-tifrontal massively parallel solver (MUMPS) [7] that are based on sparse LU and LDL^T decomposition [7, 40, 41].

Adaptive mesh refinement is used in OpenFCST during the solution process to au-tomatically refine the computational grid where the largest numerical errors are pre-dicted. This helps achieve a grid-independent solution in a computationally efficient manner, since the mesh is only refined locally where needed. Solution error is evalu-ated with the a posteriori Kelly error estimator [96] from the deal.II library. The error estimate in each cell is defined as [96, 162]

$$\eta_K^2 = \frac{h}{24} \int_\Gamma \left[[(\nabla u)n] \right]^2 d\Gamma, \tag{104}$$

where h is a measure of the cell size and $[[u]]$ denotes the jump of u at the cell bound-ary Γ. The user can specify a percentage of the cells with the largest error that are to be refined and a percentage of the cells with the smallest error that are to be coarsened in each adaptive-refinement cycle. The number of the refinement cycles and the frequency of the refinement in time can also be specified. Typically, one or two refinement cycles are recommended.

6.3.2 OpenFCST code structure

The OpenFCST source code can be downloaded from the official website in a .zip archive or as a part of a virtual machine (a preconfigured virtual linux environment that you can run without installing a new operating system on your machine). The second option is the easiest way of getting started. First, install the Oracle VM VirtualBox (see https://www.virtualbox.org/), then download the virtual machine with OpenFCST, and load that virtual machine into the virtual box. Once you start the machine and log in, you will find that all the software necessary for pre-processing (such as Gmsh [67]), OpenFCST, and all the software for post-processing (such as ParaView [5]) is already installed, so you can quickly launch simulations and analyze your results.

Prior to installation, OpenFCST contains the following main folders:

- src includes all C++ source files for OpenFCST, documentation, licenses and the users' and developers' guide;
- pre_processing includes a collection of Python scripts for processing meshes prepared with external mesh generators, such as SALOME [3, 152];
- post_processing includes a Python script for creating contour plots of the simulation results with ParaView;
- python includes a variety of additional Python scripts for mesh generation, image and microstructure analysis, data visualization, optimization, and other functionality.

After installation of OpenFCST, two additional folders are created:

- Install includes the OpenFCST binaries and copies of the documentation and simulation examples from the src folder;
- Build includes auxiliary files necessary for the compilation of OpenFCST; these files are not to be modified.

The users' workflow in OpenFCST is restricted to the Install folder, where all the files necessary to run OpenFCST and all the simulation tutorials are located. Other folders contain either the source code of OpenFCST and its supporting libraries (src and python) or auxiliary routines that are not critical for the general user.

6.3.2.1 Install directory tree

The Install directory of OpenFCST includes the following subfolders:

- bin contains binary executable files for OpenFCST, namely fuel_cell-2d.bin, fuel_cell-3d.bin, and fcst_gui; the first two files are used to run OpenFCST through the terminal to solve 2D and 3D problems; the last file is the executable for the graphical user interface;
- examples contains a set of example problems designed as tutorials for the OpenFCST users; in particular, there are examples for simulating a cathode, an anode, an MEA with macrohomogeneous and agglomerate models, and a nonisothermal MEA;
- doc contains the documentation that includes:
  - the main HTML page, index.html, that can be used to access all the code documentation in OpenFCST (in the html folder);
  - the users' and developers' guide in PDF along with its source files in LaTeX(in the RefGuide folder);

- contrib contains the external (contributing) libraries for OpenFCST, including deal.II;
- databases contains databases used to speed-up the fuel cell simulations with numerical-agglomerate submodels in catalyst layers (see [131, 194] for details); this folder is of no interest to the users who are not using numerical agglomerates;
- fcst contains the header files of the code and is not of the general user's interest;
- test contains configuration files and scripts used to run the unit tests and the example simulations in OpenFCST in order to ensure the correct installation of the code and consistency in the simulation results between the software releases;
- python contains a collection of Python scripts from the main python folder that were discussed earlier;
- my_data is an empty directory designated as the user's workspace.

Because the files in the examples folder are used in OpenFCST tutorials and testing, they should not be modified. The examples of interest can be copied to the my_data directory and modified there as necessary.

The Install directory also includes two scripts, fcst_env.sh and run_tests. The first script contains the environment-variable definitions for OpenFCST and should be sourced in the terminal by typing

```
$ . fcst_env.sh
```

every time the user opens a new terminal for executing OpenFCST in order to tell the operating system where the software's executables are located. The contents of this script can be copied directly to the user's .bash_profile so that the variables are always defined. The second script in the Install folder, run_tests, is used to run all the simulations in examples and to verify the simulation results against the precomputed reference results in the regression directory of each example. The environment variables need to be sourced before the tests are executed.

6.4 MEA simulations in OpenFCST

OpenFCST v0.3, available at www.openfcst.org, contains numerical models for cathode, anode, and full MEA. In the following section, several examples based on the non-isothermal, single-phase full-MEA model named PemfcNIThermal are discussed. The model was thoroughly validated in [30, 102]. The examples show how to solve the MEA model governed by equations (89)–(94) (liquid-water transport is not considered in PemfcNIThermal, and $s = 0$). In this section, the results obtained for the input files provided with the software are described and additional case studies are discussed that aim at illustrating the flexibility of the model for analyzing varying operating conditions and catalyst-layer submodels.

6.4.1 Base case

The tutorial and input files for this example are provided in PemfcNIThermal. In particular, in this case we will be using the input files in the subfolder polarization_curve, which are used to simulate a full polarization curve using the single-phase, nonisothermal MEA model.

The two most critical input files are main.prm, which selects the model type that the user would like to run, and the file that contains geometric and component properties, as well as meshing and solution parameters, namely data.prm. These parameters can be easily examined by means of the OpenFCST graphical user interface (GUI). In order to open the graphical user interface, go to the Install/bin folder and double-click on the GUI icon. Then load the main.prm and data.prm files from the aforementioned folder into the GUI. They will be first converted to XML format and then opened in the GUI. Figure 1 shows the graphical user interface once the example files have been loaded.

The main.prm file contains two main sections, Simulator and Logfile. The first section is used to select the application the user would like to run (meaNIT in this case, which corresponds to the PemfcNIThermal example), the nonlinear and linear solvers, the type of mesh refinement, the data file to be used to provide the properties of the cell and the type of analysis. Regarding the latter, you can select any of the following options: (a) Analysis, (b) PolarizationCurve, (c) ParametricStudy, and (d) Optimization. Analysis runs a single simulation at the operating point specified in the data tab. PolarizationCurve, ParametricStudy, and Optimization run multiple simulations; therefore, when one of those analysis types is selected, further input needs to be provided in one of the subfolders with the same name. In this example, PolarizationCurve is selected, so the initial voltage, the final voltage, and the voltage interval need to be specified in the Polarization Curve subfolder.

The data.prm file contains more detailed information regarding the simulation. The file is organized into subsections that are used to control the geometry of the cell (Grid generation), mesh adaptivity, linear- and nonlinear-solver parameters, the initial solution (initial guess), the equations that are solved, fuel cell parameters, and postprocessing options. This tutorial will focus on understanding the most critical parameters to set up the correct dimensions and properties of the fuel cells. More details on the equation and solution-strategy parameters can be found in the user guide provided with the software.

The geometry of the cell is specified in the Grid generation section. The parameter Type of mesh is used to specify whether the internal mesh generator in Open-FCST should be used or the computational grid should be read from a file prepared with an external software (such as Gmsh). In this case, the internal mesh generator of type PemfcMPL is used to create a mesh for a full MEA. The computational domain for a full-MEA simulation is shown in Figure 2. It includes gas-diffusion layers, microporous layers, catalyst layers, and the membrane. All dimensions are provided to the internal

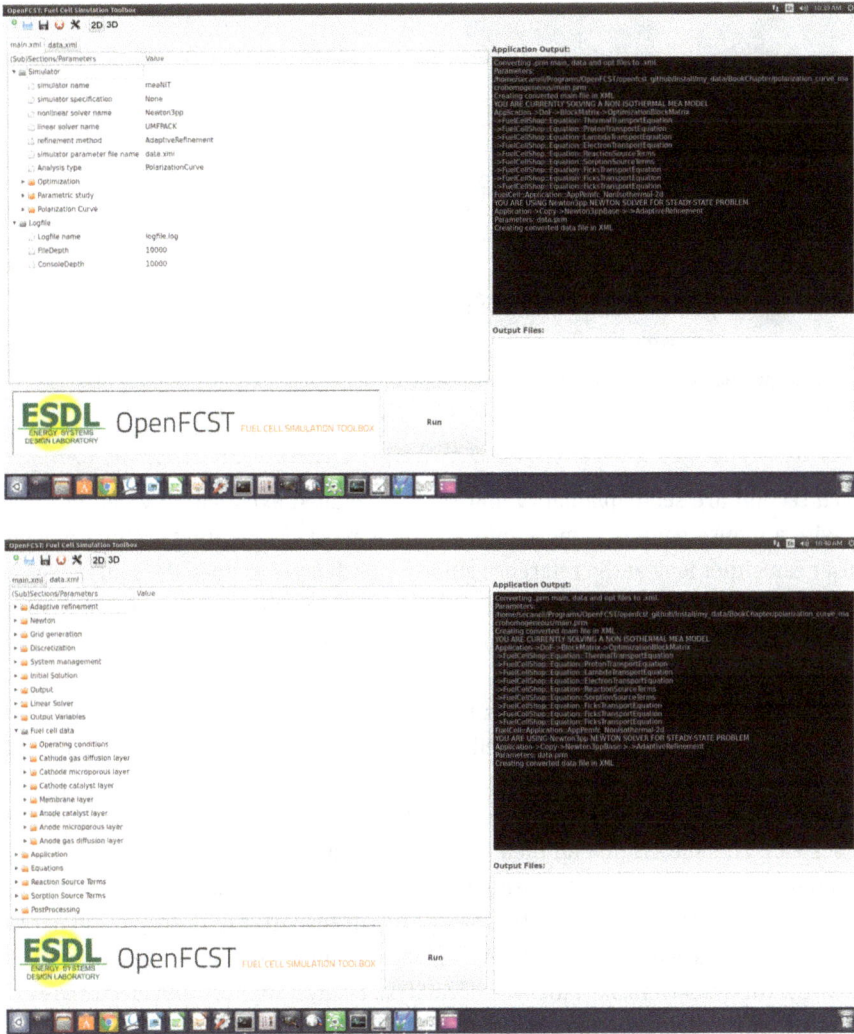

Figure 1: Screenshots of the main tab (top) and the data tab (bottom) of the GUI showing the various available options.

mesh generator through the section called `Internal mesh generator parameters`. Inside that section, the thickness of each MEA component and the channel and land widths are specified. Each material in the mesh is assigned a unique ID in this parameter section. This material ID needs to be provided to all layer-specific inputs in section `Fuel cell data`.

The other section of interest when setting up the simulation is `Fuel cell data`. This section contains subsections for specifying the operating conditions and the properties of each one of the cell components used in the simulation. Most of the parameters

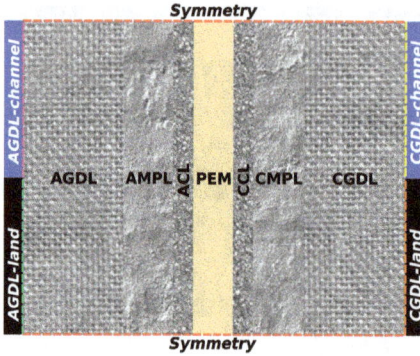

Figure 2: Computational domain used in MEA simulations. Channels and lands are treated as boundary conditions.

in each section are self-explanatory and show a detailed explanation when hovered over with a mouse pointer. If more information is needed regarding each parameter, the user can either look at the reference guide for each layer or open the source files in src/fcst/source/layers where all expressions are implemented. The OpenFCST development team aims at creating a database of available materials from which the user could pick the material of their choice.

Subsection Operating conditions inside Fuel cell data specifies that, for the default case, the temperature of the cell is set to 353.15 K and the anode and cathode pressure and relative humidity are set to 101,325 Pa and 70 %.

Be default, both anode and cathode GDLs are selected as "design" layers (Design-FibrousGDL). This means that all their effective properties will be computed based on the volume fraction of each phase using the percolation-theory equations described in Section 6.2. For example, the porosity is set to 60 %, transport is anisotropic and percolation parameters are provided for x (through-plane) and y (in-plane) directions.

For the catalyst layers, first the type of catalyst, catalyst support, and electrolyte are specified, then in subsection Materials, the specific property of each material is given. Parameter Catalyst layer type specifies the microscale model type used to approximate the local mass-transport losses, where HomogeneousCL means no microscale model (i. e., the model is purely macrohomogeneous). The last piece of information to provide is the catalyst-layer composition, which is specified in subsection ConventionalCL along with the percolation-model parameters to be used to compute the effective transport properties in the void, ionomer, and solid phases. In this case, the volumetric catalyst loading in the cathode is 400 $\mathrm{mg_{Pt}/cm_{CL}^3}$. This value is obtained by dividing the areal catalyst loading by the thickness of the electrode. The electrolyte loading is 30 wt.%, and the active area is specified to be $2 \cdot 10^5$ $\mathrm{cm_{Pt}^2/cm_{CL}^3}$. In order to compute the solid volume fraction, the amount of Pt in the platinum-carbon mix also needs to be specified; in this example, it is 46 wt.%. Note that multiple catalyst layers can be specified; the number in front of the layer's parameters is the material ID for each CL in the mesh.

After the parameters discussed above are set up, the simulation can be launched by pressing the Run button at the bottom of the screen. The simulation output will be shown in the black terminal on the right (Figure 1). After the simulation is completed, a file named polarization_curve.dat is created with tabulated data for the polarization curve. That curve only reaches 0.4 V in this example. A wider voltage range can be simulated by setting Final voltage [V] from section Simulator/Polarization Curve of the main tab to 0.15 V. The resulting polarization curve is shown in Figure 3.

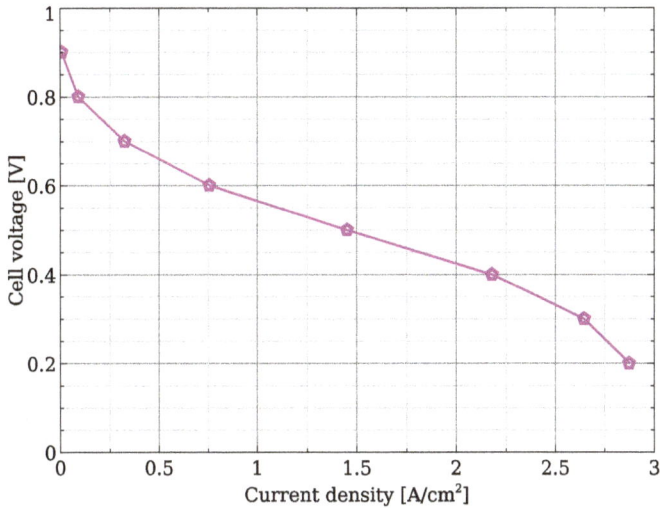

Figure 3: Polarization curve predicted by the nonisothermal MEA model. The operating conditions are 80 °C, 70% RH, 1 atm.

Further insight can be obtained by calculating additional quantities at each cell potential, such as the maximum cell temperature and relative humidity, as well as heat loss due to ohmic heating in the membrane, which can then be converted to membrane resistance by dividing the heat by the square of the current density (in accordance with equation (31)). In order to show this additional information, the number of output variables (num_output_vars) in section Output Variables of the data tab needs to be changed to 4, and then the appropriate variables need to be selected from the drop-down menus.

It might also be of user's interest to plot the solution-variable fields at each voltage to analyze, for instance, oxygen concentration or temperature variation in the cell. This can be achieved by setting Output final solution in Adaptive Refinement to true. Pressing the Run button again will launch another simulation.

Figure 4 shows the relative-humidity and temperature fields in the MEA. The relative humidity at the channel/GDL interfaces is set to 70 % and it remains relatively constant in the anode while increases to nearly 100 % under the land area in the cathode. Inside the membrane, water vapor does not exist and the relative humidity is set

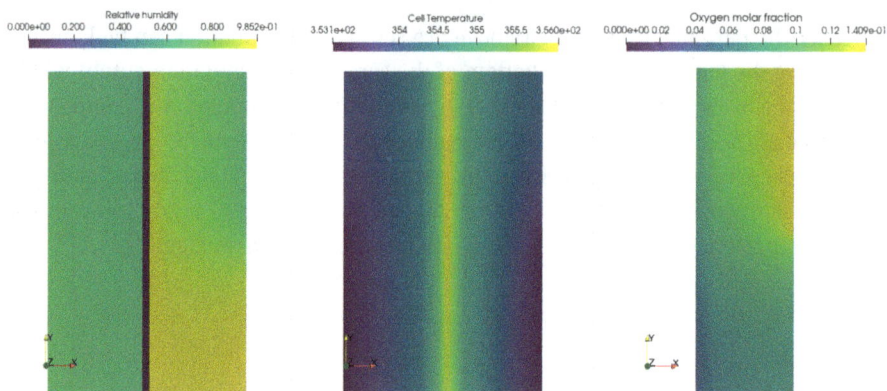

Figure 4: Relative humidity (left) and temperature in K (center) in the MEA and oxygen molar fraction (right) in the cathode catalyst layer at a current density of 1.46 A/cm², 80 °C, 70 % RH, 1 atm.

to zero. The temperature field reveals a large, three-degree difference between the catalyst layer and the plate. This temperature gradient is responsible for the heat-pipe effect that allows for efficient water removal [222]. Finally, the oxygen molar fraction shows that the area under the land is already starting to be depleted of the reactant, creating a dead zone [107].

To further investigate the effect of oxygen starvation under the land, the volumetric current density produced by the oxygen reduction reaction is shown in Figure 5. At low current densities, current is produced homogeneously inside the catalyst layer. At medium current densities, the reaction shifts toward the membrane interface (left in the image) in order to minimize voltage losses due to proton transport. Finally, at high current, a dead zone is clearly visible under the land near the membrane interface, and the reaction distribution is very nonuniform. Note these results are obtained using a macrohomogeneous model, local oxygen transport losses reduce the nonuniformity of the reaction distribution as it will be shown in the microscale analysis section.

6.4.2 Effect of operating conditions

In the previous case, a polarization curve was obtained for a cell temperature of 353.15 K, and an anode and cathode pressure and relative humidity of 101,325 Pa and 70 %, respectively. These conditions are specified in the parameter subsection Fuel cell data/Operating conditions. In this study, the effect of varying temperature and relative humidity is illustrated.

Figure 6 shows the effect of changing relative humidity from 30 % to 90 % at 80 °C. As the relative humidity increases, ohmic losses at moderate current density are reduced and the overall performance improves. However, at the higher humidity levels, the per-

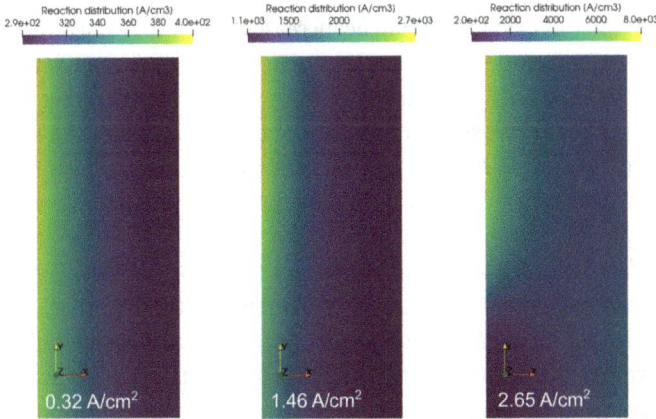

Figure 5: Volumetric current density distribution at low (left), medium (center) and high (right) current density in the cathode catalyst layer at 80 °C, 70% RH, 1 atm. The catalyst layer is stretched 40 times horizontally for clarity.

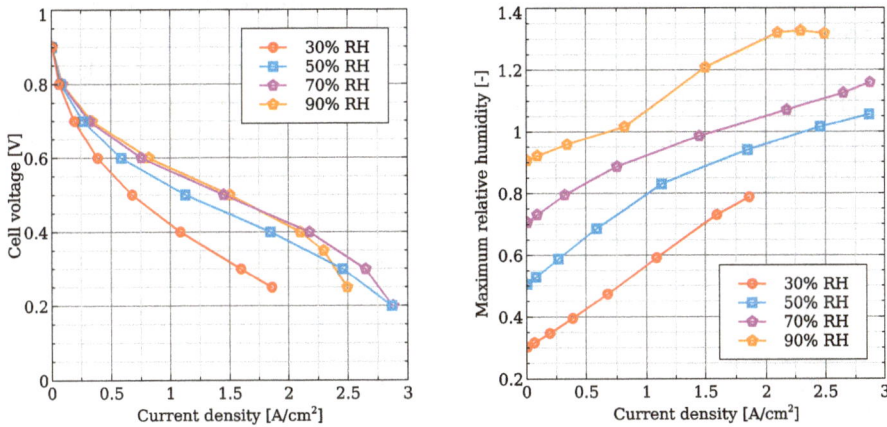

Figure 6: Effect of the operating relative humidity on the predicted polarization curve (left) and maximum relative humidity in the MEA (right). The operating conditions are 80 °C, 1 atm.

formance gain is reduced, and the limiting current density becomes lower at 90 % RH due to the higher water-vapor content in the cell that causes the oxygen levels to drop.

Because the considered model is single-phase, mass-transport effects due to electrode flooding are not observed (to observe such changes, the two-phase flow model should be used). To assess when the two-phase regime might start to occur, and hence when the single-phase model would overpredict performance, the maximum relative humidity inside the MEA can be calculated. OpenFCST can compute this value and add it to the `polarization_curve.dat` file if the user requests the maximum RH to be calculated as an output variable in subsection `Output Variables`. To achieve this goal, the number of output variables in this example should be increased from 1 to 2, and max_RH should be selected in the second entry. It can be seen in Figure 6 that the maximum relative humidity exceeds 100 % at about 0.7–2.3 A/cm^2 depending on the operating RH, which indicates the need for two-phase simulations. For the base case, relative humidity reaches 100 % at approximately 1.5 A/cm^2, in agreement with Figure 4, which also shows the exact location where water vapor is accumulating and condensation would start.

Figure 7 shows the effect of the cell temperature between 50 and 90 °C. The increase in the operating temperature positively affects the reaction rates, and thus improves the kinetic region of the polarization curve. At the current density of 0.5 A/cm^2, polarization curves start to deviate significantly. Additional information can be obtained from Open-FCST to understand the reasons for the observed change in performance. For instance, one can add PEM_proton_ohmic_heat and electron_ohmic_heat to the list of Output Variables. This will instruct OpenFCST to compute, at every point of the polarization curve, the ohmic heating (in W/cm^2) due to the proton transport in the membrane and due to the electron transport in the rest of the MEA. When added together and divided by the square of the corresponding current density, the high-frequency resistance (HFR) of the cell is obtained (in $\Omega \cdot$cm^2). For this estimate to be accurate, the effective conductivity of either the solid or the ionomer phase of the catalyst layers must be high, as discussed in [102, 103]. This is the case for the conventional carbon-supported catalyst layers considered in this example. As it can be seen in Figure 7, the HFR is generally lower for the higher operating temperature, which is in line with the observed polarization curves.

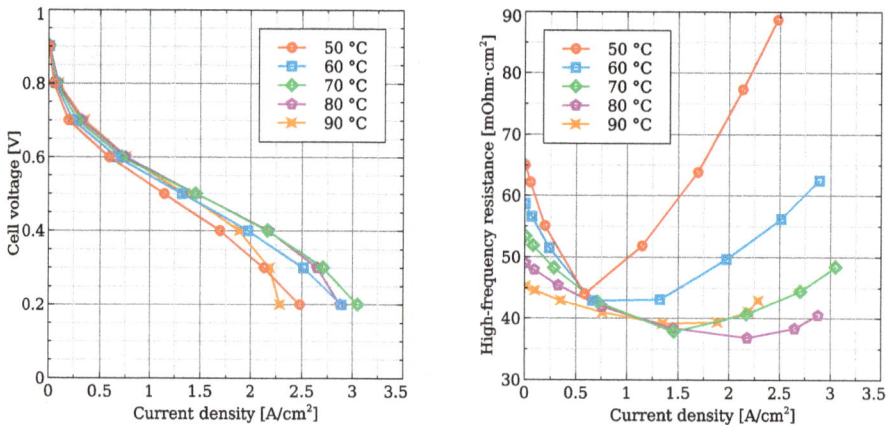

Figure 7: Effect of the operating temperature on the predicted polarization curve (left) and high-frequency resistance (right). The operating conditions are 70% RH, 1 atm.

Another important observation is that the computed HFR first decreases and then increases with the increase in current density. In this case, the HFR reduction is associated with improved hydration while its growth is associated with the anode dry-out due to the increase in the electroosmotic drag of water with current. This can be verified by visualizing the membrane water content in ParaView (the material ID of the membrane is 5). Distribution of water in the membrane is illustrated in Figure 8 for the case of 50 °C at 4.6 mA/cm^2 (the first point in the polarization curve), 0.59 A/cm^2 (the point of minimum HFR), and 2.5 A/cm^2 (the last point). Water content in the anode side of the membrane is substantially reduced at high current density.

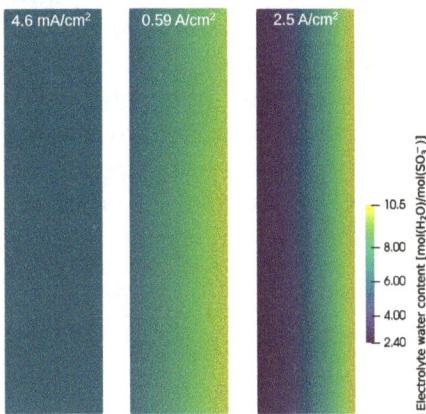

Figure 8: Distribution of absorbed water in the membrane at 50 °C at three current-density values from Figure 7. Operating conditions are 70% RH and 1 atm. The membrane is stretched 10 times horizontally for clarity.

6.4.3 Effect of microscale models

Section 6.2.3.1 highlighted that local transport losses exist at the ionomer-catalyst interface that further limit the performance of a fuel cell. In the previous examples, a macro-homogeneous model was used where all phases were assumed well-dispersed within the electrode and no local mass transport losses were accounted for. In this section, the ICCP model, where the ionomer is assumed to be covering catalyst particles and limiting the transport of oxygen, will be used and the results between the two cases will be compared. The aim of this section is to show how to modify the data file in order to set up an ICCP simulation and to compare the results to those from a macrohomogeneous CL model. Only the cathode CL will be modified.

In order to enable the ICCP in the CCL model, parameter `Fuel cell data/Cathode catalyst layer/Catalyst layer type` needs to be changed to `MultiScaleCL`; then, in the folder `MultiScaleCL`, `ICCP` needs to be selected for `Microscale type`. In the ICCP folder, users can specify the carbon-particle radius in Radius [nm] and choose whether they want to account for the finite-rate exchange of oxygen between the pore space and the ionomer (equation (85)) or to assume Henry's law. Equation (85) is enabled by setting Use non equilibrium BC to true and specifying the value for Non Equilibrium BC Rate constant (k_{O_2}). In this example, the particle radius of 39.5 nm and the nonequilibrium constant of $4.7 \cdot 10^{-4}$ m/s are assumed. It is important to note that the thickness of the ionomer film (Film Thickness [nm]) does not need to be specified, as it will be automatically computed based on the ionomer and solid-phase volume fractions and the primary-particle radius. In this case, the film thickness is 5.44 nm as can be observed in the output of the program, which is stored in logfile.log.

Figure 9 compares the polarization curves obtained with the macrohomogeneous and ICCP models. It can be clearly seen that the local transport losses do not affect the kinetic region of the polarization curve unless k_{O_2} is unrealistically small. The lower the value of this nonequilibrium rate constant, the more restrictive are the local oxygen-transport limitations in the CCL, and the lower is the observed limiting current density and the overall cell performance. Increasing the value of k_{O_2} gradually improves local transport and brings the simulated polarization curve closer to the one obtained with the macrohomogeneous model.

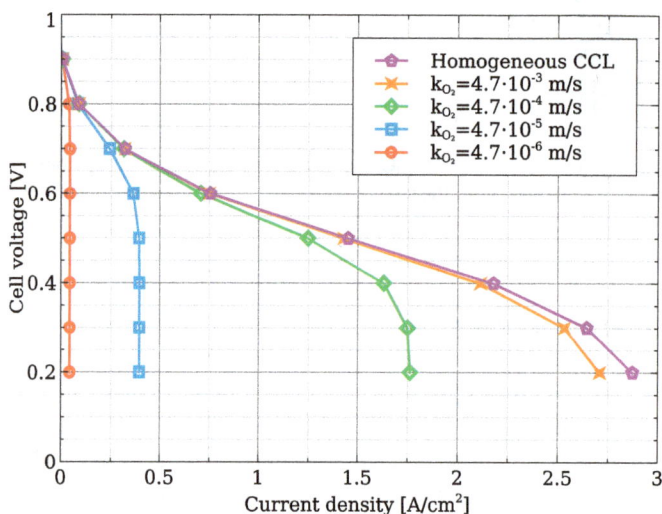

Figure 9: Comparison of macrohomogeneous and ICCP catalyst-layer models. The operating conditions are 80 °C, 70% RH, 1 atm.

Figure 10 shows the reaction distribution predicted by the ICCP model with the oxygen dissolution rate constant set to $4.7 \cdot 10^{-4}$ m/s. Comparing this current distribution to that predicted by the macrohomogeneous model in Figure 5, it can be observed that the reaction distributions follow the same trends at low and medium current density but are

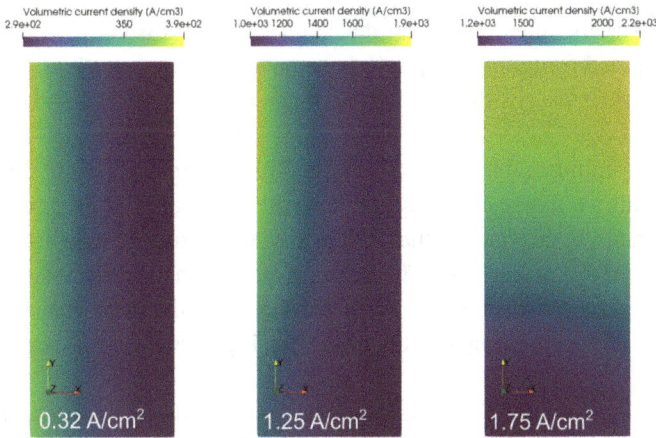

Figure 10: Volumetric current density distribution at low (left), medium (center) and high (right) current density in the cathode catalyst layer at 80 °C, 70% RH, 1 atm. The catalyst layer is stretched 40 times horizontally for clarity.

quite different at high current density, i. e., near the limiting current density, where the effect of the microscale model can be clearly observed. At low current density, the reaction distribution is uniform. At medium current densities, the reaction becomes nonuniform in both cases with higher reaction rates observed near the CL-PEM interface due to proton-transport losses in the electrolyte phase reducing the overpotential near the CL-MPL interface. At high current density, the microscale model reduces the maximum reaction rates that can be achieved. In this case, the macrohomogeneous model distribution can achieve volumetric current densities as high as 8,000 A/cm^3 while the ICCP model can only achieve 2,200 A/cm^3. From a global perspective, the limiting current density is substantially reduced. At the local scale, limiting the volumetric current density means that oxygen can easily access all areas of the catalyst layer, thereby reducing the nonuniformity of the reaction and not forming dead zones under the land, where the volumetric current density does not drop below the half of that under the channel. These results highlight the importance of including a microscale model in order to appropriately predict the catalyst-layer behavior under oxygen-limiting conditions.

6.5 Conclusions

Analyzing the performance of a fuel cell membrane-electrode assembly requires the development and solution of a coupled system of transient, nonlinear partial differential equations. In this chapter, the rationale and derivation of the equations governing mass, charge, and heat transport, as well as the electrochemical reactions in the MEA, are provided. For each governing equation, a general form is first presented, and then simplifications are applied, e. g., the assumption that the gas remains isobaric in the MEA, to arrive at a simplified transient, two-phase flow model. Closure equations are also provided to estimate the most critical transport parameters in each layer based on their composition.

The solution of the resulting system of transient, nonlinear system of partial differential equations in multiple dimensions requires in-depth knowledge of programming and numerical methods. In this chapter, the numerical solution strategy used in the open-source software OpenFCST is described. The software offers a variety of transient and nonlinear solvers. The spatial discretization is performed with the finite-element method, the functionality of which, along with linear solvers, is provided by the deal.II library [8, 17].

To introduce the reader to the software, the simulation setup and results obtained using the single-phase, nonisothermal MEA model in OpenFCST are discussed. In addition, parametric studies are performed to illustrate the effect of operating conditions (temperature and relative humidity) and microscale transport parameters. Validation of the single-phase model with experimental data can be found in [30, 102].

Since its last release, OpenFCST has been proved to be an excellent tool to accelerate progress in numerical modeling of MEAs. OpenFCST offers a unique combination of frameworks for analyzing the microstructure of MEA components [91, 92, 156–159], predicting cell flooding with pore-size-distribution-based models [101, 221, 222], investigating electrochemical impedance spectra [101–103] and open-cell voltage [130], and much more—all in one package distributed under the MIT license [1] that gives every user the freedom to adjust the software to their needs. It is the OpenFCST developers' belief that significant breakthroughs in PEMFC research are only achievable through strong collaboration within the fuel cell community and sharing of the expertise and of the tools developed.

Nomenclature

Constants
F Faraday constant, $\approx$96,485 C/mol
R universal gas constant, $\approx$8.314 J/(mol·K)

Greek
α charge-transfer coefficient
ϵ volume fraction
η overpotential, $\eta = \phi_{e^-} - \phi_{H^+} - E_{th}$, V
κ thermal conductivity, W/(cm K), or permeability, cm^2
λ absorbed-water content in the electrolyte, mol$_{H_2O}$/mol$_{SO_3^-}$
ϕ potential, V

ρ density, g/cm^3
σ electrical conductivity, S/cm
ξ portion of water produced electrochemically at the catalyst-ionomer interface

Latin

$\Delta \overline{S}_{overall}$ entropy change per mole of hydrogen, J/(mol·K)
N molar flux, mol/(cm^2 · s)
n normal unit vector
$\overline{H}$ molar enthalpy, J/mol
$\overline{H}_{lv}$ molar latent heat of water vaporization, J/mol
$\overline{H}_{sorption}$ molar enthalpy change due to water absorption/desorption by the electrolyte, J/mol
i areal current density, A/cm^2
EW   equivalent weight of the electrolyte, g/mol$_{SO_3^-}$
A_v volumetric active area, cm$^2_{cat}$/cm^3
C volumetric capacitance, F/cm^3
c concentration, mol/cm^3
D diffusion coefficient, cm^2/s
D_T thermoosmotic diffusion coefficient for water in the electrolyte, g/(cm·s·K)
E_{th} theoretical half-cell potential, V
f_{ORR} fraction of the overall entropy change due to oxygen reduction reaction
j volumetric current density, A/cm^3
k_λ rate of absorption/desorption of water by the electrolyte, 1/s
k_{O_2} rate constant for oxygen dissolution into the electrolyte, m/s
M molar mass, g/mol
n_d electroosmotic drag coefficient for water in the electrolyte, mol$_{H_2O}$/mol$_{H^+}$
T temperature, K
t time, s
x molar fraction
V voltage, V

Subscripts and superscripts

e^- electron
H^+ hydrogen proton
H_2 hydrogen
H_2O water
N_2 nitrogen
O_2 oxygen
0   property at the given operating conditions
dl   double layer

eff   effective transport property
eq   equilibrium
l   liquid water
N   electrolyte phase
S   solid, electronically conductive phase
tot   total
V   liquid-free pore (void) phase
wv   water vapor

Bibliography

[1] The MIT license, https://opensource.org/licenses/MIT. Accessed February 1, 2022.

[2] PETSc/TAO, Portable, Extensible Toolkit for Scientific Computation, Toolkit for Advanced Optimization. https://www.mcs.anl.gov/petsc/. Accessed May 31, 2021.

[3] SALOME: The Open Source Integration Platform for Numerical Simulation). https://www.salome-platform.org/. Accessed February 1, 2022.

[4] K. Adzakpa, K. Agbossou, Y. Dube, M. Dostie, M. Fournier, and A. Poulin, PEM fuel cells modeling and analysis through current and voltage transient behaviors. *IEEE Trans. Energy Convers.*, **23**(2), 581–591, 2008.

[5] J. Ahrens, B. Geveci, and C. L. Paraview, An end-user tool for large data visualization. In: *The visualization handbook*, vol. 717(8), 2005.

[6] M. B. Allen III and E. L. Isaacson, *Numerical analysis for applied science*, 2nd edition. John Wiley & Sons, 2019.

[7] P. R. Amestoy, I. S. Duff, J.-Y. L'Excellent, and J. Koster, MUMPS: A general purpose distributed memory sparse solver. In: *International Workshop on Applied Parallel Computing*, pages 121–130. Springer, 2000.

[8] D. Arndt, W. Bangerth, B. Blais, T. C. Clevenger, M. Fehling, A. V. Grayver, T. Heister, L. Heltai, M. Kronbichler, M. Maier, P. Munch, J.-P. Pelteret, R. Rastak, I. Thomas, B. Turcksin, Z. Wang, and D. Wells, The deal.II library, version 9.2. *J. Numer. Math.*, **28**(3), 131–146, 2020.

[9] U. M. Ascher and L. R. Petzold, *Computer methods for ordinary differential equations and differential-algebraic equations*, vol. 61. SIAM, 1998.

[10] K. Atkinson, W. Han, and D. E. Stewart, *Numerical solution of ordinary differential equations*. John Wiley & Sons, 2009.

[11] M. K. Baboli and M. Kermani, A two-dimensional, transient, compressible isothermal and two-phase model for the air-side electrode of PEM fuel cells. *Electrochim. Acta*, **53**(26), 7644–7654, 2008.

[12] S. K. Babu, H. T. Chung, P. Zelenay, and S. Litster, Modeling electrochemical performance of the hierarchical morphology of precious group metal-free cathode for polymer electrolyte fuel cell. *J. Electrochem. Soc.*, **164**(9), F1037, 2017.

[13] C. A. Balen, *A multi-component mass transport model for polymer electrolyte fuel cells*. Master's thesis, University of Alberta, 2016.

[14] R. Balliet, *Modeling Cold Start in a Polymer-Electrolyte Fuel Cell*. Ph. D. thesis, University of California, Berkeley, 2010.

[15] R. Balliet and J. Newman, Two-dimensional model for cold start in a polymer-electrolyte-membrane fuel cell. *ECS Trans.*, **33**(1), 1545–1559, 2010.

[16] R. Balliet and J. Newman, Cold start of a polymer-electrolyte fuel cell I. Development of a two-dimensional model. *J. Electrochem. Soc.*, **158**(8), B927–B938, 2011.

[17]   W. Bangerth, R. Hartmann, and G. Kanschat, deal.II – A general-purpose object-oriented finite element library. *ACM Trans. Math. Softw.*, **33**(4), 24, 2007.

[18]   A. J. Bard and L. R. Faulkner, *Electrochemical methods: Fundamentals and applications*. Wiley, New York, 2001.

[19]   R. Barrett, M. Berry, T. F. Chan, J. Demmel, J. Donato, J. Dongarra, V. Eijkhout, R. Pozo, C. Romine, and H. V. der Vorst, *Templates for the Solution of Linear Systems: Building Blocks for Iterative Methods*, 2nd edition. SIAM, Philadelphia, PA, 1994.

[20]   E. B. Becker, G. Carey, and J. Oden, *Finite Elements: An Introduction*, vol. 1. Prentice-Hall, Englewoods Cliffs, 1981.

[21]   T. Bednarek and G. Tsotridis, Calculation of effective transport properties of partially saturated gas diffusion layers. *J. Power Sources*, **340**, 111–120, 2017.

[22]   B. Belfort, J. Carrayrou, and F. Lehmann, Implementation of richardson extrapolation in an efficient adaptive time stepping method: applications to reactive transport and unsaturated flow in porous media. *Transp. Porous Media*, **69**(1), 123–138, 2007.

[23]   S. A. Berlinger, P. J. Dudenas, A. Bird, X. Chen, G. Freychet, B. D. McCloskey, A. Kusoglu, and A. Z. Weber, Impact of dispersion solvent on ionomer thin films and membranes. *ACS Appl. Polymer Mater.*, **2**(12), 5824–5834, 2020.

[24]   D. Bernardi and M. Verbrugge, Mathematical model of a gas diffusion electrode bonded to a polymer electrolyte. *AIChE J.*, **37**(8), 1151–1163, 1991.

[25]   D. M. Bernardi and M. W. Verbrugge, Mathematical model of the solid-polymer-electrolyte fuel cell. *J. Electrochem. Soc.*, **139**(9), 2477–2491, 1992.

[26]   T. Berning and N. Djilali, A 3d, multiphase, multicomponent model of the cathode and anode of a PEM fuel cell. *J. Electrochem. Soc.*, **150**(12), A1589–A1598, 2003.

[27]   T. Berning and N. Djilali, Three-dimensional computational analysis of transport phenomena in a pem fuel cell – a parametric study. *J. Power Sources*, **124**(2), 440–452, 2003.

[28]   T. Berning, D. Lu, and N. Djilali, Three-dimensional computational analysis of transport phenomena in a pem fuel cell. *J. Power Sources*, **106**(1–2), 284–294, 2002.

[29]   M. Bhaiya, *An open-source two-phase non-isothermal mathematical model of a polymer electrolyte membrane fuel cell*. Master's thesis, University of Alberta, 2014.

[30]   M. Bhaiya, A. Putz, and M. Secanell, Analysis of non-isothermal effects on polymer electrolyte fuel cell electrode assemblies. *Electrochim. Acta*, **147**, 294–309, 2014.

[31]   R. B. Bird, W. E. Stewart, and E. N. Lightfoot, *Transport Phenomena*, 2nd edition. Wiley international, 2002.

[32]   M. Bosomoiu, G. Tsotridis, and T. Bednarek, Study of effective transport properties of fresh and aged gas diffusion layers. *J. Power Sources*, **285**, 568–579, 2015.

[33]   W. Braff and C. K. Mittelsteadt, Electroosmotic drag coefficient of proton exchange membrane as a function of relative humidity. *ECS Trans.*, **16**(2), 309, 2008.

[34]   E. Bueler, *PETSc for Partial Differential Equations: Numerical Solutions in C and Python*. SIAM, 2020.

[35]   C. Chan, N. Zamel, X. Li, and J. Shen, Experimental measurement of effective diffusion coefficient of gas diffusion layer/microporous layer in PEM fuel cells. *Electrochim. Acta*, **65**, 13–21, 2012.

[36]   X. Chen, B. Jia, Y. Yin, and Q. Du, Numerical simulation of transient response of inlet relative humidity for high temperature PEM fuel cells with material properties. In: *Advanced Materials Research*, vol. 625, pages 226–229. Trans Tech Publ, 2013.

[37]   P. Choopanya and Z. Yang, Transient performance investigation of different flow-field designs of automotive polymer electrolyte membrane fuel cell (PEMFC) using computational fluid dynamics (CFD). In: *Proceedings of the 10th International Conference on Heat Transfer, Fluid Mechanics and Thermodynamics. International Conference on Heat Transfer, Fluid Mechanics and Thermodynamics*, 2014.

[38]   E. Cwirko and R. Carbonell, Interpretation of transport coefficients in Nafion using a parallel pore model. *J. Membr. Sci.*, **67**(2–3), 227–247, 1992.

[39] A. Damjanovic and V. Brusic, Electrode kinetics of oxygen reduction on oxide-free platinum electrodes. *Electrochim. Acta*, **12**(6), 615–628, 1967.

[40] T. A. Davis, Algorithm 832: UMFPACK V4. 3 — An unsymmetric-pattern multifrontal method. *ACM Trans. Math. Softw.*, **30**(2), 196–199, 2004.

[41] T. A. Davis, A column pre-ordering strategy for the unsymmetric-pattern multifrontal method. *ACM Trans. Math. Softw.*, **30**(2), 165–195, 2004.

[42] T. A. Davis, *Direct methods for sparse linear systems*. SIAM, 2006.

[43] E. J. Dickinson, Review of methods for modelling two-phase phenomena in a polymer electrolyte membrane fuel cell (PEMFC), 2020. https://eprintspublications.npl.co.uk/8841/1/MAT93.pdf. Accessed February 1, 2022.

[44] P. Dobson, *Investigation of the polymer electrolyte membrane fuel cell catalyst layer microstructure*. Master's thesis, University of Alberta, 2011.

[45] P. Dobson, C. Lei, T. Navessin, and M. Secanell, Characterization of the PEM fuel cell catalyst layer microstructure by nonlinear least-squares parameter estimation. *J. Electrochem. Soc.*, **159**(5), B514–B523, 2012.

[46] K. P. Domican, *Mathematical modeling and experimental validation of thin low platinum content and functionally graded cathode catalyst layers*. Master's thesis, University of Alberta, 2014.

[47] F. A. Dullien, *Porous media: Fluid transport and pore structure*. Academic press, 2012.

[48] M. Eikerling, Water management in cathode catalyst layers of PEM fuel cells: A structure-based model. *J. Electrochem. Soc.*, **153**(3), E58, 2006.

[49] M. Eikerling, A. S. Ioselevich, and A. A. Kornyshev, How good are the electrodes we use in PEFC? (Understanding structure vs. performance of membrane-electrode assemblies). *Fuel Cells*, **4**(3), 131–140, 2004.

[50] M. Eikerling and A. Kornyshev, Modelling the performance of the cathode catalyst layer of polymer electrolyte fuel cells. *J. Electroanal. Chem.*, **453**(1–2), 89–106, 1998.

[51] M. Eikerling and A. Kulikovsky, *Polymer electrolyte fuel cells: Physical principles of materials and operation*. CRC Press, 2014.

[52] D. Fairbanks and C. Wilke, Diffusion coefficients in multicomponent gas mixtures. *Ind. Eng. Chem.*, **42**(3), 471–475, 1950.

[53] R. Flückiger, S. A. Freunberger, D. Kramer, A. Wokaun, G. G. Scherer, and F. N. Büchi, Anisotropic, effective diffusivity of porous gas diffusion layer materials for PEFC. *Electrochim. Acta*, **54**(54), 551–559, 2008.

[54] G. E. Forsythe, M. A. Malcolm, and C. B. Moler, *Computer Methods for Mathematical Computations*. Prentice Hall, New Jersey, 1977.

[55] D. Froning, M. Drakselová, A. Tocháčková, R. Kodým, U. Reimer, W. Lehnert, and K. Bouzek, Anisotropic properties of gas transport in non-woven gas diffusion layers of polymer electrolyte fuel cells. *J. Power Sources*, **452**, 227828, 2020.

[56] T. F. Fuller, *Solid-polymer-electrolyte fuel cells*. Ph. D. thesis, University of California, Berkeley, 1992.

[57] G. A. Futter, P. Gazdzicki, K. A. Friedrich, A. Latz, and T. Jahnke, Physical modeling of polymer-electrolyte membrane fuel cells: Understanding water management and impedance spectra. *J. Power Sources*, **391**, 148–161, 2018.

[58] G. A. Futter, A. Latz, and T. Jahnke, Physical modeling of chemical membrane degradation in polymer electrolyte membrane fuel cells: Influence of pressure, relative humidity and cell voltage. *J. Power Sources*, **410**, 78–90, 2019.

[59] P. A. García-Salaberri, J. T. Gostick, G. Hwang, A. Z. Weber, and M. Vera, Effective diffusivity in partially-saturated carbon-fiber gas diffusion layers: Effect of local saturation and application to macroscopic continuum models. *J. Power Sources*, **296**, 440–453, 2015.

[60] P. A. García-Salaberri, G. Hwang, M. Vera, A. Z. Weber, and J. T. Gostick, Effective diffusivity in partially-saturated carbon-fiber gas diffusion layers: Effect of through-plane saturation distribution. *Int. J. Heat Mass Transf.*, **86**, 319–333, 2015.

[61]   P. A. García-Salaberri, I. V. Zenyuk, A. D. Shum, G. Hwang, M. Vera, A. Z. Weber, and J. T. Gostick,
       Analysis of representative elementary volume and through-plane regional characteristics of
       carbon-fiber papers: Diffusivity, permeability and electrical/thermal conductivity. *Int. J. Heat Mass
       Transf.*, **127**, 687–703, 2018.
[62]   T. R. Garrick, T. E. Moylan, M. K. Carpenter, and A. Kongkanand, Electrochemically active surface area
       measurement of aged Pt alloy catalysts in PEM fuel cells by CO stripping. *J. Electrochem. Soc.*, **164**(2),
       F55, 2016.
[63]   S. Ge, B. Yi, and P. Ming, Experimental determination of electro-osmotic drag coefficient in nafion
       membrane for fuel cells. *J. Electrochem. Soc.*, **153**(8), A1443–A1450, 2006.
[64]   D. Genevey, M. von Spakovsky, M. Ellis, D. Nelson, B. Olsommer, F. Topin, and N. Siegel, Transient
       model of heat, mass, and charge transfer as well as electrochemistry in the cathode catalyst layer of
       a PEMFC. In: *ASME 2002 International Mechanical Engineering Congress and Exposition*, pages 393–406.
       American Society of Mechanical Engineers, 2002.
[65]   M. R. Gerhardt, L. M. Pant, and A. Z. Weber, Along-the-channel impacts of water management and
       carbon-dioxide contamination in hydroxide-exchange-membrane fuel cells: A modeling study.
       J. Electrochem. Soc., **166**(7), F3180, 2019.
[66]   D. Gerteisen, T. Heilmann, and C. Ziegler, Modeling the phenomena of dehydration and flooding of a
       polymer electrolyte membrane fuel cell. *J. Power Sources*, **187**(1), 165–181, 2009.
[67]   C. Geuzaine and J.-F. R. Gmsh, A 3-D finite element mesh generator with built-in pre-and
       post-processing facilities. *Int. J. Numer. Methods Eng.*, **79**(11), 1309–1331, 2009.
[68]   P. Gode, F. Jaouen, G. Lindbergh, A. Lundblad, and G. Sundholm, Influence of the composition on the
       structure and electrochemical characteristics of the PEMFC cathode. *Electochim. Acta*, **48**, 4175–4187,
       2003.
[69]   A. Gomez, A. Raj, A. Sasmito, and T. Shamim, Effect of operating parameters on the transient
       performance of a polymer electrolyte membrane fuel cell stack with a dead-end anode. *Appl. Energy*,
       130, 692–701, 2014.
[70]   A. Goshtasbi, P. García-Salaberri, J. Chen, K. Talukdar, D. G. Sanchez, and T. Ersal,
       Through-the-membrane transient phenomena in PEM fuel cells: A modeling study. *J. Electrochem.
       Soc.*, **166**(7), F3154–F3179, 2019.
[71]   J. T. Gostick, M. A. Ioannidis, M. D. Pritzker, and M. W. Fowler, Impact of liquid water on reactant mass
       transfer in PEM fuel cell electrodes. *J. Electrochem. Soc.*, **157**(4), B563, 2010.
[72]   T. A. Greszler, D. Caulk, and P. Sinha, The impact of platinum loading on oxygen transport resistance.
       J. Electrochem. Soc., **159**(12), F831, 2012.
[73]   M. Grujicic and K. Chittajallu, Design and optimization of polymer electrolyte membrane (PEM) fuel
       cells. *Appl. Surf. Sci.*, **227**, 56–72, 2004.
[74]   E. Haier, S. Nørsett, and G. Wanner, *Solving Ordinary Differential Equations I: Nonstiff Problems*.
       Springer, Berlin Heidelberg, 1993.
[75]   J. T. Hinatsu, M. Mizuhata, and H. Takenaka, Water uptake of perfluorosulfonic acid membranes from
       liquid water and water vapor. *J. Electrochem. Soc.*, **141**(6), 1493–1498, 1994.
[76]   C. K. Ho and S. W. Webb, *Gas transport in porous media*, vol. 20. Springer, 2006.
[77]   S. Huo, K. Jiao, and J. W. Park, On the water transport behavior and phase transition mechanisms in
       cold start operation of PEM fuel cell. *Appl. Energy*, **233**, 776–788, 2019.
[78]   G. Hwang and A. Weber, Effective-diffusivity measurement of partially-saturated fuel-cell
       gas-diffusion layers. *J. Electrochem. Soc.*, **159**(11), F683–F692, 2012.
[79]   J. Hwang, C. Chao, C. Chang, W. Ho, and D. Wang, Modeling of two-phase temperatures in a
       two-layer porous cathode of polymer electrolyte fuel cells. *Int. J. Hydrog. Energy*, **32**(3), 405–414, 2007.
[80]   J. Hwang, C. Chao, and W. Wu, Thermal-fluid transports in a five-layer membrane-electrode assembly
       of a PEM fuel cell. *J. Power Sources*, **163**(1), 450–459, 2006.
[81]   J. Hwang and P. Chen, Heat/mass transfer in porous electrodes of fuel cells. *Int. J. Heat Mass Transf.*,
       49(13), 2315–2327, 2006.

[82]   H. Iden, K. Sato, A. Ohma, and K. Shinohara, Relationship among microstructure, ionomer property and proton transport in pseudo catalyst layers. *J. Electrochem. Soc.*, **158**(8), B987–B994, 2011.

[83]   G. Inoue, K. Yokoyama, J. Ooyama, T. Terao, T. Tokunaga, N. Kubo, and M. Kawase, Theoretical examination of effective oxygen diffusion coefficient and electrical conductivity of polymer electrolyte fuel cell porous components. *J. Power Sources*, **327**, 610–621, 2016.

[84]   M. Isichenko, Percolation, statistical topography, and transport in random media. *Rev. Mod. Phys.*, **64**(4), 961–1043, 1992.

[85]   T. Jahnke, G. A. Futter, A. Baricci, C. Rabissi, and A. Casalegno, Physical modeling of catalyst degradation in low temperature fuel cells: Platinum oxidation, dissolution, particle growth and platinum band formation. *J. Electrochem. Soc.*, **167**(1), 013523, 2019.

[86]   F. Jaouen, G. Lindbergh, and G. Sundholm, Investigation of mass-transport limitations in the solid polymer fuel cell cathode – I. Mathematical model. *J. Electrochem. Soc.*, **149**(4), A437–A447, 2002.

[87]   A. Jarauta, V. Zingan, P. Minev, and M. Secanell, A compressible fluid flow model coupling channel and porous media flows and its application to fuel cell materials. *Transp. Porous Media*, **134**(2), 351–386, 2020.

[88]   R. Jinnouchi, K. Kudo, N. Kitano, and Y. Morimoto, Molecular dynamics simulations on o2 permeation through nafion ionomer on platinum surface. *Electrochim. Acta*, **188**, 767–776, 2016.

[89]   A. Jo, S. Lee, W. Kim, J. Ko, and H. Ju, Large-scale cold-start simulations for automotive fuel cells. *Int. J. Hydrog. Energy*, **40**(2), 1305–1315, 2015.

[90]   C.-Y. Jung and S.-C. Yi, Influence of the water uptake in the catalyst layer for the proton exchange membrane fuel cells. *Electrochem. Commun.*, **35**, 34–37, 2013.

[91]   S. Jung, *Estimation of relative transport properties of fuel cell and electrolyzer transport layers by pore network and continuum based direct simulations*. Master's thesis, University of Alberta, 2020.

[92]   S. Jung, M. Sabharwal, A. Jarauta, F. Wei, M. Gingras, J. Gostick, and M. Secanell, Estimation of relative transport properties in porous transport layers using pore-scale and pore-network simulations. *J. Electrochem. Soc.*, **168**(6), 064501, 2021.

[93]   A. Kalidindi, R. Taspinar, S. Litster, and E. Kumbur, A two-phase model for studying the role of microporous layer and catalyst layer interface on polymer electrolyte fuel cell performance. *Int. J. Hydrog. Energy*, **38**(22), 9297–9309, 2013.

[94]   K. Karan, Interesting facets of surface, interfacial, and bulk characteristics of perfluorinated ionomer films. *Langmuir*, **35**(42), 13489–13520, 2019.

[95]   C. T. Kelley, *Iterative methods for linear and nonlinear equations*. SIAM, 1995.

[96]   D. Kelly, D. S. Gago, O. Zienkiewicz, I. Babuska, and et al.. A posteriori error analysis and adaptive processes in the finite element method: Part I — Error analysis. *Int. J. Numer. Methods Eng.*, **19**(11), 1593–1619, 1983.

[97]   H. Kim, S. Jeon, D. Cha, and Y. Kim, Numerical analysis of a high-temperature proton exchange membrane fuel cell under humidified operation with stepwise reactant supply. *Int. J. Hydrog. Energy*, **41**(31), 13657–13665, 2016.

[98]   S. Kim and M. Mench, Investigation of temperature-driven water transport in polymer electrolyte fuel cell: Thermo-osmosis in membranes. *J. Membr. Sci.*, **328**(1–2), 113–120, 2009.

[99]   J. Ko and H. Ju, Comparison of numerical simulation results and experimental data during cold-start of polymer electrolyte fuel cells. *Appl. Energy*, **94**, 364–374, 2012.

[100] A. Kongkanand and P. K. Sinha, Load transients of nanostructured thin film electrodes in polymer electrolyte fuel cells. *J. Electrochem. Soc.*, **158**(6), B703–B711, 2011.

[101] A. Kosakian, *Transient numerical modeling of proton-exchange-membrane fuel cells*. PhD thesis, University of Alberta, 2021.

[102] A. Kosakian, L. Padilla Urbina, A. Heaman, and M. Secanell, Understanding single-phase water-management signatures in fuel-cell impedance spectra: A numerical study. *Electrochim. Acta*, **350**, 136204, 2020.

[103] A. Kosakian and M. Secanell, Estimating charge-transport properties of fuel-cell and electrolyzer catalyst layers via electrochemical impedance spectroscopy. *Electrochim. Acta*, **367**, 137521, 2021.

[104] A. Kucernak and C. Zalitis, General models for the electrochemical hydrogen oxidation and hydrogen evolution reactions: Theoretical derivation and experimental results under near mass-transport free. *J. Phys. Chem. C*, **120**(20), 10721–10745, 2016.

[105] K. Kudo, R. Jinnouchi, and Y. Morimoto, Humidity and temperature dependences of oxygen transport resistance of Nafion thin film on platinum electrode. *Electrochim. Acta*, **209**, 682–690, 2016.

[106] K. Kudo and Y. Morimoto, Analysis of oxygen transport resistance of Nafion thin film on Pt electrode. *ECS Trans.*, **50**(2), 1487, 2013.

[107] A. Kulikovsky, J. Divisek, and A. Kornyshev, Modeling the cathode compartment of polymer electrolyte fuel cells: Dead and active reaction zones. *J. Electrochem. Soc.*, **146**(11), 3981–3991, 1999.

[108] A. A. Kulikovsky, *Analytical modelling of fuel cells*. Elsevier, 2019.

[109] A. Kusoglu, D. Kushner, D. K. Paul, K. Karan, M. A. Hickner, and A. Z. Weber, Impact of substrate and processing on confinement of Nafion thin films. *Adv. Funct. Mater.*, **24**(30), 4763–4774, 2014.

[110] A. Kusoglu, A. Kwong, K. T. Clark, H. P. Gunterman, and A. Z. Weber, Water uptake of fuel-cell catalyst layers. *J. Electrochem. Soc.*, **159**(9), F530–F535, 2012.

[111] A. Kusoglu and A. Z. Weber, New insights into perfluorinated sulfonic-acid ionomers. *Chem. Rev.*, **117**(3), 987–1104, 2017.

[112] J. M. LaManna and S. G. Kandlikar, Determination of effective water vapor diffusion coefficient in PEMFC gas diffusion layers. *Int. J. Hydrog. Energy*, **36**(8), 5021–5029, 2011.

[113] H. P. Langtangen and S. Linge, *Finite difference computing with PDEs: A modern software approach*. Springer, 2017.

[114] A. D. Le and B. Zhou, A general model of proton exchange membrane fuel cell. *J. Power Sources*, **182**(1), 197–222, 2008.

[115] H.-Y. Li, W.-C. Weng, W.-M. Yan, and X.-D. Wang, Transient characteristics of proton exchange membrane fuel cells with different flow field designs. *J. Power Sources*, **196**(1), 235–245, 2011.

[116] S. Li, J. Yuan, G. Xie, and B. Sundén, Effects of agglomerate model parameters on transport characterization and performance of PEM fuel cells. *Int. J. Hydrog. Energy*, **43**(17), 8451–8463, 2018.

[117] X. Lin, C. M. Zalitis, J. Sharman, and A. Kucernak, Electrocatalyst performance at the gas/electrolyte interface under high-mass-transport conditions: Optimization of the "floating electrode" method. *ACS Appl. Mater. Interfaces*, **12**(42), 47467–47481, 2020.

[118] S. Litster and N. Djilali, Two-phase transport in porous gas diffusion electrodes. In: *Developments in Heat Transfer*, vol. 19, pages 175–213. WIT Press, 2005.

[119] C. P. Liu, P. Saha, Y. Huang, S. Shimpalee, P. Satjaritanun, and I. V. Zenyuk, Measurement of contact angles at carbon fiber–water–air triple-phase boundaries inside gas diffusion layers using X-ray computed tomography. *ACS Appl. Mater. Interfaces*, **13**(17), 20002–20013, 2021.

[120] M. Lopez-Haro, L. Guétaz, T. Printemps, A. Morin, S. Escribano, P.-H. Jouneau, P. Bayle-Guillemaud, F. Chandezon, and G. Gebel, Three-dimensional analysis of Nafion layers in fuel cell electrodes. *Nat. Commun.*, **5**(1), 1–6, 2014.

[121] Z. Lu, M. M. Daino, C. Rath, and S. G. Kandlikar, Water management studies in PEM fuel cells, part III: Dynamic breakthrough and intermittent drainage characteristics from GDLs with and without MPLs. *Int. J. Hydrog. Energy*, **35**(9), 4222–4233, 2010.

[122] D. G. Luenberger, *Optimization by vector space methods*. John Wiley & Sons, 1997.

[123] M. Markiewicz, C. Zalitis, and A. Kucernak, Performance measurements and modelling of the ORR on fuel cell electrocatalysts–the modified double trap model. *Electrochim. Acta*, **179**, 126–136, 2015.

[124] G. A. Martynov and R. R. Salem, *Electrical double layer at a metal-dilute electrolyte solution interface*. Springer, 1983.

[125] P. A. M. Villanueva, *A mixed wettability pore size distribution model for the analysis of water transport in PEMFC materials*. Master's thesis, University of Alberta, 2013.

[126] C. K. Mittelsteadt and H. Liu, Conductivity, permeability, and ohmic shorting of ionomeric membranes. In: W. Vielstich, H. A. Gasteiger and H. Yokokawa, Eds., *Handbook of fuel cells: Fundamentals technology and applications: Advances in electrocatalysis, materials, diagnostics and durability*, pages 345–358. Springer/John Wiley & Sons, New York, 2009.

[127] M. Moore, *Investigation of the double-trap intrinsic kinetic equation for the oxygen reduction reaction and its implementation into a membrane electrode assembly model*. Master's thesis, University of Alberta, 2012.

[128] M. Moore, A. Putz, and M. Secanell, Development of a cathode electrode model using the ORR dual-trap intrinsic kinetic model. In: *International Conference on Fuel Cell Science, Engineering and Technology*, vol. 44823, pages 367–376. American Society of Mechanical Engineers, 2012.

[129] M. Moore, A. Putz, and M. Secanell, Investigation of the ORR using the double-trap intrinsic kinetic model. *J. Electrochem. Soc.*, **160**(6), F670–F681, 2013.

[130] M. Moore, S. Shukla, S. Voss, K. Karan, A. Z. Weber, I. Zenyuk, and M. Secanell, A numerical study on the impact of cathode catalyst layer loading on the open circuit voltage in a proton exchange membrane fuel cell. *J. Electrochem. Soc.*, **168**(4), 044519, 2021.

[131] M. Moore, P. Wardlaw, P. Dobson, J. Boisvert, A. Putz, R. Spiteri, and M. Secanell, Understanding the effect of kinetic and mass transport processes in cathode agglomerates. *J. Electrochem. Soc.*, **161**(8), E3125–E3137, 2014.

[132] S. Motupally, A. J. Becker, and J. W. Weidner, Diffusion of water in Nafion 115 membranes. *J. Electrochem. Soc.*, **147**(9), 3171–3177, 2000.

[133] V. Mulone and K. Karan, Analysis of capillary flow driven model for water transport in PEFC cathode catalyst layer: Consideration of mixed wettability and pore size distribution. *Int. J. Hydrog. Energy*, **38**(1), 558–569, 2013.

[134] J. H. Nam and M. Kaviany, Effective diffusivity and water-saturation distribution in single-and two-layer PEMFC diffusion medium. *Int. J. Heat Mass Transf.*, **46**(24), 4595–4611, 2003.

[135] J. Newman and K. E. Thomas-Alyea, *Electrochemical systems*. John Wiley & Sons, 2012.

[136] T. V. Nguyen and R. E. White, A water and heat management model for proton-exchange-membrane fuel cells. *J. Electrochem. Soc.*, **140**(8), 2178–2186, 1993.

[137] N. Nonoyama, S. Okazaki, A. Z. Weber, Y. Ikogi, and T. Yoshida, Analysis of oxygen-transport diffusion resistance in proton-exchange-membrane fuel cells. *J. Electrochem. Soc.*, **158**(4), B416–B423, 2011.

[138] R. O'Hayre, S.-W. Cha, W. Colella, and F. B. Prinz, *Fuel cell fundamentals*. John Wiley & Sons, 2016.

[139] M. E. Orazem and B. Tribollet, *Electrochemical impedance spectroscopy*, 2nd edition. John Wiley & Sons, 2017.

[140] J. P. Owejan, J. E. Owejan, and W. Gu, Impact of platinum loading and catalyst layer structure on PEMFC performance. *J. Electrochem. Soc.*, **160**(8), F824, 2013.

[141] L. M. Pant, M. R. Gerhardt, N. Macauley, R. Mukundan, R. L. Borup, and A. Z. Weber, Along-the-channel modeling and analysis of PEFCs at low stoichiometry: Development of a 1+2D model. *Electrochim. Acta*, **326**, 134963, 2019.

[142] A. Parthasarathy, C. R. Martin, and S. Srinivasan, Investigations of the O2 reduction reaction at the platinum/Nafion® interface using a solid-state electrochemical cell. *J. Electrochem. Soc.*, **138**(4), 916, 1991.

[143] A. Parthasarathy, S. Srinivasan, A. J. Appleby, and C. R. Martin, Pressure dependence of the oxygen reduction reaction at the platinum microelectrode/Nafion interface: Electrode kinetics and mass transport. *J. Electrochem. Soc.*, **139**(10), 2856–2862, 1992.

[144] A. Parthasarathy, S. Srinivasan, A. J. Appleby, and C. R. Martin, Temperature dependence of the electrode kinetics of oxygen reduction at the platinum/Nafion® interface—a microelectrode investigation. *J. Electrochem. Soc.*, **139**(9), 2530–2537, 1992.

[145] D. K. Paul, A. Fraser, and K. Karan, Towards the understanding of proton conduction mechanism in pemfc catalyst layer: Conductivity of adsorbed Nafion films. *Electrochem. Commun.*, **13**(8), 774–777, 2011.

[146] J. Peng, J. Shin, and T. Song, Transient response of high temperature PEM fuel cell. *J. Power Sources*, **179**(1), 220–231, 2008.

[147] J. Pharoah, On the permeability of gas diffusion media used in PEM fuel cells. *J. Power Sources*, **144**(1), 77–82, 2005.

[148] D. Pletcher, R. Greff, R. Peat, L. M. Peter, and J. Robinson, *Instrumental methods in electrochemistry*. Elsevier, 2001.

[149] W. H. Press, S. A. Teukolsky, W. T. Vetterling, and B. P. Flannery, *Numerical recipes: The art of scientific computing*, 3rd edition. Cambridge University Press, 2007.

[150] R. Rashapov and J. Gostick, In-plane effective diffusivity in PEMFC gas diffusion layers. *Transp. Porous Media*, **115**(3), 411–433, 2016.

[151] J. N. Reddy, *Introduction to the finite element method*. McGraw-Hill Education, 2019.

[152] A. Ribes and C. Caremoli, Salome platform component model for numerical simulation. In: *31st annual international computer software and applications conference (COMPSAC 2007)*, vol. 2, pages 553–564. IEEE, 2007.

[153] C. Robin, M. Gérard, M. Quinaud, J. d'Arbigny, and Y. Bultel, Proton exchange membrane fuel cell model for aging predictions: Simulated equivalent active surface area loss and comparisons with durability tests. *J. Power Sources*, **326**, 417–427, 2016.

[154] T. Rosen, J. Eller, J. Kang, N. I. Prasianakis, J. Mantzaras, and F. N. Büchi, Saturation dependent effective transport properties of PEFC gas diffusion layers. *J. Electrochem. Soc.*, **159**(9), F536, 2012.

[155] A. Roy, M. Serincan, U. Pasaogullari, M. Renfro, and B. Cetegen, Transient computational analysis of proton exchange membrane fuel cells during load change and non-isothermal start-up. In: *ASME 2009 7th International Conference on Fuel Cell Science, Engineering and Technology*, pages 429–438. American Society of Mechanical Engineers, 2009.

[156] M. Sabharwal, *Microstructural and performance analysis of fuel cell electrodes*. Ph. D. thesis, University of Alberta, 2019.

[157] M. Sabharwal, J. T. Gostick, and M. Secanell, Virtual liquid water intrusion in fuel cell gas diffusion media. *J. Electrochem. Soc.*, **165**(7), F553, 2018.

[158] M. Sabharwal, L. Pant, A. Putz, D. Susac, J. Jankovic, and M. Secanell, Analysis of catalyst layer microstructures: From imaging to performance. *Fuel Cells*, **16**(6), 734–753, 2016.

[159] M. Sabharwal, L. M. Pant, N. Patel, and M. Secanell, Computational analysis of gas transport in fuel cell catalyst layer under dry and partially saturated conditions. *J. Electrochem. Soc.*, **166**(7), F3065, 2019.

[160] R. Schlögl, Membrane permeation in systems far from equilibrium. *Ber. Bunsenges. Phys. Chem.*, **70**(4), 400–414, 1966.

[161] T. Schuler, A. Chowdhury, A. T. Freiberg, B. Sneed, F. B. Spingler, M. C. Tucker, K. L. More, C. J. Radke, and A. Z. Weber, Fuel-cell catalyst-layer resistance via hydrogen limiting-current measurements. *J. Electrochem. Soc.*, **166**(7), F3020, 2019.

[162] M. Secanell, *Computational modeling and optimization of proton exchange membrane fuel cells*. Ph. D. thesis, University of Victoria, 2007.

[163] M. Secanell, A. Jarauta, A. Kosakian, M. Sabharwal, and J. Zhou, PEM fuel cells, modeling. In: R. A. Meyers, Ed., *Encyclopedia of Sustainability Science and Technology*, pages 1–61. Springer, New York, 2017.

[164] M. Secanell, K. Karan, A. Suleman, and N. Djilali, Multi-variable optimization of PEMFC cathodes using an agglomerate model. *Electrochim. Acta*, **52**(22), 6318–6337, 2007.

[165] M. Secanell, A. Putz, S. Shukla, P. Wardlaw, M. Bhaiya, L. M. Pant, and M. Sabharwal, Mathematical modelling and experimental analysis of thin, low-loading fuel cell electrodes. *ECS Trans.*, **69**(17), 157, 2015.

[166] A. Shah, G.-S. Kim, P. Sui, and D. Harvey, Transient non-isothermal model of a polymer electrolyte fuel cell. *J. Power Sources*, **163**(2), 793–806, 2007.

[167] L. F. Shampine, Local error estimation by doubling. *Computing*, **34**(2), 179–190, 1985.

[168] Y. Shao, L. Xu, J. Li, and M. Ouyang, Numerical modeling and performance prediction of water transport for PEM fuel cell. *Energy Proc.*, **158**, 2256–2265, 2019.

[169] N. Siegel, M. Ellis, D. Nelson, and M. Von Spakovsky, Single domain PEMFC model based on agglomerate catalyst geometry. *J. Power Sources*, **115**(1), 81–89, 2003.

[170] D. Singh, D. Lu, and N. Djilali, A two-dimensional analysis of mass transport in proton exchange membrane fuel cells. *Int. J. Eng. Sci.*, **37**(4), 431–452, 1999.

[171] D. Song, Q. Wang, Z. Liu, T. Navessin, M. Eikerling, and S. Holdcroft, Numerical optimization study of the catalyst layer of PEM fuel cell cathode. *J. Power Sources*, **126**(1–2), 104–111, 2004.

[172] R. Songprakorp, *Investigation of transient phenomena of proton exchange membrane fuel cells*. Ph. D. thesis, University of Victoria, 2008.

[173] T. Springer, T. Zawodzinski, and S. Gottesfeld, Polymer electrolyte fuel cell model. *J. Electrochem. Soc.*, **138**(8), 2334–2342, 1991.

[174] D. Stauffer and A. Aharony, *Introduction to Percolation Theory*, 2nd edition. Taylor & Francis, London, 1992.

[175] W. Sun, B. A. Peppley, and K. Karan, An improved two-dimensional agglomerate cathode model to study the influence of catalyst layer structural parameters. *Electrochim. Acta*, **50**(16–17), 3347–3358, 2005.

[176] K. Sundmacher, T. Schultz, S. Zhou, K. Scott, M. Ginkel, and E. D. Gilles, Dynamics of the direct methanol fuel cell (DMFC): experiments and model-based analysis. *Chem. Eng. Sci.*, **56**(2), 333–341, 2001.

[177] T. Suzuki, K. Kudo, and Y. Morimoto, Model for investigation of oxygen transport limitation in a polymer electrolyte fuel cell. *J. Power Sources*, **222**, 379–389, 2013.

[178] R. Taylor and R. Krishna, *Multicomponent Mass Transfer. Series in Chemical Engineering*, 1st edition. John Wiley & Sons, New York, 1993.

[179] B. Tjaden, S. Cooper, D. Brett, D. Kramer, and P. Shearing, On the origin and application of the Bruggeman correlation for analysing transport phenomena in electrochemical systems. *Curr. Opin. Chem. Eng.*, **12**, 44–51, 2016.

[180] M. Tomadakis and T. Robertson, Viscous permeability of random fiber structures: comparison of electrical and diffusional estimates with experimental and analytical results. *J. Compos. Mater.*, **39**(2), 163–188, 2005.

[181] M. Tomadakis and S. Sotirchos, Ordinary and transition regime diffusion in random fiber structures. *AIChE J.*, **39**(3), 397–412, 1993.

[182] P. R. Turner, *Numerical analysis*. Macmillan, 1994.

[183] M. Verbrugge and R. Hill, Ion and solvent transport in ion-exchange membranes I. A macrohomogeneous mathematical model. *J. Electrochem. Soc.*, **137**(3), 886–893, 1990.

[184] A. Verma and R. Pitchumani, Influence of membrane properties on the transient behavior of polymer electrolyte fuel cells. *J. Power Sources*, **268**, 733–743, 2014.

[185] A. Verma and R. Pitchumani, Analysis and optimization of transient response of polymer electrolyte fuel cells. *J. Fuel Cell Sci. Technol.*, **12**(1), 011005, 2015.

[186] T. Vidaković, M. Christov, and K. Sundmacher, The use of CO stripping for in situ fuel cell catalyst characterization. *Electrochim. Acta*, **52**(18), 5606–5613, 2007.

[187] J. X. Wang, T. E. Springer, and R. R. Adzic, Dual-pathway kinetic equation for the hydrogen oxidation reaction on Pt electrodes. *J. Electrochem. Soc.*, **153**(9), A1732–A1740, 2006.

[188] J. X. Wang, J. Zhang, and R. R. Adzic, Double-trap kinetic equation for the oxygen reduction reaction on Pt (111) in acidic media. *J. Phys. Chem. A*, **111**(49), 12702–12710, 2007.

[189] Q. Wang, M. Eikerling, D. Song, and Z. Liu, Structure and performance of different types of agglomerates in cathode catalyst layers of PEM fuel cells. *J. Electroanal. Chem.*, **573**(1), 61–69, 2004.

[190] Q. Wang, F. Tang, B. Li, H. Dai, J. P. Zheng, C. Zhang, and P. Ming, Numerical analysis of static and dynamic heat transfer behaviors inside proton exchange membrane fuel cell. *J. Power Sources*, **488**, 229419, 2021.

[191] X.-D. Wang, J.-L. Xu, W.-M. Yan, D.-J. Lee, and A. Su, Transient response of PEM fuel cells with parallel and interdigitated flow field designs. *Int. J. Heat Mass Transf.*, **54**(11), 2375–2386, 2011.

[192] Z. Wang, C. Wang, and K. Chen, Two-phase flow and transport in the air cathode of proton exchange membrane fuel cells. *J. Power Sources*, **94**(1), 40–50, 2001.

[193] Z. Wang, C. Wang, and K. Chen, Two-phase flow and transport in the air cathode of proton exchange membrane fuel cells. *J. Power Sources*, **94**(1), 40–50, 2001.

[194] P. Wardlaw, *Modelling of PEMFC Catalyst Layer Mass Transport and Electro-Chemical Reactions Using Multi-scale Simulations*. Master's thesis, University of Alberta, 2014.

[195] A. Z. Weber, *Modeling water management in polymer-electrolyte fuel cells*. Ph. D. thesis, University of California, Berkeley, 2004.

[196] A. Z. Weber, Improved modeling and understanding of diffusion-media wettability on polymer-electrolyte-fuel-cell performance. *J. Power Sources*, **195**(16), 5292–5304, 2010.

[197] A. Z. Weber, R. L. Borup, R. M. Darling, P. K. Das, T. J. Dursch, W. Gu, D. Harvey, A. Kusoglu, S. Litster, M. M. Mench, R. Mukundan, J. P. Owejan, J. G. Pharoah, M. Secanell, and I. V. Zenyuk, A critical review of modeling transport phenomena in polymer-electrolyte fuel cells. *J. Electrochem. Soc.*, **161**(12), F1254, 2014.

[198] A. Z. Weber, R. L. Borup, R. M. Darling, P. K. Das, T. J. Dursch, W. Gu, D. Harvey, A. Kusoglu, S. Litster, M. M. Mench, R. Mukundan, J. P. Owejan, J. G. Pharoah, M. Secanell, and I. V. Zenyuk, A critical review of modeling transport phenomena in polymer-electrolyte fuel cells. *J. Electrochem. Soc.*, **161**(12), F1254–F1299, 2014.

[199] A. Z. Weber, R. M. Darling, and J. Newman, Modeling two-phase behavior in PEFCs. *J. Electrochem. Soc.*, **151**(10), A1715, 2004.

[200] A. Z. Weber and A. Kusoglu, Unexplained transport resistances for low-loaded fuel-cell catalyst layers. *J. Mater. Chem. A*, **2**(41), 17207–17211, 2014.

[201] A. Z. Weber and J. Newman, Modeling transport in polymer-electrolyte fuel cells. *Chem. Rev.*, **104**(10), 4679–4726, 2004.

[202] A. Z. Weber and J. Newman, Coupled thermal and water management in polymer electrolyte fuel cells. *J. Electrochem. Soc.*, **153**(12), A2205, 2006.

[203] H. Wu, P. Berg, and X. Li, Non-isothermal transient modeling of water transport in PEM fuel cells. *J. Power Sources*, **165**(1), 232–243, 2007.

[204] H. Wu, P. Berg, and X. Li, Modeling of PEMFC transients with finite-rate phase-transfer processes. *J. Electrochem. Soc.*, **157**(1), B1–B12, 2010.

[205] J. Xie, F. Xu, D. L. Wood III., K. L. More, T. A. Zawodzinski, and W. H. Smith, Influence of ionomer content on the structure and performance of pefc membrane electrode assemblies. *Electrochim. Acta*, **55**(24), 7404–7412, 2010.

[206] L. Xing, X. Liu, T. Alaje, R. Kumar, M. Mamlouk, and K. Scott, A two-phase flow and non-isothermal agglomerate model for a proton exchange membrane (PEM) fuel cell. *Energy*, **73**, 618–634, 2014.

[207] F. Xu, S. Leclerc, D. Stemmelen, J.-C. Perrin, A. Retournard, and D. Canet, Study of electro-osmotic drag coefficients in Nafion membrane in acid, sodium and potassium forms by electrophoresis NMR. *J. Membr. Sci.*, **536**, 116–122, 2017.

[208] T. C. Yau, P. Sauriol, X. T. Bi, and J. Stumper, Experimental determination of water transport in polymer electrolyte membrane fuel cells. *J. Electrochem. Soc.*, **157**(9), B1310, 2010.

[209] X. Ye and C.-Y. Wang, Measurement of water transport properties through membrane-electrode assemblies: I. Membranes. *J. Electrochem. Soc.*, **154**(7), B676, 2007.

[210] C. Zalitis, A. Kucernak, J. Sharman, and E. Wright, Design principles for platinum nanoparticles catalysing electrochemical hydrogen evolution and oxidation reactions: Edges are much more active than facets. *J. Mater. Chem. A*, **5**(44), 23328–23338, 2017.

[211] T. A. Zawodzinski Jr., M. Neeman, L. O. Sillerud, and S. Gottesfeld, Determination of water diffusion coefficients in perfluorosulfonate ionomeric membranes. *J. Phys. Chem.*, **95**(15), 6040–6044, 1991.

[212] Z. Ball, H. M. Phillips, D. Callahan, and R. Sauerbrey, Percolative Metal-Insulator Transition in Excimer Laser Irradiated Polyimide. *Phys. Rev. Lett.*, **73**(15), 2099–2103, 1994.

[213] I. Zenyuk, P. Das, and A. Weber, Understanding impacts of catalyst-layer thickness on fuel-cell performance via mathematical modeling. *J. Electrochem. Soc.*, **163**(7), F691–F703, 2016.

[214] I. Zenyuk, R. Taspinar, A. Kalidindi, E. Kumbur, and S. Litster, Computational and experimental analysis of water transport at component interfaces in polymer electrolyte fuel cells. *J. Electrochem. Soc.*, **161**(11), F3091, 2014.

[215] I. V. Zenyuk, A. Lamibrac, J. Eller, D. Y. Parkinson, F. Marone, F. N. Büchi, and A. Z. Weber, Investigating evaporation in gas diffusion layers for fuel cells with X-ray computed tomography. *J. Phys. Chem. C*, **120**(50), 28701–28711, 2016.

[216] I. V. Zenyuk, E. Medici, J. Allen, and A. Z. Weber, Coupling continuum and pore-network models for polymer-electrolyte fuel cells. *Int. J. Hydrog. Energy*, **40**(46), 16831–16845, 2015.

[217] Z. Zhan, H. Zhao, P. C. Sui, P. Jiang, M. Pan, and N. Djilali, Numerical analysis of ice-induced stresses in the membrane electrode assembly of a PEM fuel cell under sub-freezing operating conditions. *Int. J. Hydrog. Energy*, **43**(9), 4563–4582, 2018.

[218] G. Zhang, L. Fan, J. Sun, and K. Jiao, A 3D model of PEMFC considering detailed multiphase flow and anisotropic transport properties. *Int. J. Heat Mass Transf.*, **115**, 714–724, 2017.

[219] Z. Zhang, L. Jia, X. Wang, and L. Ba, Effects of inlet humidification on PEM fuel cell dynamic behaviors. *Int. J. Energy Res.*, **35**(5), 376–388, 2011.

[220] J. Zhou, *Analyzing Multiphase Flow in Membrane Electrode Assembly Using a Mixed Wettability Mathematical Model*. Ph. D. thesis, University of Alberta, 2018.

[221] J. Zhou, A. Putz, and M. Secanell, A mixed wettability pore size distribution based mathematical model for analyzing two-phase flow in porous electrodes I. Mathematical model. *J. Electrochem. Soc.*, **164**(6), F530–F539, 2017.

[222] J. Zhou, S. Shukla, A. Putz, and M. Secanell, Analysis of the role of the microporous layer in improving polymer electrolyte fuel cell performance. *Electrochim. Acta*, **268**, 366–382, 2018.

[223] J. Zhou, D. Stanier, A. Putz, and M. Secanell, A mixed wettability pore size distribution based mathematical model for analyzing two-phase flow in porous electrodes II. Model validation and analysis of micro-structural parameters. *J. Electrochem. Soc.*, **164**(6), F540–F556, 2017.

[224] D. Ziegler, *Liquid water management in gas diffusion layers (GDL) for proton exchange membrane fuel cells (PEMFC)*. Bachelor's thesis, Hochschule Mannheim/University of Alberta, 2020.

Wolfgang Olbrich, Thomas Kadyk, Ulrich Sauter, and Michael Eikerling

7 Physical models of performance

7.1 Catalyst layer modeling

The overarching objective of catalyst layer modeling is to establish physical relations between fabrication procedures and conditions, microstructure, effective properties of transport and reaction, and performance and lifetime. Figure 1 illustrates the typical layout of a catalyst layer (CL) for the oxygen reduction reaction (ORR) in the cathode catalyst layer (CCL). All species and processes that occur anywhere in the cell are also found in the CCL, including the interfacial electrochemical reaction, the supply of protons from the polymer electrolyte membrane (PEM) and through electrolyte in the CCL, electron conduction through the percolating solid phase of Pt/C, diffusion of oxygen molecules in water or gas-filled pores and the vaporization and transport of water.

Unraveling CL structure and function using theory and computation is a hierarchical challenge, as can be seen in Figure 2. At the lowest scale, the size, shape and surface atom arrangement of catalyst nanoparticles, as well as interactions between catalyst and support material, determine the ORR mechanisms and pathways, and the net specific activity. Step-by-step calculations of free energy changes along reaction pathways are nowadays routinely performed with quantum mechanical simulations at the level of density functional theory [1–4]. In combination with interface theory and microkinetic modeling, computational studies can decipher the multistep ORR mechanism and provide effective kinetic parameters, namely the intrinsic exchange current density and the Tafel slope, as functions of the electrode potential [5, 6].

The next level in Figure 2 shows a single water-filled pore with Pt nanoparticles deposited at the walls. All electrochemical reactions in polymer electrolyte fuel cells (PEFCs) proceed at Pt/water interfaces, which exist in water-filled nanopores [7]. The sizes of these pores vary from 2 to 20 nm and the confining walls must be made of an electronically conducting support material like graphitized carbon black that supports

Wolfgang Olbrich, Theory and Computation of Energy Materials (IEK-13), Institute of Energy and Climate Research, Forschungszentrum Jülich GmbH, 52428 Jülich, Germany; and Corporate Research, Robert Bosch GmbH, 71272 Renningen, Germany; and Chair of Theory and Computation of Energy Materials, Faculty of Georesources and Materials Engineering, RWTH Aachen University, 52062 Aachen, Germany
Thomas Kadyk, Theory and Computation of Energy Materials (IEK-13), Institute of Energy and Climate Research, Forschungszentrum Jülich GmbH, 52428 Jülich, Germany; and JARA Energy, Jülich Aachen Research Alliance, 52425 Jülich, Germany
Ulrich Sauter, Corporate Research, Robert Bosch GmbH, 71272 Renningen, Germany
Michael Eikerling, Theory and Computation of Energy Materials (IEK-13), Institute of Energy and Climate Research, Forschungszentrum Jülich GmbH, 52428 Jülich, Germany; and Chair of Theory and Computation of Energy Materials, Faculty of Georesources and Materials Engineering, RWTH Aachen University, 52062 Aachen, Germany; and JARA Energy, Jülich Aachen Research Alliance, 52425 Jülich, Germany

https://doi.org/10.1515/9783110622720-007

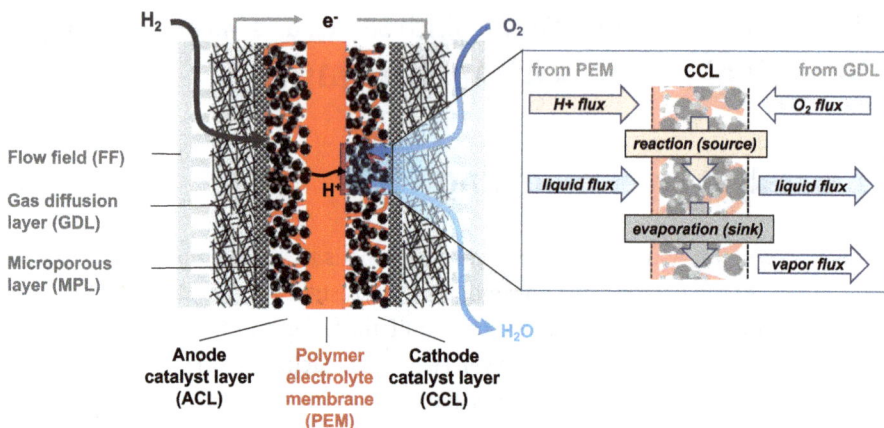

Figure 1: Layout of a polymer electrolyte fuel cell. The oxygen reduction reaction occurs in the cathode catalyst layer. Water is the reaction product and must be removed from the CCL in either vapor or liquid form.

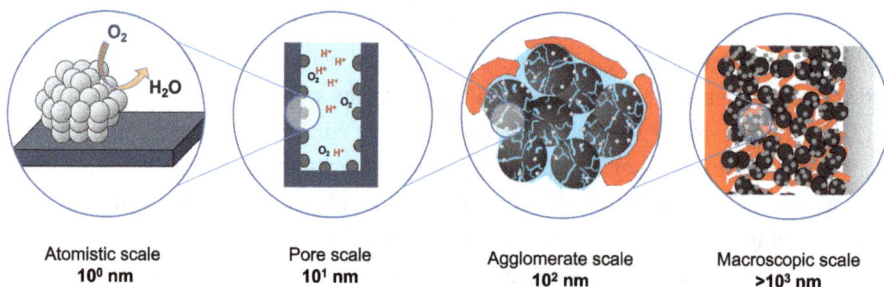

Figure 2: Main levels of the structural hierarchy in a catalyst layer: on the atomistic scale, platinum nanoparticles facilitate the oxygen reduction reaction (ORR); at the next level, water filled pores formed by the carbon support are decorated with catalyst nanoparticles; on the mesoscale, carbon support particles form agglomerates, which are covered by an ionomer film.

the Pt nanoparticles. Another type of confining wall consists of a skin-type ionomer film lined on the surface by sulfonate anions that are immobilized at the heads of ionomer sidechains. The charging properties of these walls regulate the density distribution of hydronium ions (or the local pH) in the pore, which is a key variable for the ORR activity and the Pt dissolution rate [8–12].

Agglomerates of Pt/C represent the next structural level. Agglomerates with typical sizes in the range 50 to 100 nm are partially covered by a thin ionomer film with thickness from 3 to 10 nm. It is particularly important that a large fraction of sulfonate anions at the ionomer film are oriented toward the agglomerate surface. This favorable orientation renders the interfacial region between agglomerate surface and ionomer film hydrophilic and it confers high proton concentration to the water film at the agglomerate surface, enhancing the electrocatalytic activity and the proton conductivity at the agglomerate scale [11, 13].

At the macroscopic scale, transport properties of the porous composite layer depend on percolation effects in the interpenetrating functional phases of solid Pt/C for electron transport, the network of water-filled pores for proton transport and the gaspore network for oxygen transport. The corresponding transport parameters determine the reactant distribution at the macroscale, and thus the net activity of the CCL.

Agglomerate and macroscopic scale must be coupled to determine the reaction rate distribution across the layer and inside agglomerates [14, 15]. Using a hierarchical structural model and nested mathematical descriptions to interlink transport phenomena at different scales, performance effects and catalyst utilization in CCL can be analyzed. Model-based analyses of experimental data can be used to determine parameters of transport and reaction in the layer [7], identify performance-limiting effects, generate a voltage loss breakdown [16], provide a spatially resolved activity map, and evaluate the overall effectiveness factor of Pt utilization [15].

Two basic types of CCLs exist: three-phase gas diffusion electrodes (GDE-type CCLs) and two-phase flooded electrodes (FE-type CCLs). The former represents the prevailing catalyst layer design in PEFCs. These layers, fabricated by ink-based approaches, incorporate all functional phases (metal: Pt/C; electrolyte: ionomer and liquid water; gas pores) and span all four levels of the structural hierarchy in Figure 2. The requirement for the active catalyst surface area enhancement factor and volumetric requirements for the interpenetrating phases determine the viable thickness range of this electrode design. The typical thickness of gas-diffusion-type CCL is from 2 to 10 µm.

CCLs of the second type, i. e., flooded electrodes, are ionomer-free. The thickness of these electrodes is in the range of up to ~200 nm; they are thus referred to as ultrathin CLs or UTCLs. At or below the nanopore scale, processes in UTCLs are similar to those in conventional CLs. A UTCL could be considered as an agglomerate with planar geometry. Since the thickness of a UTCL is typically about two orders of magnitude lower than that of a conventional CL, UTCLs operate in a different concentration and transport regime: the pore space of UTCLs is fully saturated with liquid water so that oxygen diffuses as a dissolved species and the transport of protons is regulated by the wall charge density of pores, which is a function of the metal phase potential [8, 9].

7.1.1 Cathode catalyst layer: important things to know

Reactive processes in PEFCs occur at metal-electrolyte interfaces in the two types of catalyst layers (CLs), as illustrated in Figure 1. The anode catalyst layer is supplied with molecular hydrogen, the fuel, to perform the hydrogen oxidation reaction (HOR). At the cathode catalyst layer (CCL), protons supplied via the polymer electrolyte membrane (PEM) and electrons arriving through the external electrical circuit react with molecular oxygen to form water in the oxygen reduction reaction (ORR).

Catalyst layer layout and basic functions

A simple cell layout with planar platinum electrodes separated by a PEM, would generate an electrode current density, j_0, in the order of ~ 10 mA cm^{-2}. Porous multiphase composite media are utilized as CLs to expand the electrode-electrolyte interface by several orders of magnitude, typically a factor in the order of 100, and proportionally bring up the current density to values around 1 A cm^{-2} per geometric area of the electrode or cell. These porous composite layers are responsible for promoting the kinetics of electrochemical reactions at the surfaces of highly dispersed catalyst nanoparticles; simultaneously, they facilitate the transport of reactant molecules (fuel and oxidant) by diffusion and of electrons and protons by migration. Moreover, the CCL plays a key role in regulating the water balance in the whole cell. This role entails retaining the necessary amount of water for optimal operation, while evacuating any excessive amount of water over adjacent layers, i. e., toward the cathode outlet via the microporous layer (MPL) and the gas diffusion layer (GDL) or toward the anode side via the PEM, as illustrated in Figure 1.

Catalyst of choice

The catalyst material of choice in PEFCs is metallic Pt, which has been known for a long time—since the invention of fuel cells—to exhibit an exceptional electrocatalytic activity for the ORR [17, 18]. The effective activity of Pt is a product of the activity per Pt surface atom (or turnover frequency) times the density of Pt surface atoms per unit volume. The latter factor depends on the surface atom density (packing density) of the catalyst, the size of Pt domains (nanoparticles or thin films) and the mass of Pt per unit CCL volume. A high Pt surface atom density can be achieved in two ways: with metallic nanoparticles of small size, typically around 2 to 5 nm, or with atomistically thin metallic catalyst films.

Among all catalyst materials being considered for PEFCs [19], Pt has remained the best in terms of stability and lifetime requirements of the cell. With proper adjustments in Pt particle sizes and under close control of operating conditions, monitoring especially the electrode potential in the cathode and its dynamic excursions during typical operating cycles that include start-up and shut-down, Pt is able to withstand harsh oxidizing conditions and various microscopic degradation mechanisms that constantly work to wear down the catalyst and cause PEFC performance to degrade [12, 20–22]. Nevertheless, catalyst stability and longevity remain foremost concerns of fuel cell developers. This is especially the case in the transportation sector that is most demanding in terms of the thirst for power density and the required tolerance of the cell to frequent and rapid load changes.

Another touted class of PEFC catalysts are organometallic compounds, which are synthesized after nature's blueprint for enzymatic catalysts [19, 23]. These catalysts could provide a high activity per active site, alas their major drawback is the low density of active sites per unit volume. Thus, they are not compatible with high power densities needed in transportation applications.

Pt mass loading
The Pt mass loading of CCLs is a crucial yardstick to determine the economic prospects of PEFCs. This property is defined as the mass of Pt per unit area of the cell and is given in units of mg cm^{-2}. Two steps have enabled the drastic reduction of this parameter, while increasing the power performance of the fuel cell. The first step involved scaling down the catalyst from bulky Pt to Pt nanoparticles and dispersing these particles on an electronically conducting carbon support. This step led to the increase of the mass-specific catalyst surface area by a factor 100 or more. Nanoparticle-dispersed Pt catalysts were introduced at the end of the 1980s by a team at the Los Alamos National Laboratory [24, 25]. The second step, introduced by the same team at Los Alamos, was the impregnation of porous Pt/C structures with a proton-conducting ionomer, which markedly increases the average proton density and the proton conductivity throughout the CCL and thereby leads to higher and more uniformly distributed reaction rates [24, 26].

Principal components and relevant sizes
The common type of CCLs is made from carbon black (mostly used: Ketjen Black or Vulcan) with particle sizes of 5 to 20 nm that support Pt nanoparticles with sizes in the range of 2 to 5 nm. Pt/C particles are mixed in the catalyst layer ink with an ionomer-based electrolyte, typically Nafion® or similar ionomer materials. In an ink solution, ionomer molecules self-aggregate into skin-type films with thickness 2 to 5 nm [27]. These films are dispersed in the resulting CCL structure and they partially encapsulate small aggregates of Pt/C particles. The resulting agglomerates of Pt/C particles and ionomer skin have sizes between 40 and 100 nm. These agglomerates are difficult to discern in microscopy studies or using other structural characterization methods. Moreover, their impact on performance remains controversial or at least difficult to quantify [28].

A strong indication that agglomeration effects play a role in the catalyst layer formation process was found originally in porosimetry studies that revealed a bimodal pore-size distribution (PSD) [29–32] and seen also in molecular dynamics simulations [33]. The first peak in such PSDs lies in the range from 5 to 20 nm and it can be identified with intraagglomerate (or primary) pores inside as well as between primary Pt/C particles. The second peak corresponds to pore sizes in the range of 20 to 50 nm and it can be assigned to interagglomerate (or secondary) pores.

Mixed wettability
CCLs exhibit mixed wetting behavior with both hydrophilic and hydrophobic pore walls. Hydrophilic surfaces can be attributed to metallic Pt, carbon, as well as the ion-lined surfaces of the ionomer. For carbons, even graphitized carbon only approaches contact angles of 90° from below but does not exceed this threshold. Hydrophobic wetting behavior is associated with surface patches of ionomer that consist of densely aggregated hydrophobic ionomer backbones. Attractive interactions of sulfonate anions with the Pt/C surface induce the preferential orientation of sidechains toward the agglomerate

surface. Hydrophobic PTFE backbones are consequently oriented toward interagglomerate spaces [33, 34]. Earlier models accounted for these wettability variations by defining two distinct types of pores with fixed contact angles [35–38]. Models are under development that will employ a continuous distribution of contact angles, similar to works addressing the mixed wettability in GDLs [39]. Understanding how modification of CCL materials, e. g., carbon support or ionomer, and the composition of CCL inks affect the contact angle distribution is subject to ongoing research [40].

Pore-size and wettability distributions determine the overall water uptake and the distribution of water in the pore volume. A crucial distinction to be made is between the water-filling of primary and secondary pores as well as between hydrophilic and hydrophobic pores. Another important difference exists between water-filled pores that are bound on all sides by walls made of Pt or C (all-metallic walls) and pores that are bound on one side by Pt/C and on the other side by an ionomer film (mixed walls). This distinction signifies a large variability of interfacial reaction conditions at the pore scale in terms of local electrolyte phase potential and pH, which has direct consequences for the spatial distributions of the rates of the ORR and of Pt dissolution.

Multiphase effect

In a well-designed CL, active catalyst nanoparticles are located at spots that are simultaneously connected to the percolating phases of transport media, namely the electrolyte phase consisting of water and ionomer domains for proton transport, the solid metal-like phase formed by Pt and C for electron conduction, and the gas-pore network for rapid diffusion of oxygen molecules. This breakdown of the composition of a CL might suggest that reactions would proceed at a triple-phase boundary. Indeed, the triple phase boundary has long been considered a useful concept to rationalize CCL operation. However, as an idealized geometrical construction the triple-phase boundary is of limited value for GDE-type CCLs in PEFCs (and obviously meaningless for FE-type CCLs). Oxygen molecules are able to diffuse a certain distance or depth in water-filled pores. This distance is the so-called reaction penetration depth or the diffusion length of dissolved oxygen. This parameter is typically on the order of 100 nm; it depends on the volumetric current density generated in the medium. Consequently, the active zone for the ORR extends away from the ideal geometric intersection that constitutes the triple phase boundary. The crucial concept of a reaction penetration depth is well known at least since Robert De Levie developed the so-called transmission line model in the 1960s to treat the problem of distributed current generation in flooded porous electrodes [17, 41, 42].

Ultrathin catalyst layers

To convince readers that the concept of a triple phase boundary is of limited practical value for describing the function of CCLs, we take a diversion from our main path (focusing on GDE-type CCLs) to briefly consider the case of ultrathin catalyst layers (UTCLs).

Realizations of this type of CCL include UTCLs fabricated by 3M's nanostructured thin-film (NSTF-) technology [43–46], layers fabricated by sputtering or ion beam-assisted deposition of Pt directly onto PEM or diffusion media [47, 48] or structured layers of a conductive nanoporous support (carbon nanotubes [49], metal foils of Au [50], metal-oxide based ordered layers [51]), into which Pt nanoislands or particles are deposited.

All of these different realizations of UTCLs enable a drastic reduction of the Pt load-ing. UTCLs are operated in a fully flooded regime, i. e., as flooded electrodes. Their typical thickness lies in the range of 200 nm, which is significantly smaller (by about a factor 20 to 50) in relation to ionomer-impregnated GDEs. This thickness range is similar to the reaction penetration depth of flooded electrodes, i. e., the typical length scale that sepa-rates a regime with uniform reaction conditions (below 200 nm) from the regime with large nonuniformity of the reaction conditions (above 200 nm) [52]. It has been demon-strated that such thin flooded electrodes, for which a triple phase boundary is nonex-istent, can exhibit very good performance as active layers for the ORR, at drastically reduced Pt loading [44].

The first theory of such layers was developed by Chan and Eikerling in [8]. It high-lighted the importance of the metal surface charging relation, i. e., the surface charge density at the walls of metallic nanopores, which is a function of electrode potential, for their operation. A refined charging relation was developed [10] and incorporated into a pore model [9].

Ionomer effect in GDEs

Coming back to GDE-type CCL, at the microscale the skin-type ionomer films create Pt-electrolyte interfaces with strongly enhanced intrinsic proton density [11, 13]. At the same time, dense aggregates of hydrophobic ionomer backbones that are oriented away from the agglomerate surface fulfill an important function as wet-proofing agents in sec-ondary pores, thereby having a beneficial effect on gaseous transport of oxygen across the layer [33].

In spite of the importance of the microscopic catalyst-ionomer interface, many ques-tions surrounding its structure and properties remain unresolved. Crucial questions persist that are related to the thickness of the water layer between Pt/C surface and ionomer film, the molecular structure (i. e., ordering) and dielectric properties of this confined water layer, the preferential orientation and interfacial density of ionomer sidechains (see Figure 3), spatial distributions of electrical potential and ion density in the water region, and the effect of the oxide coverage at Pt and C on the electrostatic properties of the interface. Recent works utilizing molecular dynamics simulations and theoretical modeling have started to address these aspects [11, 53].

Thickness effect

Transitioning from planar to porous CLs brings out the thickness l_{CL} as a crucial param-eter, which lies in the range from 2 to 10 μm for GDE-type CCL and should be <200 nm

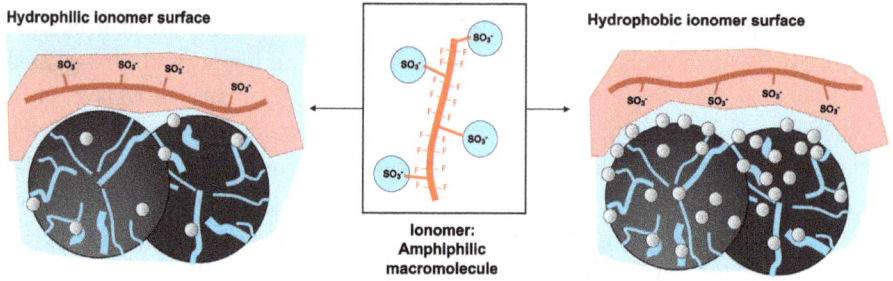

Figure 3: The ionomer film can exhibit either hydrophilic or hydrophilic wetting behavior (if viewed from the interior of the secondary pores towards the ionomer-covered surface of an agglomerate), depending on the interaction between sidechain/backbone of the amphiphilic ionomer macromolecule and the Pt/C catalyst surface. If the ionomer sidechains align toward the Pt/C surface (right), ionomer backbones are exposed to the outer surface, rendering it hydrophobic. Conversely, if ionomer backbones are oriented toward the support surface, the outer surface (toward the secondary pore space) will exhibit hydrophilic wettability (left).

for UTCLs, with a "forbidden thickness" range in between. In an operating regime with large cell current density, the local proton current density, $j(x)$, and the reaction rate distribution exhibit exponential-like decays along the thickness coordinate of the CCL. If this decay is caused by an insufficient proton conductivity, σ_p, while oxygen diffusion, with diffusion coefficient D_o, is rapid, the active sublayer, in which the exponential decay occurs, is pinned at the PEM side of the CCL. The characteristic length of this exponential decay is given by the reaction penetration depth,

$$l_p = \frac{\sigma_p b}{j_0} l_{CL},\tag{1}$$

where the operating current density of the cell is found from

$$j_0 = \sqrt{j^0 \sigma_p b}\, \exp\left(\frac{\eta_0}{2b}\right).\tag{2}$$

Here, b is the effective Tafel parameter, j^0 the corresponding effective exchange current density and η_0 the absolute value of the CCL overpotential at j_0. With the growth of j_0, the conversion domain shrinks.

For the case with rapid proton transport and slow oxygen diffusion, encountered for instance in a water-flooded CCL, the conversion domain is pinned to the boundary region at the CCL | PTL interface and the characteristic space scale of the decaying activity is given by another reaction penetration depth,

$$l_d = \frac{4FD_o c_1}{j_0} = \frac{I}{j_0} l_{CL},\tag{3}$$

with

$$j_0 = \sqrt{j^0 I} \exp\left(\frac{\eta_0}{2b}\right), \tag{4}$$

where c_1 is the oxygen concentration at the CCL | MPL boundary and $I = \frac{4FD_o c_1}{l_{CL}}$ is a characteristic current density of oxygen diffusion in the CCL.

In either case, the reaction penetration depth is proportional to the poor transport coefficient and inversely proportional to the cell current density. Highly nonuniform reaction rate distributions arise when l_{CL} is large compared to one of these two reaction penetration depths. Such conditions lead to the underutilization of Pt and large irreversible voltage losses.

Moreover, both current density vs. overpotential relations in Equations (2) and (4) exhibit a double Tafel slope behavior, as embodied in the term $2b$ in the denominator of the argument of the exponential function. This behavior is a signature of a nonuniform reaction rate distribution in the porous electrode due to a significant mass transport effect.

Practical challenges

At the current Pt mass loadings required for high cell performance, Pt is responsible for 30–70 % of the total cost of a fuel cell stack. The foremost challenge in PEFC research remains to maximize performance with a minimal amount of Pt. A significant reduction in Pt loading can translate into the economically necessary cost savings of PEFC stacks, as long as performance and lifetime targets are being met. The importance of a further Pt loading reduction is self-evident, considering that Pt is found among the least abundant elements in the upper crust of the Earth. Researchers and material developers have extensively searched for alternative catalyst materials in efforts to eliminate Pt, although even the most promising candidates still contain Pt as the catalytic base material [54–57].

7.1.2 How to evaluate the performance of porous composite electrodes

The challenge in developing a highly performing and cost-effective CCL is two-fold. The first task is to select or design an electrocatalyst material that provides high activity for the reaction of interest. This is a challenge for research in electrocatalytic materials design. Descriptor-based approaches are being developed in this field that allow for a systematic and rapid screening of the materials space [58, 59]. A convenient and widely employed tool in this field is the volcano plot [60, 61], based on the Sabatier principle [62], that could help find the most active material for a particular reaction within a certain class of materials [63, 64].

Care must be taken that the comparative assessment of catalyst activity is done under environmental and electrochemical conditions that resemble those encountered in the operating cell. Refined considerations—still dealing with the first task—should determine the optimal atomic composition and surface termination of catalyst materials as well as the size and shape of catalyst nanostructures [65]. In the case of supported catalysts, synergistic effects between catalyst and support material (in terms of their electronic and electrochemical properties), known as metal-support interaction, must be accounted for [66, 67].

Once a highly active material has been found that also fulfills the stability requirements for fuel cell electrocatalysts, the second task must be considered: how to surround the catalyst with a well-designed porous composite medium that enables high conversion rates locally and uniform reaction rate distributions globally, i. e., across the complete CCL thickness. This being a chapter on catalyst layer modeling, in what follows we will focus on the second task.

The main metric of a CCL is the overall mass activity, which can be interpreted in simple intuitive terms as the ratio of power performance to cost. The mass activity incorporates a parameter to account for the impact of a nonuniform reaction rate distribution. In this conext, we refer a compounded parameter that is the effectiveness factor of Pt utilization, which is not purely a statistical property.

In terms of the distribution of reaction rates and local conditions, CCL modeling must be able to monitor or, even better, predict the values of a minimal set of three variables: the electrical potential of the electrolyte phase, the proton density, and the oxygen concentration. In line with the majority of the literature on CCL modeling, we only consider variation of these properties in the through-plane direction (x-direction), i. e., perpendicular to the CCL | PEM interface. Deviations of any of these variables from uniformity in x-direction result in the reaction penetration depth being reduced compared to l_{CL}. Uniform conditions are thus easier to achieve for a thinner electrode, resulting in better catalyst utilization. Moreover, a thinner electrode can be built with less material and thus at reduced materials cost, especially a lower cost of Pt. However, with reduced Pt loading the peak power performance will be lower and, furthermore, the layer will be more prone to water flooding. Upon a drastic Pt loading reduction in the CCL, the nonlinear cross-component coupling involving MPL, GDL and flow field on the cathode side as well as PEM and anode side components will gain tremendously in importance.

However, before diving deeper into catalyst layer modeling, design and operation, we will discuss general aspects about the performance of nanoparticle-based composite CLs, since such layers are ubiquitous in electrochemical energy conversion devices. The main parameter of the electrode is the intrinsic exchange current density, j^{0*}. It describes the rate per unit real catalyst surface area (in units $A\,cm^{-2}$), with which electron transfer takes place across the electrified metal-electrolyte interface under a dynamic equilibrium. This parameter depends on the selection of metal and electrolyte materials and specific properties of the electrified interface formed between them.

The crucial structural parameter to be maximized in a catalytically active electrode is the electrochemically active surface area per unit volume, S_{ECSA}. This parameter is roughly proportional to the amount of catalyst, and in nanoparticle-based catalytic materials it is inversely proportional to the area-weighted mean particle radius. Using the two parameters introduced thus far, we can define the electrocatalytic activity per unit volume, $i^0 = j^{0*}S_{ECSA}$, and the ideal exchange current density per unit geometrical electrode surface area, $j^0 = j^{0*}S_{ECSA}l_{CL}$, which includes the explicit dependence on thickness.

High-performance CCLs are multiphase composites. The real accessible electrochemically active surface area, S_{ECSA}, is reduced relative to the total surface area of the catalyst, S_{tot}, due to the obstructed access to the catalyst surface of at least one of the species consumed in the ORR, namely electrons, protons or oxygen molecules. This effect is accounted for with a statistical surface utilization factor,

$$\Gamma_{stat} = \frac{S_{ECSA}}{S_{tot}}, \tag{5}$$

which can be assessed with percolation theory or other concepts in the statistical physics of heterogeneous media.

Lastly, reaction conditions will depart locally from ideal conditions, due to the limited rates of transport of protons and oxygen molecules. Nonuniform reaction conditions will reduce the effective catalyst utilization, which can be described with a specific function, i. e., the effectiveness factor, denoted as Γ. This factor depends on electrode materials and composition, which determine effective transport parameters of protons and oxygen molecules, as well as on the external or global parameters under which the cell is being operated. The effectiveness factor due to nonuniform reaction conditions entails effects that arise at different scales, i. e., in the microscopic interfacial region at the catalyst surface, in porous agglomerates of Pt/C and ionomer at the mesoscale, and at the macroscale. An overall effectiveness factor, Γ_{CL}, quantifies the compounded impacts of statistical utilization and transport phenomena.

The figure of merit of a catalyst layer is the current density that it generates at a given electrode potential and time of operation, $j_0 = f(E_{el}, t)$, divided by the mass loading of Pt at the beginning of life, m_{Pt}. This function can be defined through the following proportionalities:

$$\frac{j_0}{m_{Pt}} \propto j^{0*}\Gamma_{CL}, \tag{6}$$

where

$$\Gamma_{CL} = \Gamma_{np}\Gamma_{stat}\Gamma_{drc} \tag{7}$$

is the overall effectiveness factor. It accounts for three factors that contribute to nonideal catalyst utilization in nanoparticle-based CLs: Γ_{np} is the surface-to-volume atom ratio of catalyst nanoparticles; Γ_{stat} accounts for the simultaneous accessibility of catalyst particles to electrons, protons and reactants; and Γ_{drc}, which quantifies the uniformity of the reaction rate distribution that depends on mass transport phenomena of all species consumed in the reaction, i. e., protons, electrons and reactant gas. These transport effects occur on different length scales, namely the pore, agglomerate or CL scale, which allows to define further effectiveness factors, e. g., the effectiveness factor of agglomerate-level transport, Γ_{agg} [15], or the effectiveness factor of transport through the whole CCL based on the reaction penetration depth, Γ_δ [17].

Equations (6) and (7) can be used to rationalize and estimate an overall Pt mass activity and effectiveness factor of Pt utilization. They allow the factors that limit the performance of a CCL for a given material selection, composition, microstructure and thickness to be assessed and reserves for improvements in design and fabrication to be delineated.

The surface-to-volume atom ratio of a nanoparticle is defined as $\Gamma_{np} = N_S/N$, where N_S is the number of surface atoms and N is the total number of atoms in a particle. It is expected to be proportional to $N^{-1/3}$ for particles that are roughly spherical,

$$\Gamma_{np} = A + aN^{-1/3}, \tag{8}$$

where the parameters A and a assume different values for different particle shapes, as considered in [68]. Tetrahedral particles with four enclosing Pt(111) facets have the largest Γ_{np}, whereas cubo-octahedral particles assume the lowest Γ_{np}, about 15–20 % lower than tetrahedral particles at the same N. Another simple approximate relation to estimate Γ_{np} is

$$\Gamma_{np} \approx \frac{\sqrt{3}\bar{a}_{Pt}}{\bar{r}_{Pt}}, \tag{9}$$

where $\bar{r}_{Pt}$ is the average particle radius and $\bar{a}_{Pt} = (a_{Pt}^b)^3/(a_{Pt}^s)^2$ is the average lattice constant of the catalyst, calculated from the effective lattice constants in the bulk and at the particle surface, a_{Pt}^b and a_{Pt}^s, respectively. Particles with diameters <3 nm give $\Gamma_{np} >$ 50 %.

It is possible to determine the statistical utilization factor Γ_{stat} experimentally, since the electrochemically active area S_{ECSA} can be measured, e. g., by H-adsorption or CO-stripping voltammetry [69–72], while the total catalyst surface area can be calculated from the particle size distribution obtained by XRD measurements [69, 73–75] or from HR-TEM images [29, 74, 76–78]. For catalyst powders, a statistical utilization close to 100 % was found, with negligible losses in active surface area where the catalyst particle touches or is covered by the substrate [70]. However, the value of Γ_{stat} in an operational fuel cell was found to be significantly reduced in some cases. Values ranging from Γ_{stat} = 90 % [79] down to Γ_{stat} = 34 % [80] were found. The decrease of Γ_{stat} in an operational cell in comparison to the pure catalyst powder immersed in liquid electrolyte is caused by two effects. Firstly, a catalyst particle might be disconnected electronically from the substrate, i. e., from the percolating electron conduction pathway. Secondly, it might be disconnected from the proton transport network if it is not wetted by liquid water. As the wetting with liquid water depends on the operating conditions, the proton accessibility, and thus Γ_{stat} can change significantly under operation, even in the same material. This impact is reflected in the uncertainty observed in measured Γ_{stat}.

The third factor that lowers catalyst utilization is the nonuniform reaction rate distribution, which arises from mass and charge transport phenomena at finite current densities of fuel cell operation. This influence of the transport effects can be described by

$$\Gamma_{drc} = \frac{j_0}{j_0^{ideal}}, \tag{10}$$

i. e., the ratio of j_0 to an ideal current density, j_0^{ideal}, that would be generated at the same electrode potential if all mass and charge transport effects were negligible, i. e., if the transport rates of all species involved in the reaction were infinitely high.

Since transport effects exist on different length scales, i. e., throughout the whole electrode, inside agglomerates or inside of nanopores, Γ_{drc} can be broken down into contributions from each length scale. On the macroscale, porous electrode theory describes the interplay of reaction and transport of reactant gases, protons and electrons in the through-plane direction of the electrode. The sluggishness of each of these transport processes defines a reaction penetration depth, i. e., $\delta_{CL} = l_p$ or l_d, which is the typical length scale the species travels before being used up in reaction. If the reaction penetration depth is smaller than the thickness of the electrode, the transported species is no longer able to travel through the whole electrode, thus the far-to-reach catalyst sites are not utilized. Hence, a criterion for uniform reaction conditions on the macroscale is that the reaction penetration depth of every species involved in the reaction must be larger than the thickness of the electrode, $\delta_{CL} > l_{CL}$.

On the mesoscale, transport inside of catalyst agglomerates is considered to define a utilization factor of a single agglomerate, Γ_{agg} [15]. Since each agglomerate experiences different macroscopic reaction conditions in through-plane direction x, Γ_{agg} is a function of x. The macroscopic catalyst utilization factor Γ_{drc} can be calculated from the local utilization factors $\Gamma_{agg}(x)$,

$$\Gamma_{drc} = \frac{1}{l_{CL}} \int_0^{l_{CL}} \Gamma_{agg}(x)dx. \tag{11}$$

In a similar fashion, it is possible to break down hierarchical structures even further. For example, the utilization of an agglomerate can be broken down into contributions from different pore types, like ionomer-covered pores at the outer agglomerate surface or ionomer-free pores inside agglomerates. Inside a pore, the utilization factor can be broken down further into factors representing electrostatic and oxygen transport effects [8].

This way, the concept of catalyst utilization can be used to assess catalyst materials at all hierarchical levels and scales. The overall Pt effectiveness factor for conventional catalyst layers was found to be in the range of 4 % by model-based analyses [81] and similar values were measured at low operating currents [82]. These values decrease further at high current densities due to the higher nonuniformity of the reaction conditions, i. e., decreasing Γ_{drc} [8]. Overall, this type of assessment of the Pt effectiveness factor at the component or device level suggests that there is significant room for improvement in the design of catalyst layers and it leads to a conclusion that targets for the drastic Pt loading reduction, e. g., in the automotive industry [83], are within reach. The effectiveness factor is thus a useful concept to guide materials innovation and help tailor materials toward specific requirements.

7.2 Materials components, composition and structure-property relations

Building materials for CLs are selected by their intrinsic properties, including mechanical robustness, chemical stability, electronic conductivity, ion density and mobility, interfacial charging properties and electrocatalytic activity. The standard materials, namely carbon as the electronic conductor and support, Pt as electrocatalytically active material and the Nafion ionomer as the electrolyte provide these functions. These materials and their aggregation behavior, which can be tuned during ink fabrication, furthermore control charge density distributions and wettability of interfaces as well as the total porosity and the pore size distribution of the resulting layer.

The CCL is formed from four mixed and interpenetrating phases [17]:
- a *solid phase (S)* comprised of Pt-based catalyst and carbon-based support. Phase S determines basic electronic, electrostatic and electrocatalytic properties;
- a *liquid water phase (W)* that governs proton dynamics and controls the wetted, and thus, active Pt surface area;
- an *ionomer phase (I)* that contains water, but more importantly is a surface-activating agent that strongly interacts with the water-wetted Pt surface and determines the proton density at the microscale;
- a *gas-pore network (G)* that enables facile diffusion of oxygen gas and water vapor.

In varying amounts, any CL is built from these ingredients, which determine local reaction conditions and rates of reactive transport. A plethora of works in the literature have been focusing on different aspects of CL structure and function, such as ionomer distribution [84], catalyst support effects [29]or Pt particle dispersion [85].

Phases S and W are indispensable for CL operation as they form the microscopic electrocatalytic interface. On the other hand, a CL could function without phase I, as seen for ultrathin ionomer-free CLs [82]. However, phase I markedly enhances the local proton density, and thus the local activity, shifting these properties by up to an order of magnitude [11]. Yet, other studies have suggested an adverse impact of phase I on catalyst activity, via adsorption of sulfonate anions, i. e., the head groups of the ionomer sidechains [86–88].

Effective interactions between S and I phases, which must be separated by an interfacial water layer (or water slab), control the local interfacial proton density and the catalyst activity. Surface modification of catalyst and support modulate these interactions, and thus the interface structure and activity [84]. Furthermore, this local interface structure has an impact on the overall wetting behavior of the CCL [89].

Phase G is not desired at the microscale, as catalyst that is in contact with gas instead of water is disconnected from the proton-supplying network, and thus cannot contribute to electrochemical activity. However, the macroscale gas-pore network that exists in secondary pores is crucial for the oxygen supply.

Different operating regimes of CLs can be distinguished based on the relative volume fractions of W and G phases, with the major types being flooded electrodes and gas diffusion electrodes. The volume fractions of W and G phases depend on the pore-size distribution, the pore-space wettability and the thickness of the layer. Moreover, the water distribution depends on operating conditions (T, RH, gas pressure, current density). Upon increasing j_0, the liquid water front advances and the CCL will transition to a flooded state if vaporization of water cannot keep up with the rate of water production [90, 91].

From the point of view of statistical physics, the effective properties of random composite media depend on the volume fractions of distinct components as well structural correlations between them. At the fabrication stage, information on the composition is provided in terms of mass loadings or loading ratios. The volume fractions of distinct components in the CCL can be deduced from the mass densities of Pt, C and ionomer, viz. ρ_{Pt}, ρ_C, ρ_I, mass fractions, Y_C, Y_I, the Pt mass loading per unit geometric electrode area, m_{Pt}, and the thickness of the layer, l_{CL}, using the following relations:

$$X_{Pt} = \frac{m_{Pt}}{l_{CL}} \frac{1}{\rho_{Pt}}, \tag{12}$$

$$X_C = \frac{m_{Pt}}{l_{CL}} \frac{1 - Y_{Pt}}{Y_{Pt}\rho_C}, \tag{13}$$

$$X_I = \frac{m_{Pt}}{l_{CL}} \frac{Y_I}{(1 - Y_I)Y_{Pt}\rho_I}. \tag{14}$$

Next, the total porosity of the CCL can be calculated, $X_p = 1 - X_{Pt} - X_C - X_I$, which can be divided into porosity contributions due to primary and secondary pores, X_μ and X_M, with $X_p = X_\mu - X_M$. The volumetric filling factor of the pore space with water is defined as the liquid saturation, S_r, which depends on the pore-size distribution and the wettability of pores. The sizes and wettability distributions of pores control the build-up of the pressure distribution and liquid water saturation in the layer. Depending upon j_0, which determines the amount of water produced per unit volume of the CCL, and pressure boundary conditions, the liquid water front will advance or recede. Hydrophilic primary pores will be flooded already at j_0 below 100 mA cm^{-2}. Flooding of hydrophobic secondary pores requires liquid pressures exceeding the gas pressure, which implies that a continuous liquid water flow exists from the CCL to the flow-field—a scenario that occurs at high j_0 and that usually incurs significant voltage losses because of liquid water blocking gas diffusion pathways.

Based on Equations (12) to (14), percolation theory can be employed to determine the effective properties of CCLs [17]. Thereafter, these relations can be parameterized with experimental data from porosimetry measurements, water sorption studies, as well as imaging and tomography using x-rays and neutrons. The percolation relations for the effective proton conductivity and oxygen gas diffusivity, as well as the interfacial vaporization exchange area and the exchange current density have been presented elsewhere [17] and will not be reproduced here. However, this multiphase approach demonstrated its utility in predicting the optimal composition of the CCL in terms of ionomer loading and the optimal thickness for a targeted current density range [92, 93] and it demonstrated that functional grading of CCLs in terms of ionomer and Pt distributions could serve as a fabrication strategy to improve Pt effectiveness and power performance [94, 95].

7.2.1 The story of water in CCL: "love and hate"

Water is unavoidable as the reaction product of the ORR and without it a PEFC could not function. The membrane needs water as a pore-filling and pore-forming medium, and as the proton solvent and transport medium. The CL would not be active without water since water is the transport medium for protons. The flow field provides the supply channels for cell humidification, as well as the path for the removal of excess water, in both liquid and vapor phases. While liquid water in the flow channels can transport large amounts of water, it affects the pressure loss and the oxygen supply capability of the channels. A finely balanced approach must be taken to control the water distribution and fluxes in PEFCs. In the following, we will break down the convoluted matter of water inside PEFCs, elucidating what scientists have disentangled so far, and where gaps in understanding remain.

Water-related physics in PEFCs are closely intertwined with the role and function of the ionomer. The relationship of water and ionomer is symbiotic: perfluorosulfonic acid (PFSA) ionomers, like Nafion, have a high acid content. Upon contact with water, the acid groups dissociate, resulting in a high concentration of mobile extra protons or hydronium ions. With water as solvent, protons acquire a high mobility. The combination of high hydronium ion density and excellent proton mobility results in a high proton conductivity, both in membranes and ionomer-impregnated CCLs. But even without ionomer, water is able to sustain rates of proton transport that ensure excellent performance of ionomer-free UTCLs, as long as the oxygen diffusion path through the liquid water is short enough [8, 9]. While adding ionomer could significantly boost the proton density, and thus the activity in a CCL [11], it is not indispensable for CCL operation, as mentioned already before.

In addition to being an excellent proton solvent and medium for proton conduction, water fulfills an explicit role as cocatalyst for the ORR [3]. The activity of Pt cannot be explained considering protons and oxygen alone since water is involved in the reaction mechanism. Similarly, water is an accelerant of degradation processes: near surface water molecules facilitate the elementary process of Pt atom extraction, as a crucial step in the oxide layer growth [96]. The formation and reduction of oxide layers play key roles in the multistep process of Pt atom dissolution [12, 97]. Overall, water thus plays a double-edged role: where there is water, there will be catalytic activity for the ORR but also accelerated catalyst dissolution.

At the microscopic interface between the metal electrode surface and the electrolyte, water molecule orientations affects= the response of the surface charge density to the electrode potential [10]. This water-affected charging relation defines the local pH at the electrochemical interface that is a key variable for the ORR.

These different roles of water in a PEFC and especially in the CCL show that the story of water is a story of "love and hate." A well-functioning CCL needs both protons and oxygen molecules in high concentrations at the surface of the Pt catalyst. This translates into a design challenge for CCL materials: a benign design should afford a material with an optimal liquid water distribution that supplies protons while minimizing the transport resistance for oxygen. Protons "love" the water as their favorite solvent and they attain their highest mobility in it, relying on a relay-type mechanism that only exists in water [98–101].

Oxygen molecules, on the other hand, "hate" water. Since only a wetted Pt surface is ORR active, the only way for an oxygen molecule to get to the active Pt surface is via water—at least over the last stretch of the way. Two factors determine whether water will become a hindrance for oxygen supply in the ORR: the oxygen solubility in water and the length of the diffusion path through it. For a diffusion path shorter than ~100 nm the water pathway will not be limiting. Given this length scale, primary pores can be allowed to be flooded during PEFC operation. Flooding of secondary pores, however, would markedly increase the oxygen diffusion resistance and render oxygen transport the dominating cause of voltage loss.

The water balance becomes more delicate when the Pt loading is drastically reduced. Upon Pt loading reduction, unexpectedly high voltage losses are observed that exceed losses expected based on the proportional decrease of the ECSA. In [7], a tipping water balance was identified as the root cause of the detrimental performance effect. This interpretation has not yet been met with unanimous agreement. Indeed, the search for the origin of the dramatic Pt loading effect had become a major obsession in the field. A popular hypothesis stated that a thin ionomer layer covers the Pt catalyst sites, incurring an additional local transport loss [102–108]. However, this microscopic view is flawed, as water phenomena are usually not explicitly accounted for in analyses to support this argumentation.

A possible consequence of the delicate nature of this water balance shows in the fate of ultrathin catalyst layers (UTCL), fabriucated by the company 3M in the 2000s: they met or exceeded targets in performance and Pt loading reduction and exhibited much reduced rates of degradation at the CCL level [43]. In order to carry this performance over from the CCL to the stack level, this peculiar type of CCL needs to be enabled by well-attuned adjacent transport media, finely tuned break-in procedures and favorable operating conditions. However, ultimately, the UTCL concept failed since it required an operating window that was too narrow: under dry conditions, liquid water, and thus protons are missing in the CL. However, too much water quickly leads to flooding of the adjacent transport media. Thus, any deviation from the optimal wetting state led to drastic performance losses. This has sealed—at least for the time being—the fate of UTCLs for practical applications.

The "love and hate relationship" of the fuel cell with liquid water is not limited to the catalyst layers but spreads over the CCL boundaries into MPL, GDL and even into the flow field, as the transport through these layers and across interfaces between them determines the water fluxes, and thus the water balance in the CCL itself. In order to progress toward technological goals in performance, Pt loading reduction, and durability, the fuel cell must be designed as a whole and not as an assembly of standalone parts. Such an optimization of PEFC standalone components was predominant until roughly a decade ago, with tremendous progress in understanding [109, 110] but limited success in bringing the components to application. Nowadays, an expanded consideration of coupling to other components in the cell is central to efforts in the field [110, 111].

In principle, couplings of all transport processes need to be considered across all layers and components. While some processes might be neglected or only play an insignificant role, like electron transport, the two major coupling mechanisms are oxygen and water transport. In order to describe the local oxygen concentration at the catalyst surface, oxygen transport needs to be coupled from the flow field entrance through GDL and MPL to the CCL, where diffusive dispersion of oxygen and reaction occur. However, this oxygen transport is influenced by water phenomena, from the water production in the CCL over vaporization to the transport by vapor diffusion and convective flux of liquid water. Liquid water transport becomes complex at the transition from GDL to FF, where transport can be continuous, e. g., as film flow, or locally discontinuous due to droplet formation [112]. Water entrapment is also possible, e. g., under the ribs of the bipolar plate or flowfield channel curves and corners [113, 114]. To attain high current densities without flooding of GDL, MPL or CCL, the water transport across these layers must be managed well.

7.3 Performance modeling

A 1D model of the effects of structure and processes on steady-state performance of CLs requires a minimum of two parameters: the exchange current density j^0 and the reaction penetration depth, l_p or l_d, as introduced in Section 7.1.1. Each of these parameters is a function of structure and operating conditions. Barring issues related to the liquid water balance and other complicating effects, these two parameters uniquely define the potential loss in the CCL. Thereby, voltage efficiency, energy density, power density and effectiveness of catalyst utilization at given current density and catalyst loading can be calculated.

The present types of catalyst layers are random heterogeneous and nanoporous media. Macroscopic effective properties like transport coefficients and reaction rates are defined as averages over the corresponding microscopic quantities [115]. These averages must be taken over representative elementary volume elements (REVs), which have to be large compared to microscopic structural elements, i. e., pores and particles, and at the same time, small compared to the CCL size. The morphology of the composite layer can be related to effective properties that characterize transport and reaction, using concepts from the theory of random heterogeneous media [116]. The general ingredients of a macrohomogeneous model of CCL operation are the source term for electrochemical current conversion, the source/sink term for the interfacial phase change of water, as well as terms that account for the transport of species, that is, the migration of electrons/protons in conduction media, diffusion of dissolved oxygen and protons in water-filled pores and diffusion of oxygen and vapor in gas-filled pores.

A peculiar aspect of CCL modeling is that the structure and composition of the layer are not fixed. They undergo slow irreversible changes due to degradation (on the scale of 100s to 1000s of hours) as well as rapid reversible changes due to variations in water content (responses in surface configuration to potential). Proton transport, gas diffusion and electrocatalytic activity respond to these changes. Since the liquid water saturation S is a spatially varying function at $j_0 > 0$, physicochemical properties as well become spatially varying functions under operation. This creates a nonlinear coupling of properties and performance that demands self-consistent solution approaches.

The mathematical treatment of the problem involves a set of continuity equations to satisfy conservation of mass and charge. Transport equations and source or sink terms complete the general modeling framework. This chapter focuses on steady-state phenomena during electrode operation. The standard-type one-dimensional modeling approach will be presented. More sophisticated approaches that couple water fluxes across CCL and diffusion media with the transport of water in flow field channels require higher dimensional models. A 1D + 1D approach that accounts for this coupling is under development at the time of writing of this chapter.

In the following, we assume that species transport occurs in through-plane direction, along the x-coordinate, only, i. e., normal to the electrode plane. Moreover, isothermal conditions are considered. The widely studied macrohomogeneous model (MHM) treats the CCL as a homogeneous medium with effective parameters for electron, proton and oxygen transport. The electrostatic potentials in the solid Pt/C phase and in the electrolyte phase as well as the oxygen concentration are treated as continuous functions of x.

The general form of the continuity equations is

$$\frac{\partial \rho}{\partial t} + \nabla \cdot \boldsymbol{j} = R, \tag{15}$$

where $\rho(\boldsymbol{r}, t)$ is the density field of the considered species, $R(\boldsymbol{r}, t)$ a volumetric source or sink rate and $\boldsymbol{j}(\boldsymbol{r}, t)$ a flux density. Steady-state operation implies $\frac{\partial \rho}{\partial t} = 0$. A block of three equations suffices to describe electrochemical processes in the layer and solve for overpotential, current density and oxygen partial pressure as spatially varying functions. Proton transport follows Ohm's law,

$$\frac{d\eta}{dx} = -\frac{j(x)}{\sigma_p(S(x))}, \tag{16}$$

interfacial charge transfer is given by

$$\frac{dj}{dx} = -R_{reac}(x), \tag{17}$$

and oxygen diffusion is governed by

$$\frac{dp}{dx} = \frac{j_0 - j(x)}{4fD(S(x))}, \tag{18}$$

where $f = F/R_g T$ and D is the oxygen diffusion coefficient. Electron transport is considered fast, and thus negligible. Fluxes of oxygen and protons are related by $j_{ox}(x) = (j_0 - j(x))/4$. All fluxes are given in units of $A\,cm^{-2}$.

A second block of equations comprises the water transport and conversion processes, including water formation in the ORR and vaporization exchange,

$$\frac{dj^l}{dx} = \frac{1}{2}R_{reac}(x) - R_{lv}(x), \tag{19}$$

liquid water transport,

$$\frac{dp^l}{dx} = \frac{1}{B_0 fK^l(S(x))}\left[\left(n_d + \frac{1}{2}\right)(j(x) - j_0) + j^v(x) + n_d j_0 - j_m\right], \tag{20}$$

vaporization exchange,

$$\frac{dj^v}{dx} = R_{lv}(x), \tag{21}$$

and vapor diffusion,

$$\frac{dq}{dx} = -\frac{j^v(x)}{fD^v(S(x))}, \tag{22}$$

where $B_0 = R_g T/(V_{mw}\mu^l)$, with the molar volume V_{mw} and the dynamic viscosity of water μ^l. The solution space of the presented set of equations encompasses the partial pressure of oxygen, $p(x)$, vapor pressure, $q(x)$, liquid pressure, $p^l(x)$, electrode overpotential, $\eta(x)$, as well as fluxes (or equivalent current densities) of protons, $j(x)$, liquid water, $j^l(x)$ and vapor, $j^v(x)$. The parameter n_d is the electroosmotic drag coefficient.

The first block of equations, Equations (16) to (18), has one free parameter, namely the operating current density j_0, which is controlled by the experimentalist or the user of the cell. The second block, Equations (19) to (22) has two free parameters, i. e., the total water flux, j_m, and the proportion of the liquid water flux in the total water flux at $x = l_{CL}$. A full MEA model is needed to find a self-consistent solution for these two parameters.

The cathodic overpotential is defined

$$\eta(x) = \phi^{el}(x) - \phi^M(x) - (\phi^{el,eq}(x) - \phi^{M,eq}(x)), \tag{23}$$

where $\phi^{el}(x)$ and $\phi^M(x)$ are the local electrolyte and metal phase potentials, with the corresponding equilibrium values appearing in the bracket on the right-hand side. With the usually well-justified assumption that the metal (Pt/C) phase is equipotential, $\phi^M(x) = $ const., the overpotential incurred by the cathode is given by the local value at $x = 0$ (corresponding to the PEM|CCL boundary),

$$\eta_0 = \eta(0), \tag{24}$$

and, furthermore, the gradient of electrolyte phase potential in Ohm's law, Equation (16), can be replaced by the gradient in overpotential. In the regime of high overpotential ($\eta \gg 25\,\mathrm{mV}$), the faradaic current density can be expressed as the cathodic branch of the Butler–Volmer equation,

$$R_{reac}(x) = \frac{j^0}{l_{CL}} \frac{p(x)}{p^{FF}} \exp(a_c f\eta(x)), \tag{25}$$

where the pH, which is assumed to be constant in this type of model, is implicitly accounted for in effective parameters j^0 and a_c. It should be noted that these parameters, at best, could be considered piecewise constant over a certain potential range; for a multielectron process like the ORR, the rate-determining term will change with potential and thereby continuously modify the values of these parameters [5, 117].

The source term of water is

$$R_{lv}(x) = \frac{Fk_v}{l_{CL}} \xi^{lv}(S(x))\{q_r^s(T) - q(x)\}, \tag{26}$$

where k_v is the vaporization rate constant. The saturated vapor pressure at capillary radius $r^c(x)$ is given by the Kelvin equation,

$$q_r^s(T) = q^{s,\infty} \exp\left(-\frac{2\gamma_w \cos(\theta) V_{mw}}{R_g T r^c(x)}\right), q^{s,\infty} = q^0 \exp\left(-\frac{E_a}{k_B T}\right), \tag{27}$$

with the surface tension of water γ_w, the contact angle θ and the activation energy of vaporization at a planar water surface $E_a \approx 0.44$ eV.

The closure relation between liquid water saturation, $S(x)$, and pressure distribution is given by the Young–Laplace equation,

$$p^c(x) = \frac{2\gamma_w \cos(\theta)}{r^c(x)} = p(x) + q(x) + p^{res} - p^l(x), \tag{28}$$

with p^{res} denoting a residual gas pressure from a nonreactive gaseous component (like N_2), and integration the water-filled pore fraction in the particle radius distribution. Local values of liquid pressure and gas pressure, i. e., $p^l(x)$ and $p^g(x) = p(x) + q(x) + p^{res}$ determine the capillary radius, $r^c(x)$, which is needed to calculate the liquid saturation, $S(x)$. With $S(x)$ obtained, values of all transport coefficients can be determined, which are required as input for solving the transport equations. The presented system of equations thus represents a self-consistent loop that must be solved iteratively.

Consideration of water transport phenomena causes a highly nonlinear coupling among the complete set of equations. A much simpler model variant remains, when the liquid saturation is assumed to be uniform and constant, i. e., independent of electrode potential and local water production rate. In this case, we are left with a subset of equations for charge fluxes and conversion, Equations (16) to (18). This is the standard MHM type model of CCL operation. Different variants of this model have been developed over time [26, 93, 118]. Solutions and general capabilities of this model have been discussed in detail in [17] and the interested reader is referred to that reference. We will re-iterate only a few key aspects here.

7.3.1 Analytical solution of the macrohomogeneous model

The steady-state 1D performance model with uniform and constant liquid saturation, Equations (16) to (18), allows for an analytical solution [119] resulting in

$$\eta_0 = b \operatorname{arcsinh}\left(\frac{(\frac{j_0}{j_\sigma})^2}{2(c_h/c_{ref})(1 - \exp(-j_0/(2j_*)))}\right)$$
$$+ \frac{\sigma_t b^2}{4FDc_h}\left(\frac{j_0}{j_*} - \ln\left(1 + \frac{j_0^2}{j_*^2\beta^2}\right)\right)\left(1 - \frac{j_0}{j_{lim}^*}\right)^{-1} - b\ln\left(1 - \frac{j_0}{j_{lim}^*(c_h/c_{ref})}\right), \tag{29}$$

with the three characteristic current densities

$$j_* = \frac{\sigma_p b}{l_{CL}}, \tag{30}$$

$$j_\sigma = \sqrt{2i_* \sigma_p b}, \tag{31}$$

$$j_{lim}^* = \frac{4FD_{GDL}c_h}{l_{GDL}}, \tag{32}$$

where l_{CL} and l_{GDL} are the thickness of CL and GDL, respectively. The parameter β is the solution to the equation $\beta \tan(\beta/2) = j_0/j_*$, which can be approximated with sufficient accuracy by

$$\beta = \frac{\sqrt{2(j_0/j_*)}}{1 + \sqrt{1.12(j_0/j_*)} \exp(\sqrt{2(j_0/j_*)})} + \frac{\pi(j_0/j_*)}{2 + (j_0/j_*)}. \tag{33}$$

The equation

$$V_{cell} = V_{oc} - \eta_0 - R_\Omega j_0, \tag{34}$$

where V_{oc} is the open circuit voltage, completes the description of the polarization curve $V_{cell}(j_0)$.

In Equation (29), the first term on the right-hand side describes the overpotential losses due to the combined effect of ORR activation and proton transport. The second and third terms describe potential losses due to oxygen transport in the CCL and GDL, respectively. This separation of the contributions allows to perform a voltage loss break-down of the polarization curve as shown in Figure 4.

With increasing current density, the concerted losses due to activation and proton transport quickly increase, mainly in order to overcome the activation losses. Further increase of the current only mildly increases the losses in a weakly nonlinear fashion. The oxygen transport losses first increase in the CCL at medium current densities, while the oxygen transport in the GDL is most influential at high current densities. The ohmic losses increase linearly with current. The dependence of liquid water saturation on the current density neglected in the treatment here will worsen the oxygen transport be-havior in both CCL and GDL, ultimately leading to the strongly nonlinear breakdown of performance denoted as limiting current behavior.

7.3.2 Uses of the macrohomogeneous model

Assuming that the physical properties of the CCL, including transport and reaction pa-rameters, remain constant, the MHM could be applied to rationalize different limiting regimes and to delineate and quantify different contributions to overvoltage losses in the CCL. It is convenient to rewrite the three governing equations for this case, i. e., Equa-tions (16) to (18), in dimensionless form,

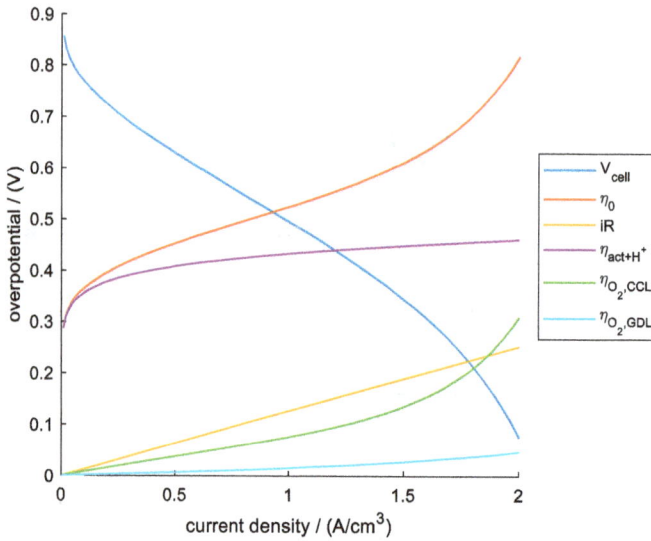

Figure 4: Voltage loss breakdown of a cell polarization curve (blue): overpotential (red) and ohmic voltage loss (iR, yellow). The total overpotential η_0 consists of losses due to reaction and proton conduction (η_{act+H^+}, purple), oxygen transport through the CCL ($\eta_{O_2,CCL}$, green) and through the GDL ($\eta_{O_2,GDL}$, cyan).

$$\frac{dP}{d\zeta} = -(1 - i), \tag{35}$$

$$\frac{d\Gamma}{d\zeta} = -g\Gamma i, \tag{36}$$

$$\frac{di}{d\zeta} = P\Gamma, \tag{37}$$

with dimensionless and normalized variables $\zeta = \frac{j_0}{I}$, $P = \frac{p}{p_L}$, $i = \frac{j}{j_0}$, and $\Gamma = \frac{j^0 I}{j_0^2} \exp(\frac{\eta}{b})$, where $I = \frac{4fD_{CL}p_L}{l_{CL}}$ is an effective oxygen diffusion parameter with units of $A\,cm^{-2}$ and b is an effective Tafel parameter. The oxygen partial pressure at the boundary of the CCL to the PTL (porous transport layer, i. e., GDL or MPL) is $p_L = p^{FF} - \frac{j_0 L_{PTL}}{4fD_{PTL}}$, with L_{PTL} and D_{PTL} being thickness and oxygen diffusion coefficient in the PTL. The solution of the dimensionless set of equations is subject to the boundary conditions

$$P(\zeta = 0) = 1, \quad i(\zeta = 0) = 0, \quad \text{and} \quad i\left(\zeta = \frac{j_0}{I}\right) = 1. \tag{38}$$

Writing the three equations of the MHM for the case of constant transport and reaction parameters in dimensionless form directly reveals that the shape of the solution is determined by a single parameter, i. e.,

$$g = \frac{4fD_{CL}p_L}{\sigma_p b}. \tag{39}$$

With this, parameter solutions of the MHM can be classified as oxygen diffusion-limited ($g \ll 1$), proton transport-limited ($g \gg 1$), and mixed ($g \sim 1$) cases, with the latter case being the most likely scenario.

The solution of the MHM can be written in general form as

$$\eta_0 = b\ln(\Gamma_0) + 2b\ln\left(\frac{j_0}{I}\right) - b\ln\left(\frac{j^0}{I}\right). \tag{40}$$

Poor oxygen diffusion will be the performance-limiting effect for CCL operation, when $j_0 \gg I$. In this limit, a sublayer of thickness $l_D = \frac{4fD_{CL}Dp_L}{j_0} \ll l_{CL}$ close to the PTL side will make the dominant contribution to the ORR current, whereas the contribution from most parts of the CCL, i.e., $l_{CL} - l_D \approx l_{CL}$ will be negligible. In this oxygen depletion regime, the performance is given by

$$\eta_0 \approx b\ln(\Gamma_c) + 2b\ln\left(\frac{j_0}{I}\right) + \frac{l_{CL}}{\sigma_p} - b\ln\left(\frac{j^0}{I}\right), \tag{41}$$

where the first term is now a constant and an explicit ohmic term has appeared that accounts for proton transport-related voltage losses incurred in the predominantly in-active part of the CCL. The solutions discussed in this section reveal a general feature of polarization curves, the appearance of a Tafel-like term with double Tafel slope (second term on the right-hand side). This doubling of the Tafel slope is a signature of any sort of transport limitation that occurs in a "thick" porous electrode.

Generally, the MHM distinguishes three performance regimes in polarization curves of CCL: a kinetic regime with simple Tafel dependence, $b\ln(\frac{j_0}{j^0})$, at $j_0 \ll I$, an intermediate regime with double Tafel slope characteristic, $2b\ln(\frac{j_0}{I})$, for $j_0 > I$ if $g < 1$ or $j_0 > \sigma_p b$ if $g > 1$, and an oxygen depletion regime with ohmic characteristic for $j_0 \gg I$ (and $l_D \ll l_{CL}$). The intermediate regime provides the best trade-off between requirements of a larger catalyst surface area that grows with the thickness of the CCL and the minimization of transport losses that also grow with the thickness of the CCL. A configuration diagram showing the three regimes was presented and discussed in [17]. It is reproduced for convenience in Figure 5.

Treating catalyst layers as random composite media, effective transport and re-action properties can be related to the thickness and the volumetric composition of the layer using percolation theory or related approaches from the theory of random heterogeneous media, as developed in [93] and refined in [90]. The resulting theory of composition-dependent CCL performance was shown to reproduce experimental trends in polarization curves. It was used to predict the composition in terms of the ionomer wt.% in the layer that yields the best power performance [92]. It was also employed to explore systematic performance improvements with functionally graded CCL [94, 95]. Compared to the normal CCL with uniform composition, a CCL with three sublayers in which the ionomer content was decreased from PEM boundary toward PTL boundary

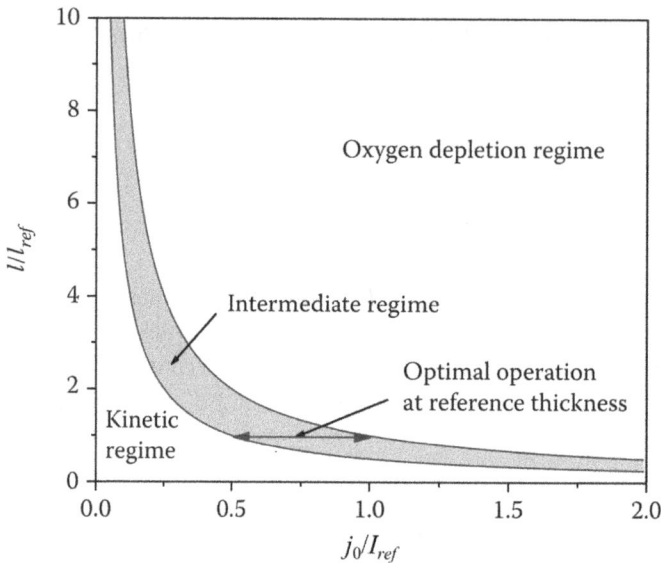

Figure 5: Configuration diagram of a cathode catalyst layer, showing the different regimes of operation assumed in dependence of catalyst layer thickness (ordinate) and fuel cell current density (abscissa). Thickness and current density are normalized to reference parameters with typical values $l_{ref} \approx 10\,\mu m$ and $I_{ref} \approx 1\,A\,cm^{-2}$. Reproduced with permission from Taylor and Francis, © 2014 [17].

in a step-like manner showed a significantly enhanced performance in experiment, as predicted by the model. Predictive capabilities of physical models of CCL operation should be improved by incorporating recent findings about structure formation during ink fabrication, micromorphology of nanoscopic ionomer films, and mixed wettability effects.

In a recent model-based study, the MHM in the form of Equations (29) to (34), was employed to analyze a wide range of fuel cell performance data from the literature [7]. Capabilities of the model-based approach in revealing trends and correlations in crucial CCL parameters, including proton conductivity, oxygen diffusion coefficients in CCL and GDL and effective exchange current density, were demonstrated. These trends were analyzed further to yield a mechanistic explanation of the impact of a Pt loading reduction on CCL performance. The logical reasoning underlying that explanation points to the important impact of the liquid water accumulation and the liquid water distribution in CCL and adjacent diffusion media that affects transport and reaction properties of these layers, and thereby influences the voltage output of the cell. The analysis presented in that work led to the definition of a correlation exponent that should find use in assessing the merit of different approaches in catalyst layer fabrication.

7.3.3 Refining the treatment of water phenomena: water retention curves

The driving force of liquid water transport in porous media is the gradient of liquid (or hydraulic) pressure. It is closely coupled to liquid saturation: depending on the pore wettability and radius, a pore will flood if a critical value of the liquid pressure is exceeded. The wettability of the pore walls and the pore-network morphology determine how water accumulates and advances in the electrode layers. The relation between liquid pressure and liquid saturation is called water retention curve (WRC). In order to achieve excellent performance, WRCs of different electrode layers have to be tuned in such way that a continuous distribution in liquid pressure is established, while local values of the liquid pressure remain low enough not to cause flooding in any layer.

For GDLs and MPLs, water retention curves are available from multiple sources and for numerous materials, but only few works provide WRCs for the CLs [120]. Assembling reliable WRC data of CLs is a laborious effort, which makes the comparison and optimization of different CL designs cumbersome. As an alternative to using a measured WRC, some models include two explicit contact angles combined with pore size distributions [36, 38, 121, 122]. First attempts to clarify which proportion of hydrophobic pores is desirable for a certain range of pore sizes led to a crude but important conclusion: micropores should be fully hydrophilic, while mesopores should exhibit a finely balanced mix of hydrophilic and hydrophobic wettability. Works capturing the wetting behavior by WRCs came to a similar result: a predominantly hydrophobic WRC helps to keep the gas-supplying pathways free of liquid water [123].

The ability to predict WRCs of CCLs on the basis of knowledge of their composition and materials selection would bypass the experimental hurdles and enable a model-based optimization of the CCL and adjacent porous media. How the CCL's material and fabrication parameter have to be tuned to match a desired wetting behavior is subject to on-going efforts in modeling [40]. Calibration of the wetting properties largely depends on ionomer structure and properties. The ionomer possesses partially hydrophobic and hydrophilic properties. Volfkovich et al. were the first to observe experimentally that the alignment of ionomer sidechains and backbones inside the ionomer thin film alters the wetting behavior of a carbon-ionomer mixture [34]. This finding was found to be in line with results of molecular dynamics simulations [27, 33, 124] and further experimental works [125]. The alignment of ionomer sidechains and orientation of backbones is imposed by the chemical nature of the Pt/C surface and the resulting interaction with ionomer sidechains or backbones.

Not only liquid water transport has to be considered in CL modeling, also the equilibrium with the gas phase and the water uptake of CL holds a few puzzles. When exposed to a water-containing atmosphere, the CL will accumulate significant amounts of water inside its microstructure where water can be absorbed in ionomer, adsorbed as a film on surfaces or behave bulk-like in sufficiently large pores. Uptake of bulk-like water by a porous medium proceeds via capillary condensation, a process that can be described by the Kelvin-equation [37, 90, 91, 126, 127]. However, water trapped at the Pt/C-ionomer interface cannot be assumed to behave as a bulk-like liquid. In the thickness range for this interfacial layer, expected to be below 2 nm, the water structure, the dielectric properties and the dynamics of water molecules and hydrated extra protons will be strongly affected by the proximity to charged and polar walls on either side of the slab [128–131]. Recent molecular dynamics studies explicitly focus on understanding the properties of water and the local reaction environment in this strongly confined interface region [13, 132].

7.4 Modeling structure-performance relations: bridging mesoscale and macroscale

Macrohomogeneous models discussed in previous sections need a parametrization for transport coefficients and kinetic parameters. These properties are largely determined by mesoscale structural features of the material. Multiple approaches have emerged to model the mesoscale structure and extract properties to be used in macrohomogeneous models. This section will briefly introduce four main types of approaches and the different aspects of the mesoscale structure they address.

The purpose of mesoscale models is to account for heterogeneities in porous composite media and capture the resulting distributions of the local reaction conditions. As depicted in Figure 6a, this can be accomplished by direct numerical simulation (DNS) of a representative elementary volume of the material. Other approaches apply different methods of abstraction, by defining a pore network or structural subunits like agglomerates or single pores.

Both DNS and pore-network models (PNM) build on 3D tomography data of the nano-to-mesoporous network, mainly employing scanning and transmission electron microscopy (SEM/TEM) and tomography [78, 121, 133–135] and nano-X-ray tomography [136]. Based on these characterization methods, DNS models discretize the structure of the medium into an ordered mesh, for which equations of transport and reaction can be solved numerically. PNMs map the information from 3D characterization data onto a network representation that consists of idealized elements, e. g., spherical pores and cylindrical throats (see Figure 6c). Overall, while DNS map the reaction and transport processes into a Cartesian mesh, PNM map them onto an idealized pore structure, e. g., spherical pores connected by cylindrical throats.

The two approaches evolved alongside modern tomography techniques. The structure of carbon and ionomer can be resolved down to 10 nm. Pt nanoparticles exhibit a tenfold better contrast in both electron and X-ray imaging due to their higher density and they can be resolved to finer structural details of 1 nm. Replacing the protons of the ionomer sulfonate head groups with Caesium ions allowed reaching a similar resolution for the ionomer film [134]. Such measurements require highly specialized and

a) Direct numerical simulation (DNS) | **b)** Agglomerate models (AM)

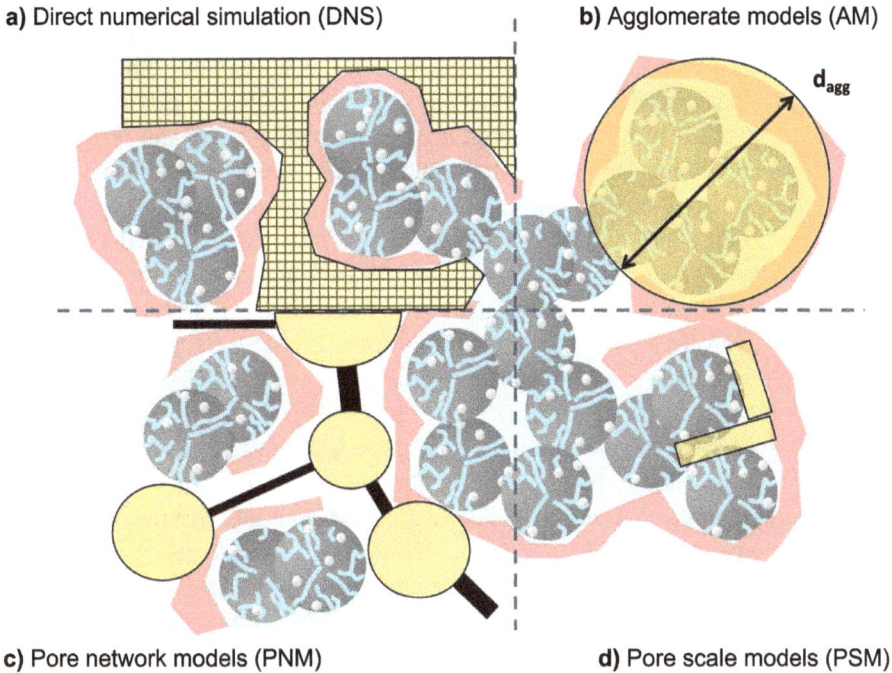

c) Pore network models (PNM) | **d)** Pore scale models (PSM)

Figure 6: Modeling approaches for the mesoscale structure of fuel cell catalyst layers. (a) direct numerical simulations, like continuum finite element methods or lattice-Boltzmann simulation, (b) agglomerate models, (c) pore-network models or (d) pores-scale models.

expensive equipment. Therefore, routines were proposed to stochastically generate 3D structures for DNS and PNMs [136, 137]. These routines can be combined with tomography data, complementing the measured 3D data with structural details below the resolution limit.

The explicit representation of mesoscale structures in DNS and PNM approaches allows studying reactive transport with a minimum of assumptions regarding pore-network geometry or topology but also entails the drawbacks of DNS and PNM approaches: besides experimental efforts to obtain tomographic images, especially DNS has high computational costs. PNMs are computationally more efficient but cannot guarantee the same accuracy as DNS [138]. DNSs and PNMs are suitable to analyze and calculate a material property for a given instance of a material class, but it would require enormous effort, both experimentally and computationally, to generalize the structure-property correlations, i. e., mapping the full parameter space of conceivable structural data. The exploration of structure-property relations calls for a different approach, to be discussed later in this chapter. Nonetheless, models analysing explicit singular representations of a class of materials, as DNS and PNMs do, might help to identify the crucial structural features shaping the macroscopic properties, such as transport properties, Pt utilization or wettability.

Different from DNS and PNM approaches, agglomerate models and pore-scale models do not model the porous network, but the subunits or building blocks of the mesoscale CCL structure. Agglomerate models were the first type of approach used to rationalize the effects of intrinsic catalyst activity, internal porosity and real catalyst surface area on reaction rate distributions and overall electrochemical performance of porous electrodes, dating back to Giner and Hunter [139] and followed by work of Iczkowski and Cutlip [140] and Björnbom [141]. These works already found the doubling of the Tafel slope as a signature of the interplay of mass transport with the distributed electrochemical activity in a porous electrode. The main value of AMs lies in rationalizing local reaction conditions and effectiveness factors [15, 142]. AMs typically consider the agglomerate to be a pseudo-homogeneous sphere (see Figure 6b), neglecting heterogeneities at the microscale. Pore-scale models (PSMs) increase the model resolution further and focus on a specific type of pore inside the agglomerate, i. e., slab-like water-filled pore that is confined by the ionomer film and Pt/C [11] or a micropore inside the Pt/C catalyst [9] (see Figure 6d). Generally, AMs and PSMs describe microstructural effects in a more analytical manner than DNS and PNMs, typically using a simplified geometry, such as spheres and cylinders. The model boundaries are clearly defined so that potential-dependent charging conditions at pore walls and the corresponding local reaction environment can be accounted for. They are an efficient tool to conduct parameter studies, e. g., on the pore geometry, the surface adsorption state, and capacitive properties of the wall material, and allow examining closely, how a particular structural modification or integration of a new material may alter the local reaction environment as well as the transport and reaction rates.

None of the mesoscale models discussed in this section provides a simple approach to account for all structural effects at the mesoscale. Whereas PNM and DNS focus on transport via the mesopore network, AM and PSM are specialized on rationalizing the local reaction environment. Few works strove to combine both aspects. For instance, Centinbas et al. [127] coupled AM and PSM with a DNS. Sadeghi [15] demonstrated that a pore-scale model accounting for local heterogeneities of proton and oxygen concentration can be combined with a MHM to resolve local effectiveness factors on the levels of agglomerates and the whole CCL. Generally, a viable approach to accomplish the bridging from mesoscale to macroscale effects should fulfill the following criteria: (1) explicitly describe mesoscopic heterogeneity of microstructure, accounting for pore-level transport, local reaction conditions ion pores and pore-space connectivity; (2) capture the structural characteristics of CCL at the different scales, as seen in Figure 2, and their impact on effective physical properties; (3) generalize structure vs. property relations from the mesoscale models and use them in MHM.

Feature (1) can be addressed by relying on statistical descriptors for structural parameters, such as pore-size distributions or averaged properties like ionomer coverage. For instance, a bimodal pore-size distribution contains information on how many micropores and mesopores exist in the the CCL, and how they vary in size. In order to fulfill (2), materials have to be characterized once by their statistical descriptors, e. g., measure a pore-size distribution, requiring a reasonable effort compared to explicit 3D tomography. Insights from DNS, PNM, AM and PSM have to be considered when identifying the most relevant structural descriptors of CCL. Besides the pore-size distribution, quantities such as ionomer coverage of Pt/C surface [143] and interior/exterior fraction (at agglomerate scale) of Pt particles [144] seem to have a significant impact. To fulfill (3), multiple approaches are conceivable. Wherever possible, mesoscale models should be scrutinized for potential generalizations to reliable structure vs. property relations, which can be readily used in MHMs. Since averaged statistical descriptors, such as porosity or ionomer coverage, will serve as the basis of a scale-bridging framework, classical random heterogeneous media theories [116], including percolation theory, will be of tremendous value in this context [145].

7.5 Dynamic phenomena and transient models

Dynamics determine the performance of a fuel cell at all stages, from the beginning of the lifecycle during ink formation and application, during MEA processing and break-in conditioning to degradation and failure. Under operation, potential variations induce changes in surface adsorption state, wettabilit, and capacitive charging at microscopic metal-electrolyte interfaces. Changes in the cell current incur transient variations in water saturation and distribution in the porous media. Dynamic processes that occur at widely different time scales shape the properties and performance of the CCLs.

To be able to control these processes and suppress associated costs and detrimental impacts for performance and lifetime, it is expedient to rationalize the underlying mechanisms. Generally, dynamic fuel cell processes can be classified into four major groups:

Electrochemical kinetics
The single most significant performance-determining electrochemical process is the oxygen reduction reaction. The ORR is a four-electron transfer that encompasses different chemisorbed oxygen species. These adsorbates play a double-edged role for the ORR, acting as vital intermediates, on the one hand, and site blockers on the other. Furthermore, interfacial effects like the orientational ordering of water molecules near the charged catalyst surface complicate the description of the ORR kinetics. Huang et al. proposed a framework that includes multiple interrelated surface phenomena and processes, shedding light on the role of the oxygen intermediates in the ORR [5, 117]. As a result, this framework gives out the apparent Tafel slope and exchange current density as continuous functions of electrode potential, which can be used as input for the MHM discussed in Section 7.3.

Dynamic mass transport phenomena
These include the dynamics of hydrogen and oxygen transport, from the cell inlet through channel and porous transport layers into the catalyst layers and through the

membrane. In addition to the educts, product water needs to be transported out, both in liquid and vapor forms. Especially on the cathode side, fluxes of gaseous components, including oxygen and water vapor, and of liquid water are coupled. The liquid saturation of the porous transport layers determines the effective transport properties for the gas. However, water uptake is slower than gas uptake, which leads to two distinct time constants of gas transport: one for the gas diffusion itself and one due to the shifting water balance. This difference in time constants can be exploited for diagnostic purposes [146], e. g., to probe the saturation as function of the applied current density.

Another form of water is encountered during freeze–start-up procedures, where phase transitions from ice to liquid occur [147, 148], and cold-start [149, 150] procedures, laying a focus on the dynamically shifting water balance. The common subject of these works is the competition between water generation and removal coupled to heat production and phase change dynamics. Transient, thermal models are emerging as essential tools to holistically understand and tailor PEM fuel cells, including design parameters of the catalyst layer [123, 151].

Ageing and degradation

This topic deals with structural changes that are irreversible and usually detrimental for fuel cell performance. The main aging mechanisms in CCL include Pt nanoparticle dissolution and Pt ion redeposition (Ostwald ripening) as well as nanoparticle aggregation, nanoparticle detachment, surface poisoning and inactivation. Available models provide kinetic expressions and reliable parametrizations for these mechanisms [20, 21, 152, 153]. Additionally, carbon support and ionomer ingredients undergo chemical and mechanical degradation. However, challenges for developing comprehensive treatments of performance and lifetime effects arise since degradation mechanisms occur on wide ranges of length and time scales. Moreover, spatially varying conditions due to the interplay of interfacial charging, pore-scale transport and electrocatalytic reactions must be accounted for. The latter aspect necessitates a coupling of degradation models with performance models.

Break-in dynamics

The dynamics of the break-in procedure are of primary importance for mass-manufacturing fuel cell stacks and a time-consuming process in the fabrication chain. Currently utilized break-in protocols rely on mostly empirical evidence but lack an explanation why certain protocols yield satisfying beginning-of-life performance of the fuel cell while others do not. Predictive modeling of structural changes during the break-in process is needed to speed up the conditioning of cells. The most relevant processes occurring during break-in and recovery procedures are the (re)distribution of ionomer, the removal of surface-contaminating agents and the formation of effective proton pathways inside the membrane [154].

7.6 Dynamic models for diagnostics: electrochemical impedance spectroscopy

Electrochemical Impedance Spectroscopy (EIS) is a well-established and highly insightful technique for fuel cell diagnostics [155–159]. In this method, the cell is excited with a harmonic signal, i. e., using a sinusoidal variation of either the fuel cell current density or voltage, and the corresponding output signal, i. e., either the time-dependent voltage or current density, is recorded. The response function is the complex time-dependent impedance that is the ratio of the voltage signal to the current density signal. Typically, this response function is analysed in the frequency domain, since in the limit of small perturbations both time-dependent variations in current density as well as voltage exhibit sinusoidal functional dependences, corresponding to a linear response. EIS models can be categorized into three classes.

Equivalent circuit models

In these empirical models, the system under study, i. e., the fuel cell is considered to be a black box. For the analysis of experimentally measured impedance spectra an equivalent electrical circuit is designed that reproduces the observed features in the EIS spectra. The model gets physical meaning when certain "hardware" elements of the circuit are interpreted as counterparts of specific physical processes in the cell. Resistors represent ohmic losses from electron or proton transport. They could represent as well a linearized charge transfer resistance in the limit of small overpotential. Capacitors represent charge or material storage capacities, re.g., of the electrochemical double layer. Inductive elements are used to describe inductive behavior that could be attributed to water transport characteristics [160], side reactions or catalyst poisoning, e. g., due to CO monoxide chemisorption [161]. Other elements are the Warburg element, describing diffusion toward or from an electrode [162] or the often discussed constant phase element introduced to describe nonhomogeneities, e. g., on rough electrodes [163] or porous electrodes with distributed reaction and relaxation processes [164]. The major drawback of equivalent circuit models is the often encountered lack of a physical basis in the interpretation of specific elements of the circuit [165] as well as ambiguities in these interpretations, meaning that different models may be able to fit a certain set of experimental data equally well [166]. These ambiguities can be further exacerbated because some parameters of the model may not be determined with sufficient accuracy [167].

Transmission line models (TLM)

In order to account for the distribution of the reaction in porous electrodes, equivalent circuit models can be extended from 0D to 1D [168]. The specific added value of these so-called transmission line models is that they describe the catalyst layer as a spatially extended electrical network [41]. The principal architecture of such a network consists of two or more parallel electrical rails that connect local equivalent circuits in parallel via transport resistances, as shown in Figure 7. In the example of the TLM for a PEFC catalyst layer in this figure, one rail represents the proton pathway starting from the membrane side, while the other rails represent the oxygen and electron pathways coming from the GDL side. Resistors along the rails represent proton, oxygen and electron transport resistances. The elements connecting the rails represent the ORR charge transfer resistance and capacitive charging of the electrochemical double layer. Transmission line models are able to simulate trough-plane processes of the CL, like proton, electron or oxygen transport and the resulting gradients in potential and concentrations [93, 169, 170]. They are especially applicable in diagnostics and characterization to detect inhomogeneities of the CL in trough-plane direction caused by fabrication or degradation [171, 172].

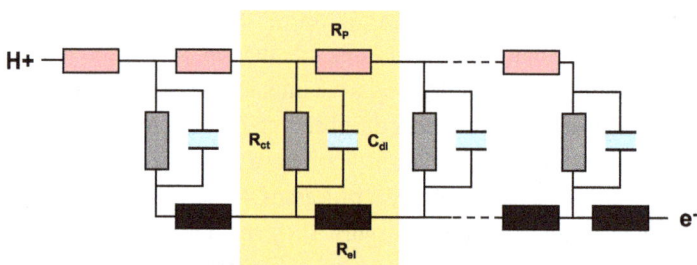

Figure 7: A one-dimensional transmission line equivalent circuit in which the elementary unit with protonic resistivity R_p, charge transfer resistivity R_{ct} and double layer capacitance C_{dl} is highlighted.

Macrohomogeneous models

Macrohomogeneous models (MHM) account for the continuous distribution of potential and concentrations of reactants and products. Macrohomogeneous electrochemical impedance models evolved from steady-state continuum models, as described in Section 7.3, which were extended by time-dependent terms in continuity equations [173, 174]. A transient MHM can be transformed to frequency space and solved for a harmonic perturbation signal, giving the EIS response as a function of the frequency of the perturbation. By fitting the parameters of the model to match experimental EIS data, parameters like proton conductivity, oxygen diffusivity or exchange current density be determined [174, 175]. A TLM can be considered the "finite difference approximation" of the transient MHM or, in return, the MHM could be seen as the continuum limit of the TLM. Also, TLM and MHM can be seamlessly combined [176].

7.7 Parameter identification, uncertainty and sensitivity analysis

Fuel cell models are becoming increasingly complex. Especially physics-based models derived from first principles contain a growing number of parameters. The predictive capabilities of these models do not only depend on the model structure, i. e., the first principles and the assumptions used to derive the model, but also on the accuracy of the parameter values. Hence, the problem of parameter identification becomes increasingly significant. The process of parameter identification encompasses three aspects: first, how accurate are the determined parameter values, i. e., how large are the parameter uncertainties? Second, how strong does the model prediction depend on the parameter values, i. e., how large are the parameter sensitivities? Third, how strongly do parameters correlate or stated differently, how independent, and thus identifiable are they?

Despite the central and increasing importance of parameter identification, this topic is still not sufficiently considered in the fuel cell community and the literature on the topic is sparse. It is an often-encountered practice that parameter values are assigned on the basis of convenience or simplicity rather than based on physical principles or understanding. Furthermore, a qualitative agreement with experimental data is often cited as sufficient evidence for model validity. Another practice is to treat parameter identification as material characterization, i. e., to determine model parameters in *ex situ* or component characterization experiments. Hereby, it is often not possible to determine all required parameters. Often, the determined parameters exhibit large uncertainties and reported parameter values vary by orders of magnitude, e. g., for thermal conductivities of various MEA layers for the same material set [177, 178].

An inherent problem in studying parameter identification is the high computational effort to solve the complex fuel cell models in combination with the typically large number of model evaluations that are required by systematic optimization-based model parametrization techniques. Hence, often only a small number of parameters is identified using optimization routines, while the remaining parameters rely on literature values. The choice, which parameters are to be identified by optimization is usually made arbitrarily or based on the researcher's intuition. Similarly, the operating conditions under which experimental data are collected, are not chosen in a systematic or optimal way. They often vary between the characterization experiments used for different parameters and consistent data sets are hard to find.

In order to close these gaps, Goshtabi et al. have recently presented a framework for parameter identification of a pseudo-2D, nonisothermal two-phase model of a PEFC [179, 180]. In the first part, they used a local sensitivity analysis carried out over several points in the parameter space to study the identifiability of the model parameters as well as the influence of the experimental conditions onto this identifiability. Based on this analysis, a parameter subset can be determined that is most suitable to be robustly identified by fitting experimental data. Then a procedure for a robust design of experiments was applied to obtain a set of experiments that are maximally informative for identification of the selected parameters. The last step is the development of a multistep parameter identification approach, which is then tested with synthetic experimental data generated by the model.

Apart from this single work in the fuel cell community, works on parameter identification are extensive and attract an increasing interest in other fields like batteries [181], fuel processing [182] or chemical engineering [183]. As fuel cell models become increasingly complex, applying these techniques in order to foster materials design, improve materials and cell characterization, as well as employing robust control of fuel cells in highly dynamic operation becomes imperative.

7.8 Summary

The global demand for highly performing and environmentally benign energy technologies is projected for a rapid growth over the next decade. This situation creates added impetus for research on polymer electrolyte fuel cells (PEFC) covering areas from the fine-tuned selection and modification of specific functional materials, to the development of advanced methods for structural characterization and property measurement, to device-level integration, testing and optimization. Among the components that are needed for a well-functioning, durable and affordable PEFC, the cathode catalyst layer (CCL), a specific type of a porous composite electrodes, is of outstanding importance. The main function to be provided by the CCL is to facilitate the oxygen reduction reaction. However, sufficiently high electrocatalytic activity of this reaction can, as of yet, only be achieved with platinum-based electrocatalysts. Given this constraint, efforts to strike an optimal balance of performance vs. cost impose the following basic rules for CCL design: incorporating the catalyst in the form of nanoparticles; providing high dispersion of catalyst nanoparticles on a carbon-based highly porous support; surrounding the catalyst with a water- and proton-friendly ionomer in order to maximize catalyst utilization and create a beneficial local reaction environment; and tuning the porosity, pore-size distribution and heterogeneous wettability of the CCL in order to facilitate oxygen diffusion and water transport. CCLs are thus hierarchically structured and multifunctional media, with an intricate interplay of transport and reaction unfolding in them.

In this realm, physical-mathematical theory, modeling and simulation provide increasingly powerful tools to understand, analyze and predict the complex relations between structure, properties and performance of active layers in PEFCs. CCL modeling has matured from the first approaches developed in the 1990s [17, 26, 93], with earlier approaches in porous electrode theory dating back even further [139–141]. CCL models can capture essential aspects of the structure and composition of materials—not only at the macroscale, but increasingly incorporating details of structure formation at meso and microscale, thanks to targeted efforts in molecular modeling that evolve alongside rapid advances in imaging and tomography. Particle and pore-size distributions as well as the structure and morphology (or film properties) of ionomer inclusions are crucial aspects that need to be accounted for [13, 27, 33, 53, 89]. Heterogeneous wettability properties and the corresponding water uptake, which can be quantified with water retention curves, are vital to link structure and composition with effective transport properties and local reaction conditions at Pt-electrolyte interfaces, and ultimately, to relate these properties to performance [40].

Reproducing polarization curves as the key performance characteristic of a PEFC is the imperative of consistent CCL modeling, albeit not an art. A model-based assessment of reaction penetration depths and effectiveness factors of Pt utilization [17], needed at a well-integrated hierarchy of structural levels [8, 9, 14, 15], can provide valuable mechanistic insights. Numerous works have demonstrated capabilities of physical modeling to deconvolute and quantify voltage loss contributions [16]; extract physico-chemical parameters like exchange current density and mass transport coefficients; and discern systematic trends and correlations among these parameters [7], structural parameters, composition, fabrication conditions and operating conditions. If applied to extensive sets of experimental performance data, crucial insights in view of solving the performance vs. cost conundrum can be gleaned from such model-based analyses. Untapped reserves for performance gains due to structural improvements at different scales can be identified.

First in a line of design considerations, macrohomogeneous modeling can propose or predict an optimal CCL thickness or macroscopic effective composition (for target operating conditions) that are straightforward to realize or monitor during MEA fabrication. Looking beyond the capabilities of macrohomogeneous CCL models, aspects like pore-size distributions and pore-network morphologies as well as wettability properties should be considered as part of the design process. However, the model-based understanding of correlations of these structural properties with viable performance metrics is still in an evolutionary phase and means for the experimental evaluation of these correlations must be developed before they can be harnessed for CCL design and MEA fabrication. Thinking further ahead, modeling and simulation must progress alongside advances in spectroscopy, microscopy, imaging, tomography and spatially localized electrochemical measurements, to decipher the fine fingerprints of structural effects at the micro and mesoscale (including pore and agglomerate levels). Questions of foremost interest in this realm pertain to (1) the role of ionomer inclusions in defining the local reaction environment, i. e., the local electrostatic potential or field and the local pH at the water-wetted interface confined by an ionomer film and the catalyst-support surface, and (2) the finely adjusted distribution or deliberate placement of Pt nanoparticles at locations with optimal reaction environment. The determination of the local reaction environment for given materials selection, microstructure, volume-averaged composition and CCL thickness as well as external operating conditions requires the modeling of macro and mesoscale transport phenomena to be self-consistently matched with interface theory and atomistic simulations.

Then of course, a CCL cannot be understood and optimized as a stand-alone component. Overarching balances at the PEFC level in terms of reactant, charge, water and heat fluxes must be considered. The development and integration (at MEA or cell level) of a new CCL design concept is likely to fail if these aspects are not given sufficient attention. Comprehensive modeling approaches must account for the coupling of the corresponding local equilibria and transport phenomena across the whole cell, including polymer electrolyte membrane, diffusion media and flow fields. Such efforts are already underway with promising results to look out for.

Bibliography

[1] A. Kulkarni, S. Siahrostami, A. Patel, and J. K. Nørskov, Understanding Catalytic Activity Trends in the Oxygen Reduction Reaction. *Chem. Rev.*, **118**, 2302–2312, 2018.

[2] M. J. Eslamibidgoli, J. Huang, T. Kadyk, A. Malek, and M. Eikerling, How theory and simulation can drive fuel cell electrocatalysis. *Nano Energy*, **29**, 334–361, 2016.

[3] M. J. Eslamibidgoli and M. H. Eikerling, Electrochemical formation of reactive oxygen species at Pt (111)—A density functional theory study. *ACS Catal.*, **5**, 6090–6098, 2015.

[4] E. Skúlason, G. S. Karlberg, J. Rossmeisl, T. Bligaard, J. Greeley, H. Jónsson, and J. K. Nørskov, Density functional theory calculations for the hydrogen evolution reaction in an electrochemical double layer on the Pt(111) electrode. *Phys. Chem. Chem. Phys.*, **9**, 3241–3250, 2007.

[5] J. Huang, J. Zhang, and M. Eikerling, Unifying theoretical framework for deciphering the oxygen reduction reaction on platinum. *Phys. Chem. Chem. Phys.*, **20**, 11776–11786, 2018.

[6] J. Huang and M. Eikerling, Modeling the oxygen reduction reaction at platinum-based catalysts: A brief review of recent developments. *Curr. Opin. Electrochem.*, **13**, 157–165, 2019.

[7] T. Muzaffar, T. Kadyk, and M. Eikerling, Tipping water balance and the Pt loading effect in polymer electrolyte fuel cells: A model-based analysis. *Sustain. Energy Fuels*, **2**, 1189–1196, 2018.

[8] K. Chan and M. Eikerling, A pore-scale model of oxygen reduction in ionomer-free catalyst layers of PEFCs. *J. Electrochem. Soc.*, **158**, B18, 2011.

[9] J. Huang, J. Zhang, and M. Eikerling, Theory of electrostatic phenomena in water-filled Pt nanopores. *Faraday Discuss.*, **193**, 427–446, 2016.

[10] J. Huang, A. Malek, J. Zhang, and M. H. Eikerling, Non-monotonic surface charging behavior of platinum: A paradigm change. *J. Phys. Chem. C*, **120**, 13587–13595, 2016.

[11] T. Muzaffar, T. Kadyk, and M. Eikerling, Physical modeling of the proton density in nanopores of PEM fuel cell catalyst layers. *Electrochim. Acta*, **245**, 1048–1058, 2017.

[12] S. G. Rinaldo, J. Stumper, and M. Eikerling, Physical theory of platinum nanoparticle dissolution in polymer electrolyte fuel cells. *J. Phys. Chem. C*, **114**, 5773–5785, 2010.

[13] A. Nouri-Khorasani, K. Malek, A. Malek, T. Mashio, D. P. Wilkinson, and M. H. Eikerling, Molecular modeling of the proton density distribution in a water-filled slab-like nanopore bounded by Pt oxide and ionomer. *Catal. Today*, **262**, 133–140, 2016.

[14] E. Sadeghi, A. Putz, and M. Eikerling, Effects of ionomer coverage on agglomerate effectiveness in catalyst layers of polymer electrolyte fuel cells. *J. Solid State Electrochem.*, **18**, 1271–1279, 2014.

[15] E. Sadeghi, A. Putz, and M. Eikerling, Hierarchical model of reaction rate distributions and effectiveness factors in catalyst layers of polymer electrolyte fuel cells. *J. Electrochem. Soc.*, **160**, F1159–F1169, 2013.

[16]   M. Baghalha, J. Stumper, and M. Eikerling, Model-based deconvolution of potential losses in a PEM fuel cell. *ECS Trans.*, **28**, 159–167, 2019.

[17]   M. Eikerling and A. Kulikovsky, *Polymer electrolyte fuel cells: Physical principles of materials and operation*. CRC Press, 2014.

[18]   W. R. Grove, On voltaic series and the combination of gases by platinum. *Philos. Mag.*, **14**, 127–130, 1839.

[19]   D. Banham and S. Ye, Current status and future development of catalyst materials and catalyst layers for proton exchange membrane fuel cells: An industrial perspective. *ACS Energy Lett.*, **2**, 629–638, 2017.

[20]   P. Urchaga, T. Kadyk, S. G. Rinaldo, A. O. Pistono, J. Hu, W. Lee, C. Richards, M. H. Eikerling, and C. A. Rice, Catalyst degradation in fuel cell electrodes: Accelerated stress tests and model-based analysis. *Electrochim. Acta*, **176**, 1500–1510, 2015.

[21]   H. A. Baroody, D. B. Stolar, and M. H. Eikerling, Modelling-based data treatment and analytics of catalyst degradation in polymer electrolyte fuel cells. *Electrochim. Acta*, **283**, 1006–1016, 2018.

[22]   J. Zhao and X. Li, A review of polymer electrolyte membrane fuel cell durability for vehicular applications: Degradation modes and experimental techniques. *Energy Convers. Manag.*, **199**, 112022, 2019.

[23]   Z. Chen, D. Higgins, A. Yu, L. Zhang, and J. Zhang, A review on non-precious metal electrocatalysts for PEM fuel cells. *Energy Environ. Sci.*, **4**, 3167, 2011.

[24]   I. D. Raistrick, Modified gas dDiffusion electrode for proton exchange. In: *Proceedings of the Symposium on Diaphragms, Separators, and Ion-Exchange Membranes*, 1986.

[25]   I. D. Raistrick, *Electrode assembly for use in a solid polymer electrolyte fuel cell*, 1988.

[26]   T. E. Springer, T. A. Zawodzinski, and S. Gottesfeld, Polymer electrolyte fuel cell model. *J. Electrochem. Soc.*, **138**, 2334–2342, 1991.

[27]   K. Malek, M. Eikerling, Q. Wang, T. Navessin, and Z. Liu, Self-organization in catalyst layers of polymer electrolyte fuel cells. *J. Phys. Chem. C*, **111**, 13627–13634, 2007.

[28]   T. Suzuki, K. Kudo, and Y. Morimoto, Model for investigation of oxygen transport limitation in a polymer electrolyte fuel cell. *J. Power Sources*, **222**, 379–389, 2013.

[29]   Y.-C. Park, H. Tokiwa, K. Kakinuma, M. Watanabe, and M. Uchida, Effects of carbon supports on Pt distribution, ionomer coverage and cathode performance for polymer electrolyte fuel cells. *J. Power Sources*, **315**, 179–191, 2016.

[30]   T. Suzuki, S. Tsushima, and S. Hirai, Effects of Nafion® ionomer and carbon particles on structure formation in a proton-exchange membrane fuel cell catalyst layer fabricated by the decal-transfer method. *Int. J. Hydrog. Energy*, **36**, 12361–12369, 2011.

[31]   T. Soboleva, X. Zhao, K. Malek, Z. Xie, T. Navessin, and S. Holdcroft, On the micro-, meso-, and macroporous structures of polymer electrolyte membrane fuel cell catalyst layers. *ACS Appl. Mater. Interfaces*, **2**, 375–384, 2010.

[32]   M. Uchida, Y. Aoyama, N. Eda, and A. Ohta, Investigation of the microstructure in the catalyst layer and effects of both perfluorosulfonate ionomer and PTFE-loaded carbon on the catalyst layer of polymer electrolyte fuel cells. *J. Electrochem. Soc.*, **142**, 4143–4149, 1995.

[33]   K. Malek, T. Mashio, and M. Eikerling, Microstructure of catalyst layers in PEM fuel cells redefined: A computational approach. *Electrocatalysis*, **2**, 141–157, 2011.

[34]   Y. M. Volfkovich, V. E. Sosenkin, and V. S. Bagotsky, Structural and wetting properties of fuel cell components. *J. Power Sources*, **195**, 5429–5441, 2010.

[35]   A. Z. Weber and M. A. Hickner, Modeling and high-resolution-imaging studies of water-content profiles in a polymer-electrolyte-fuel-cell membrane-electrode assembly. *Electrochim. Acta*, **53**, 7668–7674, 2008.

[36]   V. Mulone and K. Karan, Analysis of capillary flow driven model for water transport in PEFC cathode catalyst layer: Consideration of mixed wettability and pore size distribution. *Int. J. Hydrog. Energy*, **38**, 558–569, 2013.

[37]   T. Mashio, K. Sato, and A. Ohma, Analysis of water adsorption and condensation in catalyst layers for polymer electrolyte fuel cells. *Electrochim. Acta*, **140**, 238–249, 2014.

[38]   P. A. Mateo Villanueva, *A mixed wettability pore size distribution model for the analysis of water transport in PEMFC materials*, 2013.

[39]   A. Z. Weber, Improved modeling and understanding of diffusion-media wettability on polymer-electrolyte-fuel-cell performance. *J. Power Sources*, **195**, 5292–5304, 2010.

[40]   W. Olbrich, T. Kadyk, U. Sauter, and M. H. Eikerling, Modeling wetting phenomena in cathode catalyst layers for PEM fuel cells. *Electrochim. Acta*, **431**, 140850, 2022.

[41]   R. De Levie, On impedance measurements: The determination of the double layer capacitance in the presence of an electrode reaction. *Electrochim. Acta*, **10**, 395–402, 1965.

[42]   S. Touhami, J. Mainka, J. Dillet, S. A. H. Taleb, and O. Lottin, Transmission line impedance models considering oxygen transport limitations in polymer electrolyte membrane fuel cells. *J. Electrochem. Soc.*, **166**, F1209–F1217, 2019.

[43]   M. K. Debe, A. K. Schmoeckel, G. D. Vernstrom, and R. Atanasoski, High voltage stability of nanostructured thin film catalysts for PEM fuel cells. *J. Power Sources*, **161**, 1002–1011, 2006.

[44]   M. K. Debe, Tutorial on the fundamental characteristics and practical properties of nanostructured thin film (NSTF) catalysts. *J. Electrochem. Soc.*, **160**, F522–F534, 2013.

[45]   A. Steinbach, *Final technical report for project entitled highly active, durable, and ultra-low PGM NSTF thin film ORR catalysts and support*, 2020.

[46]   S. Du, Recent advances in electrode design based on one-dimensional nanostructure arrays for proton exchange membrane fuel cell applications. *Engineering*, **7**, 33–49, 2021.

[47]   D. Gruber and J. Müller, Enhancing PEM fuel cell performance by introducing additional thin layers to sputter-deposited Pt catalysts. *J. Power Sources*, **171**, 294–301, 2007.

[48]   M. S. Saha, A. F. Gullá, R. J. Allen, and S. Mukerjee, High performance polymer electrolyte fuel cells with ultra-low Pt loading electrodes prepared by dual ion-beam assisted deposition. *Electrochim. Acta*, **51**, 4680–4692, 2006.

[49]   P. Ramesh, M. E. Itkis, J. M. Tang, and R. C. Haddon, SWNT–MWNT hybrid architecture for proton exchange membrane fuel cell cathodes. *J. Phys. Chem. C*, **112**, 9089–9094, 2008.

[50]   R. Zeis, A. Mathur, G. Fritz, J. Lee, and J. Erlebacher, Platinum-plated nanoporous gold: An efficient, low Pt loading electrocatalyst for PEM fuel cells. *J. Power Sources*, **165**, 65–72, 2007.

[51]   B. Kinkead, J. van Drunen, M. T. Y. Paul, K. Dowling, G. Jerkiewicz, and B. D. Gates, Platinum ordered porous electrodes: Developing a platform for fundamental electrochemical characterization. *Electrocatalysis*, **4**, 179–186, 2013.

[52]   Q. Wang, M. Eikerling, D. Song, and Z.-S. Liu, Modeling of ultrathin two-phase catalyst layers in PEFCs. *J. Electrochem. Soc.*, **154**, F95, 2007.

[53]   J. A. Spooner, M. J. Eslamibidgoli, K. Malek, and M. H. Eikerling, Molecular dynamics study of the nanoscale proton density distribution at the ionomer-catalyst interface. *ECS Meet. Abstr.*, MA2020-02, 2101, 2020.

[54]   M. K. Debe, Electrocatalyst approaches and challenges for automotive fuel cells. *Nature*, **486**, 43–51, 2012.

[55]   D. F. van der Vliet, C. Wang, D. Tripkovic, D. Strmcnik, X. F. Zhang, M. K. Debe, R. T. Atanasoski, N. M. Markovic, and V. R. Stamenkovic, Mesostructured thin films as electrocatalysts with tunable composition and surface morphology. *Nat. Mater.*, **11**, 1051–1058, 2012.

[56]   X. Ren, Q. Lv, L. Liu, B. Liu, Y. Wang, A. Liu, and G. Wu, Current progress of Pt and Pt-based electrocatalysts used for fuel cells. *Sustain. Energy Fuels*, **4**, 15–30, 2020.

[57]   G. Ercolano, S. Cavaliere, J. Rozière, and D. J. Jones, Recent developments in electrocatalyst design thrifting noble metals in fuel cells. *Curr. Opin. Electrochem.*, **9**, 271–277, 2018.

[58]   J. Greeley, T. F. Jaramillo, J. Bonde, I. Chorkendorff, and J. K. Nørskov, Computational high-throughput screening of electrocatalytic materials for hydrogen evolution. *Nat. Mater.*, **5**, 909–913, 2006.

[59] O. Mamun, K. T. Winther, J. R. Boes, and T. Bligaard, High-throughput calculations of catalytic properties of bimetallic alloy surfaces. *Sci. Data*, **6**, 76, 2019.

[60] A. A. Balandin, On the theory of the selection of catalysts. *Zh. Obshchei Khimii*, **16**, 793–804, 1946.

[61] S. Trasatti, Work function, electronegativity, and electrochemical behaviour of metals. *J. Electroanal. Chem. Interfacial Electrochem.*, **39**, 163–184, 1972.

[62] P. Sabatier, Hydrogénations et déshydrogénations par catalyse. *Ber. Dtsch. Chem. Ges.*, **44**, 1984–2001, 1911.

[63] P. Quaino, F. Juarez, E. Santos, and W. Schmickler, Volcano plots in hydrogen electrocatalysis – uses and abuses. *Beilstein J. Nanotechnol.*, **5**, 846–854, 2014.

[64] J. K. Nørskov, J. Rossmeisl, A. Logadottir, L. Lindqvist, J. R. Kitchin, T. Bligaard, and H. Jónsson, Origin of the overpotential for oxygen reduction at a fuel-cell cathode. *J. Phys. Chem. B*, **108**, 17886–17892, 2004.

[65] H. Ooka, J. Huang, and K. S. Exner, The Sabatier principle in electrocatalysis: Basics, limitations, and extensions. *Front. Energy Res.*, **9**, 2021. https://doi.org/10.3389/fenrg.2021.654460.

[66] T. Binninger, T. J. Schmidt, and D. Kramer, Capacitive electronic metal-support interactions: Outer surface charging of supported catalyst particles. *Phys. Rev. B*, **96**, 165405, 2017.

[67] J. Huang, J. Zhang, and M. H. Eikerling, Particle proximity effect in nanoparticle electrocatalysis: Surface charging and electrostatic interactions. *J. Phys. Chem. C*, **121**, 4806–4815, 2017.

[68] L. Wang, A. Roudgar, and M. Eikerling, Ab initio study of stability and site-specific oxygen adsorption energies of Pt nanoparticles. *J. Phys. Chem. C*, **113**, 17989–17996, 2009.

[69] K. Shinozaki, H. Yamada, and Y. Morimoto, Relative humidity dependence of Pt utilization in polymer electrolyte fuel cell electrodes: effects of electrode thickness, ionomer-to-carbon ratio, ionomer equivalent weight, and carbon support. *J. Electrochem. Soc.*, **158**, B467, 2011.

[70] T. J. Schmidt, H. A. Gasteiger, G. D. Stäb, P. M. Urban, D. M. Kolb, and R. J. Behm, Characterization of high-surface-area electrocatalysts using a rotating disk electrode configuration. *J. Electrochem. Soc.*, **145**, 2354–2358, 1998.

[71] S. L. Gojković, S. K. Zečević, and R. F. Savinell, O2 reduction on an ink-type rotating disk electrode using Pt supported on high-area carbons. *J. Electrochem. Soc.*, **145**, 3713–3720, 1998.

[72] T. Vidaković, M. Christov, and K. Sundmacher, The use of CO stripping for in situ fuel cell catalyst characterization. *Electrochim. Acta*, **52**, 5606–5613, 2007.

[73] R. L. Borup, J. R. Davey, F. H. Garzon, D. L. Wood, and M. A. Inbody, PEM fuel cell electrocatalyst durability measurements. *J. Power Sources*, **163**, 76–81, 2006.

[74] R. Sharma and S. M. Andersen, Quantification on degradation mechanisms of polymer electrolyte membrane fuel cell catalyst layers during an accelerated stress test. *ACS Catal.*, **8**, 3424–3434, 2018.

[75] J. Shan and P. G. Pickup, Characterization of polymer supported catalysts by cyclic voltammetry and rotating disk voltammetry. *Electrochim. Acta*, **46**, 119–125, 2000.

[76] A. Colliard-Granero, M. Batool, J. Jankovic, J. Jitsev, M. H. Eikerling, K. Malek, and M. J. Eslamibidgoli, *Deep learning for the automation of particle analysis in catalyst layers for polymer electrolyte fuel cells*, 2021.

[77] S. Kabir, D. J. Myers, N. Kariuki, J. Park, G. Wang, A. Baker, N. Macauley, R. Mukundan, K. L. More, and K. C. Neyerlin, Elucidating the dynamic nature of fuel cell electrodes as a function of conditioning: An ex situ material characterization and in situ electrochemical diagnostic study. *ACS Appl. Mater. Interfaces*, **11**, 45016–45030, 2019.

[78] E. Padgett, N. Andrejevic, Z. Liu, A. Kongkanand, W. Gu, K. Moriyama, Y. Jiang, S. Kumaraguru, T. E. Moylan, R. Kukreja, and D. A. Muller, Connecting fuel cell catalyst nanostructure and accessibility using quantitative cryo-STEM tomography. *J. Electrochem. Soc.*, **165**, F173–F180, 2018.

[79] H. A. Gasteiger, J. E. Panels, and S. G. Yan, Dependence of PEM fuel cell performance on catalyst loading. *J. Power Sources*, **127**, 162–171, 2004.

[80] K. Dhathathreyan, Development of polymer electrolyte membrane fuel cell stack. *Int. J. Hydrog. Energy*, **24**, 1107–1115, 1999.

[81] Z. Xia, Q. Wang, M. Eikerling, and Z. Liu, Effectiveness factor of Pt utilization in cathode catalyst layer of polymer electrolyte fuel cells. *Can. J. Chem.*, **86**, 657–667, 2008.

[82] M. Lee, M. Uchida, H. Yano, D. A. Tryk, H. Uchida, and M. Watanabe, New evaluation method for the effectiveness of platinum/carbon electrocatalysts under operating conditions. *Electrochim. Acta*, **55**, 8504–8512, 2010.

[83] DOE, *Fuel cell technologies program multi-zear research, development, and demonstration plan*, US Department Of Energy Washington, DC, 2014.

[84] S. Ott, A. Orfanidi, H. Schmies, B. Anke, H. N. Nong, J. Hübner, U. Gernert, M. Gliech, M. Lerch, and P. Strasser, Ionomer distribution control in porous carbon-supported catalyst layers for high-power and low Pt-loaded proton exchange membrane fuel cells. *Nat. Mater.*, **19**, 77–85, 2020.

[85] G. S. Harzer, A. Orfanidi, H. El-Sayed, P. Madkikar, and H. A. Gasteiger, Tailoring catalyst morphology towards high performance for low Pt loaded PEMFC cathodes. *J. Electrochem. Soc.*, **165**, F770–F779, 2018.

[86] K. Kodama, A. Shinohara, N. Hasegawa, K. Shinozaki, R. Jinnouchi, T. Suzuki, T. Hatanaka, and Y. Morimoto, Catalyst poisoning property of sulfonimide acid ionomer on Pt (111) surface. *J. Electrochem. Soc.*, **161**, F649–F652, 2014.

[87] R. Subbaraman, D. Strmcnik, V. Stamenkovic, and N. M. Markovic, Three phase interfaces at electrified metal–solid electrolyte systems 1. Study of the Pt(h)–Nafion interface. *J. Phys. Chem. C*, **114**, 8414–8422, 2010.

[88] K. Shinozaki, Y. Morimoto, B. S. Pivovar, and S. S. Kocha, Suppression of oxygen reduction reaction activity on Pt-based electrocatalysts from ionomer incorporation. *J. Power Sources*, **325**, 745–751, 2016.

[89] T. Mashio, K. Malek, M. Eikerling, A. Ohma, H. Kanesaka, and K. Shinohara, Molecular dynamics study of ionomer and water adsorption at carbon support materials. *J. Phys. Chem. C*, **114**, 13739–13745, 2010.

[90] M. Eikerling, Water management in cathode catalyst layers of PEM fuel cells. *J. Electrochem. Soc.*, **153**, E58, 2006.

[91] J. Liu and M. Eikerling, Model of cathode catalyst layers for polymer electrolyte fuel cells: The role of porous structure and water accumulation. *Electrochim. Acta*, **53**, 4435–4446, 2008.

[92] M. Eikerling, A. S. Ioselevich, and A. A. Kornyshev, How good are the Electrodes we use in PEFC? *Fuel Cells*, **4**, 131–140, 2004.

[93] M. Eikerling and A. A. Kornyshev, Modelling the performance of the cathode catalyst layer of polymer electrolyte fuel cells. *J. Electroanal. Chem.*, **453**, 89–106, 1998.

[94] Z. Xie, T. Navessin, K. Shi, R. Chow, Q. Wang, D. Song, B. Andreaus, M. Eikerling, Z. Liu, and S. Holdcroft, Functionally Graded Cathode Catalyst Layers for Polymer Electrolyte Fuel Cells. *J. Electrochem. Soc.*, **152**, A1171, 2005.

[95] Q. Wang, M. Eikerling, D. Song, Z. Liu, T. Navessin, Z. Xie, and S. Holdcroft, Functionally graded cathode catalyst layers for polymer electrolyte fuel cells. *J. Electrochem. Soc.*, **151**, A950, 2004.

[96] M. J. Eslamibidgoli and M. H. Eikerling, Atomistic mechanism of Pt extraction at oxidized surfaces: Insights from DFT. *Electrocatalysis*, **7**, 345–354, 2016.

[97] A. A. Topalov, I. Katsounaros, M. Auinger, S. Cherevko, J. C. Meier, S. O. Klemm, and K. J. J. Mayrhofer, Dissolution of Platinum: Limits for the deployment of electrochemical energy conversion? *Angew. Chem., Int. Ed. Engl.*, **51**, 12613–12615, 2012.

[98] M. Eikerling, A. A. Kornyshev, A. M. Kuznetsov, J. Ulstrup, and S. Walbran, Mechanisms of proton conductance in polymer electrolyte membranes. *J. Phys. Chem. B*, **105**, 3646–3662, 2001.

[99] M. Eikerling, A. A. Kornyshev, and A. R. Kucernak, Water in polymer electrolyte fuel cells: Friend or foe? *Phys. Today*, **59**, 38–44, 2006.

[100] D. Marx, M. E. Tuckerman, J. Hutter, and M. Parrinello, The nature of the hydrated excess proton in water. *Nature*, **397**, 601–604, 1999.

[101] D. Marx, Proton transfer 200 Years after von Grotthuss: Insights from ab initio simulations. *ChemPhysChem*, **7**, 1848–1870, 2006.

[102] K. Sakai, K. Sato, T. Mashio, A. Ohma, K. Yamaguchi, and K. Shinohara, Analysis of reactant gas transport in catalyst layers; effect of Pt-loadings. *ECS Trans.*, **25**, 1193–1201, 2019.

[103] Y. Ono, A. Ohma, K. Shinohara, and K. Fushinobu, Influence of equivalent weight of ionomer on local oxygen transport resistance in cathode catalyst layers. *J. Electrochem. Soc.*, **160**, F779–F787, 2013.

[104] K. Kudo, R. Jinnouchi, and Y. Morimoto, Humidity and temperature dependences of oxygen transport resistance of Nafion thin film on platinum electrode. *Electrochim. Acta*, **209**, 682–690, 2016.

[105] R. Jinnouchi, K. Kudo, N. Kitano, and Y. Morimoto, Molecular dynamics simulations on O2 permeation through Nafion ionomer on platinum surface. *Electrochim. Acta*, **188**, 767–776, 2016.

[106] A. Kongkanand and M. F. Mathias, The priority and challenge of high-power performance of low-platinum proton-exchange membrane fuel cells. *J. Phys. Chem. Lett.*, **7**, 1127–1137, 2016.

[107] H. Liu, W. K. Epting, and S. Litster, Gas transport resistance in polymer electrolyte thin films on oxygen reduction reaction catalysts. *Langmuir*, **31**, 9853–9858, 2015.

[108] A. Z. Weber and A. Kusoglu, Unexplained transport resistances for low-loaded fuel-cell catalyst layers. *J. Mater. Chem. A*, **2**, 17207–17211, 2014.

[109] H. Li, Y. Tang, Z. Wang, Z. Shi, S. Wu, D. Song, J. Zhang, K. Fatih, J. Zhang, H. Wang, Z. Liu, R. Abouatallah, and A. Mazza, A review of water flooding issues in the proton exchange membrane fuel cell. *J. Power Sources*, **178**, 103–117, 2008.

[110] Y. Wang, D. F. Ruiz Diaz, K. S. Chen, Z. Wang, and X. C. Adroher, Materials, technological status, and fundamentals of PEM fuel cells – A review. *Mater. Today*, **32**, 178–203, 2020.

[111] P.-C. Sui, X. Zhu, and N. Djilali, Modeling of PEM fuel cell catalyst layers: status and outlook. *Electrochem. Energy Rev.*, **2**, 428–466, 2019.

[112] D. Kramer, J. Zhang, R. Shimoi, E. Lehmann, A. Wokaun, K. Shinohara, and G. G. Scherer, In situ diagnostic of two-phase flow phenomena in polymer electrolyte fuel cells by neutron imaging. *Electrochim. Acta*, **50**, 2603–2614, 2005.

[113] J. P. Owejan, J. J. Gagliardo, J. M. Sergi, S. G. Kandlikar, and T. A. Trabold, Water management studies in PEM fuel cells, Part I: Fuel cell design and in situ water distributions. *Int. J. Hydrog. Energy*, **34**, 3436–3444, 2009.

[114] A. Turhan, K. Heller, J. S. Brenizer, and M. M. Mench, Quantification of liquid water accumulation and distribution in a polymer electrolyte fuel cell using neutron imaging. *J. Power Sources*, **160**, 1195–1203, 2006.

[115] M. Sahimi, G. R. Gavalas, and T. T. Tsotsis, Statistical and continuum models of fluid-solid reactions in porous media. *Chem. Eng. Sci.*, **45**, 1443–1502, 1990.

[116] S. Torquato and H. Haslach, Random heterogeneous materials: Microstructure and macroscopic properties. *Appl. Mech. Rev.*, **55**, B62–B63, 2002.

[117] J. Huang, X. Zhu, and M. Eikerling, The rate-determining term of electrocatalytic reactions with first-order kinetics. *Electrochim. Acta*, **393**, 139019, 2021.

[118] M. L. Perry, Mass transport in gas-diffusion electrodes: A diagnostic tool for fuel-cell cathodes. *J. Electrochem. Soc.*, **145**, 5, 1998.

[119] A. A. Kulikovsky, A physically-based analytical polarization curve of a PEM fuel cell. *J. Electrochem. Soc.*, **161**, F263–F270, 2014.

[120] H. P. F. Gunterman, *Characterization of fuel-cell diffusion media*, 2011.

[121] T. Hutzenlaub, J. Becker, R. Zengerle, and S. Thiele, Modelling the water distribution within a hydrophilic and hydrophobic 3D reconstructed cathode catalyst layer of a proton exchange membrane fuel cell. *J. Power Sources*, **227**, 260–266, 2013.

[122] J. Zhou, A. Putz, and M. Secanell, A mixed wettability pore size distribution based mathematical model for analyzing two-phase flow in porous electrodes: I. Mathematical model. *J. Electrochem. Soc.*, **164**, F530–F539, 2017.

[123]   I. V. Zenyuk, P. K. Das, and A. Z. Weber, Understanding impacts of catalyst-layer thickness on fuel-cell performance via mathematical modeling. *J. Electrochem. Soc.*, **163**, F691, 2016.

[124]   T. Mashio, A. Ohma, and T. Tokumasu, Molecular dynamics study of ionomer adsorption at a carbon surface in catalyst ink. *Electrochim. Acta*, **202**, 14–23, 2016.

[125]   X. Li, F. Feng, K. Zhang, S. Ye, D. Y. Kwok, and V. Birss, Wettability of Nafion and Nafion/Vulcan carbon composite films. *Langmuir*, **28**, 6698–6705, 2012.

[126]   A. Chowdhury, R. M. Darling, C. J. Radke, and A. Z. Weber, Modeling water uptake and Pt utilization in high surface area carbon. *ECS Trans.*, **92**, 247–259, 2019.

[127]   F. C. Cetinbas, R. K. Ahluwalia, N. N. Kariuki, V. De Andrade, and D. J. Myers, Effects of porous carbon morphology, agglomerate structure and relative humidity on local oxygen transport resistance. *J. Electrochem. Soc.*, **167**, 013508, 2020.

[128]   N. Kavokine, R. R. Netz, and L. Bocquet, Fluids at the nanoscale: From continuum to subcontinuum transport. *Annu. Rev. Fluid Mech.*, **53**, 377–410, 2021.

[129]   A. Schlaich, A. P. dos Santos, and R. R. Netz, Simulations of nanoseparated charged surfaces reveal charge-induced water reorientation and nonadditivity of hydration and mean-field electrostatic repulsion. *Langmuir*, **35**, 551–560, 2018.

[130]   P. Loche, C. Ayaz, A. Schlaich, Y. Uematsu, and R. R. Netz, Giant axial dielectric response in water-filled nanotubes and effective electrostatic ion–ion interactions from a tensorial dielectric model. *J. Phys. Chem. B*, **123**, 10850–10857, 2019.

[131]   P. Loche, C. Ayaz, A. Schlaich, D. J. Bonthuis, and R. R. Netz, Breakdown of linear dielectric theory for the interaction between hydrated ions and graphene. *J. Phys. Chem. Lett.*, **9**, 6463–6468, 2018.

[132]   V. M. Fernández-Alvarez, K. Malek, M. H. Eikerling, A. Young, M. Dutta, and E. Kjeang, Molecular dynamics study of reaction conditions at active catalyst-ionomer interfaces in polymer electrolyte fuel cells. *J. Electrochem. Soc.*, **169**, 024506, 2022.

[133]   S. Thiele, T. Fürstenhaupt, D. Banham, T. Hutzenlaub, V. Birss, C. Ziegler, and R. Zengerle, Multiscale tomography of nanoporous carbon-supported noble metal catalyst layers. *J. Power Sources*, **228**, 185–192, 2013.

[134]   M. Lopez-Haro, L. Guétaz, T. Printemps, A. Morin, S. Escribano, P.-H. Jouneau, P. Bayle-Guillemaud, F. Chandezon, and G. Gebel, Three-dimensional analysis of Nafion layers in fuel cell electrodes. *Nat. Commun.*, **5**, 5229, 2014.

[135]   J. Jankovic, D. Susac, T. Soboleva, and J. Stumper, Electron tomography based 3D reconstruction of fuel cell catalysts. *ECS Trans.*, **50**, 353–359, 2013.

[136]   F. C. Cetinbas, R. K. Ahluwalia, N. N. Kariuki, and D. J. Myers, Agglomerates in polymer electrolyte fuel cell electrodes: Part I. Structural characterization. *J. Electrochem. Soc.*, **165**, F1051, 2018.

[137]   M. Khakbazbaboli, *Development of a micro-scale cathode catalyst layer model of polymer electrolyte membrane fuel cell*, 2013.

[138]   S. Jung, M. Sabharwal, A. Jarauta, F. Wei, M. Gingras, J. Gostick, and M. Secanell, Estimation of relative transport properties in porous transport layers using pore-scale and pore-network simulations. *J. Electrochem. Soc.*, **168**, 064501, 2021.

[139]   J. Giner and C. Hunter, The mechanism of operation of the Teflon-bonded gas diffusion electrode: A mathematical model. *J. Electrochem. Soc.*, **116**, 1124, 1969.

[140]   R. P. Iczkowski and M. B. Cutlip, Voltage losses in fuel cell cathodes. *J. Electrochem. Soc.*, **127**, 1433–1440, 1980.

[141]   P. Björnbom, Modelling of a double-layered PTFE-bonded oxygen electrode. *Electrochim. Acta*, **32**, 115–119, 1987.

[142]   Q. Wang, M. Eikerling, D. Song, and Z. Liu, Structure and performance of different types of agglomerates in cathode catalyst layers of PEM fuel cells. *J. Electroanal. Chem.*, **573**, 61–69, 2004.

[143]   N. Goswami, A. N. Mistry, J. B. Grunewald, T. F. Fuller, and P. P. Mukherjee, Corrosion-induced microstructural variability affects transport-kinetics interaction in PEM fuel cell catalyst layers. *J. Electrochem. Soc.*, **167**, 084519, 2020.

[144] M. Uchida, Y.-C. Park, K. Kakinuma, H. Yano, D. A. Tryk, T. Kamino, H. Uchida, and M. Watanabe, Effect of the state of distribution of supported Pt nanoparticles on effective Pt utilization in polymer electrolyte fuel cells. *Phys. Chem. Chem. Phys.*, **15**, 11236, 2013.

[145] D. Stauffer, *Introduction to percolation theory*. Taylor & Francis, London, Bristol, 1994.

[146] D. Bernhard, T. Kadyk, U. Krewer, and S. Kirsch, How platinum oxide affects the degradation analysis of PEM fuel cell cathodes. *Int. J. Hydrog. Energy*, **46**, 13791–13805, 2021.

[147] A. Nandy, F. Jiang, S. Ge, C.-Y. Wang, and K. S. Chen, Effect of cathode pore volume on PEM fuel cell cold start. *J. Electrochem. Soc.*, **157**, B726, 2010.

[148] T. J. Dursch, G. J. Trigub, R. Lujan, J. F. Liu, R. Mukundan, C. J. Radke, and A. Z. Weber, Ice-crystallization kinetics in the catalyst layer of a proton-exchange-membrane fuel cell. *J. Electrochem. Soc.*, **161**, F199–F207, 2014.

[149] R. J. Balliet and J. Newman, Cold-start modeling of a polymer-electrolyte fuel cell containing an ultrathin cathode. *J. Electrochem. Soc.*, **158**, B1142, 2011.

[150] S. Huo, K. Jiao, and J. W. Park, On the water transport behavior and phase transition mechanisms in cold start operation of PEM fuel cell. *Appl. Energy*, **233–234**, 776–788, 2019.

[151] H. Wu, *Mathematical modeling of transient transport phenomena in PEM fuel cells*, 2009.

[152] S. G. Rinaldo, W. Lee, J. Stumper, and M. Eikerling, Model- and theory-based evaluation of Pt dissolution for supported Pt nanoparticle distributions under potential cycling. *Electrochem. Solid-State Lett.*, **14**, B47, 2011.

[153] S. G. Rinaldo, W. Lee, J. Stumper, and M. Eikerling, Nonmonotonic dynamics in Lifshitz-Slyozov-Wagner theory: Ostwald ripening in nanoparticle catalysts. *Phys. Rev. E*, **86**, 041601, 2012.

[154] K. Christmann, K. A. Friedrich, and N. Zamel, Activation mechanisms in the catalyst coated membrane of PEM fuel cells. *Prog. Energy Combust. Sci.*, **85**, 100924, 2021.

[155] N. Fouquet, C. Doulet, C. Nouillant, G. Dauphin-Tanguy, and B. Ould-Bouamama, Model based PEM fuel cell state-of-health monitoring via ac impedance measurements. *J. Power Sources*, **159**, 905–913, 2006.

[156] X. Yuan, H. Wang, J. Colinsun, and J. Zhang, AC impedance technique in PEM fuel cell diagnosis—A review. *Int. J. Hydrog. Energy*, **32**, 4365–4380, 2007.

[157] Z. Tang, Q.-A. Huang, Y.-J. Wang, F. Zhang, W. Li, A. Li, L. Zhang, and J. Zhang, Recent progress in the use of electrochemical impedance spectroscopy for the measurement, monitoring, diagnosis and optimization of proton exchange membrane fuel cell performance. *J. Power Sources*, **468**, 228361, 2020.

[158] A. Lasia, Electrochemical impedance spectroscopy and its applications. In: *Modern aspects of electrochemistry*, pages 143–248. Springer, 2002.

[159] M. E. Orazem and B. Tribollet, In: *Electrochemical impedance spectroscopy*, pages 383–389. Wiley, 2008.

[160] I. Pivac, B. Šimić, and F. Barbir, Experimental diagnostics and modeling of inductive phenomena at low frequencies in impedance spectra of proton exchange membrane fuel cells. *J. Power Sources*, **365**, 240–248, 2017.

[161] I. Pivac and F. Barbir, Inductive phenomena at low frequencies in impedance spectra of proton exchange membrane fuel cells – A review. *J. Power Sources*, **326**, 112–119, 2016.

[162] E. Warburg, Ueber das Verhalten sogenannter unpolarisirbarer Elektroden gegen Wechselstrom. *Ann. Phys.*, **303**, 493–499, 1899.

[163] C.-H. Kim, S.-I. Pyun, and J.-H. Kim, An investigation of the capacitance dispersion on the fractal carbon electrode with edge and basal orientations. *Electrochim. Acta*, **48**, 3455–3463, 2003.

[164] M. E. Orazem, P. Shukla, and M. A. Membrino, Extension of the measurement model approach for deconvolution of underlying distributions for impedance measurements. *Electrochim. Acta*, **47**, 2027–2034, 2002.

[165] D. A. Harrington and P. van den Driessche, Mechanism and equivalent circuits in electrochemical impedance spectroscopy. *Electrochim. Acta*, **56**, 8005–8013, 2011.

[166] T. Kadyk, R. Hanke-Rauschenbach, and K. Sundmacher, Nonlinear frequency response analysis of PEM fuel cells for diagnosis of dehydration, flooding and CO-poisoning. *J. Electroanal. Chem.*, **630**, 19–27, 2009.

[167] F. Ciucci, Modeling electrochemical impedance spectroscopy. *Curr. Opin. Electrochem.*, **13**, 132–139, 2019.

[168] A. Lasia, Impedance of porous electrodes. *J. Electroanal. Chem.*, **397**, 27–33, 1995.

[169] M. C. Lefebvre, Characterization of ionic conductivity profiles within proton exchange membrane fuel cell gas diffusion electrodes by impedance spectroscopy. *Electrochem. Solid-State Lett.*, **2**, 259, 1999.

[170] R. Makharia, M. F. Mathias, and D. R. Baker, Measurement of catalyst layer electrolyte resistance in PEFCs using electrochemical impedance spectroscopy. *J. Electrochem. Soc.*, **152**, A970, 2005.

[171] D. Gerteisen, Impact of inhomogeneous catalyst layer properties on impedance spectra of polymer electrolyte membrane fuel cells. *J. Electrochem. Soc.*, **162**, F1431–F1438, 2015.

[172] T. Gaumont, G. Maranzana, O. Lottin, J. Dillet, S. Didierjean, J. Pauchet, and L. Guétaz, Measurement of protonic resistance of catalyst layers as a tool for degradation monitoring. *Int. J. Hydrog. Energy*, **42**, 1800–1812, 2017.

[173] M. Eikerling and A. A. Kornyshev, Electrochemical impedance of the cathode catalyst layer in polymer electrolyte fuel cells. *J. Electroanal. Chem.*, **475**, 107–123, 1999.

[174] G. A. Futter, P. Gazdzicki, K. A. Friedrich, A. Latz, and T. Jahnke, Physical modeling of polymer-electrolyte membrane fuel cells: Understanding water management and impedance spectra. *J. Power Sources*, **391**, 148–161, 2018.

[175] T. Reshetenko and A. Kulikovsky, Variation of PEM fuel cell physical parameters with current: Impedance spectroscopy study. *J. Electrochem. Soc.*, **163**, F1100–F1106, 2016.

[176] J. Lee, H. Salihi, J. Lee, and H. Ju, Impedance modeling for polymer electrolyte membrane fuel cells by combining the transient two-phase fuel cell and equivalent electric circuit models. *Energy*, **239**, 122294, 2021.

[177] O. S. Burheim, G. A. Crymble, R. Bock, N. Hussain, S. Pasupathi, A. du Plessis, S. le Roux, F. Seland, H. Su, and B. G. Pollet, Thermal conductivity in the three layered regions of micro porous layer coated porous transport layers for the PEM fuel cell. *Int. J. Hydrog. Energy*, **40**, 16775–16785, 2015.

[178] H. Sadeghifar, N. Djilali, and M. Bahrami, Effect of Polytetrafluoroethylene (PTFE) and micro porous layer (MPL) on thermal conductivity of fuel cell gas diffusion layers: Modeling and experiments. *J. Power Sources*, **248**, 632–641, 2014.

[179] A. Goshtasbi, J. Chen, J. R. Waldecker, S. Hirano, and T. Ersal, Effective parameterization of PEM fuel cell models—Part I: Sensitivity analysis and parameter identifiability. *J. Electrochem. Soc.*, **167**, 044504, 2020.

[180] A. Goshtasbi, J. Chen, J. R. Waldecker, S. Hirano, and T. Ersal, Effective parameterization of PEM fuel cell models—Part II: Robust parameter subset selection, robust optimal experimental design, and multi-step parameter identification algorithm. *J. Electrochem. Soc.*, **167**, 044505, 2020.

[181] N. Lin, X. Xie, R. Schenkendorf, and U. Krewer, Efficient global sensitivity analysis of 3D multiphysics model for Li-ion batteries. *J. Electrochem. Soc.*, **165**, A1169–A1183, 2018.

[182] D. G. Vlachos, A. B. Mhadeshwar, and N. S. Kaisare, Hierarchical multiscale model-based design of experiments, catalysts, and reactors for fuel processing. *Comput. Chem. Eng.*, **30**, 1712–1724, 2006.

[183] K. A. P. McLean and K. B. McAuley, Mathematical modelling of chemical processes-obtaining the best model predictions and parameter estimates using identifiability and estimability procedures. *Can. J. Chem. Eng.*, **90**, 351–366, 2012.

Yuze Hou, Qing Du, Dietmar Gerteisen, Kui Jiao, and Nada Zamel

8 Pore-scale simulation in the electrode of PEMFC – a review and tutorial

8.1 Introduction

The electrode, typically consisting of gas diffusion layer (GDL), micro-porous layer (MPL) and catalyst layer (CL), is the key component of proton exchange membrane (PEM) fuel cells, where the energy conversion and complex coupled physical-electrochemical processes occur [1]. The current two main technical issues restricting the development and commercialization of PEM fuel cells, namely cost and durability [2], are also both closely related to the electrode. Therefore, a deeper insight into the electrode is critically important for the further development of PEM fuel cells, which mostly relies on systematic designs of the electrodes.

Given the heterogeneous and anisotropic microstructures (given in Figure 1) and the complex physical-electrochemical processes in PEMFCs, it is still impossible to fully understand the complex spatial and temporal dependencies between the PEMFC electrode microstructure and performance with the state-of-the-art *in situ* equipment [3]. Numerical simulation therefore becomes a complementary approach to mitigate these experimental difficulties and obtain a deeper insight into the electrode. Divided by simulation scale, the numerical simulation methods can be categorized into macroscale simulation, which is the most mature numerical method and simplifies the porous microstructure as homogeneous medium [4], pore-scale simulation, which focuses on the frequency distribution of the molecular group [5] and microscale simulation, which tracks the behavior of each individual molecule [6]. Each simulation method has its own strengths and drawbacks. Macroscale simulation can be applied to simulate the processes in a large computation domain such as a single fuel cell or even a stack while it is incapable of studying the fundamental electrochemical mechanism due to the limited resolution scale. The microscale simulation, such as molecular dynamic method, can fundamentally resolve these processes in molecular scale, but the computation domain is restricted to the nanoscale considering the computing resources and time. With the goal of fundamentally understanding the transport mechanisms in the electrode and optimizing the microstructure design, pore-scale simulation, which is based on realistic reconstructed electrodes with acceptable computation cost, has been shown to be a reliable method to study the electrode [7–9]. It mainly focuses on the effect of the

Yuze Hou, Qing Du, Kui Jiao, State Key Laboratory of Engines, Tianjin University, 135 Yaguan Road, Tianjin 300350, China
Dietmar Gerteisen, Nada Zamel, Fraunhofer Institute for Solar Energy Systems ISE, Heidenhofstr. 2, Freiburg 79110, Germany

https://doi.org/10.1515/9783110622720-008

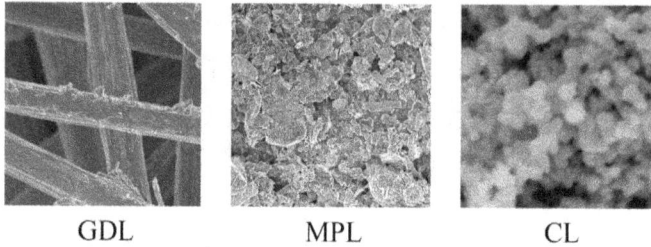

GDL   MPL   CL

Figure 1: The microstructures of the electrode of PEM fuel cells.

specific electrode microstructure on transport phenomena. In recent years, many pore-scale simulations have been conducted to study the properties of electrode, such as mass transport resistance [10], the dynamic behavior of two-phase flow [11], permeability of each component [12], etc.

As established from the discussion above, pore-scale modeling is an important tool for in-depth knowledge of PEM fuel cells and should be well understood. Hence, in this chapter, our aim is threefold:

i. Various popular pore-scale numerical tools available through the literature are introduced and compared on their advantages and disadvantages;

ii. Applications of pore-scale models in the electrode are comprehensively reviewed;

iii. A tutorial on conducting pore-scale simulations in the electrode is presented, including how to reconstruct the GDL, MPL and CL, and how to develop the single- and two-phase pore-scale models.

At the end, the key points of the chapter are summarized and the development trends of pore-scale modeling and its application are briefly discussed.

8.2 Comparison among pore-scale models

This chapter reviews the pore-scale models focused on PEM fuel cell electrode from 2010 to 2020. Visiting the literature, the popular pore-scale numerical tools include lattice Boltzmann method [5], pore-network (PN) method [13] and fine-scale conventional computational fluid dynamics (CFD) [14] and their weights are shown in Figure 2.

About 14 % of the pore-scale studies are conducted by conventional CFD method, which mainly includes the finite volume method (FVM) [15], finite element method (FEM) [16] and the volume of fluid (VOF) method [17] for the two-phase issue. The physical processes in these approaches are resolved based on governing equations, namely the Navier–Stokes (NS) equations. The conventional CFD method is mostly employed for the macroscale computation domain, such as the single fuel cell. However, the electrode microstructure would not be considered but rather it is taken as a homogeneous domain

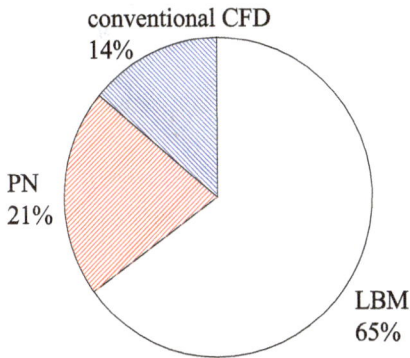

Figure 2: Categories of pore-scale models published in the literature from 2010 to 2020.

with all its properties expressed in terms of averaged properties, since there remain difficulties in discretization and boundary implementation in the porous microstructure. Recently, the discretization tool SnappyHexMesh has been widely employed and is able to automatically construct the high-quality mesh, which solves a big problem for the conventional CFD. With the powerful tool, some researchers successfully simulate the two-phase flow in the porous electrode with the VOF method [18–20]. Nevertheless, some detailed boundary conditions are still difficult to implement, such as to individually define the contact angle of carbon and PTFE. The VOF and LBM were recently compared by Deng et al. [21]. In the LB simulation, the specific distribution of PTFE in the GDL was considered and the wettability of different materials was also defined separately. The comparison results showed that numerical results by LBM agree better with experimental results than the VOF method. Additionally, the conventional CFD algorithms are less efficient in parallel computing and more complex to code than the LBM. Therefore, the conventional CFD methods are less popular in the pore-scale application.

The PN method, accounting for 21 % of the published literature, simply characterizes the porous media as connected pores and throats [22] with detailed individual pores ignored but main topology represented [23], and the water transport is based on percolation theory and determined by comparing the water/air pressure difference and the entry capillary pressure [24]. Hence, it can only give a general prediction of the phenomena. Meanwhile, the PN models save time and computation resources because the complex NS equations are not solved. As the microstructure is not comprehensively characterized, more efforts have been devoted to directly construct the pore network by extracting digital images of the microstructure by X-ray tomography [25]. However, the reconstructed pore network still represents a simplified version of the porous microstructure, since not all morphological details could be provided and nanoporosity is not reconstructed due to resolution limitations. Additionally, the PN model fails to consider the electrochemical reaction on the Pt/ionomer interface, which restricts its application in the catalyst layer. As such, it is mainly employed to give a quick prediction of the water transport in the GDL or MPL.

LBM is the most popular pore-scale numerical tool in the last decade with 65 % of the reported literature focusing on this method. It focuses on tracking representative collections of fluid molecules and describes the distribution of fluid particles and their evolution vs. time. With its underlying kinetic nature, LBM serves as a mesoscopic tool, which bridges the macroscale and microscale. It not only acts as a simple and efficient solver for the conservation equations, but also helps to fundamentally understand the underlying microscopic dynamics, while recovering the macroscopic laws. Besides, the underlying kinetic nature of LBM also makes it handy for dealing with interfacial dynamics and implementing complex boundary conditions in the microstructure. A typical example is the two-phase interface in the multiphase flow that can be automatically tracked by introducing the interaction forces, which saves much computation resources compared to the VOF in traditional CFD. Another significant advantage is that the LBM saves the meshing step, which is also complicated to conduct for porous media in conventional CFD. Compared to the conventional CFD where a complex Poisson equation needs be solved to obtain the pressure field, LB model only needs to solve a simple equation of state, which describes the relation between pressure and density. Moreover, its explicit scheme and local updated rule makes it ideal for parallel computing based on either CPUs or GPUs, which make the large-scale simulation possible. All these advantages make the LBM versatile for various applications in the electrode of PEM fuel cell. Table 1 briefly summarizes the applied scopes of these popular numerical tools.

Table 1: Comparisons among various pore-scale numerical tools.

	Conventional CFD	Pore-network model	Lattice Boltzmann method
GDL MPL	Applicable, but difficult to set boundary condition	Applicable for two-phase simulation as a fast prediction	Applicable for both single- and two-phase simulation
CL	Not suggested	Not suggested	Applicable

8.3 Application of pore-scale models to the electrode

The number of research papers focusing on pore-scale simulation in PEMFC research has been steady in the past decade, as shown in Figure 3. Most studies focus on evaluating the transport properties of the electrodes (e. g., permeability, conductivity, diffusivity), investigating the two-phase flow mechanism in these porous media and optimizing the electrode design. These studies have provided valuable data to validate the transport properties in the empirical equations and, therefore, increase the accuracy of the traditional CFD simulations. They have also been a useful resource for shedding light

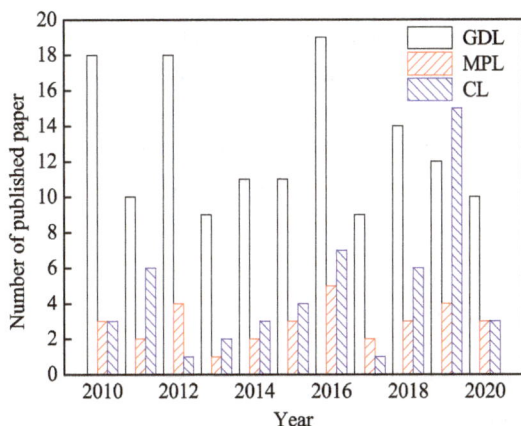

Figure 3: Pore-scale simulations published in the last decade.

onto the transport mechanisms of the two-phase flow inside the electrode. In this section, representative applications in the last decade will be reviewed according to the electrode components, namely GDL, MPL and CL. After reading this section of the chapter, the reader will have a general idea of the scope of problems that can be investigated with the current available pore-scale models.

8.3.1 Application to gas diffusion layer

In 2010, Hao and Cheng [26] calculated the relative water permeability of carbon paper GDL with LBM. It was found that water transport in the GDL was capillary pressure driven and the liquid water was likely to be trapped in the large pores in the GDL. For a low water saturation case, a hydrophobic GDL could lead to a dispersed distribution of liquid water while continuous liquid flow could be formed under the high saturation case. The relative permeabilities in both in- and through-plane were calculated and it was concluded that the flow orientation has negligible effect on relative permeabilities. In another work by Hao and Cheng [27], the water invasion in the GDL was simulated to study the effect of carbon fiber contact angle on water transport dynamics. A useful water management strategy obtained was to introduce hydrophilic passages in the GDL to help remove the water, and thus decrease the transport resistance in the GDL. The LBM was proven to be a powerful numerical tool to conduct pore-scale investigation in the porous microstructure. Note that the effects of density and viscosity ratios on the two-phase simulation in the GDL are negligible based on the nondimensional number analysis [28]. Hence, low density and viscosity ratios are initialized in most of LB two-phase simulation in GDL for a lower spurious velocity and a better numerical stability.

In 2011, Rama et al. [29] investigated the anisotropic permeability of carbon-cloth based GDL with single-phase LB model. The through- and in-plane permeabilities were calculated, respectively and the through-plane direction was about 4 times higher. The connected void regions in the GDL, playing the role of flow channels, was critically important for the gas transport. The simulation results were also used to determine the Kozeny–Carman (KC) constant K_{KC} for the widely used empirical equation KC relation (see Equation (1)), which was significant for improving the accuracy of conventional CFD modeling,

$$\frac{k}{r^2} = \frac{\varepsilon^3}{4K_{KC}(1-\varepsilon)^2} \tag{1}$$

where k is the permeability, r is the average fiber radius and ε is the porosity. The pore-scale model has also shown a strong potential for revealing the mechanisms of microscopic fluidic phenomenon in the electrode.

In the same year, Zamel et al. [30] conducted a pore-scale study focusing on the flooding effects on transport properties of a stochastic reconstructed GDL, including relative permeability, gas diffusion coefficient and thermal conductivity. The overall number of available pores in both directions made the relative permeability independent of the direction. The Leverett function was proven to underestimate the capillary pressure of the GDL. Additionally, the presence of liquid water in the GDL blocked the gas transport path and directly decreased the overall porosity of the layer, causing a larger mass transport resistance, meanwhile, increasing the overall thermal conductivity of the GDL, especially in the through-plane direction.

In 2012, Chen et al. [31] investigated the liquid water transport and distribution in the GDL with LBM. It was found that hydrophilic gas channels lead to liquid water accumulation in the GDL. The management of hydrophobic and hydrophilic fiber regions had a more significant effect on water distribution than the PTFE content. Moreover, the effect of fiber hydrophobicity on water transport, fuel or performance is also determined by the specific water distribution as the effective diffusivity is higher in hydrophilic GDL when the water is separately distributed. In a later study, they investigated the two-phase behavior in the GDL under an interdigitated flow field [32] in terms of land width and GDL contact angle. A narrower land could reduce both the water removal time and residual saturation. In the "slow creeping region," the air flow velocity was low and the water transport was capillary force dominated, while a faster air flow and shear force dominated transport in the "quick moving region." Additionally, the hydrophobic carbon fiber was only beneficial for the water transport in the shear force dominated area, and a more hydrophobic GDL could reduce the water residual saturation but prolong the water removal time.

Gostick [33] developed a pore-network model of GDL where a Voronoi tessellation was used to represent the fiber structure and a Delaunay tessellation for the pore space rather than the cubic-lattice based method in the previous pore-network models. The proposed model was employed to calculate the gas diffusivity and the results agreed well with the experimental data. Besides, the constructed pore-network could also be applied to estimate the electrical and thermal conductivity of the GDL.

In 2015, García-Salaberri et al. [34, 35] studied the effective diffusivity in a partially-saturated carbon-fiber GDL in terms of the effects of local saturation and through-plane saturation distribution with LBM. The results showed that the presence of PTFE reduced the effective diffusivity due to the decrease in pore volume and the higher tortuosity of transport paths and the local effective diffusivity follows a quadratic dependence on local water saturation. In the same year, the effects of GDL compression on single and two-phase flows were investigated by Espinoza-Andaluz et al. [36] and Jeon et al. [37], respectively. However, the simulations were limited in 2D and the compressed GDLs were simplified by changing their thickness ratio. When the GDL was compressed to 66 % of the initial thickness, the porosity would decrease around 10 % and the tortuosity increased about 6 %, which caused a larger transport resistance. Regarding the two-phase transport, the compression significantly affected the water transport behavior in the GDL and low compression rate could decrease the water saturation in the GDL and was therefore favorable for water management.

In 2016, Fazeli et al. [38] further investigated the effects of compression on water and gas transport resistance with equivalent pore-network model. The equivalent pore-networks were based on synchrotron X-ray tomography images of compressed GDLs, which was more realistic than the previous studies. The gas transport resistances under wet and dry conditions were calculated, respectively, and the result showed that the resistance under flooding was about two times higher. Besides, the optimum compression rate for Toray TGP-H-090 GDL was found to be 10 %, ensuring a low gas transport resistance in both dry and wet conditions.

In 2018, Niu et al. [18, 19] successfully employed the VOF model rooted in Open FOAM [39] to simulate the two-phase flow in the GDL. The powerful SnappyHexMesh function in the Open FOAM was used to realize the mesh discretization. The VOF model was validated by comparing the water saturation in the through-plane direction from the experimental data. The simulation results matched well under 1000 Pa while deviated from 4000 Pa. Such variation could be ascribed to the uniform distribution of wettability or the difference on reconstructed local microstructure.

Zhang et al. [40] tried to resolve the coupled two-phase flow, reactant gas transport in the GDL and CL as well as the electrochemical reaction processes in the CL. Two sets of LB models were used to realize these functions. The two-phase LB model simulated the water invasion from the CL while the single-phase LB model was used for the oxygen diffusion from GDL to CL with the electrochemical reactions coupled to this LB model. The electrochemical reaction was considered by applying a source term at the bottom layer of the lattice, which is assumed to be the CL. The presence of liquid water could affect the local concentration and diffusivity of oxygen, which connects the single and two-phase models. It was observed that the presence of a liquid water film on the CL/GDL interface reduces the performance. The effect of MPL was also investigated and proved to be capable of changing the water distribution in the GDL, thereby, increasing the performance. However, since the water is injected into the domain with an arbitrary velocity, realistic generation of the water becomes important.

In 2019, Deng et al. [21] further investigated the effect of drying method on the PTFE distribution and the layer wettability in the stochastically reconstructed GDLs. In the model, the wettability difference between carbon fiber and PTFE were considered by defining their contact angles separately. The results showed that the PTFE drying method has significant effects on water transport in the GDL. The PTFE tends to accumulate at one-layer side under the air-drying method, while the vacuum-drying method leads to a uniform PTFE distribution in the through-plane direction and a lower pressure drop.

Zhou et al. [20] studied the water transport in the compressed GDL with VOF model. The GDL was stochastically reconstructed and the finite element method was employed for the assembly pressure simulation. Such compressed GDL was more comparable to the realistic GDL than the models with simply changed thickness ratio [36]. The results showed that the compression increased the critical water breakthrough pressure but had minor effects on the pathways of water transport.

Recently, He et al. [41] simulated the ice melting process in the GDL with the coupled model of LBM and enthalpy-based method. The melting processes were comprehensively investigated in terms of melting position through the layer, porosity, temperature distribution and some nondimensional parameters, such as Rayleigh and Stefan number. The melting position was found to affect the melting process and melting process was faster in the in-plane direction. A larger Stefan number could also significantly accelerate the melting process.

As can be seen, over the last decade, substantial work has been dedicated to investigating both single and two-phase flow in the GDL. More attention should be paid on the realistic consideration of water generation as the given water inlet velocity is almost always unrealistically high, which could affect the two-phase flow pattern; hence, the real two-phase flow pattern is still unclear. However, this problem also relates to the time scale in the simulation, which will need more computational resources. More efforts should also be put on modeling the coupled phenomena in the GDL, like resolving the coupled water and gas transport in the GDL, which can give a more realistic description than the individual single or two-phase flow. Additionally, phase change processes in the GDL, such as ice melting during the cold start, pose additional challenges for the simulation which require further research.

8.3.2 Application to micro-porous layer

In 2010, Ostadi et al. [42] simulated the single-phase gas flow through the MPL reconstructed by FIB/SEM with LBM and calculated its permeability and tortuosity. The MPL sample volume was 2.7 μm^3 and resolved with a pixel resolution of 8 nm. The result showed the through-plane permeability in the MPL is 5 orders of magnitude lower than in the GDL and the tortuosity is 13 % less in the principal through-plane direction than the in-plane direction.

In 2011, Becker et al. [43] used the finite volume solver in the software package Geo-Dict [44] to estimate the binary diffusion coefficient for Knudsen number. With the same solver, Zamel et al. [45] studied the effect of MPL on thermal conductivity and diffusion coefficient in terms of thickness, porosity and penetration depth. The diffusion coefficient increased with the decrease of thickness or increase of porosity. However, the thermal conductivity decreased with the increase of porosity and was not sensitive to the thickness. Moreover, the penetration of MPL into the GDL could also increase the thermal conductivity but decrease the diffusion coefficient.

In 2013, Han and Meng [46] simulated the water transport in the MPL and GDL with pseudopotential two-phase model to investigate the effect of large, perforated pores. The results showed the perforated pores can serve as a transport pathway for liquid water and help its removal. Due to the scale difference between the GDL and the MPL, the simulation was only conducted in 2D. Following the same approach, Kim et al. [47] employed the pseudopotential LBM model to simulate 2D liquid water transport in the MPL and GDL. The simulation conditions were close to the real operation. The effects of MPL thickness and wettability were investigated. It was found that liquid water content can be reduced in the GDL by using a thicker MPL or more hydrophobic surfaces. The presence of MPL can effectively reduce the number of water breakthrough sites toward the GDL, which improves the water management.

In 2019, Deng et al. investigated the effect of the cracks in the MPL on the water transport in the electrode [21]. With the advanced parallel algorithms, for the first time, the two-phase simulations were conducted in 3D. The results showed the cracks serve as the transport path for liquid water and the drying methods of PTFE on GDL also have significant effect on water dynamic behavior.

Cetinbas et al. [48] proposed an alternate technique for modeling the liquid water transport in the porous transport layer (PTL) based on measured tomography images with momentum conservation and capillary forces accounted. The PTL was discretized into structural grid by segmenting X-ray images into pores and solid nodes. The water transport process is modeled based on capillary and viscous forces analysis. The capillary force was calculated according to the local pore size and the viscous force was determined by the local distance to the pore wall, equivalent tube radius and a calibration function. It was found that the cracks in MPL could improve water management by keeping nano-pores dry and decrease the liquid water saturation in the PTL.

Figure 3 shows that pore-scale studies of the MPL are much fewer than that of the GDL. Since the differences in terms of the numerical tools between the GDL and the MPL are negligible, the issue mainly lies in the computational cost. The reconstructed MPL needs much higher resolution than the reconstructed GDL considering the presence of nanosized pores in the MPL, 10 nanometers versus 2 micrometers, which significantly increases the difficulties in implementing the simulation, especially when the MPL and GDL are coupled. So far, the permeability and diffusion coefficient of the MPL can be well predicted and more effort should be put forward to investigate the two-phase flow in the MPL. Furthermore, since the spurious velocity around the two-phase interface may cause model instability, the minimum resolution for the two-phase simulation (10 nm) is usually at least two times higher than the single-phase flow (20 nm), making the simulation more resource-consuming.

8.3.3 Application to catalyst layer

In 2010, Lange et al. [49] simulated species transport and electrochemical reaction in a stochastically reconstructed CL at nanometer scale with FVM. They proposed a "13-direction" algorithm to determine the local pore size rather than a mean pore radius, which could consider the significant Knudsen diffusivity in small pores. The oxygen, proton and electron transports were all coupled in the model. However, since the platinum phase was not resolved in the model, the electrochemical reaction was assumed to occur uniformly at the carbon/ionomer interface. The Knudsen diffusion was proven to affect the oxygen transport in the CL. Besides, larger carbon supports could increase the effective diffusivity of oxygen. In their later work in 2012 [50], the carbon support is simplified as spheres in radii around 16 nm and the effects of some reconstruction variables like ionomer and carbon sphere distributions were studied. The results showed that the reconstruction algorithms had a significant effect on effective diffusivity, indicating the importance of CL reconstruction parameters and assumptions, and the uniform distribution of ionomer could improve the performance due to a lower transport resistance.

In 2015, Chen et al. [7] resolved the coupled physical-electrochemical processes in stochastically reconstructed CL with LBM. The platinum particles were considered in the reconstruction and the electrochemical reaction took place at the interface of the platinum, which is a more realistic description. At the same time, the presence of platinum also improved the demand on resolution and the resolution in this work was 5 nm, which is just enough to capture the Pt particles. The nonuniform distribution of ionomer increased the tortuosity, thus reducing the effective diffusivity. In their following work [51], they focused on the local gas transport around a single stochastically reconstructed Pt-based carbon support. The four-constituent microstructure of carbon support included a carbon sphere, ionomer, Pt particles and also primary pores with a resolution of 0.5 nm. The local transport resistance under various Pt and ionomer loading was investigated and it was found that the low oxygen diffusivity in ionomer is the dominant resistance for mass transport. Recently, they developed the LBM to be able to model the transport resistance and concentration drop across the ionomer/pore interface, which greatly improves the model accuracy [52].

Hou et al. [53] investigated the effect of Pt/C and I/C weight ratios on the reactive processes in a stochastically reconstructed CL. The results indicated that higher Pt/C ratio can effectively decrease the CL thickness and improve the transport. A higher I/C ratio would cause more mass transport loss, but also increase the active catalyst area ultimately increasing performance. Based on these results, an ideal CL design was proposed. The basic idea is to create relatively large (micrometer scale) oxygen transport paths to ensure a low mass transport loss under a high I/C ratio condition. This ideal design can be realized by using pore former during manufacturing. This work provided valuable insight for the CL fabrication and proved the capability of pore-scale modeling for electrode development.

Currently, pore-scale models can resolve the coupled electrochemical reaction and diffusion process in the catalyst layer with the consideration of all constituents. Most approaches place the focus on modeling the reactant transport, such as the effect of Knudsen diffusion and the reactant gas dissolution rate across the ionomer, with the electrochemical reaction source term applied on the platinum-ionomer interface. The general development trend is toward reconstructing the CL with a higher resolution in a more realistic algorithm. However, considering the complex effect of water on ionomer such as absorption and swelling, and relative studies on the flooding effect in CL are still missing, such as how the flooding could affect the reactive area and reactant transport. The review of these representative applications validates the versatility of pore-scale modeling. With a careful balance of numerical scale and computation cost, pore-scale modeling becomes suitable for in-depth investigations on structure-property-performance relationships for the PEMFC electrode. In the next section, a detailed tutorial on how to build a pore-scale model is given as a guideline for the reader.

8.4 Tutorial on pore-scale modeling

The development process of a pore-scale model can be briefly summarized as:
a. Reconstruct the electrode microstructure as a computation domain;
b. Build numerical model to resolve the complex transport and reaction processes

This section first introduces various approaches to reconstruct the electrode and the instructions for numerical reconstruction algorithms are given. Then a tutorial is given on the LBM for its popularity and the emphasis is placed on its model modification under different application conditions. In general, the reader will have a basic idea on how to develop a pore-scale model after reading this section.

8.4.1 Electrode reconstruction

Realistic reconstruction of the electrode is a prerequisite for the pore-scale simulation since it serves as the computation domain and determines the simulation results. The reconstruction methods can be categorized into experimental imaging approaches and numerical stochastic approaches [9]. The experimental approaches rely on the numerical descriptors of the sample microstructure obtained via an imaging technique, such as X-ray computed tomography (X-CT) [54] or focused ion beam/scanning electron microscopy (FIB/SEM) tomography [55]. Since the reconstruction is based on the real electrode sample, the obtained numerical microstructure is authentic. However, this method places heavy demands on experimental facilities, and its reconstructed resolution is limited (typically ~10 μm for X-ray microtomography, ~50 nm for X-ray nanotomography and ~20 nm for FIB/SEM [56]). In addition, parametric analysis of the electrode's structural properties, such as porosity, pore size distribution and tortuosity, cannot be carried out easily and accurately. As for the numerical stochastic approach, the reconstructed electrode is artificially generated based on a "reconstruction algorithm." Therefore, it is highly flexible to control the resolution and macro properties of the reconstructed electrode and the presence of every constituent in the electrode can be considered during the reconstruction. The issue with such an approach, however, is that the assumptions in the reconstruction algorithms may affect the authenticity of the reconstructed structure. For this reason, the reconstructed structure should be compared with the real electrode quantitatively for validation. The pore size distribution is often used to carry out this validation. Nevertheless, the obvious advantages on cost, implementation and flexibility still make it a more popular option than the experimental imaging approach. In the next subsections, the introduction and development of these two approaches are presented.

8.4.1.1 Experimental imaging approach

The details about the PEMFC electrode characterization using imaging approaches are given in Chapter 3. In this chapter, a short overview of the methods is given with respect to application in modeling.

X-ray tomography method

X-ray computed tomography (XCT) method is a noninvasive technique for the 3D visualization of the microstructure without extensive requirements on sample preparation. A schematic of its experiment setup is shown in Figure 4(a) and the reconstruction process can be simplified as:

i.   Obtaining 2D gray scale images via XCT;
ii. Image thresholding process to obtain binary images;
iii. Reconstructing 3D binary microstructures.

Koido et al. [57] and Becker et al. [58] carried out the pioneering work of reconstructing carbon paper GDLs via X-ray tomography with a resolution of 1 μm and 0.7 μm, respectively. Later, Rama et al. [12, 29] reconstructed the carbon paper with micro and nanotomography, respectively. Under the same reconstruction process, the resolution

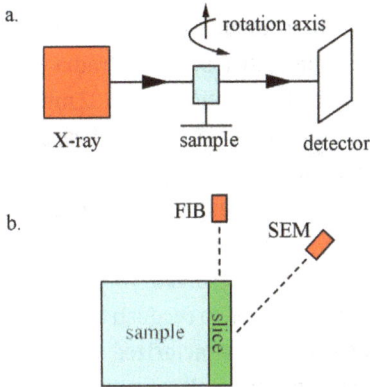

Figure 4: Experimental schematics of (a) X-ray tomography and (b) FIB/SEM method.

of the microtomography images by Skyscan 1072 system was 1.76 µm while the resolution of nanotomography images by Skyscan 2011 system was 680 nm, which shows that the reconstruction accuracy highly relies on the experimental equipment. Ostadi et al. [59] investigated the influence of small alterations in the threshold value on reconstruction properties and found that the porosity, anisotropy and the mean pore radius are sensitive to the threshold variation. Hence, the importance of determining the correct threshold value when using the X-ray tomography technique was established. However, this process often relies on the user and it is as subjective process. Rama et al. [60] successfully reconstructed a compressed GDL by XCT. To permanently maintain the compressed state of GDL, the sample was saturated with polydimethylsiloxane (PDMS) and then heated to 333 K while under compression. The PDMS is an optically clear silicone-based polymer with similar density to air and low X-ray attenuation coefficient. Eller et al. [61] developed the XCT to realize the *in situ* observation of liquid water distribution in the GDL and found the XCT shows a better spatial and temporal resolution than the neutron imaging while sacrificing the contrast for liquid water. Regarding the liquid contrast issue, the liquid water phase can be segmented by subtracting the dry dataset from the water containing data set. García-Salaberri et al. [34, 35] also segmented the XCT reconstructed saturated GDL into three phases: GDL solid structure, water and air with a custom-made code. Inoue et al. [62] reconstructed the compressed GDL with phase segmentation into fibers, binder and pores with a resolution of 2.4 µm per pixel.

From the above brief survey, it can be seen that the XCT is already a mature approach to reconstruct the GDL and can distinguish between carbon fibers, binder, water and pores. The reconstruction resolution is also sufficient to resolve the microstructure of the GDL.

As for its application to the MPL reconstruction, nano XCT is used to achieve a higher spatial resolution. Basically, the implementation process is the same as that for the GDL reconstruction. In 2013, Wargo et al. [56] reconstructed the MPL with nano-XCT and quantitatively validated the reconstructed structure in terms of porosity, pore connectivity, tortuosity and chord length distribution by comparing with the experimentally measured value. Kotaka et al. [63] used the nano-XCT system (Xradia Inc. UltraXRm-L200) to reconstruct an MPL sample with a pixel resolution about 32 nm. In 2016, Andisheh-Tadbir et al. [64] reconstructed the MPL with the segregation of carbon, PTFE and pore phases. It was found that PTFE is distributed in conglomerated regions and acts as a binder for carbon particles. The resolution could reach 50 nm per pixel, which is also sufficient to resolve the MPL. Therefore, the state-of-the-art nano-XCT equipment can reconstruct the MPL with both carbon and PTFE resolved.

Regarding the CL reconstruction, in 2012, Epting et al. [65] tried to reconstruct the CL via nano-XCT (UltraXRM L200). A general carbon structure was characterized as well as the large pores. However, the microstructure details, such as the smaller primary pores within the carbon support and the thin ionomer films (around 10 nm thick) coated on the agglomerate, were lost with the 50 nm nano-CT resolution, not to mention the Pt particles whose diameter was about 3–5 nm. In 2013, with the same equipment, Litster et al. [66] imaged the CL microstructure and verified the reconstructed CL against mercury intrusion porosimetry and TEM micrographs of the same CL. Since the reconstruction resolution is highly determined by the equipment, the state-of-the-art X-ray tomography is not able to resolve the detailed microstructure of CL for now. On the other hand, since the XCT is noninvasive, it is widely applied for the *in situ* electrode observation.

FIB/SEM methods

Compared to the X-ray tomography technique, FIB/SEM is a destructive technique and its schematic is shown in Figure 4(b), and discussed in Section 3.2. The sample is milled away slice by slice via the FIB to reveal successive cross-sections, which are imaged by the SEM. It usually requires some preparation process like impregnation of sample pores for a better result and the implementation should be conducted in a vacuum environment. FIB/SEM cannot be applied in GDL reconstruction due to the discontinuity in solid microstructure.

Regarding the application in the MPL, Ostadi et al. [42] used a dual-beam FIB/SEM system to obtain the 3D microstructure data of MPL sample with a resolution of 14 nm. A 100 nm-thick layer of platinum was coated on the surface prior to side-wall milling to protect the soft surface from ion bombardment and to reduce the redeposition effect. The general process can be described as:

i. Mill away a thin slice (~5–20 nm) of the side-wall of a trench with FIB;
ii. Record an SEM image of the new surface;
iii. Repeat steps (I) and (II) to produce a stack of SEM images.

Like the XCT, the obtained 2D aligned images are then converted to binary and used for the 3D microstructuring. Inoue et al. [62] used the SMF-1000 (Hitachi High-tech Science) reconstructing the MPL with a 4 nm resolution. The traditional preprocessing would fill all pores with uniform materials via embedded resin or metal evaporation to present clear contrast differentials in SEM images. However, this method has some drawbacks:

i.   The delicate microstructure may be destroyed during the preprocessing;
ii.   It is hard to fill all the micropores;
iii. It is difficult to distinguish the resin binder from the pore-filling material in SEM observation.

To overcome these issues, they proposed a new technique eliminating the need for pre-processing. The general idea was when the cross-sectional image is captured, both secondary-electron images and back scattered electron images should also be captured. The cross-sectional structure was identified from the composite image to reconstruct the MPL. The advantages offered by the synthesis of the secondary and reflected electron images were utilized to optimize SEM observation. Terao et al. [67] presented a novel segmentation technique for preparing binary images by using the composite image of secondary electron and energy-selective backscattered (EsB) electron images. The dark deep solid was removed by contrast stretching using the accumulated SE images and the advantage of EsB on uniform and anomalous contrast can make up the hollow low-intensity area. The resolution could reach 4 nm per pixel and the reconstructed structure was validated by comparing the data to experimental results.

Regarding the catalyst layer reconstruction, Zils et al. [68] were among the first to visualize the PEM fuel cell electrode using dual FIB/SEM tomography with only carbon phase presented. Later, Lange et al. [69] found that the sputtered material from the FIB could contaminate the sample chamber and affect the SEM resolution in the dual FIB/SEM system, which can be resolved by using a standalone FIB/SEM system. The proposed standalone FIB/SEM system is realized by fixing the sample on a holder to ensure the constant position and swap between FIB and SEM and can provide a much finer porous interior structure. Gao et al. [70] carefully conducted the sample preparation including, deposition of protective Pt layer; rough milling; recognizing a fiducial mark to reach a high resolution of 15 nm. Thiele et al. [71] successfully extracted the digital CL structures data with a resolution of 2.5 nm per pixel in the SEM imaging direction (x-y), which is the best resolution achieved so far. However, the imaging process still cannot distinguish the ionomer since the density of carbon and ionomer are too close to give a good contrast in the SEM device. Later, the same team [72] proposed a new approach for integrating reconstruction information. After the CL was reconstructed by FIB/SEM, transmission electron microscopy (TEM) was used to analyze the distribution of platinum particles. Then the Pt volume distribution obtained from TEM information and the SEM were combined to visualize the Pt nanoparticles in the porous CL, yielding knowledge of reaction pathways and the number of reactions sites. Klingele et al. [73] further presented a model to fix the errors in FIB/SEM caused by the FIB cutting distance and the reconstruction accuracy was significantly improved.

The above brief literature overview indicates that FIB/SEM can be used to reconstruct the MPL with enough resolution, but still needs to be adapted for the reconstruction of the CL. Although the resolutions of FIB/SEM and nano-XCT may be near or even higher than the feature size of the material phases, more efforts should be devoted to accurately resolve the distributions of various phases. As mentioned, whether the XCT or FIB/SEM techniques, they are both based on the real electrode sample, and hence the authenticity of the reconstruction can be ensured, which makes these approaches more suitable for the analysis of the electrode properties and providing quantitative data for the validation of stochastic reconstruction. Note that electron tomography (ET) can also be applied to reconstruct the CL in a higher resolution (~1 nm) [74], however, its reconstruction scale is very limited (approximate size 0.5×0.5×0.1 μm), making it not suitable for the pore-scale investigation. Hence, ET is not emphasized in this chapter.

8.4.1.2 Numerical stochastic approach

Due to the advantages of easy implementation, flexibility and fast reconstruction, the stochastic reconstruction approach is more popular in the pore-scale simulation with the stochastic reconstruction algorithm being the key for this method. For each layer of the electrode, the evolution and detailed instructions of the stochastic algorithms will be presented below.

Gas diffusion layer
The nonwoven carbon paper is the most common material for GDL fabrication due to its porous structure, good mechanical properties and its high electrical and thermal conductivity. For the carbon paper GDL, the numerical stochastic algorithms have not changed very much through the years due to the relatively simple geometry of the carbon fibers.

Schladitz et al. [75] were the first to reconstruct the nonwoven fiber numerically. The microstructure was modeled by a macroscopically homogeneous random system of straight cylinders. The fibers were generated by a spatially stationary random system of lines, dilated by a sphere. The anisotropy was described by a parametric distribution of the fiber directions. Many of reconstruction algorithms found in literature are based on this work. Schulz et al. [76] used the same method to reconstruct the nonwoven carbon fiber GDL. The reconstructed microstructure was determined by some structural inputs, such as fiber diameter, orientation and porosity and based on the following assumptions for simplification:
i.   The fibers are long compared to the sample size and the crimp is negligible;
ii. The fibers are allowed to overlap;
iii. The fiber system is macroscopically homogeneous and isotropic in the in-plane direction, defined as x-y plane.

In this way, the density distribution $p(\theta, \varphi)$ can be described as a function of the altitude θ and the through-plane thickness φ, given by

$$p(\theta, \varphi) = \frac{1}{4\pi} \frac{\beta \sin \theta}{(1 + (\beta^2 - 1) \cos^2 \theta)^{1.5}} \qquad (2)$$

β is the anisotropy parameter. $\beta = 1$ describes an isotropic cylinder system and increasing β makes the fibers more parallel to the x-y plane.

Hao and Cheng [77] further considered the compression and existence of PTFE. However, in their compression assumption, only the thickness of the reconstructed GDL was changed and the fibers shape remained. One more step was added after building the GDL when the PTFE is considered. The void cells near the fiber were randomly converted into PTFE cells until the prescribed PTFE content is achieved. Becker et al. [45] further considered the binder and numerically reconstructed an assembled MPL/GDL microstructure. For the binder generation process, first the pore size of the 3D geometry was calculated and then the pore nodes with lower pore size had higher possibility to change to binder nodes.

Daino and Kandlikar [78] developed a phase-differentiated GDL reconstruction model with the consideration of void, fiber, binder and PTFE. The fiber skeleton of GDL was modeled as a random collection of intersecting cylinders. General process was similar to the previous work. The binder material was added to the substrate in regions of small crevasses near the intersection of fibers and proportional to the spherical structuring element radius. The PTFE was then added using a closing with a spherical structuring element of a larger radius than was used for the binder addition.

Deng et al. [21] optimized the GDL reconstruction by using the experimentally measured data from the literature. It was assumed that carbon fibers were straight cylinders with fixed radius, no fiber was generated in the through-plane direction and the carbon fibers were randomly generated in the in-plane direction. The layers were numerically assembled to construct the GDL and the porosity of each layer was checked to meet the experimental data. In the GDL manufacture process, the PTFE distribution was determined by the drying method (air-drying or vacuum-drying). To reconstruct PTFE numerically, first, the volume fraction of PTFE in each layer was calculated based on the total PTFE content and experimentally measured relative content along the thickness direction. Second, every pore node was identified and the "converting probability," P, was based on the solid node number around the pore node. Third, comparing P with the critical value, the pore node changed to PTFE node if P was larger. The last step was to check the volume fraction of PTFE in this layer and repeated such process until the target fraction was achieved.

Note that the carbon cloth, as GDL material, was found to suffer mechanical and mass transport issues, and hence showed poor performance in comparison with carbon paper [79–81]. Therefore, it is not widely adopted in the state-of-the-art fuel cell and here we only review one representative work by Salomov et al. [82]. The assumptions for carbon cloth based GDL are:

i. GDL fibers are two pairs of mutually orthogonal bundles;
ii. The cross profile of bundle is elliptic;

iii. Carbon fibers are homogeneously distributed in the bundle;
iv. The fiber is a sinusoidal-shaped cylinder;

The algorithm can be briefly described as:
i. Generate an elliptic bundle of fibers with circular sections and a given radius;
ii. Merge two elliptic bundles to create a 2D slice;
iii. Create all 2D slices by shifting the centers of fiber sections along the sinusoidal direction. These two directions are shifted relative to each other by half the wavelength;
iv. Create the orthogonal fibers in a similar way and assemble the two pairs of fibers;

These representative examples of numerical stochastic work point out that, for the carbon paper GDL, the advanced stochastic models can realistically reconstruct the GDL with PTFE and binder amounts and distributions considered. The macro properties like porosity, pore size distribution can also be identical or very close to the experimental data. To further realistically describe the carbon paper GDL, more attention is suggested to be paid on the PTFE and binder generation, such as its specific size or thickness, which may need higher reconstruction resolution. As for the carbon cloth, the present algorithm can already properly describe its structure. In conclusion, the numerical reconstruction of the GDL is very mature. A GDL reconstruction tutorial from [21] and its validation test are presented below.

The illustration of the reconstruction process is clarified in Figure 5(a) and the process follows:
i. Set the GDL thickness, target porosity distribution and the radius of the carbon fiber.
ii. Randomly generate two points A and B in a fiber layer as shown.
iii. Construct the line connected by points A and B.
iv. Calculate the distance of each node in the fiber layer to the line and change the node from void to solid if the distance is within the fiber radius. Then one carbon fiber can be constructed.
v. Repeat process ii–iv until the target porosity of this layer is reached and the reconstruction of one fiber layer is finished.
vi. Keep reconstructing fiber layers until the GDL thickness is reached.

The PTFE reconstruction process can be described as:
i. The total volume fraction of PTFE is calculated based on the PTFE content in the GDL.
ii. Every pore node is given a "generation probability" P, which is proportional to the solid node number adjacent to the pore node.
iii. For each pore node, the value P is compared with a critical value (self-defined), if P is larger, the pore node converts to PTFE node.
iv. Repeat such process until the total PTFE volume is reached.

This algorithm can accurately meet the target porosity distribution and PTFE volume fraction as shown in Figure 5(b). Additionally, it is flexible to adjust the fiber parameter, GDL thickness, porosity distribution, PTFE content and, therefore, recommended.

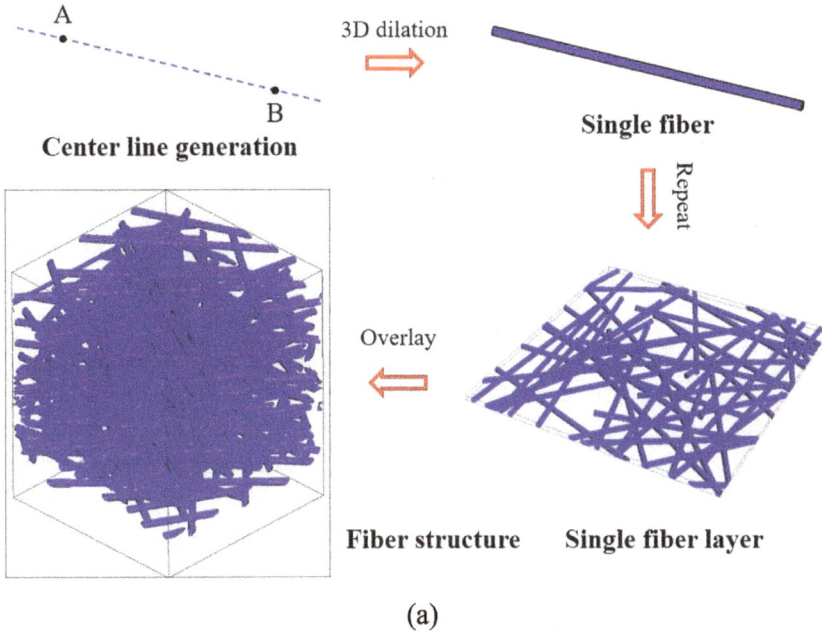

(a)

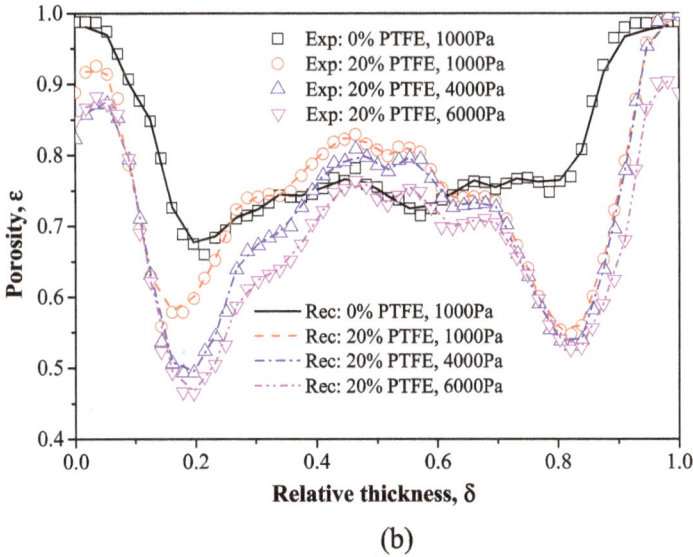

(b)

Figure 5: (a) Illustration of GDL reconstruction; (b) Comparison of GDL porosities between reconstructed and the experimentally measured data [21].

Micro-porous layer

Compared to the GDL, the microstructure of the MPL is much more complex and stochastic as it not only has the nanoscale pores, but also microscale cracks. Therefore, researchers have presented various different approaches to reconstruct the microstructure of the MPL.

Becker et al. [43] is one of the first groups to reconstruct the MPL and validate the reconstruction by comparing its pore size distribution and cross-section profile to the real MPL sample. The MPL was assumed to be constructed with agglomerates of N number of spheres. The distance between each two centers of the spheres should be more than a minimum value, while less than a maximum value (user defined). In this way, the carbon agglomerates built by connected carbon spheres were reconstructed. Then the carbon agglomerates were added to the domain one by one until the target porosity was reached. In 2012, Zamel et al. [45] reconstructed the MPL by another method. First, large spheres were randomly distributed to define the large pores. Then small spheres were initialized as carbon particles, which could overlap, but not be inside the large spheres. Last step, the small particles were glued together by filling the void nodes in between. This method was flexible with 5 changeable parameters: the diameter and the volume fraction of the large sphere, the diameter and the volume fraction of the small spheres and the maximum pore diameter to be filled. A comparison between the reconstructed pore size distribution and that measured from an MPL sample showed good agreement between model and experiment. In 2015, Hannach et al. [83] presented another reconstruction algorithm with the consideration of PTFE. In this method, the fraction of the seed particles, the overlap rate between each particle and the connectivity between particles were changeable parameters and could be tuned to reach the realistic microstructure of the MPL. The PTFE was generated in the narrow spaces between carbon particles taking into consideration the pore size distribution. Recently, Deng et al. [52] reconstructed a simplified MPL with the consideration of microcracks. Since the work focused on the two-phase flow in the MPL, the nanoscale pores were not resolved. The morphology and width of the crack was set according to those observed in SEM images of an MPL.

In this subsection, we can see that the pore size distribution and cross-section profiles are usually used to validate the reconstructed structure. More quantitative comparisons will further optimize the MPL reconstruction. However, the question still arises as how to realistically consider the presence of cracks in the reconstruction of the MPL. Due to its multiscale character, the reconstruction of the MPL with consideration of both nano-particles and microcracks is still lacking, and hence more effort for a realistic construction should be put forward.

For the MPL reconstruction tutorial, the algorithm proposed by Hannach et al. [83] is suggested and illustrated in Figure 6(a).

i. Set the target volume fraction of carbon and PTFE and the range of carbon sphere size.

MPL:

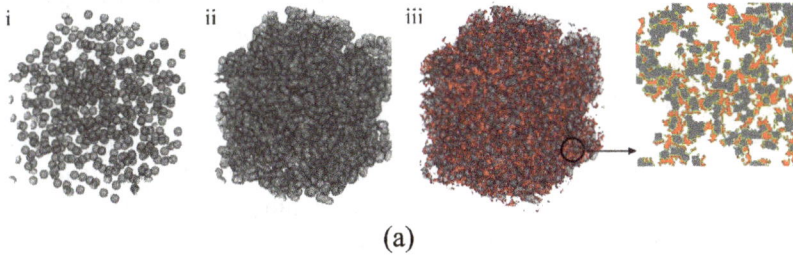

(a)

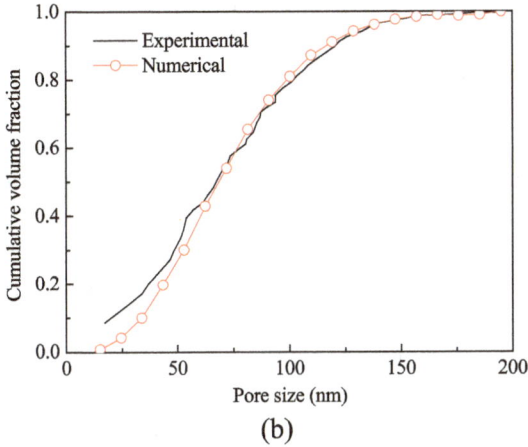

(b)

Figure 6: (a) Illustration of MPL reconstruction; (b) Comparison of the pore size distribution between the reconstructed MPL and the experimental data [11].

ii. Seed carbon spheres are randomly initialized in the computation domain. The initial spheres are not allowed to overlap.

iii. The rest carbon spheres are added in the domain one-by-one until reaching the target porosity. Each time a carbon sphere is added, it must overlap with at least one preexisting sphere within a specified overlap range.

iv. Calculate the local pore size of the domain. The void node with a smaller pore size will change to PTFE node in a higher possibility. Repeat this process until the volume fraction of PTFE is reached. Note that the changed void node has to be next to the carbon or PTFE node.

The pore size distribution of the reconstructed MPL can be adjusted by varying the initial seed number and the overlap rate. The algorithm is validated by the quantitative comparison with the experimental data as shown in Figure 6(b).

Catalyst layer

The CL is the most complex component in the electrode with ionomer to provide a path for proton conduction, Pt to provide the electrochemical reaction site, carbon to support the structure and provide a path for electron conduction and pores for the transport of reactants and products. Due to its nanometer scale and complex microstructure, the stochastic reconstruction is currently the more viable approach to reconstruct the CL with detailed description of every ingredient.

Mukherjee et al. [28] reconstructed the CL with a Gaussian random field method employed. Only two-point correlation functions could be used to characterize the reconstructed structure and the CL was simply segmented into pore and solid phase under a very low resolution, which is inadequate to describe the CL morphology. Kim et al. [84] optimized the flexibility of the stochastic reconstruction by using a sphere-based simulated annealing method. The ionomer phase was also considered and uniformly generated on the carbon support.

A classic CL reconstruction was proposed by Lange et al. [49]. The general process is as follows:

i.   An initial carbon sphere center is chosen in the computational domain using a random number generator.
ii. A random number generator is used to determine whether or not the next carbon sphere is required to overlap with spheres, which are already in the computational mesh.
iii. A random number generator is used to generate trial sphere centers.
iv. Continue this process until the specified number of spheres has been placed in the computational mesh or no more spheres can fit in the computational mesh without exceeding the specified overlap tolerance.
v.   Loop over each cell in the computational domain. If the cell center is within the radius of any of the carbon spheres, tag the cell as a carbon cell.
vi. Loop over each cell in the computational domain. If the cell center has a distance from the sphere center, which is greater than the radius of any of the carbon spheres, but less than the sum of the radius and the ionomer thickness for the sphere, tag the cell as ionomer.
vii. Tag the remaining cells in the computational domain as pore cells.

Chen et al. [7] reconstructed the CL with a resolution of 2 nm. The reconstruction method was based on the synthesis process proposed by Siddique et al. [85] with the quartet structure generation set (QSGS) method [86]. The CL was reconstructed based on the fabrication process including carbon, platinum and ionomer. These ingredients were all generated via the QSGS method. In this way, the carbon phase was not agglomerative spheres based, but it was able to consider the nonuniform distribution of the ionomer phase.

Hou et al. [53] combined the Lange [49] and QSGS method to reconstruct the CL. The carbon support was built by connected carbon spheres defined by Lange's method, while the platinum and ionomer were generated using the QSGS method. Such method could more realistically describe the CL morphology.

Ishikawa et al. [87] recently presented a novel CL reconstruction algorithm. First, carbon aggregates consisted of 25 primary carbon spheres with the same size. The aggregates were constructed by a pseudo-attractive force between each sphere based on the probability density function and there were 100 types of aggregate shapes. Then the aggregates were randomly placed in the domain with random shape from the 100 conditions until the target carbon fraction was reached. Following that, the platinum particles were randomly loaded on the carbon surface. Lastly, the nonuniformly distributed ionomer was generated according to the adhesion model, where the ionomer prefers to coat the smaller pores between the carbon spheres.

The algorithm proposed by Ishikawa is so far the most advanced method among the reviewed approaches. Although the CL is still stochastically reconstructed, the generation processes of these constituents are based on physical mechanisms, which make it more reasonable and realistic. It also provides a trend for the development of the general stochastic method, which makes the stochastic algorithm more physically reasonable. More experimental quantitative data describing the microstructure of the electrode would be of great help for the development and validation of numerical stochastic approaches, like the total reactive area under different ionomer and platinum loading. Jankovic et al. [74] for the first time reconstructed a complete catalyst layer and provided a realistic microstructure, however, the method is time consuming and labor-intensive. Compared to the experimental approach, the numerical approach needs further development to increase its reliability, especially for the MPL and CL. Nevertheless, it is still a useful tool for pore-scale simulation, especially for studying the effects of various microstructure characteristics and designing novel electrode structures.

Considering the simplicity and authenticity of the reconstruction algorithm, the method employed in [53] is suggested and illustrated in Figure 7(a).

i. Calculate the thickness of the reconstructed CL (l) and the volume fraction of platinum, ionomer and carbon (ε_{Pt}, ε_I and ε_{carbon}) according to the platinum loading (γ_{Pt}), Pt/C ratio (θ_{Pt}) and I/C ratio (θ_I), following:

$$\varepsilon_{Pt} = (1-\varepsilon)\frac{\theta_{Pt}/\rho_{Pt}}{\theta_I/\rho_I + (1-\theta_{Pt})/\rho_{carbon} + \theta_{Pt}/\rho_{Pt}} \tag{3}$$

$$\varepsilon_I = (1-\varepsilon)\frac{\theta_I/\rho_I}{\theta_I/\rho_I + (1-\theta_{Pt})/\rho_{carbon} + \theta_{Pt}/\rho_{Pt}} \tag{4}$$

$$\varepsilon_{carbon} = (1-\varepsilon)\frac{(1-\theta_{Pt})/\rho_{carbon}}{\theta_I/\rho_I + (1-\theta_{Pt})/\rho_{carbon} + \theta_{Pt}/\rho_{Pt}} \tag{5}$$

$$l = \frac{\gamma_{Pt}}{\varepsilon_{Pt}\rho_{Pt}} \tag{6}$$

where ρ_{Pt}, ρ_I and ρ_{carbon} are the density of Pt, ionomer and carbon, respectively.

1. Generate Carbon 2. Generate Platinum 3. Generate Ionomer

(a)

(b)

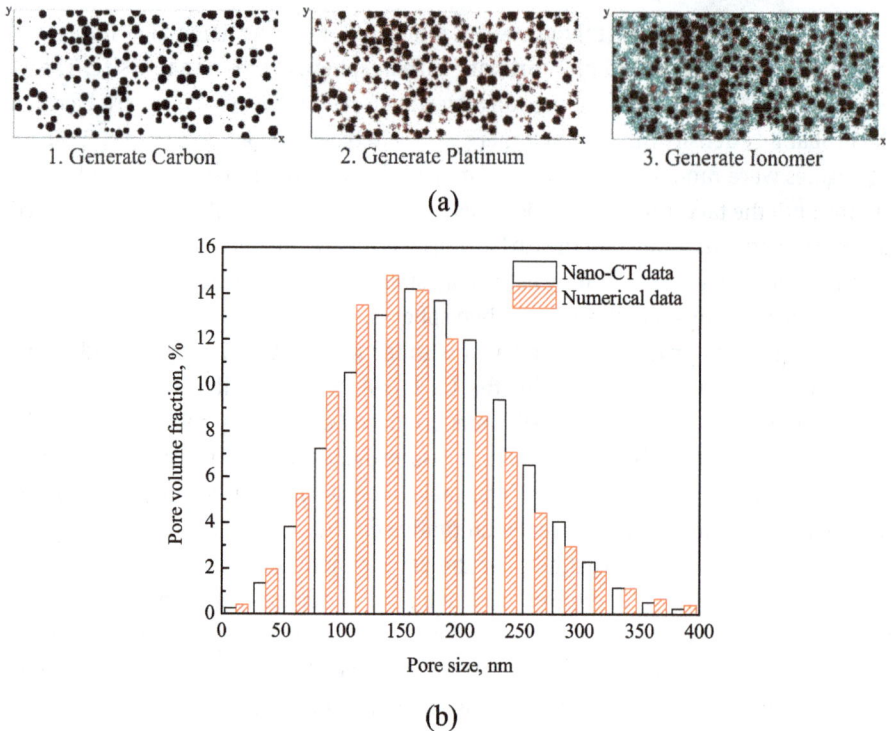

Figure 7: (a) Illustration of CL reconstruction; (b) Comparison of the pore size distribution between the reconstructed CL and the experimental data [53].

ii. Generate the carbon support. It is generally the same process as that for the generation of carbon in the MPL reconstruction.

iii. Generate Pt particles. The reconstruction resolution is 4 nm, which is the side length of the Pt cube. Each potential Pt node (the void cell next to the carbon cell) is identified and assigned a probability function, which determines if the void node will be converted to Pt node or not. The converting possibility is set sufficiently low (1/100,000) to ensure the uniform distribution of Pt. Repeat such loop until the target Pt fraction is reached.

iv. Generate ionomer phase. This process is similar to the Pt generation. Each potential ionomer node (pore node adjacent to carbon, Pt, or ionomer) is identified and assigned a probability function. The converting possibility is 1/10,000. Repeat such loop until the target volume fraction of ionomer is reached.

The pore-size distribution of the reconstructed CL is compared against the experimental data to validate the algorithm as presented in Figure 7(b). This model can also be modified to realize the gradient distribution of ionomer.

8.4.2 Numerical model development

So far, we have introduced how to numerically reconstruct the GDL, MPL and CL structures, respectively. The next step is to build the numerical model to simulate the complex transport processes in each layer. As discussed in Section 8.2, LBM is the most popular numerical tool with advantages on versatility, simplicity and computing efficiency. In this subsection, we will give a detailed tutorial on how to apply the LBM to simulate single- and two-phase flow in the electrode.

8.4.2.1 Introduction of LBM

The LBM is basically the calculation of the discrete-velocity distribution function, $f_i(x, t)$, which represents the density distribution of particles along the ith direction at lattice site x and time t. The lattice spacing and time step are defined as Δx and Δt, respectively, commonly set as 1 in the lattice unit. The mass density ρ and fluid velocity u can be obtained through weighted sums of f_i, calculated as

$$\rho = \sum_i f_i \tag{7}$$

$$\rho u = \sum_i c_i f_i \tag{8}$$

c_i is the discrete velocity. The discrete velocities and lattice sound speed c_s are determined by the lattice scheme, which is typically denoted by $D_n Q_m$, where n and m represent the numbers of dimensions and discrete velocities, respectively. Note that c_s connects the density and pressure via $p = c_s^2 \rho$. The D2Q5, D2Q9, D3Q7 and D3Q19 are the most common lattice schemes (presented in Figure 8) for the pore-scale simulation and should be chosen according to the specific simulation condition as more discrete directions bring higher accuracy but are also much more demanding computationally. Categorized by the collision operator, the LB models can be generally classified into single-relaxation time (SRT) model [88] and multiple-relaxation time (MRT) model [89].

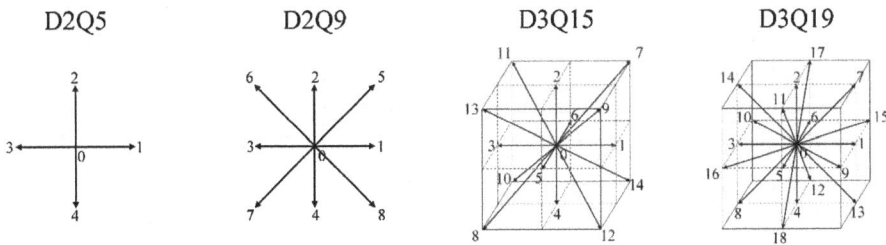

Figure 8: D2Q5, D2Q9, D3Q15, D3Q19 velocity direction sets.

The most common SRT LB equation reads

$$f_i(x + c_i \Delta t, t + \Delta t) = f_i(x, t) - \frac{1}{\tau}(f_i(x, t) - f_i^{eq}(x, t)). \tag{9}$$

The equation can be divided into the so-called "collision" process,

$$f_i^*(x, t) = f_i(x, t) - \frac{1}{\tau}(f_i(x, t) - f_i^{eq}(x, t)) \tag{10}$$

and the "streaming" process,

$$f_i(x + c_i \Delta t, t + \Delta t) = f_i^*(x, t) \tag{11}$$

where the distribution functions $f_i(x, t)$ are locally updated and the streaming process is totally linear, which are ideal for parallel computing. The equilibrium distribution $f_i^{eq}(x, t)$ is given by

$$f_i^{eq}(x, t) = w_i \rho \left[1 + \frac{c_i \cdot u}{c_s^2} + \frac{1}{2} \frac{(c_i \cdot u)^2}{c_s^4} - \frac{1}{2} \frac{u \cdot u}{c_s^2} \right] \tag{12}$$

where w_i is the corresponding weighting factors. The relaxation time τ determines the rate approaching the equilibrium state and the kinematic viscosity of the fluid as

$$v = c_s^2 \left(\tau - \frac{1}{2} \right) \frac{\Delta x^2}{\Delta t} \tag{13}$$

The advantages of SRT scheme are its simplicity and computation savings. However, the advantages come with the sacrifices of accuray and stability. SRT scheme is not versatile on model adjustment with the one and the only tunable parameter τ. It should be mentioned that the choice of τ has a significant effect on the computation efficiency and accuracy. It cannot be too close to 0.5 for the consideration of numerical stability, while it also cannot be much greater than 1.0 for a larger numerical error. When it is applied in the CL where the diffusivity usually varies in a wide range (e. g., oxygen diffusivity in pore and ionomer), the relaxation time will also change dramatically and the model stability and accuracy become difficult to ensure. Therefore, regarding the application in the electrode, though MRT model is a bit more complex and computationally expensive, it is still quite necessary, considering a better numerical stability and accuracy.

The difference between SRT and MRT operator is the collision term. The general collision process of the MRT operator is:

i. Convert the distribution functions into moment space via a transformation matrix;
ii. Individualy relax the moments to equilibrium via a relaxation matrix;
iii. Convert the updated moments to the distribution fuctions via an inverse matrix.

The kernel idea of the MRT operator is to individually relax different moments rather than distribution functions to optimize the model stability and accuracy. The evolution equation of MRT model can be presented as

$$f_i(x + c_i\Delta t, t + \Delta t) = f_i(x, t) - \mathbf{M}^{-1}\Lambda\mathbf{M}(f_i(x, t) - f_i^{eq}(x, t)), \tag{14}$$

where Λ is the relaxation matrix, $\mathbf{M}$ is the transformation matrix and $\mathbf{M}^{-1}$ is its inverse matrix. The specific transformation matrices of the mentioned lattice schemes can be found in [5].

Now, let us take D2Q9 model as an example. The distribution functions can be converted into the moment space with the transformation matrix, given by

$$\mathbf{m}^{eq} = \mathbf{M}f^{eq} = \rho(1, -2 + 3|u|^2, 1 - 3|u|^2, u_x, -u_x, u_y, -u_y, u_x^2 - u_y^2, u_x u_y)^{\mathrm{T}} \tag{15}$$

with the corresponding relaxation matrix presented as

$$\Lambda = \mathrm{diag}(\tau_\rho^{-1}, \tau_e^{-1}, \tau_\zeta^{-1}, \tau_j^{-1}, \tau_q^{-1}, \tau_j^{-1}, \tau_q^{-1}, \tau_v^{-1}, \tau_v^{-1}) \tag{16}$$

where τ_ρ^{-1} and τ_j^{-1} are the relaxation times of conserved moments and usually set as: 1. τ_v^{-1} is related to kinematic viscosity via $v = c_s^2(\tau_v - \frac{1}{2})\frac{\Delta x^2}{\Delta t}$. For the D2Q9 mode, MRT has three more tunable relaxation factors than the SRT scheme to achieve better stability, which is why MRT is superior over SRT [90]. Compared to the benefits it brings, the extra 15–20 % computation cost is worth it [5].

The most significant advantage for applying LBM in the electrode is the simplicity in boundary conditions implementation. For the nonslip wall, LBM has the so-called bounce-back method, which can easily handle the complex stochastic computation domain. The general idea is during the streaming process, if a particle meets the wall, it will be reflected back to its original location with its velocity direction reversed, schematically presented in Figure 9. More details about the LBM boundary conditions can also be found in [91].

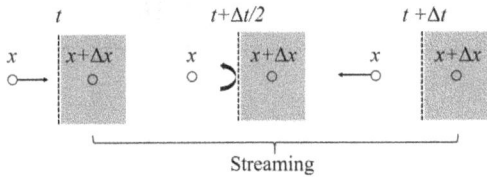

Figure 9: Schematic of halfway bounce-back boundary condition.

In the next section, we will introduce how to implement the LBM to simulate single- and multiphase flow in the electrode.

8.4.2.2 Development of single-phase flow model

Advection-diffusion problem is a classic topic in the electrode study. The governing equation of the advection-diffusion problem can be presented as below:

$$\frac{\partial C}{\partial t} + \nabla \cdot (C\boldsymbol{u}) = \nabla \cdot (D\nabla C) + q, \tag{17}$$

where the scalar field C could be concentration or temperature, $\boldsymbol{u}$ is the fluid velocity, D is the diffusivity and q is the source term. The advection-diffusion equation can be transformed to LB equation through the Chapman–Enskog analysis:

$$g_i(x + c_i\Delta t, t + \Delta t) - g_i(x, t) = \Omega(x, t) + Q(x, t), \tag{18}$$

where $\Omega(x, t)$ is the collision operator. The concentration field $C = \sum_i g_i$ and the diffusivity is determined by the relaxation time τ by $D = c_s^2(\tau - \frac{1}{2})\frac{\Delta x^2}{\Delta t}$. Note that the velocity field is obtained from another standard lattice set resolving the flow field and imposed externally. The source term $Q(x, t)$ is added according to the specific problem and the Dirichlet boundary condition is normally employed to simulate the reactant consumption. To further investigate the transport of multicomponent, we only need to add one more lattice set to represent the new component. Here is the basic single phase LB model, which can be directly applied in the GDL investigation. To be further applied in nanometer scale MPL or CL, the Knudsen diffusion effect should be considered and usually coupled as below:

$$D_p = \left(\frac{1}{D_b} + \frac{1}{D_{kn}}\right)^{-1} \tag{19}$$

$$D_{Kn} = 48.5d_p\sqrt{\frac{T}{32}} \tag{20}$$

where the local diffusivity D_p is determined by the bulk diffusivity D_b and the Knudsen diffusivity D_{kn}, which relates to the local pore size d_p and temperature T. Therefore, the diffusivity is different on each cell, which needs MRT operator considering the numerical stability and accuracy as mentioned. When it comes to the catalyst layer, there is a concentration gap between the pore cell and ionomer cell due to the Henry law [92] and a cross-interface dissolution rate k_{dis}, shown as

$$D_p\frac{\partial C_p}{\partial n} = -k_{dis}\left(\frac{C_p}{H} - C_I\right) = D_I\frac{\partial C_I}{\partial n}, \tag{21}$$

where C_p and C_I are the concentration in pore cell and ionomer cell, respectively, H is the Henry constant and k_{dis} is the dissolution rate constant, which has been proved to be the main cause for the extra transport resistance [52]. The following scheme is employed to resolve the concentration gap across the interface:

$$g_i(x_p, x + \Delta t) = \frac{1}{-\frac{w_i \Delta x}{k_{dis} \Delta t} - \frac{1}{H} - 1} \left[\left(-\frac{w_i \Delta x}{k_{dis} \Delta t} + \frac{1}{H} - 1 \right) \widehat{g}_{i*}(x_p, x) - 2\widehat{g}_i(x_I, t) \right]. \tag{22}$$

The detailed mathematical derivation can be found in [57]. In this way, we can simulate the cross-ionomer diffusion process. In the CL, the Dirichlet boundary condition is usually applied on the Pt-ionomer interface to model the electrochemical reaction, presented as

$$D_I \frac{\partial C_I}{\partial n} = k_{elec}. \tag{23}$$

k_{elec} is the electrochemical reaction rate calculated by the Butler–Volmer equation.

$$k_{elec} = \frac{1}{nF} i_{ref} \frac{C_I}{C_{ref}} \left[\exp\left(-\frac{\alpha F}{RT}\eta \right) - \exp\left(\frac{(1-\alpha)F}{RT}\eta \right) \right], \tag{24}$$

where F is the Faraday constant, n is the number of transferred electrons, i_{ref} and C_{ref} are the reference current density and concentration, respectively, α is the transfer coefficient and η is the overpotential.

8.4.2.3 Development of two-phase flow model

The two-phase flow issue is another important subject in the electrode since the flooding affects the mass transport resistance and performance. The porous microstructure and its inner micropores increase the difficulty on capturing the gas-liquid interface. As shown in Section 3.2, LBM is widely applied for simulating the two-phase flow. The popular multiphase models include the color gradient model proposed by Gunstensen et al. [93], the pseudopotential model proposed by Shan and Chen [94] and the free energy model proposed by Swift et al. [95]. The pseudopotential model is the most popular LB multiphase model applied in the electrode due to its simple implementation and automatic phase separation. This section will give a brief introduction of the pseudopotential model and its implementation in the electrode.

In nature, the inner attractive force among the molecules determines the state, boiling point and vapor pressure of the liquid. Inspired by this phenomenon, the pseudopotential model realizes the automatic phase separation by introducing an interaction force between each fluid lattice. The interaction strength is determined by the distance between the two fluid lattices and their local density. Usually, it is assumed that only the neighboring lattice within $c_i \Delta t$ can apply the interaction force. The total interaction force action on a fluid element at lattice site x can be presented as

$$F_{int}(x) = -G\varphi(x) \sum_i w_i \varphi(x + c_i \Delta t) c_i \Delta t, \tag{25}$$

which is the integral force from the neighboring lattices. G is the interaction strength factor and can adjust the surface tension. Positive G applies repulsive force while negative for attractive. w_i is the lattice weighted factor. The local density is replaced by local pseudopotential φ for the numerical stability. The pseudopotential is basically a function of local density and can be calculated as

$$\varphi(x) = 1 - \exp(-\rho(x)). \tag{26}$$

For the solid boundary, the pseudopotential is usually set as the average of the neighboring fluid pseudopotentials for model stability.

To adjust the wettability of porous microstructure, the fluid-solid interaction force is further introduced. The effective mass-based method proposed by Li et al. [96] is used to achieve a wide range of contact angles, expressed as

$$F_{ads}(x) = -G_w\varphi(x) \sum_i w_i\varphi(x)s(x + c_i\Delta t)c_i\Delta t \tag{27}$$

where $s(x + c_i\Delta t)$ acts as an indicator, set as 0 for fluid lattice and 1 for solid. The wall wettability phenomenon is determined by the interaction force between fluid and solid, which therefore can be adjusted by changing the interaction strength parameter G_w. For water component, positive G_w implements hydrophobic effect and negative for hydrophilic.

This is the basic concept on how to realize phase separation and wettability within the pseudopotential model. Regarding the method to couple these phenomena, the MRT lattice scheme with the forcing scheme proposed by Li et al. [97] is highly recommended. The spurious velocity around the gas–liquid interface can cause numerical instability, especial in the MPL due to a smaller average pore size. The MRT operator and advanced forcing scheme could effectively reduce the spurious velocity by adjusting the relaxation parameters. This model can be directly applied in the GDL simulation. Regarding the MPL, more attention should be focused on controlling the spurious velocity.

For more information and details about the pseudopotential models, the following representative literatures are strongly recommended [98–100]. Chen et al. [98] gave a critical review of the pseudopotential model where more details of the model can be found. Li et al. [99] provided an extensive overview of the development and application of pseudopotential model and a book by Huang et al. [100] was dedicated to the introduction of the theory and application of the multiphase method. Hou et al. [101] successfully simulated the realistic multicomponent two-phase flow with the latest pseudopotential model developments.

8.5 Summary and outlook

It is clear that pore-scale modeling is a powerful and viable tool to simulate the complex phenomena in the electrode with detailed consideration of its microstructure. Pore-scale investigations are significant for the development of PEM fuel cells since they can reveal the transport mechanisms in the micro or even nanoscale, where state-of-the-art experimental observation cannot be easily conducted. This chapter analyzes the advantages and disadvantages of different popular pore-scale numerical methods, offers a representative review on the pore-scale studies applied in the electrode and includes a tutorial on how to develop the pore-scale model, comprising electrode reconstruction and the development of single- and two-phase LB models.

LBM, PN and the traditional CFD methods are most popular with employed rates of 65 %, 21 % and 14 %, respectively. LBM occupies the dominant position due to its distinctive strengths over macro and microscopic models on computation efficiency and ease of implementation. Many validation tests have been conducted to prove its viability and accuracy and the results agree well with both the experimental data and analytical models. More focus is suggested to be placed on solving the coupled processes. With the increasing demand for computation, the parallelization scheme should also be carefully optimized to achieve a more efficient performance, which is usually neglected but critical for large-scale simulation. The advanced PN model is mainly applied to give a fast prediction of the water transport in the TPL. With the advent of powerful meshing tools, researchers are also starting to use the traditional CFD methods to model single- and two-phase flow. However, it is still difficult to implement the detailed boundary conditions, such as the electrochemical reaction at the Pt/Ionomer interface.

A review of representative pore-scale studies in the electrode found abundant studies for the GDL focusing on transport properties and the two-phase flow mechanism. The transport properties are well understood, while the mechanism of two-phase flow still needs more comprehensive analyses. The published studies mostly initialize liquid water as a droplet or set the water inlet boundary with a fixed velocity. Such initialization methods cannot realistically represent the real operating condition. Therefore, realistic representation of the breakthrough of water is important. For the MPL and CL, high difficulty is associated with the reconstruction and more computing resources are required; hence, pore-scale-based works are still scarce, especially for the simulation of two-phase phenomena. Also, the simulation should be conducted on the premise of a validated reconstructed microstructure. Moreover, most of the reviewed work focuses on investigating the phenomena in a single layer excluding the effects of the adjacent layers and the transport phenomena at the interface of two layers. A coupling of two layers together is more computationally expensive, due to the resolution gap between the two components, which is why efficient, parallelized code is crucial.

For electrode reconstruction, the methods can be categorized into two groups based on (i) experimental data from structural imaging/characterization methods and (ii) artificial creation using stochastic methods. Experimental imaging methods use X-ray tomography or FIB/SEM facilities for the reconstruction and are generally a good approach to capture the details of the structure. However, it is still challenging to reconstruct the CL due to its various ingredients, complex structure and nanometer scale, which may be solved with the advancement of the experimental techniques. Numerical stochastic methods are more popular for its easy-implementation and flexibility. The reconstructed electrode can be quantitatively validated in terms of structural properties such as porosity, tortuosity and pore-size distribution. The most advanced reconstruction can even incorporate the basic physical rules into the stochastic algorithms, which makes the process more realistic. Although the reconstruction algorithm for a specific electrode component is approaching maturity, there is still no work dedicated to reconstruct the assembled components and characterize the interface morphology [102] where many interesting phenomena are still to be investigated.

With the development of LBM in the last decades, the current LB model can already solve most of the physicochemical processes in the electrode, including the transports of multispecies, two-phase flow, electrochemical reaction at Pt sites, phase-change, etc. However, due to its transient nature and small time-step, the current LB model still cannot realize a time-scale analysis, thus having issues on modeling some long lasting phenomenon, such as the CL contamination. Additionally, although LBM is efficient using parallel computing, its memory requirements limit the size of the computation domain. Hence, as future work it is suggested to focus on the extension of LBM in time- and space-coordinates, which could be supported by coupling with the artificial intelligence (AI) [103].

Bibliography

[1]   K. Jiao and X. Li, Water transport in polymer electrolyte membrane fuel cells. *Prog. Energy Combust. Sci.*, **37**(3), 221–291, 2011.

[2]   J. Wang, Barriers of scaling-up fuel cells: Cost, durability and reliability. *Energy*, **80**, 509–521, 2015.

[3]   K. Jiao and M. Ni, Challenges and opportunities in modelling of proton exchange membrane fuel cells (PEMFC). *Int. J. Energy Res.*, **41**, 1793–1797, 2017.

[4]   G. Zhang and K. Jiao, Multi-phase models for water and thermal management of proton exchange membrane fuel cell: A review. *J. Power Sources*, **391**, 120–133, 2018.

[5]   K. Timm, H. Kusumaatmaja, A. Kuzmin et al., *The lattice Boltzmann method: principles and practice*. Springer, Switzerland, 2016. ISSN 1868-4521.

[6]   T. Hansson, C. Oostenbrink, and W. F. van Gunsteren, Molecular dynamics simulations. *Curr. Opin. Struct. Biol.*, **12**(2), 190–196, 2002.

[7]   L. Chen, G. Wu, E. F. Holby et al., Lattice Boltzmann pore-scale investigation of coupled physical-electrochemical processes in C/Pt and non-precious metal cathode catalyst layers in proton exchange membrane fuel cells. *Electrochim. Acta*, **158**, 175–186, 2015.

[8]   G. R. Molaeimanesh, H. S. Googarchin, and A. Q. Moqaddam, Lattice Boltzmann simulation of proton exchange membrane fuel cells—A review on opportunities and challenges. *Int. J. Hydrog. Energy*, **41**(47), 22221–22245, 2016.

[9]   M. H. Shojaeefard, G. R. Molaeimanesh, M. Nazemian et al., A review on microstructure reconstruction of PEM fuel cells porous electrodes for pore scale simulation. *Int. J. Hydrog. Energy*, **41**(44), 20276–20293, 2016.

[10]   J. Liu, S. Shin, and S. Um, Comprehensive statistical analysis of heterogeneous transport characteristics in multifunctional porous gas diffusion layers using lattice Boltzmann method for fuel cell applications. *Renew. Energy*, **139**, 279–291, 2019.

[11]   Y. Hou, X. Li, Q. Du et al., Pore-Scale Investigation of the Effect of Micro-Porous Layer on Water Transport in Proton Exchange Membrane Fuel Cell. *J. Electrochem. Soc.*, **167**(14), 144504, 2020.

[12]   P. Rama, Y. Liu, R. Chen, H. Ostadi, K. Jiang, X. Zhang, R. Fisher, and M. Jeschke, An X-ray tomography based lattice Boltzmann simulation study on gas diffusion layers of polymer electrolyte fuel cells. *J. Fuel Cell Sci. Technol.*, **7**(3), 031015, 2010.

[13]   I. Fatt, The network model of porous media. *Trans. AIME*, **207**(01), 144–181, 1956.

[14]   T. J. Chung, *Computational fluid dynamics*. Cambridge university press, 2010.

[15]   R. Eymard, T. Gallouët, and R. Herbin, Finite volume methods. *Handb. Numer. Anal.*, **7**, 713–1018, 2000.

[16]   K. H. Huebner, D. L. Dewhirst, D. E. Smith, and T. G. Byrom, *The finite element method for engineers*. John Wiley & Sons, 2001.

[17]   C. W. Hirt and B. D. Nichols, Volume of fluid (VOF) method for the dynamics of free boundaries. *J. Comput. Phys.*, **39**(1), 201–225, 1981.

[18]   Z. Niu, Z. Bao, J. Wu, Y. Wang, and K. Jiao, Two-phase flow in the mixed-wettability gas diffusion layer of proton exchange membrane fuel cells. *Appl. Energy*, **232**, 443–450, 2018.

[19]   Z. Niu, Y. Wang, K. Jiao, and J. Wu, Two-phase flow dynamics in the gas diffusion layer of proton exchange membrane fuel cells: volume of fluid modeling and comparison with experiment. *J. Electrochem. Soc.*, **165**(9), F613, 2018.

[20]   X. Zhou, Z. Niu, Z. Bao, J. Wang, Z. Liu, Y. Yin, Q. Du, and K. Jiao, Two-phase flow in compressed gas diffusion layer: Finite element and volume of fluid modeling. *J. Power Sources*, **437**, 226933, 2019.

[21]   H. Deng, Y. Hou, and K. Jiao, Lattice Boltzmann simulation of liquid water transport inside and at interface of gas diffusion and micro-porous layers of PEM fuel cells. *Int. J. Heat Mass Transf.*, **140**, 1074–1090, 2019.

[22]   M. J. Blunt, *Multiphase flow in permeable media: A pore-scale perspective*. Cambridge University Press, 2017.

[23]   X. Yang, Y. Mehmani, W. A. Perkins, A. Pasquali, M. Schönherr, K. Kim, M. Perego, M. L. Parks, N. Trask, M. T. Balhoff, M. C. Richmond, M. Geier, M. Krafczyk, L.-S. Luo, A. M. Tartakovsky, and T. D. Scheibe, Intercomparison of 3D pore-scale flow and solute transport simulation methods. *Adv. Water Resour.*, **95**, 176–189, 2016.

[24]   M. Aghighi and J. Gostick, Pore network modeling of phase change in PEM fuel cell fibrous cathode. *J. Appl. Electrochem.*, **47**(12), 1323–1338, 2017.

[25]   T. Agaesse, A. Lamibrac, F. N. Büchi, J. Pauchet, and M. Prat, Validation of pore network simulations of ex-situ water distributions in a gas diffusion layer of proton exchange membrane fuel cells with X-ray tomographic images. *J. Power Sources*, **331**, 462–474, 2016.

[26]   L. Hao and P. Cheng, Pore-scale simulations on relative permeabilities of porous media by lattice Boltzmann method. *Int. J. Heat Mass Transf.*, **53**(9–10), 1908–1913, 2010.

[27]   L. Hao and P. Cheng, Lattice Boltzmann simulations of water transport in gas diffusion layer of a polymer electrolyte membrane fuel cell. *J. Power Sources*, **195**(12), 3870–3881, 2010.

[28]   P. P. Mukherjee, C. Y. Wang, and Q. Kang, Mesoscopic modeling of two-phase behavior and flooding phenomena in polymer electrolyte fuel cells. *Electrochim. Acta*, **54**(27), 6861–6875, 2009.

[29]   P. Rama, Y. Liu, R. Chen, H. Ostadi, K. Jiang, X. Zhang, Y. Gao, P. Grassini, and D. Brivio, Determination of the anisotropic permeability of a carbon cloth gas diffusion layer through X-ray computer micro-tomography and single-phase lattice Boltzmann simulation. *Int. J. Numer. Methods Fluids*, **67**(4), 518–530, 2011.

[30]   N. Zamel, X. Li, J. Becker, and A. Wiegmann, Effect of liquid water on transport properties of the gas diffusion layer of polymer electrolyte membrane fuel cells. *Int. J. Hydrog. Energy*, **36**(9), 5466–5478, 2011.

[31]   L. Chen, H. B. Luan, Y. L. He, and W. Q. Tao, Numerical investigation of liquid water transport and distribution in porous gas diffusion layer of a proton exchange membrane fuel cell using lattice Boltzmann method. *Russ. J. Electrochem.*, **48**(7), 712–726, 2012.

[32]   L. Chen, H. B. Luan, Y. L. He, and W. Q. Tao, Pore-scale flow and mass transport in gas diffusion layer of proton exchange membrane fuel cell with interdigitated flow fields. *Int. J. Therm. Sci.*, **51**, 132–144, 2012.

[33] J. T. Gostick, Random pore network modeling of fibrous PEMFC gas diffusion media using Voronoi and Delaunay tessellations. *J. Electrochem. Soc.*, **160**(8), F731, 2013.

[34] P. A. García-Salaberri, G. Hwang, M. Vera et al., Effective diffusivity in partially-saturated carbon-fiber gas diffusion layers: Effect of through-plane saturation distribution. *Int. J. Heat Mass Transf.*, **86**, 319–333, 2015.

[35] P. A. García-Salaberri, J. T. Gostick, G. Hwang et al., Effective diffusivity in partially-saturated carbon-fiber gas diffusion layers: Effect of local saturation and application to macroscopic continuum models. *J. Power Sources*, **296**, 440–453, 2015.

[36] M. Espinoza, M. Andersson, J. Yuan, and B. Sundén, Compress effects on porosity, gas-phase tortuosity, and gas permeability in a simulated PEM gas diffusion layer. *Int. J. Energy Res.*, **39**(11), 1528–1536, 2015.

[37] D. H. Jeon and H. Kim, Effect of compression on water transport in gas diffusion layer of polymer electrolyte membrane fuel cell using lattice Boltzmann method. *J. Power Sources*, **294**, 393–405, 2015.

[38] M. Fazeli, J. Hinebaugh, Z. Fishman, C. Tötzke, W. Lehnert, I. Manke, and A. Bazylak, Pore network modeling to explore the effects of compression on multiphase transport in polymer electrolyte membrane fuel cell gas diffusion layers. *J. Power Sources*, **335**, 162–171, 2016.

[39] H. Jasak, OpenFOAM: open source CFD in research and industry. *Int. J. Nav. Archit. Ocean Eng.*, **1**(2), 89–94, 2009.

[40] D. Zhang, Q. Cai, and S. Gu, Three-dimensional lattice-Boltzmann model for liquid water transport and oxygen diffusion in cathode of polymer electrolyte membrane fuel cell with electrochemical reaction. *Electrochim. Acta*, **262**, 282–296, 2018.

[41] P. He, L. Chen, Y. T. Mu, and W. Q. Tao, Lattice Boltzmann method simulation of ice melting process in the gas diffusion layer of fuel cell. *Int. J. Heat Mass Transf.*, **149**, 119121, 2020.

[42] H. Ostadi, P. Rama, Y. Liu et al., 3D reconstruction of a gas diffusion layer and a microporous layer. *J. Membr. Sci.*, **351**(1–2), 69–74, 2010.

[43] J. Becker, C. Wieser, S. Fell et al., A multi-scale approach to material modeling of fuel cell diffusion media. *Int. J. Heat Mass Transf.*, **54**(7–8), 1360–1368, 2011.

[44] A. Wiegmann, S. Rief, and A. Latz, Geodict and filterdict: Software for the virtual material design of new filter media. In: *Proc. New Developments in Filtration Technology*, Loughborough, Angleterre, 2006.

[45] N. Zamel, J. Becker, and A. Wiegmann, Estimating the thermal conductivity and diffusion coefficient of the microporous layer of polymer electrolyte membrane fuel cells. *J. Power Sources*, **207**, 70–80, 2012.

[46] B. Han and H. Meng, Numerical studies of interfacial phenomena in liquid water transport in polymer electrolyte membrane fuel cells using the lattice Boltzmann method. *Int. J. Hydrog. Energy*, **38**(12), 5053–5059, 2013.

[47] K. N. Kim, J. H. Kang, S. G. Lee et al., Lattice Boltzmann simulation of liquid water transport in microporous and gas diffusion layers of polymer electrolyte membrane fuel cells. *J. Power Sources*, **278**, 703–717, 2015.

[48] F. C. Cetinbas, R. K. Ahluwalia, A. D. Shum, and I. V. Zenyuk, Direct simulations of pore-scale water transport through diffusion media. *J. Electrochem. Soc.*, **166**(7), F3001, 2019.

[49] K. J. Lange, P. C. Sui, and N. Djilali, Pore scale simulation of transport and electrochemical reactions in reconstructed PEMFC catalyst layers. *J. Electrochem. Soc.*, **157**(10), B1434–B1442, 2010.

[50] K. J. Lange, P. C. Sui, and N. Djilali, Determination of effective transport properties in a PEMFC catalyst layer using different reconstruction algorithms. *J. Power Sources*, **208**, 354–365, 2012.

[51] L. Chen, R. Zhang, P. He et al., Nanoscale simulation of local gas transport in catalyst layers of proton exchange membrane fuel cells. *J. Power Sources*, **400**, 114–125, 2018.

[52] L. Chen, Q. Kang, and W. Tao, Pore-scale study of reactive transport processes in catalyst layer agglomerates of proton exchange membrane fuel cells. *Electrochim. Acta*, **306**, 454–465, 2019.

[53] Y. Hou, H. Deng, F. Pan et al., Pore-scale investigation of catalyst layer ingredient and structure effect in proton exchange membrane fuel cell. *Appl. Energy*, **253**, 113561, 2019.

[54]   C. R. Parkinson and A. Sasov, High-resolution non-destructive 3D interrogation of dentin using X-ray nanotomography. *Dent. Mater.*, **24**(6), 773–777, 2008.

[55]   M. A. Groeber, B. K. Haley, M. D. Uchic et al., 3D reconstruction and characterization of polycrystalline microstructures using a FIB–SEM system. *Mater. Charact.*, **57**(4–5), 259–273, 2006.

[56]   E. A. Wargo, T. Kotaka, Y. Tabuchi et al., Comparison of focused ion beam versus nano-scale X-ray computed tomography for resolving 3-D microstructures of porous fuel cell materials. *J. Power Sources*, **241**, 608–618, 2013.

[57]   T. Koido, T. Furusawa, and K. Moriyama, An approach to modeling two-phase transport in the gas diffusion layer of a proton exchange membrane fuel cell. *J. Power Sources*, **175**(1), 127–136, 2008.

[58]   J. Becker, V. Schulz, and A. Wiegmann, Numerical determination of two-phase material parameters of a gas diffusion layer using tomography images. *J. Fuel Cell Sci. Technol.*, **5**(2), 021006, 2008.

[59]   H. Ostadi, P. Rama, Y. Liu et al., Influence of threshold variation on determining the properties of a polymer electrolyte fuel cell gas diffusion layer in X-ray nano-tomography. *Chem. Eng. Sci.*, **65**(6), 2213–2217, 2010.

[60]   P. Rama, Y. Liu, R. Chen et al., A numerical study of structural change and anisotropic permeability in compressed carbon cloth polymer electrolyte fuel cell gas diffusion layers. *Fuel Cells*, **11**(2), 274–285, 2011.

[61]   J. Eller, T. Rosén, F. Marone et al., Progress in in situ X-ray tomographic microscopy of liquid water in gas diffusion layers of PEFC. *J. Electrochem. Soc.*, **158**(8), B963–B970, 2011.

[62]   G. Inoue, K. Yokoyama, J. Ooyama et al., Theoretical examination of effective oxygen diffusion coefficient and electrical conductivity of polymer electrolyte fuel cell porous components. *J. Power Sources*, **327**, 610–621, 2016.

[63]   T. Kotaka, Y. Tabuchi, and P. P. Mukherjee, Microstructural analysis of mass transport phenomena in gas diffusion media for high current density operation in PEM fuel cells. *J. Power Sources*, **280**, 231–239, 2015.

[64]   M. Andisheh-Tadbir, F. P. Orfino, and E. Kjeang, Three-dimensional phase segregation of micro-porous layers for fuel cells by nano-scale X-ray computed tomography. *J. Power Sources*, **310**, 61–69, 2016.

[65]   W. K. Epting, J. Gelb, and S. Litster, Resolving the Three-Dimensional Microstructure of Polymer Electrolyte Fuel Cell Electrodes using Nanometer-Scale X-ray Computed Tomography. *Adv. Funct. Mater.*, **22**(3), 555–560, 2012.

[66]   S. Litster, W. K. Epting, E. A. Wargo et al., Morphological analyses of polymer electrolyte fuel cell electrodes with nano-scale computed tomography imaging. *Fuel Cells*, **13**(5), 935–945, 2013.

[67]   T. Terao, G. Inoue, M. Kawase et al., Development of novel three-dimensional reconstruction method for porous media for polymer electrolyte fuel cells using focused ion beam-scanning electron microscope tomography. *J. Power Sources*, **347**, 108–113, 2017.

[68]   S. Zils, M. Timpel, T. Arlt et al., 3D visualisation of PEMFC electrode structures using FIB nanotomography. *Fuel Cells*, **10**(6), 966–972, 2010.

[69]   K. J. Lange, H. Carlsson, I. Stewart et al., PEM fuel cell CL characterization using a standalone FIB and SEM: Experiments and simulation. *Electrochim. Acta*, **85**, 322–331, 2012.

[70]   Y. Gao, Using MRT lattice Boltzmann method to simulate gas flow in simplified catalyst layer for different inlet–outlet pressure ratio. *Int. J. Heat Mass Transf.*, **88**, 122–132, 2015.

[71]   S. Thiele, R. Zengerle, and C. Ziegler, Nano-morphology of a polymer electrolyte fuel cell catalyst layer—imaging, reconstruction and analysis. *Nano Res.*, **4**(9), 849–860, 2011.

[72]   S. Thiele, T. Fürstenhaupt, D. Banham et al., Multiscale tomography of nanoporous carbon-supported noble metal catalyst layers. *J. Power Sources*, **228**, 185–192, 2013.

[73]   M. Klingele, R. Zengerle, and S. Thiele, Quantification of artifacts in scanning electron microscopy tomography: Improving the reliability of calculated transport parameters in energy applications such as fuel cell and battery electrodes. *J. Power Sources*, **275**, 852–859, 2015.

[74] J. Jankovic, S. Zhang, A. Putz, M. S. Saha, and D. Susac, Multi-scale imaging and transport modeling for fuel cell electrodes. *J. Mater. Res.*, **34**(4), 579–591, 2019.

[75] K. Schladitz, S. Peters, D. Reinel-Bitzer et al., Design of acoustic trim based on geometric modeling and flow simulation for non-woven. *Comput. Mater. Sci.*, **38**(1), 56–66, 2006.

[76] V. P. Schulz, J. Becker, A. Wiegmann et al., Modeling of two-phase behavior in the gas diffusion medium of PEFCs via full morphology approach. *J. Electrochem. Soc.*, **154**(4), B419–B426, 2007.

[77] L. Hao and P. Cheng, Lattice Boltzmann simulations of anisotropic permeabilities in carbon paper gas diffusion layers. *J. Power Sources*, **186**(1), 104–114, 2009.

[78] M. M. Daino and S. G. Kandlikar, 3D phase-differentiated GDL microstructure generation with binder and PTFE distributions. *Int. J. Hydrog. Energy*, **37**(6), 5180–5189, 2012.

[79] V. Radhakrishnan and P. Haridoss, Differences in structure and property of carbon paper and carbon cloth diffusion media and their impact on proton exchange membrane fuel cell flow field design. *Mater. Des.*, **32**(2), 861–868, 2011.

[80] P. G. Stampino, L. Omati, and G. Dotelli, Electrical performance of PEM fuel cells with different gas diffusion layers. *J. Fuel Cell Sci. Technol.*, **8**(4), 041005, 2011.

[81] S. Park and B. N. Popov, Effect of a GDL based on carbon paper or carbon cloth on PEM fuel cell performance. *Fuel*, **90**(1), 436–440, 2011.

[82] U. R. Salomov, E. Chiavazzo, and P. Asinari, Pore-scale modeling of fluid flow through gas diffusion and catalyst layers for high temperature proton exchange membrane (HT-PEM) fuel cells. *Comput. Math. Appl.*, **67**(2), 393–411, 2014.

[83] M. El Hannach, R. Singh, N. Djilali et al., Micro-porous layer stochastic reconstruction and transport parameter determination. *J. Power Sources*, **282**, 58–64, 2015.

[84] S. H. Kim and H. Pitsch, Reconstruction and effective transport properties of the catalyst layer in PEM fuel cells. *J. Electrochem. Soc.*, **156**(6), B673–B681, 2009.

[85] N. A. Siddique and F. Liu, Process based reconstruction and simulation of a three-dimensional fuel cell catalyst layer. *Electrochim. Acta*, **55**(19), 5357–5366, 2010.

[86] D. Guan, J. H. Wu, and L. Jing, A statistical method for predicting sound absorbing property of porous metal materials by using quartet structure generation set. *J. Alloys Compd.*, **626**, 29–34, 2015.

[87] H. Ishikawa, Y. Sugawara, G. Inoue et al., Effects of Pt and ionomer ratios on the structure of catalyst layer: A theoretical model for polymer electrolyte fuel cells. *J. Power Sources*, **374**, 196–204, 2018.

[88] X. He and L. S. Luo, Theory of the lattice Boltzmann method: From the Boltzmann equation to the lattice Boltzmann equation. *Phys. Rev. E*, **56**(6), 6811, 1997.

[89] D. d'Humieres, Multiple–relaxation–time lattice Boltzmann models in three dimensions. *Philos. Trans. R. Soc., Math. Phys. Eng. Sci.*, **360**(1792), 437–451, 2002.

[90] Z. Yu and L. S. Fan, Multirelaxation-time interaction-potential-based lattice Boltzmann model for two-phase flow. *Phys. Rev. E*, **82**(4), 046708, 2010.

[91] Z. Guo, C. Zheng, and B. Shi, An extrapolation method for boundary conditions in lattice Boltzmann method. *Phys. Fluids*, **14**(6), 2007–2010, 2002.

[92] J. Jaffré and A. Sboui, Henry'law and gas phase disappearance. *Transp. Porous Media*, **82**(3), 521–526, 2010.

[93] A. K. Gunstensen and D. H. Rothman, Lattice-Boltzmann studies of immiscible two-phase flow through porous media. *J. Geophys. Res., Solid Earth*, **98**(B4), 6431–6441, 1993.

[94] X. Shan and H. Chen, Lattice Boltzmann model for simulating flows with multiple phases and components. *Phys. Rev. E*, **47**(3), 1815, 1993.

[95] M. R. Swift, E. Orlandini, W. R. Osborn et al., Lattice Boltzmann simulations of liquid-gas and binary fluid systems. *Phys. Rev. E*, **54**(5), 5041, 1996.

[96] Q. Li, K. H. Luo, Q. J. Kang, and Q. Chen, Contact angles in the pseudopotential lattice Boltzmann modeling of wetting. *Phys. Rev. E*, **90**(5), 053301, 2014.

[97] Q. Li, K. H. Luo, and X. J. Li, Lattice Boltzmann modeling of multiphase flows at large density ratio with an improved pseudopotential model. *Phys. Rev. E*, **87**(5), 053301, 2013.

[98] L. Chen, Q. Kang, Y. Mu et al., A critical review of the pseudopotential multiphase lattice Boltzmann model: Methods and applications. *Int. J. Heat Mass Transf.*, **76**, 210–236, 2014.

[99] Q. Li, K. H. Luo, Q. J. Kang et al., Lattice Boltzmann methods for multiphase flow and phase-change heat transfer. *Prog. Energy Combust. Sci.*, **52**, 62–105, 2016.

[100] H. Huang, M. Sukop, and X. Lu, *Multiphase lattice Boltzmann methods: Theory and application.* John Wiley & Sons, 2015.

[101] Y. Hou, H. Deng, Q. Du, and K. Jiao, Multi-component multi-phase lattice Boltzmann modeling of droplet coalescence in flow channel of fuel cell. *J. Power Sources*, **393**, 83–91, 2018.

[102] F. E. Hizir, S. O. Ural, E. C. Kumbur et al., Characterization of interfacial morphology in polymer electrolyte fuel cells: Micro-porous layer and catalyst layer surfaces. *J. Power Sources*, **195**(11), 3463–3471, 2010.

[103] Y. Wang, B. Seo, B. Wang, N. Zamel, and X. C. Adroher, Fundamentals, materials, and machine learning of polymer electrolyte membrane fuel cell technology. *Energy AI*, **1**, 100014, 2020.

Max Cimenti and Jürgen Stumper

9 Voltage loss breakdown in PEM fuel cells

9.1 Introduction

In the field of applied electrochemistry, Voltage Loss Breakdown (VLB), Voltage Loss Analysis or simply Loss Analysis all refer to the same process: the quantification of the thermodynamic inefficiencies associated with an operating electrochemical cell. In general, these inefficiencies can be broken down by type and by component. From a fundamental understanding viewpoint, the breakdown by type is of interest, whereas the breakdown by component is more interesting from an engineering perspective. The goal for VLB is to characterize and quantify all major loss sources and identify the fuel cell components responsible. In proton exchange membrane fuel cells (PEMFC) in particular, loss analysis is necessary for the optimization of the polarization curve to maximize power output.

In fact, loss analysis is a fundamental engineering tool to advance PEMFC technology, because when successfully applied, it allows for the identification and quantification of loss-sources in Membrane Electrode Assemblies (MEA), such as kinetic, ohmic and transport losses associated with the individual subcomponents. Furthermore, VLB results are quantitative in nature (units of V or mV) allowing for an objective ranking of losses. In essence, loss analysis is the tool/procedure to answer the following questions: where in the fuel cell are voltage losses originating, and what are the processes causing these?

The most common use of loss analysis is in diagnosing low-performing or failing MEAs, where the goal is to identify the root cause of performance-loss, and subsequently to guide effective corrective actions regarding operating conditions and/or component design. Another common use is in MEA design and engineering, where the focus is on understanding the interactions between MEA-components and trade-offs between the various choices. The selection of membranes (PEM) and gas diffusion layers (GDL) for optimal MEA water management is one example of such application. Loss analysis can also be used as a tool in MEA-component design, where the focus is on understanding interactions for chosen materials or processes. This application is particularly useful for cathode design, where most losses are located (see Chapter 7). Finally, VLB has been used since the early days for PEMFC modeling for fundamental understanding purposes.

In summary, loss analysis is a "must have utensil" in the toolbox of applied researchers and engineers working on MEA, MEA-components as well as fuel cell materials research and design.

Max Cimenti, Ballard Power Systems Inc., Burnaby, BC V5J 5J9, Canada
Jürgen Stumper, Juergen Stumper & Assoc., Vancouver, BC V5M 1H5, Canada

https://doi.org/10.1515/9783110622720-009

This chapter is intended for scientists and engineers entering the field of PEM fuel cell MEA performance diagnostics. Although the topics are presented at an introductory level, an understanding of fuel cell and electrochemistry basics is assumed.

The first section of this chapter provides a summary of the main ingredients required for loss analysis and the various approaches are introduced. Next, these approaches are described in greater detail using examples for the published literature. Then the experimental procedures used to measure the most impactful properties of MEAs are discussed. Strategies to select an appropriate VLB approach as introduced next. Finally, a generalized approach for loss breakdown analysis is proposed.

9.2 Approaches to diagnose PEMFC performance losses

Since the early days of PEMFC, development attempts to deconstruct the measured polarization curves and identify the sources of voltage losses have been undertaken by researchers. In 1988, Srinivasan et al. proposed an empirical model of PEMFCs [1]. The model consisted of a constant and three variable terms that were associated to hypothetical loss sources, namely to the cathode kinetics, ohmic resistances; in a later publication [2], the authors proposed a refinement of the same model including an exponential mass-transfer term that was assumed to represent the combined effect of proton and oxygen transfer. This simple empirical model was found to fit accurately polarization data obtained under a variety of operating conditions (Figure 1). In addition, the trends of the fitting parameters were found to behave consistently with the physicochemical processes hypothesized. Although the model was too simple to fully describe MEA behavior, it quickly became the preferred approach to explain and even predict PEMFC performance mostly due to its simplicity and analytical format. This 0-D empirical MEA model has often been used in simplified unit cell and stack models used in the industry.

A few years later, Bernardi and Verbrugge presented the first physics-based model of a PEMFC [3], where all MEA components were considered (including anode), and processes normally hard to predict, such as liquid water transport in the porous phase, were mathematically represented and quantified. The authors proposed a calculation of the voltage loss contributions in the MEA that exemplifies what is meant by voltage loss breakdown (Figure 2).

Perry, Newman and Cairns proposed a model of fuel cell cathodes that clearly illustrates the combined effects of proton and oxygen transport limitation in the porous cathode [4]. Using this model, the authors showed the effects of the interplay between proton and oxygen transport, resulting in distinct double Tafel slopes for the apparent cathode overpotential, and also proposed a diagnostic tool to unveil the dominant transport process in PEMFC cathodes operating at high current densities that is based on the

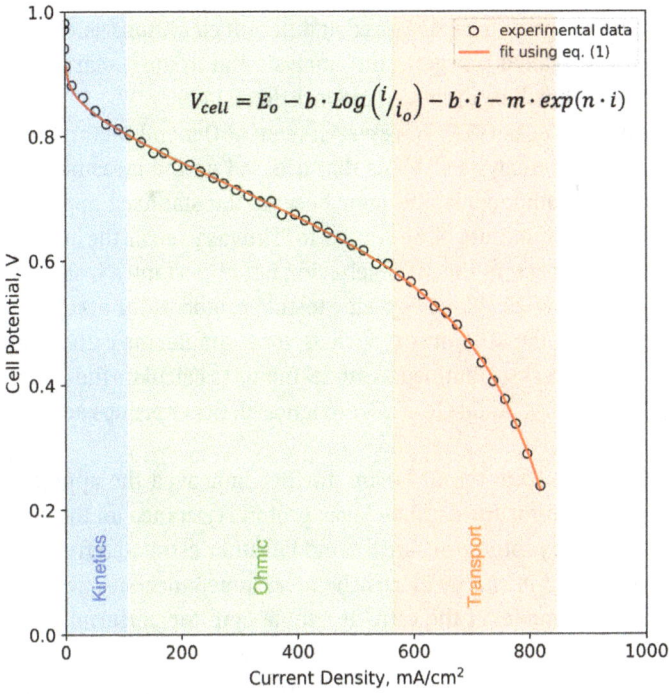

Figure 1: Example of empirical fitting of PEM fuel cell polarization data from Kim et al. [2]. Shaded regions represent the dominant overpotential type.

The equation shown in Figure 1:

$$V_{cell} = E_o - b \cdot Log\left(\frac{i}{i_o}\right) - b \cdot i - m \cdot exp(n \cdot i)$$

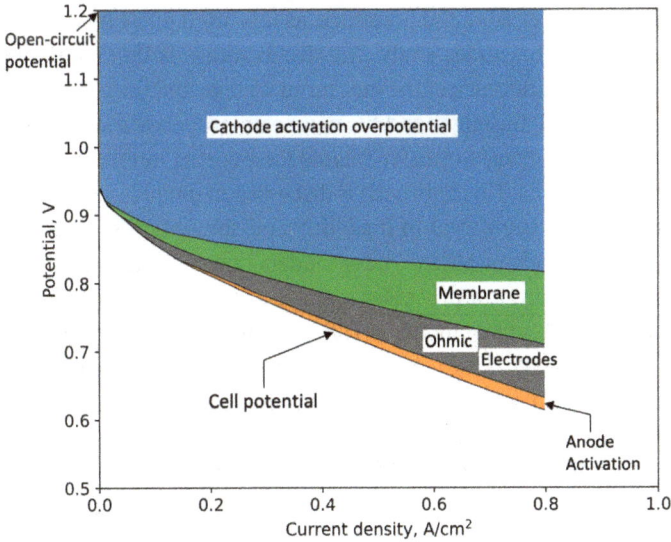

Figure 2: Example of voltage loss adapted from [3].

comparison of the electrode overpotentials measured at different conditions. Such a diagnostic tool then evolved in so-called "oxygen gain" analysis and in the "polarization change curve" (ΔV curve), which will be discussed in more detail later.

In the early 2000s, Gasteiger et al. (PEM FC research team at General Motors) proposed a methodology for the loss analysis of MEAs that is based on the measurement of effective properties of MEA components and later became the standard approach adopted by the broader PEMFC community [5, 6]. Similar to Srinivasan et al., the authors adopted a simplified model to represent the MEA behavior, but importantly, instead of using it to fit polarization data they developed specific testing protocols for accurately measuring effective properties of the MEA *in situ*, such as, for example, the cathode kinetics, and then used the parameters so obtained as model inputs to calculate the contribution of the individual loss sources. Anode losses were ignored. Other groups adopted a similar approach [7].

Later on, Neyerlin et al., still from the GM team, further improved the approach by developing a model to account for the cathode layer proton resistance on the electrode utilization [8]. The effective proton resistance could be either estimated from the cathode composition or measured *in situ* by Electrochemical Impedance Spectroscopy (EIS), providing a 1-D analytical model of the cathode usable both for performing the Tafel analysis, and also to quantify the cathode overpotential during normal operation. Although very simplified the MEA model proposed by these authors is not purely empirical; it is physics-based but simplified to include only the essential processes, as highlighted by Perry et al., so that the analysis of MEA data is very approachable and only an essential set of inputs is needed to carry on the voltage loss analysis. The advantage of this approach is in removing the ambiguities that arise when the inputs must be taken either from *ex situ* measurements or from the literature, as it is often unclear whether these are representative of the MEA under study. The disadvantage is that transport losses are simply obtained as the difference of the measured voltage and the sum of the estimated losses. More recently [9], Gasteiger and his new team proposed a simple analytical expression to quantify the oxygen transfer-related losses using inputs obtained from limiting current measurements. This approach is discussed in detail later.

Voltage loss analysis has also been used in modeling and simulation studies utilizing complex physics-based models of MEAs. These models typically include all the established electrochemical and physical processes applied to a macro-homogeneous framework representing the MEA subcomponents [10]. Recently, Gerhardt et al. provided guidelines for voltage loss analysis focusing on detailed physics-based models [11]. These complex numerical models are typically two-dimensional to properly represent the reactants and product distributions within realistic cell geometries and need over fifty input parameters, which are often obtained from the literature. Typically, a calibration of the model is performed to refine the certain input parameters using selected polarization data for a specific MEA. The calibration is normally applied to the least certain and most sensitive inputs; then loss breakdown is obtained from post-processing of the model results. Similarly, to the empirical case, this approach relies heavily on some sort of fitting to polarization data while using a complex and physically rigorous model.

As an alternative to the complete loss analysis for PEMFC, two simple diagnostic methods have been proposed, which are based on the comparisons of polarization curves obtained at different operating conditions or different stages in life for the same MEA. The first technique is called "oxygen gain analysis" and can be used to differentiate between mass-transport resistance within the catalyst layer and within the gas-diffusion layer [12]. The second technique is named "polarization change" analysis and is useful to determine the causes of performance loss changes during the lifetime of a MEA [13]. These techniques are very valuable due to their simplicity and rapidity of application and can provide rapid qualitative insight on the type of losses; however, they do not provide a quantitative breakdown of losses.

In summary, different approaches have been proposed to analyze losses in PEMFC. As it is not possible to directly measure/quantify individual losses in real MEAs, it becomes necessary to extract these loss-components using an MEA model in combination with model inputs derived from experiments as specific operating conditions. While this need is common to all approaches, the selection of the model (dimensionality and complexity) and of the experimental source of inputs varies. Three basic cases can be identified:

- Empirical VLB – A purely empirical approach where a simple analytical MEA model is selected and fitted to polarization data.
- Effective-property based VLB (EPB VLB) – An approach based on inputs derived from measuring *in situ* effective properties of the MEA and using a 1-D analytical or semi-analytical model to calculated overpotentials.
- Numerical VLB – An approach based on selecting a comprehensive physics-based numerical model that uses as inputs detailed information on the structure and composition of the MEA, plus parameters obtained from the literature; the model typically requires a "calibration" based on polarization data.

The schematic shown in Figure 3 exemplifies the relationship between model, experimental data and type of loss analysis for each VLB approach.

The "empirical VLB" approach utilizes a simple 0D model of the MEA that is fitted to a polarization curve, after which the type of loss (kinetics, ohmic and mass transfer) associated to the model terms are estimated.

The "EPB VLB" utilizes a different strategy: a simplified 1D model, typically isothermal and single phase, is selected; inputs are drawn mostly from *in situ* measurements of the MEA effective properties; next, the model is used to estimate individual losses (overpotentials) by type and component; finally, the predicted cell voltage, obtained subtracting the sum of overpotentials from the equilibrium potential, is compared to the measured polarization for the purpose of verification.

For the "numerical VLB," a comprehensive physics-based model is chosen, which draws inputs from material properties and *ex situ* characterizations. The model is cali-

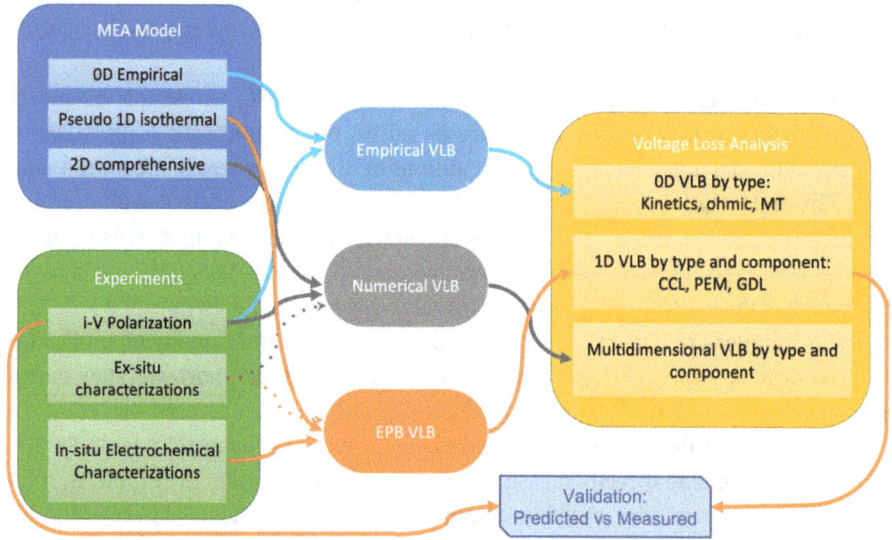

Figure 3: Schematic showing the model type (blue), experimental inputs (green) and outputs (yellow) for each VLB approach. Empirical VLB: curve fitting of polarization data using 0D analytical model. Numerical VLB: prediction of performance and overpotentials based on fundamental properties after model-calibration against polarization data. EPB VLB: prediction of overpotentials and cell performance based on inputs derived from specific *in situ* (and *ex situ*) measurements of the MEA effective properties. Validation for all cases based on the comparison of predicted versus measured cell voltage.

brated to better represent the specific MEA under investigation, and finally the numerical results are post-processed to derive losses assigned to individual processes/type in each component.

The approaches identified above will be discussed in more detail in the following sections.

9.3 Empirical approach for MEA loss analysis

By the "empirical MEA model," we refer to 0D analytical models including one or more empirical terms that are not unequivocally associated with a specific physical or chemical process nor to a specific MEA component. One of the earliest empirical models proposed to represent PEMFC polarization is given by equation:

$$V_{\text{Cell}}(i) = E_0 - b \cdot \text{Log}\left(\frac{i}{i_0}\right) - R \cdot i - m \cdot \exp(m \cdot i), \tag{1}$$

where $V_{\text{cell}}(i)$ is the cell voltage as function of the current density i; All other parameters in the equation are empirical and can be obtained by fitting to a polarization curve [2].

It was recognized early on that cathode electrode kinetics plays a vital role in determining the overall performance of PEMFC. Srinivasan et al. proposed that overpotentials are the result of steps occurring in series from the electrode surface, while losses from other sources were considered negligible. They identified five loss-regions in the voltage vs. current plot, as shown in Figure 1. In their analysis, three types of losses are encountered: (i) activation overpotential mainly due to slow kinetics of electro reduction of oxygen, which is predominant at low current densities; (ii) ohmic overpotential at intermediated currents due to resistance to proton transfer in the electrolyte (membrane); and (iii) mass transport losses at higher currents due to lower concentrations of reactants at the electrode/electrolyte interfaces, which is typically dominant at high current densities. The authors provided the following physical interpretation of the model of Equation (1): the first term corresponds to the open circuit potential, also equal to the reversible potential minus a current-independent loss due to hydrogen crossover and the mixed potential at the cathode due to platinum oxide formation; the second term accounts for the cathode activation overpotential, where b is an apparent Tafel slope and i_o is the apparent exchange current of the cathode; the third term includes all linear losses of the MEA and only minor contributions from anode activation overpotential, thus the parameter R is assumed to represent the ohmic resistance of membrane and GDLs; the last term represents transport limitations in the cathode, generally including oxygen and proton transport losses within the layer.

Srinivasan and coworkers provided a theoretical interpretation of results obtained from Equation (1) based on the analysis of the variation of the differential resistance of the cell, defined as the derivative of Equation (1) with respect to the current density. By plotting the differential resistance vs. the current density one can more easily differentiate the characteristic loss-regions (activation, ohmic and mass-transport) and attempt to explain the fitting results in terms of physical processes. In general, while this model can fit quite accurately, polarization curves obtained at various operating conditions, it cannot be used to assign unequivocally the MEA loss sources. In absence of fuel leaks, the activation parameters (i_o and b) obtained from fitting can in fact approximate well to the effective ORR parameters of the cathode layer obtainable through dedicated and controlled measurements. However, even in ideal conditions, both ohmic and activation terms in Equation (1) are "lumping" effects coming from multiple loss sources. For example, the ohmic resistance obtained using this approach also includes a portion of the proton transfer resistance of the cathode layer, as discussed in detail later. On the other hand, no theoretical interpretation was provided for the mass-transfer parameters, therefore, the model of Equation (1) cannot provide a reliable breakdown of losses at intermediate and high current densities.

Various attempts have been made to modify the above empirical model and make it more mechanistic. Squadrito et al. [14] proposed to substitute the last term of Equation (1) with a modification of a concentration overpotential:

$$V_{\text{Cell}}(i) = E_0 - b \cdot \text{Log}\left(\frac{i}{i_0}\right) - R \cdot i - a \cdot i^k \cdot \text{Log}\left(1 - \frac{i}{i_{\text{LIM}}}\right), \tag{2}$$

where a and k are empirical parameters. When k is zero, the third term of Equation (2) is equivalent to a concentration overpotential due to a concentration gradient of oxygen arising from diffusion within the gas diffusion layer of the cathode [15]. The pre-logarithmic term acts as an "amplification term" and is assumed to capture deviations from the 1-dimensional diffusion film caused by the complex structure of porous electrodes. The authors tested this modified model at various conditions using MEAs with different membranes and found that it could fit quite accurately most experimental data; by comparing the fitting results against those obtained using the model of Equation (1), they also proposed that their model allows for a better separation of ohmic contributions. In conclusion, the mass-transfer term in Equation (2) has a clear physical origin and contains as a parameter the electrode limiting current, which can be related to the effective properties of the MEA as discussed later and can be measured with dedicated experiments.

Pisani et al. [16] proposed a mass-transfer term to substitute the third term of Equation (1) that was derived assuming that the diffusion region in the cathode is made by a network of pores capable of transporting gas and liquid mixtures:

$$V_{\text{Cell}}(i) = E_0 - b \cdot \text{Log}\left(\frac{i}{i_0}\right) - R \cdot i - \beta \cdot \text{Log}\left(1 - \frac{i}{i_{\text{LIM}}} \cdot S^{-\mu \cdot (1 - \frac{i}{i_{\text{LIM}}})}\right), \tag{3}$$

where β and μ are parameters with a specified physical meaning, while S is a flooding empirical parameter related to the equilibrium saturation of the porous media. Although quite similar to Srinivasan's model of Equation (1), the authors proposed that this model better represents the phenomenology that is applicable in the cathode as the current density is increased.

The main strength of empirical models is that they can fit quite accurately polarization curves measured over largely different operating conditions and for a variety of MEA designs. These types of models are still extensively used in PEM FC stack development and for control purposes. Typically, maps or look-up tables of the model parameters are obtained by fitting polarization curves measured at different operating conditions over relevant ranges and using adjustment factors to account for MEA aging and degradation. Despite being intensive in terms of testing time, this approach can still provide sufficiently accurate estimates of the FC performance at different stages of the stack lifetime.

Regarding the physical meaning of the model parameters obtained from fitting i-V data, simple empirical models such as Equation (2) can be used to obtain some MEA effective properties under controlled conditions. For example, the cathode kinetics constants can be obtained by fitting the activation portion (i. e., low current density) of H_2/O_2 and H_2/Air polarization curves [17]. Similarly, the model of Equation (2) can be used to obtain the apparent limiting current of a cathode for polarization curves measured by controlling the potential down to 100 mV [18]. However, caution should be exercised in the interpretation of the regression results, particularly if unweighted regression is performed.

The main limitation of empirical models is that while they can accurately fit polarization data the resulting parameters cannot be associated univocally to the physical and electrochemical phenomena and processes occurring within the MEA. This is particularly true at high current densities (i. e., transition between ohmic and mass-transfer region of the polarization curve) where cross-contamination between ohmic and transfer losses is likely. Various studies have been published to evaluate the consistency and performance of empirical models. For example, Haji [19] used the model of Equation (2) (with fixed $k = 0$) to obtain Tafel constant, ohmic resistance and limiting current through nonlinear regression, and also by performing multiple linear regression analysis in combination to dedicated measurement to obtain the limiting current. The author concluded that although the model fits accurately the data when using nonlinear regression, the obtained parameters were not reasonable as MEA effective properties, while linear regression analysis results were more consistent. Another recent study [20] compared results obtained using different empirical models on the same data set and concluded that different interpretations (of the parameters) can be provided for the different models. Curiously, the authors also found that the model of Equation (1) performed better than the others as for fitting accuracy, despite using a purely empirical term to fit the mass-transfer region. This result indicates that while semiempirical models such as in Equations (2) and (3) use coefficients with precise physical meaning, they do not necessarily guarantee a better representation of the complexity of MEA processes.

In summary, empirical models can fit PEMFC polarization curves quite accurately under various operating conditions and are still extensively used for modeling stack performance particularly for system control purposes; as part of the empirical approach, these models can also be used to extract some effective properties of MEA when combined with controlled experiments. However, the usefulness of these models is very limited for voltage loss analysis because of their inability to univocally segregate loss sources.

9.4 Loss analysis based on measuring effective MEA properties

Gasteiger and his team proposed a methodology that became popular for analyzing voltage losses in PEM FC [5]. In this approach, dedicated electrochemical measurements (e. g., cyclic-voltammetry, H_2–O_2 polarizations, limiting current measurements and electrochemical impedance spectroscopy) are performed to extract effective properties of the MEA under investigation, which in turn are the inputs for a MEA model used to predict individual losses. The MEA model is simplified but physically meaningful and consistent, as it includes the most impactful physical processes without being overly complicated. Model assumptions and formulation are such that the final model is expressible in analytical form and the overpotentials of each MEA component are easily separable. The model inputs are measured ad hoc and individual losses are calculated using the model. Finally, the predicted cell performance is compared to the measured cell voltage to verify the consistency of the loss breakdown.

This section includes a description of the model and VLB examples, together with the analysis of the relative magnitude and importance of each overpotential term. A description of the methodologies used to measure effective properties usable as inputs in the approach is provided later. Finally, strengths and limitations are discussed.

A 1D MEA model should provide a sufficiently accurate description of the MEA behavior if all in-plane gradients in the cell are negligible. This assumption implies that both independent and dependent variables are almost constant both along-the-channel and in the channels/ribs directions, while gradients are still present in the through-plane direction (perpendicular to the membrane). Under this assumption the MEA can be represented as in Figure 4 and the cell voltage is equal to the reversible voltage minus the various losses of each component:

$$V_{\text{cell}} = E_{\text{rev}} - |\eta_{\text{Anode}}| - |\eta_{\text{Cathode}}| - |\eta_{\Omega,\text{PEM}}| - |\eta_{\Omega,\text{other}}| \tag{4}$$

$$E_{\text{rev}} = E_{\text{rev}}^0 + \frac{R \cdot T}{2 \cdot F} \cdot \text{Ln}\left[\frac{(\frac{pH_2}{\text{pref}}) \cdot (\frac{pO_2}{\text{pref}})^{\frac{1}{2}}}{(\frac{pH_2O}{\text{pref}})} \right], \tag{5}$$

where E_{rev} is the thermodynamic equilibrium potential, η_{Cathode} and η_{Anode} are the total overpotentials of anode and cathode, respectively, $\eta_{\Omega,\text{PEM}}$ is the ohmic loss due to proton transport in the membrane and $\eta_{\Omega,\text{other}}$ are the ohmic losses in GDLs, plates and contacts, including every interface of the MEA sandwich (see Chapter 7 for a detailed description of the chemical and physical processes involved).

Anode and cathode total overpotentials are defined as

$$\eta_{\text{Anode}} = \eta_{\text{HOR}} + \eta_{\text{H}^+-\text{An}} + \eta_{\text{MT}-\text{An}} \tag{6}$$

and

$$\eta_{\text{Cathode}} = \eta_{\text{ORR}} + \eta_{\text{H}^+-\text{Cath}} + \eta_{\text{MT}-\text{Cath}}, \tag{7}$$

where η_{HOR} and η_{ORR} are the activation overpotentials of anode and cathode electrodes evaluated at a reference condition chosen as the operating conditions in the flow channels at the inlet of the cell, where the reactants have the highest enthalpy, $\eta_{\text{H}^+-\text{An}}$ and $\eta_{\text{H}^+-\text{Cath}}$ are the ohmic losses due to proton transport within the porous electrodes and $\eta_{\text{MT}-\text{An}}$ and $\eta_{\text{MT}-\text{Cath}}$ are the H_2 and O_2 transport losses for each electrode.

(a) MEA Schematic

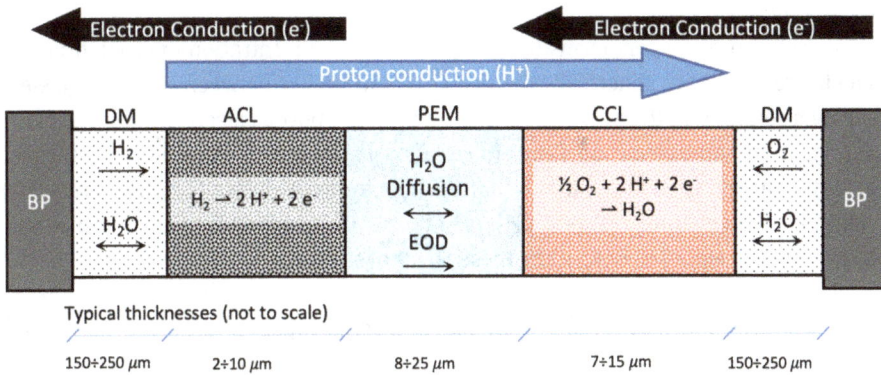

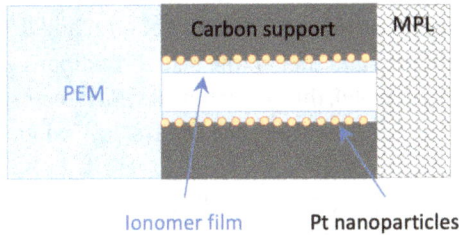

Figure 4: (a) 1D schematic of a MEA showing the various components: diffusion media (DM), bipolar plates (BP), anode and cathode electrodes (ACL, CCL) and proton exchange membrane (PEM). Reactions and processes occurring in each component/domain are also shown (modified from [21]). (b) Schematics representing an idealized single pore for the CCL representative of the average electrode structure.

The ohmic loss in each porous catalyst layer (CL) can be expressed as the potential difference in the ionomer-phase of the catalyst layer and is also equivalent to the integral of the ohmic losses within the porous electrode according to the following equation:

$$\eta_{\text{H}^+-\text{CL}} = \left|\phi_{\text{H}^+}(\delta_{\text{CL}}) - \phi_{\text{H}^+}(0)\right| = \int_0^{\delta_{\text{CL}}} \frac{1}{\kappa_{\text{H}^+}^{\text{eff}}(z)} \cdot i_{\text{H}^+}(z) \cdot dz, \tag{8}$$

where CL is either the anode or cathode electrode, Φ_{H^+} is the proton potential within the CL, $\kappa_{\text{H}^+}^{\text{eff}}$ is the effective proton conductivity and i_{H^+} is the protonic current, both of which are functions of the position in the layer and depend on the local hydration and temperature within the CL. Effective properties, such as the electrode proton conductivity, result from the combined effect of material properties and structural properties of the electrode.

The reactant transport terms (η_{MT}) represent the concentration overpotentials of each electrode and are generically defined as the difference between the average electrode overpotential at the actual conditions within the porous electrode and the overpotential of the same electrode if the entire catalyst surface was exposed to the highest reactants concentration found in the channel inlet for each electrode [15]. The reactant conditions in the channels (T, pO_2, pH_2, pH_2O_{an}, pH_2O_{cath}) are chosen as reference values. The generic equation describing these terms are

$$\eta_{MT-An} = \left(E_{rev}^{An} - \eta_{HOR}\right)_{Ref-An} - \left(E_{rev}^{An} - \eta_{HOR}\right)_{Pt-An} \tag{9}$$

and

$$\eta_{MT-Cath} = \left(E_{rev}^{Cath} - \eta_{ORR}\right)_{Ref-Cath} - \left(E_{rev}^{Cath} - \eta_{ORR}\right)_{Pt-Cath}, \tag{10}$$

where the activation overpotentials were defined earlier, while E_{rev}^{An} and E_{rev}^{Cath} are the Nernst potentials for the anode and cathode half-reactions, respectively. These terms are evaluated first at reference conditions, where Ref – An and Ref – Cath are chosen as the channel inlet conditions, and then at the actual conditions within the electrode (Pt – An and Pt – Cath, respectively), the evaluation of which requires the knowledge of the local MEA dependent variables within each catalyst layer, obtainable in turn using the 1D MEA model.

The individual losses of Equations (6) to (10) can be derived in analytical form if thermal gradients through plane and effect of liquid water saturation are negligible, that is, under the assumption of isothermal single-phase conditions. These assumptions are chosen here because they allow an intuitive understanding of each overpotential term and the VLB approach. However, one or more of these assumptions are often not satisfied at realistic conditions, as discussed later (see Section 9.9). The derivation that follows is based on Gu et al. [21], using the same governing equations recommended by leading PEMFC model developers [10]. The model does not contain any empirical terms, that is, all terms have a well-defined physical meaning associated to the MEA components and to the related processes. In addition to the three assumptions stated earlier, the following assumptions are also necessary: (1) porous electrode theory is applied to anode and cathode, that is, both electrodes are uniform in thickness, porosity and composition, and their properties are describable by effective properties; (2) the water content in the membrane and ionomer of the CLs is close to equilibrium; (3) oxygen transport resistance through the pores of the electrode is negligible if compared to the transport resistance resulting from dissolution and diffusion in the ionomer film. As shown in Figure 4b, under these assumptions the electrode reduces to an ensemble of identical straight pores, the membrane behaves as a pure ohmic conductor; the GDLs are the medium through which reactant- and product-diffusion occurs.

Starting from the cathode, the activation overpotential is given by the high-field approximation of the Butler–Volmer equation, also known as Tafel equation [22]:

$$\eta_{ORR} = \frac{Ln(10) \cdot R \cdot T}{\alpha_C \cdot F} \cdot Log\left(\frac{i_{cell} + i_{xover}}{rf_{CCL} \cdot i_{0,ORR}(T, pO_2)}\right), \tag{11}$$

where i_{cell} is the input variable, i_{xover} is a current equivalent to the rate of permeation of hydrogen through the membrane, rf_{CCL} is the Pt roughness factor of the electrode, α_C is the apparent charge transfer coefficient for the ORR reaction [22] and $i_{0,ORR}$ is the apparent exchange current density for the ORR reaction, which is property specific of the catalyst type used and depends on temperature and O_2 partial pressure, as in the following equation [23]:

$$i_{0,ORR}(T, pO_2) = i_{0,ORR}^{Ref} \cdot \left(\frac{pO_2}{pRef}\right)^{\gamma_{O_2}} \cdot \exp\left[-\frac{E_{a,ORR}^{rev}}{R \cdot T} \cdot \left(1 - \frac{T}{T^{Ref}}\right)\right], \tag{12}$$

where $i_{0,ORR}^{Ref}$, α_C, γ_{O_2}, and $E_{a,ORR}^{rev}$ are the four kinetics parameters for the cathode ORR, that is, exchange current density, charge transfer coefficient (related to Tafel slope), reaction order and activation energy, respectively.

The ohmic loss due to proton resistance in the cathode layer is given by Equation (8). Neyerlin et al. obtained a semiexplicit analytical equation for the ionomer potential [8] by solving the cathode governing equation, and then derived an effective proton resistance of the layer that allows to express the loss as an ohmic loss (see Figure 5a):

$$\eta_{H^+-Cath} = |\phi_{H^+}(\delta_{CCL}) - \phi_{H^+}(0)| = R_{H^+-CCL}^{eff} \cdot i_{cell} \tag{13}$$

$$R_{H^+-CCL}^{eff} = \frac{R_{H^+-CCL}}{3 + \zeta}. \tag{14}$$

The effective proton resistance of the CCL ($R_{H^+-CCL}^{Eff}$) is related to the apparent proton resistance of the layer (R_{H^+-CCL}) as shown in Equation (14), where ζ is a dimensionless correction factor that varies between 0 and 3 as function of the cell current and is approximately equal to

$$\zeta = 0.68777 \cdot \beta_{CCL} - 0.0061569 \cdot \beta_{CCL}^2 + 0.000051952 \cdot \beta_{CCL}^3, \tag{15}$$

where $\beta_{CCL} = \frac{i_{cell} \cdot R_{H^+-CCL}}{b_C}$ is a nondimensional ratio of the proton transport vs. kinetic losses within the layer, and b_C is the Tafel slope given by the first term in Equations (11). Figure 5b shows the dependence of ζ as function of β_{CCL}.

The oxygen transport losses for the cathode are obtained by applying Equation (10), resulting in the following:

$$\eta_{MT-Cath} = \frac{Ln(10) \cdot R \cdot T}{F} \cdot Log\left[\frac{pO_{2Ref}}{pO_{2Pt}}\right] \cdot \left\{\frac{1}{4} \cdot Log\left[\left(\frac{pH_2O_{Pt}}{pH_2O_{Ref}}\right)^2\right] + \frac{\gamma_{O_2}}{\alpha_C}\right\}, \tag{16}$$

(a)

(b)

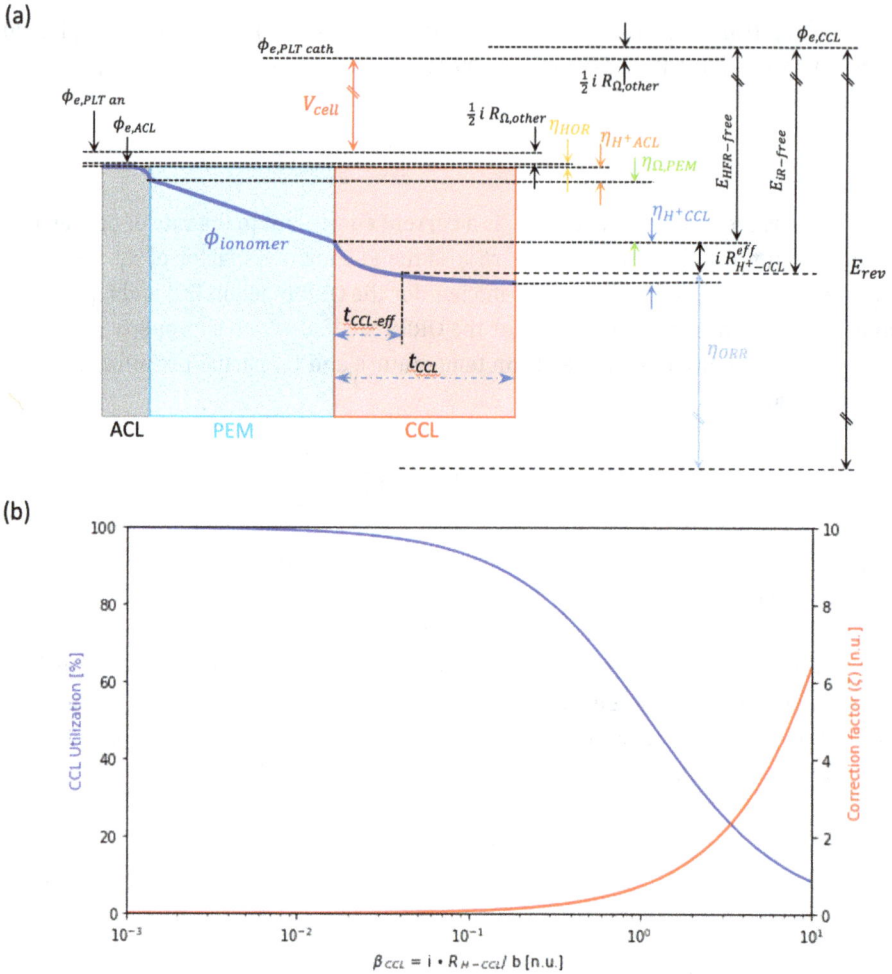

Figure 5: (a) Schematic showing a typical profile of the protonic potential within the electrodes and membrane of a PEM FC when a galvanic current is present, (b) the catalyst utilization and the correction factor (ζ) used to correlate layer utilization to the effective proton resistance of the electrode (R_{H^+-CCL}) as a function of the dimensionless factor (β_{CCL}). (Adapted and modified from [8].)

where pO_{2Pt} and pO_{2Ref} are the oxygen partial pressures at the catalyst surface and in the cathode gas channel, respectively. The first term in the right-hand side of Equation (16) represents the Nernstian voltage loss due to the concentration drop between channel and electrode surface, while the second term is the drop associated with the change in electrode kinetics. The combined terms represent a concentration overpotential of the cathode. The oxygen partial pressure at the catalyst surface is obtained as

$$pO_{2Pt} = pO_{2Ref} - \frac{R \cdot T}{4 \cdot F} \cdot R_{MT-O_2} \cdot i_{cell}. \tag{17}$$

Here, $R_{\text{MT-O}_2}$ is the oxygen transfer resistance in the GDL and cathode electrode. Similarly, the vapor partial pressure at the catalyst surface is obtained as

$$pH_2O_{\text{Pt}} = pH_2O_{\text{Ref}} - \frac{R \cdot T}{2 \cdot F} \cdot R_{\text{MT-H}_2\text{O}} \cdot i_{\text{cell}}. \tag{18}$$

The concentration of reactants at the Pt surfaces is uniquely identified if the oxygen diffusion rate through the pores of the CCL is much larger compared to the combined dissolution and diffusion rates in the ionomer layer, as assumed for the model under consideration. In general, oxygen presents a concentration gradient in the cathode layer. Therefore, the concentration at the Pt surface is not unique. However, the concept of Equation (10) is still applicable if the average overpotential of the electrode is considered, or equivalently an effective concentration.

The concentration overpotential of Equation (16) can be further broken down if the total mass transfer resistance can be separated in components. For example, the concentration overpotential of the GDL can be obtained by substituting the concentration of oxygen and vapour at the Pt surface with the concentrations at GDL/CCL interface. This concept is applicable to any intermediate interface within the cathode side of the MEA sandwich.

Moving to the anode, the activation overpotential is given by the low-field approximation of the Butler–Volmer equation for the HOR reaction [24]:

$$\eta_{\text{HOR}} = \frac{R \cdot T}{(\alpha_{Aa} + \alpha_{Ac}) \cdot F \cdot \text{rf}_{\text{ACL}} \cdot i_{o,\text{HOR}}(T, pH_2)} \cdot i_{\text{cell}} = R_{K,\text{HOR}} \cdot i_{\text{cell}}, \tag{19}$$

where $i_{o,\text{HOR}}$ is the exchange current density for the HOR reaction, which is specific of the catalyst type used and depends on temperature and H_2 partial pressure, as in the following equation:

$$i_{o,\text{HOR}}(T, pH_2) = i_{o,\text{HOR}}^{\text{Ref}} \cdot \left(\frac{pH_2}{p\text{Ref}}\right)^{\gamma_{\text{H}_2}} \cdot \exp\left[-\frac{E_{a,\text{HOR}}^{\text{rev}}}{R \cdot T} \cdot \left(1 - \frac{T}{T^{\text{Ref}}}\right)\right], \tag{20}$$

where $i_{o,\text{HOR}}^{\text{Ref}} \gamma_{\text{H}_2}$ and $E_{a,\text{HOR}}^{\text{rev}}$ are the three kinetics parameters for the anode HOR.

In a similar manner as discussed for the cathode, the ohmic loss due to proton resistance in the anode layer is given by

$$\eta_{\text{H}^+-\text{An}} = |\phi_{\text{H}^+}(\delta_{\text{ACL}}) - \phi_{\text{H}^+}(0)| = R_{\text{H}^+-\text{ACL}}^{\text{eff}} \cdot i_{\text{cell}}, \tag{21}$$

where $R_{\text{H}^+-\text{ACL}}^{\text{eff}}$ is the effective proton resistance of the ACL, which is related to the apparent proton resistance of the layer ($R_{\text{H}^+-\text{ACL}}$) as follows:

$$R_{\text{H}^+-\text{ACL}}^{\text{eff}} = R_{\text{H}^+-\text{ACL}} \cdot \frac{\left(\frac{e^{\beta_{\text{ACL}}} + e^{-\beta_{\text{ACL}}}}{e^{\beta_{\text{ACL}}} - e^{-\beta_{\text{ACL}}}} - \frac{1}{\beta_{\text{ACL}}}\right)}{\beta_{\text{ACL}}}, \tag{22}$$

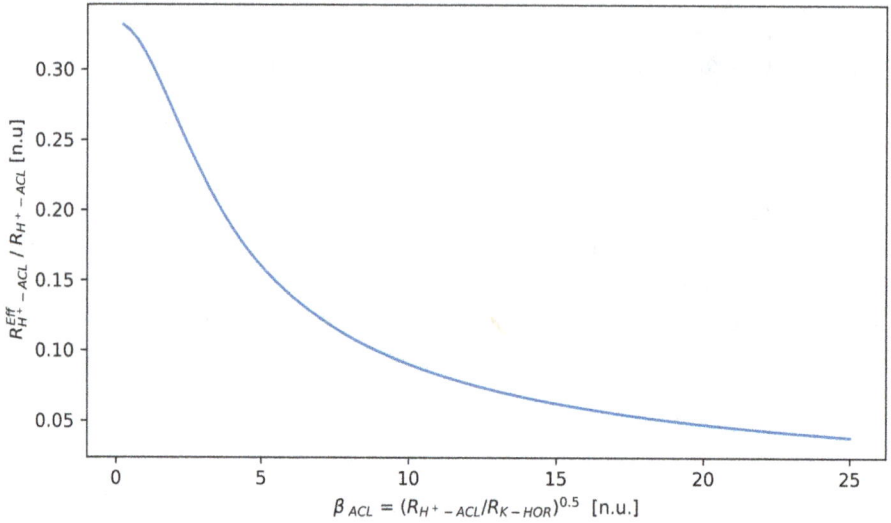

Figure 6: Effective proton resistance of the anode electrode as a function of the dimensionless factor (β_{ACL}).

where β_{ACL} is a nondimensional factor given by the square of the ratio of proton resistance of the ACL and charge transfer resistance for HOR:

$$\beta_{ACL} = \sqrt{\frac{R_{H^+-ACL}}{R_{K,HOR}}} \tag{23}$$

Equations (21) and (22) were obtained from the analytical solution of the anode governing equation by Thomson [25] under the simplifying assumptions stated earlier.

The hydrogen transport losses for the anode are obtained applying Equations (9), which gives

$$\eta_{MT-An} = \frac{R \cdot T}{2 \cdot F} \cdot \text{Ln}\left(\frac{pH_{2Ref}}{pH_{2Pt}}\right) - \frac{R \cdot T \cdot i_{cell}}{(\alpha_{A-an} + \alpha_{C-an}) \cdot F \cdot \text{rf}_{ACL} \cdot i_{0,HOR}(T, P^{Ref})}$$
$$\cdot \left(1 - \left(\frac{pH_{2Ref}}{pH_{2Pt}}\right)^{\gamma_{H_2}}\right) \tag{24}$$

where pH_{2Pt} and pH_{2Ref} are the hydrogen partial pressures at the catalyst surface and in the anode gas channel, respectively. The two terms of Equation (24) represent the "thermodynamic" and "kinetics" part of the anode concentration overpotential. The hydrogen partial pressure at the catalyst surface is obtained as

$$pH_{2Pt} = pH_{2Ref} - \frac{R \cdot T}{2 \cdot F} \cdot R_{MT-H_2} \cdot i_{cell}, \tag{25}$$

where R_{MT-H_2} is the hydrogen transfer resistance in the GDL and anode catalyst layer.

Finally, ohmic losses in the membrane and GDL plus contacts are given by

$$\eta_{\Omega,\text{PEM}} = R_{\text{PEM}} \cdot i_{\text{cell}} \tag{26}$$

$$\eta_{\Omega,\text{other}} = R_{\Omega,\text{other}} \cdot i_{\text{cell}} = (R_{e-\text{GDL}} + R_{e-\text{PLTs}} + R_{e-\text{Contacts}}) \cdot i_{\text{cell}} \tag{27}$$

The input parameters of Equations (10) to (27) can be measured using the experimental procedures described later. Some of these inputs can vary significantly depending on the MEA design and state-of-health, and can have a large impact on the losses, while other parameters vary less or are less impactful and their measurement might be avoided if estimates are available. The relative importance of these inputs can be understood by analyzing the magnitude of each loss term. Next, the model described above will be used to simulate two hypothetical operating conditions as an example of voltage loss analysis.

The predicted performance is shown in Figure 7a, while Figure 7b and c represent the magnitudes of each loss as function of the current densities for the simulated two cases. Finally, in Figure 7d the ranking of the losses as percentages of the total is provided. From this analysis, it can be inferred that the cathode activation losses are always dominant, accounting for more than 50 % of the total losses until the limiting current is approached, while anode losses are always minor. The ohmic losses in PEM, GDL, plates and contacts rank second among all losses, while the ohmic loss due to proton resistance in the cathode layer ranks third in the low to medium current density range.

Oxygen transport losses in the cathode and GDL become more relevant than the ohmic losses only close to the limiting current, while they are larger than the CCL proton transport losses already at intermediate current densities in case of low oxygen concentration.

As shown in Figure 8, analysis at different hydrogen dilutions shows that anode losses are always the least prominent, and the magnitude of anode losses exceeds 20 mV only above $2\,\text{A/cm}^2$ for this specific example. Based on this observation, one can conclude that it is often reasonable to neglect anodic losses for the purpose of loss analysis.

According to this simplified model and for the MEA design considered in this analysis, we can conclude that the three dominant losses are the ORR activation losses, the total ohmic losses of the MEAs and the proton transport losses in the CCL, accounting for more than 80 % of the total losses up to approximately limiting current (~ 98 % of I_{LIM}). Oxygen transport losses in the cathode become dominant only when approaching limiting current. The ranking of losses as well as this preliminary conclusion will be reevaluated later (see Section 9.9) for the case in which the assumptions of isothermal conditions, uniform hydration and negligible saturation effects (single phase) no longer hold.

Interestingly, when anodic overpotentials are neglected Equation (4) will have only three terms dependent on the current that can be broadly grouped in cathode activation, ohmic and mass-transfer terms similar to the empirical models described in the

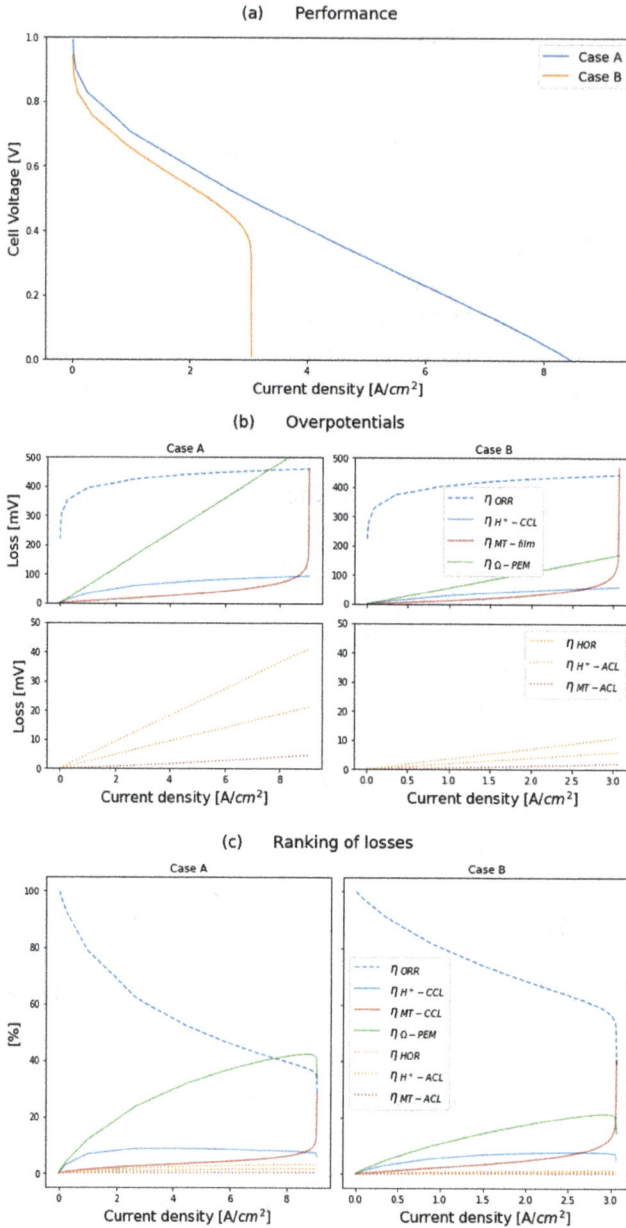

Figure 7: Simulated performance (a) and individual overpotentials (b) for Case A [80 °C, 2.5 bar and 60 % RH in H_2/Air] and Case B [95 °C, 2.3 bar and 95 % RH in H_2/10 % O_2]; (c) Relative ranking of the losses.

previous section. Although simplified and analytical, the model just described is not empirical but rather derived from physical principle under a set of assumptions that are reasonable and applicable at least in the case of low to moderate current densities.

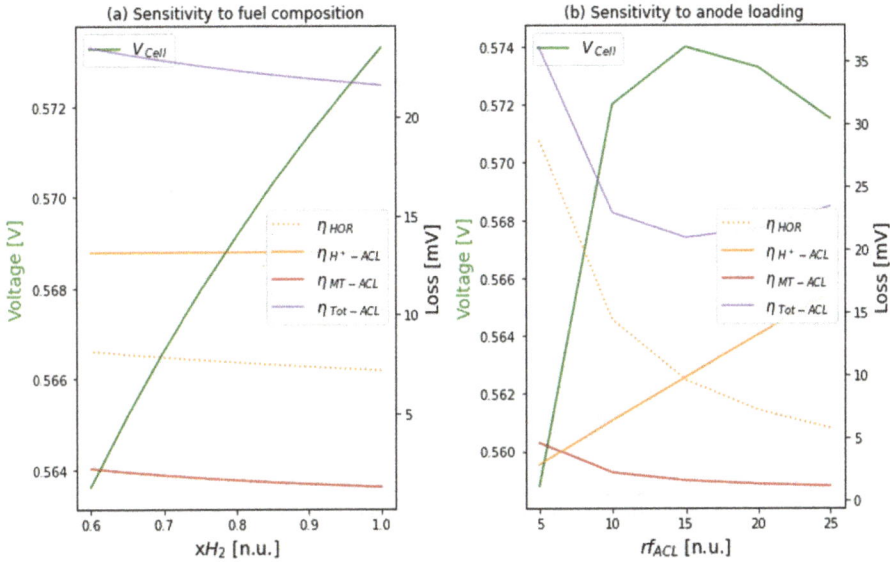

Figure 8: Performance sensitivity at constant current (3 A/cm²) to (a) hydrogen concentration and (b) anode Pt roughness factors simulated using the same model of Figure 7.

An analogous analytical model of the MEA is provided in Chapter 7, where the case of nonnegligible O_2 transport through the primary pores of the cathode layer is considered. As shown in Figure 4 of Chapter 7, although the losses predicted by that model differ in magnitude if compared to what presented previously, the ranking of losses does not change, even after including mass transport within the CCL. In his book, Kulikovsky [26] provides an exhaustive collection of analytical models of PEM FC, including asymptotic cases that are equivalent to the one described here. Another very good review of analytically treatable PEM FC models is the book by Eikerling and Kulikovsky [27]. In these references, the losses related to proton and oxygen transport in the CCL are quantified as "utilization" terms, which allow for a loss analysis analogous to what described above.

Next, we review a few examples from the literature where loss analysis based on measuring effective MEA properties was applied.

As mentioned previously, Gasteiger et al. [5] realized the limitation of the empirical VLB approach previously used in PEMFC research and proposed to combine dedicated measurement to obtain MEA ohmic resistance and cathode kinetics losses. The authors analyzed the expected anode losses as function of Pt loading and concluded that these should be negligible at practical loads ($\sim$ 10 mV at 1.2 A/cm²). Then they performed a loss analysis for varying cathode loadings by performing a Tafel analysis of H_2–O_2 polarization data after removing the ohmic losses, which in turn were obtained from High Frequency Resistance (HFR) measurements (iR-free cell potential, $E_{iR\text{-}free}$ in Figure 9). By plotting the $E_{iR\text{-}free}$ as a function of the Pt-mass specific current, they observed the

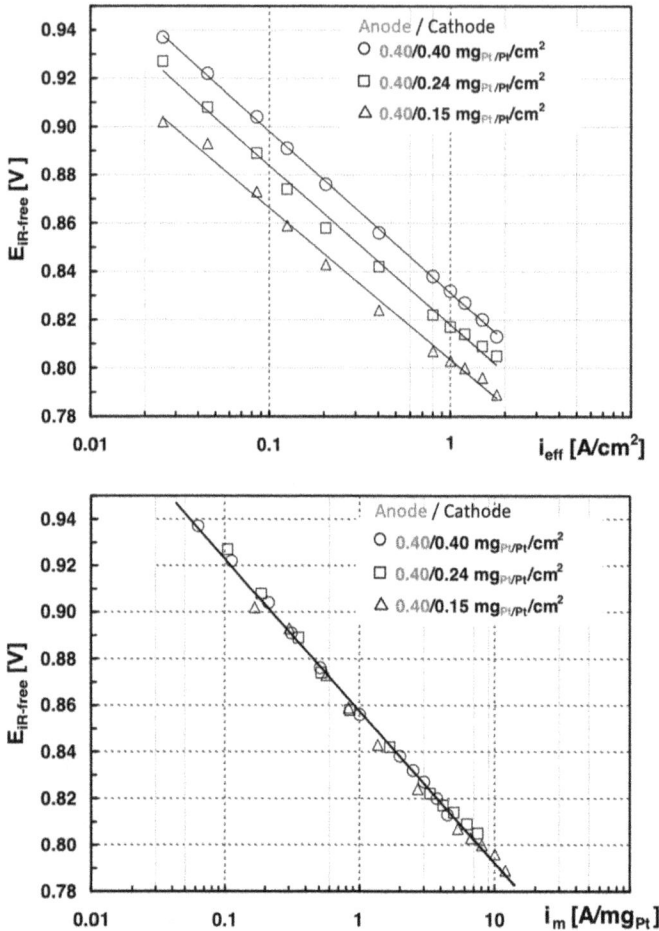

Figure 9: (a) Effect of cathode loading of the resistance-corrected cell voltage for H_2/O_2 operation, and (b) mass-specific current density showing Tafel slope of 65 mV/dec. Reproduced with permission from Elsevier, © (2004) [5].

"collapse" of the curves measured for the various loadings into one, confirming that the "true" ORR could be measured using this approach.

In a later publication [6], the same authors obtained a loss breakdown of an MEA running in H_2/Air. They used a 0D model similar to Equation (1) where the activation and ohmic overpotential were obtained from specific measurement, while mass transfer losses were obtained from the difference between the predicted (η_{tx} – free E_{cell} in Figure 10) and the measured voltages.

Neyerlin et al. developed a 1D model of the cathode already described in Equation (13) and Figure 6 [8] that relates the loss due to proton transfer in the porous cathode to the layer sheet resistance, which is in turn measurable by EIS. This model

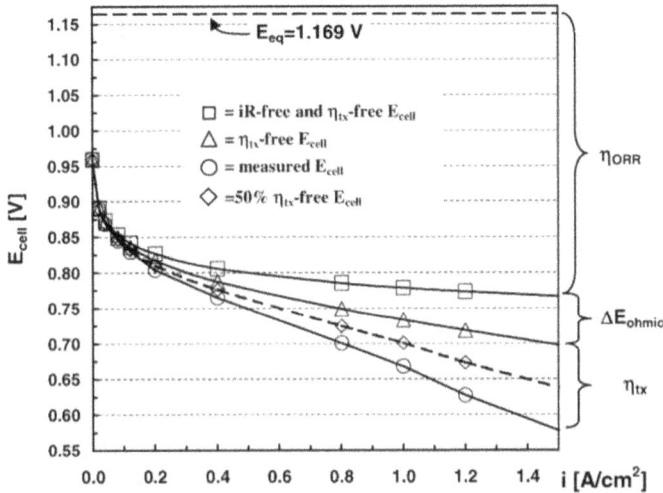

Figure 10: VLB example applied to H_2/Air polarization data measured in H_2/Air at 80 °C and 1.5 bar. The square symbols represent the mass-transport and ohmic free voltage obtained using *in situ* measurement of HFR. Reproduced with permission from Elsevier © (2005) [6].

can be used both to improve the accuracy for obtaining ORR kinetics parameters and for loss analysis of H_2/Air data. For example, Papadias [28] investigated the durability of PtCo alloy catalysts by combining accelerated stress tests with voltage loss analysis. The ORR activity was obtained from H_2/O_2 data after iR, and H_2 cross-over corrected currents, while proton resistance was estimated from electrode compositions. Oxygen mass transport losses were from the difference between measured and predicted voltage.

Recently, this loss approach further evolved with the inclusion of oxygen transport losses estimated from mass-transport resistance measurements obtained by limiting current measurements. Orfanidi et al. [9] derived an expression for oxygen transport losses that is analogous to Equation (16) and estimated losses from mass-transport resistance measurements described in Section 9.8.4. By doing so, all losses were estimated from dedicated measurement done for the same MEA under scrutiny. When comparing the predicted to the measured voltage, the authors still found up to ~ 40 mV difference at the higher current densities for some MEA designs. These were termed "unaccounted losses" and attributed to hypothetical causes such as deviation in ORR kinetics due to oxide coverage or possible variations in mass-transfer resistance. This study exemplifies the power of this approach: all individual losses are obtained from dedicated measurement, and finally the comparison between predicted and measured voltage is used to validate the breakdown and, in case a difference larger than an established tolerance is observed, it provides a magnitude for not yet explained/accounted losses.

More recent examples from the same team [29] demonstrate that this approach is applicable to a wide variety of electrocatalyst and cathode designs, and importantly, it can be combined with AST to understand changes during degradation (see Figure 11).

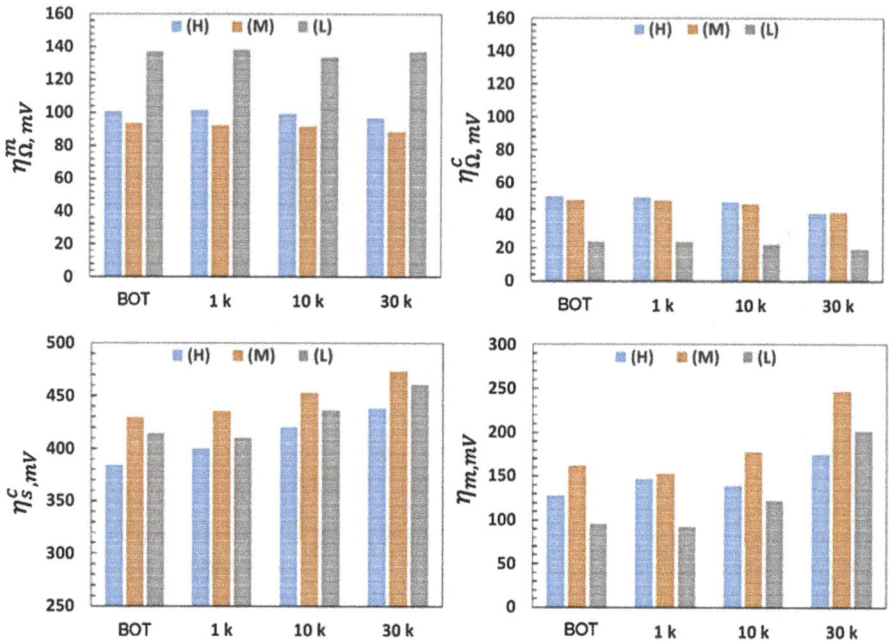

Figure 11: Change in VLB with electrode aging where H, M and L represent MEAs with cathodes containing PtCo catalyst with different composition from various suppliers. Reproduced with permission from IOP Publishing © (2018) [28].

Ramaswany et al. applied this VLB approach to characterize the impact on performance of different carbon supports ([30] in Figure 12a), and ionomer of different side-chain-length and EW ([31] in Figure 12b) using an agglomerate model to distinguish between the exterior and interior of the Pt/C particles' reaction.

The main strength of the VLB approach (à la Gasteiger) described here is its ability to quantify individual losses with a high degree of certainty. Another less evident benefit is that this approach is inherently more robust and accurate if compared to non-linear regression of polarization data. In fact, since dedicated/controlled experiments are performed to obtain MEA effective properties, it is more likely that measurement variabilities are detected and controlled for each individual property that is measured, and that deviations between data and models used to extract these properties are spotted early on preventing cross contamination between various losses, while in the case of regression for polarization data the measurement uncertainty is unknown a priory. The risk of cross contamination is also minimized by including dedicated measurements for all the dominant sources of losses. Finally, the approach offers a way to validate its correctness in the final comparison of predicted versus measured cell voltage, while the quantification of "unaccounted" losses offers initial evidence for formulating further refinements, as it will be discussed in later sections.

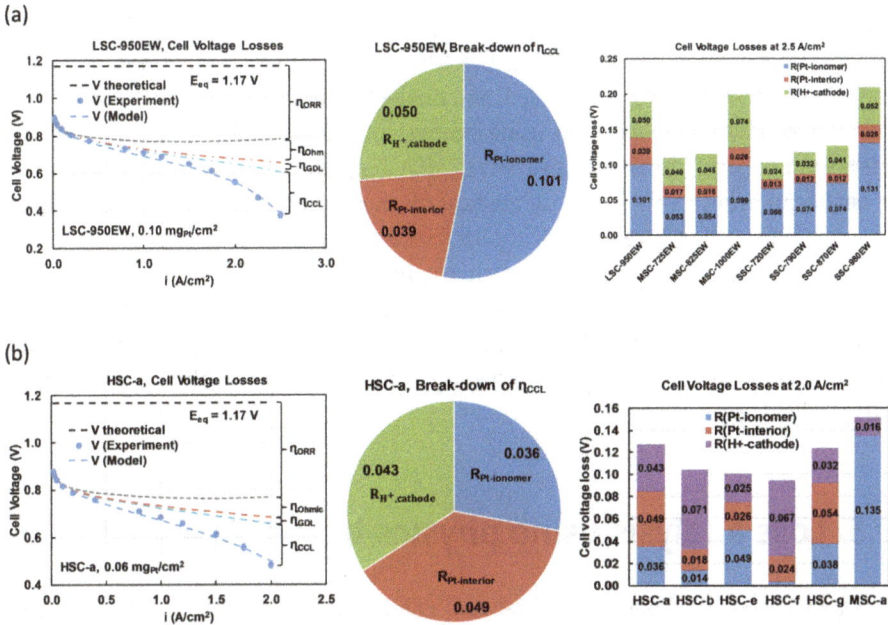

Figure 12: Examples of EPB VLB: (a) Comparison vs predicted polarization for MEA with long-side chain ionomer (right) and VLB at 2.5 A/cm^2 (left) (reproduced from [31] with permission from IOP Publishing); (b) Comparison of measured and predicted polarization curve for MEA using the catalyst high surface area carbon supported catalyst (right) and VLB at 2.0 A/cm^2 (left) (reproduced with permission from IOP Publishing © (2018) [30]).

On the other hand, this approach has some limitations. First, it requires inputs and data from differential cell measurements, and the condition of having negligible along-the-channel gradients is hard to satisfy at high current densities. Second, in certain situations there is a significant risk of mismatch between the inputs and MEA behavior at a specific point in time. For example, the MEA might be experiencing temporary degradation due to some form of contamination. Some of the protocols applied to measure the effective properties required as inputs for the VLB might have some conditioning effect, resulting in a change in MEA state. Last, but not least, the high current density region is often the most interesting for the purpose of analyzing losses and at many of the assumptions adopted for the MEA model are likely to fail, in particular there might be nonnegligible temperature and hydration gradients in the through plane direction, or there might be large differences also in-plane between the under-the-channel and -rib areas, or the effects of liquid water saturation in the diffusion media and cathode might become dominant. All these cases require the adoption of MEA models of higher dimensionality (2D) and including all physical processes.

In summary, the loss analysis approach described in this section evolved from the empirical approach by progressively incorporating dedicated measurements to improve the diagnostic accuracy. The first measurements to be included were the HFR and the Tafel analysis of the ORR kinetics. Then the effective proton resistance of the cathode obtained by EIS was included, and finally the oxygen mass-transfer resistance broken down into components by using limiting current measurements. This approach has been successfully applied to a variety of MEA designs and to characterize MEA state-of-health. To draw an analogy with medical diagnosis, this approach does not rely on the symptoms alone but requires specialized tests to produce a diagnosis, which is time consuming and could be limiting, particularly in the case of "temporary" degradation. Some deviations have been observed between predicted and measured voltage ("unassigned losses") particularly at high current density, indicating that a more sophisticated MEA modeling approach might be necessary as the technology evolves toward higher current densities.

9.5 Loss analysis using physics-based MEA models

The development of detailed physics-based PEMFC models started over 30 years ago. A break-down analysis of voltage losses was proposed already in the model developed early on by Bernardi and Verburgge [3], but interestingly VLB based on detailed physics-based models has been very rarely discussed after this case, until recently.

There is consensus within the research community on the best approaches to represent and model electrochemical and transport phenomena in PEM fuel cells [10, 27]. While the representation of some processes is well established, it has also been recognized that the macro homogeneous modeling (MHM) approach might not be adequate to accurately represent and predict the complex behavior of vapor and liquid water in porous diffusion media due to the impossibility to define a representative elementary volume for the diffusion media [32, 33]. Moreover, it is evident from experimental work that the correlation between material properties, electrode composition and structure and the effective properties required as inputs in MHM is not yet fully understood. Specifically, the ORR properties of PEM FC cathodes, which, as seen earlier, is responsible for the major losses at most operating conditions, are governed by processes occurring at the platinum/ionomer interface, which are yet fully understood [34, 35]. As a result, researchers still aspire to predict the behavior of MEAs starting from the fundamental principles, basic material properties and structural information of components, and experimental verification of new MEA concepts and designs remains necessary.

An overview of PEMFC modeling is given in Chapters 6, 7 and in dedicated publications [10, 27]. Analytical solutions are not typically available for multidimensional multiphysics fuel cell models; therefore, numerical solutions are required. Weber and Newman proposed one of the most popular multiphysics models of an MEA [36]. The model applies the MHM approach on multiple domains representing the MEA subcomponents; proton and water transport in the membrane are described using the concentrated solution theory, thermal and two-phase water behavior are included. Secanell proposed a two-dimensional multiphysics model including the spatial effects of channel and ribs of bipolar plates [37]. The common features of these steady-state models are the full coupling between the seven or so dependent variables, and requirement of many input parameters (up to 50), typically obtained from published literature or estimated using various approaches. Weber's team recently expanded their model in the along-the-channel direction, adding the capability to predict current and reactant distribution over the active area [38].

Various transient multiphysics models of MEA have also been published. Zenyuk et al. [39] developed a two-dimensional, multiphase, transient model to investigate the effect of cathode thickness on water transport focusing on low temperature. Goshtasbi et al. [40] proposed a 2D fully coupled and comprehensive transient model to study the effects of cathode loading and operating conditions, again focusing on the water transport effect. Such detailed models represent in great detail the behavior of electrodes and membranes and provide an effective tool to unravel the intricate and often non-intuitive effects brought about by MEA design changes. Despite this great capability of resolving individual MEA processes and interactions, these models offer modest accuracy in predicting performance. As it can be seen from the published validation results, the predicted performance often differs by more than 50 mV versus experiments, particularly at high current densities, limiting the applicability of such numerical models for the analysis of voltage losses.

One of the reasons for this apparent lack of accuracy in predicting performance is the already mentioned high number of input parameters required for these detailed multiphysics models and their associated uncertainty. Model sensitivity studies have been presented recently. Vetter and Schumacher [41, 42] performed a detailed study on the uncertainty of constitutive model parameters from published experimental works and on the effect of this uncertainty on performance modeling. The authors concluded that the most sensitive parameters are membrane hydration, electroosmotic drag, water diffusivity in ionomer and proton conductivity. An analogous study by Goshtasbi et al. [43, 44] using sensitivities based on voltage, HFR and water crossover rate predictions came to similar conclusions although using a different model: ionomer and cathode properties are dominant. Furthermore, the authors concluded that many of the model parameters are not identifiable from cell-level measurement. The latter conclusion is particularly important because it demonstrates the limits of nonlinear regression of polarization curves as an approach to derive model inputs and for model calibration.

A recent publication by Weber's team [45] presented an alternative sensitivity study using a 2D multiphysics model. After analyzing local sensitivities, the authors concluded that sensitivities change greatly depending on the operating conditions. While at low humidity cathode kinetics and ionomer conductivities dominate, at cold and wet conditions gas transport and water saturation in porous media become dominant.

The important learning from these sensitivities studies is that for the purpose of accurately analyzing losses it is paramount to select a MEA model that best captures the dominant electrochemical and physical processes without need of an excessive number of input parameters. The MEA researcher needs to strike a balance between the "resolution" capacity of the models, the minimum number of required experiments (in addition to performance curves), and the capacity of the approach to accurately predict performance.

In a recent publication, Gerhardt et al. [11] provided an analysis of practices and pitfalls in voltage breakdown analysis of electrochemical energy-conversion systems, focusing on physics-based numerical models. The authors described two possible approaches applicable to PEMFC models: `
- Sequential limiting-case analysis, which can be further subdivided into "standalone" and "cumulative"
- Power-loss method

The "sequential limiting-case" analysis consists in computing the polarization curve twice: once for the 'baseline" case including all mechanisms, and then for a "limiting-case" where the individual mechanisms are sequentially removed by setting the corresponding transport or kinetics coefficient to an arbitrarily high value. The overpotential attributed to the mechanism in question is given by

$$\Delta V_k(i_{cell}) = V_{cell}(i_{cell})|_{(...,q_k=\infty,...)} - V_{cell}(i_{cell})|_{(...,q_k \neq \infty,...)}, \tag{28}$$

where q_k is the coefficient associated to the process k, and $\Delta V_k(i_{cell})$ is the overpotential associated with process k and is given by the difference in predicted voltage between the "baseline" case (all q finite) and the "limiting case" (q_k = inf, all other q finite).

In the "standalone limiting-case" approach each process is considered individually, and the associated overpotential is obtained as in Equations (28). The total loss is given by the summation of all overpotentials.

In the "cumulative limiting-case," the processes are removed sequentially and the associated overpotential is obtained according to the following:

$$\Delta V_k(i_{cell}) = V_{cell}(i_{cell})|_{(q_1=\infty,q_2=\infty,...q_k=\infty,q_{k+1},...,q_n)} - V_{cell}(i_{cell})|_{(q_1=\infty,q_2=\infty,...q_{k-1}=\infty,q_k,...,q_n)} \tag{29}$$

In the second term on the left-hand side of Equation (29), parameters q_1 to q_{k-1} have been set to infinity (in fact an arbitrary large values) and the other are the same as in the "baseline" case (i. e., finite), while in the first term also parameter q_k has been set to infinity. In this case, the sum of the overpotentials should correspond to the difference between OCV, where all sources of overpotential have been removed, and the baseline case, where all overpotentials are present.

The "power-loss" approach aims to obtain a voltage breakdown from the output of a single simulation by using a post-processing approach based on calculating the power loss within a domain (i. e., volume of one of the MEA sub-subcomponents) according to the following definition:

$$\Delta V_k(i_{\text{cell}}) = \frac{\oiint_{S_k} i \cdot n\phi dS_k}{I_{\text{cell}}}, \tag{30}$$

where i is the current density vector, n is the unit vector normal to the surface S_k, and ϕ is the potential associated with process k. The total overpotential is obtained from the summation of the individual terms obtained applying Equation (30) for each process.

Figure 13 shows examples of loss analysis obtained using the above approaches to the same case, corresponding to the polarization in H_2/Air at 80 °C, 100 % RH.

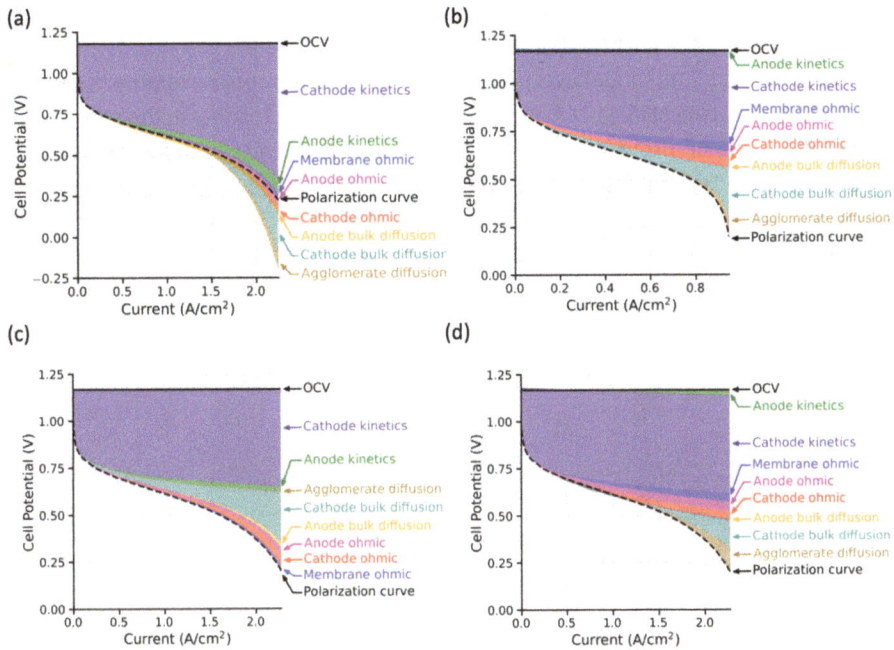

Figure 13: Comparison of (a) Standalone limiting-case, (b) Cumulative limiting-case order 1, (c) Cumulative limiting-case order 2 and (d) Power loss analysis estimated for 80 °C, 100 % RH, H_2/air. (Reproduced with permission from IOP Publishing © (2021) [11]).

Figure 13(a) shows the VLB obtained using the "sequential limiting-case" approach. It can be observed that "cathode kinetics" losses dominate at all loads, followed by "cathode bulk diffusion" related to diffusion in secondary pores of the CCL, and "cathode ohmic" related mainly to proton resistance with the CCL. Anode kinetics is also sizable, while other sources of loss are relatively smaller. The total loss (i. e., difference between OCV and the summation of all overpotentials) does not match the predicted performance (i. e., polarization curve, Figure 13) by more than 400 mV at the highest load. Gerhardt et al. attributed this effect to the "decoupling" between processes when applying the procedure of Equations (28).

Figure 13(b) shows the VLB obtained using the "cumulative limiting-case" approach for a given sequence of parameters (order 1), while Figure 13(c) shows the analogous case using a difference sequence (order 2). In both examples, the total loss matches the predicted performance, while the magnitude of the individual overpotentials changes depending on the order chosen. The authors attributed this "ambiguity" to differences in coupling between different processes.

Finally, Figure 13(d) represents the VLB obtained by applying the 'Power Loss' analysis. This approach results in matching between the sum of losses and the predicted total loss, and the magnitude of overpotentials is unique.

Figure 14 shows the comparison of the VLB results obtained using these approaches for the same case for a current density of $2\,A/cm^2$. It can be observed that both magnitude and ranking of individual overpotentials vary depending on the approach chosen. This ambiguity is concerning from a practical perspective. According to the authors the "power loss" method provides a more robust definition of overpotentials and is faster compared to the "sequential limiting-case" analysis. Nevertheless, they also argue that the latter is "predictive," meaning that it can show what would happen if a MEA property was altered, while the power loss method is "descriptive" in nature.

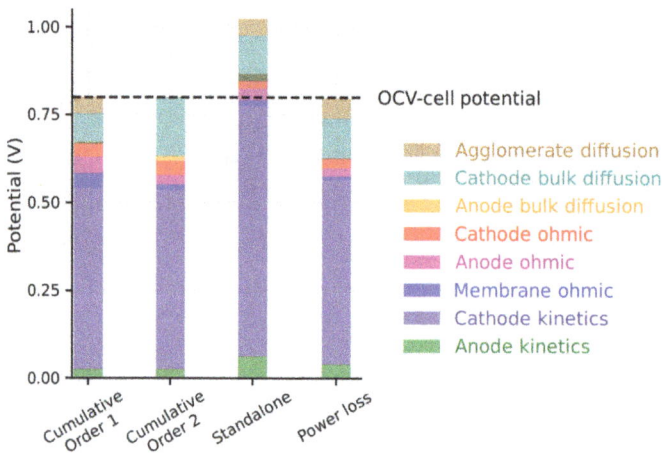

Figure 14: Comparison of different voltage loss breakdown methods in a PEMFC model at $2\,A/cm^2$. Standalone method does not match cell potential as interactions between different losses are excluded (see text for more details). (Reproduced with permission from IOP Publishing © (2021) [11].)

In summary, in this important publication the authors provided VLB methodologies applicable to physics-based numerical models highlighting risks and benefits, but did not elaborate on the possible ways of relating these VLB estimates to experimental results, nor did they report validation data, comparing, for example, predicted versus measured polarization or other measurable properties (e. g., HFR, performance sensitivities, water crossover rate).

To complete the evaluation of model-based VLB approaches, we now consider possible ways of applying these methodologies to experimental results, where to the best of our knowledge no representative study has yet been published on this aspect.

The simplest possible example of application of a model-based VLB approach consists in directly comparing model estimates to a measured polarization curve after adjusting the model inputs to match the design and composition of the MEA under investigation (e. g., PGM loading of the electrodes, ECSA of catalyst, ionomer EW and I/C, membrane EW and thickness and GDL thickness and properties). The VLB obtained for example applying the "power loss" approach can be obtained right after the simulation, and a comparison of the predicted to measured performance provides validation to the results obtained if within a preestablished tolerance.

Another example that can be considered if the previous procedure does not satisfy the desired accuracy criteria consists in gathering more experimental data, such as multiple performances and sensitivities together, if possible, with additional measurements (e. g. HFR, Cathode CVs and EIS data) and use some of these for "calibrating" the model and others for validation. The calibration of some critical model inputs should improve the predictive capacity of the model.

Finally, one could consider performing a rigorous calibration the model using specific MEA measurement (e. g., HFR, roughness factors for each electrode, CL proton resistance by EIS, H_2-O_2 polarizations and ILIMs for the diffusion properties, as done in the VLB à la Gasteiger) for the purpose of maximizing the accuracy and representativeness of the critical model inputs. After this calibration, the model can be used to predict the cell performance and VLB, and the residual between predicted and measured performance used to validate the results as usual. Noncritical model inputs can be obtained either from previous investigations or from the literature on similar MEAs.

A global regression of the model against multiple data set (including polarizations, sensitivities and other electrochemical measurements) is also conceivable. However, as discussed earlier an appropriate strategy for assigning regression weights should be identified a priori to avoid cross-contamination of the model inputs, given the heteroskedastic nature of these types of measurements.

The main strength of the model-based VLB approach described here is the increased capability in resolving sources of losses. This is a direct result of the greater complexity of the physics-based models. At the same time, the main limitation is the large number of inputs required and the apparently limited accuracy of these models in predicting performance, judging from the published literature. Another limitation is related to the ambiguity of VLB results if using the limiting case approach described earlier.

In summary, the loss analysis approach described in this section is still in an early stage of development. Only a few publications described these VLB methods and validation data is still lacking. It has been shown that the "power loss" method is more robust and less calculation intensive compared to the "limiting-case" approaches. It has been proposed that the latter is "predictive" and should be preferred for MEA engineering, although one could argue that by combining a parameter-sensitivity studies to the "power loss" method this can be made both "descriptive" and "predictive" without ambiguities in magnitude and ranking of the individual overpotentials.

A possible strategy for future improvements of the VLB procedure using physics-based numerical models might consist in using dedicated experiments to obtain more accurate inputs for the critical processes, and possibly to simplify the models based on model sensitivity analysis to capture the most impactful MEA processes.

9.6 Rapid diagnostic methods

The objectives of PEMFC durability studies are to quantify degradation and to understand what phenomena are contributing to losses; therefore, loss analysis is a fundamental tool for PEMFC durability investigations. Several accelerated stress tests (AST) have been proposed [42] targeting specific degradation modes. For practical reasons, it is highly desirable to obtain the maximum possible information from the minimum number of measurements, and while ASTs are designed to amplify only one degradation factor, there are often interactions between processes leading to the need of additional diagnostics. As discussed earlier, accurate loss analysis requires dedicated tests that are time intensive, thus it is generally not practical to add many VLB diagnostics checks during an AST other than at the beginning (BOL) and at the end (EOL).

Fortunately, there is a diagnostic technique for PEMFC based on the comparisons of polarization curves obtained at different points in the MEA lifetime called "polarization change" analysis [13, 46] that can resolve what type of overpotential (kinetics, ohmic or transport) is dominant during degradation.

To understand how this technique works, it is useful to consider again the simplified analytical MEA model described in Section 9.4. If we exclude anode overpotential and assume that the effective properties of a MEA changed between times t_A and t_B, we can derive the following changes in overpotential:

$$\Delta\eta_{H_2-xover} = \frac{\text{Ln}(10) \cdot R \cdot T}{a_C \cdot F} \cdot \text{Log}\left(1 + \frac{\Delta i_{xover}}{i_{cell} + i_{xover,init}}\right) \tag{31}$$

$$\Delta\eta_{ORR} = -\frac{\text{Ln}(10) \cdot R \cdot T}{a_C \cdot F} \cdot \text{Log}\left(1 - \frac{\Delta(\text{rf}_{CCL} \cdot i_{o,ORR})}{(\text{rf}_{CCL} \cdot i_{o,ORR})_{init}}\right) \tag{32}$$

$$\Delta\eta_{\Omega,tot} = \Delta R_{\Omega,tot} \cdot i_{cell} \tag{33}$$

$$\Delta\eta_{MT} = \frac{\text{Ln}(10) \cdot R \cdot T}{F} \cdot \left(\frac{1}{4} + \frac{\gamma_{O_2}}{a_C}\right) \cdot \text{Log}\left(\frac{1 - \theta_A \cdot i_{cell}}{1 - \theta_B \cdot i_{cell}}\right) \tag{34}$$

$$\theta_i = \frac{R \cdot T}{4 \cdot F \cdot pO_{2Ref}} \cdot R_{MT-O_2,i} \tag{35}$$

The changes in overpotential with current are plotted individually in Figure 15(a). The change kinetic overpotential is independent of the current, the ohmic the change in-

(a)

(b)

(c)

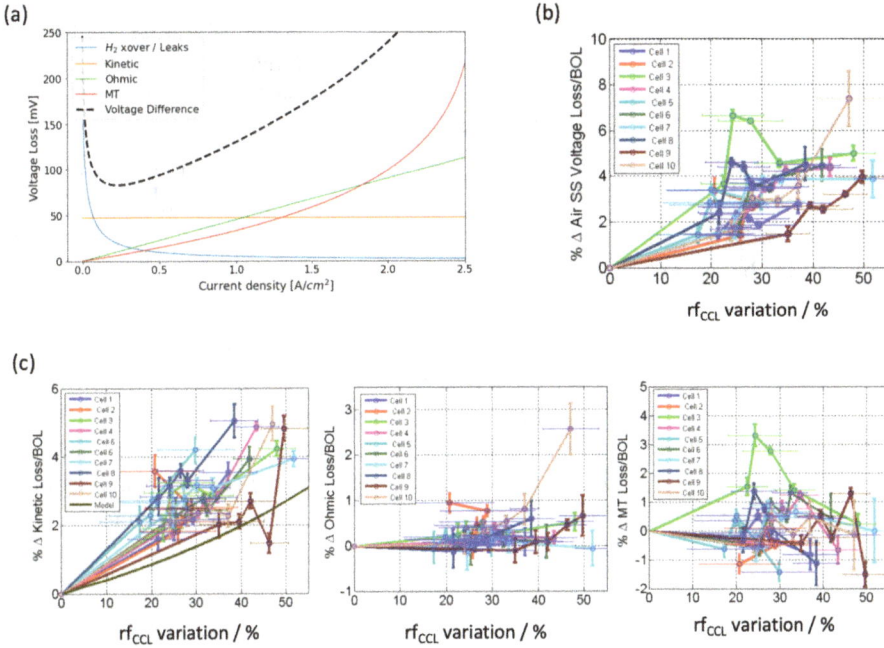

Figure 15: (a) Simulated "voltage change" vs. current density; Example of voltage degradation (b) and the related degradation fingerprints (c) as function of the variation in cathode roughness factor (rf$_{CCL}$) obtained in 10-cell stack by Stumper et al. [47]. Reproduced with permission from IOP Publishing © (2009).

creases linearly with the current, while for the oxygen transport the change increases in exponential manner. Finally, an increase in leaks should result in an apparent increase on kinetics overpotential that disappears with increasing the current. When kinetics, ohmic and oxygen transport properties change simultaneously the resulting potential difference as function of the current exhibits a shape that depends on the relative magnitudes of these MEA losses. In Figure 15(b) kinetic, ohmic and mass transport losses are plotted against the relative change in cathode roughness factor during lifetime testing, constituting a "degradation fingerprint" [47, 48]. As a result, the characteristic shape of the potential difference vs time or current density can be used as a "fingerprint" of degradation and used as a pointer toward the dominant degradation process.

The "polarization change" methodology is routinely used in the fuel cell industry. The approach is effective in pointing out if a dominant degradation process exists, in which case the voltage change also provides a direct quantification of the increase for the dominant overpotential. In the case of simultaneous kinetics, ohmic and transport degradation, the deconvolution of losses is not straightforward, and this approach can lead to ambiguities unless complemented by other measurements.

Another useful technique based on the comparison of polarization curves is called "oxygen gain analysis"; this can be used to identify the dominant source of oxygen transport resistance in a cathode at a fixed point in time. The oxygen gain is the voltage difference when operating a fuel cell on an oxygen-depleted and an oxygen-rich stream. If the performance measurements are done at a constant level of membrane hydration, the voltage gain is resulting only from changes in kinetics and mass transport overpotentials. O'Neil et al. [12] developed an analytical cathode model using the Thiele modulus approach to account for the effects of oxygen diffusion within the platinum agglomerates. Their analysis demonstrated that a decisively different oxygen gain behavior is observed in the asymptotic cases of internal (i. e., agglomerate) versus external (i. e., GDL) diffusion limitation. In the case of internal transport limitations, the oxygen gain is bound within a finite range, while in the case of external transport limitation, a change in limiting current is observed and the oxygen gain is not bonded. By plotting the oxygen gain versus the air potential in dimensionless form, a characteristic curve should be obtained, which can be compared to the asymptotic fingerprints of Figure 16. Although not quantitative, this methodology can be very useful for the comparison of cathode designs, particularly in the design of MEAs for high current density performance.

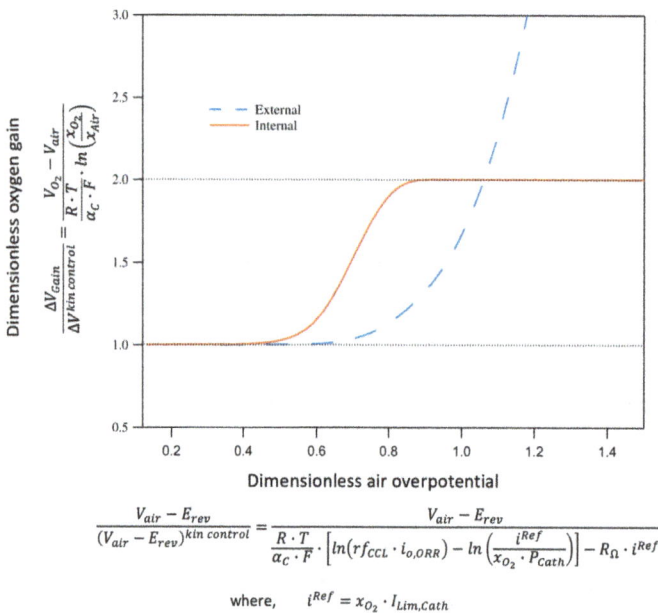

$$\frac{V_{air} - E_{rev}}{(V_{air} - E_{rev})^{kin\ control}} = \frac{V_{air} - E_{rev}}{\frac{R \cdot T}{\alpha_c \cdot F} \cdot \left[ln\left(rf_{CCL} \cdot i_{o,ORR} \right) - ln\left(\frac{i^{Ref}}{x_{O_2} \cdot P_{Cath}} \right) \right] - R_\Omega \cdot i^{Ref}}$$

where, $i^{Ref} = x_{O_2} \cdot I_{Lim,Cath}$

Figure 16: Dimensionless oxygen gain vs. air potential as defined by O'Neil in [12]. Reproduced with permission from Elsevier © (2012).

The basic concepts of the "polarization change" and "oxygen gain" analysis were adopted by Pant et al. [49] who used a detailed 2D numerical model of a cell in place of simplified analytical models. The authors found that the shape of the "polarization change" curve and its sensitivity to oxygen concentration are characteristics for each degradation pathway. The uniqueness of the degradation fingerprint, which is easily measurable, combined with the higher capacity of physics-based models to resolve for loss sources provides a powerful tool not only to determine the primary degradation mechanism, but also to untangle the interactions between the changes in MEA properties and their effects on performance. The model used by the authors accounted for reactant diffusion, liquid and gas convection, electronic conduction, water and proton transport according to the concentrated solution theory and energy transport; for the ORR in the cathode, a Pt-oxide dependent Tafel model was used combined with a new agglomerate model based on fractional reaction order. Besides simulating a variety of scenarios and obtaining a database of "fingerprints" (i. e., "voltage loss" vs. current curves), the authors also validated the approach using experimental data obtained from a differential cell using a Pt dissolution AST protocol (see Figure 17).

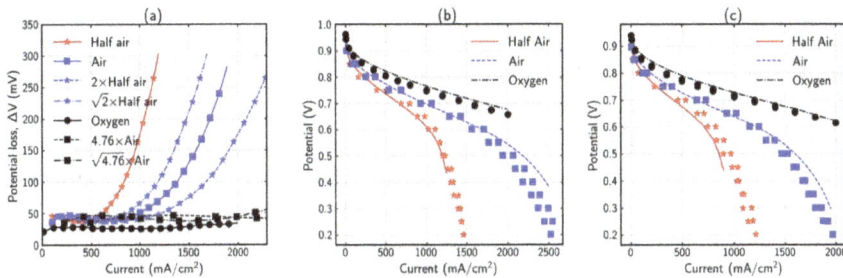

Figure 17: Analysis and validation of cell polarization curves using the "polarization change" approach: (a) polarization change at different O_2 concentrations; (b) BOL polarization curves; (c) EOT polarization curves after Pt-dissolution AST. Reproduced with permission from IOP Publishing © (2019) [49].

The results of the example shown in Figure 17 confirmed that degradation was predominantly kinetic and ohmic, as expected for the type of AST applied. The simulation procedure required a calibration of the input parameter for both BOL and EOL data using a trial-and-error approach; it can be observed that while the model captures the experimental trends there is still a relatively large deviation between predicted and measured performances. Despite the relatively onerous simulation effort required, this approach might provide unique insight on MEA degradation phenomena in all the cases that are not already obvious from the standard analysis of AST results.

The main strength of the rapid diagnostic methods described in this section is their simplicity, rapidity and cost effectiveness. The disadvantage is that they might be ineffective in resolving the source of loss or the simulation steps might become cumbersome, depending on the complexity of the model chosen.

In summary, the diagnostic approach based on comparing polarization data at different points in the lifetime of a cell and/or at different oxygen concentrations offers a valuable tool to unveil dominant degradation modes, which is indispensable within the toolbox of MEA researchers and designers.

9.7 Voltage loss analysis based on EIS

Electrochemical Impedance Spectroscopy (EIS) is a powerful technique used in many application fields. It consists in applying a small sinusoidal perturbation in current or voltage and measuring the resulting changes in magnitude and phase for the other electrical signal over a wide range of frequencies. Linear EIS data can be represented as complex values containing information of all MEA processes within the range of time-constants probed. The response of a specific process is in principle separable from the others based on differences in time constants. The extraction of process-related parameters from EIS data is typically done by nonlinear complex fitting using an impedance model representing the system under study. Details on EIS and its application to PEM fuel cells are provided elsewhere [50–52].

A common way to analyze PEM fuel cell impedance is by using the "equivalent circuit approach," which consists in using a system of ideal electrical circuit elements to model the MEA impedance. A very popular model used for PEM fuel cells consists of two Randles elements, a resistor and an inductor combined in series. Often constant phase elements (CPE) replace the ideal capacitors to account for the capacitive dispersion phenomena, while other lumped parameters such as Warburg impedance might be added to represent diffusion in the GDLs. In most cases, these equivalent circuits are empirical in nature and often contain two or more identical elements (e. g., resistors and capacitor for each electrode) that cannot be unequivocally assigned to a specific MEA process when EIS is measured at a unique condition. Therefore, a strategy to link model parameters to MEA physical processes is necessary, which requires performing a sensitivity analysis, whereby gas compositions and RH are varied individually, and EIS is measured. Next, the trends of individual fitting parameters are monitored versus the controlled variable, and finally the parameters are assigned to MEA components and to processes based on the agreement of trends versus the expected behavior. Examples of this procedure have been discussed in various publications [52, 53], and in several cases a loss analysis based on overpotentials calculated from the fitted parameters has also been described.

The limitations of the equivalent circuit approach and of VLB are analogous to those already discussed earlier for the empirical fitting of polarization curves (see Section 9.3): the models are often too simplistic to fully describe the complex MEA behavior (e. g., coupling of proton and oxygen transport in porous electrode), and a weighted regression should be used account for the heteroscedasticity of the measurements. In addition, there is debate about the validity of some of the lumped parameters commonly used. For example, while using CPEs in the equivalent circuit model can give excellent fitting, it has been argued that the dispersive behavior of electrodes is true only for limited frequency ranges; therefore, the CPE might give a physically inconsistent representation of the real electrode interface [54]. Furthermore, the Randles circuit does not represent porous electrodes therefore it can be argued that CPEs function in part as fudge-factors to improve the apparent fitting accuracy. Finally, there is often ambiguity on which equivalent circuit is more representative for a given data set, and selection criteria are often based on goodness of fit rather than on physicochemical consistency of the model with the specific cell design and conditions [20].

An alternative analysis approach to interpret EIS data is the Distribution of Relaxation Times (DRT) technique (see Figure 18). As indicated by the name, this approach provides a distribution of the dominant time-constants from the mathematical manipulation of EIS data. DRT is a well-established approach in the characterization of solid oxide fuel cells [55, 56], and more recently, it has been adopted and adapted also for PEMFC [56]. Once the DRT is available, one can associate the peaks to the time-constant of a selected MEA process and deduce which ones might be dominant for the case under test. The analysis of the shifts in the DRT peaks measured at different currents and for different electrode designs (e. g., Pt loading) can also be used to quantify the resistances associated to the MEA processes and from this obtain important estimates for properties otherwise inaccessible, such as, for example, in a study by Reshetenko and Kulikovsky, who proposed this approach to quantify the magnitude the ionomer film diffusivity over Pt [57]. Furthermore, an appropriate impedance model can be selected based on DRT results, which can then be used for data fitting.

EIS data are also commonly analyzed by nonlinear complex fitting of physics-based models. A compendium of analytical solutions of the impedance of PEMFC cathodes and cells under various limiting cases was published by Kulikovky [26]. In addition, more complex physics-based numerical models can also be used for the deconvolution of impedance data. Reshetenko and Kulikovsky extracted kinetic and transport parameters for PEM fuel cell cathodes (see Figure 19) by fitting a numerical impedance model based on the transient macro homogeneous equations for a PEM fuel cell cathode [58].

Jaouen and Lindbergh provided another excellent example for the use of transient techniques to investigate PEM fuel cell electrodes [59, 60]. In this, the authors combined the current-interrupt technique, EIS and steady-state polarization measurements to obtain kinetic, proton transport and oxygen transport properties of the cathode using a numerical impedance model including the agglomerate properties of the cathode. The parameters obtained using this procedure were consistent with those obtained from other techniques. In general, when using detailed physics-based models the fitting procedure is not straightforward and requires detailed understanding of the model and of the physics included.

As shown in the previous examples, EIS can be used to directly obtain all relevant electrode properties; however, this technique is more commonly used to measure individual MEA properties such as the total ohmic resistance (HFR) and the effective proton resistance of the cathode, sometimes named "sheet resistance" of a porous electrode. EIS and HFR measurements are fundamental for the VLB approach described in Section 9.4 and will be discussed more in detail in the following sections.

Several studies have been published on the use of EIS to monitor individual MEA properties, particularly in relation to cathode degradation. A notable example of this

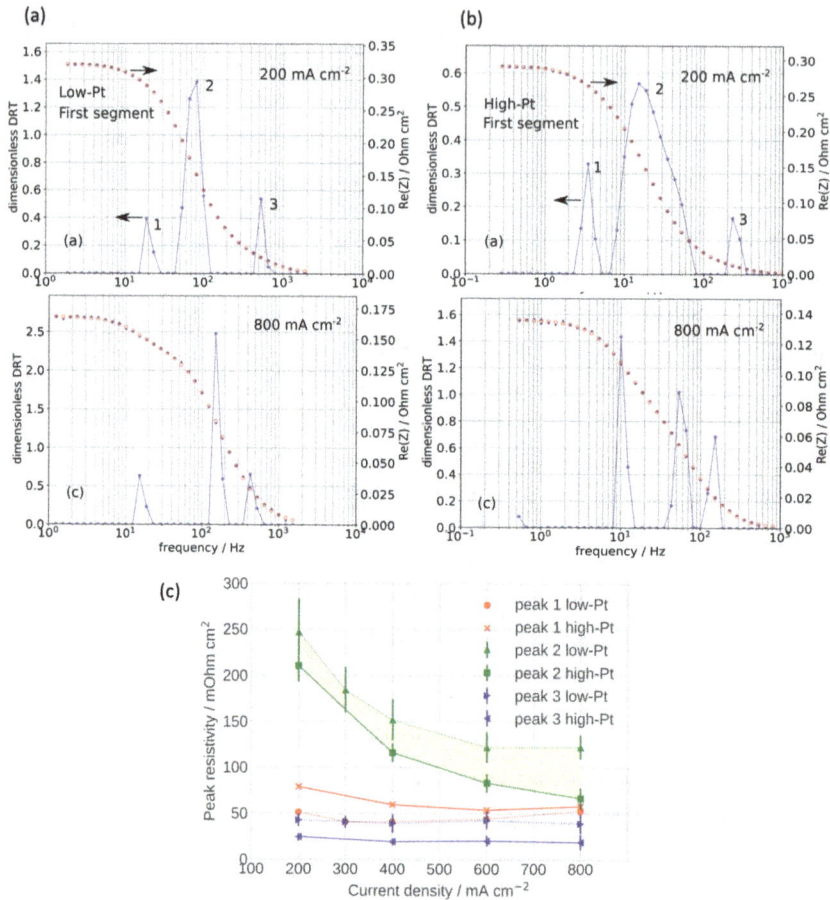

Figure 18: DRT of an MEA with (a) low Pt-loaded and (b) high Pt-loaded cathode at different current densities; (c) Resistivity corresponding to the three frequency peaks identified by DRT analysis as function of the current density. Reproduced with permission from Elsevier © (2021) [57].

approach is the work published by Young et al. [61] where EIS measurements were combined to SEM and CO_2 release analysis during a carbon corrosion AST protocol to identify correlations between structural changes and effective transport properties of the cathode. The same authors published another study [62] focusing on the effect of ionomer degradation on the MEA properties and included a loss analysis based on estimates of the cathode utilization obtained from the combined measurements. HFR results indicated a clear increase in membrane resistance during the OCV AST, while the effective layer proton resistance showed initially a decrease and later an increase versus the cumulative fluoride release, which was attributed to the combined effects of ionomer degradation and other structural changes of the cathode. Other more recent studies followed an analogous approach to monitor changes of cathode properties during accelerated degradation tests as mentioned previously [28].

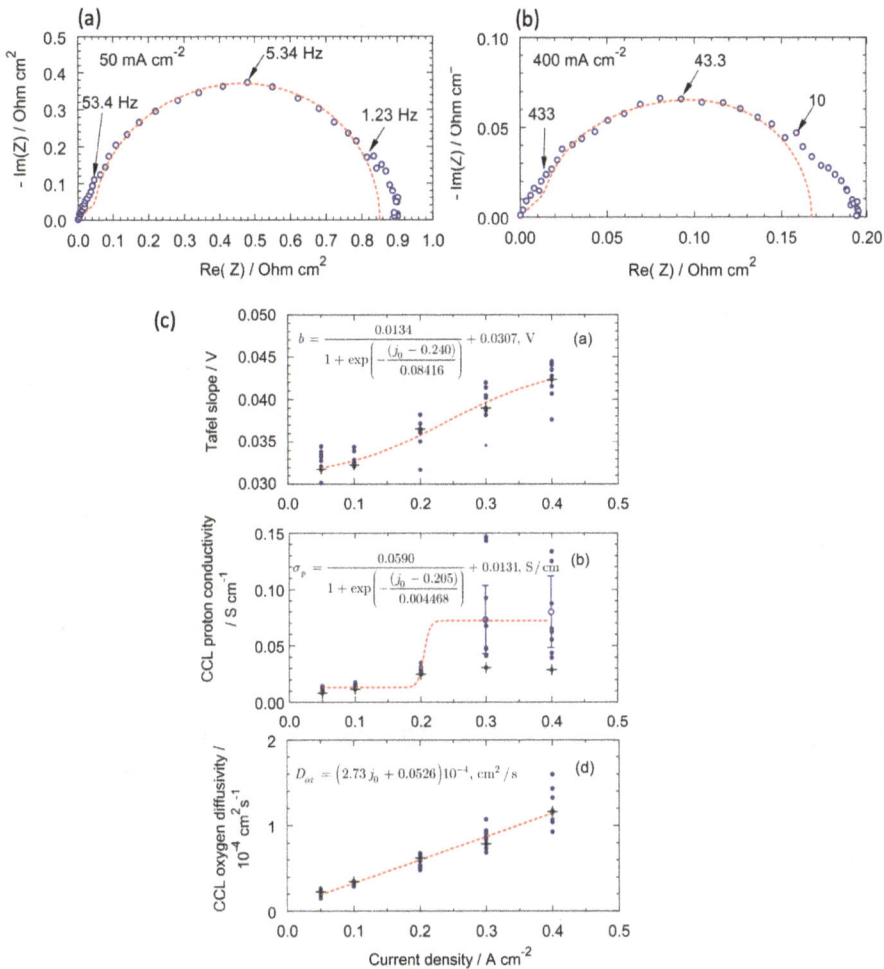

Figure 19: (a) Example of experimental EIS measured for a single PEMFC at different current densities. The arrows indicate specific frequencies, while the dashed lines are the results from model fitting. (b) Trends for the EIS model parameters of the MEA after fitting to the experiment data in the frequency range between the rightmost and leftmost arrow. Reproduced with permission from IOP Publishing © (2016) [58].

As for every other approach, voltage loss analysis based on EIS has strengths and limitations. The opportunity of resolving/separating MEA processes with different time-constants from an individual spectrum makes this technique very attractive. The fact that EIS can be measured at normal operating conditions should lead upon data analysis to properties that are more representative of realistic operating modes if compared to those obtained from limiting cases (e. g., limiting current).

On the other hand, obtaining high quality EIS data is challenging for PEM fuel cells. High- frequency measurements can be affected by various experimental artifacts [63], while high-accuracy, low-frequency measurements, which are needed to derive oxygen transport properties, can take a significant amount of time (>20 min) during which there might be changes in the MEA behavior, particularly for degraded or poisoned electrodes [64].

The interpretation strategy of EIS data is also challenging. Complex nonlinear regression implies the selection of an appropriate impedance model and knowledge of the error structure of the measurement for the purpose of weighted regression [51, 65].

Finally, the same EIS trends can be caused by different processes or combinations of processes [66], which could lead to ambiguity in interpretation. Subsequently, it is often advisable to complement EIS with results from other measurement techniques to confirm the validity of the hypothesized model.

In summary, EIS is an indispensable tool for voltage loss analysis, both combined with other techniques for measuring effective MEA properties and also as a standalone approach. Various strategies can be adopted for the interpretation of EIS results. For the purpose of loss analysis, we believe that the use of empirical equivalent circuits has limited values, and we do not recommend this approach. On the other hand, it has been demonstrated that when using physics-based impedance models for data regression it is possible to obtain meaningful results that are representative of the actual conditions of the MEA.

9.8 Measuring MEA performance and effective properties

In this section, experimental aspects related to loss analysis of PEM fuel cells are discussed. First, an overview of key properties of the MEA components is presented. Next, the important topic of cell design and operating mode for MEA testing in single cells is discussed in detail. Then the techniques for measuring cell performance and key MEA effective properties for loss analysis are presented. Finally, additional MEA properties, often less accessible but still relevant for VLB at high current density, are briefly reviewed.

9.8.1 Basic properties of MEA components

The properties and functionalities of MEA components have been discussed extensively in the literature and also within this book. Kusoglu and Weber presented a detailed review of the properties of PFSA ionomers and membranes [35]. Gostick et al. provided an overview of GDL properties and characterization methods [67]. Sun et al. reviewed the current state of development for catalyst materials, cathode layer design and characterization techniques [68]. Another recent review on cathode design for automotive applications was published by Wang et al. [69].

Finally, a comprehensive discussion of functions, properties and measurement methods for PEMFC components focusing on quality aspects was recently published by a German-Canadian consortium [70].

MEA components consist of complex structures composed of different materials; these composites present a heterogeneous structure at the microscale (see Chapter 2). While it is desirable to investigate structures and processes at the microscale for fundamental understanding purposes, to model an MEA it is often sufficient to account only for the essential features of each component, without going down to the exact geometric detail; this is possible by adopting the macro-homogeneous approximation mentioned earlier [71]. From this macroscopic perspective, the volume-averaged value of a property of a component is defined as *"effective property."* It is often possible to define a correlation between the fundamental properties of the materials and the structure of the composite (see p. 245 in Eikerling and Kulikovsky [27]). An example is the effective electrical conductivity of the cathode electrode; from a macroscopic view, the effective conductivity of the layer is equivalent to the conductivity of carbon (material property) times the volume fraction of carbon within the layer, divided by a tortuosity factor (empirical factor) to account for the path increase between connected carbon particles. Other effective properties can be defined in an analogous way.

When modeling an electrochemical system using the macro-homogeneous approach, there are always at least two layers within the model: the first, not always explicit, is a set of models and functions to translate the complex structure of each MEA component in terms of simple effective properties; the second is given by the governing equations and boundary conditions applied to the various domains/components constituting the macro-homogenous model.

While there is good agreement on the basic governing equations, there are still differences in the approach to model effective properties. Some effective properties can be measured using *ex situ* or *in situ* methods, others need to be estimated from compositional and structural information. Finally, a few others, particularly those related to the catalyst/ionomer interface, are still poorly understood and not yet accessible.

A list of the basic MEA properties required for voltage loss analysis is given in Table A in Appendix.

In general, these properties contain information about the size, design and composition and transport properties of each component and are related to specific sources of loss.

9.8.2 Cell designs, operating modes and measurement errors in MEA testing

MEA testing is typically done using a *single cell* set-up. This is composed by a set of monopolar flow field plates and the compression hardware, interfacing the cell to the test station. The flow field plates consist of an arrangement of channels and ribs designed to provide an adequate amount of reactants to the entire electrode surface. Other important functionalities of the plates are transport of electrical current and heat, and removal of product water from the MEA. The most popular channel configurations for MEA testing are single-channel serpentine, multiple-channel serpentine and straight parallel flow. The interdigitated arrangement of channels is also possible but less common.

An example of single-channel serpentine flow field design is shown in Figure 20(a). The flow path is continuous from start to end providing an efficient distribution of reactants over the entire electrode surface. However, the pressure drop along the channel can be large due to the path length, resulting in a nonuniform distribution of pressure and concentrations over the active area, which, in turn, results in differences in current density distribution. While single-channel serpentine is likely the most common flow configuration design for subscale MEA testing, it is also the least appropriate for voltage loss analysis, particularly at high current densities because of the high likelihood of uneven distributions.

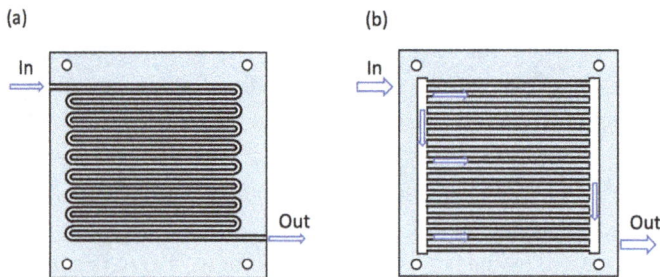

Figure 20: Types of plate designs commonly used in subscale PEM fuel cell testing: (a) serpentine and (b) parallel flow field.

The parallel flow field design is shown in Figure 20(b). In this design, the reactants are distributed among multiple channels, which are typically short resulting in a reduced pressure drop and more uniform distributions of reactants and current. This flow configuration is preferable for VLB. The disadvantage of the parallel flow configuration is that an obstruction in any of the channels results in flow redistribution among the remaining channels, often referred as "in cell flow sharing," which leads to uneven distribution of reactants and load. A sufficiently high gas velocity is required for the removal of liquid droplets accumulating in each channel. As a result, the required gas pressure gradient is larger, and a robust design of the transition regions between the gas manifold and the channels is also needed to minimize flow sharing effects.

PEMFC performance is usually referenced at the inlet conditions for reactant concentrations (xH_2, xO_2), pressures (P_{an}, P_{cath}), relative humidity (RH_{an}, RH_{cath}) and temperature (T). These should be considered as "nominal operating conditions." In fact, significant gradients are normally present in real cells. Considering a single cell with straight flow field design, we can intuitively identify three main directions for these gradients:

- along-the-channel direction (x) for coolant/plate temperature, gas pressure, molar composition and flow of reactants and products (including both vapor and liquid phase)
- channel/rib direction (y) mainly for T, xH_2, xO_2, RH and liquid saturation
- Through-plane direction (z) for the same variables as above (already encountered in Sections 9.4 and 9.5).

For the purpose of performance comparison, it is acceptable to test MEAs using a cell where gradients in all three directions are present. In fact, the Joint Research Center of the EU suggested using a single cell with counter-flow configuration operated at low fuel and oxidant stoichiometries for testing MEAs [72]; this configuration necessarily leads to large gradients along-the-channel and, subsequently, nonuniform current distribution over the active area. On the other hand, for the purpose of voltage loss analysis it is preferable to reduce or possibly eliminate along-the-channel gradients by using the "differential" approach. The concept (and terminology) of "differential cell" is likely derived from the theory of chemical reactor engineering [73] and was first adopted by Benziger et al. [74]. In general, a "differential" fuel cell should provide a uniform concentration of reactants, temperature and pressure, and subsequently, uniform current density over the entire active area. This operating mode is achievable by reducing the active area and simultaneously increasing the flow rate of hydrogen and oxygen at any given load (i. e., running at high stoichiometries). The flow configuration is also important: parallel flow with short channels, resulting in low pressure drop, is preferable. A common approach to achieve differential conditions in a regular single cell consists in masking a portion of the active area. Alternatively, dedicated cell designs for high stoichiometry testing have been proposed and commercialized (see Figures 21, 22).

A recent initiative funded by the European Community, under the Fuel Cells and Hydrogen second Joint Undertaking (FCH2JU) framework, developed a reference hardware for harmonized testing of PEM single cell fuel cells named JRC ZEROVCELL [75, 76]. The design documentation of the hardware is publicly offered under CERN Open-hardware license.

A similar initiative to develop a standardized differential cell to be used by the research community was undertaken in Japan by FC-Cubic under a NEDO project. The new *FC-Cubic small fuel cell* has an active area of 1 cm^2 and is operated at stoichiometries up to 20 (<5 % reactant utilization). Further information on the cell design is available [77].

To provide a quantitative definition of *differential cell* and *differential operating mode*, the following questions must be answered:

1. How much variability is tolerable?
2. What key operating parameters and MEA variables should be considered?

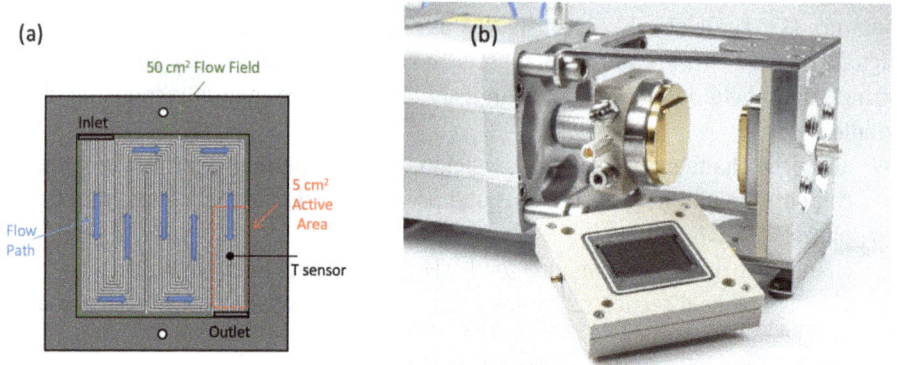

Figure 21: Examples of differential cell: (a) Masked concept used by GM (redrawn from ref. [21]); (b) High load differential cell developed by Baltic FuelCell in collaboration with the Fraunhofer Institute of Solar Energy Systems (reproduced with permission from copyright owner Baltic Fuel Cells GmbH [78]).

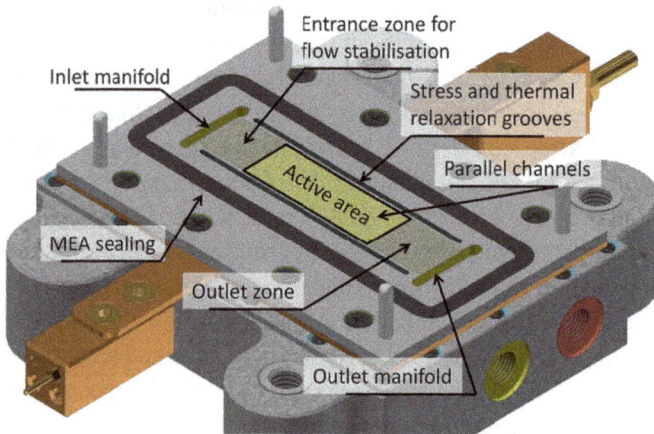

Figure 22: Examples of differential cell: Features for the JRC ZEROVCELL single cell testing hardware (reproduced with permission under Creative Commons License CC BY 4.0 [79]).

3. What are the minimum stoichiometries and maximum pressure drops allowed?
4. What cell design—including flow-field profile, active area size and aspect ratio—should be used to achieve the targeted tolerances?

The answer to the first question is arbitrary and depends on what one is trying to measure, while others are somehow interrelated. As mentioned earlier, the important channel variables in the x-direction are T, P_{an} and P_{cath}, xH_2 and xO_2, RH_{an} and RH_{cath}, for anode and cathode, respectively. The minimum set of MEA (dependent) variables that should be considered are the local current density, $i(x)$, and local ohmic resistance, HFR(x), which are measurable/verifiable by using a segmented cell. Other important

MEA variables are the local membrane hydration, $\lambda(x)$ and water crossover rate $\beta(x)$, which are not yet measurable.

Once a tolerance for one or more MEA variables has been specified, an experimental approach can be adopted to measure the change in variability as a function of the operating conditions. This approach requires accurate mapping capabilities that are not often available and might not be applicable to a variety of plate concepts. Alternatively, a validated model can be used to predict the same variabilities. For example, if considering a straight flow field plate design, a MEA model extended along-the-channel should be sufficient to predict the effects of different inlet conditions on the local MEA variables. Here, the pseudo 2D model developed and validated by Guo et al. [21] was used to perform a sensitivity analysis of a hypothetical differential cell with parallel flow field design considering both co- and counter-flow configurations. This model uses the same MEA sandwich model as described in Section 9.4 and calculates changes in variables down-the-channel using stoichiometries, total pressure drops and coolant temperature increase as additional model-inputs. An example of the model predictions is given in Figure 23, which shows the distributions along-the-channel of a hypothetical cell operate at 1.5, and 3.0 A/cm^2 according to the EU Harmonized Test Protocols [72]. The coolant temperature change for the coolant is assumed to be less than 1 °C, while the pressure drop is assumed to be 7.5 and 15 kPa for the anode, and 15 and 30 kPa for the cathode at 1.5, and 3.0 A/cm^2, respectively. The MEA properties used in this simulation were the same as in Figure 7.

According to these simulations, the local current density deviation versus the average is up to 0.5 and 0.7 A/cm^2 for the two set point values. The local HFR is about 35 % higher than average at the inlet at 1.5 A/cm^2, while it is more uniform at higher load because of increased water production. The temperature profiles at the electrodes vary by about 1 °C, analogously to the channel variation, with higher temperature at higher loads as expected. The mole fractions of O_2 and vapor in the cathode show little variability with the load, while for the H_2 and vapor in the anode a peak toward the center is observed, as expected for the counter-flow configuration. The channel relative humidity shows similar trends at both loads, reaching saturation with the central portion of the cell for the anode channels, and saturation in the second half of the cell for the cathode channels. The membrane hydration (λ) shows a peak around the center of the cell, with the lower point at the cathode inlet for both loads. The average lambda is lower for the higher load because of the higher temperature gradient in the through plane direction. In general, these results demonstrate that along-the-channel gradients are not negligible when operating a single cell at relatively low and constant stoichiometries, irrespective of the active area. A comparison of the average voltage loss breakdown for the example of Figure 23 and for a hypothetical differential cell operated at the same inlet conditions is shown in Figure 24. The counter-flow low stoich case shows higher performance due to the higher average hydration of membrane and electrodes. According to the VLB results, the low-stoich case has slightly higher ORR loss compared to the differential case (10 mV), and lower membrane resistance (20 and 40 mV at 1.5 and 3.0 A/cm^2,

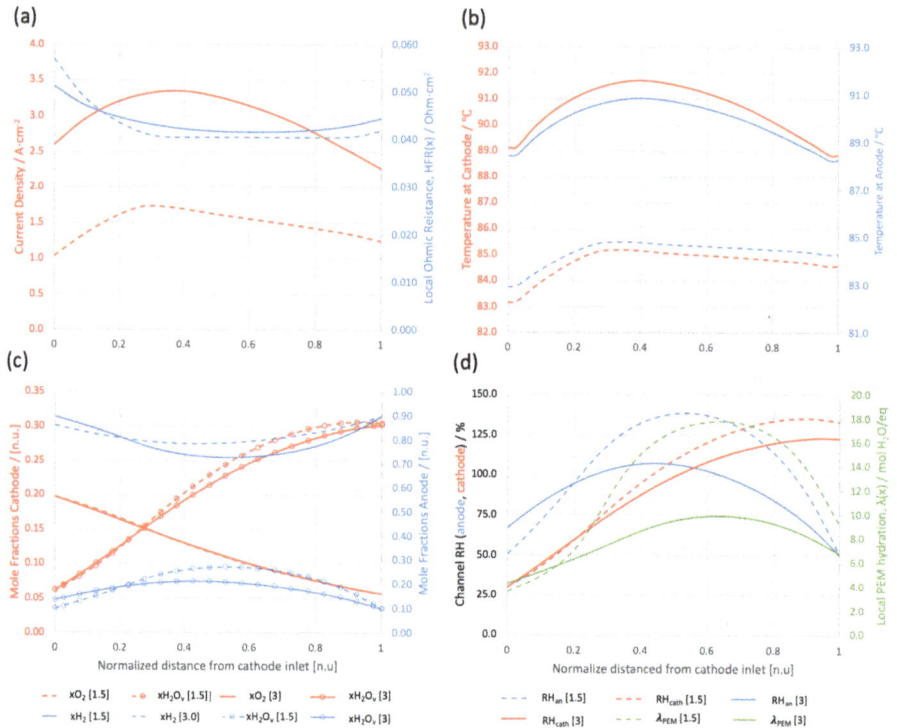

Figure 23: MEA variables calculated along-the-channel for single cell operating in counter-flow at stoichiometries of 1.3 and 1.5 for anode and cathode, respectively, and at inlet operating conditions as in Table 1 in [70]; Distributions of (a) Current density and HFR, (b) Temperature, (c) Reactants and (d) RH and local membrane hydration. Dotted and solid lines correspond to 1.5 and 3.0 A/cm^2 average current density, respectively.

respectively). The loss associated with proton transport in the cathode is also higher in the differential case, while oxygen transport losses are lower if compared to the low stoich case. In general, this example illustrates the necessity to achieve *true* differential conditions for the loss breakdown to be properly linked to the referenced operating conditions. As we will discuss in Section 9.10, a local VLB should be performed whenever the variation of channel conditions is not small, as in the low stoich example discussed above. However, high stoichiometries require high volumetric flow or reactants, which subsequently lead to increased pressure drops.

A sensitivity study was performed using the same model as above to quantify the variations in local conditions as a function of stoichiometry and pressure drop for both co- and counter-flow configurations. The reference inlet conditions for the simulations were 80 °C, H$_2$/Air at balance pressure of 2.5 bar and average current of 3 A/cm^2. Two cases were considered for the humidification: 100 % and 50 % RH, balanced between anode and cathode.

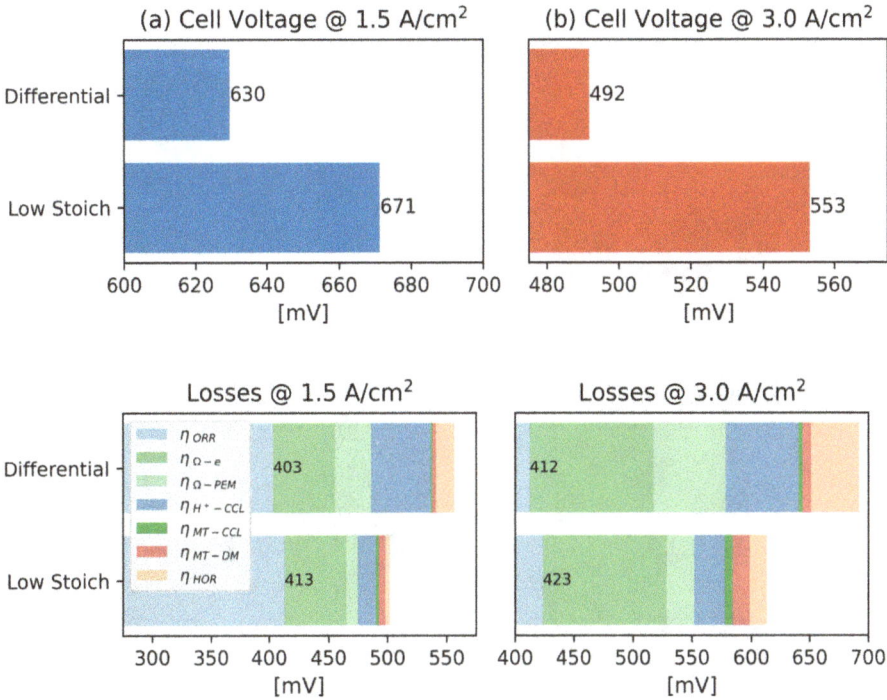

Figure 24: Simulated VLB for the same MEA operated as in Figure 23 (i. e., counterflow with stoichiometries of 1.3 and 1.5 for anode and cathode, respectively) in differential mode at the same inlet conditions at an average current density of (a) 1.5 A/cm^2 and (b) 3.0 A/cm^2.

The variation in the outlet O_2 concentration versus the stoichiometry is shown in Figure 25(a). The difference between inlet and outlet is independent of operating conditions since this is simply related to the average current density for each stoichiometry value. A stoichiometry higher than 8 is required to prevent the concentration from dropping more than 2 %. Figure 25(b) and (c) show examples of the predicted current distribution for three different stoichiometries at fixed pressure drop and humidification. In general, the current density decreases from inlet to outlet for both flow configurations when the inlet gas streams are well humidified. This variation is less steep as the stoichiometry is increased because the drop in O_2 concentration is smaller, while the slope increases as the pressure drop is increased. In the case of lower inlet humidification (Figure 25(c)), the co-flow configuration shows an almost linear increasing trend in current density, due to the increasing humification along the channel, while the profile for the counter-flow configuration is parabolic, showing a clear benefit in limiting the variability.

Figures 26 and 27 summarize the results of the sensitivity study. These show the trends in cell voltage, current density, HFR and membrane hydration variation as a function of stoichiometry. Again, co- and counter-flow configurations are considered for the

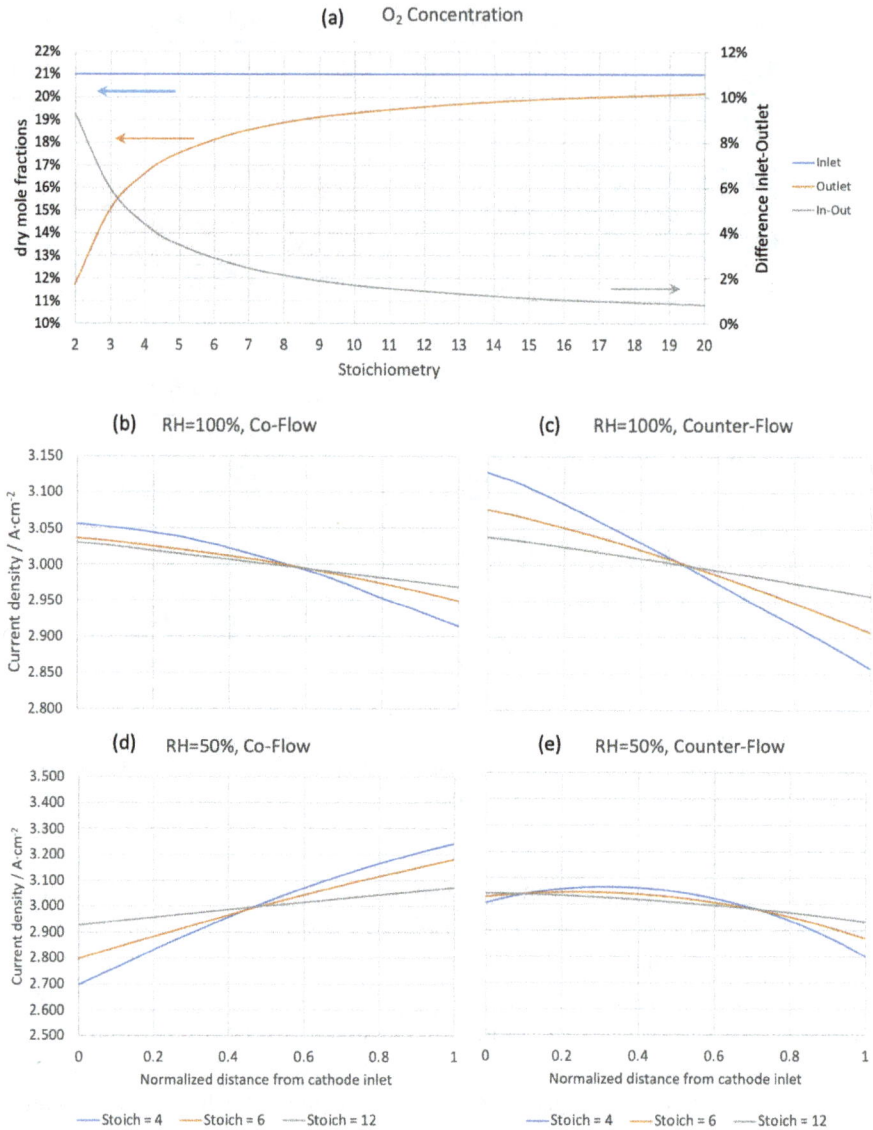

Figure 25: (a) Variation in outlet O_2 concentration versus the stoichiometry; Comparison of simulated current density distribution for co- and counter-flow at three different stoichiometries for inlet gases at (a) 100 % RH and (b) 50 % RH. The pressure drop is 50 mbar for both cases.

fully humidified and dry cases. The "variations" in the plots correspond to the range (min and max) for the along-the-channel distributions of each parameter; the different gas pressure drops (inlet–outlet) used for the simulations are indicated as values in mbar. The dash-dot lines added to the plots represent the values for the *ideal* differential cell (case of zero along-the-channel gradients).

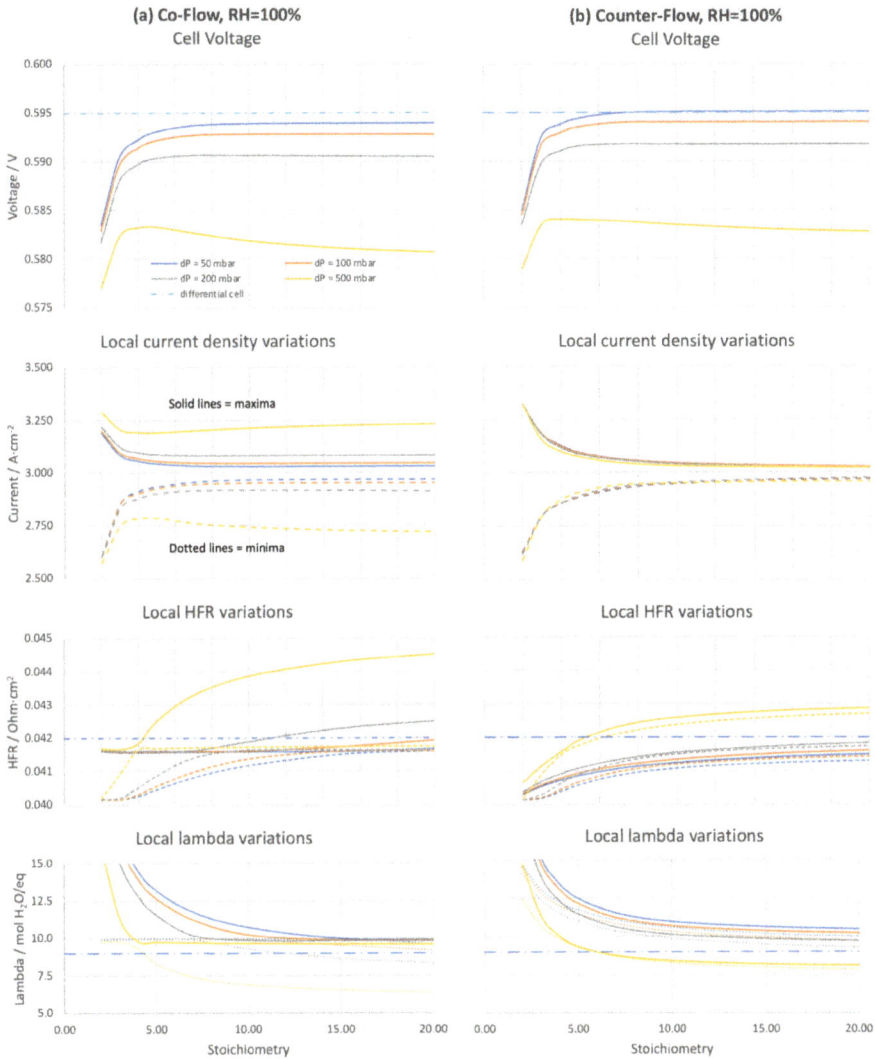

Figure 26: Variation of local conditions as function of reactants stoichiometry for a cell with straight flow field operated in (a) co-flow and (b) counter-flow at an average current density of 3 A/cm^2 and 100 % RH (fully humidified inlet conditions). Y-axes are shared by row; X-axes are shared by column. The solid and dotted lines represent the minimum and maximum values for each variable from the along-the-channel distributions. "dP" represents the pressure drop (inlet–outlet) in mbar; legend labels and colors apply to all graphs.

The cell voltage, which is uniform over the plate by assumption, shows similar trends versus the stoichiometry for co- and counter-flow cases: increasing with stoichiometry for the humidified case (Figure 26) and decreasing for the dry case (Figure 27). The voltages level off for stoichiometries greater than 10 in the humidified case, while

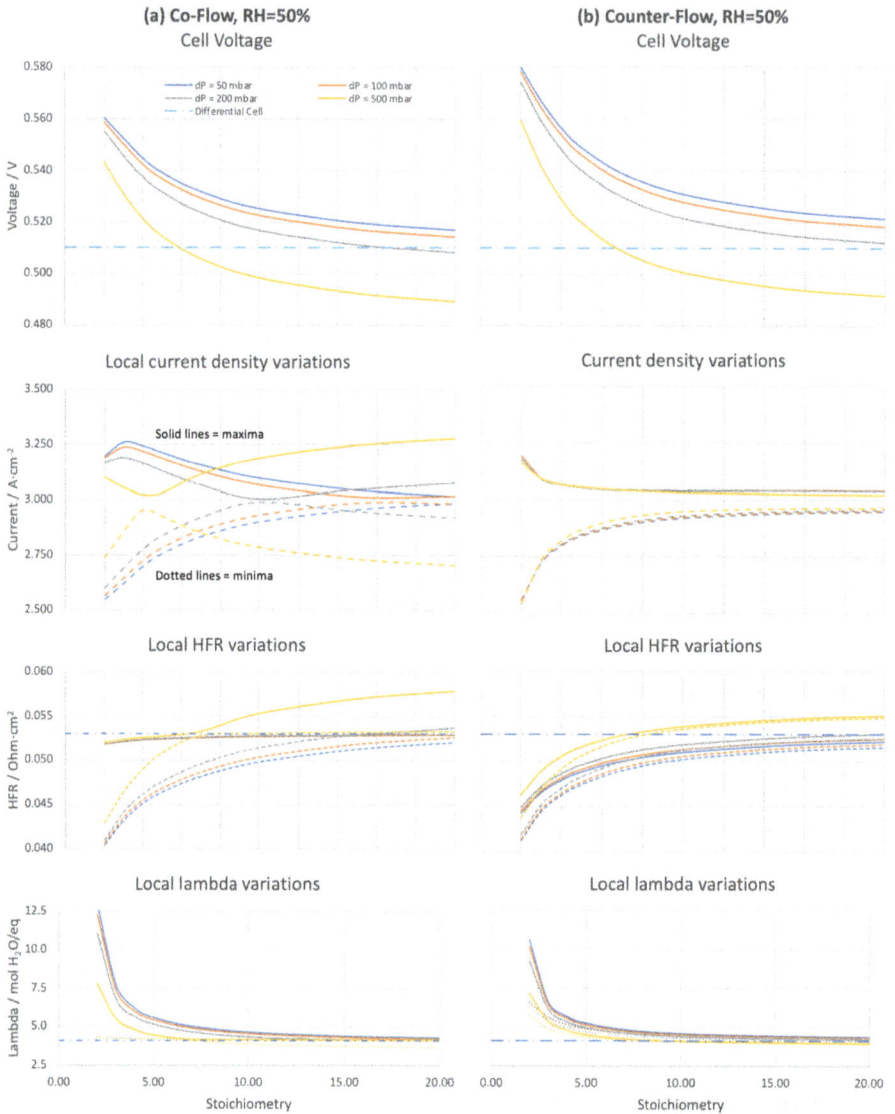

Figure 27: Variation of local conditions as function of reactant stoichiometry for a cell with straight flow field operated in (a) co-flow and (b) counter-flow at an average current density of 3 A/cm² and 50 % RH (dry inlet conditions). Y-axes are shared by row; X-axes are shared by column. The solid and dotted lines represent the minimum and maximum values for each variable from the along-the-channel distributions. "dP" represents the pressure drop (inlet–outlet) in mbar; legend labels and colors apply to all graphs.

there is still a down-trend for stoichiometries greater than 15 in the dry case. Increasing pressure drop results in lowering the cell voltage for both flow configurations, as expected.

The difference in current density (i. e., maximum and minimum values along-the-channel for each stoichiometry) decreases with increasing stoichiometries and is almost stable for values higher than 10. As noted in Figure 25(c), counter-flow results in less variability at the dry conditions, and the difference in current is also not sensitive to the pressure drop. A benefit is also observed in the variation of membrane hydration (lambda) and HFR, which results from the combined effects of the opposite direction for RH and pressure gradients in the anode and cathode channels.

Other simulation results show that counter-flow is also preferable if different stoichiometries and pressure drops are used between cathode and anode. For example, for stoichiometries of 8 and 4 and pressure drops of 100 and 50 mbar for cathode and anode, respectively, the variability in current density, HFR and lambda (not shown here) is much smaller compared to the co-flow case.

In conclusion, counter-flow configuration is preferable and, if operated at stoichiometries of at least 8 and 4, and with a maximum pressure drop of 100 mbar should give less than 5 % variation in current density, lambda and other MEA variables at an average current density of $3\,A/cm^2$ and for the inlet conditions and MEA design considered here.

While slightly higher variability is predicted when the inlet concentrations of O_2 and/or H_2 are reduced, the *differential-cell* behavior is still approached when using sufficiently high stoichiometries (>10) and low pressure drops (< 100 mbar). The pressure drop in the channels increases almost linearly with increasing gas velocity, and the stoichiometry is directly proportional to the volumetric flow rate for a given current. At a fixed stoichiometry, the volumetric flow rate is inversely proportional to the O_2 concentration at the inlet; for example, the flow rate needs to be doubled when the inlet concentration goes from air to half-air. As a result, it is challenging to simultaneously satisfy the conditions of high stoichiometry and low pressure drop required by the differential cell approach. Furthermore, it is desirable to achieve gas velocities in the channels of a differential cell that are comparable to those occurring in a real stack, to achieve a similar flow regime and the same degree of convective flow on the GDL surface. Furthermore, a sufficiently low Reynolds number is required to prevent the onset of turbulent flow and higher pressure drops [80–82]. When testing at very low concentrations, it is often challenging or impossible to simultaneously satisfy these conditions.

The analysis reported here is not representative for cold and wet conditions since the model used is single phase and does not account for the presence of liquid water. Channel flooding is not expected to be significant when operating at high stoichiometry (i. e., channel blockage is not likely), whereas saturation of diffusion media and electrode is possible. As a result, when operating at high stoichiometries, the GDL and CCL saturation are not expected to change drastically in the channel direction. Therefore, differential conditions should be achievable even for the cold and wet case, although the expected variability is not quantifiable by this analysis.

Although a complete analysis for the optimization of a differential cell is beyond the scope of this review, we can conclude with the following observations:

- Using counter-flow configuration is preferable because it gives more uniform current density distributions at both humidified and dry conditions.
- The cathode pressure drop should be minimized, ideally within 100 mbar.
- The anode pressure drop causes less variability, unless diluted hydrogen is used, and should not be larger than for the cathode.
- Using a low aspect ratio for the active area (length:width) should minimize pressure drop for a given active area, and is likely preferable.
- Stoichiometries of at least 8 and 4 for cathode and anode, respectively, should be sufficient to approach differential behavior for the conditions simulated here.
- Finally, an approach similar to the one described here can be used to verify the validity of "differential cell" assumption both for the design of a new cell, and also for the interpretation of existing results.

Lastly, we consider the topics of measurement error and uncertainty in MEA testing. While a detailed analysis is beyond our scope, here we want to reflect on a few specific aspects of fuel cell testing that have important consequences on VLB, and more generally data quality:
- Each fuel cell measurement is usually referenced to the *nominal inlet conditions*.
- There is normally a difference between these nominal conditions and the *actual conditions* in the channels over the active area; this is due in part to the presence of along-the-channel gradients, as discussed above, and also to the specific location and configuration of the sensors and meters in the test station.
- The variability of the actual inlet conditions for the fluids entering the cell depends on the station design and components used.
- While the accuracy of the electrical measurements (voltage and current) is known and typically high, the same cannot be said for all other parameters, and the variability of these depends on the propagation of uncertainties associated to several variables and components for the station used.
- The measurement uncertainty in fuel cell testing is heteroskedastic; that is, both accuracy and precision of fuel cell measurement vary depending on the operating conditions.

To illustrate the complexity of this topic, let us consider the uncertainty for the cathode inlet RH. The RH of the gas stream to the cell is typically controlled by setting the temperature of the humidifier (dew temperature) relative to the cell temperature. The cell temperature can be monitored and controlled at the plate or at the coolant inlet (in case of liquid cooled cells). The uncertainty of these two temperatures depends on the specifications for the temperature sensors used, while the accuracy of the cathode channel temperature depends also on the location of the temperature probe relative to the channel, as there can be a nonnegligible temperature difference between the probe and the channel inlet. Considering the propagation of errors and uncertainties, it should be possible to quantify the variability for the cathode inlet RH (for approach refer to [83]), which can be surprisingly different depending on the testing setup chosen, even when components of comparable specifications are used.

In conclusion, it is important to consider measurement errors and uncertainties in fuel cell testing, particularly for comparing results obtained with different cells, test stations and laboratories, and in the validation of models and in hypothesis testing for MEA diagnosis. The resolution and tolerance applied to voltage loss analysis should not exceed the variability of the measurements.

9.8.3 MEA performance

The most basic measure of a fuel cell performance is the polarization curve, as already discussed in previous chapters. Cell polarization can be measured in a variety of ways, such as:
–   Load or voltage control
–   Constant stoichiometry or constant flow rate
–   Load or voltage step sequence (Increasing or decreasing)
–   Duration of each step/condition (Steady state or constant interval)

For example, the protocol proposed by the US-DOE in 2007 DOE for measuring PEMFC polarization prescribes load control at constant stoichiometries (1.2/2.0; anode/cathode) with a sequence of increasing current density steps of 20 minutes each [84]. The reported performance is obtained by averaging the last minute for each step. The protocols proposed in 2013 again by the US-DOE [85] requires load control at constant stoichiometries (1.5/1.8) with increasing and then decreasing load sequences of 3 minutes, while the European JRC [72] proposes a similar protocol with different stoichiometries (1.3/1.5) using sequences of 2 minutes. The average current for the descending and ascending polarization curves at each voltage is the reported performance. Obviously, the polarization curve will differ depending on the protocol used to measure it.

For the purpose of loss analysis, it is necessary that performance is measured at high stoichiometries (differential cell) at steady-state conditions (i. e., at least 10–15 minutes at a given condition) and the performance is relatively stable. The stability condition is particularly important for VLB and, more generally, in MEA diagnostics.

Before testing a fuel cell, it is necessary to *break-in* the MEA, and then to *condition* the MEA at the specific operating set points. The purpose of the *break-in procedure* is to bring the MEA to a state where maximum performance is achieved; the procedure typically includes a phase (or more) to hydrate the membrane and maximize the conductivity by helping with the generation of protonic-pathways, a phase to clean and maximize the active area of the electrodes, cathode in particular and a phase to wash out possible contaminants and by-products from the manufacturing process. These phases require different operating conditions and are often arranged in repeated sequences/cycles, with intermittent checks to verify when a stable performance is achieved. The purpose of the *conditioning procedure* is to prepare the fuel cell for testing by bringing the operating conditions to a stable setpoint, and also to "resetting" the state of electrodes by removing what can be causing temporary degradation (e. g., removal of Pt-oxide and/or anions adsorbed on Pt surface). The conditioning procedure is not always completely effective in bringing back the MEA to a given state, depending on the sample history and other factors. This can be a problem when additional measurements are done on the MEA to perform VLB. In fact, some of the protocols described below can have a "conditioning" effect and change the state of the MEA, increasing performance after applying the diagnostic measurements. A useful strategy to monitor performance stability is to insert performance checks interspersed in the testing sequence planned for VLB.

It is advisable to measure the high-frequency resistance (HFR) of the cell together with performance. Most commercial test stations offer this capability. The HFR is a direct measurement of the ohmic losses and can also be used to estimate the average hydration of membrane and cathode when operating at dry conditions.

Measuring performance sensitivities is useful to obtain a deeper characterization of MEAs. In this case, the cell current or voltage is held constant and one of the other parameters (e. g., O_2 model fraction, RH, T) is swept over an appropriate range. Performance and HFR sensitivities are very useful for diagnostics.

9.8.4 Cathode kinetics parameters

As discussed earlier, cathode kinetic losses are predominant in PEM fuel cells. Therefore, the ORR parameters are among the most important for VLB. The number and meaning of these parameters depend on the kinetic expression used. In the cases of the Tafel expression of Equations (11) and (12), the parameters are five: Pt roughness factor, exchange current density, charge transfer coefficient, reaction order and activation energy. These can be obtained using the dedicated measurement described below.

It should be clarified that the oxygen reduction reaction over platinum is a complex reaction involving intermediates that are not experimentally accessible in typical MEA testing. The ORR rate is represented by a set of differential equations, involving reactants and all intermediates, that is not reducible to a single expression, such as Equation (11). This topic is discussed in detail elsewhere [22]. For the purpose of VLB, we should consider the ORR rate as an empirical equation of the cathode kinetics rather than as a fundamental property of the catalyst. Subsequently, the ORR parameters discussed here can be considered as effective properties of the cathode.

The roughness factor of the cathode (rf_{CCL}) quantifies the Pt-surface available for electrochemical reaction, it is a nondimensional value [cm^2_{Pt}/cm^2_{Geom}], and when multiplied to the Pt-loading (L_{Pt}) it yields the electrochemical active surface area (ECSA) of the catalyst in the electrode in m^2/g_{Pt}. The cathode roughness factor can be measured from hydrogen adsorption/desorption (HAD) or CO stripping of the electrode by cyclic voltammetry. Details on the protocols are provided elsewhere [86].

The roughness factor is typically measured with fully humidified gases. It is useful to measure this parameter also as a function of RH to better understand how the platinum ionomer interface modifies with different atmospheres.

The standard procedure to determine the ORR parameters of MEAs is by polarization in saturated H_2–O_2 at low to moderate loads (Figure 28), where the ohmic resistance is known to be constant and mass-transfer losses are negligible [23]. The reference conditions for *in situ* measurement of ORR activity are 80 °C, 1 bara H_2 and O_2 partial pressure saturated with water vapor, which correspond to 1.5 bar total pressure. The cathode kinetics is assumed to follow the Tafel equation (Equation (11)), and the main kinetics constants are obtained from the linear trend in the Tafel plot. The analysis requires some corrections for ohmic and proton transport losses and for the parasitic currents. Therefore, additional measurements are required. Ohmic and proton-transport resistances can be measured using the EIS method described later; alternatively, the HFR values measured during the polarization can be used to obtain the total and membrane resistances, and from the latter the proton-resistance can be estimated if Pt-loading and I/C of the electrode are known. The H_2 crossover current can be obtained by cyclic voltammetry or by linear scanning voltammetry, as described by Kocha et al. [87].

The *Tafel analysis* consists in plotting either the ORR overpotential (η_{ORR}) or the cell voltage corrected from ohmic and proton transport losses ($E_{iR\text{-free}}$) in a semilog chart, where the x-axis is the decimal logarithm of the current density, and the y-axis is the overpotential or potential. ORR overpotential and iR-free voltage can be obtained from Equations (4) and (7), resulting in the following:

$$\eta_{ORR} = E_{rev} - V_{cell} - \eta_{\Omega,PEM} - \eta_{\Omega,other} - \eta_{H^+ - Cath} \tag{36}$$

$$E_{iR-free} = V_{cell} + R_{\Omega,tot} \cdot i_{cell} + R^{eff}_{H^+ - CCL} \cdot i_{cell} \tag{37}$$

The measured current is also corrected by adding the H_2 crossover current, and finally a linear fit (in the semilog plot) is applied to the corrected data, and from slope and intercept the kinetics parameters are derived. The *apparent exchange current density* ($i_{o\text{-App}}$) is defined as the intercept between the line best fitting the data and the reversible potential (E_{rev}), while from the slope (*Tafel slope*) the charge transfer coefficient is obtained (see Equation (11)). The current density at 900 mV $E_{iR\text{-free}}$ is also used to quantify ORR activity; it is more convenient to use this quantity in place of the exchange current because it can be determined with higher precision, given that the latter is extrapolated several hundred of mV above the measured voltage range.

After obtaining $i_{o\text{-App}}$, the ORR Specific (i_{SA}) and Mass activities (i_{MA}) can be calculated if roughness factor and Pt loading of the cathode are known.

Reaction order and activation energy are obtained by repeating the measurements and analysis at different O_2 partial pressures and temperatures, respectively. Alternatively, a global fitting procedure has also been applied, which produces all the kinetics parameters at once [23].

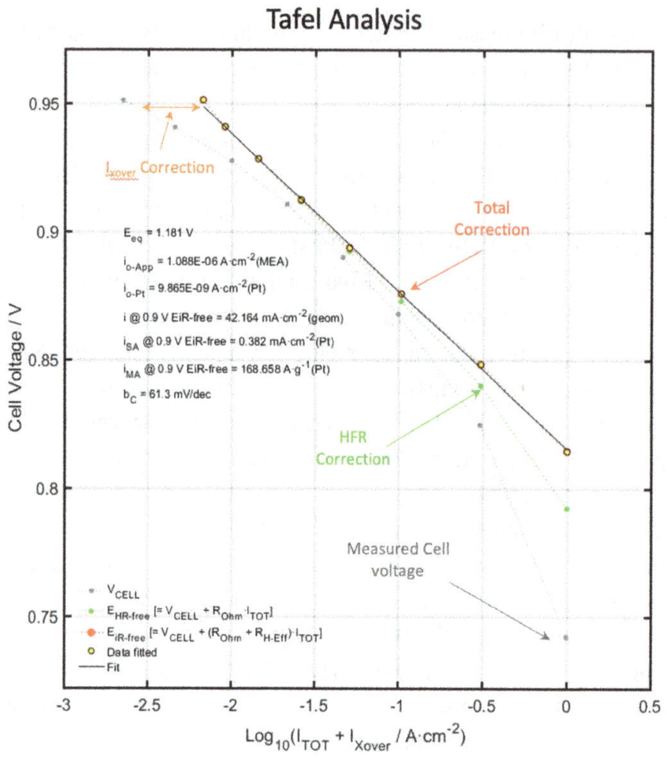
Tafel Analysis
0.95
0.9
0.85
0.8
0.75
Cell Voltage / V
Ixover Correction
Total Correction
HFR Correction
Measured Cell voltage
Eeq = 1.181 V
io-App = 1.088E-06 A·cm-2(MEA)
io-Pt = 9.865E-09 A·cm-2(Pt)
i @0.9 V EiR-free = 42.164 mA·cm-2(geom)
iSA @0.9 V EiR-free = 0.382 mA·cm-2(Pt)
iMA @0.9 V EiR-free = 168.658 A·g-1(Pt)
bC = 61.3 mV/dec
VCELL
EHR-free [= VCELL + ROhm·ITOT]
EiR-free [= VCELL + (ROhm + RH-Eff)·ITOT]
Data fitted
Fit
-3
-2.5
-2
-1.5
-1
-0.5
0
0.5
Log10(ITOT + IXover / A·cm-2)

Figure 28: Example of Tafel analysis for the extraction of kinetics parameters from H_2–O_2 polarization curves after correcting for ohmic and CCL proton transport losses.

As mentioned previously for the roughness factor, it is also useful to measure the ORR activity at different relative humidity. Quantifying the changes of ORR specific activity with RH is fundamentally important to understand the characteristic behavior of the Pt/ionomer interface for the electrode under test.

For VLB, it is not always necessary to measure all ORR parameters. To a minimum, one should measure the cathode roughness factor and check the magnitude of the exchange current density for comparing the specific ORR activity against a baseline or values reported in the literature. A single measurement in voltage control at 0.9 V $E_{\text{iR-free}}$ and reference conditions is sufficient for this relative comparison. The other parameters can be assumed to be equal as in previous measurements done for similar electrodes. A complete analysis is required when high diagnostic accuracy is desired.

As discussed earlier, the Butler–Volmer equation, and subsequently the Tafel equation, from which this is derived, does not always represent the ORR accurately due to the formation of a passivation layer on the platinum surface. For a more accurate representation of the cathode kinetics, it is necessary to consider the effect of oxide-coverage on the ORR kinetics. This is particularly relevant for cathodes with low Pt-loading or low roughness factor. Subramanian et al. [88] proposed the following coverage dependent expression for MEA testing:

$$i_{ORR}(\eta) = i_{o,ORR}^{Ref} \cdot \left(\frac{pO_2}{p^{Ref}} \right)^{\gamma_{O_2}} \cdot (1 - \theta) \cdot \exp\left(-\frac{a \cdot F \cdot \eta}{R \cdot T} \right) \cdot \exp\left(-\frac{\omega \cdot \theta}{R \cdot T} \right)$$
$$\cdot \exp\left[-\frac{E_{a,ORR}^{rev}}{R \cdot T} \cdot \left(1 - \frac{T}{T^{Ref}} \right) \right], \tag{38}$$

where θ is the fractional oxide coverage on Pt, ω is an energy parameter for the oxide coverage, while the other parameters were already defined for Equations (11) and (12). The determination of these new parameters requires the measurement of the oxide coverage as a function of the cathode potential. A procedure to quantify oxide coverage was published by Liu et al. [89]. An example of the oxide coverage changes with cathode potential is shown in Figure 29 (a); above 0.8 V, the coverage exceeds 50 % and increases with the potential following a sigmoidal shape. Figure 29(b) illustrates the relative importance of the oxide coverage effects; the deviation between the Tafel kinetics of Equation (11) and the coverage dependent of Equation (38) is much more evident for low loaded electrodes. Figure 29(c) and (d) show the same deviation as function of the cathode roughness factor. At rf_{CCL} = 100, a typical value for state-of-the art PEM fuel cell electrodes for automotive applications, the simple Tafel kinetics underpredicts the cathode overpotential by 5 mV at 1 A/cm^2, while at 3.0 A/cm^2 the gap grows to 20 mV. At high current densities and lower cathode roughness (r_{CCL}), this effect is not negligible; subsequently, the coverage dependent ORR should be used in VLB as shown in recent publications [28].

Finally, the ORR parameters can also be measured using fast transient current measurements. The protocol consists in measuring the voltage response of the cell during a rapid current pulse, typically an exponential-waveform of period 5 to 10 s. Alternatively, a triangular wave in voltage control can be used. The current (or voltage) pulse is performed after the cathode has been exposed to a fixed potential ($E_{iR-free}$) for a given time interval to control the amount of oxide formed. The transient polarization is then corrected for the capacitive current, in addition to corrections for H$_2$ crossover current and ohmic and proton-transport losses as done in the Tafel analysis previously described. The apparent double-layer capacitance of the cathode can be measured by EIS or cyclic voltammetry experiments. The advantage of this procedure is its rapidity, which makes it suitable to cases where the MEA is showing performance instability or poisoning [90].

9.8.5 Electrical and proton resistances of MEA-layers

High frequency resistance (HFR) measurements are routinely done during performance testing. The HFR is an impedance measurement at fixed frequency, typically between 2 and 3 kHz, obtained using a small current perturbation superimposed to the load (AC <

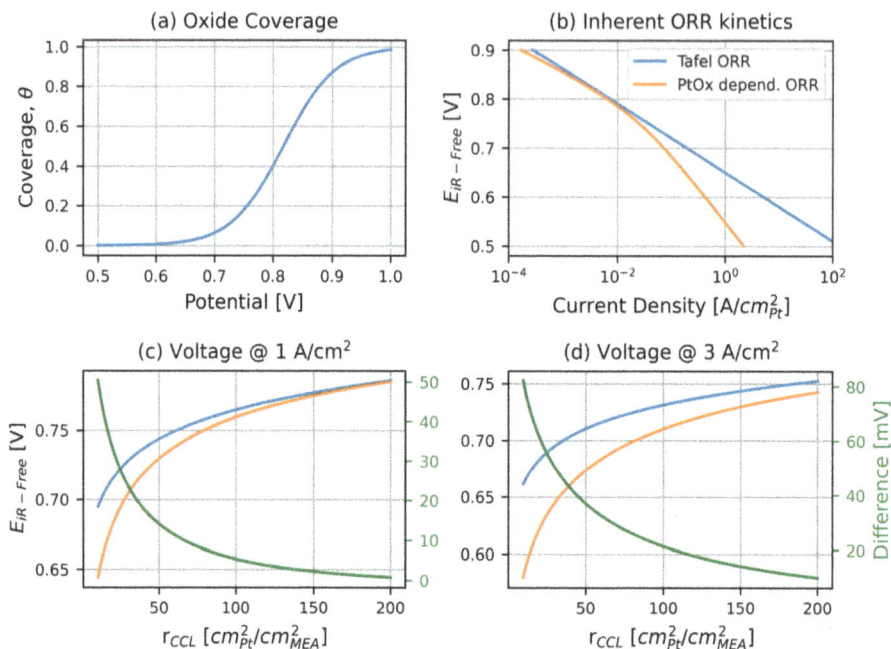

Figure 29: (a) Pt oxide coverage as a function of the cathode potential; (b) examples of simple Tafel versus coverage-dependent ORR kinetics for cathodes of different loadings; difference in ORR overpotential prediction between the simple Tafel and coverage-dependent ORR rate as a function of the cathode roughness factor at (c) 1.0 A/cm² and (d) 2.5 A/cm².

5 % DC). The HFR is a measure of the *total ohmic resistance* ($R_{\Omega\text{-tot}}$) of the MEA, including membrane, GDL, plates and GDL/Plate contact resistances. To separate the electrical ($R_{\Omega\text{-other}}$) from the membrane ($R_{\Omega\text{-PEM}}$) resistances, it is necessary to obtain the electrical resistance from a dedicated experiment. A common procedure consists in using a "dummy" MEA made using only the two GDL assembled with the MPLs facing each other and mounted to the same cell used for performance measurements. A fixed load is applied, and the voltage drop is measured at the same probing point normally used over the temperature range of interest. The electrical resistance so measured includes the GDL/Plate contact resistance, while the resistance between MPL/CCL are assumed to be negligible. For a break-down of all electrical resistance components, it is necessary to rely on dedicated *ex situ* experiments, or estimates based on material properties and modeling. Jiang et al. proposed a more rigorous procedure to separate all through-plane electrical resistances in a MEA; this procedure is based on combining *in situ* and *ex situ* measurements including EIS under varying compression forces [91].

The effective proton resistance of the cathode is measured by electrochemical impedance spectroscopy (EIS) in H_2/N_2, as described by Liu et al. [92–94]. The procedure consists in applying a sinusoidal voltage perturbation to the electrode at a constant bias between 0.2 and 0.45 V in a frequency range between 15 kHz and 0.1 Hz while the electrode is exposed to N_2. At these conditions, the electrode behaves like a polarizable electrode (i. e., double-layer capacitor). The impedance behavior of PEMFC electrodes is analogous to that of porous electrodes in aqueous electrochemistry. In PEMFC electrodes, proton transport occurs within the ionomer network through which platinum nanoparticles embedded in the porous structure and further away from the membrane can be reached. The measured impedance of a PEMFC electrode with uniform ionomer distribution and thickness shows identical features as that of a cylindrical pore with ohmic drop in a liquid electrolyte solution, which is described mathematically by the De Levie's model [95]. The model commonly used to extract the effective proton resistance of the cathode (R_{H^+-CCL}) from EIS data is the following:

$$Z_{CL}(f) = R_{\Omega,tot} + \sqrt{\frac{R_{H^+-CCL}}{2 \cdot \pi \cdot j \cdot C_{CCL} \cdot f}} \coth(\sqrt{2 \cdot \pi \cdot j \cdot R_{H^+-CCL} \cdot C_{CCL} \cdot f}) \qquad (39)$$

where $Z_{CL}(f)$ is a complex-function representing the total MEA impedance when the cathode is exposed to N_2, C_{CCL} is the apparent double-layer capacitance of the electrode, and j is the complex unit. Effective proton conductivity and specific capacitance of the electrode can be extracted from R_{H^+-CCL} and C_{CCL}. A typical example of the cathode impedance is shown in the complex-plot of Figure 30(a). The intercept with the real axis represents the total ohmic resistance ($R_{\Omega,tot}$); in the high to mid frequency range (10 kHz to 10 Hz), the complex spectrum increases linearly with 45 °slope, while in the low frequency range (<1 Hz) the spectrum becomes vertical. This characteristic change in slope is typical of uniform porous electrodes. For relative humidity higher than 50 %, the model of Equation (39) can represent quite accurately impedance data in the high to intermediate frequency range, while it shows large deviations at lower frequencies (<1 Hz, see of Figure 30(a)). At lower relative humidities, this model deviates from the data even at mid frequencies (> 1 kHz) and the complex spectrum increases gradually rather than showing the characteristic change in slope. To extract R_{H^+-CCL}, it is sufficient to fit only the data in high and mid frequency range, although there is some ambiguity in data fitting caused by the absence of the vertical portion of the spectrum (see of Figure 30(b)).

During EIS measurement in H_2/N_2, there is no water production at the cathode. If the relative humidity in the anode and cathode channels is equal, the hydration in the membrane and ionomer-phase of the electrodes can be considered at equilibrium, provided that sufficient conditioning was applied. As a result, the measured effective properties ($R_{\Omega,tot}$, R_{H^+-CCL} and C_{CCL}) are directly related to an equilibrium hydration value, by means of the hydration isotherms of the MEA components (i. e., $\lambda_{PEM-eq}(RH)$ and $\lambda_{CCL-eq}(RH)$) that can be measured *ex situ*.

Repeating EIS H_2/N_2 measurements over a range of RH provides valuable information on the behavior of the electrode. Figure 31 shows an example of the trends versus RH for effective properties obtained by EIS measurement in H_2/N_2. Each point was measured after conditioning the MEA at constant RH for 20 minutes. A hysteresis is observed

(a)

(b)

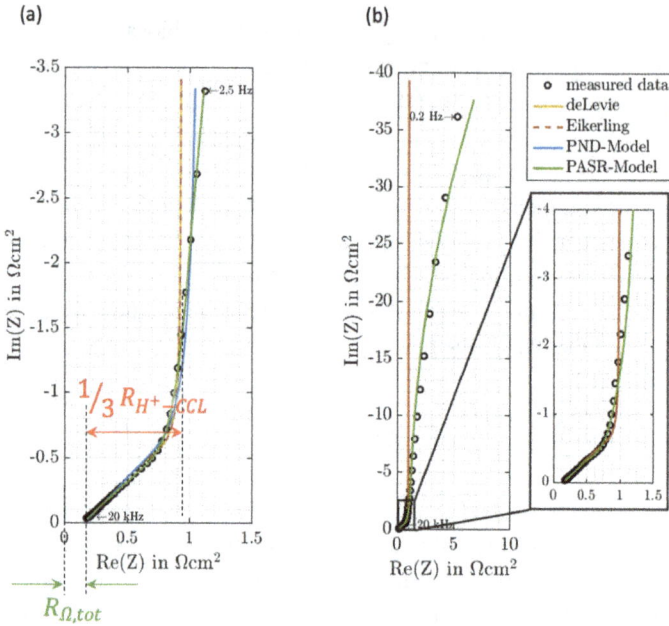

Figure 30: (a) Typical complex impedance plot for PEM fuel cell electrodes showing the characteristic change in slope and (b) typical deviations from an ideal transmission line model (Eikerling model in red) in the low frequency range. Adapted under Creative Commons License CC BY 4.0 from [97].

for RH lower than 50 %; the amount of hysteresis is larger for the cathode properties (R_{H^+-CCL}, and C_{CCL}) providing indirect evidence of changes at the Pt/ionomer interface at very low relative humidities, as already observed using other methods [64, 96]. The amount of hysteresis depends on the RH value and duration of exposure of the MEA at lower RH values.

During normal cell operation, water is produced at the cathode, the hydration within the MEA is not at equilibrium and deviations from equilibrium are expected particularly at lower channel RH and high current densities. Thus, the MEA properties measured by EIS H_2/N_2 are not necessarily representative of the actual conditions under load. For VLB purposes, two strategies are possible. First, the average membrane and CCL hydration under load can be estimated if the HFR was also measured at the same load. The estimate is obtained from the plot of the total ohmic resistance measured at equilibrium, as shown in Figure 28(a). The estimated RH value can then be used to obtain the effective proton resistance "corrected" to account for the water production under load. The measured HFR and the "corrected" R_{H^+-CCL} can be used to estimate the related losses. Alternatively, the effective conductivities as a function of hydration can be derived from the EIS H_2/N_2 measurements and a MEA model including membrane water management is used to estimate the loss associated with proton transport. In this case, additional measurements are required as described in Section 9.8.7. Finally,

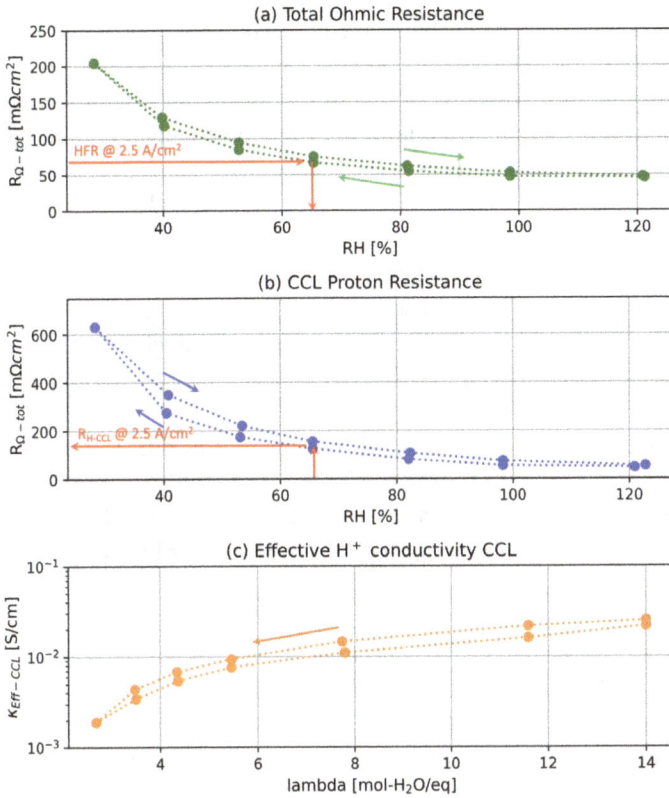

Figure 31: (a) Total ohmic resistance and CCL proton resistance versus RH; (b) Effective proton conductivity vs. RH. The arrows indicate the measurement sequence, and a hysteresis is observed. The HFR in green corresponds to the measure HFR under load at 50 % channel RH; using this the average hydration of the MEA, expressed as RH-equivalent, is estimated and from this a 'corrected' value of the CCL proton resistance at the specific condition is obtained.

it is possible to obtain the proton conductivity during normal operation directly from EIS measurements performed in H_2/Air under load, as described by Reshetenko and Kulikovsky [58], although the relative error in the values obtained appear to be high at higher loads (see Figure 19), possibly because of the multiparametric fitting procedure used.

Errors associated with fitting are also relevant for EIS measured in H_2/N_2. As discussed above, the relatively simple model of Equation (39) shows deviations versus experimental data, particularly at low RH. These deviations have been attributed to various causes. Obermaier et al. [97] proposed a new model, called PND (Pore number distribution), to account for the effects due to the presence of distribution of porosities within real electrodes. The PND model takes as input the pore size distribution of the electrode measured by mercury porosimetry. By using the PND model, a more accurate fit is obtained in the mid frequency range, when vertical data points are absent (see of Figure 30(b)). Furthermore, the same authors propose a model, called PASR (Parallel adsorption and side reaction), to account for the adsorption of sulfonic acid groups on platinum. The PASR-model provides a more accurate fit for the low frequency range (see of Figure 30(b)). The effect of H_2 crossover was also considered to explain deviations at high frequencies.

Deviation from the ideal pore model of Equation (39) can be caused also by other factors. For example, variations in electrode thickness and/or in through-plane distribution of ionomer would also result in the absence of vertical data points in the mid frequency range, and in slopes higher than 45°, while the presence of electrical shorting in the MEA causes a curved trend in the very low frequencies similar to the ionomer adsorption proposed by Obermeir.

As discussed earlier, the same EIS spectrum can be produced by different models and additional independent measurements are often required to identify the most likely physical processes causing the observed behavior. Unfortunately, the complex nature and structure of PEMFC electrodes does not allow for a straightforward interpretation of EIS data.

In a recent publication, Kulikovsky provided a detailed theoretical analysis on the effects of proton and electron transport on the impedance of cathodes during H_2/N_2 testing. Kulikovsky derived a novel analytical model of the cathode impedance and, using this, he demonstrated that the model of Equation (39)—in which the electron conductivity is assumed to be infinite—does not provide an accurate value of the true proton resistance of the electrode when the ratio between effective electron and proton conductivities of the layer is lower than 3 [98]. The effective electron conductivity of typical PEMFC electrodes can vary between 0.1 to 1 S/cm, while the proton conductivity can reach 0.02 S/cm; therefore, the assumption of negligible electron resistance if justified. While the model of Equation (39) is applicable to standard electrode designs, it is important to assess the magnitude of the effective electron conductivity of the layer in novel design using low Pt loading or PGM-free electrodes. Kulikovsky's model should be used for a more rigorous analysis.

The effective electron conductivity of CCL can be measured *ex situ* using various approaches; for example, Suzuki et al. proposed a relatively straightforward procedure based on in-plane four-probe resistance measurements done on the electrode combined with image analysis of the electrode to quantify cracks and nonuniformities of the layer, and from this the in-plane connectivity of the layer is estimated; subsequently, the in-plane electrical conductivity is corrected and from this effective through-plane electron conductivity of the CCL assuming isotropy [99].

For VLB, it is desirable to measure this quantity with a sufficiently high level of accuracy and precision since the proton transport loss ranks 3rd for most operating conditions.

9.8.6 Oxygen transfer resistances of porous layers

For moderate current densities and under normal or dry operating conditions, the oxygen transport losses of an MEA are negligible; however, as the current density increases these losses grow exponentially and end up dominating over all other losses as the cathode limiting current is approached. Furthermore, at cold and wet conditions liquid water is expected to partially saturate the pore volume in the diffusion media and electrode, causing local flooding and inhomogeneous current densities over the active area (e. g., saturation under the ribs, resulting in difference in current density between channel and rib areas) that can significantly decrease cell performance. For these reasons, it is fundamental to quantify oxygen transfer losses, which is doable by dedicated experiments at limiting conditions. The parameter commonly used to quantify oxygen transport losses is the mass transfer resistance (R_{MT}), which is defined as the resistance to the normal molar flux of a oxygen (N_{O_2}) driven by a given difference in oxygen concentration (ΔC_{O_2}), according to the equation:

$$R_{MT} = \frac{\Delta C_{O_2}}{N_{O_2}} = 4 \cdot F \cdot \frac{\Delta C_{O_2}}{i}, \tag{40}$$

where i is the current density and F is the Faraday's constant.

Different types and sources of transport resistance occur in the cathode. Starting from the channel, oxygen transfer is limited by molecular diffusion within a gas boundary layer that forms over the GDL surface in the channel; the associated resistance is defined as flow channel resistance (R^{Ch}). Within the GDL thickness and in part of the MPL intermolecular diffusion within the pores is still the dominant source of transport limitations. Once the oxygen molecules reach the electrode, which is characterized by much smaller pore sizes if compared to the GDL, they will start interacting with the pore walls and Knudsen-type diffusion will become more important. Figure 32 shows that a transition from molecular to Knudsen diffusion regimes occurs within a specific pore-size range that depends on the conditions. Typical pore sizes are between 10 and 50 μm for GDL, 0.5 and 5 μm for MPL, while electrodes are characterized by bimodal pore size distributions, where the primary pores are between 25 and 50 nm in diameter and the secondary pores are larger than 100 nm and comparable to the smaller pores in the MPL. Based on Figure 32, molecular diffusion is the primary transport mechanism in GDL in absence of convective flows; in the MPL, both molecular and Knudsen diffusion are important, while for the cathode a combination of molecular and Knudsen diffusion is expected within the secondary pores of the cathode and predominantly Knudsen diffusion in the primary pores. The last transport phase in the cathode consists with the dissolution and diffusion of oxygen in the hydrated ionomer film covering the platinum nanoparticles. This type of diffusion, often referred to as oxygen diffusion in the ionomer film, is analogous to the diffusion in liquid water, although ionomer films

O₂ diffusion type vs pore size

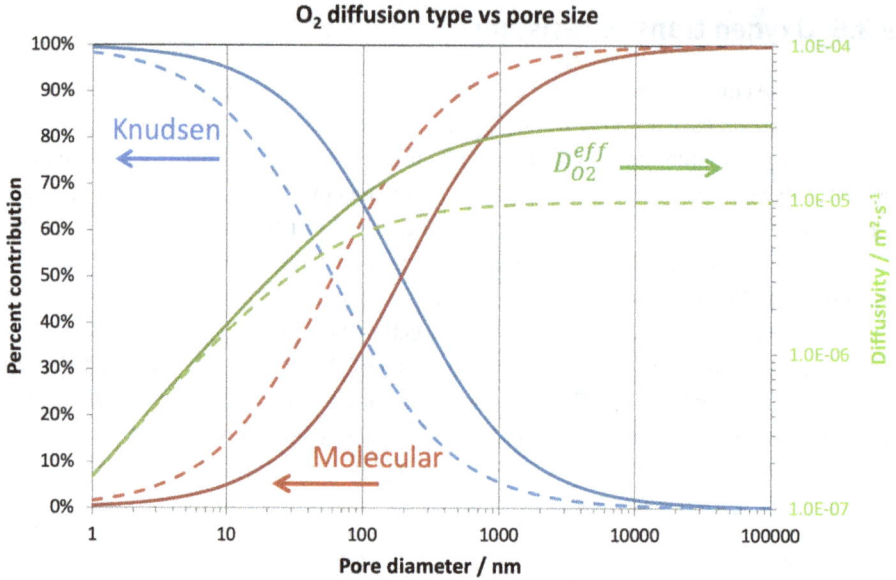

Figure 32: Theoretical variation of the diffusion regime (molecular vs. Knudsen) as a function of the pore size for oxygen at 1 bar pressure (solid) or 3 bar pressure (dashed) and 80 °C. $D_{O_2}^{eff}$ is the effective diffusivity of the pore.

are not homogeneous at microscopic level (nm), and oxygen transport can occur both through the hydrophobic and hydrophilic volume fractions of the ionomer film.

The different dependencies of three types of diffusion described (molecular, Knudsen and liquid/ionomer) on temperature and gas pressure, as well as the interaction of oxygen with the other diluting gas species, have been used to break down the total mass transfer resistance of the cathode into its various constituents.

Baker et al. exploited the dependencies on the gas pressure to separate the pressure-dependent (R_P) from the pressure-independent (R_{NP}) mass-transport resistances [100]. To intuitively understand the dependence on the gas pressure, it is useful to recall that R_{MT} can be defined in terms of the effective diffusivity as follows:

$$R_{MT} = \frac{f_{ch} \cdot h}{D_{O_2}^{eff}},$$ (41)

where f_{ch} is a nondimensional parameter to convert the mass flux in the diffusion layer from anisotropic two-dimensional to an equivalent isotropic one-dimensional, and h is the thickness of the diffusion layer, both of which are geometric parameters. The effective diffusivity is a physical property that changes with temperature and pressure. The change in effective diffusivity with pressure as function of the pore diameter in which transfer occurs is shown in Figure 32, where the values for two different pressures are plotted. For pore sizes lower than 10 nm, the effective diffusivity is independent of the pressure, while for pores larger than 1 μm the effective diffusivity behaves as molecular diffusivity, that is, it increases as $T^{1.5}$ with temperature and as 1/P with pressure. Therefore, the portion of R_{MT} associated with GDL and to the connected network of larger pores in the MPL is pressure dependent, while the portion associated with the cathode is essentially pressure independent. The portion of R_{MT} related to oxygen dissolution and transfer in the ionomer-film is also pressure independent, with solubility and diffusivity following an Arrhenius relationship.

The method to separate R_{NP} from R_P consists in several steps, described in detail in [100]. In brief, the limiting current of a cell is measured at different oxygen concentrations and fixed temperature and pressure, and the (total) oxygen transport resistance is obtained from the slope of the limiting current versus the dry oxygen mole fraction plot. The procedure is then repeated at different pressures (at least three), and finally the transfer resistance is plotted versus the gas pressure in the cathode channel, resulting in a linear trend, and R_{NP} and R_P are obtained from the intercept and the reminder, respectively.

Baker et al. proposed a breakdown of R_{MT} as terms in series:

$$R_{MT} = R_P + R_{NP} = R^{Ch} + R^{DM} + R^{MPL} + R^{Other}, \tag{42}$$

where R_P is the pressure dependent mass-transfer resistance and RNP is the pressure independent term. R^{Ch} is the channel mass-transfer resistance, which due to molecular diffusion within a gas boundary-layer generated over the GDL surface, R^{DM} is the term corresponding to the GDL paper and is also due to molecular diffusion. R^{MPL} is the term associated with the MPL layer in the GDL. R^{Other} combines the transfer resistance effects of the electrode and is assumed to equate R_{NP}, possibly with some contribution from Knudsen diffusivity in the MPL. The breakdown of the pressure dependent term into channel, DM and MPL is possible knowing the thicknesses and porosity of the layers. Figure 33 shows the breakdown of R_{MT} into pressure dependent R_P and pressure independent R_{NP} components.

Caulk and Baker further evolved this limiting-current procedure to distinguish dry and wet conditions in the diffusion medium [101]. The procedure consists in measuring the limiting current successively for increasing dry mole fractions of oxygen from less than 1 % up to air concentration and repeating the sequence for different pressures and temperatures. The limiting-current measurements are essentially the same as in the previous procedure, where the maximum oxygen mole fraction was limited to 4 %. As the xO_2 increases, so does the limiting current and at some operating conditions water vapor condensation occurs inside the MPL and/or GDL. The same authors also proposed a related model to predict flooding conditions in a MEA based on operating conditions and GDL properties [102]. The model is simplified and decoupled from the other MEA processes (cell performance and water crossover through the membrane as input to the model) yet it allows predicting flooding conditions for a GDL type of known thermal and

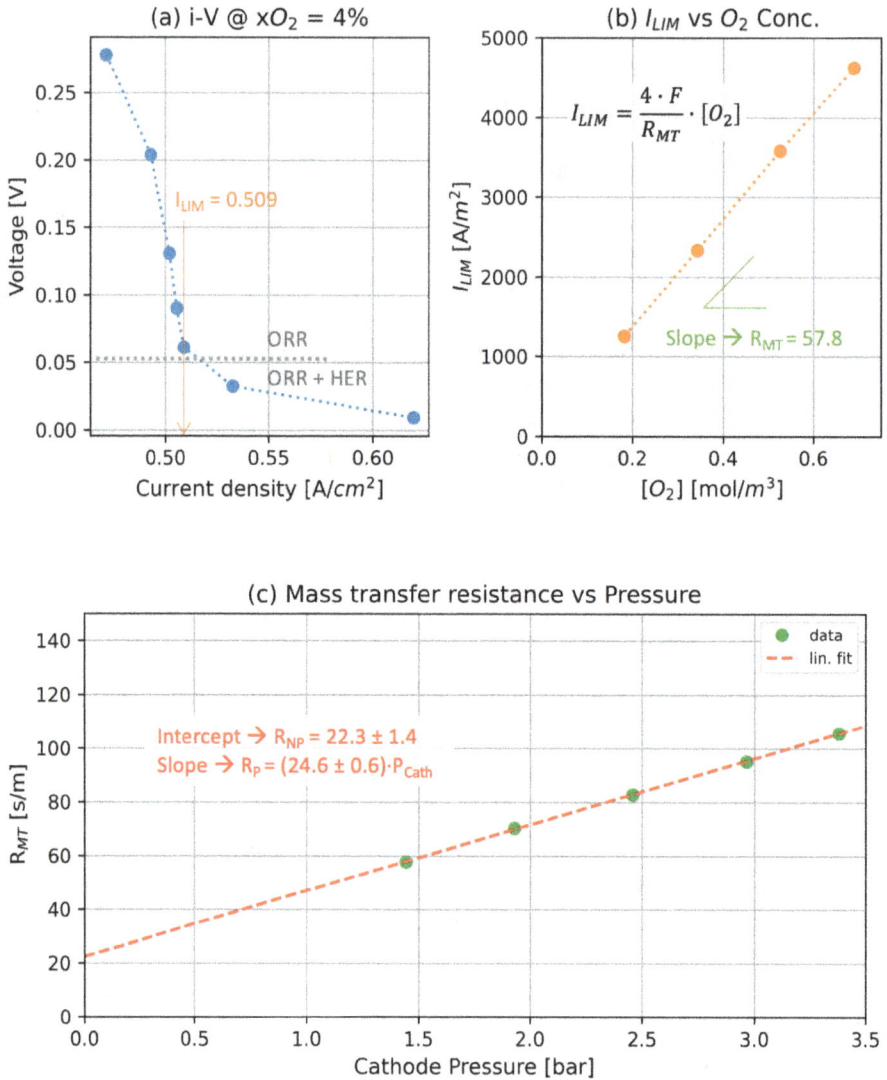

(a) i-V @ xO$_2$ = 4%

I$_{LIM}$ = 0.509

ORR
ORR + HER

Voltage [V]
Current density [A/cm^2]

(b) I$_{LIM}$ vs O$_2$ Conc.

$$I_{LIM} = \frac{4 \cdot F}{R_{MT}} \cdot [O_2]$$

Slope → R$_{MT}$ = 57.8

I$_{LIM}$ [A/m^2]
[O$_2$] [mol/m^3]

(c) Mass transfer resistance vs Pressure

Intercept → R$_{NP}$ = 22.3 ± 1.4
Slope → R$_P$ = (24.6 ± 0.6)·P$_{Cath}$

data
lin. fit

R$_{MT}$ [s/m]
Cathode Pressure [bar]

Figure 33: Example of the analysis sequence performed to separate the cathode mass-transfer resistance (R$_{MT}$) into its pressure dependent and independent components: (a) Identification of the limiting current (I$_{LIM}$) one condition; (b) Linear trend of I$_{LIM}$ versus oxygen concentration obtained from repeating step (a) at different dry mole fractions; extraction of R$_{MT}$ from the slope; (c) Linear trend of R$_{MT}$ versus the channel pressure; pressure independent and dependent terms are obtained from the intercept and slope, respectively.

oxygen-transfer properties (see Figure 34(b)). The transfer properties of the GDL are obtained using the extended limiting current procedure just described. In a different study, Owejan et al. performed similar limiting-current combined with neutron imagining to measure the local water content in the diffusion media [103] and found that water sat-

(a)

(b)

Cases	A	B	C
$\mathcal{D}^o_{H_2O-N_2}/\mathcal{D}^{eff}_{DM(dry)}$	3.6	3.6	6.0
κ^{DM}_T	0.1	0.5	0.5

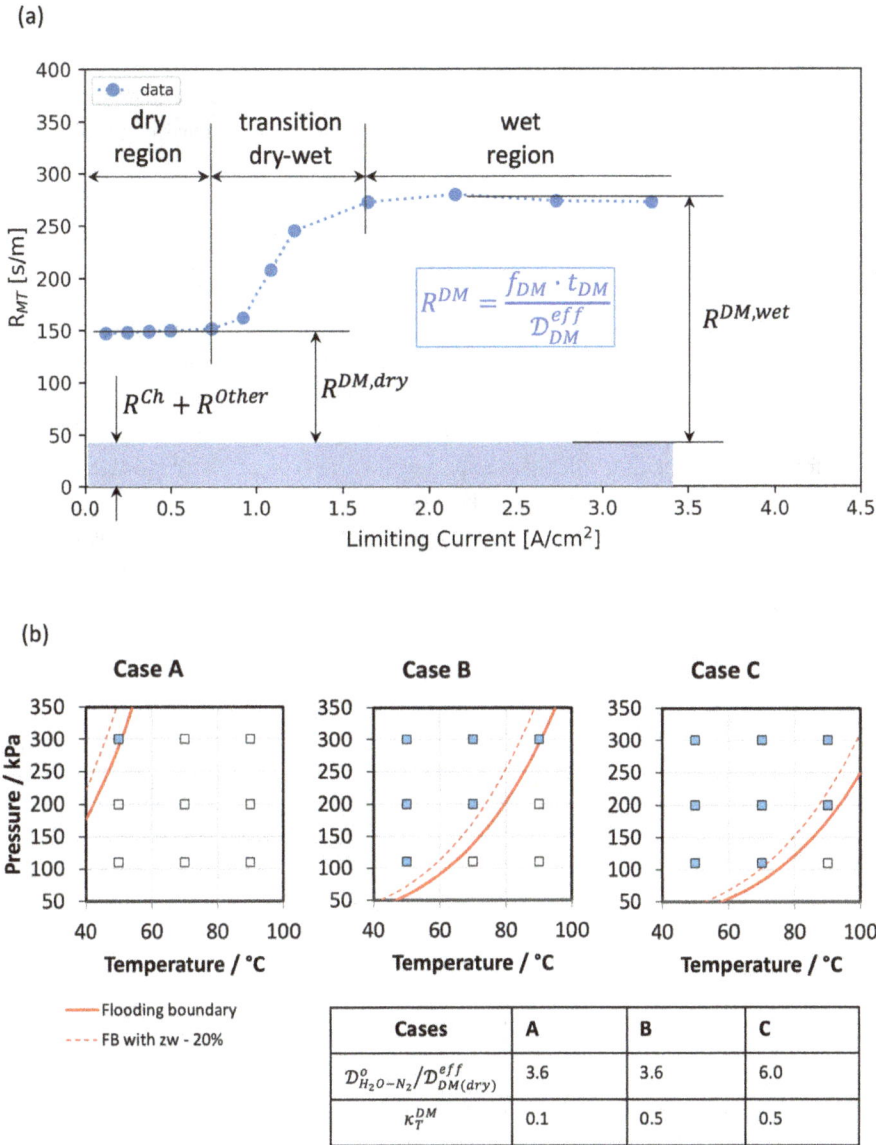

Figure 34: (a) Variation of R_{MT} of a cathode for increasing limiting currents, resulting from increasing oxygen mole fractions; a dry region is observed for low mole fractions (< 4 %) while, depending on the operating temperature and pressure, a transition region and a plateau attributed to GDL flooding is observed as the mole fractions increase. Depending on the GDL properties and operating conditions a "wet region" might appear or not (adapted from [102]); (b) Maps of dry and wet behavior of three hypothetical diffusion layers predicted using the model of [102].

uration occurred mainly under the ribs in the GDL, confirming the interpretation proposed by Caulk.

Greszler et al. further explored the physical meaning of the pressure independent terms (R^{Other}) discovered earlier by Baker [104]. The authors measured the effect of varying Pt-loading in the cathode on the oxygen transport resistance and found an inverse relation between R^{Other} and the electrode roughness factor (rf_{CCL}), according to this equation:

$$R_{MT} = 4 \cdot F \cdot \frac{C_{O_2}^{ch}}{I_{Lim}} = R^{Ch} + R^{DM} + R^{MPL} + R^{CL,Knud} + \frac{R_{O_2}^{Pt}}{rf_{CCL}}, \tag{43}$$

where $R_{O_2}^{Pt}$ is a transport-like resistance on the platinum surface and obtained from the slope in the plot of R_{MT} versus the inverse of the cathode roughness factor ($1/rf_{CCL}$) measured for electrodes of different loadings but similar thickness. The value obtained for $R_{O_2}^{Pt}$ was approximately 12 s/cm (see Figure 35(a)), which exceeded the value estimated for the ionomer film thickness expected for the type of electrodes tested. In a different study, Owejan performed similar experiments showing that the larger than expected values for $R_{O_2}^{Pt}$ might be related to the presence of an interfacial resistance on the ionomer surface, consistently with a model of platinum/ionomer interface [105].

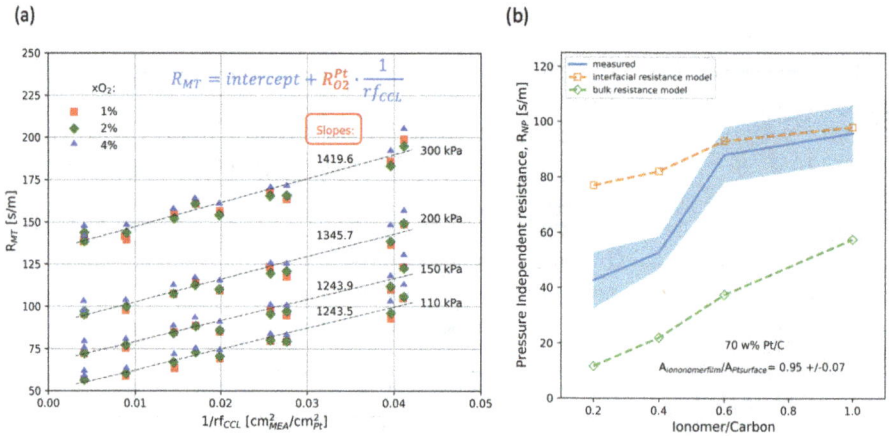

Figure 35: (a) Plot of the oxygen transfer resistance (R_{MT}) as function of the inverse of the cathode Pt roughness factor at different pressures for electrodes of similar thickness. Similar slope of approximately 12 s/cm*(cm^2(MEA)/cm^2(Pt)) was observed. (Redrawn from [104]). (b) Pressure independent transport resistance versus I/C for different electrodes of similar roughness factor (~25) compared to two different models of the limiting current over Pt nanoparticles. (Redrawn from [105]).

Nonoyama et al. proposed an alternative approach for separating oxygen transfer resistance associated to GDL and electrode by performing limiting current measurement in low concentration oxygen diluted in nitrogen or helium repeated at different temperatures to exploit the known dependencies on temperature for the different diffusion type [106]. Using this approach, the authors could separate the transfer resistance terms associated to Knudsen diffusion and to diffusion in an ionomer film and showed that diffusion in the ionomer-film dominates over Knudsen for the electrodes tested.

Reshetenko and St-Pierre developed yet another procedure to separate oxygen transfer resistance in gas and ionomer phases of PEMFC electrode based again on using diluent gases of different molecular weight [107]. In this study, the limiting current of a commercial CCM was measured at a single condition (5 % O_2, 60 °C and 100 % RH) using a segmented cell for nine different diluent gases. The presence of a along-the-channel concentration gradient was accounted for to derive the transport resistance. The oxygen transfer resistance associated with the ionomer obtained with this method was between 0.4 and 0.8 s/cm, based on which $R_{O_2}^{Pt}$ values between 25 s/cm and 50 s/cm can be estimated (assuming a Pt roughness of 70), which is 2 to 5 times higher if compared to the results obtained by Owejan and Greszler.

Springler et al. proposed the use of the hydrogen limiting current to investigate and break down the transport resistances of PEMFC electrodes [108]. Using limiting currents measured in hydrogen pumping experiments allows the assessment of the effective diffusivity as a function of liquid saturation at equilibrium, unlike all other techniques based on using dilute-oxygen limiting current where water and heat production complicate this analysis. The separation of pressure dependent and independent terms is still possible using the hydrogen pumping method, based on the same rationale. The authors controlled the saturation level of the GDL by first submerging these in water followed by exposure to controlled RH in a desiccator for several hours prior to assembling the layer in the cell. The oxygen transport resistance can be obtained knowing the ratio of the oxygen and hydrogen diffusivity, which varies between 0.250 and 0.275 for Knudsen and molecular diffusivity, respectively. Chowdhury et al. used the same technique to study more in detail the pressure independent transport resistance for electrodes of different I/C [109]. In agreement with the previously mentioned studies by Greszler and Owejan, the existence of a mass-independent interfacial resistance, which contributed to 20 to 30 % of the total CL resistance was hypothesized to explain the results. Schuler et al., from the same team in LBNL, recently published a comprehensive study where the possible sources of CL mass-transport resistance were identified using H_2-pump experiments combined with changes in electrode composition, operating conditions and H_2 or D_2 are reacting gas to probe the impact of reactant MW [110]. The results of this study (see Figure 36) revealed that CL resistance is dominated by Knudsen- and ionomer-film transport resistance (40 to 70 %) followed by the interfacial resistance, and that the breakdown of the CL transport-resistance parts strongly depend on the materials used, electrode design and also operating conditions. The equivalent $R_{O_2}^{Pt}$ estimated from the values measured in H_2 for the known ratios of diffusivity is approximately 65 s/cm, approximately 6 times higher compared to the values measured by Greszler.

Recognizing the possible limitations of the approach proposed by Caulk and Baker [101] in producing transport resistance results representative of realistic operating conditions, Göbel et al. proposed a novel transient limiting current (TLC) method whereby the oxygen transfer resistance of the MEA is measured from the initial portion of the

(a)

(b)

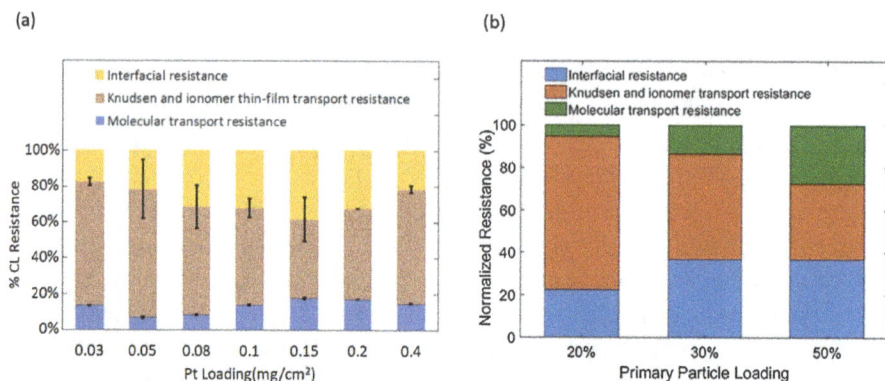

Figure 36: Breakdown of H_2 CL resistance types for (a) electrodes of different Pt-loadings using 20 wt% Pt/C and (b) for electrodes using Pt/C with different metal wt.%. Reproduced from IPO Publishing © (2019) [110].

load transient after a sudden voltage step voltage to 0.2 V after a conditioning period at steady load for a realistic operation [111]. The change in oxygen transfer resistance when condensation occurs is understood to be the result of local saturation of the diffusion medium, which in turn is related to the temperature gradients established in the MEA during the measurement. The hypothesis is that the temperature gradients and subsequently the saturation distribution in the MEA is different in realistic conditions compared to limiting conditions, such as during stationary limiting current (SLC) measurements, where the heat generation is maximum, and a quasi steady-state condition is maintained. The TLC method consists of a sequence of arbitrary conditioned load points followed by a hold at 0.2 V for 10 s, as shown in Figure 37(a). The limiting current corresponding to the specific conditioned load is the current values measured 1 s after the beginning of the voltage hold at 0.2 V. In Göbel's study, results for TLC were compared to SLC, showing that the wet region of Figure 34(a) occurs at lower current densities suggesting a difference in the distribution of liquid water in the diffusion medium. The authors also presented synchrotron X-ray radiographic and tomographic results to quantify and localize liquid water with the MEA during and after TLC testing. Although a one-to-one comparison with SLC was not shown, the results appear to be consistent with the hypothesis; that is, in MEAs using GDL more prone to flooding liquid water was first observed under the cathode ribs, consistently with previous observations (e. g., by Owejan et al. using the SLC method). For the MEA conditioned in air, liquid water was observed also over the cathode and, interestingly, anode electrode and GDL.

EIS under moderate load has been proposed to extract the MEA oxygen transport properties at realistic operating conditions. Kulikovsky developed an analytical model of the PEMFC cathode operating a low-current and low-stoichiometry using which the effective diffusivities of GDL and CCL were obtained [112]. In more recent publications, Reshetenko and Kulikovsky could also extract the oxygen transport resistance within

(a) Polarization points including transient limiting current (TLC)

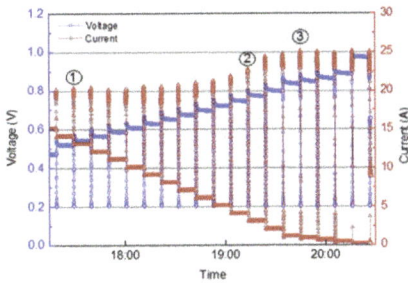

TLC for high current (wet)

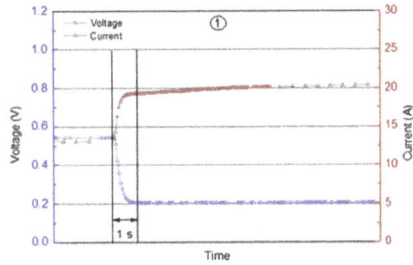

TLC for transition (wet/dry)

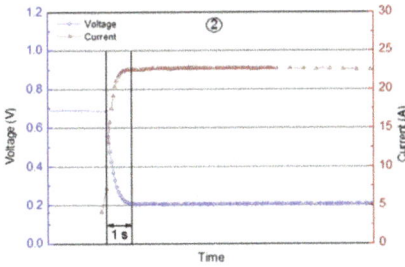

TLC for low current (dry)

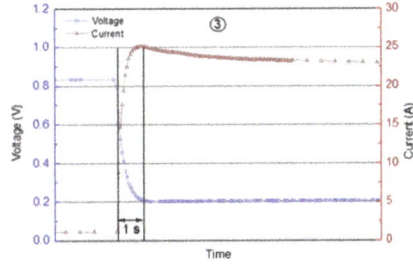

(b) Avg saturation a limiting current for different GDL types (SGL 28BC and Freudenberg H14C7)

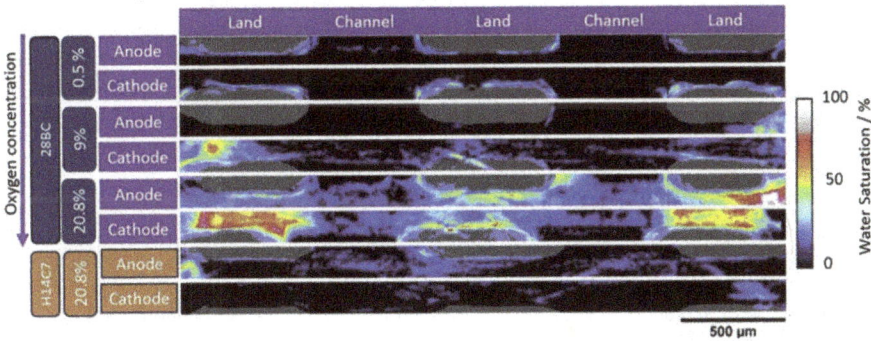

Figure 37: (a) Overview of the TLC sequence with voltage drops after different condition points (polarization curve) and (b) liquid saturation at limiting current for SGL and Freudenberg GDLs. Reproduced with permission from Elsevier © (2018) [111].

the ionomer film resistance by using a modified version of the previous impedance model and impedance data obtained using a segmented cell [113, 114]. The information contained in the impedance spectrum that allows the separation between pore and ionomer-film transport type is contained in the frequency range between 10 and 100 s Hz. The $R_{O_2}^{Pt}$ estimated from the values read from Figure 38(b) and for the reported

(a) Experimental EIS measured in H_2/air at 80°C, 1.5 bar, 100 and 50% RH for cathode and anode, respectively.

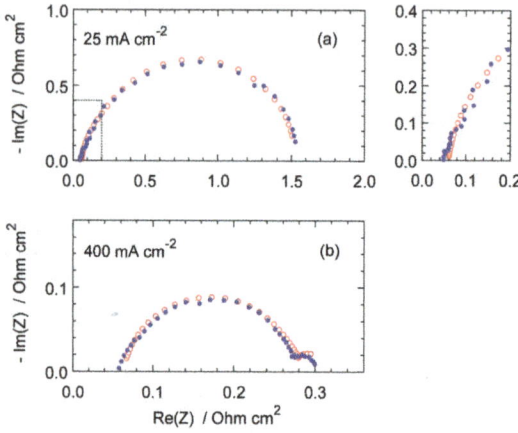

(b) Parameters obtained from model fitting.

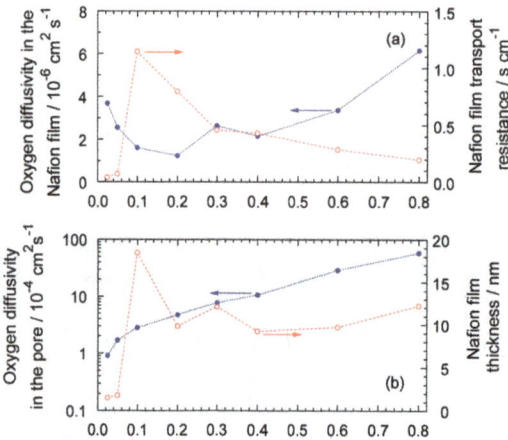

Figure 38: (a) Complex impedance and (b) breakdown of the oxygen transfer resistance into ionomer and molecular components vs. current density obtained by complex fitting using the impedance model proposed by Kulikovsky. Reproduced with permission from Royal Society of Chemistry © (2019) [114].

Pt-loading is approximately 50 s/cm at low loads decreasing to about 10 s/cm at high loads, consistent with the values measured by Greszler by limiting current.

In summary, various approaches have been developed to measure oxygen transfer resistance and to break it down into parts associated with MEA components and types. Independently on the approach adopted, the O_2 transfer resistance of the cathode electrode was found to be higher than what estimated theoretically for an ionomer-film of thickness characteristic of the CCL design, where the majority of the O_2 transfer losses were hypothesized. The existence of an interfacial resistance has been suggested to explain this observation, but the physical processes involved have not yet been proved. Since we do not have a clear understanding of what this interfacial resistance represents, it is difficult to assess if any of the approaches proposed to measure it is preferable. For now, we can compare the magnitude of $R_{O_2}^{Pt}$ obtained using the different methods to assess for consistency, although we should be aware that the electrodes tested were not identical.

Limiting current measurements in diluted O_2 pose a challenge for the interpretation of flooding conditions possibly because of differences in local saturation of the pores between normal operation and the limiting conditions at which the measurements are performed. H_2 limiting current measurements have been proposed to overcome this issue. However, even with this approach the MEA conditions are quite remote normal operation, and a way to translate the H_2 transfer into O_2 transfer resistance is still required for the processes occurring in the ionomer and interface.

Recently proposed transient and EIS methods might offer a valid approach for the measurement and breakdown of oxygen transfer resistances at realistic conditions, possibly also reducing the testing time. However, the reproducibility and robustness of these methods has not been verified yet.

In conclusion, reliable quantification and breakdown of these transfer resistances is fundamental for maximising accuracy and resolution in voltage loss analysis.

9.8.7 Other properties relevant for advanced VLB

So far, our focus was on measuring effective properties of the MEA related to the dominant losses (ohmic, cathode kinetics, proton and oxygen transfer losses).

At high current densities (> $3\,A/cm^2$) and/or at extreme operating conditions (dry and wet) also other losses, otherwise neglected, need to be accounted for, and the coupling between cell performance, and thermal- and water-management in the MEA need to be considered.

The MEA water management—intended as the combined effects of water, proton and heat transfer—has important effects on cell performance. At dry conditions, the product water can back diffuse from the cathode increasing the cathode and membrane hydration, while at cold and wet conditions, water crossover from cathode to anode can alleviate the effects of cathode flooding.

Based on the current understanding of transport phenomena in PEM [10], the key properties affecting water management are the effective water diffusivity, the electroosmotic drag coefficient and the hydration isotherm of the PEM membrane, as well as their dependencies on temperature.

Membrane water diffusivity is measurable by *ex situ* techniques, such as dynamic vapor sorption and nuclear magnetic resonance, and it has also been measured with high accuracy in-situ on MEAs using a steady permeation method [115].

The electroosmotic drag coefficient (EOD) of a membrane is a measure of the number of water molecules that are carried with each proton, and, as such, it couples proton-to water-transfer effects. The EOD coefficient is not expected to vary widely between membranes, and it is not measured routinely; however, this property can be quantified *in situ* by hydrogen-pump measurements combined with water crossover [116, 117]. Hydration isotherms related the number of moles of water adsorbed by each equivalent in the membrane (λ) at equilibrium with the vapor surrounding the membrane. These isotherms are obtained by dynamic vapor sorption as described elsewhere [35], and are measurable for membranes and for electrodes.

For the purpose of loss analysis of nonaged samples (BOL MEA), it is not necessary to measure these properties, as they are usually available from suppliers or publications. However, in case of aged samples it is relevant either to verify indirectly if there are changes in water management of the MEA, or even to try measuring some of these properties directly using *in situ* or *ex situ* techniques.

A convenient way to monitor the water management behavior of MEA is by measuring the water crossover rate, which is defined as the fraction of water produced that is crossing over to the anode. This rate, which combines the effects of all the properties discussed above, is a function of the operating conditions and depends quite strongly on the MEA design used and on its state-of-health. Water crossover rates of a cell can be measured using a gravimetric approach, based on quantifying the amount of condensed water at cathode and anode outlets, or, for more rapid measurements, using dedicated test-stations equipped with high accuracy RH-sensors, and fluid flow meters. Yau et al. developed a testing bench for the rapid measurement of water crossover rates [118–120]. The tool was then used to study the behavior of MEAs using different GDL designs and configurations [120]. With this tool, the effective water diffusivity of the membrane can be obtained directly by measuring the crossover rates at zero load for different anode and cathode RHs, while the electroosmotic drag can be extracted using a model to fit the results measured for different loads and operating conditions [121].

The properties of gas diffusion layers have important effects on water management, by the combined effect of mass- and heat-transfer. Mass-transfer properties of the GDL, namely the dry effective diffusivities for oxygen and water vapor, are implicitly included in the mass transfer resistance measurable by limiting-current. The effect of liquid saturation on the effective diffusivities of the GDL is also included in R_{MT} measured at flooding conditions using one of the techniques described earlier.

When applying a VLB approach based on a detailed physics-based model (see Section 9.5), it is also necessary to obtain capillary and permeation properties as function of saturation for the GDL used [67]. These properties are measurable *ex situ*, but they are not collected routinely and cannot be easily measured in aged samples.

The electrical and thermal properties of GDL are also measured using *ex situ* techniques [122] and are usually available from suppliers.

For VLB at high current density, it is relevant to quantify the contribution of electrical and thermal contacts. High-frequency resistance (HFR) provides a direct measurement of the total ohmic resistance of an operating cell, that as seen in Section 9.8.5, can be separated into electrical and proton contributions. Electrical contact resistance constitutes a significant portion of the total ohmic resistance, depending on the flow-field design and materials used, it typically exceeds 30 % of the HFR in a fully humidified MEA (i. e., > 15 mOhm $\cdot$ cm^2). A method for measuring the electrical contact resistance *in situ* was already discussed earlier [91]. An alternative *in situ* approach based on adding voltage probes to the plate has been recently proposed by Lædre et al. [123]. Kleemann et al. developed a suite of *ex situ* techniques to measure GDL properties, including the contact resistance at the GDL/Plate and MPL/CCL interfaces [124]. Using these *ex situ* techniques, it is possible to break down the total electrical resistance into components, and by applying the Ohm's law quantify the associated losses. Thermal contact resistance at the GDL/Plate interface has been measured *ex situ* by Bahrami, Stumper and team [125–128].

When considering the breakdown of electrical and thermal resistances, it is important to consider their dependence on the compressive force applied to the GDL, as these quantities can vary significantly with the applied force and can show hysteresis upon cycling.

Lastly, we consider the losses associated with the anode. As seen in Section 9.4, the HOR loss is typically negligible; however, at high current densities and/or in diluted hydrogen the anode overpotentials can easily exceed 20 mV even for a healthy MEA.

HOR and HER kinetics parameters have been measured for various electrocatalysts by Sheng et al. [129] using the rotating disk electrode technique combined with EIS. Considering the low sensitivity of cell performance versus the HOR parameters, it is not necessary to remeasure these parameters for the purpose of VLB. It should suffice to measure the platinum roughness factor for the anode using the same methods described earlier (see Section 9.8.4). The effective proton resistance of the anode should be also measured (see Section 9.8.5) or estimated if the electrode composition is known, since the loss associated with the proton-transport in the anode layer is typically the largest contributor of the total anode loss. Nevertheless, there might be specific situations where the anode overpotentials must be measured *in situ* for the specific MEA, as, for example, in case of degraded/damaged electrodes or poisoning. The HOR activity and the associated kinetics parameters (see Equation (19)) are measured in a MEA by hydrogen-pump measurements combined with EIS. The procedure and analysis were described by Neyerlin et al. [24], Wiezell et al. [130, 131] and Durst et al. [132].

9.9 Choosing the appropriate VLB approach

Voltage loss analysis can arise from various motivations, which can be broadly grouped into three main categories:

-   Diagnosis – where the goal is to identify the causes of MEA performance loss, instability or more generally malfunctioning; Papadias et al. provided an excellent example of loss analysis applied to samples undergoing durability testing [28].
-   Engineering – where the objective is to verify that changes in the design of the MEA, or of one of its components, provide hypothesized/expected performance improvements, and to unveil interactions and/or trade-offs between design changes, performance and durability; Murata et al. [133] provided an excellent example of the use of VLB for MEA engineering and optimization; more recently publications from Gasteiger's team also demonstrate the use of VLB for electrode engineering purposes [9, 29].
-   Fundamental understanding – where the focus is on isolating sources of losses to identify correlations between the various MEA effective properties and performance; in this case, VLB can be used to guide and validate physics-based models correlating material properties, electrode (or other component) design and microstructure, to the resulting overpotentials; Ramaswamy et al. provide an example of using VLB to understand the impact of different ionomers on the effective properties of the cathode layer and the resulting performance changes [31].

The selection of the most appropriate VLB approach is based on the nature of the problem or issue being addressed. Two important factors to consider are whether a complete or a partial breakdown is required and what resolution level (in mV) is desired for the identifiable loss sources.

The *completeness* of the VLB typically depends on the scope of the investigation. For example, in case of electrode design and engineering the focus is typically on optimizing the ionomer content (I/C) and electrode structure, for which it is relevant to measure only a few effective properties of the layer, such as the effective kinetics parameters and the proton- and oxygen-transfer resistances. Furthermore, new electrodes are typically compared versus a baseline design and *in situ* measurement can be limited to what is relevant for the cathode. On the other hand, when dealing with an unexpected performance degradation without a clear hypothesis on the causes it is preferable to opt for a complete VLB. The results of the VLB should reveal where the losses originate, suggesting directional hypotheses.

The *resolution* of the VLB approach can be defined as the minimum overpotential (or overpotential difference) that can be distinguished. This resolution is necessarily related to the variability of the measurements. In the case of a complete VLB, where all overpotentials are estimated and summed up to predict the cell performance, the resolution cannot be higher than the variability in measuring cell performance, since this is eventually used to validate the breakdown analysis. For example, overpotentials smaller than 10 mV cannot be simultaneously distinguished and validated if the uncertainty in measuring performance at a given current density is ±5 mV. On the other hand, when focusing on a specific loss the resolution can be assumed to be a function of the uncertainty for the measurement of the effective properties associated to that loss (after accounting for the propagation of uncertainties). Evidently, the VLB resolution will show different variabilities depending on the operating conditions analogous to performance measurements, as already discussed in Section 9.8.2.

The selection of the most appropriate VLB strategy also depends on how much/what kind of information is available. Typically, performance measurements and details on the MEA composition and manufacturing process are known. During the validation and verification stages of a new FC product (DV Testing), a complete history of the MEA—including variations in performance and properties measured on differential cells—is normally available. In the case of products returning from the field, there might be only rather generic indications of out-of-spec behaviors (i. e., lower than expected power output) requiring a root-cause investigation, which is typically comprised of several analysis steps, including the VLB of one or more MEAs pulled out from the stack.

Eventually, the selection of the approach might also depend on the state-of-health of the MEA and on the accessible diagnostic tools and methods. Based on these limitations and on the key questions, one should choose the VLB approach that has the highest probability of providing reliable results.

In conclusion, selecting a VLB approach involves selecting the "right tool for the job," i. e., one should choose the appropriate resolution level for the issue being addressed as in sculpting one should use a point chisel for coarse surfaces and a rasp for refining.

Based on our experience on voltage loss analysis, we recommend the use of an *increasing complexity approach*, which consists in progressively refining the resolution of the analysis until overpotential types and sources can be distinguished down to a desired/required level. This is achievable by integrating into the analysis additional results from dedicated measurements, and by selecting progressively more comprehensive MEA models, compatible with the new measurements.

In the following example, we demonstrate a possible VLB workflow allowing insight into the principal MEA losses and associated processes by exploiting the *increasing complexity approach* mentioned above using four levels of refinement. The schematic of Figure 39 illustrates how the approach starts from simple analysis of the polarization curves and incrementally becomes more sophisticated, requiring specific *in situ* measurements and involving more comprehensive MEA models, which—while consistent with simpler versions used in the earlier levels—progressively include processes/effects previously ignored or assumed negligible.

This VLB example is applied to two polarization curves: the first was measured on a fresh MEA (BOL), while the second was obtained for the same MEA after applying the accelerated stress test (EOT) to induce Pt dissolution proposed by the US Drive Fuel Cell Technical Team [84]. Performance was measured in voltage control mode using a differential cell at 85 °C in H_2/Air at constant flowrate, 2.9 and 2.7 bar gas pressure (an/cath) and at 60 % RH; the stoichiometries were greater than 7 for all loads. The same experimental data are represented in Figure 40 to Figure 44. Information on the MEA composition and design was available from the start (e. g., electrode PGM loading and I/C, mem-

Measurements/Information	VLB Approach	MEA Model
Polarization Curves	**Level 1:** Polarization Change Analysis	0D empirical
+		
MEA design information (Composition, literature data)	**Level 2:** Attempt VLB using 1D model and inputs estimated from MEA info	1D isothermal single-phase
+		
In-situ measurements of MEA effective properties	**Level 3:** Attempt VLB using 1D model and inputs from dedicated in-situ measurements	1D isothermal single-phase
+		
Additional in-situ measurements, ex-situ measurements of specific components	**Level 4:** Attempt VLB using comprehensive model and inputs from both in-situ and ex-situ measurements	Comprehensive (various options)

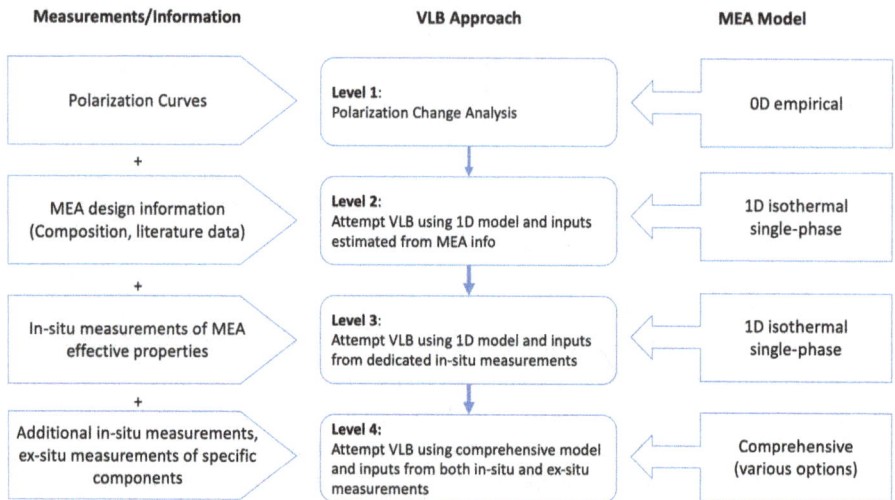

Figure 39: Schematic of the increasing complexity approach to voltage loss analysis. Top to bottom: VLB Levels 1–4 with increasing complexity.

brane and GDL types and properties), in addition MEA effective properties described in Section 9.8—such as cathode and anode roughness factors, ORR kinetics parameters, proton resistances and oxygen transfer resistance of the cathode—were measured both at BOL and EOT. However, here these inputs were introduced in the analysis only gradually through the workflow for illustrative purposes.

Level 1 in Figure 39 represents a simplified loss analysis that one would attempt when only polarization curves are available. In this case, we applied a direct fit of Equation (1), as described in Section 9.3, as well as the polarization change analysis shown in Section 9.6 to identify what loss types are dominant. The model fitting results and corresponding loss breakdown are shown in Figure 40, and also summarized in Table 1 and Figure 46. In general, the empirical model fits accurately both polarization curves, with residual within ±5 mV over the entire curve.

The polarization change in Figure 41 shows a linear trend with intercept at approximately 40 mV, and a slope corresponding to 100 m$\Omega \cdot$ cm^2. Based on the principles illustrated in Section 9.6, the "fingerprint" observed is indicative of combined catalyst and proton- and/or electron-transport degradation. According to (32), a drop in ORR of 40 mV should correspond to 70 % reduction in cathode roughness factor (rf$_{CCL}$). The same can also be deduced by comparing the ORR kinetics fitting parameter (i @ 0.9 $E_{iR\text{-free}}$). No further insight can be gained from this analysis unless additional information is integrated.

Level 2 in Figure 39 represents a VLB approach that one would attempt when in addition to the polarization curve some basic information on the MEA is available. In this case, we assume that the MEA composition was known, and that some properties of the components were obtained from literature or from previous work on similar materials.

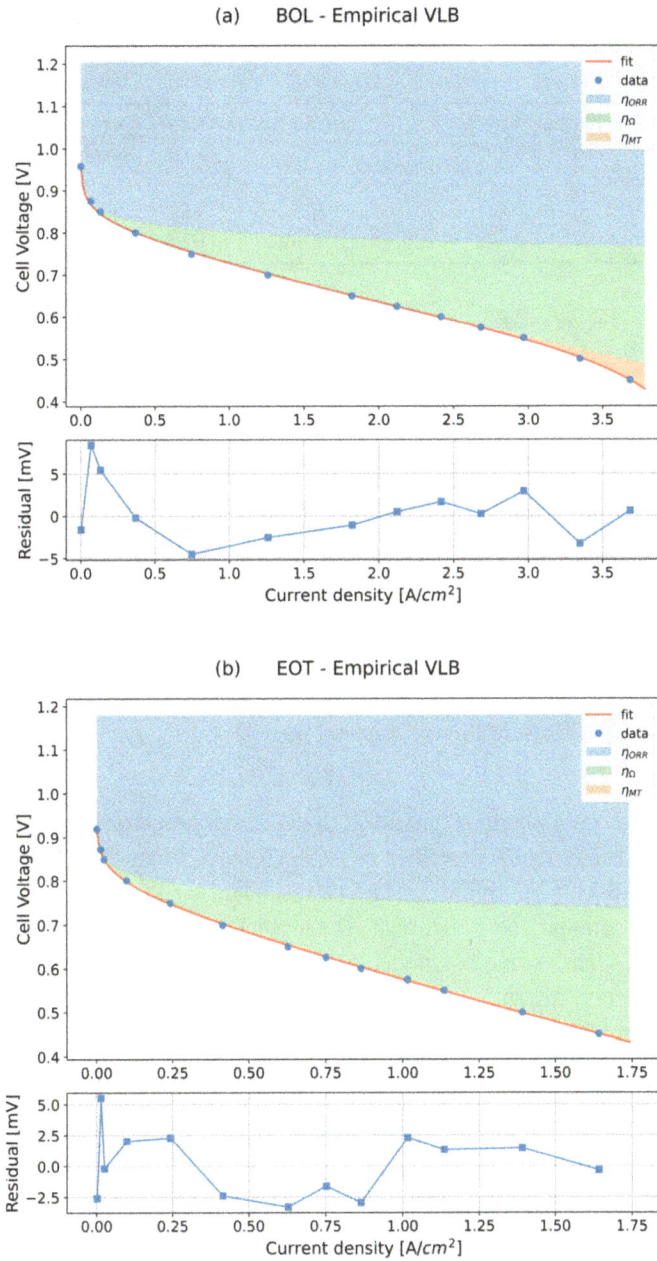

Figure 40: VLB Level 1: Application of the empirical model of Equation (1) for the voltage breakdown of MEA performance measured in H_2/Air at 85 °C, 60 % RH at (a) BOL and (b) EOT, after application of a Pt dissolution AST; (c) Voltage-difference analysis applied to the same data set.

Table 1: Outputs of VLB Level 1: Results obtained from fitting Equation (1) to the data of Figure 40.

Parameter	Description	Value BOL	Value EOT	Units
i_0	apparent ORR exchange current	5.968E-07	4.129E-7	A/cm^2(MEA)
i @ 0.9 V $E_{iR\text{-free}}$	apparent ORR current at 0.9 V cathode potential	23.724	6.412	A/cm^2(MEA)
b	Tafel Slope	60.9	66.8	mV/dec
R_Ω	Resistance	73.5	172.6	m$\Omega \cdot$ cm^2(MEA)

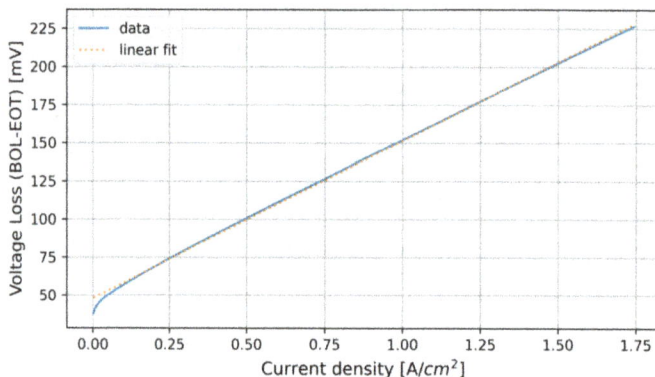

Figure 41: Voltage-difference analysis applied to the same data set of Figure 40.

From this information, we could obtain estimates of the effective properties listed in Table 2, and attempt using the 1D model described in Section 9.4 to estimate MEA losses. Anode losses were excluded from this analysis. The VLB results for the BOL sample are shown in Figure 42(a) and summarized in Figure 46. The predicted performance agrees well with the data up to 1 A/cm^2, while the model (VLB prediction) overestimates the losses at higher currents up to 3 A/cm^2, also showing a deviation in curvature of the polarization curves at even higher loads.

Figure 42(b) shows the VLB results applied to EOT. Two different predictions were attempted:

A. only the effect of a change in cathode roughness factor was considered (as per polarization change analysis).

B. Ohmic effects caused by the appearance of a Pt-depletion band in the cathode electrode at the membrane-electrode interface were considered in combination to the rf$_{CCL}$ reduction of case A. The formation of a Pt-rich band within the membrane and of an adjacent Pt-depleted band in the electrode after Pt-dissolution AST has been documented elsewhere [134, 135]. In this analysis, we assumed an extreme situation with complete removal of Pt from the depleted band in the CCL; this should cause an increase in ohmic resistance proportional to the thickness of the depletion band, which in turn, was estimated from the rf$_{CCL}$ drop.

Table 2: Inputs for VLB Level 2: MEA information and estimated effective properties used as inputs for the loss analysis of the data in Figure 42.

Parameter	Description	Value	Units	Source
Design Info				
L_{Pt}	Cathode Pt loading	0.25	[mg/cm^2(MEA)]	–
ECA	ORR catalyst electrochemical surface area	50	[m^2/g]	–
t_{PEM}	Thickness	15	[μm]	Gore Select
EW$_{PEM}$	Equivalent weight	850	[g/eq]	–
t_{GDL}	Thickness	170	[μm]	Freudenberg HC14C10
Estimated MEA effective properties (VLB Inputs)				
rf$_{CCL}$	Cathode roughness factor	125	[cm^2(Pt)/cm^2(MEA)]	estimated
$R_{\Omega,tot}$	Ohmic resistance of MEA	62	mΩ · cm^2(MEA)	estimated
R_{H^+-CCL}	Proton resistance of cathode layer	175	mΩ · cm^2(MEA)	estimated
R_{NP}	P-independent O$_2$ transport resistance	10	[s/m]	estimated
R_P	P-dependent O$_2$ transport resistance	70	[s/m]	estimated
$i_{o,ORR}^{Ref}$	ORR Exchange current density	2.47E-8	[A/cm^2(Pt)]	from ref. [23]
a_C	ORR Charge transfer coefficient	1.0	NA	[23]
γ_{O_2}	ORR Reaction order	0.54	NA	[23]
$E_{a,ORR}^{rev}$	ORR Activation energy	49	[kJ/mol]	[23]

In the first case (prediction A), the results show that the predicted voltage is accurate only for the first couple of data points dominated by ORR overpotentials. On the other hand, in the second case (prediction B) cell voltage matches the measurements up to 0.5 A/cm^2, but the VLB prediction is still inaccurate at higher loads suggesting that more measurements—including the structural characterization of the electrode (see Part I)—and dedicated MEA models are necessary.

The next level of the analysis (Level 3 in Figure 39) is representative of a VLB based on measuring selected effective properties, as discussed in Section 9.4. The effects of both anode and cathode were included. As for the effective properties, these were measured on the specific MEA both at BOL and EOT and included: cathode and anode roughness factors; ORR kinetics parameters; total ohmic resistance and effective proton resistance of the cathode as function of RH; oxygen transfer resistance of the cathode as function of gas pressure and HFR for loads > 0.5 A/cm^2. The values are summarized in Table 3. For the EOT results, the same cases described for the analysis of Figure 42(b) were applied. The results of the loss analysis are shown in Figure 43.

The VLB results for BOL (Figure 43(a)) show that cell voltage is predicted with sufficient accuracy (±5 mV) up to 3.0 A/cm^2, after which the 1D model overpredicts performance.

The VLB results for EOT Figure 43(b)) show some analogies with those obtained in the previous level of analysis (Figure 42(b)). In this case, the cathode roughness was measured both at BOL and EOT, and the loss observed was approximately 50 %, in contrast to the 70 % predicted by voltage-difference analysis (Figure 40(c)). A comparison of the ORR parameters (Table 3) shows that in addition to the reduction in Pt roughness factor

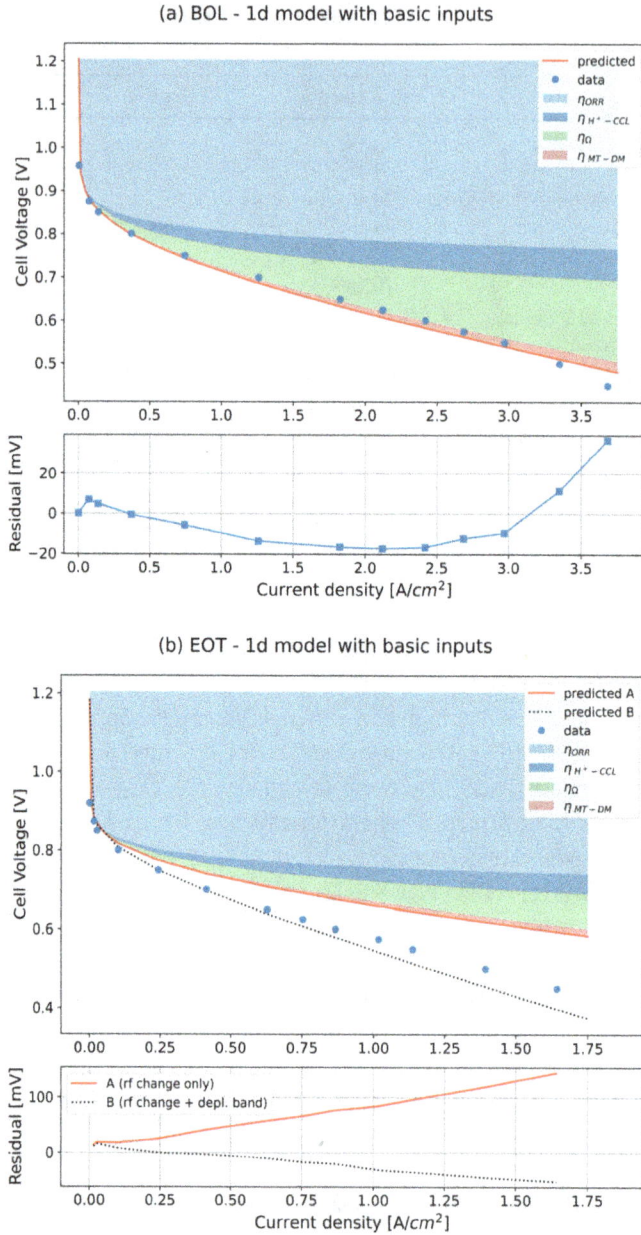

Figure 42: VLB Level 2: 1D with basic inputs: Application of the 1D isothermal model (see Section 9.4) for the loss breakdown analysis of MEA performances measured at (a) BOL and (b) EOT, as described in Figure 40, using only basic inputs obtained from the MEA supplier and from the literature. In the VLB analysis of EOT "Prediction A" was obtained using the same input parameters as for the BOL case except for the cathode roughness factor, which was reduced according to the voltage-difference analysis of Figure 40(c); "Prediction B" was obtained assuming the formation of a Pt-depletion band in the cathode layer leading to an increase in apparent ohmic resistance.

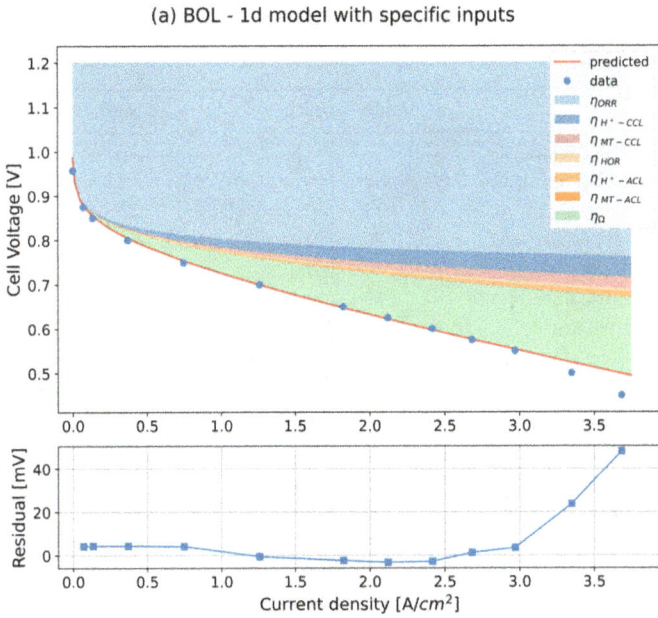

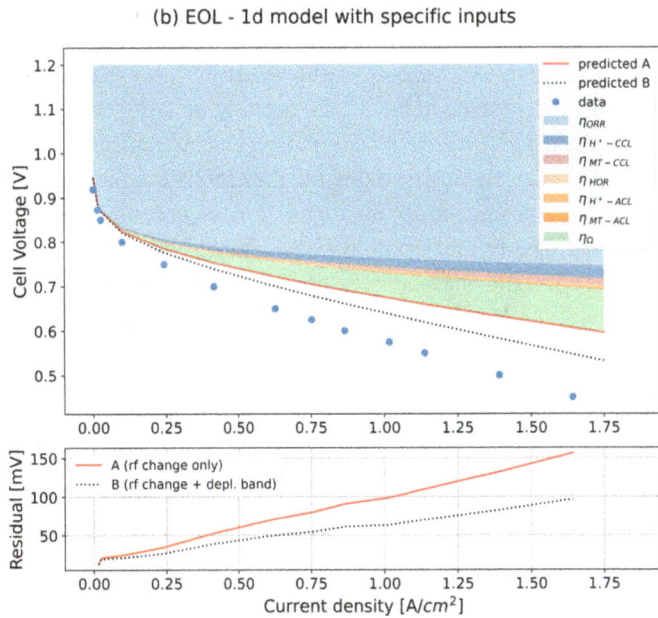

Figure 43: VLB Level 3: 1D with specific inputs: Application of the "Effective-Property Based VLB" approach (EPB VLB of Section 9.4 (a) BOL and (b) EOT using specific VLB inputs obtained from *in situ* measurement of effective properties for the same MEA. "Prediction A" and "Prediction B" were obtained using the same assumptions as in Figure 42(b).

Table 3: Inputs for VLB Level 3: Effective properties used as inputs for the loss analysis of the data in Figure 43.

Parameter	Description	Value	Units	Method/Source
Measured MEA effective properties (VLB Inputs)				
rf_{CCL}	Cathode roughness factor	110	[cm^2(Pt)/cm^2(MEA)]	§ 9.8.4
$R_{\Omega,tot}$	Ohmic resistance of MEA	47	m$\Omega \cdot$ cm^2(MEA)	§ 9.8.5
R_{H^+-CCL}	Proton resistance of cathode layer	80	m$\Omega \cdot$ cm^2(MEA)	§ 9.8.5
R_{NP}	P-independent O$_2$ transport resistance	11	[s/m]	§ 9.8.6
R_P	P-dependent O$_2$ transport resistance	60	[s/m]	§ 9.8.6
$i_{o,ORR}^{Ref}$	ORR Exchange current density	3.21E-8	[A/cm^2(Pt)]	§ 9.8.4
a_C	ORR Charge transfer coefficient	1.03	NA	§ 9.8.4
y_{O_2}	ORR Reaction order	0.75	NA	§ 9.8.4
$E_{a,ORR}^{rev}$	ORR Activation energy	54	[kJ/mol]	§ 9.8.4
rf_{ACL}	Anode roughness factor	25	[cm^2(Pt)/cm^2(MEA)]	§ 9.8.4
$i_{o,HOR}^{Ref}$	HOR Exchange current density	0.535	[A/cm^2(Pt)]	lit. value [24]
$a_{Aa} + a_{Ac}$	HOR Charge transfer coefficients	1.0	NA	lit. value [24]
y_{H_2}	HOR Reaction order	0.75	NA	lit. value [24]
$E_{a,HOR}^{rev}$	HOR Activation energy	25	[kJ/mol]	lit. value [24]
R_{H^+-ACL}	Proton resistance of anode layer	10	[m$\Omega \cdot$ cm^2(MEA)]	§ 9.8.5

also a decrease in specific ORR activity was present although even this was not sufficient to justify the observed drop in kinetics at low loads, possibly caused by effects on MEA conditioning occurring during AST discussed elsewhere [136]. The estimated cell voltage was significantly higher than the measured even after including the possible ohmic effects due to Pt depletion (prediction B), highlighting once more the need for a deeper investigation and/or of a more representative MEA model. Nevertheless, this VLB attempt still provides some insight. For example, the trend of the residuals in Figure 42(b) increases linearly with the current density. This observation can be used to narrow down hypotheses for improvement.

In the last level of analysis (Level 4 in Figure 39), a 1D numerical model of the MEA including temperature and water management effects was considered. The model reproduces the non-isothermal single phase sandwich model proposed by Gu et al. [21] and takes as inputs material properties (see Table 4), which in this case were adjusted to match the same effective properties introduced in Level 3 plus other properties measured *ex situ* for the same type of components used here.

The VLB results for BOL are shown in Figure 44. Similarly, to the previous case (Figure 43(a)), the model accurately predicts performance up to 3 A/cm^2, and even at higher loads the residual is reduced to 20 mV, which is close to the measurement error. However, it can be observed by comparing Figure 43(a) and Figure 44 that the relative distribution of losses varies at high current density. Figure 45 represents the predicted cathode temperature, HFR and average water uptake of the membrane. The cathode layer temperature increases almost linearly, reaching 100 °C at 3.5 A/cm^2, which illustrates the

Table 4: Inputs for VLB Level 4: Additional ME properties used as inputs for the loss analysis of the data in Figure 44.

Parameter	Description	Value	Units	Method/ Source
$R_{\Omega,other}$	Electrical resistance MEA	32	$[m\Omega \cdot cm^2]$	§ 9.8.7
$\kappa^T_{Eff-GDL}$	Thermal conductivity GDL	0.17	$[W/(m \cdot K)]$	§ 9.8.7
$\kappa^T_{Eff-PEM}$	Thermal conductivity PEM	0.13	$[W/(m \cdot K)]$	§ 9.8.7
$R^T_{GDL|PLT}$	Thermal contact resistance	5.0E-5	$[m^2 \cdot K/W]$	§ 9.8.7
$\lambda_{PEM}(a = RH)$	PEM water uptake	$0.3 + 10.8 \cdot a - 16.0 \cdot a^2 + 14.1 \cdot a^3$	$[mol\ H_2O/eq]$	
$\kappa^{H^+}_{PEM}(\lambda)$	Proton conductivity PEM	$32.1 \cdot (\frac{18 \cdot \lambda}{400 + 18 \cdot \lambda} - 0.075)^{0.97}$ $\cdot \exp(\frac{1.5e4}{R} \cdot (\frac{1}{303} - \frac{1}{T}))$	$[S/m]$	§ 9.8.5
$\kappa^{H^+}_{Eff-CCL}(\lambda)$	Proton conductivity CL	$5.8 \cdot (\frac{18 \cdot \lambda}{400 + 18 \cdot \lambda} - 0.077)^{0.94}$ $\cdot \exp(\frac{1.5e4}{R} \cdot (\frac{1}{303} - \frac{1}{T}))$	$[S/m]$	§ 9.8.5
$\mathcal{D}^{H_2O}_{PEM}(\lambda)$	Water diffusivity PEM	$5.39 \cdot 10^{-6} \cdot (1 + 2.7 \cdot 10^{-3} \cdot \lambda^2)$ $\cdot [1 + \tanh(\frac{\lambda - 2.6225}{0.8728})] \cdot \exp(\frac{-3343}{T})$	$[m^2/s]$	
$n_D(\lambda)$	Electro-osmotic drag coefficient	$1.2 \cdot \tanh(\lambda/2.5)$	$[n.u.]$	

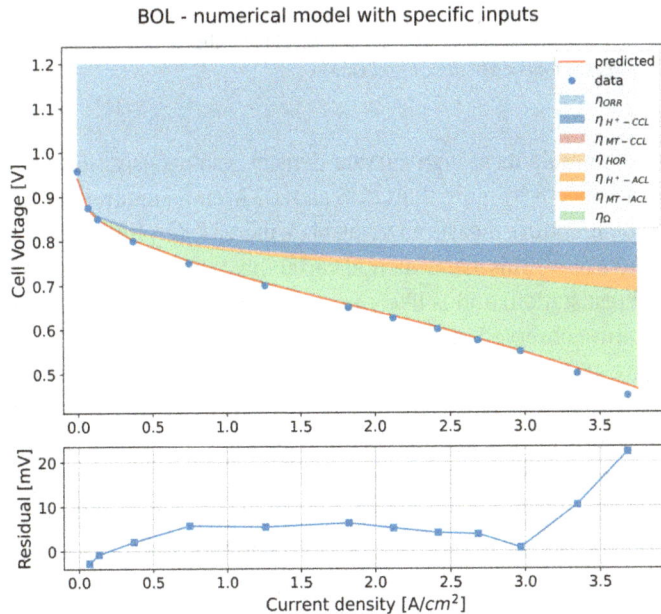

Figure 44: VLB Level 4: Numerical with specific inputs: Application of a 1 D numerical MEA model including temperature and water management effect for the VLB of BOL data.

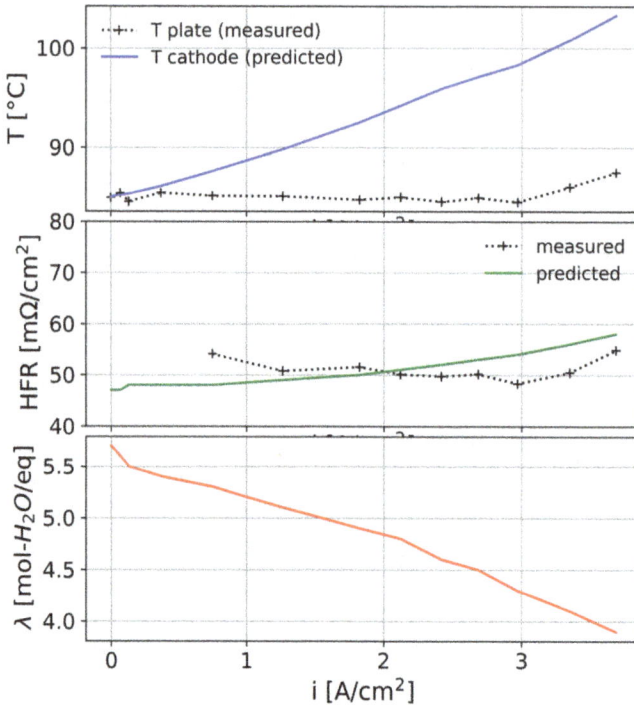

Figure 45: VLB Level 4: Numerical with specific inputs: model predictions of the cathode temperature, HFR and average membrane water uptake as function of the current density.

importance of through-plane gradients at high current density. Analogously, the average membrane hydration also drops by more than 20 % above $3\,A/cm^2$. Finally, it is also important to note that the temperature measured at the plate increased by 2 or more degrees for loads higher than $3\,A/cm^2$. This experimental variability is likely related to the hardware design (cell and test station) and is likely to contribute to the measurement variability and to the curvature observed in the measured polarization.

The bar charts of Figure 46 show a comparison of the overpotentials estimated using VLB Level 1–4 for the analysis of BOL MEA performance. In the kinetic region of the polarization curve (Figure 46(a), the four methods are in reasonable agreement, with a small deviation for the empirical approach, which underpredicted performance due to fitting effects. In the ohmic region (Figure 46(b), it is apparent that the empirical (Level 1) VLB incorrectly attributes to ohmic losses the overpotentials pertaining to the cathode- and anode-layer proton resistances. Finally, for the high current density region (Figure 46(c) it can be observed that the relative magnitude of the losses varies depending on the type of model used. For example, VLB Levels 3 and 4 coincidentally predict the same total overpotential at this current density but also show a different distribution of the losses due to the combined effect of thermal and water management. Furthermore, the difference between predicted and measured cell voltage is within ±5 mV for both

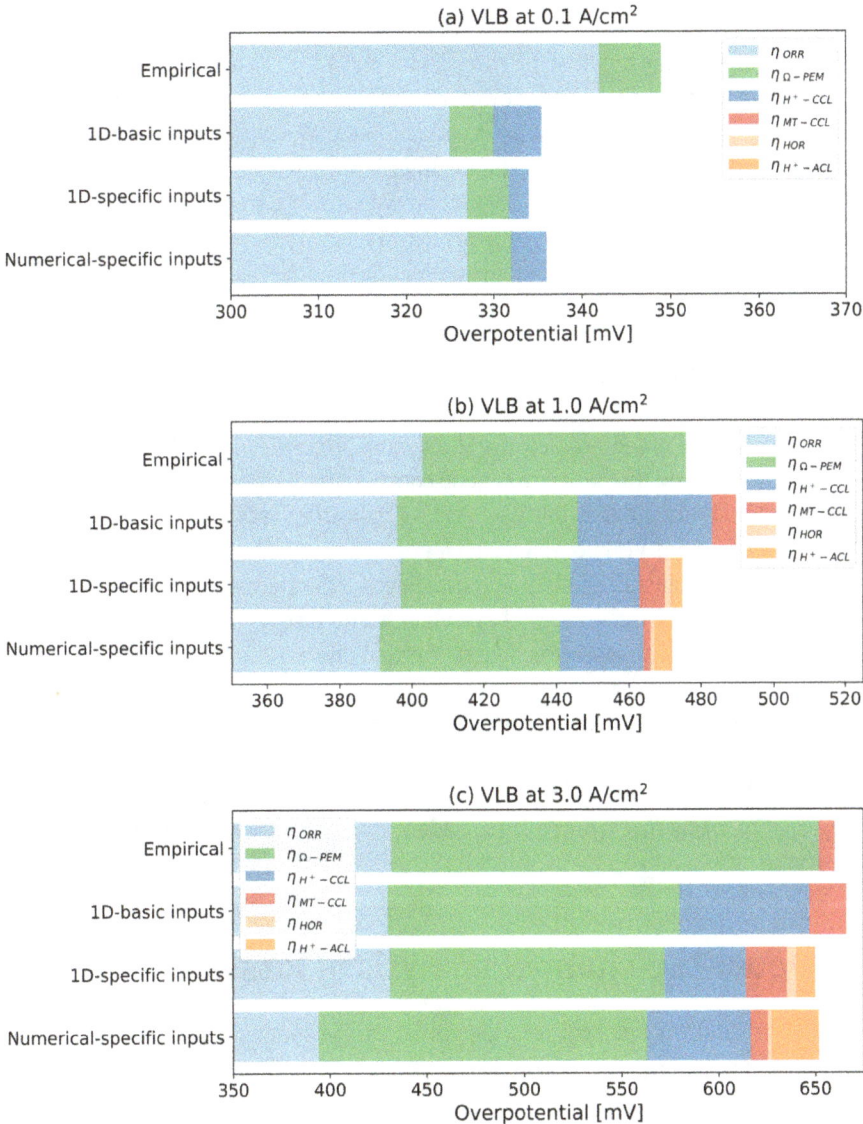

Figure 46: Estimated overpotentials at (a) 0.1, (b) 1.0 and (c) 3.0 A/cm² using VLB Levels 1–4 for the MEA performance at BOL.

VLB approaches (compare Figure 43(a) to Figure 44(a)) highlighting that this comparison (i. e., predicted vs. measured voltage) is a necessary but not sufficient condition for the validity of the VLB results. In this case, it is more likely that the results obtained at Level 4 are correct given that these integrate additional properties and processes of the MEA and the larger residual for Level 3 at > 3 A/cm².

In summary, from this example we can conclude the following:

- The empirical VLB approach (VLB Level 1) provides reasonable estimates in the kinetic region and, for this specific case, useful insight when applied to compare polarization results using the voltage difference approach. However, at higher current densities it does not allow for a correct attribution of the type of loss and of the component where this is originated. On the other hand, the voltage different approach can provide useful insight if the fingerprint type shows a simple trend (e. g., linear) as for this example.
- The VLB using a 1D isothermal model with basic input derivable from MEA composition (VLV Level 2) can provide a reasonable estimate of the main overpotentials up to moderate current densities, although representative properties (e. g., ORR kinetics and transport resistances) might not be readily available other than for commercial catalysts.
- The loss analysis based on measuring effective properties of the same MEA (VLB Level 3) provides in this case a very good compromise between accuracy in predicting overpotentials and feasibility in terms of measurements.
- The loss analysis of VLB Level 4, using a numerical model including the dominant processes (most sensitive) and input parameters calibrated using the effective properties measured for the same MEA, provides useful insight on how the distribution of losses can shift at high loads, where thermal and water management effects in the through-plane direction cannot be ignored.
- Finally, none of the approaches attempted here could resolve the loss sources in the degraded sample (EOT), showing how specific measurements, analysis of the microstructural changes within the MEA, and models adaptable to the type of degradation occurring need to be considered in order to resolve the degradation mechanisms.

9.10 Down-the-channel analysis of losses

The voltage breakdown concepts described so far can be adapted to the case of high reactant utilization, that is, when gradients are present along-the-channel, and the cell is no longer operating in differential mode. In this case, the loss analysis should be applied locally and a *VLB distribution* or *VLB map* over the active area is obtainable. At a local level, that is at a defined point over the active area, the same procedure discussed earlier for the MEA sandwich in a differential cell is valid. Local MEA properties together with local channel conditions, including current density, are required as inputs for this analysis. In general, this implies the use of advanced diagnostic tools, such as a segmented cell, capable of simultaneously mapping current density, temperature and HFR/EIS. In the case of aged samples, local CV (potentiodynamic) capabilities are also desirable. In the case of fresh MEA samples (BOL) with uniform compositions over the entire active area, it is appropriate to assume that effective properties are uniform and to measure the same for entire MEA or for a representative sample (or a cut-out) using a differential cell.

The use of segmented cells for current mapping both in steady-state and transient conditions has been used extensively for MEA and unit cell diagnostics. Stumper et al. [121, 137, 138] developed a methodology named MRED, which provides insight into the distribution of liquid water in a fuel cell under operation and is based on the measurement of current transients using a segmented unit. Gu et al. used a segmented cell capable of simultaneously measuring current density and HFR combined with water crossover measurements for the purpose model validation [21].

An analysis of such mapping tools is beyond the scope of this review; more details can be found elsewhere [139–141].

The change in local conditions down-the-channel implies that the equilibrium potential changes from point to point, therefore, an additional loss is expected. The difference in equilibrium potential between a point down-the-channel and a reference location, conveniently chosen as the cell inlet, can be interpreted as a "concentration/thermodynamic-state" loss, which combines the effects of reactant-concentration and RH changes, local channel temperature and pressure drop down-the-channel. For a co-flow configuration, the reference conditions (gas inlet) corresponding to the state of maximum energy for both anode and cathode reactants coincide, while in the counter-flow case they are located at opposite ends of the plate.

If we consider an infinitesimal segmentation of the active area in the along-the-channel direction, the channel conditions of each segment become the "local reference" for the corresponding segment and the local VLB (i. e., pertaining to that specific segment) can be carried out for the 1D sandwich model as discussed previously.

When cell performance is measured together with current- and temperature-distributions, and inlet and outlet pressures are known, it is possible to estimate the local channel conditions everywhere on the cell by applying the VLB analysis (i. e., solving the MEA sandwich model) and the mass balances for each segment sequentially. Furthermore, the VLB verification previously discussed is also doable. If nonrandom, the distribution of voltage residuals (i. e., trend of local "unassigned losses") could provide further diagnostic insights related to the unit-cell behavior in addition to showing limitations for the assumptions made.

When only the total cell current is measured/controlled (i. e., mapping is not available), it is not possible to apply the VLB analysis sequentially to the cell segments. However, the local operating conditions and overpotentials can be estimated by using a down-the-channel model of the cell (e. g., 3-dimensional or "1 + 1D" model). A VLB map down-the-channel can be estimated, but only the validation based on the average performance is possible, which offers only limited insight if a sizable average residual is found.

Below we provide an example of VLB distribution down-the-channel using simulation results obtained using the down-the-channel model proposed by Gu et al. [21].

The simulation results of Figure 47 and Figure 48 illustrate the well-known benefits of the counter-flow: cell performance is higher on average (17 mV at 3 A/cm^2); the membrane hydration (i. e., average lambda) is higher; and more water is removed from the anode side (i. e., beta is positive toward the cathode outlet, indicating presence of water

(a)   Distributions along-the-channel

(b)   Loss Distributions along-the-channel

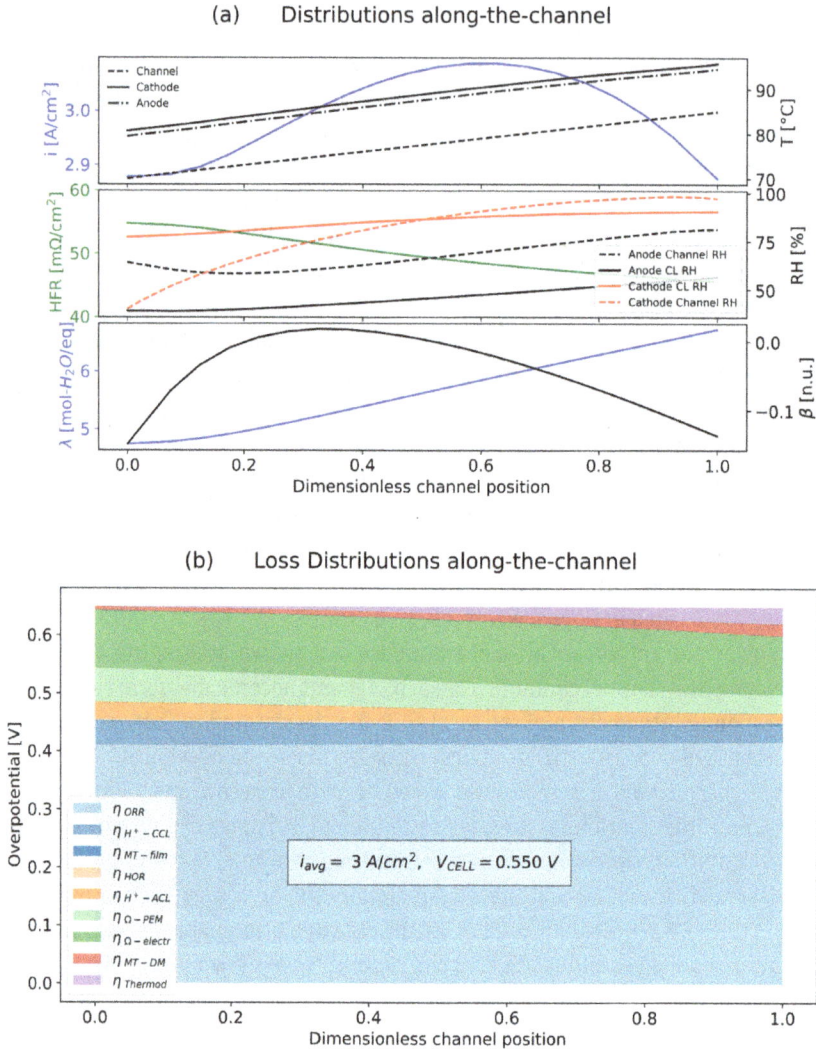

Figure 47: Co-Flow Case: Modeling results showing (a) Local current density, temperatures, relative humidities (left axis), average water uptake in the PEM (lambda), and normalized water flux through the membrane (beta, which represents the fraction of water produced at the cathode crossing over to the anode) and (b) breakdown of losses down-the-channel for a PEM fuel cell operated at 3 A/cm² average, 80 °C inlet temperature, at 2.7 and 2.5 bar, 1.2 and 1.3 for anode and cathode, respectively.

back diffusion to the anode whenever it becomes negative). In addition, the VLB distributions provide some insight on the key differences in MEA behavior between the two flow configurations: the average ohmic losses (PEM + electronic components) is lower for counter-flow due to the improved membrane hydration; mass-transfer and "thermodynamic" losses are lower for counter-flow due to the larger back diffusion and to the reduced pressure differential (anode to cathode), respectively.

(a) Distributions along-the-channel

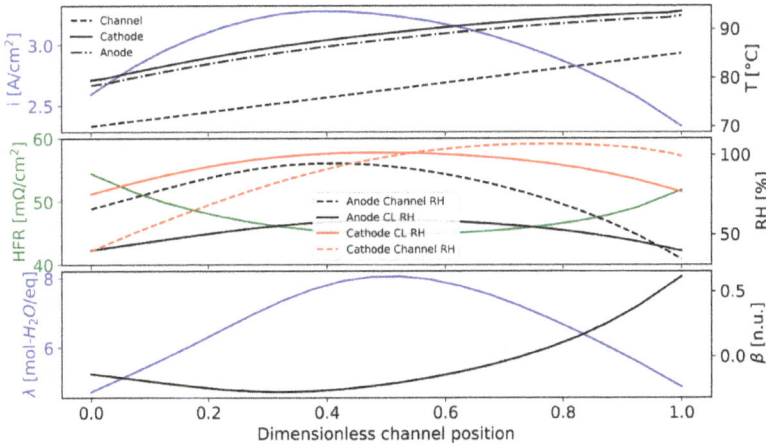

(b) Loss Distributions along-the-channel

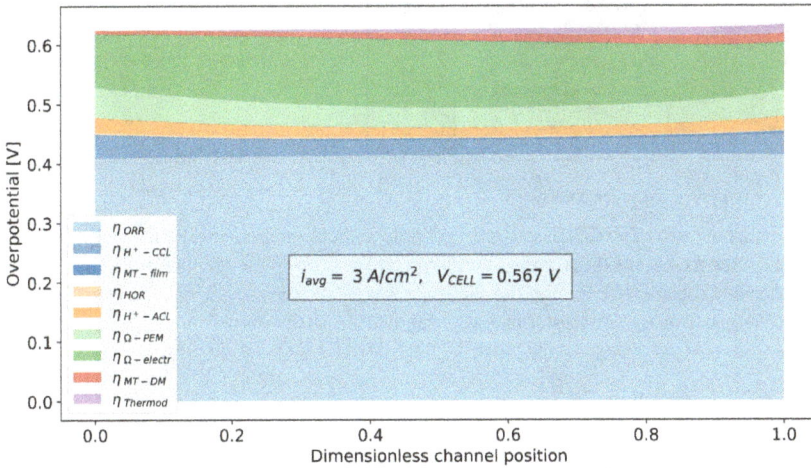

Figure 48: Counter-Flow Case: Modeling results showing (a) local current density, temperatures, relative humidity (left axis), average water uptake in the PEM (lambda) and normalized water flux through the membrane (beta) and (b) breakdown of losses down-the-channel for a PEM fuel cell operated at 3 A/cm^2 average, 80 °C inlet temperature, at 2.7 and 2.5 bar, 1.2 and 1.3 for anode and cathode, respectively.

A similar type of analysis can be very insightful when the effects of liquid water saturation are included. Gerhardt et al. provided an excellent example of VLB applied along-the-channel in a modeling study on the impacts of water management and carbon-dioxide contamination for a hydroxide-exchange-membrane fuel cell (HEMFC) [142]. In this study, VLB showed how different portions of the cell are limited by different mechanisms. Figure 49 shows an example of the simulations presented by Gerhardt, where at

(a)

Anode: $H_2 + 2\,OH^- \rightarrow 2\,H_2O + 2\,e^-$

Cathode: $\tfrac{1}{2}\,O_2 + H_2O + 2\,e^- \rightarrow 2\,OH^-$

(b)

(c)

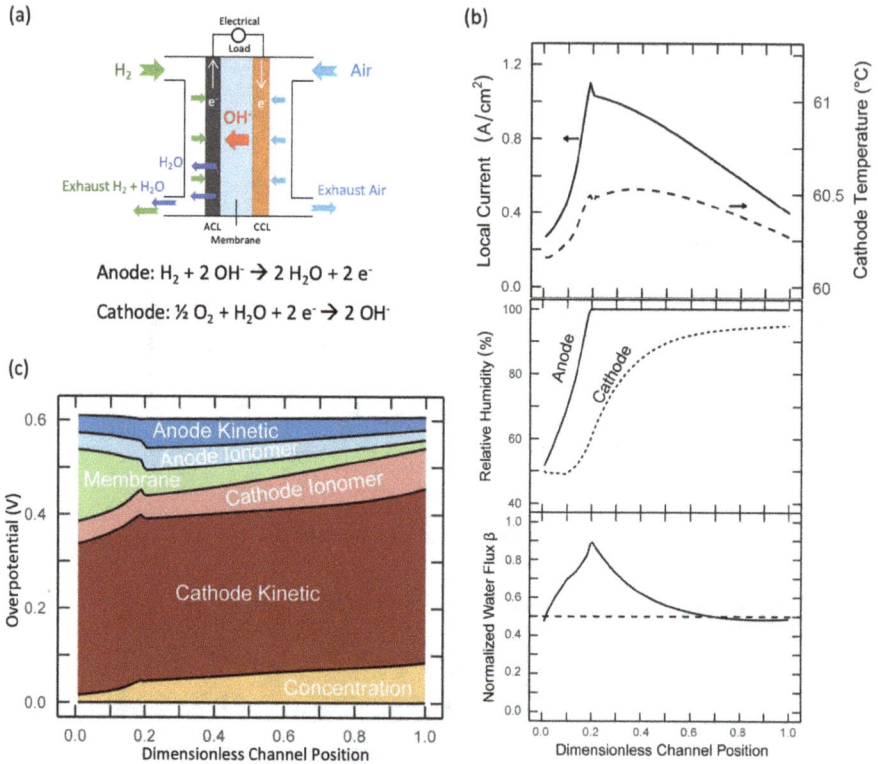

Figure 49: (a) Schematic of a HEMFC, (b) Local current density, cathode temperature, RH and normalized water flux through the membrane (beta) and (c) breakdown of losses down-the-channel for a HEMFC operated at 0.6 V and 60 °C coolant temperature. The anode gas was H_2 and the cathode gas was CO_2-free air at 1.21 bar and 50 % RH (balanced), and 1.2 stoichiometry. Reproduced with permission from IOP Publishing © (2019) [142].

the channel inlet the membrane loss is significant, but as the membrane hydrates down the channel the ohmic loss related to hydrogen transfer decreases in magnitude, while transport losses due to lower oxygen concentration increase. As for the PEMFCs cathode, kinetics is the dominant loss also in HEMFC, although anode losses are significantly higher than in PEMFC, as expected.

The examples described above show the potential of voltage loss analysis extended along-the-channel to study interactions and trade-offs between MEA design, unit cell concept (intended as the combination of plate and flow-field geometry, and of the distribution arrangement of gases and coolant), and operating conditions strategy.

A natural evolution of the VLB concepts applied to differential cells described in this chapter consists in combining voltage loss analysis to mapping tools for a single cell to gain insight on the interactions between MEA design, unit cell concepts and operating strategy for the development of more cost effective and durable commercial products.

9.11 Conclusions

This chapter provides a detailed review of the approaches used to analyze voltage losses in PEM fuel cells. Voltage loss analysis evolved starting from the early development stages of PEMFCs; it was first based on simple empirical models fitted to polarization curves to capture the major loss-types (kinetic, ohmic and mass transport), and it then evolved to include more complex behaviors resulting from the simultaneous transport of protons and reactants through the porous structure of the electrodes and the still elusive effects of liquid water saturation in the porous media.

According to the preceding discussion, VLB based on 0-D analytical models containing one or more empirical terms has limited validity and it is not recommended other than for a preliminary assessment of the type of loss affecting degraded MEAs (e. g., polarization change analysis).

VLB based on detailed physics-based numerical models offer the highest capacity in terms of loss resolution; however, performance prediction using these models is still not sufficiently accurate and extensive calibration efforts are necessary for each MEA design, limiting the practicality of such approach.

The VLB approached pioneered by Gasteiger et al., based on the measurement of key effective-properties of the MEA (VLB Level 3) and on a simplified 1-dimensional MEA model to predict the associated overpotentials, currently provides the best compromise between accuracy of the final breakdown and practicality in terms of testing and modeling effort. The validation procedure, based on the comparison of predicted versus measured performance, is a powerful and practical way to assess whether the VLB analysis achieved the desired level of resolution, or if further refinements of the approach are necessary due to the presence of "unassigned" residual losses. This validation is conceptually identical to the one typically done in MEA modeling. However, the VLB Level 3 integrates input from dedicated in-situ measurements done on the same MEA under test, resulting in a more accurate and reliable breakdown.

VLB at high current densities ($> 3\,\text{A/cm}^2$) is challenging due to (i) the difficulties of operating in differential mode and (ii) the presence of nonnegligible through-plane temperature and hydration gradients in the MEA, requiring the use of more detailed numerical models.

As the current density of operation is expected to increase over time in order to increase stack power density, it will become necessary to use increasingly detailed physics-based models in VLB, possibly integrating a model-calibration approach based on the measurement of additional key effective properties for the same MEA under test as currently done in the VLB Level 3.

The extension of voltage loss analysis in the along-the-channel direction is another emerging trend, where the combination of VLB and current-, temperature-, and HFR-mapping could expand the scope of this diagnostic method beyond the MEA level to include unit cell development as well as optimization of operating conditions.

Appendix A

Table A: List of MEA properties and parameters relevant for voltage loss analysis.

Property Symbol	Name	Units	Correlations
L_{Pt}	Pt loading of electrode	[mg(Pt)/cm^2]	
rf$_{CL}$	Electrode roughness	[cm^2(Pt)/cm^2]	$10 \cdot ECSA \cdot L_{Pt}$
t_{CL}	Electrode thickness	[μm]	
Pt w%	Metal content of electrocatalyst	[g(Pt)/g(Catalyst) in %]	
ECSA	Electrochemical surface of electrocatalyst	[m^2(Pt)/g(Catalyst)]	
SA$_{ecat}$	Surface area of electrocatalyst	[m^2/g(Catalyst)]	
$r_{Pt/C}$	Pt-to-C weight ratio	[g(Pt)/g(C)]	$(Pt\,w\%/100^{-1} - 1)^{-1}$
Ionom. w%	Ionomer content of electrode	[g(Ion)/g(CL) in %]	$r_{I/CL} \cdot 100$
$r_{I/C}$	Ionomer-to-C weight ratio	[g(Ion)/g(C)]	$(r_{Pt/C} + 1)/$ $(100/Ionom\,w\% - 1)$
$r_{Pt/CL}$	Pt-to-CL weight ratio	[g(Pt)/g(CL)]	$(1 + r_{Pt/C}^{-1} \cdot (1 + r_{I/C}))^{-1}$
$r_{I/CL}$	Ionomer-to-CL weight ratio	[g(Ion)/g(CL)]	$(1 + r_{I/C}^{-1} \cdot (1 + r_{Pt/C}))^{-1}$
$r_{C/CL}$	Carbon-to-CL weight ratio	[g(C)/g(CL)]	$(1 + r_{I/C} + r_{Pt/C})^{-1}$
L_C	Carbon loading of electrode	[mg(C)/cm^2]	$\frac{L_{Pt}}{r_{Pt/C}}$
L_{Ion}	Ionomer loading of electrode	[mg(Ion)/cm^2]	$\frac{L_{Pt} \cdot r_{I/C}}{r_{Pt/C}}$
L_{ecat}	Catalyst loading of electrode	[mg(Catalyst)/cm^2]	$L_{Pt} + L_C,\ \frac{100 \cdot L_{Pt}}{Pt\,w\%}$
L_{solid}	Solid loading of electrode	[mg/cm^2]	$L_{Pt} + L_C + L_{Ion}$
ρ_k	Density (k = Pt, C, Ionomer)	[g(k)/cm^3(CL)]	21.45, 2.09, 1.9
ε_{Pt}	Volume fraction of Pt in CL	[cm^3(Pt)/cm^3(CL)]	$\frac{10 \cdot L_{Pt}}{\rho_{Pt} \cdot t_{CL}}$
ε_C	Volume fraction of C in CL	[cm^3(C)/cm^3(CL)]	$\frac{10 \cdot L_{Pt}}{\rho_C \cdot r_{Pt/C} \cdot t_{CL}}$
ε_{ion}	Volume fraction of Ionomer in CL	[cm^3(Ion)/cm^3(CL)]	$\frac{10 \cdot r_{I/C} \cdot L_{Pt}}{\rho_{ion} \cdot r_{Pt/C} \cdot t_{CL}}$
ε_{CL}	Average porosity of CL	[cm^3(void)/cm^3(CL)]	$1 - \varepsilon_{Pt} - \varepsilon_C - \varepsilon_{ion}$
τ_{ion}	Tortuosity of Ionomer in CL	[n. u.]	
τ_{CL}	Gas tortuosity in CL	[n. u.]	
$t_{ion-film}$	Theoretical ionomer film thickness	[nm]	$\frac{10 \cdot Pt\,w\% \cdot r_{I/C}}{r_{Pt/C} \cdot \rho_{ion} \cdot SA_{ecat}}$
$\kappa_{H^+-CL}^{eff}$	Effective proton conductivity of CL	[S/m]	$\kappa_{H^+-Ionomer}^{bulk} \cdot \frac{\varepsilon_{ion}}{\tau_{ion}}$
$R_{H^+-CL}^{eff}$	Effective proton resistance of the electrode	[Ohm $\cdot$ cm^2]	$\frac{t_{CL}}{\kappa_{H^+-CL}^{eff}}$
$\mathcal{D}_k^\circ$	Diffusion coefficient of species k (O_2, H_2, H_2O) in gas mix	[m^2/s]	$\frac{1-x_k}{\sum_{i \neq k}^{j=m} x_{ij} \mathcal{D}_{i-j}^\circ}$
$\mathcal{D}_{k-CL}^{eff}$	Effective diffusivity of species k (O_2, H_2, H_2O) in CL	[m^2/s]	$\mathcal{D}_{k-CL}^{Knud} \cdot \frac{\varepsilon_{CL}}{\tau_{CL}}$
t_{L-p}	Thickness of component p (DM, MPL)	[μm]	
ε_{L-p}	Average porosity of layer p	[cm^3(void)/cm^3(p)]	
$\mathcal{D}_{k-p}^{eff}$	Effective diffusivity of species k (O_2, H_2, H_2O) in p (DM, MPL)	[m^2/s]	$\mathcal{D}_k^\circ \cdot \frac{\varepsilon_p}{\tau_p} = \frac{t_{CL}}{R_{MT}^p} \cdot \frac{\mathcal{D}_k^\circ}{\mathcal{D}_{O_2}^\circ}$
R_{MT}^p	Mass transfer resistance of species k in p	[s/m]	$\frac{t_l}{\mathcal{D}_{k-o}^{eff}}$
t_{PEM}	Membrane thickness	[μm]	
κ_{H^+-PEM}	Proton conductivity of ionomer in PEM	[S/m]	$a \cdot \left(\frac{18 \cdot \lambda}{EW/\rho_{ion}+18 \cdot \lambda} - b\right)^c$
R_{PEM}	Membrane resistance	[Ohm $\cdot$ cm^2]	$\frac{t_{PEM}}{\kappa_{H^+-PEM}}$

Bibliography

[1]   S. Srinivasan, E. A. Ticianelli, C. R. Derouin, and A. Redondo, Advances in solid polymer electrolyte fuel cell technology with low platinum loading electrodes. *J. Power Sources*, **22**, 359–375, 1988. https://doi.org/10.1016/0378-7753(88)80030-2.

[2]   J. Kim, S. Lee, S. Srinivasan, and C. E. Chamberlin, Modeling of Proton Exchange Membrane Fuel Cell Performance with an Empirical Equation. *J. Electrochem. Soc.*, **142**, 2670, 1995. https://doi.org/10.1149/1.2050072.

[3]   D. M. Bernardi and M. W. Verbrugge, A Mathematical Model of the Solid-Polymer-Electrolyte Fuel Cell. *J. Electrochem. Soc.*, **139**, 2477–2491, 1992. https://doi.org/10.1149/1.2221251.

[4]   M. L. Perry, J. Newman, and E. J. Cairns, Mass Transport in Gas-Diffusion Electrodes: A Diagnostic Tool for Fuel-Cell Cathodes. *J. Electrochem. Soc.*, **145**, 5, 1998. https://doi.org/10.1149/1.1838202.

[5]   H. A. Gasteiger, J. E. Panels, and S. G. Yan, Dependence of PEM fuel cell performance on catalyst loading. *J. Power Sources*, **127**, 162–171, 2004. https://doi.org/10.1016/j.jpowsour.2003.09.013.

[6]   H. A. Gasteiger, S. S. Kocha, B. Sompalli, and F. T. Wagner, Activity benchmarks and requirements for Pt, Pt-alloy, and non-Pt oxygen reduction catalysts for PEMFCs. *Appl. Catal. B, Environ.*, **56**, 9–35, 2005. https://doi.org/10.1016/j.apcatb.2004.06.021.

[7]   M. V. Williams, H. R. Kunz, and J. M. Fenton, Analysis of Polarization Curves to Evaluate Polarization Sources in Hydrogen/Air PEM Fuel Cells. *J. Electrochem. Soc.*, **152**, A635, 2005. https://doi.org/10.1149/1.1860034.

[8]   K. C. Neyerlin, W. Gu, J. Jorne, A. Clark, and H. A. Gasteiger, Cathode catalyst utilization for the ORR in a PEMFC – Analytical model and experimental validation. *J. Electrochem. Soc.*, **154**, B279, 2007.

[9]   A. Orfanidi, P. Madkikar, H. A. El-Sayed, G. S. Harzer, T. Kratky, and H. A. Gasteiger, The Key to High Performance Low Pt Loaded Electrodes. *J. Electrochem. Soc.*, **164**, F418–F426, 2017. https://doi.org/10.1149/2.1621704jes.

[10] A. Z. Weber, R. L. Borup, R. M. Darling, P. K. Das, T. J. Dursch, W. Gu, D. Harvey, A. Kusoglu, S. Litster, M. M. Mench, R. Mukundan, J. P. Owejan, J. G. Pharoah, M. Secanell, and I. V. Zenyuk, A Critical Review of Modeling Transport Phenomena in Polymer-Electrolyte Fuel Cells. *J. Electrochem. Soc.*, **161**, F1254, 2014. https://doi.org/10.1149/2.0751412jes.

[11] M. R. Gerhardt, L. M. Pant, J. C. Bui, A. R. Crothers, V. M. Ehlinger, J. C. Fornaciari, J. Liu, and A. Z. Weber, Method—Practices and Pitfalls in Voltage Breakdown Analysis of Electrochemical Energy-Conversion Systems. *J. Electrochem. Soc.*, **168**, 074503, 2021. https://doi.org/10.1149/1945-7111/abf061.

[12] K. O'Neil, J. P. Meyers, R. M. Darling, and M. L. Perry, Oxygen gain analysis for proton exchange membrane fuel cells. *Int. J. Hydrog. Energy*, **37**, 373–382, 2012. https://doi.org/10.1016/j.ijhydene.2011.08.085.

[13] M. L. Perry, R. Balliet, and R. M. Darling, Chapter 7 – Experimental Diagnostics and Durability Testing Protocols. In: M. M. Mench, E. C. Kumbur and T. N. Veziroglu, Eds., *Polymer Electrolyte Fuel Cell Degradation*, pages 335–364. Academic Press, Boston, 2012. https://doi.org/10.1016/B978-0-12-386936-4.10007-7.

[14] G. Squadrito, G. Maggio, E. Passalacqua, F. Lufrano, and A. Patti, An empirical equation for polymer electrolyte fuel cell (PEFC) behaviour. *J. Appl. Electrochem.*, **29**, 1449–1455, 1999. https://doi.org/10.1023/A:1003890219394.

[15] J. Newman and K. E. Thomas-Alyea, *Electrochemical Systems*. Wiley-Interscience 3rd edition, 2004.

[16] L. Pisani, G. Murgia, M. Valentini, and B. D'aguanno, A new semi-empirical approach to performance curves of polymer electrolyte fuel cells. *J. Power Sources*, **108**, 192–203, 2002. https://doi.org/10.1016/S0378-7753(02)00014-9.

[17] S. Srinivasan, D. J. Manko, H. Koch, M. A. Enayetullah, and A. J. Appleby, Recent advances in solid polymer electrolyte fuel cell technology with low platinum loading electrodes. *J. Power Sources*, **29**, 367–387, 1990. https://doi.org/10.1016/0378-7753(90)85011-Z.

[18]   U. Beuscher, Experimental Method to Determine the Mass Transport Resistance of a Polymer Electrolyte Fuel Cell. *J. Electrochem. Soc.*, **153**, A1788, 2006. https://doi.org/10.1149/1.2218760.

[19]   S. Haji, Analytical modeling of PEM fuel cell i–V curve. *Renew. Energy*, **36**, 451–458, 2011. https://doi.org/10.1016/j.renene.2010.07.007.

[20]   L. Zhao, H. Dai, F. Pei, P. Ming, X. Wei, and J. Zhou, A Comparative Study of Equivalent Circuit Models for Electro-Chemical Impedance Spectroscopy Analysis of Proton Exchange Membrane Fuel Cells. *Energies*, **15**, 386, 2022. https://doi.org/10.3390/en15010386.

[21]   W. Gu, D. R. Baker, Y. Liu, and H. A. Gasteiger, Proton exchange membrane fuel cell (PEMFC) down-the-channel performance model. In: W. Vielstich, H. A. Gasteiger and H. Yokokawa, Eds., *Handbook of Fuel Cells*. Wiley, Hoboken, 2009.

[22]   J. Newman, Electrode Kinetics. In: J. Newman and K. E. Thomas-Alyea, Eds., *Electrochemical Systems*, 3rd edition. Wiley 2004.

[23]   K. C. Neyerlin, W. Gu, J. Jorne, and H. A. Gasteiger, Determination of Catalyst Unique Parameters for the Oxygen Reduction Reaction in a PEMFC. *J. Electrochem. Soc.*, **153**, A1955, 2006. https://doi.org/10.1149/1.2266294.

[24]   K. C. Neyerlin, W. Gu, J. Jorne, and H. A. Gasteiger, Study of the Exchange Current Density for the Hydrogen Oxidation and Evolution Reactions. *J. Electrochem. Soc.*, **154**, B631, 2007. https://doi.org/10.1149/1.2733987.

[25]   E. L. Thompson. *Behavior of Proton Exchange Membrane Fuel Cells at Sub-Freezing Temperatures*. Ph. D., University of Rochester, 2007.

[26]   A. Kulikovsky. *Analytical models for PEM fuel cell impedance*. Eisma, 2017.

[27]   M. Eikerling and A. Kulikovsky, Modeling of Catalyst Layer Performance. In: *Polymer Electrolyte Fuel Cells Physical Principles of Materials and Operation*. CRC Press, 2017.

[28]   D. D. Papadias, R. K. Ahluwalia, N. Kariuki, D. Myers, K. L. More, D. A. Cullen, B. T. Sneed, K. C. Neyerlin, R. Mukundan, and R. L. Borup, Durability of Pt-Co Alloy Polymer Electrolyte Fuel Cell Cathode Catalysts under Accelerated Stress Tests. *J. Electrochem. Soc.*, **165**, F3166–F3177, 2018. https://doi.org/10.1149/2.0171806jes.

[29]   G. S. Harzer, A. Orfanidi, H. El-Sayed, P. Madkikar, and H. A. Gasteiger, Tailoring Catalyst Morphology towards High Performance for Low Pt Loaded PEMFC Cathodes. *J. Electrochem. Soc.*, **165**, F770–F779, 2018. https://doi.org/10.1149/2.0311810jes.

[30]   N. Ramaswamy, W. Gu, J. M. Ziegelbauer, and S. Kumaraguru, Carbon Support Microstructure Impact on High Current Density Transport Resistances in PEMFC Cathode. *J. Electrochem. Soc.*, **167**, 064515, 2020. https://doi.org/10.1149/1945-7111/ab819c.

[31]   N. Ramaswamy, S. Kumaraguru, R. Koestner, T. Fuller, W. Gu, N. Kariuki, D. Myers, P. J. Dudenas, and A. Kusoglu, Editors' Choice—Ionomer Side Chain Length and Equivalent Weight Impact on High Current Density Transport Resistances in PEMFC Cathodes. *J. Electrochem. Soc.*, **168**, 024518, 2021. https://doi.org/10.1149/1945-7111/abe5eb.

[32]   P. A. García-Salaberri, J. T. Gostick, I. v. Zenyuk, G. Hwang, M. Vera, and A. Z. Weber, On the Limitations of Volume-Averaged Descriptions of Gas Diffusion Layers in the Modeling of Polymer Electrolyte Fuel Cells. *ECS Trans.*, **80**, 133, 2017. https://doi.org/10.1149/08008.0133ecst.

[33]   P. A. García-Salaberri, I. V. Zenyuk, A. D. Shum, G. Hwang, M. Vera, A. Z. Weber, and J. T. Gostick, Analysis of representative elementary volume and through-plane regional characteristics of carbon-fiber papers: diffusivity, permeability and electrical/thermal conductivity. *Int. J. Heat Mass Transf.*, **127**, 687–703, 2018. https://doi.org/10.1016/j.ijheatmasstransfer.2018.07.030.

[34]   K. Karan, Interesting Facets of Surface, Interfacial, and Bulk Characteristics of Perfluorinated Ionomer Films. *Langmuir*, **35**, 13489–13520, 2019. https://doi.org/10.1021/acs.langmuir.8b03721.

[35]   A. Kusoglu and A. Z. Weber, New Insights into Perfluorinated Sulfonic-Acid Ionomers. *Chem. Rev.*, **117**, 987–1104, 2017. https://doi.org/10.1021/acs.chemrev.6b00159.

[36]   A. Z. Weber and J. Newman, Coupled Thermal and Water Management in Polymer Electrolyte Fuel Cells. *J. Electrochem. Soc.*, **153**, A2205, 2006. https://doi.org/10.1149/1.2352039.

[37]   M. Secanell, N. Djilali, and A. Suleman, Optimization of Membrane Electrode Assemblies for PEMFC. In: *49th AIAA/ASME/ASCE/AHS/ASC Structures, Structural Dynamics, and Materials Conference*, 2008.

[38]   L. M. Pant, M. R. Gerhardt, N. Macauley, R. Mukundan, R. L. Borup, and A. Z. Weber, Along-the-channel modeling and analysis of PEFCs at low stoichiometry: Development of a 1+2D model. *Electrochim. Acta*, **326**, 134963, 2019. https://doi.org/10.1016/J.ELECTACTA.2019.134963.

[39]   I. V. Zenyuk, P. K. Das, and A. Z. Weber, Understanding Impacts of Catalyst-Layer Thickness on Fuel-Cell Performance via Mathematical Modeling. *J. Electrochem. Soc.*, **163**, F691–F703, 2016. https://doi.org/10.1149/2.1161607jes.

[40]   A. Goshtasbi, P. García-Salaberri, J. Chen, K. Talukdar, D. G. Sanchez, and T. Ersal, Through-the-Membrane Transient Phenomena in PEM Fuel Cells: A Modeling Study. *J. Electrochem. Soc.*, **166**, F3154–F3179, 2019. https://doi.org/10.1149/2.0181907jes.

[41]   R. Vetter and J. O. Schumacher, Experimental parameter uncertainty in proton exchange membrane fuel cell modeling. Part I: Scatter in material parameterization. *J. Power Sources*, **438**, 227018, 2019. https://doi.org/10.1016/j.jpowsour.2019.227018.

[42]   R. Vetter and J. O. Schumacher, Experimental parameter uncertainty in proton exchange membrane fuel cell modeling. Part II: Sensitivity analysis and importance ranking. *J. Power Sources*, **439**, 126529, 2019. https://doi.org/10.1016/j.jpowsour.2019.04.057.

[43]   A. Goshtasbi, J. Chen, J. R. Waldecker, S. Hirano, and T. Ersal, Effective Parameterization of PEM Fuel Cell Models—Part I: Sensitivity Analysis and Parameter Identifiability. *J. Electrochem. Soc.*, **167**, 044504, 2020. https://doi.org/10.1149/1945-7111/ab7091.

[44]   A. Goshtasbi, J. Chen, J. R. Waldecker, S. Hirano, and T. Ersal, Effective Parameterization of PEM Fuel Cell Models—Part II: Robust Parameter Subset Selection, Robust Optimal Experimental Design, and Multi-Step Parameter Identification Algorithm. *J. Electrochem. Soc.*, **167**, 044505, 2020. https://doi.org/10.1149/1945-7111/ab7092.

[45]   L. M. Pant, S. Stewart, N. Craig, and A. Z. Weber, Critical Parameter Identification of Fuel-Cell Models Using Sensitivity Analysis. *J. Electrochem. Soc.*, **168**, 074501, 2021. https://doi.org/10.1149/1945-7111/ac0d68.

[46]   M. Mench, E. C. Kumbur, and T. N. Veziroglu, *Polymer Electrolyte Fuel Cell Degradation*. Elsevier Inc., 2011. https://doi.org/10.1016/C2010-0-67819-9.

[47]   J. Stumper, R. Rahmani, and F. Fuss, In-situ Diagnostics for Cell Performance and Degradation. *ECS Trans.*, **25**, 1605, 2009. https://doi.org/10.1149/1.3210716.

[48]   S. G. Rinaldo, J. Stumper, and M. Eikerling, Physical theory of platinum nanoparticle dissolution in polymer electrolyte fuel cells. *J. Phys. Chem. C*, **114**, 5773–5785, 2010. https://doi.org/10.1021/jp9101509.

[49]   L. M. Pant, Z. Yang, M. L. Perry, and A. Z. Weber, Development of a Simple and Rapid Diagnostic Method for Polymer-Electrolyte Fuel Cells. *J. Electrochem. Soc.*, **165**, F3007, 2018. https://doi.org/10.1149/2.0011806jes.

[50]   H. Wang, X. Z. Yuan, and H. Li, *PEM fuel cell diagnostic tools*, 2011. https://doi.org/10.1201/b11100.

[51]   M. E. Orazem and B. Tribollet, *Electrochemical Impedance Spectroscopy*, 2nd edition. Wiley, 2017.

[52]   N. Wagner, Electrochemical Impedance Spectroscopy. In: H. Wang, X.-Z. Yuan and H. Li, Eds., *PEM Fuel Cell Diagnostic Tools*. CRC Press, 2012.

[53]   N. Wagner, T. Kaz, and K. A. Friedrich, Investigation of electrode composition of polymer fuel cells by electrochemical impedance spectroscopy. *Electrochim. Acta*, **53**, 7475–7482, 2008. https://doi.org/10.1016/j.electacta.2008.01.084.

[54]   A. Lasia, *Electrochemical Impedance Spectroscopy and its Applications*. Springer, New York, 2014.

[55]   S. Dierickx, A. Weber, and E. Ivers-Tiffée, How the distribution of relaxation times enhances complex equivalent circuit models for fuel cells. *Electrochim. Acta*, **355**, 136764, 2020. https://doi.org/10.1016/j.electacta.2020.136764.

[56]   A. Kulikovsky, PEM fuel cell distribution of relaxation times: A method for the calculation and behavior of an oxygen transport peak. *Phys. Chem. Chem. Phys.*, **22**, 19131–19138, 2020. https://doi.org/10.1039/d0cp02094j.

[57] T. Reshetenko and A. Kulikovsky, Understanding the distribution of relaxation times of a low–Pt PEM fuel cell. *Electrochim. Acta*, **391**, 138954, 2021. https://doi.org/10.1016/j.electacta.2021.138954.

[58] T. Reshetenko and A. Kulikovsky, Variation of PEM Fuel Cell Physical Parameters with Current: Impedance Spectroscopy Study. *J. Electrochem. Soc.*, **163**, F1100–F1106, 2016. https://doi.org/10.1149/2.0981609jes.

[59] F. Jaouen, G. Lindbergh, and K. Wiezell, Transient Techniques for Investigating Mass-Transport Limitations in Gas Diffusion Electrodes. *J. Electrochem. Soc.*, **150**, A1711, 2003. https://doi.org/10.1149/1.1624295.

[60] F. Jaouen and G. Lindbergh, Transient Techniques for Investigating Mass-Transport Limitations in Gas Diffusion Electrodes. *J. Electrochem. Soc.*, **150**, A1699, 2003. https://doi.org/10.1149/1.1624294.

[61] A. P. Young, J. Stumper, and E. Gyenge, Characterizing the Structural Degradation in a PEMFC Cathode Catalyst Layer: Carbon Corrosion. *J. Electrochem. Soc.*, **156**, B913, 2009. https://doi.org/10.1149/1.3139963.

[62] A. P. Young, J. Stumper, S. Knights, and E. Gyenge, Ionomer Degradation in Polymer Electrolyte Membrane Fuel Cells. *J. Electrochem. Soc.*, **157**, B425, 2010. https://doi.org/10.1149/1.3281899.

[63] M. Cimenti, M. Tam, and J. Stumper, High Frequency Artifacts in Electrochemical Impedance Spectroscopy Measurements on PEM Fuel Cells. *Electrochem. Solid-State Lett.*, **12**, B131, 2009. https://doi.org/10.1149/1.3162829.

[64] S. Jomori, K. Komatsubara, N. Nonoyama, M. Kato, and T. Yoshida, An Experimental Study of the Effects of Operational History on Activity Changes in a PEMFC. *J. Electrochem. Soc.*, **160**, F1067–F1073, 2013. https://doi.org/10.1149/2.103309jes.

[65] S. Roy and M. E. Orazem, Guidelines for Evaluation of Error Structure for Impedance Response of Polymer Electrolyte Membrane (PEM) Fuel Cells. *ECS Trans.*, **13**, 153–169, 2008. https://doi.org/10.1149/1.3004035.

[66] S. K. Roy, M. E. Orazem, and B. Tribollet, Interpretation of Low-Frequency Inductive Loops in PEM Fuel Cells. *J. Electrochem. Soc.*, **154**, B1378, 2007. https://doi.org/10.1149/1.2789377.

[67] J. T. Gostick, M. A. Ioannidis, M. W. Fowler, and M. D. Pritzker, Characterization of the Capillary Properties of Gas Diffusion Media. In: U. W. Chao-Yang and Pasaogullari, editors, *Modeling and Diagnostics of Polymer Electrolyte Fuel Cells*, pages 225–254. Springer, New York, 2010. https://doi.org/10.1007/978-0-387-98068-3_7.

[68] Y. Sun, S. Polani, F. Luo, S. Ott, P. Strasser, and F. Dionigi, Advancements in cathode catalyst and cathode layer design for proton exchange membrane fuel cells. *Nat. Commun.*, **12**, 5984, 2021. https://doi.org/10.1038/s41467-021-25911-x.

[69] H. Wang, R. Wang, S. Sui, T. Sun, Y. Yan, and S. Du, Cathode Design for Proton Exchange Membrane Fuel Cells in Automotive Applications. *Automot. Innov.*, **4**, 144–164, 2021. https://doi.org/10.1007/s42154-021-00148-y.

[70] X. Z. Yuan, C. Nayoze-Coynel, N. Shaigan, D. Fisher, N. Zhao, N. Zamel, P. Gazdzicki, M. Ulsh, K. A. Friedrich, F. Girard, and U. Groos, A review of functions, attributes, properties and measurements for the quality control of proton exchange membrane fuel cell components. *J. Power Sources*, **491**, 229540, 2021. https://doi.org/10.1016/j.jpowsour.2021.229540.

[71] J. Newman, Porous Electrodes. In: *Electrochemical Systems*, 3rd edition. Wiley-Interscience, 2004.

[72] G. Tsotridis, A. Pilenga, G. Marco, and T. Malkow, *EU harmonised test protocols for PEMFC MEA testing in single cell configuration for automotive applications*, 2015.

[73] O. Levenspiel, *Chemical Reaction Engineering*, 3rd edition. Wiley, 1998.

[74] J. Benziger, E. Chia, E. Karnas, J. Moxley, C. Teuscher, and I. G. Kevrekidis, The stirred tank reactor polymer electrolyte membrane fuel cell. *AIChE J.*, **50**, 1889–1900, 2004. https://doi.org/10.1002/aic.10158.

[75] T. Bednarek. The JRC ZERO∇CELL design documentation, 2021. https://op.europa.eu/en/publication-detail/-/publication/57794ea3-7585-11eb-9ac9-01aa75ed71a1/language-en (accessed June 19, 2022).

[76]   E. Commission, J. R. Centre, T. Bednarek, and G. Tsotridis, *Development of reference hardware for a harmonised testing of PEM single cell fuel cells*. Publications Office, 2021. https://doi.org/10.2760/83818.

[77]   FC-Cubic Small Fuel Cell – Technical Information (2022). https://www.fc-cubic.or.jp/technical-info/ (accessed June 19, 2022).

[78]   P. D. Schneider. QuBK – Qualification of Fuel Cell Components, 2020. https://www.ise.fraunhofer.de/en/research-projects/qubk.html (accessed June 19, 2022).

[79]   T. Bednarek and G. Tsotridis, Assessment of the electrochemical characteristics of a Polymer Electrolyte Membrane in a reference single fuel cell testing hardware. *J. Power Sources*, **473**, 228319, 2020. https://doi.org/10.1016/J.JPOWSOUR.2020.228319.

[80]   M. Sano and K. Tamai, A universal transition to turbulence in channel flow. *Nat. Phys.*, **12**, 249–253, 2016. https://doi.org/10.1038/nphys3659.

[81]   S. G. Kandlikar, Transition to Turbulence in Microchannels. In: D. Li, Ed., *Encyclopedia of Microfluidics and Nanofluidics*, pages 2093–2095. Springer, Boston, 2008. https://doi.org/10.1007/978-0-387-48998-8_1622.

[82]   R. W. Hanks and H.-C. Ruo, Laminar-Turbulent Transition in Ducts of Rectangular Cross Section. *Ind. Eng. Chem. Fundam.*, **5**, 558–561, 1966. https://doi.org/10.1021/i160020a022.

[83]   J. R. Taylor, *An Introduction to Error Analysis: The Study of Uncertainties in Physical Measurements*, 2nd edition. University Science Books, Sausalito, California, 1997.

[84]   DOE CELL Component Accelerated Stress Test Protocols For Pem Fuel Cells, US DOE (2007). https://www1.eere.energy.gov/hydrogenandfuelcells/fuelcells/pdfs/component_durability_profile.pdf (accessed June 21, 2022).

[85]   U. S. DRIVE Fuel Cell Tech Team, Appendix A: FCTT AST and Polarization Curve Protocols for PEMFCs, 2013. https://www.energy.gov/sites/default/files/2015/08/f25/fcto_dwg_usdrive_fctt_accelerated_stress_tests_jan2013.pdf (accessed June 21, 2022).

[86]   T. R. Garrick, T. E. Moylan, M. K. Carpenter, and A. Kongkanand, Editors' Choice—Electrochemically Active Surface Area Measurement of Aged Pt Alloy Catalysts in PEM Fuel Cells by CO Stripping. *J. Electrochem. Soc.*, **164**, F55–F59, 2017. https://doi.org/10.1149/2.0381702jes.

[87]   S. S. Kocha, J. Deliang Yang, and J. S. Yi, Characterization of gas crossover and its implications in PEM fuel cells. *AIChE J.*, **52**, 1916–1925, 2006. https://doi.org/10.1002/aic.10780.

[88]   N. P. Subramanian, T. A. Greszler, J. Zhang, W. Gu, and R. Makharia, Pt-Oxide Coverage-Dependent Oxygen Reduction Reaction (ORR) Kinetics. *J. Electrochem. Soc.*, **159**, B531–B540, 2012. https://doi.org/10.1149/2.088205jes.

[89]   Y. Liu, M. Mathias, and J. Zhang, Measurement of Platinum Oxide Coverage in a Proton Exchange Membrane Fuel Cell. *Electrochem. Solid-State Lett.*, **13**, B1, 2010. https://doi.org/10.1149/1.3257595.

[90]   M. Cimenti, R. Kehoe, and J. Stumper, Effect of oxide formation on the oxygen reduction kinetics on Pt and PtCo PEMFC cathodes. In: *European Fuel Cell Forum 2011*, pages 42–51, 2011.

[91]   R. Jiang, C. K. Mittelsteadt, and C. S. Gittleman, Through-Plane Proton Transport Resistance of Membrane and Ohmic Resistance Distribution in Fuel Cells. *J. Electrochem. Soc.*, **156**, B1440, 2009. https://doi.org/10.1149/1.3240877.

[92]   Y. Liu, C. Ji, W. Gu, J. Jorne, and H. A. Gasteiger, Effects of Catalyst Carbon Support on Proton Conduction and Cathode Performance in PEM Fuel Cells. *J. Electrochem. Soc.*, **158**, B614–B621, 2011. https://doi.org/10.1149/1.3562945.

[93]   Y. Liu, C. Ji, W. Gu, D. R. Baker, J. Jorne, and H. A. Gasteiger, Proton Conduction in PEM Fuel Cell Cathodes: Effects of Electrode Thickness and Ionomer Equivalent Weight. *J. Electrochem. Soc.*, **157**, B1154, 2010. https://doi.org/10.1149/1.3435323.

[94]   Y. Liu, M. W. Murphy, D. R. Baker, W. Gu, C. Ji, J. Jorne, and H. A. Gasteiger, Proton Conduction and Oxygen Reduction Kinetics in PEM Fuel Cell Cathodes: Effects of Ionomer-to-Carbon Ratio and Relative Humidity. *J. Electrochem. Soc.*, **156**, B970, 2009. https://doi.org/10.1149/1.3143965.

[95] A. Lasia, Impedance of Porous Electrodes. In: *Electrochemical Impedance Spectroscopy and Its Applications*, 1st edition. Springer, New York, 2014.

[96] K. Kodama, K. Motobayashi, A. Shinohara, N. Hasegawa, K. Kudo, R. Jinnouchi, M. Osawa, and Y. Morimoto, Effect of the Side-Chain Structure of Perfluoro-Sulfonic Acid Ionomers on the Oxygen Reduction Reaction on the Surface of Pt. *ACS Catal.*, **8**, 694–700, 2018. https://doi.org/10.1021/acscatal.7b03571.

[97] M. Obermaier, A. S. Bandarenka, and C. Lohri-Tymozhynsky, A Comprehensive Physical Impedance Model of Polymer Electrolyte Fuel Cell Cathodes in Oxygen-free Atmosphere. *Sci. Rep.*, **8**(1), 1–9, 2018. https://doi.org/10.1038/s41598-018-23071-5.

[98] A. Kulikovsky, Analysis of proton and electron transport impedance of a PEM fuel cell in H2/N2 regime. *Electrochem. Sci. Adv.*, **1**, e2000023, 2021. https://doi.org/10.1002/elsa.202000023.

[99] T. Suzuki, H. Murata, T. Hatanaka, and Y. Morimoto. Analysis of the Catalyst Layer of Polymer Electrolyte Fuel Cells. R&D Review of Toyota CRDL 39, 2008.

[100] D. R. Baker, D. A. Caulk, K. C. Neyerlin, and M. W. Murphy, Measurement of Oxygen Transport Resistance in PEM Fuel Cells by Limiting Current Methods. *J. Electrochem. Soc.*, **156**, B991, 2009. https://doi.org/10.1149/1.3152226.

[101] D. A. Caulk and D. R. Baker, Heat and Water Transport in Hydrophobic Diffusion Media of PEM Fuel Cells. *J. Electrochem. Soc.*, **157**, B1237, 2010. https://doi.org/10.1149/1.3454721.

[102] D. A. Caulk and D. R. Baker, Modeling Two-Phase Water Transport in Hydrophobic Diffusion Media for PEM Fuel Cells. *J. Electrochem. Soc.*, **158**, B384, 2011. https://doi.org/10.1149/1.3551504.

[103] J. P. Owejan, T. A. Trabold, and M. M. Mench, Oxygen transport resistance correlated to liquid water saturation in the gas diffusion layer of PEM fuel cells. *Int. J. Heat Mass Transf.*, **71**, 585–592, 2014. https://doi.org/10.1016/J.IJHEATMASSTRANSFER.2013.12.059.

[104] T. A. Greszler, D. Caulk, and P. Sinha, The Impact of Platinum Loading on Oxygen Transport Resistance. *J. Electrochem. Soc.*, **159**, F831–F840, 2012. https://doi.org/10.1149/2.061212jes.

[105] J. P. Owejan. *Transport Resistance in Resistance in Polymer Electrolyte Fuel Cells*. Ph. D., The University of Tennessee, 2014.

[106] N. Nonoyama, S. Okazaki, A. Z. Weber, Y. Ikogi, and T. Yoshida, Analysis of Oxygen-Transport Diffusion Resistance in Proton-Exchange-Membrane Fuel Cells. *J. Electrochem. Soc.*, **158**, B416, 2011. https://doi.org/10.1149/1.3546038.

[107] T. V. Reshetenko and J. St-Pierre, Separation Method for Oxygen Mass Transport Coefficient in Gas and Ionomer Phases in PEMFC GDE. *J. Electrochem. Soc.*, **161**, F1089–F1100, 2014. https://doi.org/10.1149/2.1021410jes.

[108] F. B. Spingler, A. Phillips, T. Schuler, M. C. Tucker, and A. Z. Weber, Investigating fuel-cell transport limitations using hydrogen limiting current. *Int. J. Hydrog. Energy*, **42**, 13960–13969, 2017. https://doi.org/10.1016/J.IJHYDENE.2017.01.036.

[109] A. Chowdhury, C. J. Radke, and A. Z. Weber, Transport Resistances in Fuel-Cell Catalyst Layers. *ECS Trans.*, **80**, 321–333, 2017. https://doi.org/10.1149/08008.0321ecst.

[110] T. Schuler, A. Chowdhury, A. T. Freiberg, B. Sneed, F. B. Spingler, M. C. Tucker, K. L. More, C. J. Radke, and A. Z. Weber, Fuel-Cell Catalyst-Layer Resistance via Hydrogen Limiting-Current Measurements. *J. Electrochem. Soc.*, **166**, F3020–F3031, 2019. https://doi.org/10.1149/2.0031907jes.

[111] M. Göbel, S. Kirsch, L. Schwarze, L. Schmidt, H. Scholz, J. Haußmann, M. Klages, J. Scholta, H. Markötter, S. Alrwashdeh, I. Manke, and B. R. Müller, Transient limiting current measurements for characterization of gas diffusion layers. *J. Power Sources*, **402**, 237–245, 2018. https://doi.org/10.1016/J.JPOWSOUR.2018.09.003.

[112] A. Kulikovsky, A Fast Low-Current Model for Impedance of a PEM Fuel Cell Cathode at Low Air Stoichiometry. *J. Electrochem. Soc.*, **164**, F911–F915, 2017. https://doi.org/10.1149/2.0561709jes.

[113] T. V. Reshetenko and A. Kulikovsky, Impedance Spectroscopy Measurements of Ionomer Film Oxygen Transport Resistivity in Operating Low-Pt PEM Fuel Cell. *Membranes*, **11**, 985, 2021. https://doi.org/10.3390/membranes11120985.

[114] T. Reshetenko and A. Kulikovsky, Nafion film transport properties in a low-Pt PEM fuel cell: impedance spectroscopy study. *RSC Adv.*, **9**, 38797–38806, 2019. https://doi.org/10.1039/C9RA07794D.

[115] D. A. Caulk, A. M. Brenner, and S. M. Clapham, A Steady Permeation Method for Measuring Water Transport Properties of Fuel Cell Membranes. *J. Electrochem. Soc.*, **159**, F518–F529, 2012. https://doi.org/10.1149/2.016209jes.

[116] X. Ye and C.-Y. Wang, Measurement of Water Transport Properties Through Membrane Electrode Assemblies. *J. Electrochem. Soc.*, **154**, B683, 2007. https://doi.org/10.1149/1.2737384.

[117] X. Ye and C.-Y. Wang, Measurement of Water Transport Properties Through Membrane-Electrode Assemblies. *J. Electrochem. Soc.*, **154**, B676, 2007. https://doi.org/10.1149/1.2737379.

[118] P. Sauriol, D. S. Nobes, X. T. Bi, J. Stumper, D. Jones, and D. Kiel, Design and Validation of a Water Transfer Factor Measurement Apparatus for Proton Exchange Membrane Fuel Cells. *J. Fuel Cell Sci. Technol.*, **6**, 041014, 2009. https://doi.org/10.1115/1.3007900.

[119] T. C. Yau, P. Sauriol, X. Bi, and J. Stumper, Water transfer factor measurement. In: *PEM Fuel Cell Diagnostic Tools*, pages 149–179, 2011.

[120] T. C. Yau, P. Sauriol, X. T. Bi, and J. Stumper, Experimental Determination of Water Transport in Polymer Electrolyte Membrane Fuel Cells. *J. Electrochem. Soc.*, **157**, B1310, 2010. https://doi.org/10.1149/1.3456621.

[121] J. Stumper, H. Haas, and A. Granados, In Situ Determination of MEA Resistance and Electrode Diffusivity of a Fuel Cell. *J. Electrochem. Soc.*, **152**, A837, 2005. https://doi.org/10.1149/1.1867673.

[122] M. E. Mathias, J. Roth, J. B. Fleming, and W. Lehnert, Diffusion media materials and characterization. In: *Handbook of Fuel Cells – Fundamentals, Technology and Applications*. John Wiley & Sons, Ltd, 2003.

[123] S. Lædre, O. E. Kongstein, A. Oedegaard, F. Seland, and H. Karoliussen, Measuring In Situ Interfacial Contact Resistance in a Proton Exchange Membrane Fuel Cell. *J. Electrochem. Soc.*, **166**, F853–F859, 2019. https://doi.org/10.1149/2.1511912jes.

[124] J. Kleemann, F. Finsterwalder, and W. Tillmetz, Characterisation of mechanical behaviour and coupled electrical properties of polymer electrolyte membrane fuel cell gas diffusion layers. *J. Power Sources*, **190**, 92–102, 2009. https://doi.org/10.1016/j.jpowsour.2008.09.026.

[125] H. Sadeghifar, N. Djilali, and M. Bahrami, A new model for thermal contact resistance between fuel cell gas diffusion layers and bipolar plates. *J. Power Sources*, **266**, 51–59, 2014. https://doi.org/10.1016/J.JPOWSOUR.2014.04.149.

[126] M. Andisheh-Tadbir, E. Kjeang, and M. Bahrami, Thermal conductivity of microporous layers: Analytical modeling and experimental validation. *J. Power Sources*, **296**, 344–351, 2015. https://doi.org/10.1016/J.JPOWSOUR.2015.07.054.

[127] M. Ahadi, M. Tam, M. S. Saha, J. Stumper, and M. Bahrami, Thermal conductivity of catalyst layer of polymer electrolyte membrane fuel cells: Part 1 – Experimental study. *J. Power Sources*, **354**, 207–214, 2017. https://doi.org/10.1016/J.JPOWSOUR.2017.02.016.

[128] M. Ahadi, A. Putz, J. Stumper, and M. Bahrami, Thermal conductivity of catalyst layer of polymer electrolyte membrane fuel cells: Part 2 – Analytical modeling. *J. Power Sources*, **354**, 215–228, 2017. https://doi.org/10.1016/J.JPOWSOUR.2017.03.100.

[129] W. Sheng, H. A. Gasteiger, and Y. Shao-Horn, Hydrogen Oxidation and Evolution Reaction Kinetics on Platinum: Acid vs Alkaline Electrolytes. *J. Electrochem. Soc.*, **157**, B1529, 2010. https://doi.org/10.1149/1.3483106.

[130] K. Wiezell, P. Gode, and G. Lindbergh, Steady-State and EIS Investigations of Hydrogen Electrodes and Membranes in Polymer Electrolyte Fuel Cells. *J. Electrochem. Soc.*, **153**, A759, 2006. https://doi.org/10.1149/1.2172561.

[131] K. Wiezell, P. Gode, and G. Lindbergh, Steady-State and EIS Investigations of Hydrogen Electrodes and Membranes in Polymer Electrolyte Fuel Cells. *J. Electrochem. Soc.*, **153**, A749, 2006. https://doi.org/10.1149/1.2172559.

[132] J. Durst, C. Simon, F. Hasché, and H. A. Gasteiger, Hydrogen Oxidation and Evolution Reaction Kinetics on Carbon Supported Pt, Ir, Rh, and Pd Electrocatalysts in Acidic Media. *J. Electrochem. Soc.*, **162**, F190–F203, 2015. https://doi.org/10.1149/2.0981501jes.

[133] S. Murata, M. Imanishi, S. Hasegawa, and R. Namba, Vertically aligned carbon nanotube electrodes for high current density operating proton exchange membrane fuel cells. *J. Power Sources*, **253**, 104–113, 2014. https://doi.org/10.1016/j.jpowsour.2013.11.073.

[134] S. Kundu, M. Cimenti, S. Lee, and D. Bessarabov, Fingerprint of automotive fuel cell cathode catalyst degradation: Pt band in PEMs. *Membr. Technol.*, **2009**, 7–10, 2009. https://doi.org/10.1016/S0958-2118(09)70212-5.

[135] V. Berejnov, Z. Martin, M. West, S. Kundu, D. Bessarabov, J. Stumper, D. Susac, and A. P. Hitchcock, Probing platinum degradation in polymer electrolyte membrane fuel cells by synchrotron X-ray microscopy. *Phys. Chem. Chem. Phys.*, **14**, 4835–4843, 2012. https://doi.org/10.1039/C2CP40338B.

[136] S. Kabir, D. J. Myers, N. Kariuki, J. Park, G. Wang, A. Baker, N. Macauley, R. Mukundan, K. L. More, and K. C. Neyerlin, Elucidating the Dynamic Nature of Fuel Cell Electrodes as a Function of Conditioning: An ex Situ Material Characterization and in Situ Electrochemical Diagnostic Study. *ACS Appl. Mater. Interfaces*, **11**, 45016–45030, 2019. https://doi.org/10.1021/acsami.9b11365.

[137] J. Stumper, M. Löhr, and S. Hamada, Diagnostic tools for liquid water in PEM fuel cells. *J. Power Sources*, **143**, 150–157, 2005. https://doi.org/10.1016/J.JPOWSOUR.2004.11.036.

[138] P. Berg, K. Promislow, J. Stumper, and B. Wetton, Discharge of a Segmented Polymer Electrolyte Membrane Fuel Cell. *J. Fuel Cell Sci. Technol.*, **2**, 111–120, 2004. https://doi.org/10.1115/1.1867977.

[139] S. A. Freunberger, M. Reum, and F. N. Buechi, Design approaches for determining local current and membrane resistance in polymer electrolyte fuel cells (PEFCs). In: W. Vielstich, H. A. Gasteiger, A. Lamm and H. Yokokawa, Eds., *Handbook of Fuel Cells*, 1st edition. Wiley, New York, 2009.

[140] M. Geske, M. Heuer, G. Heideck, and Z. A. Styczynski, Current Density Distribution Mapping in PEM Fuel Cells as An Instrument for Operational Measurements. *Energies*, **3**, 770–783, 2010. https://doi.org/10.3390/en3040770.

[141] D. Natarajan and T. van Nguyen, Current Mapping. In: *PEM Fuel Cell Diagnostic Tools*. CRC Press, 2011. https://doi.org/10.1201/b11100-11.

[142] M. R. Gerhardt, L. M. Pant, and A. Z. Weber, Along-the-Channel Impacts of Water Management and Carbon-Dioxide Contamination in Hydroxide-Exchange-Membrane Fuel Cells: A Modeling Study. *J. Electrochem. Soc.*, **166**, F3180–F3192, 2019. https://doi.org/10.1149/2.0171907jes.

Jake deVaal and Jürgen Stumper

10 Relationships between stack performance, transfer leaks and H_2-emissions

10.1 Introduction

In order to enable widespread application of PEMFCs in transportation applications, certain cost, performance and durability targets have to be met as laid out in the US Drive Fuel Cell Technical Team Roadmap 2017 [1], which defines levels of capability necessary for commercial viability. The durability of a PEM fuel cell stack is determined by the durability of its components, i. e., Flowfield Plates, Seals and Membrane-Electrode Assembly (MEA). MEA durability in turn depends on the properties of the electrodes and, most importantly, the membrane. One of the key functions of the membrane, in addition to providing proton conduction, is to separate the fuel from the oxidant on the cathode side. Any degradation mechanisms leading to an increased leakage of gases between anode and cathode will negatively impact stack-durability, performance and safety. Consequently, membrane degradation is of particular importance for gas crossover. Membrane degradation can be of (i) mechanical, (ii) chemical, (iii) thermal nature and lead to thinning and/or pinhole formation, which can cause internal transfer leaks. Under conditions where the anode gas pressure is larger than the gas pressure on the cathode, the resulting pressure driven H_2 cross-over increase directly enhances the rate of the last two degradation mechanisms, thereby leading to positive feedback that makes internal transfer leaks a dominant failure mode [2–4], both during accelerated lifetime testing [5, 6] as well as during normal operation.

Due to the importance of this failure mode, a variety of methods have been developed to locate and/or quantify MEA internal transfer leaks in fuel cell stacks. Fundamentally, they can be divided into *ex situ* and *in situ* methods. Whereas *ex situ* methods may require stack disassembly, this is not the case for *in situ* methods. The latter can be further divided into those with or without power production by the stack.

In situ methods with power production
Air starvation method (see Section 10.3.2). This method is based on the measurement of individual cell voltages during the reduction of the air stoichiometry at a fixed current while maintaining a positive anode-cathode pressure difference $\Delta p > 0$. Due to the recombination of H_2 and O_2 to water on the cathode (recombination effect), those cells

Jake deVaal, Ballard Power Systems Inc., Burnaby, BC V5J 5J9, Canada
Jürgen Stumper, Juergen Stumper & Assoc., Vancouver, BC V5M 1H5, Canada

https://doi.org/10.1515/9783110622720-010

with internal transfer leaks will show O_2 starvation first, their cell voltages will drop to $\approx 0\,V$ and the corresponding cells can be identified.

In situ methods without power production

The OCV method (see Section 10.3.1) relies on the measurement of the open-circuit voltage (OCV) of the individual cells of a stack with flows of H_2 and N_2 to anode and cathode, respectively, while maintaining a positive pressure differential from anode to cathode. Under these conditions H_2 will transfer from anode to cathode in a leaky cell and the OCV is a measure for the ratio of the H_2 concentrations on anode and cathode. Using measured flows, total and partial gas pressures and OCV, the H_2 leak rate Q^{H_2} [mol/s] can be calculated [7, 8].

The localized OCV method (see Section 10.2.2.1) is a variation of the OCV method, where the OCV is measured at several locations along the length/perimeter of the cell while supplying H_2 and air to anode and cathode, respectively, and maintaining a positive pressure differential from anode to cathode. Local OCV minima correspond to transfer leak locations.

The applied voltage method (see Section 10.2.2.2). Similar to above, flows of H_2 and N_2 are supplied to anode and cathode, respectively, while maintaining a positive pressure differential from anode to cathode. Instead of measuring the OCV, a potential difference is applied to an individual cell, where the cathode potential is positive relative to the anode. The observed current is then a direct measure for the amount of H_2 transferring from anode to cathode [9].

Tracer method (see Section 10.2.2.4). Here, an inert gas such as He is added to one gas stream in one fluid manifold and detected in another fluid manifold [10].

Ex situ methods

Bubble test (see Section 10.2.2.3). For this test, the anode outlet and cathode inlet are sealed, and pressurized air is supplied to the anode ($\Delta p \approx 0.3\,bar$). Air leaking from anode to cathode is collected and measured using a graduated cylinder submerged in a water tank. This test can be performed both on single cells and stacks.

Transfer leak location within the active area. For this test, the cell/stack is disassembled, and the MEA of interest is placed in a special fixture exposing one electrode to H_2 and the other side to ambient air. Upon supply with H_2, any H_2 leaking through pinholes will recombine with the O_2 in the air leading to localized heating, which is detectable using an IR camera [11].

Since the development of internal transfers is highly likely during the operational lifetime of a fuel cell stack, there is strong demand for advanced diagnostic tools to locate and quantify internal transfer leaks in fuel cell stacks both during service/refurbishment as well as *in situ* during normal operation (on-board diagnostics).

Therefore, this chapter provides an overview of methods of transfer leak quantification and -location as well as more advanced methods suitable for on-board health monitoring of PEM fuel cell stacks during operation.

10.2 H$_2$ crossover through an MEA—fundamentals

For the following discussion, we consider a PFSA membrane of thickness t_{mem}, with a certain number of pinholes of diameter d and a pinhole density N_p [1/cm^2] (for simplicity we assume that all pinholes have the same size).

Because for typical fuel cell operating conditions $\lambda/d < 10^{-2}$ where λ is the mean free path length for the gas, the convective molar H$_2$- flux $Q_{conv}^{H_2}$ [mol/s] through a pinhole can be treated as viscous flow through a tube:

$$Q_{conv}^{H_2} = \frac{A^2}{8\pi \cdot \mu} \cdot \frac{p_A}{R \cdot T} \cdot \frac{\Delta p}{t_{mem}} \tag{1}$$

where $A = \pi \cdot d^2/4$ is the pinhole area, p_A, p_C are the upstream (anode) and downstream (cathode) pressures, respectively, $\Delta p = p_A - p_C \ll p_A$ the pressure differential across the pinhole, μ the dynamic gas viscosity, R: gas constant, T: temperature.

Even without pinholes, H$_2$ can still cross over from anode to cathode through permeation [12]. The molar H$_2$-flux permeating through the membrane per cm^2 can be written as

$$Q_{perm}^{\prime H_2} = \psi^{H_2} \cdot \frac{\Delta p_{H_2}}{t_{mem}}, \tag{2}$$

where ψ^{H_2} is the H$_2$-permeability and $\Delta p_{H_2}/t_{mem}$ the H$_2$ partial pressure gradient across the membrane. ψ^{H_2} depends on both temperature and RH and shows characteristic Arrhenius behavior with an activation energy of $\approx$20 kJ/mol. With increasing RH, the permeability curves shift to higher values as the higher permeable aqueous phase increases in volume. At 60 °C, ψ^{H_2} ranges from 1 to 2 $\cdot$ 10^{-13} mol $\cdot$(cm $\cdot$ kPa $\cdot$s)$^{-1}$ [13]. With (1), (2) the total H$_2$ crossover flux $Q_{tot}^{\prime H_2}$ [mol s^{-1} cm^{-2}] can be written as

$$Q_{tot}^{\prime H_2} = Q_{perm}^{\prime H_2} + Q_{conv}^{H_2} \cdot N_p, \tag{3}$$

where N_p denotes the number of pinholes/cm^2. The red curve in Figure 1 shows $Q_{tot}^{\prime H_2}$ for a 25 μm membrane, 10 μm pinhole diameter and $N_p = 1$/cm^2 whereas the blue curve shows the contribution of $Q_{perm}^{\prime H_2}$ alone. Only when the anode pressure is larger than the cathode pressure ($p_C = 150$ kPa) is $\phi_{conv}^{H_2} > 0$ and there is a contribution of the pinhole to the total H$_2$ crossover flux from anode to cathode.

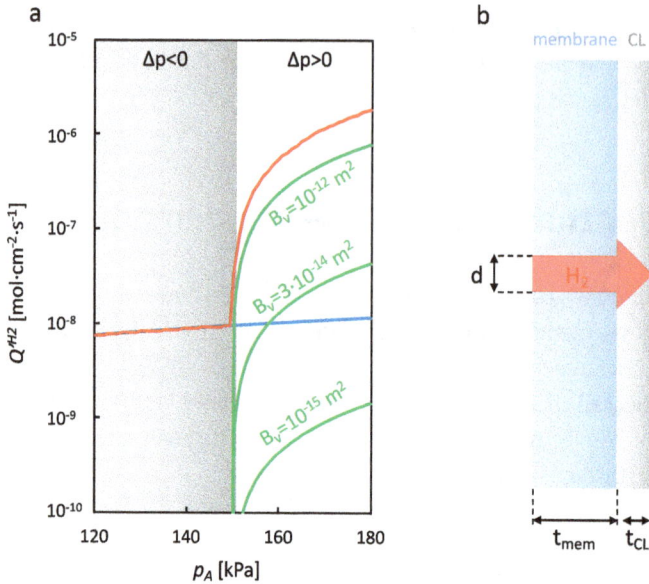

Figure 1: (a) Molar H_2 crossover-flux per cm^2 vs. anode pressure p_A for a pinhole density $N_p = 1/cm^2$. $T = 60\,°C$, RH $= 40\,\%$, Cathode pressure 150 kPa, membrane thickness $t_{mem} = 25\,\mu m$, CL thickness $t_{CL} = 10\,\mu m$, pinhole diameter $d = 10\,\mu m$. Blue: H_2 flux due to membrane permeation alone, red: total H_2 flux (H_2 permeation + flux through pinhole). H_2 crossover through pinhole contributes only for $Dp > 0$. Green: effect of a 10 µm CL with a permeability $B_v = 10^{-12}, 3 \cdot 10^{-14}, 10^{-15}\,m^2$, respectively, on H_2 flux through pinhole. Y-axis scale: $10^{-8}\,mol\,cm^{-2}\,s^{-1}$ corresponds to a current density of $\approx 2\,mA/cm^2$. (b) schematic diagram of a membrane with pinhole of diameter d bonded to a CL.

10.2.1 H_2 crossover through an MEA with pinholes

When integrated in an MEA, the membrane is bonded to the catalyst layers (CL), so the question arises as to the effect of the permeability of the CL's on $Q_{tot}^{\prime H_2}$. The permeability B_v [m^2] of a porous medium is defined according to Darcy's law:

$$\frac{\Delta p}{\Delta x} = -\frac{\mu}{B_v} \cdot v \tag{4}$$

where $\frac{\Delta p}{\Delta x}$ is the pressure gradient, μ [Pa s] the dynamic gas viscosity and v the gas-velocity. Writing (4) as flux yields

$$Q_{darcy}^{\prime H_2} = \frac{B_v}{\mu} \cdot \frac{p_A}{R \cdot T} \cdot \frac{(p_A - p_C)}{t_{CL}}, \tag{5}$$

where p_A, p_C are the upstream and downstream pressures, respectively, and t_{CL} the CL thickness. In order to calculate the flux through a membrane pinhole bonded to a CL (see Figure 1b), one needs to solve

$$Q'^{\,H_2}_{darcy} \cdot A_{darcy} = Q^{H_2}_{conv} \tag{6}$$

Figure 1a shows the solution for a range of permeabilities assuming the pinhole area $A_{darcy} = A$ for the Darcy flow. There are only very few measurements of CL permeability in the literature, with Zhao et al. [14] reporting $B_v = 1.5 - 3.7 \cdot 10^{-15}$ m^2 and Xu [15] $B_v = 3 - 15 \cdot 10^{-16}$ m^2.

Figure 1a shows that for values $B_v \leq \cdot 10^{-13}$ m^2 the convective H_2-flux through the pinhole is in effect determined by the CL permeability and that for $B_v \leq \cdot 10^{-15}$ m^2 it becomes so small that it can be neglected compared to H_2 permeation. The H_2 permeation flux (blue curve in Figure 1) was calculated using a membrane permeability $\psi^{H_2} = 1.65 \cdot 10^{-13}$ mol $\cdot$ (cm $\cdot$ kPa $\cdot$s)$^{-1}$ for $T = 60\,°C$, RH $= 40\,\%$.

Kreitmeier et al. [4] studied convective and diffusive He permeabilities for an MEA with membrane pinholes of $N_p = 9$ cm^{-2} and found a ratio of convective/diffusive He crossover of ≈ 10 (i. e., convection/permeation ratio of ≈ 1 for $N_p = 1$ cm^{-2}) at a pressure differential of 10 kPa. Figure 1 indicates that this ratio would correspond to a Darcy permeability $B^{CL}_v \leq 10^{-14}$ m^2 (see green curves in Figure 1) for a single CL in contact with the membrane. This is in reasonable agreement with the experimental values for B_v when considering that for Figure 1 the area for permeation through the CL was assumed to be the same as the pinhole area. In reality, there is likely to be some divergence of the flux as H_2 diffuses through the CL thickness, which would reduce the impact of the CL and reduce the permeability B^{CL}_v at which the convection/permeation H_2-flux ratio would be about 1 even further.

10.2.2 H_2-crossover quantification

10.2.2.1 OCV method

The open circuit voltage (OCV) of a fuel cell is the difference between the electrochemical potentials of the cathode and the anode. For H_2/O_2 operation, it can be written as the difference between the equilibrium potentials for the HOR/HER and ORR/OER reactions for cathode and anode, respectively,

$$V_{oc} = E^{eq}_c - E^{eq}_a = E^0_c - E^0_a + \frac{R \cdot T}{2 \cdot F} \ln\left(\sqrt{\frac{c_{O_2}}{c^{ref}_{O_2}}} \cdot \frac{c_{H_2}}{c^{ref}_{H_2}} \right), \tag{7}$$

where E^0_c, E^0_a are the standard cathode, anode electrochemical potentials at p^{ref}, T^{ref}, F is Faraday's constant, c_{O_2}, c_{H_2} are the gas concentrations and $c^{ref}_{O_2}$, $c^{ref}_{O_2}$ the corresponding reference values. With $T^{ref} = 298.15 \cdot K$ and $p^{ref} = 101$ kPa,

$$E^0_c - E^0_a = 1.23V - 8.5 \cdot 10^{-4} V/K \cdot (T - T^{ref}) \tag{8}$$

In the case of H_2 crossing over from the anode to the cathode, E_c^{eq} becomes a *Mischpotential* [16] E_c^{mix} with $E_c^{mix} < E_c^{eq}$, i. e., the open circuit voltage is reduced by ΔV_{oc}. In order to calculate $\Delta V_{oc} = E_c^{eq} - E_c^{mix}$ we write the Butler–Volmer equation for the OER/ORR reaction

$$j_c = j_c^0 \cdot \left\{ \frac{c_{H_2O}}{c_{H_2O}^{ref}} \cdot \exp\left(\frac{E - E_c^{eq}}{\frac{R \cdot T}{a \cdot z \cdot F}} \right) - \frac{c_{O_2}}{c_{O_2}^{ref}} \cdot \exp\left(-\frac{E - E_c^{eq}}{\frac{R \cdot T}{(1-a) \cdot z \cdot F}} \right) \right\}, \qquad (9)$$

where j_c^0 is the exchange current density, a the symmetry factor and z the number of charges transferred in the rate-determining step. Typical H_2-crossover current densities $j_{H_2} = 2 \cdot F \cdot \phi_{tot}^{H_2}$ are in the mA/cm^2 range, which is large compared to typical values for j_c^0 and we can neglect the first term in Equation (9) (Tafel approximation) to obtain (assuming complete oxidation of the H_2 crossover flux):

$$\Delta V_{oc} = b_{ORR} \cdot \log\left(\frac{j_{H_2}}{j_c^0} \right) \qquad (10)$$

with $b_{ORR} = RT / (1-a) \cdot z \cdot F$. Using the data from Figure 3 in Gasteiger et al. [17]. Figure 2 shows ΔV_{oc} as function of H_2 crossover j_{H_2} for three MEA's with different cathode Pt-loading calculated according to Equation (8) with $b_{ORR} = 65$ mV for the Tafel slope.

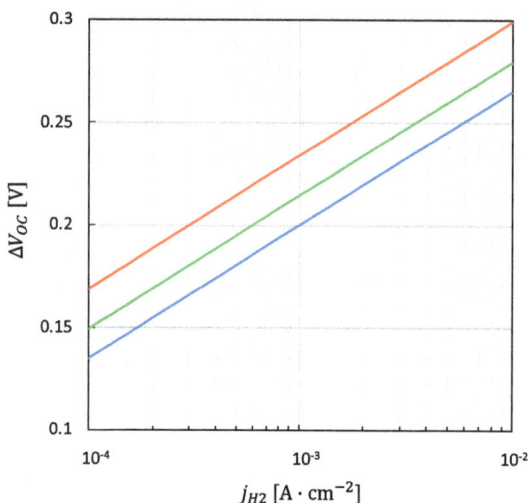

Figure 2: Reduction ΔV_{oc} in OCV [V] for H_2/O_2 operation (80 C, 100 % RH, 270 kPa) as function of H_2 crossover [A/cm^2] for different cathode Pt loadings: Red: $L_{Pt} = 0.15$ mg/cm^2, green: $L_{Pt} = 0.24$ mg/cm^2, blue: $L_{Pt} = 0.4$ mg/cm^2 (see Figure 3 in Gasteiger et al. [17]).

For every order of magnitude, change in H_2 crossover V_{oc} changes by one Tafel slope. Therefore, by measuring the deviation ΔV_{oc} from the theoretical value (see Equa-

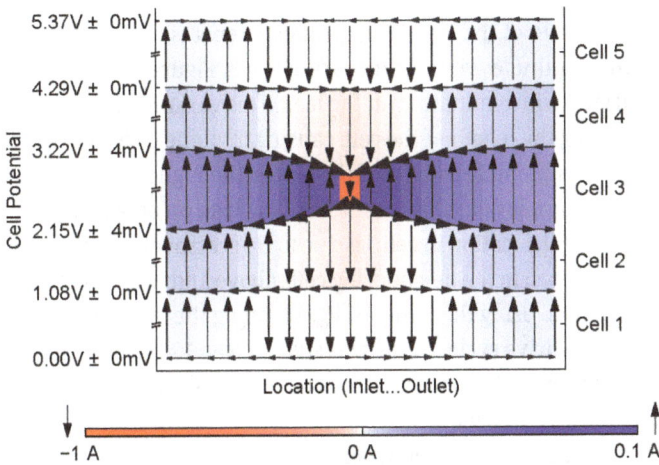

Figure 3: Theoretical profiles of OCV and internal currents vs. location along cell length calculated for a 5-cell Ballard Mk 9 stack operated with H$_2$/O$_2$ exhibiting a H$_2$ crossover leak in the middle of cell 3. Leak size: leakage current I_L = 1 A. Through plane currents: color intensity codes magnitude [A], blue: current in same direction as during normal operation (proton movement from A → C), red: proton movement from C → A (current reversal). Y-axis labels indicate the cell voltages E measured against the anode bus plate (bottom) and are represented as $E \pm \Delta E$: where $E = (E_{max} + E_{min})/2$, $\Delta E = (E_{max} - E_{min})/2$ and E_{max}, E_{min} refer to the maximum voltages, respectively (reprinted from [18] with permission form Elsevier).

tions (7)–(8)), the H$_2$ crossover flux can be determined (neglecting any O$_2$ crossover or electrical shorts).

For a fuel cell stack, the individual cells will in general exhibit different H$_2$ crossover fluxes, and correspondingly will show different open circuit voltages. Consequently, by monitoring the distribution of open circuit voltages across the stack, leaky cells with elevated H$_2$ crossover can be identified [7].

Not only can leaky cells be identified within a stack, the location of transfer leaks within an MEA is also possible if V_{oc} is measured with spatial resolution on an individual cell (V_{oc}-mapping). Using this method, Stumper et al. [18] investigated transfer leak formation during lifetime testing of a Ballard 10-cell Mk 9 stack. The stack was equipped with a current mapping setup and cell voltage monitoring at eight locations along the cell length. Using this setup spatially resolved information could be obtained both under load and under open-circuit conditions. An electrical network model was then used to calculate local currents. As illustration of the method, Figure 3 shows a theoretical calculation of the effect of a transfer leak equivalent to a current of 1 A in the middle of cell 3 in a 5-cell stack. The vertical arrows symbolize the direction and size of the through plane currents corresponding to the direction of proton movement. At the location of the leak, V_{oc} shows a minimum and the 1 A arrow is pointing downward, symbolizing proton movement in the opposite direction compared to normal operation (red color). Since the total cell current is zero at OCV, the sum of all through plane currents is zero for each cell. This leads to current loops circulating in cell 3 in clockwise/counterclockwise direction on the left/right side of the leak. This behavior repeats in the adjacent cells, but with decreasing in magnitude with distance from cell 3. Figure 3 also clearly shows the reduction in local OCV with a minimum at the leak. This variation in OCV also leads to potential gradients along the length of the cell, causing in-plane currents in the flow-field plates. The lower the in-plane resistivity of the plates, the higher the in-plane currents and the stronger and more localized is the effect of the leak on adjacent cells.

The current reversal at the leak location in cell 3 is due to H_2-crossover, whereas the current reversal (downward arrows) in the neighboring cells above and below is caused by the increased local cell voltages. Using this setup of V_{oc}-mapping the development of transfer leaks during automotive drive cycle testing was studied on a 10-cell stack and it was found that (i) transfer leak formation started already at 66 % test duration and (ii) that the leaks were located at the cell inlets.

Another variation of this method is realized when flowing inert gas such as N_2 on the cathode (see Section 10.3.1). In this case, the H_2 crossing over does not react and the fuel cell acts as a H_2-concentration-cell. The open-circuit voltage in this case is given by the Nernst equation:

$$V_{oc} = \frac{R \cdot T}{2 \cdot F} \ln\left(\frac{p_{aH_2}}{p_{cH_2}}\right), \tag{11}$$

where p_{aH_2}, p_{cH_2} are the H_2 partial pressures on anode and cathode side, respectively. Consequently, by measuring V_{oc} the H_2 partial pressure on the cathode can be determined and, using the operating parameters of the fuel cell, the H_2 crossover flux $Q_{tot}^{H_2}$ [mol/s] can be calculated [8]:

$$Q_{tot}^{H_2} = Q^{N_2} \cdot \frac{\frac{p_{cH_2}^f}{p_c^f} \cdot \left(1 + \frac{p_{cH_2O}^i}{p_c^i - p_{cH_2O}^i}\right)}{1 - \frac{p_{cH_2}^f}{p_c^f} \cdot \left(1 + \frac{p_{aH_2O}^i}{p_{aH_2}^i}\right)}, \tag{12}$$

where Q^{N_2} is the N_2 flow on the cathode, $p_{cH_2}^f$, $p_{cH_2O}^i$ are cathode H_2 and H_2O partial pressures at outlet and inlet, respectively, p_c^f is the cathode outlet pressure and $p_{aH_2O}^i$, $p_{aH_2}^i$ the anode H_2O- and H_2 inlet partial pressures. The H_2 partial pressure $p_{cH_2}^f$ can be calculated using Equation (11) from the known H_2 partial pressure at the anode and the open circuit voltage.

The inert gas method can also be used to investigate coolant/fuel or coolant/oxidant leaks. In this case, inert gas is supplied to both the anode and cathode and H_2 is supplied to the coolant manifold. Furthermore, the H_2 pressure in the coolant manifold maybe higher than the inert gas pressure in both the anode and cathode manifolds. Depending on the relative magnitudes of the coolant/fuel and coolant/oxidant leaks, the open circuit voltage maybe positive or negative, allowing determination of the relative sizes of coolant/fuel and coolant/oxidant leaks (see Equation (11)).

10.2.2.2 Applied voltage method

Here, the operating conditions of the cell are identical to the case of an inert gas such as N_2 on the cathode discussed in the previous section, the only difference being that instead of measuring V_{oc}, a potential difference of 0.35–0.6 V is applied to the cathode vs the anode [9], and the current I_{H_2} is measured. This current corresponds to the oxidation of all H_2 that crosses over from anode to cathode (assuming complete oxidation):

$$I_{H_2} = 2 \cdot F \cdot Q_{tot}^{H_2},$$ (13)

which is directly proportional to the total H_2 crossover flux. Depending on the pressure difference between anode and cathode, both convective and diffusive components of the H_2-crossover flux can be determined.

10.2.2.3 Bubble test

This test can be performed on both single cells and stacks under nonoperational conditions (i. e., all compartments purged with inert gas) and can be used to determine leak rates between any two of the three compartments fuel, oxidant and coolant. In the case of determining the leak rate from compartment 1 to compartment 2, outlet 1 is closed and compartment 1 is fed with pressurised gas such as air, N_2. Furthermore, inlet 2 is closed and outlet 2 is connected to a tube feeding into a measurement cylinder submerged in a water tank in order to capture any gas bubbles. Inlet and outlet are closed for the third compartment not of interest. Typically, the test is performed at ambient conditions so that the leak rate can be directly determined in slpm (std l/min.).

10.2.2.4 Tracer method

This method detects fluid leaks within a fuel cell assembly by introducing a tracer into a first fluid supply and monitoring the concentration of the tracer in the exhaust of a second fluid stream [10]. The tracer is preferably inert and not one of the fuel or oxidant reactants nor a reaction product that may be present in the monitored fluid stream without the presence of a leak such as, for example, He, Ar, N_2 or CO_2, and need only be present in minute amounts compared to the reactants. The method can also be used to detect fluid leaks to the external environment if the fuel cell is placed into a chamber and the tracer is detected in the chamber. Kreitmeier et al. [4] used He as tracer gas and mass-spectrometry as detection method to study crossover through an MEA with 10 μm pin-holes. With 10 %/90 % He/H_2 mix on the anode, N_2 on the cathode and an anode-cathode pressure difference of 10 kPa they found a H_2/He crossover flux ratio of 6.5:1, indicating an increased crossover for He. This is in accordance with the ratio of dynamic viscosities $\mu_{He}/\mu_{H_2} \approx 2$. When switching from N_2 to O_2 on the cathode, the H_2,He crossover dropped to 10 %, 50 %, respectively. This means that the recombination rate for H_2 is only 90 %, as indicated by presence of unreacted H_2 on the cathode. The 50 % reduction observed for He crossover was attributed to a secondary "sealing" effect caused by the liquid H_2O production from H_2 recombination.

10.3 Relationship between stack performance, transfer leak and hydrogen emissions

In the following section, research assessing the capability of EIS (Electrochemical Impedance Spectroscopy) to diagnose a stack performance decrease due to transfer leaks is presented. This work began as part of a collaborative Canadian federal government NSERC-funded (Natural Sciences and Engineering Research Council of Canada) research project between Ballard Power Systems, Simon Fraser University (SFU) and University of Victoria (UVic), managed through NSERC's Automotive Partnership Canada (APC) program for Automotive Research Innovation. Due to its complementarity to the primary goal of improving MEA durability, this project examined ways to detect, extend the operating window and mitigate the impacts of internal transfer leaks on hydrogen emissions in a system framework. An ultimate goal would be the development of on-board diagnostics systems with similar capabilities to those currently used in IC-engine vehicles. As part of this activity, (a) the effect of operating conditions on H_2 crossover, (b) on-board transfer leak characterization, (c) on-board hydrogen sensor calibration and fault detection and (d) development of an on-board health monitoring system was investigated. While this is a rapidly evolving area with significant hardware and signal analysis patents developed over the past few years, it is useful here to review the early work as well as more recent research done on both EIS spectra and hardware to understand how this new tool will improve serviceability, and eventually, control of PEM fuel cell systems.

10.3.1 H_2 transfer leak detection using electrochemical impedance spectroscopy

As discussed in greater detail in papers by Mousa et al. [19, 20], the original intent of studying single cells and short stacks using Electrochemical Impedance Spectroscopy (EIS) was to determine whether EIS could be used to detect internal transfer leaks. EIS typically employs a frequency response analyzer (FRA) to apply either a small AC voltage or current perturbation signal to the cell or stack, and then measures the current or voltage response over a wide frequency range. The impedance is then calculated by dividing the voltage by current, in the form of a magnitude and phase angle, at each specific frequency tested.

As shown in Figure 4a, the most basic equivalent circuit used to represent fuel cell operation is the Randles circuit [21]; where C_{dl} is the double layer capacitance of the catalyst surfaces, R_{HF} is the high frequency resistance of the cell components including the contribution from contact resistance between components and movement within the conducting media and R_{ct} is the charge-transfer resistance representing electrochemical reaction at the electrode/electrolyte interface.

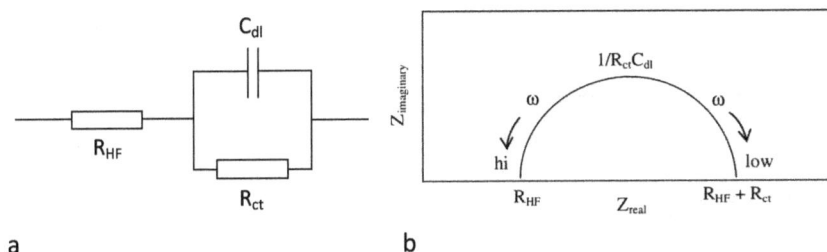

a   b

Figure 4: (a) Randles equivalent circuit (b) Nyquist plot for the Randles circuit (reprinted from [19] with permission from Elsevier).

The Nyquist plot in Figure 4b presents the resistance versus reactance for a Randles-type circuit measured over the range of frequencies scanned by the EIS measurements. This plot is typically broken down into three frequency regions: high, medium and low. The high frequency limit of the impedance spectrum represents the high frequency resistance R_{HF} of the cell(s), whereas the low frequency region represents the sum of high frequency resistance and charge transfer resistance: at high frequency, the C_{dl} is short circuited and the R_{HF} is measured. As frequency decreases, the impedance becomes a combination of resistance and reactance from the capacitive element. At low frequency, C_{dl} acts like a blocking diode and the total resistance is equal to R_{HF} and R_{ct}.

Impedance spectroscopy has the ability to characterize many of the electrical properties of materials and their interfaces, and of cell operating conditions on impedance. This ability has made EIS widely used in modeling and diagnostics of PEM fuel cells, where individual contributions affecting cell performance can be isolated by fitting the impedance spectrum into parameters of an equivalent circuit model. As such, electrical circuits with different configurations, components and degrees of complexity have been proposed and used in the literature based on behaviors observed in the EIS spectra.

In the initial single-cell tests performed by Mousa et al. [19] full-size FCvelocity™-1100 fuel cells (used in Ballard HD6 bus modules) were tested at Ballard Power Systems at 20, 50 and 125 A loads. Cells with small, medium and large hydrogen leaks were tested with a differential pressure ($\Delta p = p_A - p_C$) of 2, 4, 6 and 8 psig (13.8, 27.6, 41.4 and 55.2 kPa) using air as oxidant, and the normal and leaky cells were also tested at different cathode oxygen (O_2/N_2) concentrations. Figure 5a shows the effects of different operating currents on measured impedance using in-vehicle air stoichiometries for a normal cell

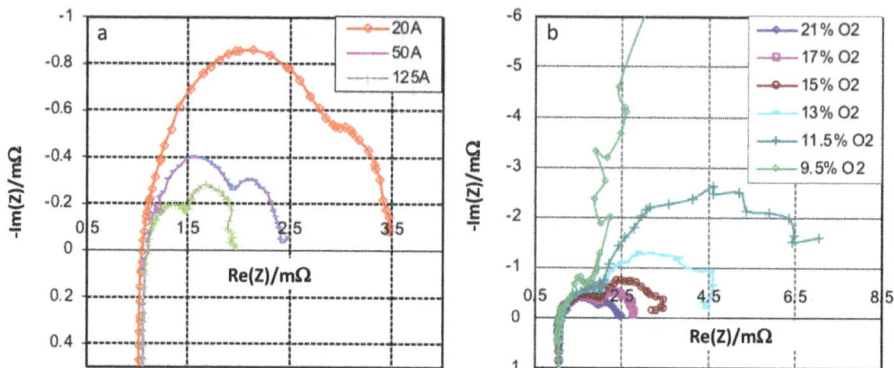

Figure 5: (a) Impedance of a normal (leak free) FCvelocity™-1100 single cell vs. current under H$_2$/air operation. (b): Impedance normal FCvelocity™-1100 single cell @50 A vs. O$_2$ concentration (reprinted from [19] with permission from Elsevier).

(where higher-current responses nest as smaller impedances while sharing the same R_{HF}), while Figure 5b shows the effect of reduced oxygen concentration in the same in-vehicle air flow on a normal cell operating at 50 A, causing significant increases in low-frequency impedance.

In initial single-cell tests with different MEA leak sizes (i. e., small, medium and large from MEAs recovered from field-returned stacks) a behavior similar to that observed with reduced O$_2$ concentration (of an increased impedance in the lower-frequency mass transport arc) was seen, but somewhat less dramatic; as only small flows of hydrogen can be forced through the MEA holes at up to $\Delta p = 8$ psig (55.2 kPa); where pressures were intentionally kept low to prevent hole growth during testing. From these tests, it was concluded that EIS is able to detect hydrogen leaks and oxygen concentrations in a single cell very effectively; especially during full air-starvation, where the EIS signature was very distinctive and impedance values were high. With increasing hydrogen leak rates, it was observed that the impedance signatures changed mostly in the mass transport region. Furthermore, in tests where the flow direction through the cathode side was reversed, putting the observed inlet leak locations near/at the cell outlets, the EIS signatures did not change much with higher Δp testing, indicating that hydrogen was able to leave the cell without the near-total recombination seen with inlet-side transfer leaks. Switching of cathode flow direction is therefore another diagnostic method to roughly locate internal transfer leaks (see Section 10.2.2.1).

As EIS can identify *single-cells* with near-inlet transfer leaks, the question arises whether individual cells with transfer leaks can also be detected within automotive *stacks*. This was investigated by Mousa et al. [20] in their second paper. Figure 6 shows a schematic of the test arrangement used for this FCvelocity™-1100 short stack testing, which was used to test 5, 9 and 19 cell stacks with a leaky MEA installed in the middle cell position (i. e., the single anode/cathode shown was actually a multicell stack).

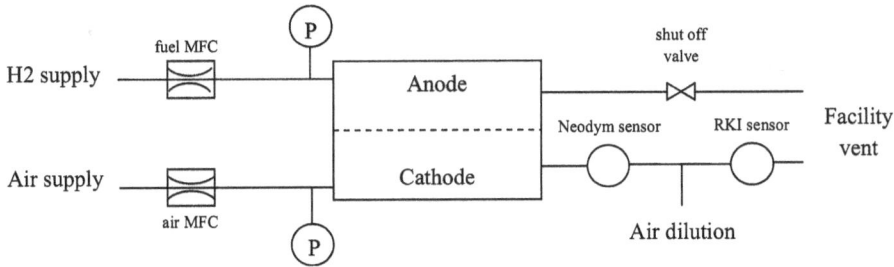

Figure 6: Schematic diagram of experimental setup for short stack testing. Neodym, RKI: H$_2$ sensors (reprinted from [8, 20, 22] with permission from Elsevier).

Two sets of data were obtained in this study, where the first set of data was measured at reduced oxygen concentration rates on the cathode side while keeping the differential pressure (Δp) across the membrane at the inlet of the stack at $\Delta p = 0$ (so the leaky cell would not leak). The second set of data were measured at increased $\Delta p > 0$ between the anode and cathode while keeping the oxygen concentration in the input flow air at normal level, i. e., 21 %. The two impedance data sets differed thus by the absence/presence of leaking hydrogen through holes in the leaky MEA. However, in the first case ($\Delta p = 0$) the oxygen concentration was reduced by an equal amount in *all cells*, whereas in the second case ($\Delta p > 0$) the impedance was obtained with a reduction of oxygen *only in the single leaky cell* in the n-cell stack.

In order to compare the EIS results for the two cases, a Neural Network (NN) approach [23] was chosen to correlate the two data sets. Mapping was done by training the network using oxygen concentration impedance with its associated concentrations see Figure 7 for approach.

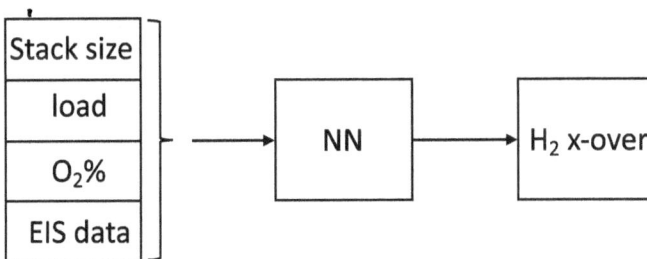

Figure 7: A schematic diagram of neural network (NN) training.

Using this NN, an empirical relationship can be developed that allows determination of the hydrogen crossover through its effect on oxygen concentration for $\Delta p > 0$ from the data for reduced O$_2$ concentration at $\Delta p = 0$. Increasing the amount of Δp at the inlet of the MEA increases hydrogen leakage through the pinholes, also leading to reduced O$_2$ concentration through recombination. Having no access to the upstream of the MEA where the pinholes are, the leak rate was estimated by using oxygen concentrations simulated by the NN at different Δp's. Oxygen stoichiometry λ_{O_2} (utilization) was calculated by using the following equation:

$$\lambda_{O_2} = \frac{O_2 \text{ provided}}{O_2 \text{ consumed}} = \frac{Q^i \cdot x_{O_2}}{C_{O_2} \cdot I} \tag{14}$$

where

λ_{O_2} is 4.7/4.02 for 20 A/50 A load, respectively

Q^i is the inlet oxidant flow rate [slpm]

x_{O_2} is the O_2 concentration in the oxidant flow

$C_{air} = 0.0167 \text{ slpm} \cdot A^{-1}$ is the air consumption

$C_{O_2} = 3.36 \cdot 10^{-3} \text{ slpm} \cdot A^{-1}$ is the O_2 consumption

I is the load current [A]

The oxygen flux $Q^f_{O_2}$ at the outlet is
Case (1) $\Delta p = 0$

$$Q^f_{O_2}(x^1_{O_2}) = Q^i \cdot x^1_{O_2} - C_{O_2} \cdot I \tag{15}$$

Case (2) $\Delta p > 0$

$$Q^f_{O_2}(\Delta p) = Q^i \cdot x^2_{O_2} - C_{O_2} \cdot (I + I_{H_2}(\Delta p)), \tag{16}$$

where $x^1_{O_2}$, $x^2_{O_2}$ refer to the inlet O_2 concentrations for case 1, 2, respectively, for which matching EIS signatures were observed. Therefore, by matching the EIS signatures at fixed current I for the two cases using the neural network (NN) the H_2 crossover equivalent current $I_{H_2}(\Delta p)$ can be calculated by equating (15), (16):

$$I_{H_2} = \frac{x^2_{O_2} - x^1_{O_2}}{C_{O_2}} \cdot Q^i \tag{17}$$

and the consumed hydrogen due to leak by

$$Q_{H_2} = \frac{I_{H_2}}{2 \cdot F} \tag{18}$$

A discussion of the details of the role of the NN fitting of the observed impedance signatures is beyond the scope of this review; but it is illustrative to examine the measured impedance signatures themselves. Here, similar to the previous single-cell tests, impedance magnitudes were observed to increase with decreasing amounts of oxygen concentration. This pattern was consistent for different stack sizes; see Figure 8a, b. However, in either case, whenever the amount of oxygen was insufficient to deliver enough power to the stack, the impedance increased sharply; see Figure 8b.

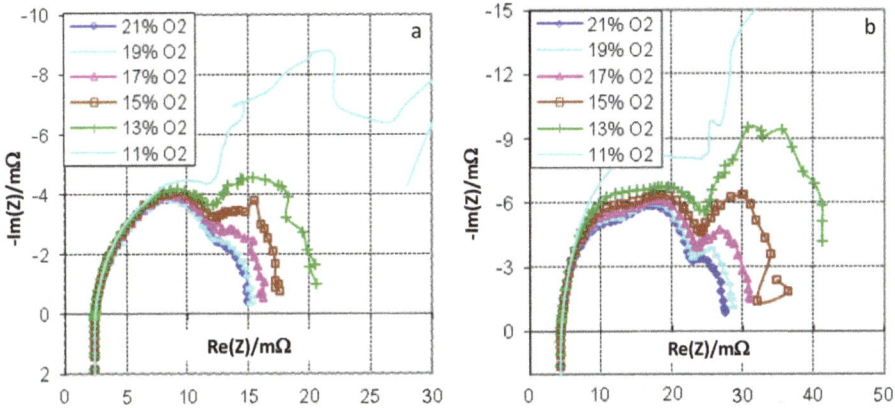

Figure 8: Impedance signatures at 20A load as function of O_2 concentration for a 5-cell (a) and 9-cell (b) FCvelocity™-1100 short stack with a leaky MEA in the middle of the stack.

A similar, but less pronounced behavior was observed with increasing Δp across the stack but O_2 concentration fixed at 21 %, where impedance signatures increased but not as significantly, especially for smaller leaks (see Figure 9a,b: medium/large-sized leak). Compared to Figure 8, here the increase in impedance was due to the recombination of oxygen in the leaking cell only resulting in a smaller effect. Also, with increasing Δp (see Figures 9b, 10; large-leaky cell), the impedance magnitude actually reduced again. This reduction in impedance is believed to be due to the nearly complete loss of power of the leaky cell. In other words, the leaky cell was not contributing to the stack impedance at high Δp because its voltage had already declined to near-zero due to air starvation.

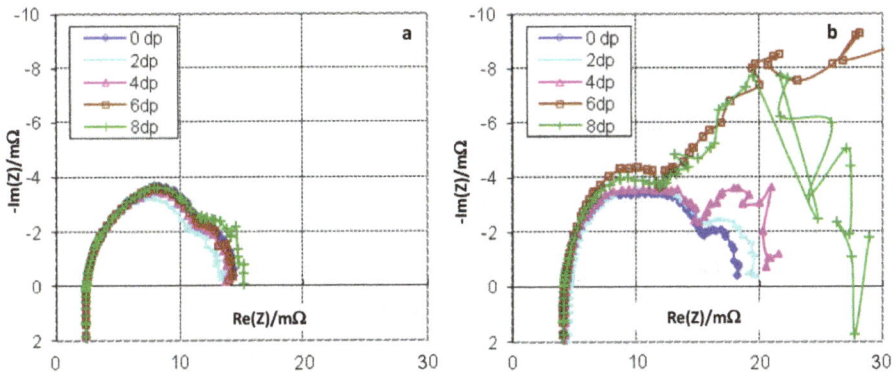

Figure 9: Impedance as function of anode-cathode pressure differential $\Delta p = p_A - p_C$ for a 5-cell FCveloc-ity™-1100 short stack at 20A with medium leaky cell (a) and large leaky cell (b).

In this condition with further increase in leak rates, impedance was saturated and no reduction in the signatures was noticed. The increase in impedance signatures were

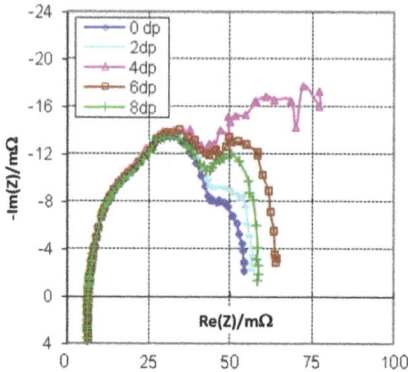

Figure 10: Impedance of a 19-cell FCvelocity™-1100 short stack at 20A load with same large leaky cell as in Figure 8b as a function of anode-cathode pressure differential $\Delta p = p_A - p_C$.

more significant for the smaller sized stacks, especially with the large leaky cell; see Figure 8b. With an increase in stack size (see Figure 10 with a single large leaky cell in a 19-cell stack), the significance of signatures reduced, but were still quite detectable for a situation where about 5 % of the cells had significant leaks. In tests performed at higher loads, similar impedance behavior was noticed, but with smaller overall low-frequency reactance arcs (see Figure 5).

This research work showed that, while the measured impedance signatures showed a clear relationship between oxygen consumption and hydrogen leak-rate, impedance signatures alone could not be used to quantify leak rates because the impedance changes with the number of cells present in the stack (e. g., compare Figure 10 with Figure 9b, where the leak rate is the same but the number of cells in the stack is different (5 vs. 19). Thus, in an attempt to understand the nature of multicell data in test stacks of different sizes, NN analysis was used to try to quantify the recombined amount of oxygen in the cathode due to leak(s).

The use of NN enabled comparison of a large collection of short stack data, recognizing that the impedance signatures of reduced oxygen concentration at $\Delta p = 0$ and $\Delta p > 0$ were interchangeable, because increasing Δp resulted in more hydrogen leak and less oxygen concentration in the cathode. This observation was then used to establish the amount of hydrogen leak in the n-cell stack using NN. The network training was conducted using the impedance data of particular stack sizes and leak rates. The network simulation was then used along with Equations (15)–(18) to calculate the hydrogen leak rate in the stack.

Using this diagnostic method, the leak rate could be inferred effectively from the impedance signature during operation of the fuel cell system [20]. This estimation was found to differ from the off-line *ex situ* leak measurements, however, most likely because water produced in operation blocked or partially-blocked the pinholes, as also observed by Kreitmeier et al. [4]. At lower Δp, the off-line and inferred leak rates are almost identical, while at higher Δp the off-line measured leak rates are typically higher. This showed the feasibility of the diagnostic method, as Δp's across the MEAs under operation are usually kept small in most fuel cell systems.

In the next subsection, a practical implementation of a setup enabling AC impedance spectroscopy on full-size fuel cell stacks is presented.

10.3.1.1 Voltage reduction technique for use with AC impedance spectroscopy on fuel cell stacks

While single cells or short stacks (< 20 cells) are well studied using electrochemical impedance spectroscopy ((EIS), see previous section), the high DC voltages of larger stacks pose interface challenges due to the need to couple small AC voltages between the power and measurement circuits. In this section, we show that it is possible to reduce the stack DC voltage for the EIS measurement using *DC Level-Reducing (DLR) circuitry*, without altering any relevant AC information. This DLR approach for reducing the DC voltage level of high-voltage stacks permits EIS measurement on the entire high-voltage stack or battery, and safer and more reliable processing of this information using lower-voltage measurement equipment [24]. Figure 11 shows the schematic diagram of the setup for AC impedance measurements on a full-size fuel cell (or battery) stack.

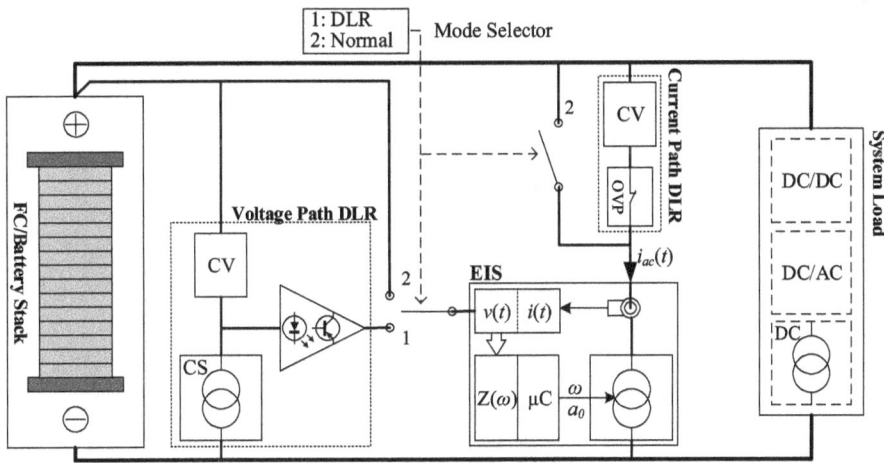

Figure 11: An electrochemical stack setup for EIS testing. The DC Level Reduction (DLR) circuit is shown in two blocks: voltage path DLR and current path DLR. With the mode selector, the setup can be used with DLR interface or for direct testing to compare results with and w/o DLR. CS: Current Source, CV: Constant Voltage module, OVP: Overvoltage Protection (reprinted from [24] with permission from IEEE).

EIS measurements on a FC or battery stack require both current and voltage measurement and the DLR should reduce the DC voltage for both measurements, preferably to a level of <60 V [25]. Figure 11 shows independent voltage reduction circuits for both Voltage- and Current paths. The DLR circuit for the voltage path comprises a constant voltage (CV) module connected in series with a DC constant current source (CS). A linear opto-isolator is used for galvanic isolation of the EIS unit. In addition, the DLR circuit has a unity AC gain and near-zero phase shift for frequencies up to 10 kHz and operates in the linear region to avoid introducing harmonic distortion. For more details on the DC Level-Reducing (DLR) circuitry, the reader is referred to [24, 26].

A DLR setup according to Figure 11 was tested at lower voltage on a 9-cell fuel cell stack, allowing for direct comparison between measurement with and without DLR.

Figure 12a shows the results for a direct measurement without DLR and for a measurement with a DLR according to Figure 11. The impedance $Z^{DLR}(\omega_i)$ matches that of the direct measurement with high precision over the entire frequency spectrum, as can be seen in Figure 12b, c. Subsequently, a high-voltage version of the DLR setup was tested on a 110-cell fuel cell stack, reducing the stack voltage to <10 V for the impedance measurement and it was found that $Z^{DLR}(\omega_i)$ closely matched results from direct measurement.

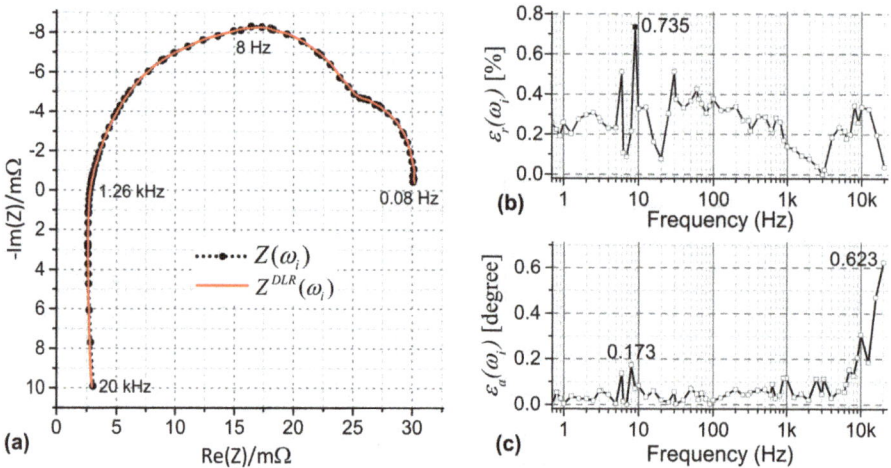

Figure 12: Experimental results of the low-voltage DLR setup according to Figure 11 using 9-cell stack. (a) Comparison of Nyquist plots for tests with DLR and without it. (b) The relative error of the magnitude of the impedance. (c) Absolute phase error (reprinted from [24] with permission from IEEE).

These results show that it is possible to utilize AC impedance instrumentation common for low voltage applications also on full-size fuel cell stacks by implementing a DC level reduction (DLR) strategy that only affects the DC component of the stack voltage, while leaving the AC components unchanged. This also results in an electrically safer operation and can save time and cost through use of existing equipment.

10.3.2 Quantification of H_2 transfer leaks using cell voltages

In Section 10.3.1, it was shown that while crossover leaks can be detected *in situ* using EIS, they are often not particularly well correlated to *ex situ* air-based leak measurements. Consequently, additional research was performed to develop techniques to more accurately measure the hydrogen crossover rate in individual cells based on (i) OCV measurements [8] (see Section 10.2.2.1) while controlling the fuel/air pressure difference Δp and (ii) by reducing the cell/stack air flow rate at a fixed current while maintaining a fixed fuel/air pressure difference [22] (oxidant starvation method).

Both of these approaches have their respective advantages and disadvantages; the first method means the stack does not produce power and is best suited for R&D-type applications, where Cell Voltage Monitoring (CVM) and nitrogen are typically available in stack test station environments, and access to individual cell voltages can be used to accurately estimate individual cell H_2 crossover flows. The second method (using measured stack voltages and reduced cathode air flows) can be performed while the stack is producing power and is better suited for system-level operational testing, where reducing the measured air flow is often the only practical way of identifying and characterizing transfer. As such this second method, while not as accurate as the first, is well suited to field-type or on-board diagnostics tests to determine if transfer is developing to become a life-limiting condition, or has not yet initiated.

10.3.2.1 H_2 leak diagnostics without power production

A Ballard Power Systems distributed generation FCgen™-1300 cell architecture in a 10-cell short stack configuration was used to develop and test the OCV diagnostic tool, while Ballard FCvelocity™-1100 fuel cells (of the type used in HD6 bus modules) were used for single-cell testing of the OCV diagnostic tool and for the development of the reduced air-flow transfer test. The test station used in both developments was similar to that shown schematically in Figure 13, where 3 mass flow controllers, MFC1, MFC2 and MFC3, with respective maximum flow rates of 20, 200 and 2000 sccm, were used to inject hydrogen into the cathode stream in order to simulate transfer leaks of a known rate; ball valves V2 and V3, V4 and V5 were used to isolate the unused MFCs.

In the single-cell testing, as shown in Figure 13, the anode and cathode input stream pressures were measured using PT-Ain and PT-Cin with a maximum pressure of 100 psi and accuracy of 1 % Full Scale (FS), and controlled using manual backpressure control valves to 13 and 10 psig, respectively. Anode and cathode input Mass Flow Controllers (MFCs) with a maximum flow rate of 3 and 5 slpm and accuracy of 1 % FS were used for the supply gases, respectively. A nitrogen MFC with a maximum flow rate of 30 slpm was used to supply the cathode with pure nitrogen through three-way valve V1. In this case, the nitrogen flow rate was set above the requirement for the cathode stream, and the excess nitrogen flow rate was vented to the outlet manifold in order to ensure that suf-

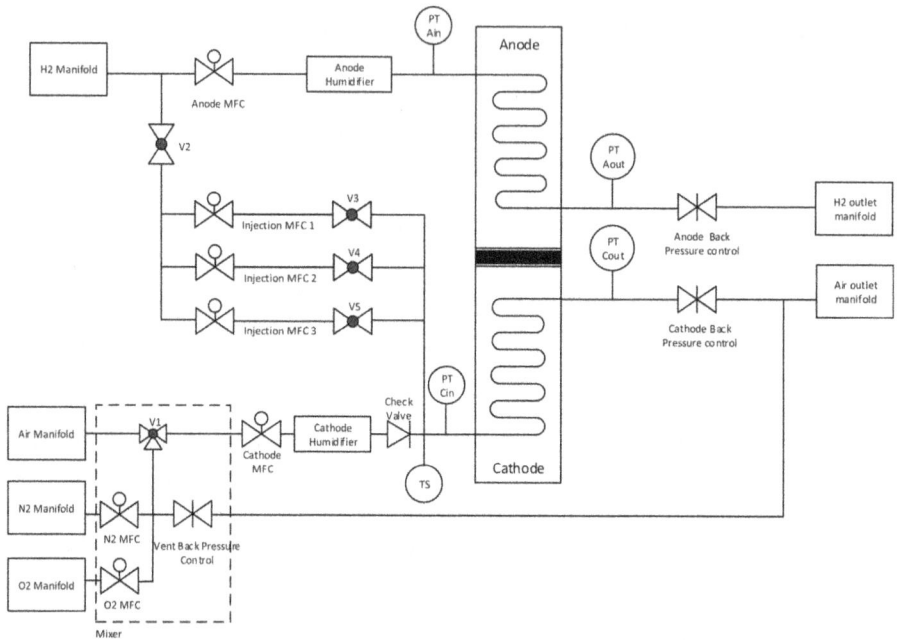

Figure 13: Schematic of Test Station used for Transfer Leak Method Development (reprinted from [22] with permission from Elsevier).

ficient flow was always available at the cathode. The vent (cathode) backpressure controller maintained a high nitrogen pressure to ensure nitrogen flow through the stack.

As protection against flammable concentrations of hydrogen caused by injection of hydrogen into the cathode (air) flow back-flowing into the upstream air mixing equipment, a check valve was added on the cathode line to prevent hydrogen ingress, where thermocouple TS was further installed at the hydrogen/air mixture junction to signal station shutdown if temperatures above 120 °C occurred, indicating hydrogen combustion occurring in the pipes. Anode and cathode streams were humidified to 60 °C using bubbling humidifiers and heated to 70 °C by heating the supply tubes. Stack temperature was controlled at 65 °C by circulating deionized hot water. A Princeton Applied Research Model 263A potentiostat with a 10A rating was also used to calibrate the diagnostic tool for hydrogen permeation through the membrane. All experimental conditions in this work were as explained above, unless mentioned otherwise.

When hydrogen and nitrogen are supplied to the anode and cathode of a PEM fuel cell, hydrogen crosses over from the anode to cathode, resulting in a hydrogen partial pressure in the cathode stream. In this case, anode and cathode hydrogen pressures, p_{aH_2} and p_{cH_2} respectively, and the open circuit voltage, V_{oc}, are related by the Nernst Equation (11). Using this equation, the partial pressure of hydrogen in the cathode can be calculated from the cell voltage, temperature and anode pressure, where the partial pressure of water in the anode, cathode can also be neglected because we are typically supplying enough hydrogen to the leaky cell to ignore the presence of water vapor (i. e., $p_{aH_2} = p_a$ and $p_{cN_2} = p_c$). Under these conditions, the ratio of the cathode partial pressures is given by

$$\frac{p_{cH_2}}{p_{cN_2}} = \frac{Q_{H_2}}{Q_{N_2}} \tag{19}$$

where Q_{N_2}, Q_{H_2} are the nitrogen flow, H_2 crossover rates and p_{cN_2} is the nitrogen partial pressure in the cathode.

The partial pressure of water in the cathode can also be calculated from the cathode dew temperature using the empirical relationship in [27]. From Equation (12) follows for dry operation ($p^i_{aH_2O} = p^i_{cH_2O} = 0$),

$$Q_{H_2} = \frac{Q_{N_2}}{\frac{p_c}{p_a} \cdot e^{\frac{2F \cdot V_{OC}}{R \cdot T}} - 1} \tag{20}$$

From Equation (20), the rate of hydrogen crossover from anode to cathode, Q_{H_2}, can be estimated using *in situ* measurements of cell voltage, temperature, pressure and cathode nitrogen flow rate. Next, the above equation was used with known hydrogen injection rates to verify the accuracy of the technique for crossover measurement, as well as separation of permeation and convection components, followed by implementation of the technique on a single leaky cell, and finally, on a full stack.

In order to develop a diagnostic tool for accurate estimation of hydrogen transfer leak through membrane pinholes, hydrogen permeation through the membrane was treated as a disturbance that needed to be accounted for separately. One technique for measuring the rate of hydrogen permeation is to supply the anode and cathode with hydrogen and nitrogen, respectively, and to measure the cell current in the double layer region ($\approx$350–600 mV) [7, 28], where the hydrogen crossover rate can be accurately estimated from the cell current using Faraday's law. Alternatively, hydrogen and nitrogen can be supplied to the anode and cathode of a fuel cell without pinholes, and the OCV measured to estimate the permeation rate using Equation (20). To test and compare the proposed OCV-based technique, a nonleaky MEA was supplied with different hydrogen and nitrogen flow rates on anode and cathode, respectively. At each nitrogen flow rate, a cyclic voltammogram was acquired from OCV to 500 mV with a sweep rate of 1 mV/s, using the potentiostat described previously. The result of this experiment is shown in Figure 14a. The figure shows that the OCV (x axis) increases with an increase in nitrogen flow rate, because higher nitrogen flow rates result in lower hydrogen partial pressure in the cathode, which increases the hydrogen partial pressure gradient between the anode and cathode and causes an increase in the OCV in agreement with Equation (11). The figure also shows that the cell current at 350 mV is similar for all nitrogen flow rates, as it represents oxidation of the permeated hydrogen, which is independent of the nitrogen flow rate (see Section 10.2.2.2).

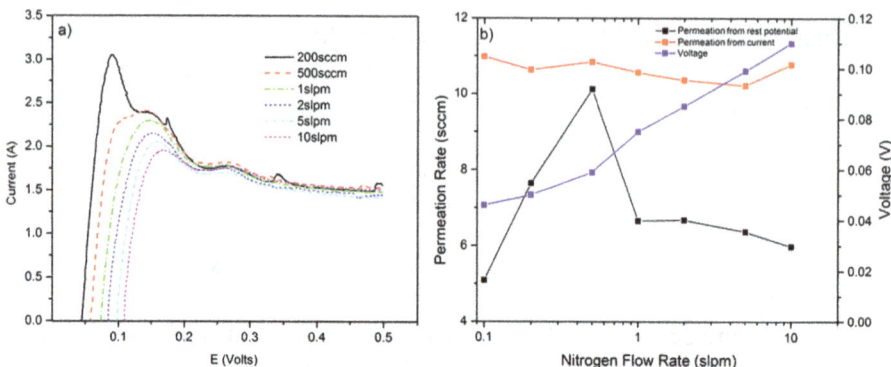

Figure 14: Permeation measurement on a leak-free MEA. (a) Positive voltage sweep at 1 mV/s with H_2/N_2 on anode/cathode with different N_2 flows on the cathode, (b) black: H_2 Permeation rate from OCV, red: H_2 permeation rate form applied voltage method (current at 350 mV) vs. different N_2 flow rate on the cathode (reprinted from [8, 22] with permission from Elsevier).

Figure 14b shows the permeation rate at different nitrogen flow rates, calculated using the two techniques explained above. We can see that the applied voltage technique is robust to changes in nitrogen flow rate and provides a consistent estimate of the permeation rate. The OCV-based technique yields a permeation rate dependent on N_2 flow, which is close to the Faradaic method at ~0.5 slpm N_2 flow, while underestimating it for other N_2 flows. This might be explained by the fact that Equation (20) only holds for dry gases, where fuel humidification increases viscosity reducing crossover flow, and hydrogen permeation is also a function of membrane hydration. In the rest of this section, the permeation rate estimated at a nitrogen flow rate of 0.5 slpm from OCV is used to calibrate small hydrogen transfer leaks (and 1 slpm for larger leaks), as the permeation estimation at this nitrogen flow rate has the smallest error. For leak rates significantly larger than the permeation rate such calibration is not required, as shown in the next subsection.

To examine the accuracy of the proposed diagnostic tool for estimating the rate of transfer, hydrogen and nitrogen was supplied to the anode and cathode of a fuel cell with a nonleaky MEA and used MFCs 1–3 of Figure 1 to inject 2 to 500 sccm hydrogen into the cathode stream. This technique allows simulating a known transfer leak rate in order to examine the accuracy of the proposed diagnostic technique. Figure 15 shows the measured and estimated OCVs for the hydrogen injection rates mentioned above, as the OCV can be calculated from the nitrogen and hydrogen flow rates by rearranging Equation (20) as

$$V_{oc} = \frac{RT}{2F} \cdot \ln\left\{ \frac{p_a}{p_c} \cdot \frac{Q_{H_2} + Q_{N_2}}{Q_{H_2}} \right\} \qquad (21)$$

Figure 15a shows the OCVs measured (blue), estimated w/o H_2 permeation (red) and calculated with H_2 permeation (black) for 2 to 10 sccm hydrogen injection into a 500 sccm

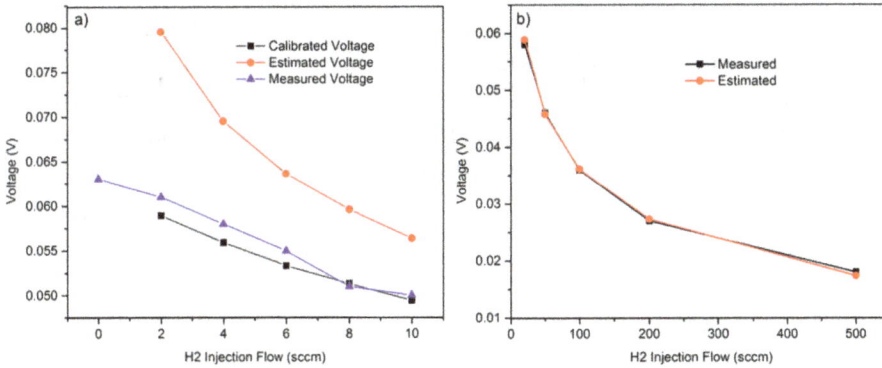

Figure 15: Measured and estimated OCV with cathode hydrogen injection Ballard for a FCvelocity™-1100 single cell. (a) 2–10 sccm H₂ injection into 500 sccm N₂ flow on anode, red: OCV calculated w/o H₂ permeation, black: OCV calculated with H₂ permeation; (b) 20–500 sccm hydrogen injection into 1 slpm N₂, red: OCV calculated w/o H₂ permeation, black: measured OCV (reprinted from [8, 22] with permission from Elsevier).

nitrogen flow in the cathode side. The measured and estimated OCVs are in relatively good agreement, with ~30 % error. This error is partially due to the effect of hydrogen permeation on the estimated OCV, as it changes the hydrogen flow rate in the cathode. We can also see that the error is reduced at higher injection rates, as permeation becomes less significant and the effect of humidification of the fuel becomes less at higher leak flows.

To calibrate the technique for the permeation error, the H₂ permeation rate was calculated from OCV at no hydrogen injection using Equation (4) for a single nonleaky Ballard FCvelocity™-1100 fuel cell to be ~7 sccm. Next, this hydrogen permeation rate was added to the 2–10 sccm injected hydrogen and calculated the calibrated OCV from Equation (21). As shown in Figure 15a, the calibrated OCV (black) closely matches the measured OCV (blue) for the hydrogen injection rates examined.

Figure 15b shows the measured and estimated cell potential for 20 to 500 sccm hydrogen injections into a 1 slpm nitrogen flow in the cathode. At these higher injection rates, the cell OCV estimation is very close to the measured value without correcting for the permeation rate. In summary, Figure 15 shows that the relationship between the cell OCV and the rate of hydrogen transfer leak can be represented accurately by the proposed diagnostic technique.

In [8], the relationship between *in situ* hydrogen crossover and *ex situ* leak measurements with air is also examined in some detail, where at the same overpressure, the flow rate ratio is given by

$$\frac{Q_{H_2}}{Q_{air}} = \frac{\mu_{air}}{\mu_{H_2}} \tag{22}$$

where Q_{air} is the measured air flow rate, and μ_{air} and μ_{H_2} are air and hydrogen dynamic viscosities, respectively; the ratio of which is ~2. Figure 16 shows the leak rate, measured in a single leaky cell versus overpressure by the *ex situ* technique and adjusted for hydrogen using Equation (22); see blue data. The hydrogen transfer leak rate estimated by the proposed *in situ* diagnostic technique closely matches the *ex situ* measurement (see pink data). This shows that the proposed diagnostic tool can accurately estimate the leak rate.

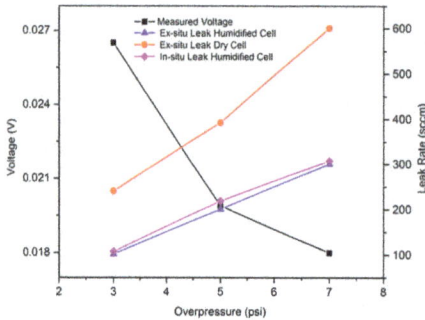

Figure 16: Hydrogen transfer leak rate at different anode overpressures for a Ballard FCvelocity™-1100 single cell: Estimated *in situ* from cell potential and measured *ex situ* at different membrane humidification conditions. Blue: *ex situ* H$_2$ leak rate calculated from *ex situ* air leak rate using Equation (22), purple: *in situ* H$_2$ leak rate calculated from measured voltage using Equation (20), red: *ex situ* H$_2$ leak rate calculated from *ex situ* air leak rate using Equation (22) after drying for ≈1000 h at room temperature (reprinted from [8, 22] with permission from Elsevier).

Figure 16 also shows that exposing an MEA to an extended period of drying in room conditions (≈1000 h) caused a significant increase in the hydrogen transfer leak rate. This is particularly important when using the *ex situ* technique for measuring the leak rate of individual cells in a fuel cell stack, as stack disassembly affects membrane humidification conditions and the measured leak rate.

In conclusion, an *in situ* diagnostic method has been developed for determining the hydrogen transfer leak rate based on supplying hydrogen and nitrogen to the anode and cathode of a PEM fuel cell stack, respectively, and measuring the cell open-circuit voltages while maintaining an anode overpressure. It was further shown that this tool can accurately estimate the hydrogen transfer leak rate in individual cells of a PEM fuel cell stack using temperature, pressure, humidity, N$_2$ flow and cell OCV measurements.

The accuracy of the diagnostic tool was demonstrated by injecting known amounts of hydrogen into the cathode, simulating a transfer leak. The method was shown to accurately estimate small leaks after correcting for hydrogen permeation, whereas for larger leaks, the correction was not necessary. The method was tested on a fuel cell with pinholes and was found to closely match *ex situ* measurements. Its reproducibility was then tested on a different stack with 10 cells, and the estimated individual leak rates summed to the *ex situ* measurement of the overall stack leak rate.

10.3.2.2 H_2 leak diagnostics with power production

While the OCV diagnostic technique is useful in R&D-type situations, in operation lack of access to individual cell voltages and of nitrogen flow on the cathode limits this technique to laboratory use. In the next cycle of transfer technique development at Ballard, a diagnostic tool that enables characterizing the rate and distribution of transfer leaks in an operational fuel cell stack was developed [22]. In the following, first the effects of oxygen concentration and hydrogen transfer leak on a single-cell voltage are characterized; then it is shown that once the air flow reaches a value close to oxygen starvation, further reduction in air flow drops the cell voltage sharply to zero volts. Furthermore, it is shown that at a fixed current, hydrogen transfer increases the air flow rate at which the cathode starves, as the crossover hydrogen consumes the oxygen in the cathode. Therefore, the air flow rate at which the cell voltage drops to zero can be used to estimate the rate of hydrogen transfer leak in each cell. Next, this methodology is expanded to characterize the leak rates and distribution of leaky cells for a fuel cell stack. The calculations are based on the residual between the expected linear decrease in stack voltage and the measured stack voltage as the air flow rate is reduced from a high to low value. This residual represents the number of leaky cells that are oxygen starved at the lower air flow rate and can be used to estimate the rate and distribution of hydrogen transfer leak in the fuel cell stack. Since the technique requires only measurements of air flow and stack voltage, it is suitable for integration in an operational fuel cell system, as these measurements are readily available in these systems (on-board diagnostics).

For the single-cell experiments, a Ballard FCvelocity™-1100 fuel cell configuration was used to study the impact of different cathode flows (see Figure 13 for test schematics); oxygen and nitrogen mass flow meters were added to the station cathode system so that different oxygen/nitrogen mixture concentrations could be supplied to the stack. Operating conditions used in this experiment are shown in Table 1, unless mentioned otherwise.

For stack-level testing, a 408 cell Ballard FCvelocity™-1100 fuel cell stack with pinholes was used. The anode inlet pressure was controlled with a pressure regulator, and a Hydrogen Recirculation Blower (HRB) was used to circulate hydrogen. Anode purging was enabled through a solenoid valve at the outlet. The cathode air flow rate was controlled with an MFC and humidified with a water-spray humidifier. The gas streams were heated with electric elements and the stack temperature was controlled with deionized water as a coolant. The stack load was controlled with an electronic switching load, and individual cell voltages were measured with CVM hardware. The temperature and humidity set points used in the experiment are also listed in Table 1, unless otherwise specified.

Table 1: Single-cell operating conditions.

Parameter	Value
Anode pressure	13 psig
Anode temperature	70 °C
Anode humidity	63 °C dewpoint
Anode flow rate	2 slpm
Cathode pressure	10 psig
Cathode temperature	70 °C
Cathode humidity	63 °C dewpoint
Cell temperature	65 °C
Load	20 A

When operating a leaky MEA with anode overpressure, the resulting hydrogen transfer leak can affect cell voltage through kinetic- and mass transport effects. This is due to recombination which reduces oxygen concentration, resulting in a reduction in cell voltage. At low air flow rates, this can result in cathode oxygen starvation and a drop in cell voltage down to zero volts. This property can be used to quantify the hydrogen transfer leak Q^{H_2} from cell voltage measurement through reduction of the air flow rate at a fixed current until the cell voltage drops to zero. Under these conditions, the leak rate can be estimated by subtracting faradaic oxygen consumption from the total oxygen supplied (see Equation (23)).

In order to quantify hydrogen crossover using oxygen starvation, it is preferred to operate at low current densities in order to minimize the effect of water production on hydrogen transfer leaks through, for example, the sealing of pinholes. Figure 17 shows the effect of load and hydrogen injection into the cathode for a Ballard FCvelocity™-1100 single-cell.

Figure 17 shows that injecting hydrogen into the cathode at OCV up to ≈100 sccm results in a drop in the cell voltage (see red curve). However, higher injection rates do not lead to further reduction in cell voltage suggesting that the recombination rate reaches a saturation value. It appears that at this injection rate the cathode is fully activated w. r. t. recombination. The high air stoich of ≈15 ensures that even at 200 sccm H_2 injection, equivalent to about 30 A, the air stoich is still about $\lambda_{air} \approx 6$, i. e., the cell is not oxygen starved. This suggests that a load equivalent to 100 sccm hydrogen consumption, which is ≈14 A, fully activates the cathode. Therefore, we used a load of 20 A for the H_2 injection experiments (see black data) in Figure 17. Note that due to the large active area of the Ballard heavy-duty bus stack architecture, 20 A results in a relatively small current density ($<0.1\,A/cm^2$) that corresponds to the beginning of the ohmic region. To illustrate this, we have also graphed the cell voltage at the currents equivalent to the hydrogen injection flow rates, as calculated by Faraday's law. We can see that the cell voltage that results from load (see blue data in Figure 17) closely follows that during hydrogen injection at low currents, while the deviation increases at higher currents. This is due to the fact that hydrogen injection results only in cathode kinetic and mass transport losses,

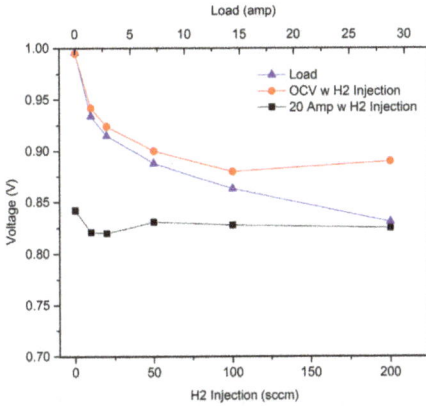

Figure 17: Effect of hydrogen injection into cathode on Tafel kinetics with anode(H_2) and cathode (air) flow rates at 2 and 5 slpm, respectively, for a Ballard FCvelocity™-1100 single cell. Blue: effect of load on cell voltage w/o H_2 injection Red: effect of H_2 injection on cell voltage w/o current (OCV), black: effect of H_2 injection on cell voltage at 20 A and air flow rate of 5 slpm (air stoich ≈15). Top/bottom x-axis: Load and H_2 injection scales are equivalent according to Faraday's law for H_2 (Equation (13)), i. e., 10 A load current corresponds to 67 sccm H_2 (reprinted from [22] with permission from Elsevier).

while the load current imposes additional losses due to anode kinetics as well as ohmic resistances. Furthermore, the H_2 recombination rate may saturate, so that the recombination efficiency drops with higher H_2 injection rates. These observations suggest that the 20 A point is a suitable point to be relatively insensitive to kinetic losses caused by hydrogen transfer leak, while being small enough to not produce significant amounts of water, which could lead to sealing of transfer leaks.

Figure 18 shows the cell voltage when reducing the cathode air flow rate from a high to low value, while keeping the load constant at 20 A. Under these conditions, the cell voltage vs. air flow graph exhibits two distinct regions. At higher air flow rates, above 0.6 slpm ($\lambda_{air} \geq 2$), changes in the air flow rate have a relatively small and linear effect on the cell voltage. However, once the air flow rate is reduced below 0.6 slpm, the cell voltage drops sharply. The cell voltage hits zero volts at about 0.32 slpm ($\lambda_{air} \approx 1$), which is equal to the rate of oxygen consumption at 20 A. This suggests that, at the chosen low current density, the flow where the cell voltage drops to zero corresponds to $\lambda_{air} \approx 1$ and can therefore be used to estimate oxygen consumption on the cathode.

Figure 18 also shows that reduction in oxygen flow results in only a small drop in cell voltage at flow rates well above oxygen starvation. This indicates only small changes in the oxygen concentration profile through the cathode at these conditions. In order to investigate the effect of changes in oxygen concentration on the equilibrium potential, Figure 19 shows the measured (see black data) and estimated (see red data) cell OCV at different oxygen concentrations. For the estimated OCV, we used the measured cell voltage at 21 % oxygen concentration as the reference value and deducted the drop in cell voltage as captured by the Nernst equation (see Equation (7)).

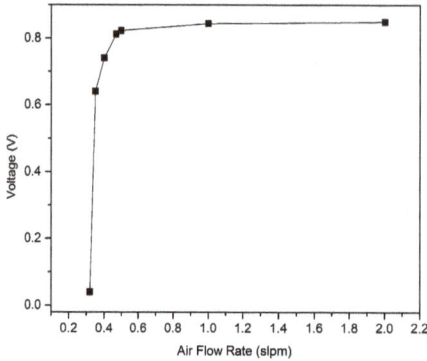

Figure 18: Effect of air flow on cell voltage at a fixed current of 20 A for a Ballard FCvelocity™-1100 single cell (reprinted from [22] with permission from Elsevier).

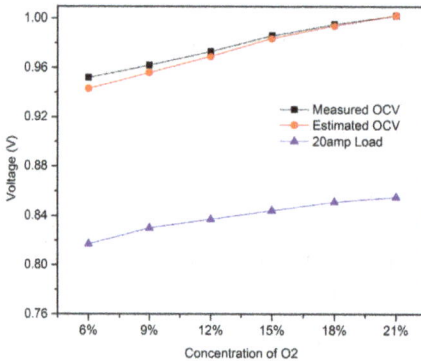

Figure 19: Effect of oxygen concentration on cell voltage at OCV and 20 A load with anode for a Ballard FCvelocity™-1100 single-cell and cathode flow rates of 2 and 5 slpm, respectively. Black: measured OCV, red: OCV estimated by subtracting corresponding change in cathode equilibrium potential (Nernst equation), blue: cell voltage at 20 A load (reprinted from [22] with permission from Elsevier).

Figure 19 shows that there is a small discrepancy between the measured and esti-mated OCV, as the Nernst equation overestimates the effect of oxygen concentration on cell voltage [29]. Also graphed is the effect of reduction in oxygen concentration on the cell voltage at 20 A, showing a similar slope compared to the measured OCV. Therefore, the cell voltage exhibits essentially a linear behavior w. r. t. oxygen concentration, espe-cially at higher oxygen concentrations. This property will be used later in this section for the estimation of size and distribution of transfer leaks in PEM fuel cell stacks.

To characterize the effect of hydrogen transfer leak on oxygen starvation, a known amount of hydrogen was injected into the cathode of a single cell while the cell voltage was measured at a fixed current of 20 A and different air flow rates. The result of this experiment is shown in Figure 20. It can be seen that at high air flow rates of 1 slpm and above, hydrogen injection has a small effect on cell voltage. This is because hydrogen

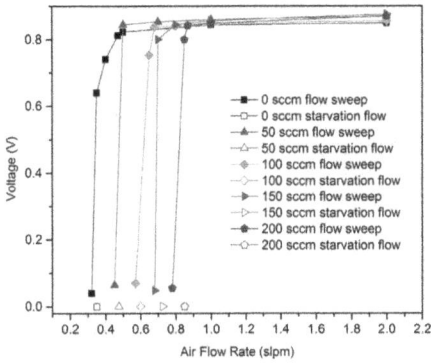

Figure 20: Effect of hydrogen injection on cell voltage drop at 20A (the oxygen starvation flow rate is marked using open symbols as calculated from the injection rate using Equation (23)), (reprinted from [22] with permission from Elsevier).

injection results in no mass transport losses at high air flow rates, and only small effect on activation losses at 20 A. However, the figure also shows that with the reduction of air flow to lower values, the cell voltage drops to zero as a result of oxygen starvation. In Figure 20, the starvation flow rates are also shown (open symbols), as calculated from the load and hydrogen injection rate using the following relationship:

$$Q_{air}^s = \frac{1}{x_{O_2}} \left(\frac{I}{4 \cdot F} + \frac{Q_{H_2}}{2} \right) \tag{23}$$

where Q_{air}^s is the stoichiometric air flow required to drive the reactions in slpm, I is the load current, x_{O_2} is the oxygen mole fraction and Q_{H_2} is the flow rate of injected hydrogen. The first term on the right side of Equation (23) represents the Faradaic oxygen consumption, and the second represents oxygen consumption due to recombination with the injected hydrogen at the CCL. Equation (23) shows that the air flow rates at which the cell oxygen starves increase with higher hydrogen injection rates (see open symbols in Figure 20). Therefore, the air flow at which point the fuel cell oxygen starves, along with the cell current, can be used to estimate the hydrogen transfer leak rate Q_{H_2} in Equation (23).

To examine the accuracy of the diagnostic technique for characterizing hydrogen transfer leaks in PEM fuel cells, a commercial MEA was used which was returned from the field with a pinhole at the cathode inlet. The leak rate was characterized by measuring the cell OCV while hydrogen and nitrogen was supplied to the anode and cathode, respectively, as explained in Section 10.3.2.1. The red curve in Figure 21a shows the corresponding hydrogen transfer leak rates obtained from this technique at different anode overpressures. For comparison, the blue curve in Figure 21a shows the estimated the hydrogen transfer leak rates from the oxygen starvation air flow rates using Equation (23). Figure 21b shows the cell voltage under operation at 20A while reducing the air flow rate

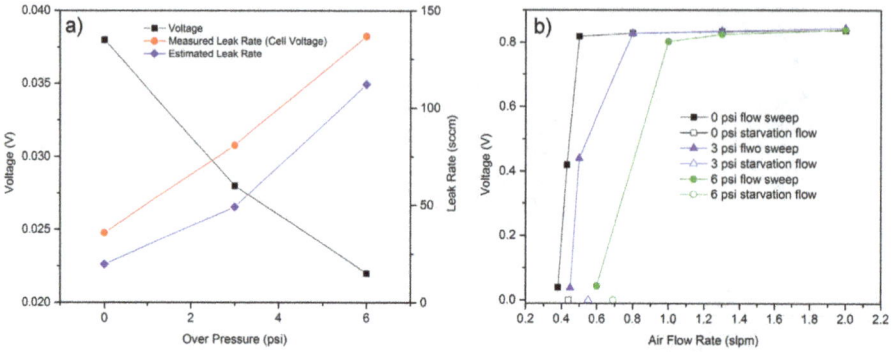

Figure 21: Rate of hydrogen transfer leak vs. anode overpressure estimated using hydrogen/nitrogen and oxygen starvation techniques for a single fuel cell with a pinhole leak. (a) Hydrogen and nitrogen flow rates of 2 and 0.5 slpm used to quantify hydrogen crossover from open circuit voltage. Red: H_2 Leak rate calculated using OCV method; see Equation (20). Blue: H_2 leak rates calculated from oxygen starvation flow rates (see Figure 21b) using Equation (23). (b) Effect of air flow on cell voltage at different anode overpressures at 20A. Open symbols: starvation flows calculated using Equation (23) using H_2 leak rates from OCV method (see Figure 21a), (reprinted from [22] with permission from Elsevier).

from a high to low value at the same anode-cathode overpressures. Similar to Figure 20, Equation (23) was used to calculate the air flow rate at which the cell oxygen starves at different overpressures (open symbols in Figure 21b).

The leak rates in Figure 21a show a similar trend, however, the hydrogen/nitrogen technique yields values which are larger by 20 to 30 sccm. In Figure 21b, as in Figure 20, the calculated flow rates at which the cell oxygen starves (using Equation (23)) are slightly higher than the actual values; however, they still fall within the sharp voltage drop region. This suggests that the assumption of zero cell voltage to identify the starvation flows may not be correct and may need to be modified to "near zero." This is a source of error in the proposed leak measurement technique, but the overall accuracy is reasonable for larger leaks.

Based on the starvation method, in the following a method to determine leak rates and leak rate distribution of leaky cells in a stack is developed. As discussed previously, two distinct operating regions for each cell are assumed. The first region is not oxygen starved, with a linear dependence of cell voltage upon air flow rate. In the second region close to oxygen starvation, the cell voltage drops sharply to near zero values. Consequently, the voltage drop in the stack will have two components: one resulting from reduced oxygen concentration and one due to oxygen starvation of leaky cells, as illustrated in Figure 22:

$$V_{\text{ref}} - V(Q) = n_{\text{leak}}(Q) \cdot \left(V_{\text{ref}}^1 - V_{\text{leak}}^1(Q)\right) + \left(n - n_{\text{leak}}(Q)\right) \cdot \Delta V^1(Q) \tag{24}$$

And with $V_{\text{leak}}^1(Q) \approx 0$ and $V_{\text{ref}}^1 = V_{\text{ref}}/n$:

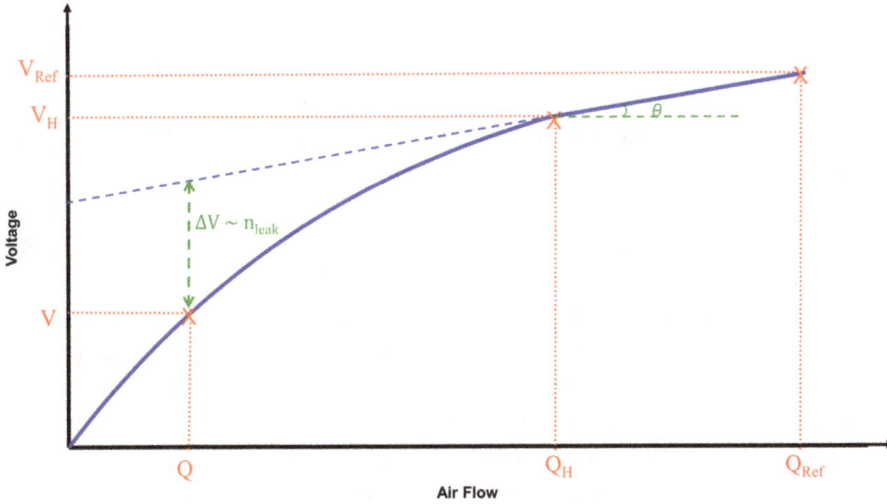

Figure 22: Illustration on the effect of reduction in air flow on cell voltage. At high air flows, represented by Q_{ref} and Q_H, the effect of reduction in air flow on stack voltage is linear, with a slope represented by $\tan(\theta)$. The divergence from the linear drop at lower air flows is due to oxygen starvation of the cells with hydrogen transfer leak (reprinted from [22] with permission from Elsevier).

$$V_{ref} - V(Q) = \frac{n_{leak}(Q)}{n} \cdot V_{ref} + (n - n_{leak}(Q)) \cdot \Delta V^1(Q), \tag{25}$$

where V_{ref}, V_H are the stack voltages at a high air flow rate where we assume no oxygen starvation, $V(Q)$ is the stack voltage at air flow Q where $n_{leak}(Q)$ cells have entered starvation, V_{ref}^1 refers to V_{ref} for a single cell, $V_{leak}^1(Q) \approx 0$ is the single-cell voltage under starvation for a leaky cell, n is the total number of cells in the fuel cell stack and $\Delta V^1(Q)$ represents the linear drop in single-cell voltage when the air flow is reduced from the high to a low value due to changes in oxygen concentration (see blue dashed line in Figure 22). The left-hand side of Equation (25) represents the total changes in the stack voltage when the air flow is reduced from a high value where there is no oxygen-starved cells, to a lower flow rate. The first term on the right-hand side of Equation (25) captures drop in stack voltage as a result of oxygen starvation in the leaky cells (where $V_{leak}^1(Q) \approx 0$ for the individual leaky cells), and the second term represents the linear drop in stack voltage in the remaining nonleaking part of the stack.

In this derivation, it is assumed that all cells in the stack have the same voltage at high air flow rates, with the same linear rate of voltage drop due to changes in oxygen concentration. This drop in cell voltage as a result of reduction in air flow rate Q can be calculated as

$$\Delta V^1(Q) = \tan(\theta) \cdot \frac{Q_{ref} - Q}{n} \tag{26}$$

with

$$\tan(\theta) = \frac{V_{\text{ref}} - V_H}{Q_{\text{ref}} - Q_H},\tag{27}$$

where $\tan(\theta)$ is the slope of stack voltage drop in the linear range at high air flow rates due to reduction in oxygen concentration (see Figure 22). By substituting Equation (26) and Equation (27) in Equation (25), and solving for the number of leaky cells, it follows:

$$n_{\text{leak}}(Q) = n \cdot \frac{V_{\text{ref}} - V(Q) - \tan(\theta) \cdot (Q_{\text{ref}} - Q)}{V_{\text{ref}} - \tan(\theta) \cdot (Q_{\text{ref}} - Q)}\tag{28}$$

Equation (28) allows the calculation of the total number of starved cells at any air flow from the stack voltage, air flow, the total number of cells and the stack voltage slope at high air flow rates, all of which can be measured in an operational fuel cell system. Therefore, by systematically reducing the stack air flow rate Q, the additional oxygen-starved cells can be associated with a leak rate $Q_{H_2}(Q)$ corresponding to the respective air flow. This single-cell leak rate can be calculated from Equation (23) as

$$Q_{H_2}(Q) = 2 \cdot \left(x_{O_2} \cdot \frac{Q}{n} - \frac{I}{4 \cdot F} \right)\tag{29}$$

To examine the proposed diagnostic technique for determining the leak rates and distribution in a fuel cell stack, an experimental setup capable of running a full-size stack with air and hydrogen similar to that presented in Figure 15 was used. The stack current was fixed at 20A and the air flow reduced from 649 slpm (stoich of ~4) to 222 slpm (stoich of 1.6). Under these conditions, stack and individual cell voltages were measured. Figure 23 shows two snapshots of the CVM data for high and low flow rates. At a high flow rate, most cells have a high voltage of about 0.85 V, whereas the voltage of about 15 cells drops significantly (below 0.4 V) at low flow, indicating oxygen starvation in the respective cells as a result of hydrogen transfer leaks.

Figure 24 shows the stack voltage and the cumulative number of leaky cells as calculated from Equation (28). Also graphed is the number of oxygen-starved cells at each flow rate, as counted directly from the CVM data (red curve). It can be seen that the counted and estimated number of oxygen-starved cells match closely, validating the present method. Also shown is the lower bound on the leak rate as calculated from air flow and a constant current of 20A from Equation (29). The results show that the diagnostic method detects oxygen-starved cells when air flow is reduced to about 350 slpm, corresponding to a leak rate of about 270 sccm. This information provided by the method could be obtained on-board of a vehicle, for example, and could be used to estimate remaining lifetime and make service plans. Figure 25 shows the differential distribution of oxygen-starved cells vs. single-cell leak rate as calculated through Equation (29).

The results show that the proposed diagnostic technique can be used to detect the rate and distribution of hydrogen transfer leaks in a PEM fuel cell stack. The technique is suitable for leak characterization of operational fuel cell systems and may be useful for on-board diagnostics, as it requires only stack voltage and air flow measurements.

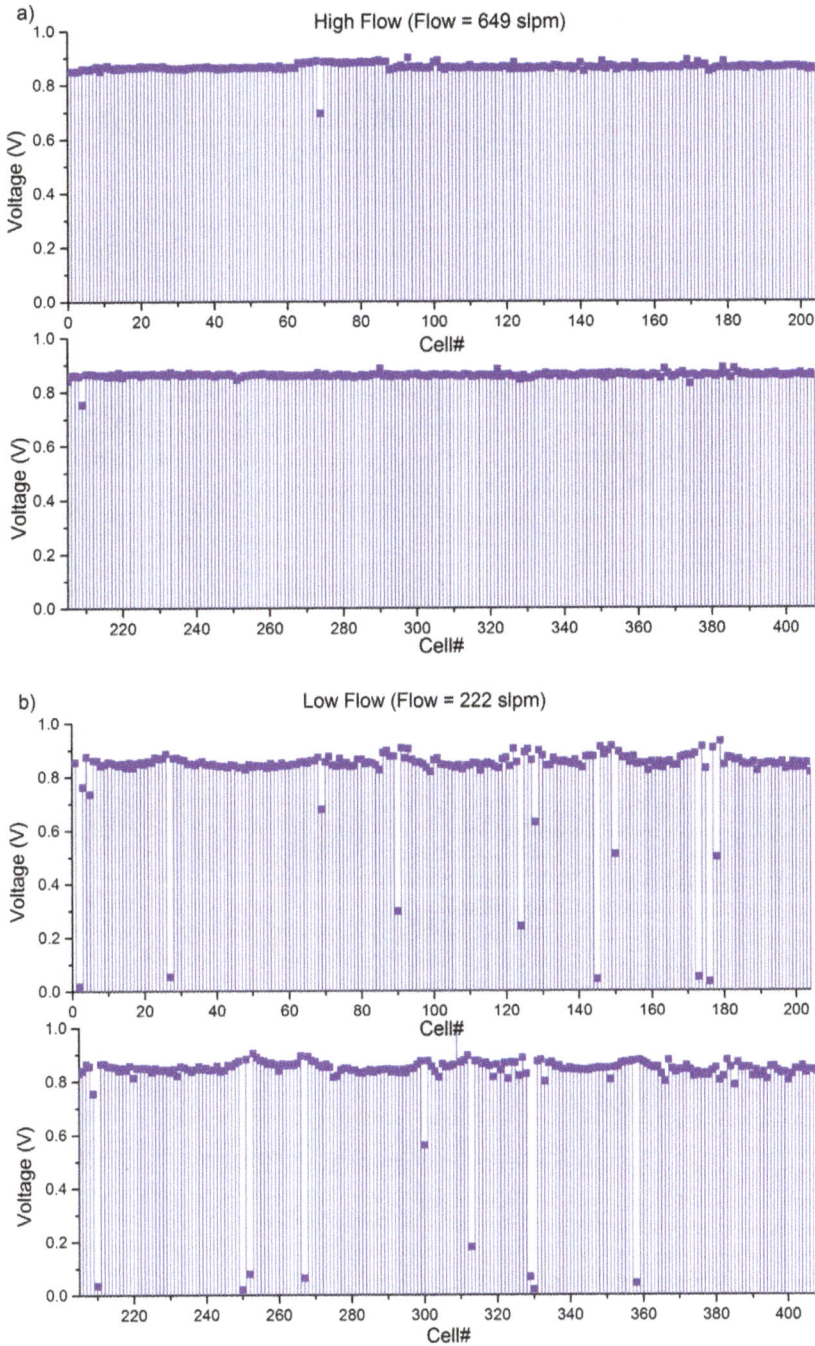

Figure 23: Snapshot of the HD6 stack CVM data at 20 A and Δp = 0.2 bar. (a) At high air flow rate of 648 slpm (stoich 4). (b) At low air flow rate of 222 slpm (stocih 1.6) (reprinted from [22] with permission from Elsevier).

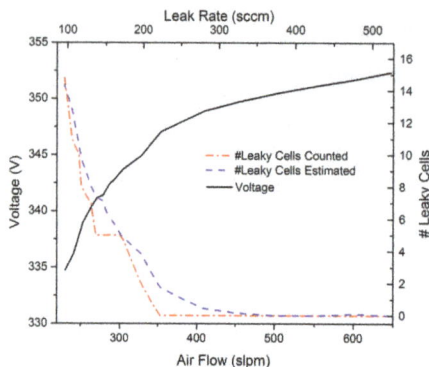

Figure 24: Stack voltage and number of leaky cells, as calculated from Equation (28) as well as direct counting of oxygen-starved cells, at different air flow rates (fixed current of 20 A and over pressure of 3 psi). The top x-axis is the lower bound on the hydrogen transfer leak rate, calculated from the air flow using Equation (29) (reprinted from [22] with permission from Elsevier).

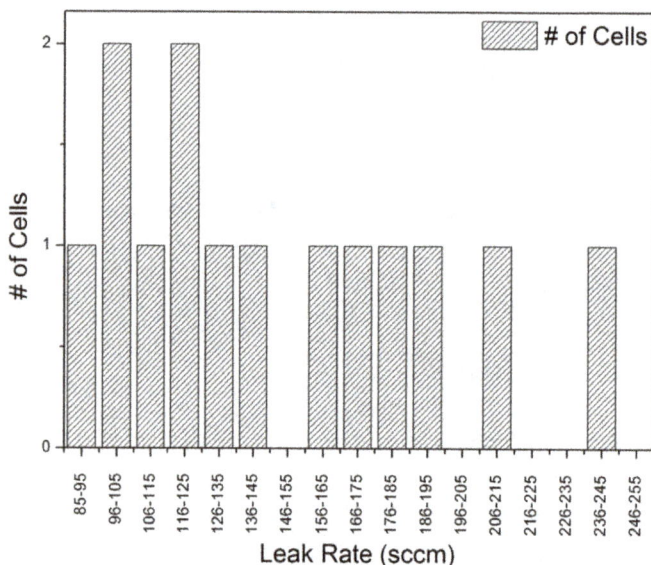

Figure 25: Number of leaky cells and associated H_2 leak rates in PEM fuel cell stack (reprinted from [22] with permission from Elsevier).

10.3.3 H_2 injection and emissions due to internal transfer leaks

For a fuel cell with either (i) an internal H_2 transfer leak or (ii) a H_2 injection flux $Q_{H_2}^i$ into the cathode stream the H_2 concentration $x_{H_2\_ox}$ in the oxidant outlet can be expressed as

$$x_{H_2\_ox} = \frac{Q^f_{H_2}}{Q^f_{tot}} \tag{30}$$

To calculate $x_{H_2\_ox}$, we first calculate the O_2 flux at the cathode outlet $Q^f_{O_2}$:

$$Q^f_{O_2} = Q^i_{O_2} - \frac{I}{4F} - Q^{rec}_{O_2} = \frac{I}{4F} \cdot \left(\lambda_{O_2} - 1 - \eta_r \cdot \frac{Q^i_{H_2}}{\frac{I}{2F}} \right), \tag{31}$$

where $Q^i_{O_2}$, $Q^{rec}_{O_2} = \frac{1}{2}Q^{rec}_{H_2} = \frac{1}{2}\eta_r \cdot Q^i_{H_2}$ refer to the inlet, recombining O_2 flux, respectively, λ_{O_2} to the oxygen stoichiometry, η_r to the fraction of the H_2 flux recombining with O_2 to form water (recombination efficiency) and the term $r_l = Q^i_{H_2}/\frac{I}{2F}$ to the H_2 *leak ratio* (*ratio of H_2 leak to H_2 consumption*; it is useful here to normalize the H_2 injection flux with respect to the H_2 consumption in order to nondimensionalize the bracketed term in Equation (31)). For increasing $Q^i_{H_2}$, the O_2 flux at the cathode outlet decreases until it reaches zero at the point of starvation for $r_l = r_{l\_0}$:

$$r_{l\_0} = \frac{\lambda_{O_2} - 1}{\eta_r} \tag{32}$$

For $r_l < r_{l\_0}$, the cathode outlet H_2 concentration $Q^f_{H_2}$ and total flux at the cathode outlet Q^f_{tot} can be written as

$$Q^f_{H_2} = Q^i_{H_2}(1 - \eta_r) \tag{33}$$

and

$$\begin{aligned}
Q^f_{tot} &= Q^f_{N_2} + Q^f_{O_2} + Q^f_{H_2O} + Q^f_{H_2} \\
&= \frac{I}{4F} \left(\frac{\lambda_{O_2}}{r_{ON}} + \{\lambda_{O_2} - (1 + \eta_r \cdot r_l)\} + \lambda_{O_2} \left(\frac{RH \cdot p_{sat}(T)}{p_{tot} - RH \cdot p_{sat}(T)} \right) \left(1 + \frac{1}{r_{ON}} \right) + 2(1 + r_l) \right)
\end{aligned} \tag{34}$$

Here, RH denotes the cathode inlet relative humidity, p_{tot} the total cathode pressure, $p_{sat}(T)$ the saturation vapor pressure at operating temperature T and $r_{ON} = 0.21/0.79$ the oxygen/nitrogen molar ratio of the oxidant air. All water is assumed to be in gaseous form.

Figure 26 shows the H_2 concentration in the cathode outlet according to Equation (33). If there is no recombination, the concentration increases nonlinearly, reaching about 10 % for $r_l = 1$. With increasing recombination efficiency, the H_2 concentration is reduced, and a distinct transition is observed at the point of starvation $r_l = r_{l\_0}$, after which the H_2 concentration increases at a faster rate. In general, the recombination efficiency for any H_2-leakage into the cathode gas stream is dependent on the position of

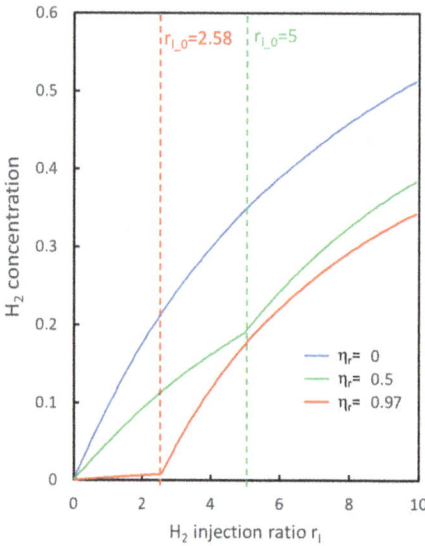

Figure 26: H_2 concentration $x_{H_2\_ox}$ in the cathode outlet of a fuel cell upon H_2 injection into the cathode gas stream at p_{tot} = 1.5 bar, RH = 50 %, T = 60 °C, air stoich λ_{O_2} = 3.5 as a function of the H_2 leak ratio r_l. blue: no recombination η_r = 0, green: η_r = 0.5, red: η_r = 0.97.

the leak: as the leak position moves toward the cell outlet, the recombination efficiency η_r drops to zero.

Using the reduced-voltage circuitry (see Section 10.3.1.1) for performing EIS on relatively large stacks, the relationship between H_2 crossover and EIS response was studied in more detail by injecting H_2 into the cathode stream. This testing, as shown in Figure 27, was performed on a 30 kW Ballard fuel cell module containing 180 FCvelocity™-9SSL cells, where hydrogen was injected just downstream of the air compressor, and stack voltage, cathode outlet O_2 concentrations and H_2 emissions were measured.

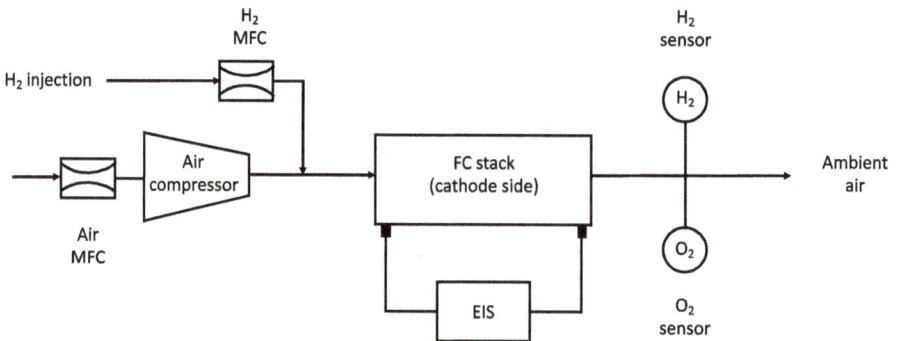

Figure 27: Test setup schematic for 30kW Module cathode H_2 injection testing. H_2 sensor: RKI Model FHD-752, O_2 sensor Nova Model 5301-A.

This arrangement permitted simulating increasing H_2 crossover by incrementally increasing the H_2 injection flow delivered by MFC into the cathode inlet air flow, simulating a uniform transfer leak in the cathode inlet of every cell in the stack. As noted in Table 2, a series of tests was performed with increasing H_2 flows, with return-to-zero injection flows between individual injections. The EIS response returned to its baseline condition, showing that no additional transfer was created via this testing (in fact, the H_2 injections tended to "seal" small previous transfer leaks detectable by EIS prior to the testing). In the first of these injection tests (i. e., Test 3), the module was set up to purge from the anode loop into the cathode exhaust just upstream of a dedicated 0–4 % H_2 RKI Model FHD-752 cathode exhaust hydrogen sensor, but in the later tests (Tests 8–10), the module was modified to purge downstream of this sensor to give more stable measured hydrogen emissions. Another change introduced in Test 8 was the use of a separate added air-flow sensor installed upstream of the main air-flow sensor used in the product to provide more accurate air flows. Additional instrumentation used in this testing included a Nova Model 5301-A oxygen sensor, and a 0–100 % Neodym Panterra thermal-conductivity-based hydrogen sensor, although all emissions reported in this work were measured with the RKI H_2 sensor. For testing safety, the hydrogen was injected through a stainless-steel line into a metal spool piece with a downstream thermocouple used to detect combustion, and an interlocked upstream solenoid valve on the H_2 injection line that would immediately shut off if/when the module shuts down.

Table 2: Tests performed using H_2 injection to simulate transfer.

Test	Description	Additional information	H_2 emission
Test 3	20A; 0,2,4,6,8,10 slpm H_2	Purged upstream of H_2 sensor	No
Test 8	20A; 10,15,20 slpm H_2 injections	1st use added air-flow sensor	No
Test 9	20A; 20,25,20 slpm H_2 injections	Some H_2 emissions, but small	Yes
Test 10	20A; 25,30,35,40,45,48 slpm H_2	Small H_2 emit on large H_2 inject	Yes

Figure 28 presents the results of EIS scans performed at 20 A load (idle), as measured at the module DC power outlet, for increasing hydrogen injection flows. The left-hand figure shows 0 to 25 slpm H_2 injections, while the right-hand figure shows 25 to 48 slpm H_2 injections (with returns to baseline/zero injection flows). Attempts to perform EIS scans at even higher injection flows were not possible as the module shut down due to low stack voltage (i. e., < 0.55 V/cell).

As seen previously Figure 28 shows that increasing the hydrogen injection rate increases the size of the third (low frequency) arc in the EIS scans, which represents the mass transport losses resulting in stack voltage drop. The graph on the left suggests a possible sensitivity of about 6 slpm H_2 for distributed transfer detection in this module (assuming all cells are exhibiting an equal leak rate), corresponding to 6 slpm/180 ≈ 33 sccm H_2 per cell.

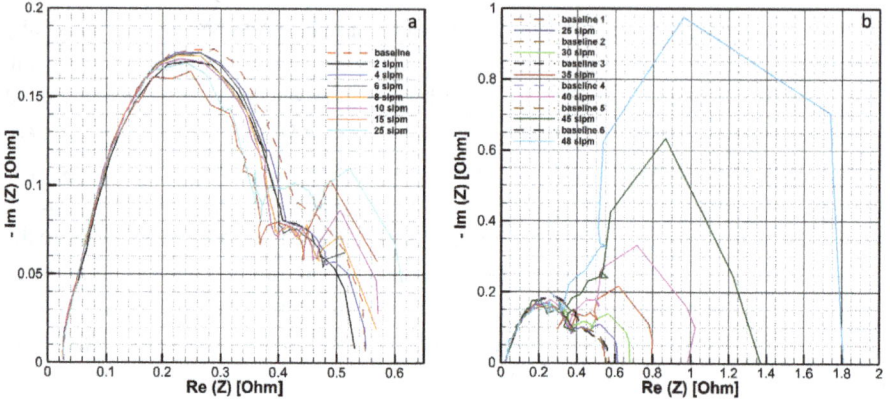

Figure 28: (a, b) EIS Scans at 20A (idle) Load on Ballard 30 kW Module with increasing hydrogen injection flows.

Hydrogen injection and air flow reduction both result in decreasing oxygen stoichiometry defined by

$$\lambda_{O_2} = \frac{O_2 \text{ provided}}{O_2 \text{ consumed}} = 4 \cdot F \cdot \frac{Q^i_{O_2} - \frac{Q^i_{H_2} + Q^{leak}_{H_2}}{2}}{n_{cell} \cdot I} \tag{35}$$

where

λ_{O_2} is the oxygen stoichiometry

$Q^i_{O_2}$ is the dry input oxygen flow rate

$Q^i_{H_2}$ is the cathode hydrogen injection flow rate

$Q^{leak}_{H_2}$ is the hydrogen transfer leak rate

n_{cell} is the number of cells in the stack module

I is the stack current, and

F is Faraday's constant

and where it is assumed that all H_2 entering into the cathode undergoes complete recombination.

Using Equation (35) to determine the oxygen stoichiometry for the various hydrogen injection tests in Table 2 (and assuming that the underlying transfer leak rate, $Q^{leak}_{H_2}$, is zero due to the H_2 injection recombining to water and sealing up any small preexisting transfer leaks) results in a linear dependence of stack voltage versus oxygen stoichiometry (see Figure 28).

Figure 29 shows that, over a very wide range of operation (i. e., at 20A idle with ~0.8V/cell down to 0.55 V/cell) the stack voltage-loss is linear, even if the hydrogen delivered to the cells via cathode injection may not be completely uniform. In order to understand this phenomenon based on its impact on fuel economy, or versus the amount

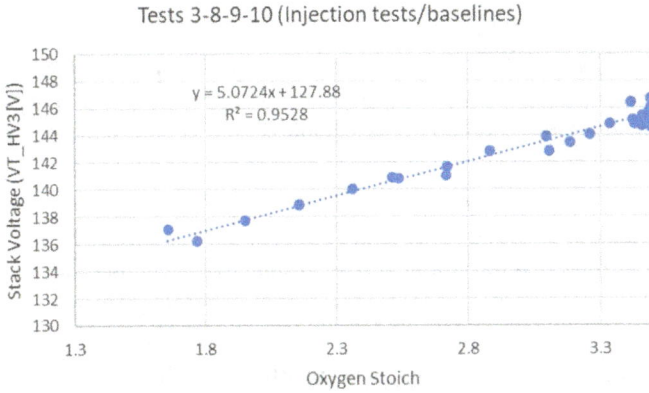

Figure 29: Stack voltage of Ballard 30 kW Module at 20A versus O_2 stoichiometry for H_2 injection tests 3, 8, 9 and 10 combined. Dotted line: linear fit.

of fuel lost to transfer relative to that consumed without transfer. Figure 29 shows the results from Figure 28 versus the hydrogen injection rate.

Figure 30 also shows the idle-level fuel consumption of ≈26 slpm (without transfer) on the x-axis. The results illustrate that the module can operate at idle with additive injection/transfer flows almost double the idle fuel consumption, i. e., a H_2 leak ratio of $r_l \leq 2$. As a consequence, in the performance limit, a ≈3x increased fuel use compared to beginning-of-life may be possible at idle conditions due to transfer leaks. On a per cell basis, the linear fit in Figure 29 indicates that a transfer leak rate of 100 sccm H_2 per cell causes a 19.4 mV drop in the cell voltage at 20A current draw.

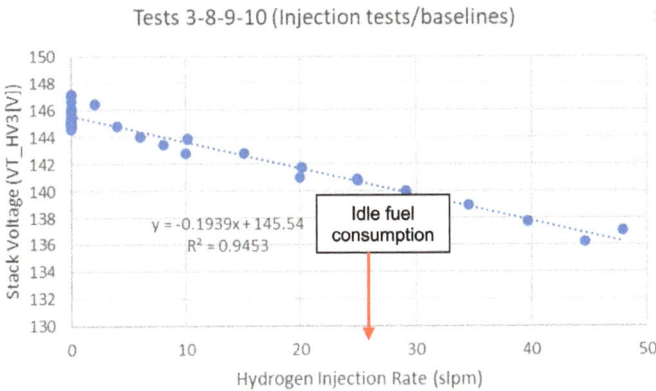

Figure 30: Stack voltage at 20A versus Hydrogen Injection Rate for H_2 injection tests 3, 8, 9 and 10 combined. Idle H_2 consumption: ≈26 slpm.

In reality, in a leaky stack, most cells will not exhibit transfer, therefore the observed linear relationship between stack voltage and hydrogen injection rate represents rather the *maximum voltage drop expected by transfer*. The voltage drop expected for a stack with nonuniform transfer (e. g., at end-of-life) should be less than calculated by this worst-case scenario.

As a final examination on the impact of cathode hydrogen injection on H_2 emissions, Figure 31 presents the measured cathode outlet hydrogen concentration from the four tests in Table 2. The figure shows that the results scatter around the calculated continuous line, which is plotted according to Equation (30) (using Equation (33), (34)), using an oxygen stoich λ_{O_2} = 3.5 and a recombination efficiency η_r = 0.97 leading to a starvation ratio r_{l_0} = 2.58. The high recombination efficiency is expected since the H_2 is injected at the oxygen inlet of the stack, maximizing the exposure to O_2 and catalyst.

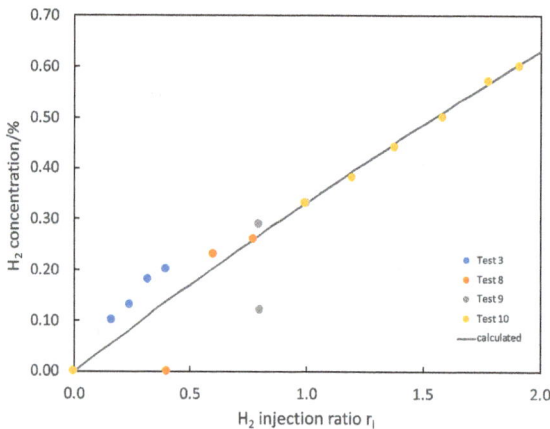

Figure 31: Measured (see Table 2) and calculated cathode-outlet H_2 concentration versus cathode inlet H_2 injection ratio for a 30 kW Ballard fuel cell module at idle conditions (20A), calculated curve corresponds to a recombination efficiency η_r = 0.97.

Figure 26 shows that for leak ratios $r_l < r_{l\_0}$ the H_2 concentration in the cathode outlet depends strongly on the recombination efficiency η_r, ranging from 0-≈10 % at an injection/leak rate of r_l = 1. This means that there exists a region at lower leakage/injection rates $r_l < r_{l\_0}$, where H_2 emissions may not occur. However, H_2 emission concentrations >0 should always be expected in the case of significant nonuniform transfers for a multicell stack.

For $r_l > r_{l\_0}$ on the other hand, there will always be H_2 emissions, which as Figure 26 shows, increase monotonously with decreasing recombination efficiency η_r so that under these conditions, the calculated curves in Figure 26 can be taken as a lower limit for the cathode outlet H_2 concentration.

Figure 31 demonstrates that H_2-emissions can be kept far less than flammable (≈4 % H_2 in air) (even out to life-terminating voltage loss) if MEAs are designed to delay transfer leaks, or to develop transfer leaks with higher recombination efficiency than is seen in today's designs.

10.4 Summary and conclusions

This chapter provides an overview over diagnostics and quantification for one of the leading failure modes for fuel cell stacks: fuel-to-air (transfer) leak formation. Such leaks can lead to H$_2$ emissions in the exhaust gas, posing potential safety risks. While utilization of an H$_2$ sensor in the exhaust is possible, the main challenges of this approach are sensor degradation/lifetime of the sensor and preventing the sensor from being wetted by excessive liquid water in the cathode exhaust. As an alternative, two different approaches of monitoring stack leaks by characterizing operating parameters of the entire stack are presented. Both offer the potential for on-board application (on-board diagnostics): (i) stack (total) voltage measurement during air starvation at a fixed current, while maintaining a fixed anode-cathode gas pressure difference, and (ii) stack AC impedance measurements. The air starvation method can even provide the leak-rate distribution of all the cells, which can act as indicator of overall stack health.

For stack AC impedance measurements, a DC Level-Reducing (DLR) technique is demonstrated to reduce the DC voltage-level of the stack without altering any relevant AC information. This DLR approach permits performing EIS measurements over the entire high-voltage stack of fuel cells (or batteries), enabling safer and more reliable processing of this information using standard low-voltage measurement equipment. While individual-cell voltage profiling can also be used to identify leaky cells, in general this approach is not practical for on-board diagnostics.

Stack AC impedance measurements during H$_2$ injection into the cathode gas stream to simulate H$_2$ transfer leaks showed that such uniform "transfer leaks" are detectable down to a leak rate of ≈33 sccm/cell at idle conditions. On the other hand, for AC measurements on fuel cell stacks with nonuniform transfers, i. e., with individual leaky cells, the sensitivity of transfer detection depends on the number of leaky cells as well as their H$_2$ leak rates. As a rough estimate about 5 % leaky cells in a stack should be detectable. For the nonuniform case, H$_2$ emissions are expected to be larger than for the "uniform" case with H$_2$ injection, as the latter represents the case of optimum recombination.

Bibliography

[1]   T. Abdel-Baset et al., US Drive Fuel Cell Technical Team Roadmap, 2017.
[2]   N. Macauley, M. Watson, M. Lauritzen, S. Knights, G. G. Wang, and E. Kjeang, Empirical membrane lifetime model for heavy duty fuel cell systems. *J. Power Sources*, **336**, 240–250, 2016.
[3]   S. Kreitmeier, P. Lerch, A. Wokaun, and F. N. Büchi, Local Degradation at Membrane Defects in Polymer Electrolyte Fuel Cells. *J. Electrochem. Soc.*, **160**(4), F456–F463, 2013.

[4]   S. Kreitmeier, M. Michiardi, A. Wokaun, and F. N. Büchi, Factors determining the gas crossover through pinholes in polymer electrolyte fuel cell membranes. *Electrochim. Acta*, **80**, 240–247, 2012.

[5]   R. Mukundan et al., Membrane Accelerated Stress Test Development for Polymer Electrolyte Fuel Cell Durability Validated Using Field and Drive Cycle Testing. *J. Electrochem. Soc.*, **165**(6), F3085–F3093, 2018.

[6]   S. Stariha et al., Recent Advances in Catalyst Accelerated Stress Tests for Polymer Electrolyte Membrane Fuel Cells. *J. Electrochem. Soc.*, **165**(7), F492–F501, 2018.

[7]   R. W. J. Bailey, D. P. Wilkinson, and S. A. Campbell, US 6,638,650 Method and apparatus for detecting transfer leaks in fuel cells. US 6,638,650 (2003).

[8]   A. M. Niroumand, O. Pooyanfar, N. Macauley, J. Devaal, and F. Golnaraghi, In-situ diagnostic tools for hydrogen transfer leak characterization in PEM fuel cell stacks part I: R&D applications. *J. Power Sources*, **278**, 652–659, 2015.

[9]   D. P. Wilkinson and S. A. Campbell, US 6,475,651 Method and apparatus for detecting transfer leaks in fuel cells. US 6,475,651 (2002).

[10] S. Knights, D. P. Wilkinson, R. Nebelsiek, and J. Stumper, US 6,492,043 Method and apparatus for detecting a leak within a fuel cell. US 6,492,043 (2002).

[11] G. J. Lamont and D. P. Wilkinson, US 5,763,765 Method and apparatus for detecting and locating perforations in membranes employed in electrochemical cells. US 5,763,765 (1998).

[12] A. Z. Weber, Gas-Crossover and Membrane-Pinhole Effects in Polymer-Electrolyte Fuel Cells. *J. Electrochem. Soc.*, **155**(6), B521, 2008.

[13] C. K. Mittelstaedt and H. Liu, Conductivity, permeability, and ohmic shorting of ionomeric membranes. In: W. Vielstich, H. Yokokawa and H. A. Gasteiger, Eds., *Fuel Cell Handbook*, vol. 5, page 345. John Wiley & Sons, 2009.

[14] J. Zhao, S. Shahgaldi, I. Alaefour, Q. Xu, and X. Li, Gas permeability of catalyzed electrodes in polymer electrolyte membrane fuel cells. *Appl. Energy*, **209**, 203–210, 2018.

[15] H. Xu. *Experimental Measurement of Mass Transport Parameters of Gas Diffusion Layer and Catalyst Layer in PEM Fuel Cell*. University of Alberta, 2019.

[16] K. J. Vetter, *Elektrochemische Kinetik*. Springer, Berlin, 1961.

[17] H. A. Gasteiger, W. Gu, R. Makharia, M. F. Mathias, and B. Sompalli, BOL MEA performance – Efficiency loss contributions. In: H. A. Gasteiger, W. Vielstich and A. Lamm, Eds., *Fuel Cell Handbook*, vol. 3, page 593. John Wiley&Sons, 2009.

[18] J. Stumper, R. Rahmani, and F. Fuss, Open circuit voltage profiling as diagnostic tool during stack lifetime testing. *J. Power Sources*, **195**(15), 4928–4934, 2010.

[19] G. Mousa, F. Golnaraghi, J. DeVaal, and A. Young, Detecting proton exchange membrane fuel cell hydrogen leak using electrochemical impedance spectroscopy method. *J. Power Sources*, **246**, 110–116, 2014.

[20] G. Mousa, J. Devaal, and F. Golnaraghi, Diagnosis of hydrogen crossover and emission in proton exchange membrane fuel cells. *Int. J. Hydrog. Energy*, **39**(35), 20116–20126, 2014.

[21] T. E. Springer, T. A. Zawodzinski, M. S. Wilson, and S. Gottesfeld, Characterization of Polymer Electrolyte Fuel Cells Using AC Impedance Spectroscopy. *J. Electrochem. Soc.*, **143**(2), 587–599, 1996.

[22] A. M. Niroumand, H. Homayouni, J. DeVaal, F. Golnaraghi, and E. Kjeang, In-situ diagnostic tools for hydrogen transfer leak characterization in PEM fuel cell stacks part II: Operational applications. *J. Power Sources*, **322**, 147–154, 2016.

[23] H. G. Mousa, J. W. DeVaal, and F. Golnaraghi. US 10581099 Use of neural network and EIS signal analysis to quantify H_2 crossover in operating PEMFC. US 10,581,099 (2020).

[24] H. Homayouni, J. Devaal, F. Golnaraghi, and J. Wang, Voltage reduction technique for use with electrochemical impedance spectroscopy in high-voltage fuel cell and battery systems. *IEEE Trans. Transp. Electrif.*, **4**(2), 418–431, 2018.

[25] EU Regulation No 100 rev 2 BEV requirements. United Nations Commission for Europe, p. L 45/21, 2009.

[26] S. Franco, *Design with operational amplifiers and analog integrated circuits*. McGraw-Hill series in electrical and computer engineering, 2015.

[27] A. M. Niroumand and M. Saif, Two-phase flow measurement system for polymer electrolyte fuel cells. *J. Power Sources*, **195**(10), 3250–3255, 2010.

[28] M. Inaba, T. Kinumoto, M. Kiriake, R. Umebayashi, A. Tasaka, and Z. Ogumi, Gas crossover and membrane degradation in polymer electrolyte fuel cells. *Electrochim. Acta*, **51**(26), 5746–5753, 2006.

[29] J. Larminie and A. Dicks, *Fuel Cell Systems Explained*. John Wiley & Sons, 2000.

Part III: **Catalysis/CL components**

Lei Zhang, Xulei Sui, Junjie Li, Kieran Doyle-Davis, and Xueliang Sun

11 Electrocatalysts: from nanostructures to single atoms

11.1 Introduction

As one of the high efficiency energy-conversion devices, polymer electrolyte membrane fuel cells (PEMFCs) have attracted more and more attention. During the operation of PEMFCs, the rate determining reaction is the slow oxygen reduction reaction (ORR) at the cathodic side. Two different pathways have been explored in different types of catalysts for the ORR process: two-electron pathway ($O_2 + 2H^+ + 2e^- \rightarrow H_2O_2; E = 0.70$ V) and four-electron pathway ($O_2 + 4H^+ + 4e^- \rightarrow H_2O; E = 1.23$ V). The final product is significantly determined by the selectivity of catalysts and the coordination environment. For instance, the ORR on Pt nanoparticles usually undergo the four-electron pathway [1]. The single Pt atoms on different substrates might induce a two-electron pathway [2] or a four-electron pathway according to their coordinated atoms.

The commercialization of PEMFCs is limited by the cost of various components of a fuel cell. The catalysts used is the highest contributor to the overall cost of a fuel cell. In the PEMFCs, Pt has been commonly used as the active electrocatalyst. However, commercial Pt catalysts in the PEMFCs exhibited low durability during the operation under severe conditions. In addition, the low abundance, limited supplies and ever-increasing price of Pt-based catalysts have inspired researchers to further improve its performance [3]. Improvement of the atom utilization efficiency, exposed surface structure and the composition are three strategies to improve the mass activity of Pt catalysts [4].

In addition to Pt-based noble metal catalysts, extensive efforts have been devoted to developing alternative nonnoble metal electrocatalysts for ORRs. In recent years, metal and nitrogen co-doped carbon (M–N–C) catalysts nanomaterials have been widely investigated as alternative electrocatalysts for ORRs. Recent research work revealed that Metal-Organic Framework (MOF)-derived nanomaterials exhibit excellent catalytic activity and durability during electrocatalysis for ORR. For example, Chen et al. prepared a highly reactive and stable isolated single-atom Fe/N-doped porous carbon (ISA Fe/CN) catalyst with Fe loadings up to 2.16 wt.% and reported that the catalyst showed excellent ORR performances with a half-wave potential ($E_{1/2}$) of 0.900 V that outperformed

Lei Zhang, College of Chemistry and Environmental Engineering, Shenzhen University, Shenzhen, Guangdong 518060, P.R. China; and Department of Mechanical and Materials Engineering, The University of Western Ontario, London, ON N6A 5B9, Canada
Xulei Sui, Junjie Li, Kieran Doyle-Davis, Xueliang Sun, Department of Mechanical and Materials Engineering, The University of Western Ontario, London, ON N6A 5B9, Canada, e-mail: xsun@eng.uwo.ca

https://doi.org/10.1515/9783110622720-011

commercial Pt/C [5]. Therefore, carbon supported nonnoble metal catalysts are another promising catalyst applied in PEMFCs in the future.

In this chapter, we will describe recent progress on the development of Pt-based catalysts and nonprecious metal catalysts. First, we will introduce effective routes for improving the activity of Pt catalysts, such as reducing the size, surface control and alloy formation. In addition, Pt single atoms will also be introduced as a pathway to high performance with low mass loading. Then we will systematically summarize nonprecious metal catalysts according to synthesis methods, single atom catalysts, dimer catalysts and catalyst durability. The authors hope that this summary will provide a useful guide for the design of PEMFC catalysts in the future.

11.2 Pt-based catalysts

11.2.1 Pt catalysts with tuned sizes

The specific activity and mass activity are the two most important performance characteristics for the evaluation of Pt catalysts. Whereas the specific activity is normalized by the electrochemical active surface area (ECSA), which is highly determined by the surface structure and the composition of the catalyst, the mass activity is greatly dependent on the atom utilization efficiency (AUE). Reducing the particle size of Pt catalysts is one of the most direct ways to improve the AUE. Consequently, studies on the relationship between the Pt catalyst particle size and catalytic performance have been widely reported.

For example, Shao et al. synthesized the Pt catalysts with various particle sizes by replacing the Cu under-potential-deposition (UPD) layer with Pt atoms [6]. The Pt catalysts with the size of 1.3, 1.84, 2.46 and 4.65 nm can be controlled as shown in Figure 1. The truncated octahedral particles were simulated by atomic models, and the detailed atom number in each particle was calculated. During the electrochemical testing, the ECSA increases gradually with the decrease in particle size due to the increase of AUE. The catalysts larger than 2.2 nm exhibited the best specific activity, which was much better than that of 1.3 nm catalysts. The {111} active sites caused the observed increase in specific activity. For the mass activity, the 2.2 nm Pt catalysts exhibited the highest mass activity, which is different from the trend for the specific activity. This result indicated that the best ORR mass activity results from the combination of a high ratio of {111} surface sites and a high AUE. Arenz et al. studied the ORR activities of four high surface area (HSA) carbon supported catalysts with different average sizes [7]. The result indicated that the AUE performs a significant role for improving the mass activity of the catalysts. Shao-Horn et al. examined the ORR activity of carbon supported Pt catalysts with particle sizes ranging from 1.6 nm to 4.7 nm in $HClO_4$ and H_2SO_4 [8]. The Pt/C catalyst containing 46 % Pt nanoparticles was treated in Ar at 900 °C for 1 minute or 2 hours to generate particles with larger diameters. In the ORR test, the Pt/C catalysts exhibited

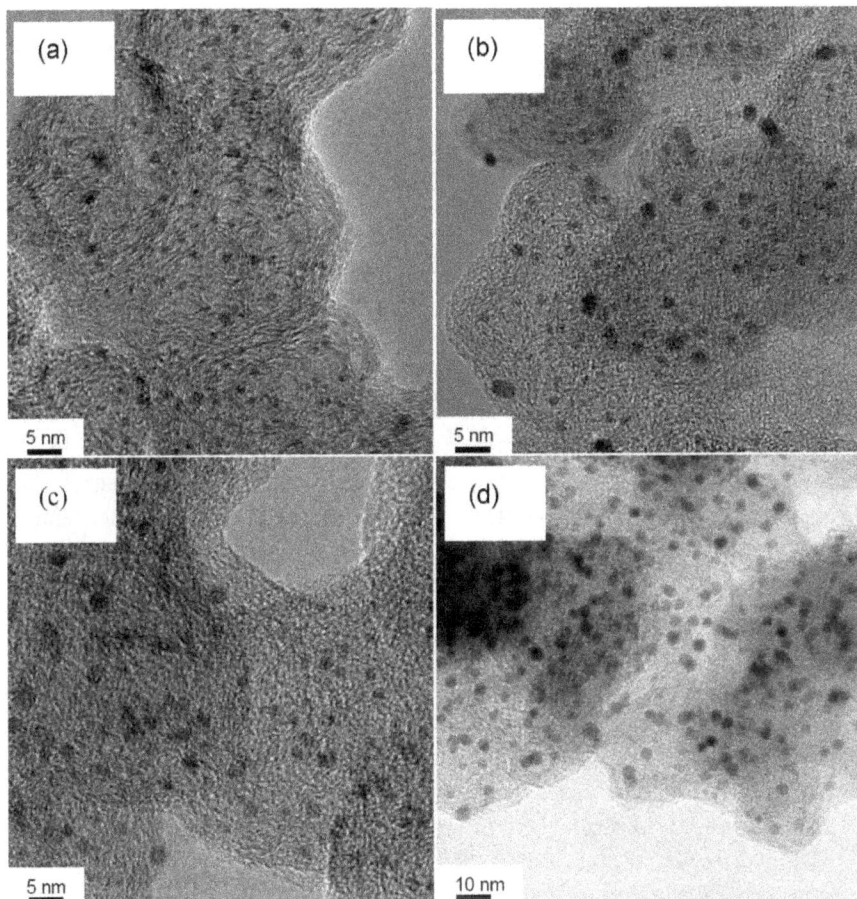

Figure 1: TEM images of Pt particles supported on carbon black with the size of 1.3, 1.84, 2.46 and 4.65 nm, respectively. Reprinted with permission [6], Copyright 2011 American Chemical Society.

a five-fold reduction in the specific ORR activity in H_2SO_4 relative to $HClO_4$, due to the strong adsorption of sulfate. In addition, the specific activities of the catalysts were similar at 0.9 V RHE (reversible hydrogen electrode). The authors mentioned that the specific ORR activity on majority terrace sites is much greater than those of minority undercoordinated sites, explaining the size-independent activity. They also found that the mass activity increased from 0.07 to 0.3 A mg_{Pt}^{-1}, with the size decreasing from 4.7 nm to 1.7 nm due to the greatly increased AUE.

11.2.2 Pt-based alloys

The formation of multicompositional Pt-based catalysts can effectively improve the specific activity of Pt due to synergy effects between Pt and other metals (Ni, Co, Fe, etc.). The introduction of Ni was reported to significantly enhance the activity of Pt. Stamenkovic et al. demonstrated that extended single crystal surfaces of Pt3Ni(111) exhibit an enhanced ORR activity, which is 10-fold higher than Pt(111) and 90-fold higher than the current state-of-the-art Pt/C catalysts [9]. The formation of the PtNi alloy can weaken the OH adsorption due to the decrease of the d-band center on the Pt skin formed by surface segregation. Fang and coworkers synthesized nanometer-sized single crystal Pt_3Ni particles by a wet-chemical approach and prepared monodisperse Pt_3Ni nanooctahedra and nanocubes terminated with {111} and {100} facets (Figure 2) [10]. They found that in the ORR the mass activity at 0.9 V of the Pt_3Ni nanooctahedra is ~2.8 times of that of Pt_3Ni nanocubes and ~3.6 times of that of Pt nanocubes. Xia et al. reported the synthesis of uniform 9 nm Pt-Ni octahedra with the use of oleylamine and oleic acid as surfactants and $W(CO)_6$ as a source of CO that can promote the formation of {111} facets in the presence of Ni [11]. The coverage of surfactant on the surface of resultant Pt-Ni octahedra was significantly reduced while the octahedral shape was still attained by using benzyl ether as a solvent. The obtained PtNi catalysts were treated with acetic acid to further remove the surfactants. The ORR activity of the PtNi catalysts was significantly improved with a specific activity 51-fold higher than that of the state-of-the-art Pt/C catalyst for the ORR at 0.93 V, together with a record high mass activity of 3.3 A mg_{Pt}^{-1} at 0.9 V. To improve the stability and activity of PtNi octahedra, Huang and coworkers tried to introduce transition metals on the surface. $Mo-Pt_3Ni/C$ showed the best ORR performance, with a specific activity of 10.3 mA/cm^2 and mass activity of 6.98 A/mg$_{Pt}$, resulting in 81- and 73-fold enhancements, respectively, compared to a commercial Pt/C catalyst (0.127 mA/cm^2 and 0.096 A/mg$_{Pt}$) [12]. Theoretical calculations suggest that Mo prefers subsurface positions near the particle edges in vacuum and surface vertex/edge sites in oxidizing conditions, where it enhances both the performance and the stability of the Pt_3Ni catalyst.

Besides PtNi alloy catalysts, Pt-Co systems have received considerable attention because of their relatively high ORR activity and stability in acidic environments [13]. For example, Abruña et al. synthesized a new class of Pt-Co nanocatalysts composed of ordered Pt_3Co intermetallic cores with a 2–3 atomic-layer-thick platinum shell by pretreatment of the Pt_3Co/C at different temperatures under a flowing H_2/N_2 mixed gas atmosphere [14]. These nanocatalysts exhibited over 200 % increase in mass activity and over 300 % increase in specific activity when compared with the disordered Pt_3Co alloy nanoparticles as well as Pt/C. In another case, Huang et al. reported the synthesis of a class of hierarchical Pt-Co NWs enclosed with high-density high-index facts through a robust large-scalable wet-chemical approach as electrocatalysts for fuel cell reactions [15]. These hierarchical Pt-Co NWs with highly uneven surfaces are tailored to have an ordered intermetallic structure, high-index facets and a Pt-rich surface. In electrochemical tests, they showed an ORR performance with specific and mass activities of 7.12 mA cm^{-2} and 3.71 A mg$_{Pt}^{-1}$, respectively, which were 39.6 and 33.7 times higher than those of a state-of-the-art Pt/C catalyst.

In addition to the typical PtNi and PtCo bimetallic catalysts, several other Pt-M alloy catalysts are developed, including Pt-Pd, Pt-Fe, Pt-Cu, Pt-Pb, etc. For example, Strasser

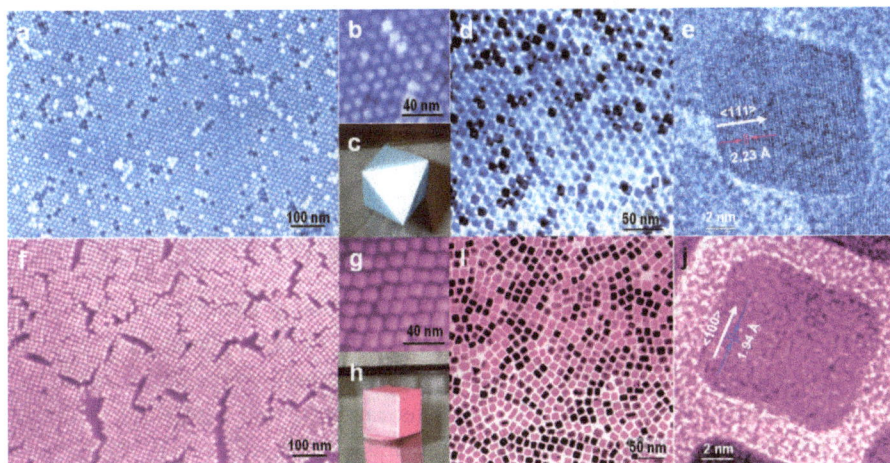

Figure 2: The morphology of Pt$_3$Ni nanooctahedra and nanocubes. (a–e) SEM and TEM images for Pt3Ni nanooctahedra. (f–j) SEM and TEM images for Pt3Ni nanocubes. Reprinted with permission [10], Copyright 2010 American Chemical Society.

et al. investigated the ORR activity for carbon supported PtCu$_3$, PtCu and Pt$_3$Cu alloy nanoparticles in different electrolytes and pH values [16]. The PtCu$_3$, PtCu and Pt$_3$Cu alloy nanoparticle electrocatalysts were prepared using a liquid precursor impregnation-freeze-drying-annealing method. They found that the voltage cycling in 0.1 M HClO$_4$ immediately causes the dissolution of Cu and results in Pt-enriched particle surface. The PtCu$_3$ catalyst exhibited the best mass activity compared to PtCu and Pt$_3$Cu, which increases by 3–4 times compared to the benchmark pure Pt. Bao and Zeng prepared Pd–Pt hollow frame structures composed of double-shell cubes linked by body diagonals [17]. The formation of Pd–Pt tesseracts via etching is highly dependent on the Pd/Pt ratio in the initial Pd–Pt nanocubes. The oxidative etching of Pd–Pt nanocubes with the Pd/Pt ratio of 6.8, 3.5 and 2.0 led to the formation of octapods, tesseracts and cubic nanoframes, respectively. During ORR, Pd–Pt tesseracts exhibited the highest mass activity of 1.86 A mg^{-1} and specific activity of 2.09 mA cm^{-2} among Pd–Pt nanocrystals, which were 11.6 times and 8.4 times greater than those of Pt/C. Cho and Lee synthesized Pt-Fe alloy nanoparticles and employed carbon coating to protect the nanoparticles from agglomeration during heat treatment [18]. The Pt$_3$Fe/C exhibited a mass activity (0.454 A mg$_{Pt}^{-1}$) and good stability with superior resistances to nanoparticle agglomeration and iron leaching. It showed 11 times higher mass activity (0.343 A mg$_{Pt}^{-1}$) than Pt/C (0.030 A mg$_{Pt}^{-1}$) after ADT testing of 8,000 cycles. Recently, several studies found compressive surface strains can boost ORR activity for alloys. Huang et al. reported a class of platinum-lead/platinum (PtPb/Pt) core/shell nanoplate catalysts with large biaxial strains [19]. This Pt-Pb alloy nanoplate catalyst exhibited high ORR specific and mass activities that reach 7.8 mA cm^{-2} and 4.3 A mg^{-1} at 0.9 V. Density functional theory calculations reveal that the edge-Pt and top (bottom)–Pt (110) facets undergo large tensile strains that help optimize the Pt-O bond strength. The intermetallic core and uniform four layers of Pt shell of the PtPb/Pt nanoplates appear to cause the high durability of these catalysts, which can undergo 50,000 voltage cycles with negligible activity decay and no apparent structure and composition changes.

11.2.3 Highly branched Pt nanostructures

The surface energy of Pt will increase with the reduction of particle size, which will cause aggregation of Pt particles. To avoid aggregation, Pt-based porous structures are fabricated. Thanks to the channels in the porous structure, the interior atoms of the porous structures can pontificate in the reactions. Yamauchi et al. reported the Pluronic F127 block copolymer assisted synthesis of dendritic Pt nanoparticles [20]. The Pluronic chains can form cavities and induce the formation of Pt porous structure. The surface area of the dendritic Pt nanoparticles is 56 m^2 g^{-1}, which is much higher than that of Pt black. In addition, Lee and coworkers prepared porous Pt dendrites by simply tuning the reaction temperature. The activities of dendritic Pt/C showed much better mass activity and specific activity compared to a commercial Pt/C catalyst [21].

In addition to pure Pt metals, Pd–Pt bimetallic porous structures were fabricated to obtain further enhanced performance. Yang and coworkers prepared the Pt-on-Pd bimetallic hetero-nanostructures using a sequential synthetic method (Figure 3) [22]. Due to the very small lattice mismatch of 0.77 % existed between Pt and Pd, Pt nanoparticles can grow along the (111) crystal planes on Pd supports without formation of grain boundaries or defects. The as-prepared Pt on Pd dendrites exhibited improved stability, which could be due to a favorable interface between Pt and Pd supports, as well as the larger than usual overall particle size of Pt-on-Pd nanostructure. Pd–Pt dendrites can also be prepared through a seed-mediated growth method [23]. During the growth step of Pt, Pt atoms can be attached onto the Pd surfaces due to the multiple nucleation sites. In addition, a highly ordered orientation of Pd and Pt atoms was observed and the whole particle showed a single crystalline structure. Thanks to the high surface area and the exposed {100}, {111}, {110} and some high-index facets, the mass activity of the porous structure is 2.5 times that of commercial Pt/C.

11.2.4 Rational designed surface-controlled structures

11.2.4.1 Core-shell structures

To obtain a high AUE, core-shell structures have been developed due to their bimetallic composition and unique structure. Although many studies have been successful in preparing X-Pt (X = Cu, Au, Pd) core-shell structures, most of the as-prepared Pt shells were too thick (>5 nm). It is of great importance to reduce the thickness of shell to less

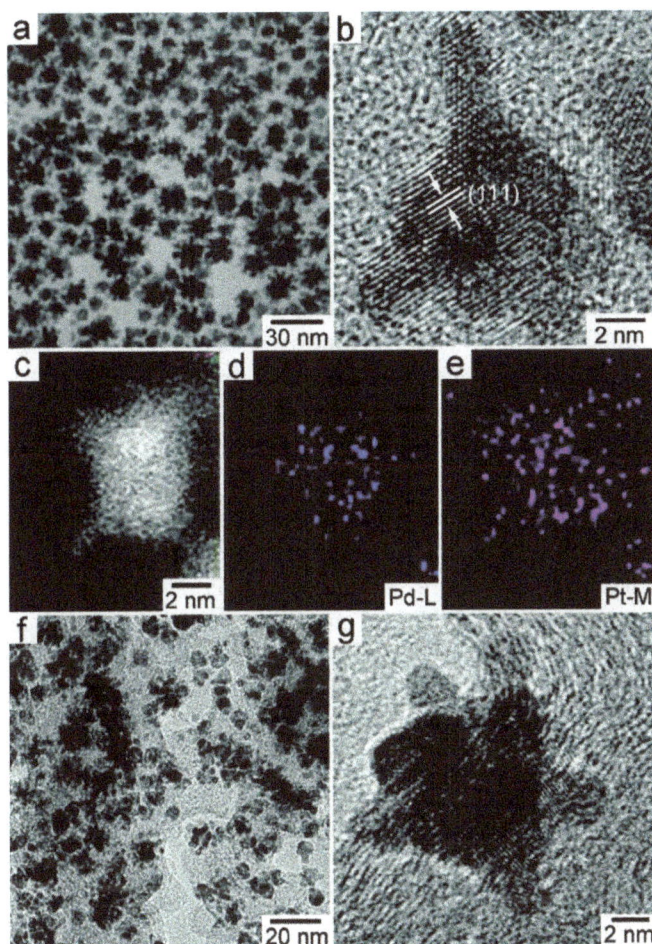

Figure 3: The morphology of Pt-on-Pd bimetallic nanodendrites. (a) TEM, (b) HR-TEM, (c) HAADF-STEM images and (d, e) elemental maps for Pd and Pt metals of Pt-on-Pd bimetallic nanoparticles; (f) TEM and (g) HR-TEM images of carbon-supported Pt-on-Pd bimetallic catalysts after the thermal treatments. Reprinted with permission [22], Copyright 2009 American Chemical Society.

than 4 atomic layers. To achieve such an ultrathin Pt shell, Adzic and coworkers first obtained a Cu monolayer on a Pd substrate by an underpotentially deposited (UPD) method. Then the Cu monolayer was replaced by Pt atoms through a galvanic replacement reaction. Interestingly, a smooth monolayer of Pt was formed on Pd due to the slow deposition rate of Pt and the surface diffusion of Pt adatoms [24]. When the core was changed from Pd to Pd_9Au_1, the stability of Pt monolayer catalysts was significantly enhanced [25]. In addition, with the introduction of Ni and Ru into the core can further improve the activity of the Pt monolayer core-shell catalysts obtained by a chemical reduction method [26].

Besides, a pulse electro-deposition method was developed for the deposition of ultrathin Pt layers on TiNiN substrates without galvanic replacement reaction [27]. The vacuum deposition method can also fabricate a Pt thin film (1.5 nm) on a TaB_2 (0001) single-crystal substrate at room temperature [28]. Another widely employed method for the preparation of uniform ultrathin Pt shells is the wet-chemical reduction method. The conformal deposition of ultrathin shells of Pt on Pd nanocrystals can be achieved by slowly injecting Pt precursors at a high temperature [29]. The thickness of the Pt shell can be precisely controlled from 1 to 8 layers by adjusting the amount of Pt precursor. As shown in Figure 4, the Pt atoms are evenly deposited on Pd nanocubes, maintaining the initial core surface structure. Although the particle size of Pd@Pt core-shell structures is large, they exhibited higher mass activity compared with commercial Pt/C. The growth method can be extended from cubic Pd to octahedral Pd by using a weak reducing agent [30]. In addition, the mass activity of Pd@Pt core-shell structures can be further increased by using Pd seeds with twin defects, because the Pt overlayers are forced to evolve into a compressed, corrugated structure due to the lateral confinement imposed by the boundaries of twin defects [31]. Simulation results showed that the binding energy of OH was weakened by the compressive strain. When Pt was deposited onto Pd decahedra, the as-prepared particles exhibited a concave structure due to the strong adsorption of Pt adatoms on edge/ridge sites [32]. Yang et al. investigated the deposition of Pt shell onto Au decahedral cores with a starlike geometry [33]. Similar to the Pd core, the as-prepared catalysts exhibited enhanced ORR performance than Pt/C, due to the strong electron interaction arising from a lattice-matched interface.

11.2.4.2 Nanocage structures

Generally, Pt-based nanocage structures can be fabricated by galvanic replacement reaction with Ag nanocubes as the seeds. The sizes and shapes of the Pt nanocages can be controlled by the morphologies and sizes of Ag particles [34]. In addition to Ag, Pd nanoparticles are another ideal substrate for the preparation of Pt-based nanocages [35]. To obtain the Pt nanocages with subnanometer-thick walls, Zhang et al. selectively etched Pd cores of Pd@Pt core-shell nanostructures with $FeCl_3$ and HCl as the etchants [36]. The thickness of the nanocages can be controlled to 4 atomic layers. DFT results indicated that some Pd atoms have already be incorporated into the Pt shell with the formation of Pd@Pt structure, which provided a channel to dissolve the Pd atoms. Thanks to the formation of nanocages with subnanometer-thick walls, the octahedral nanocages with {111} facets exhibited a mass activity of 0.75 A mg_{Pt}^{-1} at 0.9 V. Furthermore, the activity of Pt octahedral nanocages reduced by 36 % after 10,000 cycles, still showing 3.4-fold enhancement relative to the pristine Pt/C. To further increase the activity of the nanocage structure, the Pt-based icosahedral nanocages were prepared using the Pd@Pt icosahedra as seeds (Figure 5) [37, 38]. During the selective etching of Pd cores, the Pt atoms on the shell reconstruct and the lateral confinement imposed by the twin boundaries

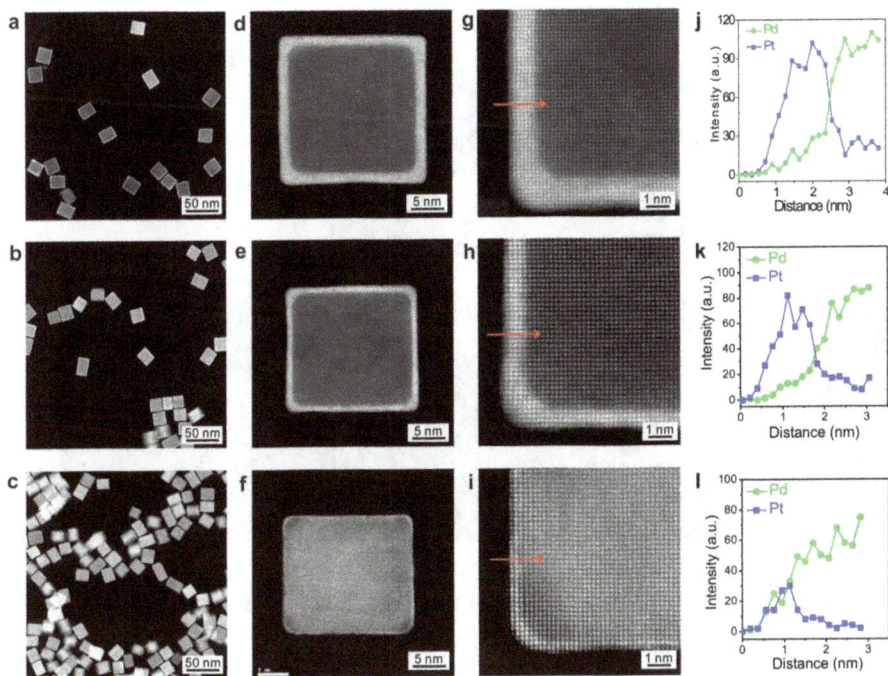

Figure 4: The precisely controlled Pt atomic layers on Pd nanocubes. (a–c) Low-magnification HAADF-STEM images Pd@Pt with (a) 6 Pt layers, (b) 4 Pt layers and (c) 1 Pt layers; (d–f) HAADF-STEM images of individual Pd@Pt nanocubes with (a) 6 Pt layers, (b) 4 Pt layers and (c) 1 Pt layers; (g–l) atomic-resolution HAADF-STEM images and EDX line scan analyses taken from the Pd@Pt nanocubes with (a) 6 Pt layers, (b) 4 Pt layers and (c) 1 Pt layers. Reprinted with permission [29], Copyright 2014 American Chemical Society.

no longer exists in the nanocage structure (Figure 5e). Thanks to the synergistic effect of {111} facets, twin defects and ultrathin walls, the icosahedral nanocages exhibited a mass activity of $1.28 \, A \, mg^{-1}$.

11.2.4.3 Nanoframe structures

Contrasting the fabrication of ultrathin nanocages, most of the Pt-based nanoframes were obtained through a one-pot synthesis. However, the edges of nanoframes were thick, which are not good for the improvement of mass activity [39]. Recently, Pt-Ni nanoframes with ultrathin frames were prepared from the $PtNi_3$ rhombic dodecahedron [40, 41]. The edge thickness of the Pt skin was only 2 nm and the whole nanoframe retained the high crystallinity. The Pt_3Ni nanoframes exhibited an extremely high mass activity and good durability. Li et al. synthesized Pt-Ni nanoframes by transforming truncated octahedral Pt-Ni alloy particles through a priority-related chemical etching process. They found that the etching mechanism followed the Kirkendall effect stemming from distinct diffusion coefficients of Ni and Pt [42].

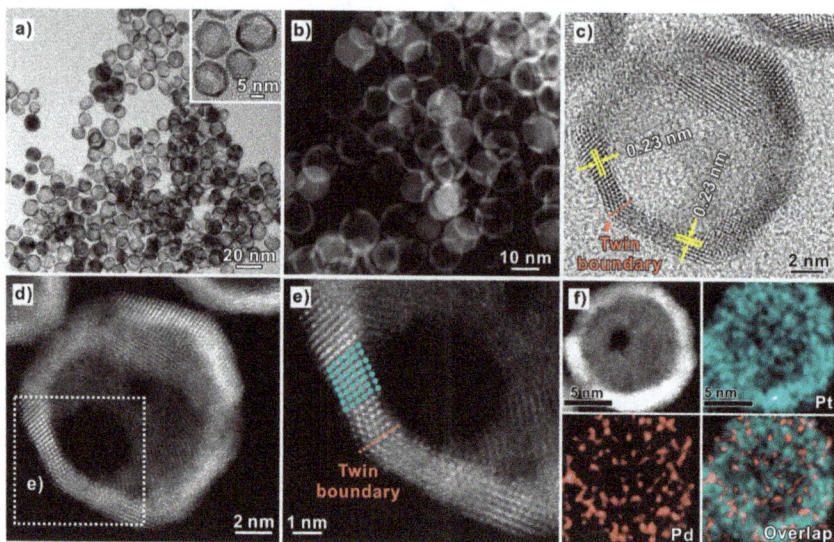

Figure 5: (a) TEM and (b) low-magnification HAADF-STEM images of the Pt icosahedral nanocages. (c) Bright-field and (d, e) atomic-resolution HAADF-STEM images taken from a single nanocrystal along a 2-fold symmetry axis. (f) HAADF-STEM image of an icosahedral nanocage and the corresponding EDX mapping of Pd and Pt. Reprinted with permission [37], Copyright 2016 American Chemical Society.

11.2.5 Pt single atom catalysts

To further improve the utilization efficiency of Pt and reduce the cost of catalysts, Liu and coworkers fabricated Pt single atom electrocatalysts on N-doped carbon black (Pt_1-N/BP) as shown in Figure 6 [43]. The loading amount of Pt is 0.4 wt.% according to the inductively coupled plasma mass spectrometry (ICP-MS) test. Aberration-corrected scanning transmission electron microscopy (STEM) reveals that only individual Pt atoms are present in Pt_1–N/BP (Figure 6a). When this catalyst is applied in an acidic single cell, it exhibited high performance, with the power density up to 680 mW cm^{-2} at 80 °C with an ultralow platinum loading of 0.09 mg$_{Pt}$ cm^{-2}. Furthermore, after a 200 h durability test, the current of the fuel cell still remains 74 % of the fresh at 80 °C, and 90 % remains at 70 °C, indicating good stability of Pt_1-N/BP as cathode in acidic fuel cells. Theoretical calculations reveal that the main effective sites on the Pt single-atom electrocatalysts are single-pyridinic-nitrogen-atom-anchored Pt centers, which are tolerant to carbon monoxide/methanol, but highly active for the oxygen reduction reaction. Cho et al. developed a hydrothermal ethanol reduction method and the loading amount of Pt single atoms on N/BP can reach about 5 wt.% [44]. By further heat-treating Pt_1/NBP, the coordination environment of isolated Pt atoms was reconstructed to produce the uniquely nitrogen-anchored Pt single atoms with 47.8 % retention of the isolated Pt atoms and well-dispersed Pt nanoparticles on the carbon support (Pt_1@Pt/NBP). The maximum power density of the acidic H_2/O_2 fuel cell with the Pt_1@Pt/NBP cathodes is

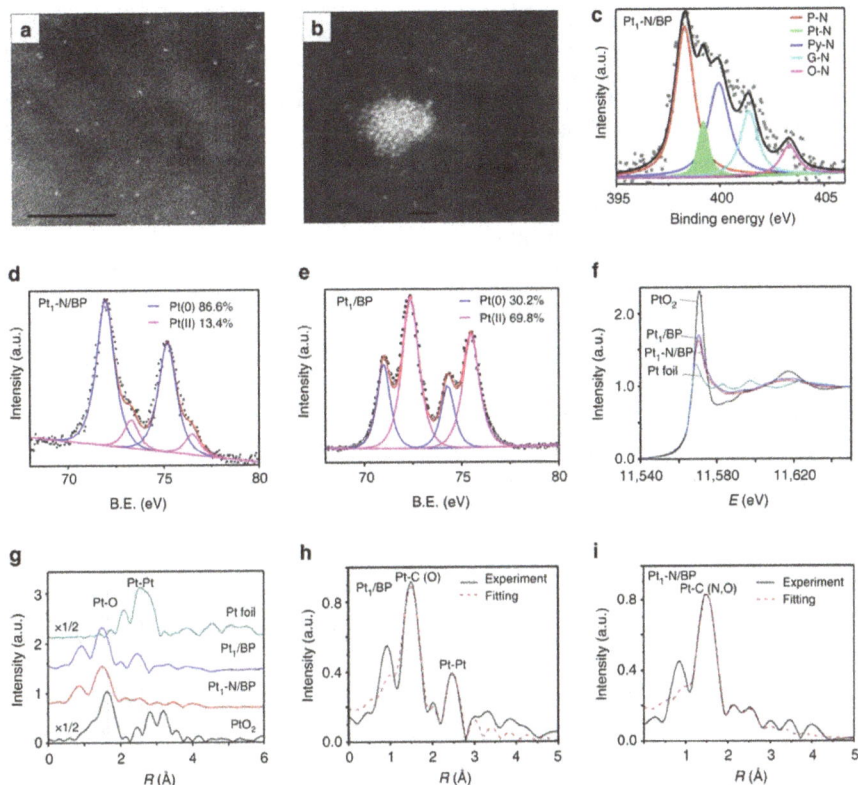

Figure 6: HAADF-STEM images of (a) Pt1-N/BP and (b) Pt1/BP (scale bar, 5 nm (a) and 1 nm (b)). (c) XPS spectra for N 1s in Pt_1-N/BP. (d, e) XPS spectra for Pt 4f in (d) Pt_1-N/BP and (e) Pt1/BP. (f–i) XAS analysis for Pt_1-N/BP and Pt_1/BP. Reprinted with permission [43], Copyright 2017 Nature Publishing Group.

$0.844\,W\,cm^{-2}$. Liu et al. synthesized carbon black with defects as the support for Pt single atoms deposition by treating hydrothermally in sealed ethanol solution containing a certain amount of H_2O_2. The loading amount of Pt in the optimal catalysts for ORR is 1.1 wt.%. In an acidic single-cell test, the maximum power density of the Pt1.1/BP defect-based PEMFC reaches $520\,mW\,cm^{-2}$ at 80 °C, corresponding to the highest Pt utilization $(0.09\,g_{Pt}\,kW^{-1})$ [45].

In addition to carbon black support, the Pt single atoms can also be deposited onto metal-organic frameworks. Song et al. reported the atomic layer deposition (ALD) strategy for synthesis of Pt SACs on the metal–organic framework (MOF)-derived N-doped carbon (NC) [46]. The Pt catalysts can be precisely controlled from single atoms to subclusters and NPs by adjusting the ALD exposure time of Pt precursor. The EXAFS spectra results indicate that the coordination numbers of Pt-Pt in the Pt single atoms and Pt subclusters are 0.1 and 0.8, respectively, indicating the formation of Pt isolated atoms and subclusters instead of Pt crystals. The mass activity of $1.17\,A\,mg_{Pt}^{-1}$ is achieved for Pt SAs-ZIF, which is 6.5 times greater than that of $0.18\,A\,mg_{Pt}^{-1}$ for Pt NPs-ZIF catalysts.

In addition, the average electron transfer number of Pt SAs-ZIF is 3.79, approaching to a four-electron oxygen reduction pathway for the ORR. The H_2-O_2 polarization curves before and after the 5000 triangle potential cycles for the single-cell MEA composed with Pt SAs-ZIF cathode shows a 24 mV voltage loss at 0.8 mA cm^{-2}, which suggests the excellent stability of Pt SAs-ZIF under practical fuel cell operation. Density functional theory calculations confirmed that Pt SAs prefer to be anchored by the pyridinic N sites from ZIF-NC support. With coadsorption of both hydroxyl and oxygen, the unoccupied orbitals of Pt SAs could be modulated with enhanced activity in the ORR. The ORR on Pt single-atom catalyst can occur in multichannel instead of single pathway as the Pt nanoparticle, which greatly decreases the free energy change for the rate-determining step and enhances the activity of Pt SAs for ORR.

In recent years, nonprecious metal catalysts (NPMCs) Fe–N–C catalysts have emerged as promising alternatives to Pt as the oxygen reduction reaction (ORR) catalysts for proton exchange-membrane fuel cells. Zeng and coworkers deposited Pt single atoms on the Fe centers of Fe–N–C, creating a new active moiety of Pt_1-O_2-Fe_1-N_4. A loading of 2.1 wt.% Pt single-atoms can be achieved on Fe–N–C [47]. The modulated Fe–N_4 moiety cannot only increase the PEMFC power density, but also display considerably improved ORR stability in acidic media, attributed to a possible protective effect of the grafted Pt_1-O_2-cap on the slightly reduced Fe^{3+} atoms. The linear sweep voltammetry curve in ORR shows an onset potential of 0.93 V and a half-wave potential ($E_{1/2}$) of 0.80 V for Pt_1@Fe–N–C, which is much better than Fe–N–C and Pt single atoms. In addition, the $E_{1/2}$ of Pt_1@Fe–N–C decayed only 12 mV after 10000 cycles, much less than the 36 mV degradation of Fe–N–C, suggesting that grafting Pt_1-O_2-Fe_1-N_4 moiety is an effective strategy to improve the durability of Fe–N–C electrocatalysts. In a single cell test, the power density curve of Pt_1@Fe–N–C shows a peak power density of 0.86 W cm^{-2} at 0.49 V and a high-power density of 0.79 W cm^{-2} at 0.6 V. With the formation of Pt single atoms on Fe–N–C, the Pt_1-O_2-cap could prevent or disturb the Fe center from catalyzing the Fenton's reaction, resulting in the good stability.

11.3 Nonprecious metal catalysts

11.3.1 Brief introduction of nonprecious metal catalysts

To make PEMFCs more competitive with other batteries, PEMFCs need to meet an ultimate cost target of US$30 per kW according to the requirements of the U. S. Department of Energy. The key point is to significantly reduce the usage of Pt-based catalysts but not to drastically reduce its performance including activity, durability and power density. In recent years, nonprecious metal catalysts for ORR have attracted more and more researchers' attention due to their cost advantages [48, 49]. The related research has made great progress, and a series of nonprecious metal catalysts sprung up like mushrooms. The ORR catalytic performance has also been greatly improved [50, 51].

However, most of the research results are limited to alkaline electrolyte systems, and ORR activity in acidic media is still challenging. As we all know, the currently practical PEMFC is used in acidic systems, so the development of nonprecious metal catalysts suitable for acidic systems has great practical significance [52]. So far, the most studied ORR nonprecious metal catalysts in acidic systems can be roughly divided into three categories: (1) nonprecious metal oxides, nitrides, oxynitrides and carbonitrides; (2) nonprecious metal chalcogenides; (3) nitrogen-coordinated $3d$ transition metal (M: Fe, Co, Mn…) macrocyclic compounds, polymers and their pyrolysis derivatives (metal and nitrogen co-doped carbon: M–N–C nonprecious metal catalysts (NPMCs)) [53–56]. The first type of catalyst has high acid-corrosion resistance and good stability, but its ORR activity is too poor to practical application [57]. Although the second type of catalyst has better activity than the first type, its activity is still far from meeting the practical application [58, 59]. For the third type of catalyst, nitrogen-coordinated transition metal macrocyclic compounds or polymers have limited activity and stability and are difficult to be applied in practice, but their derivatives (M–N–C) obtained by heat treatment have been verified by the early work of Yeager and coworkers to display much better ORR activity [53, 54]. Subsequently, various pyrolysis precursors are developed to replace the expensive macrocyclic compounds, and remarkable progress in exploring M–N–C catalysts has been made through the unremitting efforts of scientists in recent years. Therefore, M–N–C catalysts have been identified as the most promising alternative to Pt-based catalysts for ORR [60].

Generally, the atomically dispersed $3d$ transition metals stabilized via bonding with nitrogen atoms in M–N–C catalysts are considered to be responsible for the catalytic performance in the whole ORR process [50]. The carbon-based conductive substrates are also necessary to maintain the high electron transmission capacity. In the past few years, in order to improve the catalytic performance of M–N–C catalysts, different synthesis strategies have been explored and different types of M–N–C catalysts for ORR have been developed. In addition, the mechanism of catalyst degradation has been well studied, and the improvement strategies have also been proposed. In the following discussion, the related progress in these areas will be separately described in detail.

11.3.2 Synthesis methods

Due to the atom-level dispersion of transition metals in M–N–C catalysts, there is a strong tendency of aggregation and cluster/nanocrystal growth in order to reduce the specific surface energy. Therefore, it is a huge challenge to successfully prepare the well-defined atomically dispersed M–N–C catalysts, which has attracted great attention around the world [61, 62]. A variety of synthesis methods have already been developed, including coordinative pyrolysis strategy, adsorption anchoring strategy, spatial confinement strategy and thermal transfer and capture strategy [63, 64]. For all the synthesis strategies, nitrogen doping on a carbon substrate plays a very important role in the anchoring of transition mental atoms, resulting in the formation of atomic dispersion. In addition, the uniformity of the introduced metal precursors is crucial to their anchoring and atomic dispersion in the later stage [65]. Consequently, the synthesis strategies mainly focus on the achievement of uniform dispersion of transition metal in the precursors. Here, the classification of synthesis strategies is based on the different approaches of introducing transition metals.

11.3.2.1 Coordinative pyrolysis strategy

$3d$ transition metals in general contain empty d-sublayer electron orbitals, which can accept lone pairs of electrons. As a result, they can easily act as central ions to form coordination bonds with the unsaturated coordination atoms in organic functional groups, such as N, O and S [64]. This coordination effect can automatically make the transition metal ions achieve atomic dispersion in the precursors, because the presence of organic groups hinders their agglomeration. After pyrolysis, the transition metal atoms can be readily anchored by doped nitrogen around them, leading to the atomically dispersed M–N–C catalysts. Due to its highly effective synthesis, much effort has been devoted to exploring coordinative pyrolysis strategy, and different coordinative precursors have been successfully developed. Metal–organic frameworks (MOFs) are one of the most commonly used coordinative precursors, which have a high, specific surface area, well-defined porous structure and structural diversity [66, 67]. The abundant metal nodes and coordinative organic ligands in MOFs allow for the incorporation of the large amount of transition metal atoms. Meanwhile, the unique porous structure endows them with great potential to anchor the transition metal atoms [68]. Another common coordination precursor are polymers with abundant heteroatoms, which can generate the strong coordination interaction with transition metal atoms. A variety of polymers have already been applied to coordinate with transitional metal atoms. The structure of polymer precursors is mainly responsible for the anchoring of transition metal atoms during the process of pyrolysis [69]. Therefore, the construction of suitable polymers is a hot direction in this area. In addition, the carbon-based materials modified with organic functional groups/molecules/oligomer can serve as the coordinative precursors [70]. The modification methods and the structure of carbon-based materials have great influence on the final dispersion of transition metal atoms. Based on the above materials, great advances have been made to promote the development of coordinative pyrolysis strategy.

11.3.2.2 Adsorption anchoring strategy

The adsorption anchoring strategy is the foundation of another class of typical methods of the preparation of M–N–C catalysts [65]. To realize the atomic dispersion of transition metal, controlling the metal content of adsorption is generally the most direct approach. Also, the number of defects on substrates is very crucial to the uniform adsorption of transition metal because defects usually have high surface energy and can serve as the adsorption sites [50, 64]. Hence, it is of great importance to select suitable substrates with high specific surface area, which is beneficial for providing the sufficient defect sites. In addition, the introduction of surface defects is an effective approach to create more adsorption sites on substrates. The common methods generally include heteroatom doping and the construction of intrinsic defects, such as vacancies, edge-corner sites and porous structure. It is worth noting that the adsorption methods also have great influence on the dispersion of transition metal precursors. Impregnation strategy is usually adopted to achieve the adsorption of metal ions [71]. Subsequently, filtration or evaporation is used to separate the adsorption product. Sometimes a freeze-drying method is employed to protect the adsorbed metal ions from the agglomeration [72]. The ball-milling method is also a facial and scalable approach to make the metal precursors uniformly adsorb on substrates [73]. In the process of ball milling, in addition to mixing metal precursors with substrates uniformly, it can cause the partial damage of substrates and the formation of some defects, thereby further improving the dispersion of metal precursors. After adsorption, the heat treatment is necessary to make transition metal bond with surrounding N atoms.

11.3.2.3 Spatial confinement strategy

The spatial confinement strategy can lead to atomic dispersion of transition metal precursors by encapsulating them into molecular-scale cages of substrates [64]. In order to ensure their atomic dispersion, two conditions have to be fulfilled: (1) the transition metal precursors have to maintain the atomic dispersion during the encapsulation process. Metal–organic coordination complexes are generally used as the metal precursors because of their highly monodispersed metal atoms, such as metal acetylacetonate, phthalocyanine and so on. (2) the size of single transition metal precursor needs to be compatible with the molecular-scale cages of substrates. On the other hand, it should be larger than the diameter of the holes in the cage to ensure that the metal precursor will not escape from the cage. MOFs are the commonly used substrates for the spatial confinement strategy [66, 74]. The regular porous structure of MOFs provides great potential to spatially confine and encapsulate the transition metal precursors. Additionally, MOFs are carbon-based materials and have high electronic conductivity after pyrolysis, which is the basic requirement for electrocatalysts. After the encapsulation of monodispersed metal precursors, the ligands of metal precursors should be removed, and then the single metal atom could be fixed by the skeletons of substrates to form the M–N–C catalysts.

11.3.2.4 Thermal transfer and capture strategy

Other than through the above-mentioned strategies, transition metal atoms can be introduced into substrates though the thermal transfer and capture strategy [75]. The thermal transfer process is crucial to obtain the atomically dispersed metal species through breaking the original chemical binding of metal precursors. In addition, the substrates need to provide sufficient anchoring sites to capture the metal precursors. The two processes could occur in succession or simultaneously. In the case of successive occurrences, it is generally necessary to add additives to atomize the precursor; in the case of simultaneous occurrence, the strong trapping effect of substrates enables metal atoms to directly detach from metal precursors and form a new bond with substrates. For the thermal transfer and capture strategy, usually M–N–C catalysts can be directly acquired, and no further post-treatment is required.

11.3.3 Nonprecious metal single-atom catalysts

Atomically dispersed M–N–C catalysts have been developed as the most promising alternative to Pt-based catalysts for ORR. Many efforts have been devoted to exploring the different types of M–N–C catalysts. A series of nonprecious transition metals have been successfully introduced to prepare single atom catalysts such as Fe, Co, Mn, Cr, Cu, Ni, Mo, W, etc. Although the nature of the catalytic active sites for ORR is still on debate, there is no doubt that the metal centers of M–N–C catalysts have profound influence on their ORR activity. According to the recent research results, the most promising M–N–C catalysts include Fe–N–C, Co–N–C, Mn–N–C and Cr–N–C due to their high ORR performance.

11.3.3.1 Fe–N–C single-atom catalysts

Compared with other transition metals, Fe–N–C catalysts exhibit the best ORR catalytic activity and are considered to be the most promising nonnoble metal catalysts for acid systems [76, 77]. Therefore, many researchers have paid their attention to the study of Fe–N–C catalysts in recent years. Herein, recent advances on Fe–N–C catalysts are summarized.

Early research work by the Dodelet group forms the foundation for this field. In 2009, they developed a ball-milling pyrolysis method to prepare Fe–N–C catalyst with improved ORR activity in PEMFCs [78]. Based on the previous work, micropores are mainly responsible for hosting catalytic sites [79]. To overcome the limitations resulting from the difficulty of N-doping in some micropores, they innovatively filled a mixture of pore filler and iron precursor into these micropores through a ball-milling method. The pore filler could fix the Fe atoms through the N dopants during the pyrolysis process, serving as a bridge between porous carbon substrate and Fe atoms. This strategy increased the density of active sites and improved the ORR performance. Further work by Wu et al. explored polyaniline as a promising template compound to prepare Fe–N–C catalysts with high performance, which opened a new era of research [80]. Recently, Fe-based single atom catalyst has been greatly studied, and its ORR performance has been significantly improved. MOFs-based materials as Fe substrates have been well developed and deliver an excellent ORR performance. The Wu group conducted a systematic study to prepare Fe–N–C catalysts for PEMFCs application. Zeolitic imidazolate frameworks (ZIF-8) were employed as a host to catch Fe single atoms [67, 71, 81]. Because Fe ions can coordinate with dimethylimidazole, they can be chemically doped into ZIF-8 by replacing partial Zn ions. After pyrolysis, atomically dispersed Fe–N–C catalysts were obtained by anchoring the separate Fe atoms on surrounding nitrogen dopants. The impact of ZIF-8 size and thermal activation temperature on the ORR performance were studied in depth (Figure 7) [67]. It was found that the Fe–N–C catalyst derived from 50 nm ZIF-8 at 800 °C delivered the better ORR performance with a half-wave potential ($E_{1/2}$) of 0.85 V in 0.5 M H_2SO_4. To further increase the density of Fe active sites, the effect of Fe content was studied on the Fe–N–C catalyst [81]. The highest density of active sites was achieved by 1.5 at% Fe-ZIF-8 precursors, which gave rise to a about $E_{1/2}$ of 0.88 V in 0.5 M H_2SO_4. The formation mechanism of Fe active sites was further elucidated based on the ZIF-8-derived carbon by a controlled thermal activation [71]. The result indicated that the Fe active sites could be formed at relatively low temperature (400 °C), while the Fe active sites obtained at 700 °C possessed the ideal coordination structure, exhibiting the best ORR performance.

11.3.3.2 Co–N–C single-atom catalysts

Although many important advances have been made in the research on Fe–N–C catalysts, durability still needs further improvement. Fe species are easy to corrode and dissolve in an acidic system, leading to loss of active sites. Furthermore, the produced iron ions can catalyze the formation of active free radical from the ORR by-product H_2O_2 (Fenton reaction), which further leads to the catalytic performance degradation [82]. As a result, exploring Fe-free metal catalysts is a very active area of research for nonprecious metal catalysts. The Co–N–C catalyst gives the most promising alternative to replace Fe–N–C catalyst due to their similar high activity for ORR [83–85]. Recently, many excellent Co–N–C catalysts have been developed and reported.

Like Fe–N–C catalysts, the density of Co active sites has significant impact on the ORR performance of Co–N–C catalysts. Wang and coworkers prepared a high-performance atomically dispersed Co–N–C catalyst through thermal activation of Co-doped MOFs (ZIF-8) [83]. The Co-doping content and thermal activation temperature were systematically investigated. The optimized Co–N–C catalyst shows a superior ORR performance with an $E_{1/2}$ of 0.80 V in 0.5 M H_2SO_4. More importantly, its related durability in both acidic media and real fuel cell was also significantly improved. Based on the Co-MOFs derived Co–N–C catalyst, an effective surfactant-assisted confinement strategy was developed to further increase the density of Co active sites through introducing

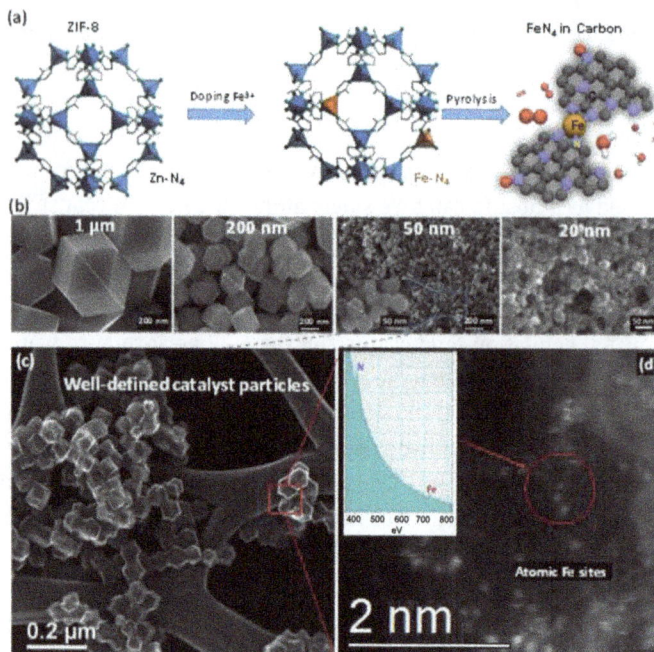

Figure 7: (a) Schematic of the synthesis of Fe–N–C catalyst and (b–d) the related morphologies of samples with different particle sizes and thermal activation temperatures [67]. Reproduced with permission, Copyright 2017, American Chemical Society.

F127 on the surface of Co-ZIF-8 [84]. The carbon shell derived from F127 could retain the complete carbon matrix on the surface to anchor the more Co active sites. This strategy could greatly improve the ORR activity of the Co–N–C catalyst compared with conventional surfactant-free method. The performance of Co–N–C catalyst in PEMFCs was further carefully studied by Xie and coworkers [85]. Compared with the Fe–N–C catalyst, the Co–N–C catalyst could achieve a similar catalytic activity in PEMFCs but with a four-time enhanced durability. The high durability could be ascribed to the lower activity of Co-ions for the Fenton reaction.

11.3.3.3 Other nonprecious single-atom catalysts

Apart from Co–N–C catalysts, some Fe-free nonprecious catalysts have been developed aiming to achieve high activity and durability for ORR. The Mn–N–C catalyst is one of the most promising nonprecious catalysts. Li et al. used ZIF-8 as anchoring matrix to prepare Mn–N–C catalyst through a two-step process involving coordination and adsorption strategies, which effectively increased the density of Mn active sites [86]. The as-prepared Mn–N–C catalyst exhibited a high catalytic activity with an $E_{1/2}$ of 0.80 V in

0.5 M H_2SO_4, along with an improved durability compared with Fe–N–C catalyst. Additionally, Luo and coworkers developed atomically dispersed Cr–N–C catalyst to replace the Fe–N–C catalyst with excellent ORR performance in acidic media [87]. The obtained Cr–N–C catalyst displayed an $E_{1/2}$ of 0.773 V in 0.5 M H_2SO_4. Despite of its lower activity, the durability of the Cr–N–C catalyst was significantly superior to the Fe–N–C catalyst. For the Mn–N–C and Cr–N–C catalysts, the research results show that their high stability can be ascribed to the substantially restrained Fenton reaction. These studies indicate that it is promising to develop new Fe-free M–N–C catalysts with high stability for ORR, which should be attract more attention moving forward.

11.3.4 Nonprecious dimer catalysts

Single atom catalysts (SACs) with unique isolated active sites attracted significant attention and have been widely applied across many catalysis fields. However, due to their unique local structures, SACs are unable to catalyze some chemical reactions and achieve satisfied results [88, 89]. Therefore, further improvement of SACs to meet the high requirement of catalysis is necessary. Because of the unique local structure of SACs, their electronic properties are strongly affected by their neighboring atoms. Thus, modifying their most closely neighboring sites by introducing another single atom may lead to favorable geometric and electronic effects. This synergistic effect may modulate charge transfer between two single atoms and lead to a high catalytic activity by optimization of the adsorption of reactants. Recently, several dimeric catalysts with homo and hetero pairs have been successfully achieved and applied in different research fields, such as electrocatalysis [90], hydrogenation [91] and other reactions [92]. The rising star of dimeric catalysts showed improved catalytic activity compared to the SACs, which would be very promising for future practical applications. Compared with noble metal-based catalysts, the nonprecious dimer catalysts are more promising because of the lower expense and can be a practical alternative to Pt in the electrochemical reactions.

Lu and coworkers used a competitive complexation strategy to design a Zn/Co bimetallic sites catalyst [93]. In the synthesis method, chitosan was used as the C and N source, zinc chloride and cobalt acetate were used as Zn and Co sources, respectively. Because of the functional groups on the chitosan chains, the Co^{2+} or Zn^{2+} would be coordinated with the $-NH_2$ and $-OH$ groups through a simultaneous competitive complexation process, which could form the homogeneous dispersion of Zn and Co (Figure 8a). The bimetallic Zn/Co atomic catalyst was achieved after pyrolysis at 750 °C. The Zn/Co atomic pairs were confirmed by the HAADF-STEM and EXAFS results (Figure 8b–d). In the ORR under 0.1 M KOH with rotating disk electrode, the Zn/CoN–C dimers exhibited an E_{onset} at 1.004 V, which is much more positive than that of Pt/C (0.970 V). Moreover, the $E_{1/2}$ on Zn/CoN–C is as high as 0.861 V, which is also more positive than that on Pt/C (0.826 V), CoN–C (0.793 V) and ZnN–C (0.706 V), indicating its best catalytic activity in ORR. The electron transfer number of Zn/CoN–C was evaluated about four through

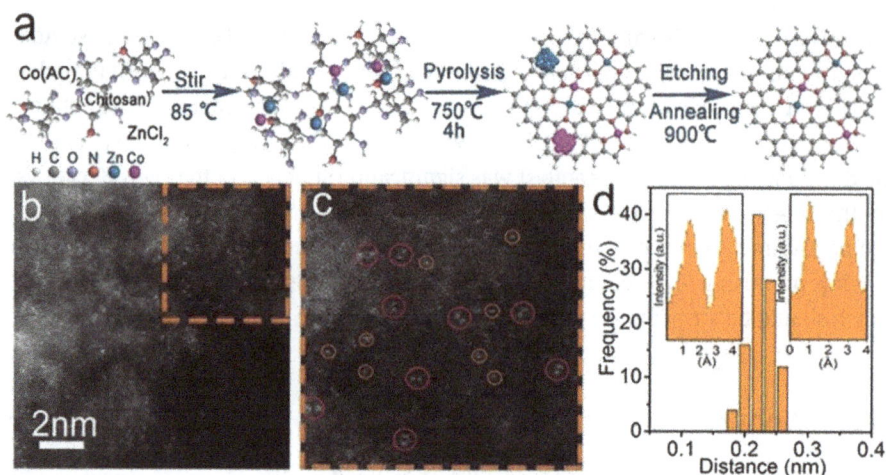

Figure 8: (a) Illustration of the formation of Zn/CoN–C, (b) Aberration-corrected HAADF-STEM image and (c) enlarged images of the Zn/CoNC. (d) Statistical Zn-Co distance in the observed diatomic pairs, in which the two insets are intensity profiles obtained on two bimetallic Co/Zn sites [93]. Reproduced with permission, Copyright 2019, John Wiley and Sons.

the Koutecky–Levich equation and rotating ring-disk electrode. More importantly, the Zn/CoN–C showed remarkable stability, which was only 9 mV changes on the $E_{1/2}$ after the 10,000 cycles durability test. The high catalytic performance and stability on Zn/CoN–C was in line with the DFT predication.

Metal-organic frameworks (MOFs) based materials, which has special confinement porous structure and high surface area, show promising application in the synthesis of non-precious dimer catalysts [94–96]. Wang et al. first reported the FeCo dimers can be achieved using a MOFs derived strategy [94]. In their work, a ZIF-8 structure like Co^{2+} and Zn^{2+} nodes with 2-methylimidazole was used as the host to adsorb the Fe ions, which acted as the guest ion. The Fe^{3+} ions are encapsulated in the cavity. The obtained materials were further treated in a pyrolysis process, in which the adsorbed Fe species catalyze the formation of N-doped graphene and simultaneously reduced by the neighboring Co atoms. In the high temperature pyrolysis process, the Zn species are evaporated and the N-coordinated Fe-Co duals sites formed. The Fe-Co pairs were confirmed by HAADF-STEM. In the ORR under 0.1 M $HClO_4$ solution, the (Fe, Co)/N-C catalyst exhibited a high ORR activity with $E_{1/2}$ = 0.863 V and E_{onset} = 1.06 V, which are much more positive than the Fe SAs/N-C, Co SAs/N-C and Pt/C catalysts. These (Fe, Co)/N-C dimers also yield a high stability, without showing obvious activity decline even after 50,000 cycles. The maximum power density in H_2/O_2 fuel cell could reach to about 0.85 and 0.98 mW cm^{-2} at back pressures of 0.1 and 0.2 MPa, respectively. In a more practical H_2/air fuel cell, its peak power density exceeded 505 mW cm^{-2} at 0.42 V, which achieved about 76 % of commercial Pt/C. Its ultrahigh stability was also impressive with the negligible voltage loss after 100 h at operation conditions of 600 mA cm^{-2} and 1000 mA cm^{-2}.

Apart from the hetero-paired dimers, some single element dimers also show remarkable catalytic performance in electrochemical reactions. Recently, Xie et al. developed the planar-like Fe_2N_6 structure with dual-metal atom sites by coupling isolated FeN_4 single sites through thermal migration strategy for ORR [97]. The Fe_2N_6 exhibited a positive $E_{1/2}$ of 0.84 V, which is much more positive than that of single site FeN_4 (0.76 V) and Fe-N nanoparticles (0.81 V). The Fe_2N_6 also shows the highest specific surface area activity of 26.2 mA m^{-2} at 0.8 V. More importantly, the mass activity on Fe_2N_6 is over 700 % higher than FeN_4 In the MEA performance testing in PEMFCs, the Fe_2 dimer catalyst achieved outstanding power densities (845 and 360 mW cm^{-2}) under 0.2 MPa H_2/O_2 and H_2/air conditions, respectively, which is much higher than a Fe single-atom catalyst.

11.3.5 Catalysts durability improvement strategies

The electrocatalysts used in both anode and cathode of PEMFCs are Pt-based materials on carbon supports. Because the harsh working conditions such as high overpotentials, acidic electrolyte and oxidation environment, the degradation of electrocatalysts occur, which severely reduced the working life of catalysts. The potential degradation mechanisms of Pt-based carbon electrocatalysts can be suggested: (a) platinum dissolution— some of the Pt atoms; (b) Ostwald ripening; (c) agglomeration; (d) particle detachment; (e) carbon corrosion [98–100]. Due to the harsh working conditions, some of the Pt metal will dissolve into the electrolyte, leading to the decreased total amount of Pt. If the interaction between Pt NPs and the carbon support is not strong enough, Oswald ripening, agglomeration and particle detachment will sharply reduce the number of active Pt sites. More seriously, carbon corrosion will not only decrease the active sites but also reduce the conductivity of materials. Therefore, rational design highly efficient and stable electrocatalysts are critical important.

Design of shape controlled, and alloyed Pt nano catalysts is one of the promising strategies to address these issues. Chen et al. synthesized $PtNi_3$ polyhedra in oleylamine. The capped $PtNi_3$ polyhedra were dispersed in nonpolar solvent and kept for 2 weeks. Further increasing the temperature to 120 °C, the Pt_3N_i nanoframes form the desired Pt-skin surface structure. This unique nanoframe catalyst with Pt-skin surface shows remarkable catalytic activity in the ORR. At 0.95 V, the Pt_3Ni nanoframes exhibited more than 16-fold higher specific activity than the commercial Pt/C. In the durability test, the Pt_3Ni nanoframes showed extremely high stability, while the commercial Pt/C lost 40 % of its initial specific activity due to the dissolution of Pt surface atoms and agglomeration. Such a high stability of Pt_3Ni is attributed to the unique Pt-skin surface, which lower the coverage of oxygenated intermediates from weaker oxygen binding strength, thus reducing the Pt dissolution [101]. Li and coworkers successfully achieved the PtNi alloy nanowires through the thermal annealing of Pt/NiO core/shell nanowires. The PtNi alloy nanowires can further transformed into jogged Pt nanowires under electrochemical dealloying [102]. After fully activation, the j-PtNWs reached the ECSA as high as

$118 \, m^2/g_{Pt}$. In ORR evaluation, the j-PtNWs showed the high mass activity of $13.6 \, A/mg_{Pt}$ at 0.9 V, which is 52 times higher than that of the Pt/C. The j-PtNWs also showed superior ORR durability, with only a minor 7 % ECSA and 12 % mass activity loss after 6000 cycles. The high catalytic performance of j-PtNWs is attributed to the unique geometry of 1-D structure of Pt nanowires, which has highly stressed surface configurations.

Apart from the modification on the Pt NPs, developing highly stable support is also important to promote the stability of electrocatalysts. Using N-doped carbon materials as the support for electrocatalyst is one of the possible choices. N-doping not only promotes ORR activity but also strengthens the resistance to corrosion. Geng and coworkers synthesized the N-doped graphene by the treatment of graphene under ammonia atmosphere [103]. In ORR, the achieved N-graphene (900) exhibited higher onset potential and more positive $E_{1/2}$ than the commercial Pt/C. More importantly, the N-graphene (900) showed great stability under ORR, without showing any obvious activity decline after 5000 cycles durability test. Rather than traditional carbon-based materials, some novel noncarbon supports are also applicable. For example, Cheng and coworkers used the ZrC as the substrate to support Pt NPs using the ALD method [104]. The achieved Pt-on-ZrC catalyst exhibited remarkable catalytic activity in ORR, showing the mass activity of $0.122 \, A/mg_{Pt}$ at 0.9 V, which is much higher than the commercial Pt/C ($0.074 \, A/mg_{Pt}$). During the durability test, the ALD-Pt/ZrC catalyst was five times more stable than the Pt/C, in which the ALD-Pt/ZrC had the activity loss of 26 % but Pt/C lost 76 % of its initial activity. The X-ray absorption near edge structures results proved the strong interaction between Pt NPs and ZrC, which could be the main reason for its high stability.

The weak interaction between Pt NPs and the substrate is one of the reasons that lead to the low stability of the Pt-based electrocatalysis. Therefore, increasing the interactions, several effective strategies have been developed to increase the interactions thus inhibiting the migration and coalescence of Pt NPs. In Cheng and coworkers' work, the oleylamine was used as a blocking agent to the surface of Pt NPs and then deposited ZrO_2 using ALD (Figure 9) [105]. Due to the blocking agent, the ZrO_2 were not directly deposited on the surface of Pt NPs but the immediate surrounding surface. As a result, with the increasing cycles of ZrO_2 ALD, the ZrO_2 NPs functioned as the nanocages, which strongly increases the sinter resistance of the Pt NPs. In the ORR, the ZrO_2 coated Pt NPs catalyst showed a mass activity of 0.28 A/mg at 0.9 V, which was 6.4 times higher than the Pt/C. More importantly, the ZrO_2 coated Pt NPs catalyst showed outstanding stability, without showing obvious activity loss after ADT, which is much more stable than the commercial Pt/C. Based on the TEM characterization results, the size of Pt NPs on the ZrO_2 coated sample almost unchanged, while the size of Pt NPs on Pt/C sharply increased from 3.9 to 8.1 nm. These results unambiguously show that ZrO_2 promotes the stability of Pt NPs. It is worthy to note that this area selective ALD method, which created the nanocages to stabilize Pt NPs can also be applied to other oxide coatings. For instance, Song et al. used TaO_x coated on oleylamine decorated Pt/C and further thermal treated under H_2 to remove the oleylamine [106]. This TaO_x coated Pt/C sample showed remarkable stability in ORR. In the MEA stability test, it only lost 12 % of its initial power

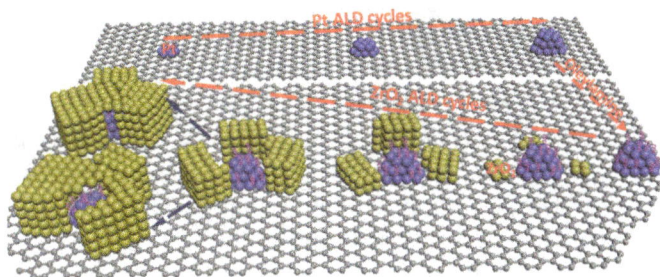

Figure 9: Schematic diagram of platinum encapsulated in zirconia nanocages structure fabricated by area-selective ALD [105]. Reproduced with permission, Copyright 2015, John Wiley and Sons.

density after 120 h ADT, which is much more stable than the commercial Pt/C. These high stabilities are attributed to the strong interaction between Pt NPs and the oxide.

11.4 Outlook and summary

In this chapter, we summarized the recent development of the Pt-based catalysts and nonprecious metal catalysts, as well as their applications in PEMFCs. By tuning particle size, surface structure and composition, the ORR activity of Pt-based catalysts can be significantly enhanced compared with commercial Pt/C catalysts. The AUE of Pt catalysts can be significantly improved by rational design of porous, core-shell, nanocage and nanoframe structures. Furthermore, the synthesis of Pt single atoms can promote the AUE to even 100 %. MOF-derived single atom catalysts are promising alternative non-noble metal electrocatalysts for ORR. With the remarkable advantages of a high surface area, well-tuned coordination environments and high AUE, MOF-derived transition metal catalysts can be used as the most promising materials to substitute Pt catalysts. Currently, several synthesis methods have been developed, such as coordinative pyrolysis strategy, adsorption anchoring strategy and others. In addition to single atom catalysts, dual metal sites are also fabricated to improve the performance even further.

It has been widely proved that the activity of nanostructured Pt catalysts is highly dependent on their AUE, surface structure and composition. The composition adjustment can be used simultaneously with high AUE or surface control to improve the performance of Pt-based catalysts. However, with the decrease of the particle size, the ratio of surface/edge atoms decrease at the same time, which brings great challenges to accomplish surface control, and reduce particle size at the same time. In the future, the novel Pt structures, such as nanowires, core-shell and hollow structures should be developed to combine the three factors to achieve further enhanced activities and durability.

Another problem is synthesis scale-up to commercial quantities. The commercial Pt/C with 1–5 nm size can be obtained through industrial production. The large-scale fabrication of well-controlled Pt nanostructures might be achieved through continuous flow synthesis based on droplets. The pyrolysis method has been proved a potential large-scale preparation method for the single atom catalysts. In addition, increasing the loading amount of Pt on substrates is particularly important, as most of the Pt wt.% in the Pt single atom catalysts are very low. The achievement of Pt single atom catalysts with higher Pt loadings might accelerate their applications in industry.

For nonnoble metal electrocatalysts, most electrocatalysts are based on transition metals and carbon, which provide remarkable electrocatalytic activity in alkaline media. However, most of these M–N–C electrocatalysts still suffer from severe stability problems in acidic conditions, inhibiting their commercial application in polymer electrolyte membrane fuel cells. Therefore, pursuing a M–N–C electrocatalyst that can work in acidic media with both high activity and durability is highly desirable. One of the future directions in the development of M–N–C electrocatalysts should therefore focus on the design of stable electrocatalysts under fuel cell working conditions. Moreover, the enhanced mechanism should be systematically investigated. Meanwhile, multiple transition metal SACs are worth to be developed due to their synergistic effect.

Bibliography

[1]   M. K. Debe, *Nature*, **486**, 43–51, 2012.
[2]   C. H. Choi, M. Kim, H. C. Kwon, S. J. Cho, S. Yun, H. T. Kim, K. J. Mayrhofer, H. Kim, and M. Choi, *Nat. Commun.*, **7**, 10922, 2016.
[3]   L. Zhang, L. Wang, C. M. B. Holt, B. Zahiri, Z. Li, K. Malek, T. Navessin, M. H. Eikerling, and D. Mitlin, *Energy Environ. Sci.*, **5**, 6156–6172, 2012.
[4]   S. Sui, X. Wang, X. Zhou, Y. Su, S. Riffat, and C.-J. Liu, *J. Mater. Chem. A*, **5**, 1808–1825, 2017.
[5]   Y. Chen, S. Ji, Y. Wang, J. Dong, W. Chen, Z. Li, R. Shen, L. Zheng, Z. Zhuang, D. Wang, and Y. Li, *Angew. Chem., Int. Ed. Engl.*, **56**, 6937–6941, 2017.
[6]   M. Shao, A. Peles, and K. Shoemaker, *Nano Lett.*, **11**, 3714–3719, 2011.
[7]   M. Nesselberger, S. Ashton, J. C. Meier, I. Katsounaros, K. J. Mayrhofer, and M. Arenz, *J. Am. Chem. Soc.*, **133**, 17428–17433, 2011.
[8]   W. Sheng, S. Chen, E. Vescovo, and Y. Shao-Horn, *J. Electrochem. Soc.*, **159**, B96–B103, 2011.
[9]   V. R. Stamenkovic, B. Fowler, B. S. Mun, G. Wang, P. N. Ross, C. A. Lucas, and N. M. Marković, *Science*, **315**, 493, 2007.
[10]   J. Zhang, J. Yang, J. Fang, and S. Zou, *Nano Lett.*, **10**, 638–644, 2010.
[11]   S. I. Choi, S. Xie, M. Shao, J. H. Odell, N. Lu, H. C. Peng, L. Protsailo, S. Guerrero, J. Park, X. Xia, J. Wang, M. J. Kim, and Y. Xia, *Nano Lett.*, **13**, 3420–3425, 2013.
[12]   X. Huang, Z. Zhao, L. Cao, Y. Chen, E. Zhu, Z. Lin, M. Li, A. Yan, A. Zettl, Y. M. Wang, X. Duan, T. Mueller, and Y. Huang, *Science*, **348**, 1230, 2015.
[13]   M. Oezaslan, F. Hasché, and P. Strasser, *J. Electrochem. Soc.*, **159**, B394–B405, 2012.
[14]   D. Wang, H. L. Xin, R. Hovden, H. Wang, Y. Yu, D. A. Muller, F. J. DiSalvo, and H. D. Abruna, *Nat. Mater.*, **12**, 81–87, 2013.
[15]   L. Bu, S. Guo, X. Zhang, X. Shen, D. Su, G. Lu, X. Zhu, J. Yao, J. Guo, and X. Huang, *Nat. Commun.*, **7**, 11850, 2016.
[16]   M. Oezaslan, F. Hasché, and P. Strasser, *J. Electrochem. Soc.*, **159**, B444–B454, 2012.

[17]   S. Chen, J. Zhao, H. Su, H. Li, H. Wang, Z. Hu, J. Bao, and J. Zeng, *J. Am. Chem. Soc.*, **143**, 496–503, 2021.
[18]   C. Jung, C. Lee, K. Bang, J. Lim, H. Lee, H. J. Ryu, E. Cho, and H. M. Lee, *ACS Appl. Mater. Interfaces*, **9**, 31806–31815, 2017.
[19]   L. Bu, N. Zhang, S. Guo, X. Zhang, J. Li, J. Yao, T. Wu, G. Lu, J.-Y. Ma, D. Su, and X. Huang, *Science*, **354**, 1410, 2016.
[20]   L. Wang and Y. Yamauchi, *J. Am. Chem. Soc.*, **131**, 9152–9153, 2009.
[21]   C. Kim, J.-G. Oh, Y.-T. Kim, H. Kim, and H. Lee, *Electrochem. Commun.*, **12**, 1596–1599, 2010.
[22]   Z. Peng and H. Yang, *J. Am. Chem. Soc.*, **131**, 7542–7543, 2009.
[23]   B. Lim, M. Jiang, P. H. C. Camargo, E. C. Cho, J. Tao, X. Lu, Y. Zhu, and Y. Xia, *Science*, **324**, 1302, 2009.
[24]   J. X. Wang, H. Inada, L. Wu, Y. Zhu, Y. Choi, P. Liu, W.-P. Zhou, and R. R. Adzic, *J. Am. Chem. Soc.*, **131**, 17298–17302, 2009.
[25]   K. Sasaki, H. Naohara, Y. Choi, Y. Cai, W. F. Chen, P. Liu, and R. R. Adzic, *Nat. Commun.*, **3**, 1115, 2012.
[26]   H. Nan, X. Tian, J. Luo, D. Dang, R. Chen, L. Liu, X. Li, J. Zeng, and S. Liao, *J. Mater. Chem. A*, **4**, 847–855, 2016.
[27]   X. Tian, J. Luo, H. Nan, H. Zou, R. Chen, T. Shu, X. Li, Y. Li, H. Song, S. Liao, and R. R. Adzic, *J. Am. Chem. Soc.*, **138**, 1575–1583, 2016.
[28]   E. Toyoda, R. Jinnouchi, T. Ohsuna, T. Hatanaka, T. Aizawa, S. Otani, Y. Kido, and Y. Morimoto, *Angew. Chem., Int. Ed. Engl.*, **52**, 4137–4140, 2013.
[29]   S. Xie, S. I. Choi, N. Lu, L. T. Roling, J. A. Herron, L. Zhang, J. Park, J. Wang, M. J. Kim, Z. Xie, M. Mavrikakis, and Y. Xia, *Nano Lett.*, **14**, 3570–3576, 2014.
[30]   J. Park, L. Zhang, S.-I. Choi, L. T. Roling, N. Lu, J. A. Herron, S. Xie, J. Wang, M. J. Kim, M. Mavrikakis, and Y. Xia, *ACS Nano*, **9**, 2635–2647, 2015.
[31]   X. Wang, S. I. Choi, L. T. Roling, M. Luo, C. Ma, L. Zhang, M. Chi, J. Liu, Z. Xie, J. A. Herron, M. Mavrikakis, and Y. Xia, *Nat. Commun.*, **6**, 7594, 2015.
[32]   X. Wang, M. Vara, M. Luo, H. Huang, A. Ruditskiy, J. Park, S. Bao, J. Liu, J. Howe, M. Chi, Z. Xie, and Y. Xia, *J. Am. Chem. Soc.*, **137**, 15036–15042, 2015.
[33]   T. Bian, H. Zhang, Y. Jiang, C. Jin, J. Wu, H. Yang, and D. Yang, *Nano Lett.*, **15**, 7808–7815, 2015.
[34]   J. Chen, B. Wiley, J. McLellan, Y. Xiong, Z.-Y. Li, and Y. Xia, *Nano Lett.*, **5**, 2058–2062, 2005.
[35]   R. Wu, Q.-C. Kong, C. Fu, S.-Q. Lai, C. Ye, J.-Y. Liu, Y. Chen, and J.-Q. Hu, *RSC Adv.*, **3**, 12577–12580, 2013.
[36]   L. Zhang, L. T. Roling, X. Wang, M. Vara, M. Chi, J. Liu, S.-I. Choi, J. Park, J. A. Herron, Z. Xie, M. Mavrikakis, and Y. Xia, *Science*, **349**, 412, 2015.
[37]   X. Wang, L. Figueroa-Cosme, X. Yang, M. Luo, J. Liu, Z. Xie, and Y. Xia, *Nano Lett.*, **16**, 1467–1471, 2016.
[38]   D. S. He, D. He, J. Wang, Y. Lin, P. Yin, X. Hong, Y. Wu, and Y. Li, *J. Am. Chem. Soc.*, **138**, 1494–1497, 2016.
[39]   Z. Fang, Y. Wang, C. Liu, S. Chen, W. Sang, C. Wang, and J. Zeng, *Small*, **11**, 2593–2605, 2015.
[40]   C. Chen, Y. Kang, Z. Huo, Z. Zhu, W. Huang, H. L. Xin, J. D. Snyder, D. Li, J. A. Herron, M. Mavrikakis, M. Chi, K. L. More, Y. Li, N. M. Markovic, G. A. Somorjai, P. Yang, and V. R. Stamenkovic, *Science*, **343**, 1339, 2014.
[41]   N. Becknell, Y. Kang, C. Chen, J. Resasco, N. Kornienko, J. Guo, N. M. Markovic, G. A. Somorjai, V. R. Stamenkovic, and P. Yang, *J. Am. Chem. Soc.*, **137**, 15817–15824, 2015.
[42]   Y. Wu, D. Wang, G. Zhou, R. Yu, C. Chen, and Y. Li, *J. Am. Chem. Soc.*, **136**, 11594–11597, 2014.
[43]   J. Liu, M. Jiao, L. Lu, H. M. Barkholtz, Y. Li, Y. Wang, L. Jiang, Z. Wu, D. J. Liu, L. Zhuang, C. Ma, J. Zeng, B. Zhang, D. Su, P. Song, W. Xing, W. Xu, Y. Wang, Z. Jiang, and G. Sun, *Nat. Commun.*, **8**, 15938, 2017.
[44]   J. Liu, J. Bak, J. Roh, K.-S. Lee, A. Cho, J. W. Han, and E. Cho, *ACS Catal.*, **11**, 466–475, 2020.
[45]   W. Zhang, Y. Liu, L. Zhang, and J. Chen, *Nanomaterials*, **9**, 1402, 2019.
[46]   Z. Song, Y. N. Zhu, H. Liu, M. N. Banis, L. Zhang, J. Li, K. Doyle-Davis, R. Li, T. K. Sham, L. Yang, A. Young, G. A. Botton, L. M. Liu, and X. Sun, *Small*, **16**, e2003096, 2020.
[47]   X. Zeng, J. Shui, X. Liu, Q. Liu, Y. Li, J. Shang, L. Zheng, and R. Yu, *Adv. Energy Mater.*, **8**, 1701345, 2018.
[48]   S. T. Thompson and D. Papageorgopoulos, *Nat. Catal.*, **2**, 558–561, 2019.
[49]   Y. Shao, J.-P. Dodelet, G. Wu, and P. Zelenay, *Adv. Mater.*, **31**, 1807615, 2019.
[50]   Y. He, S. Liu, C. Priest, Q. Shi, and G. Wu, *Chem. Soc. Rev.*, **49**, 3484–3524, 2020.

[51] Y. He, Q. Tan, L. Lu, J. Sokolowski, and G. Wu, *Electrochem. Energy Rev.*, **2**, 231–251, 2019.

[52] H. A. Gasteiger and N. M. Markovic, *Science*, **324**, 48–49, 2009.

[53] R. Jasinski, *Nature*, **201**, 1212–1213, 1964.

[54] S. Gupta, D. Tryk, I. Bae, W. Aldred, and E. Yeager, *J. Appl. Electrochem.*, **19**, 19–27, 1989.

[55] Y. Nie, L. Li, and Z. Wei, *Chem. Soc. Rev.*, **44**, 2168–2201, 2015.

[56] M. Shao, Q. Chang, J.-P. Dodelet, and R. Chenitz, *Chem. Rev.*, **116**, 3594–3657, 2016.

[57] Y. Takasu, M. Suzuki, H. Yang, T. Ohashi, and W. Sugimoto, *Electrochim. Acta*, **55**, 8220–8229, 2010.

[58] M.-R. Gao, J. Jiang, and S.-H. Yu, *Small*, **8**, 13–27, 2012.

[59] M.-R. Gao, Y.-F. Xu, J. Jiang, and S.-H. Yu, *Chem. Soc. Rev.*, **42**, 2986–3017, 2013.

[60] G. Wu, *Front. Energy*, **11**, 286–298, 2017.

[61] J. Liu, *ACS Catal.*, **7**, 34–59, 2016.

[62] A. Wang, J. Li, and T. Zhang, *Nat. Rev. Chem.*, **2**, 65–81, 2018.

[63] N. Cheng, L. Zhang, K. Doyle-Davis, and X. Sun, *Electrochem. Energy Rev.*, **2**, 539–573, 2019.

[64] Y. Chen, S. Ji, C. Chen, Q. Peng, D. Wang, and Y. Li, *Joule*, **2**, 1242–1264, 2018.

[65] S. Ji, Y. Chen, X. Wang, Z. Zhang, D. Wang, and Y. Li, *Chem. Rev.*, 2020. https://doi.org/10.1021/acs.chemrev.9b00818.

[66] Z. Song, L. Zhang, K. Doyle-Davis, X. Fu, J. L. Luo, and X. Sun, *Adv. Energy Mater.*, **10**, 2001561, 2020.

[67] H. Zhang, S. Hwang, M. Wang, Z. Feng, S. Karakalos, L. Luo, Z. Qiao, X. Xie, C. Wang, D. Su, Y. Shao, and G. Wu, *J. Am. Chem. Soc.*, **139**, 14143–14149, 2017.

[68] P. Yin, T. Yao, Y. Wu, L. Zheng, W. Lin, W. Liu, H. Ju, J. Zhu, X. Hong, Z. Deng, G. Zhou, S. Wei, and Y. Li, *Angew. Chem., Int. Ed. Engl.*, **55**, 10800–10805, 2016.

[69] Y. Han, Y.-G. Wang, W. Chen, R. Xu, L. Zheng, J. Zhang, J. Luo, R.-A. Shen, Y. Zhu, W.-C. Cheong, C. Chen, Q. Peng, D. Wang, and Y. Li, *J. Am. Chem. Soc.*, **139**, 17269–17272, 2017.

[70] W. C. Cheong, W. Yang, J. Zhang, Y. Li, D. Zhao, S. Liu, K. Wu, Q. Liu, C. Zhang, D. Wang, Q. Peng, C. Chen, and Y. Li, *ACS Appl. Mater. Interfaces*, **11**, 33819–33824, 2019.

[71] P. Li, M. Wang, X. Duan, L. Zheng, X. Cheng, Y. Zhang, Y. Kuang, Y. Li, Q. Ma, Z. Feng, W. Liu, and X. Sun, *Nat. Commun.*, **10**, 1711, 2019.

[72] S. Zhou, L. Shang, Y. Zhao, R. Shi, G. I. N. Waterhouse, Y. C. Huang, L. Zheng, and T. Zhang, *Adv. Mater.*, **31**, e1900509, 2019.

[73] D. Deng, X. Chen, L. Yu, X. Wu, Q. Liu, Y. Liu, H. Yang, H. Tian, Y. Hu, P. Du, R. Si, J. Wang, X. Cui, H. Li, J. Xiao, T. Xu, J. Deng, F. Yang, P. N. Duchesne, P. Zhang, J. Zhou, L. Sun, J. Li, X. Pan, and X. Bao, *Sci. Adv.*, **1**, e1500462, 2015.

[74] Y. Chen, S. Ji, Y. Wang, J. Dong, W. Chen, Z. Li, R. Shen, L. Zheng, Z. Zhuang, D. Wang, and Y. Li, *Angew. Chem., Int. Ed. Engl.*, **56**, 6937–6941, 2017.

[75] Y. Qu, Z. Li, W. Chen, Y. Lin, T. Yuan, Z. Yang, C. Zhao, J. Wang, C. Zhao, X. Wang, F. Zhou, Z. Zhuang, Y. Wu, and Y. Li, *Nat. Catal.*, **1**, 781–786, 2018.

[76] N. D. Leonard, S. Wagner, F. Luo, J. Steinberg, W. Ju, N. Weidler, H. Wang, U. I. Kramm, and P. Strasser, *ACS Catal.*, **8**, 1640–1647, 2018.

[77] Q. Li, W. Chen, H. Xiao, Y. Gong, Z. Li, L. Zheng, X. Zheng, W. Yan, W.-C. Cheong, R. Shen, N. Fu, L. Gu, Z. Zhuang, C. Chen, D. Wang, Q. Peng, J. Li, and Y. Li, *Adv. Mater.*, **30**, 1800588, 2018.

[78] M. Lefevre, E. Proietti, F. Jaouen, and J.-P. Dodelet, *Science*, **324**, 71–74, 2009.

[79] M. Lefèvre and J.-P. Dodelet, *Electrochim. Acta*, **53**, 8269–8276, 2008.

[80] G. Wu, K. L. More, C. M. Johnston, and P. Zelenay, *Science*, **332**, 443–447, 2011.

[81] H. Zhang, H. T. Chung, D. A. Cullen, S. Wagner, U. I. Kramm, K. L. More, P. Zelenay, and G. Wu, *Energy Environ. Sci.*, **12**, 2548–2558, 2019.

[82] C. H. Choi, H.-K. Lim, M. W. Chung, G. Chon, N. Ranjbar Sahraie, A. Altin, M.-T. Sougrati, L. Stievano, H. S. Oh, E. S. Park, F. Luo, P. Strasser, G. Dražić, K. J. J. Mayrhofer, H. Kim, and F. Jaouen, *Energy Environ. Sci.*, **11**, 3176–3182, 2018.

[83] X. X. Wang, V. Prabhakaran, Y. He, Y. Shao, and G. Wu, *Adv. Mater.*, **31**, 1805126, 2019.

[84] Y. He, S. Hwang, D. A. Cullen, M. A. Uddin, L. Langhorst, B. Li, S. Karakalos, A. J. Kropf, E. C. Wegener, J. Sokolowski, M. Chen, D. Myers, D. Su, K. L. More, G. Wang, S. Litster, and G. Wu, *Energy Environ. Sci.*, **12**, 250–260, 2019.

[85] X. Xie, C. He, B. Li, Y. He, D. A. Cullen, E. C. Wegener, A. J. Kropf, U. Martinez, Y. Cheng, M. H. Engelhard, M. E. Bowden, M. Song, T. Lemmon, X. S. Li, Z. Nie, J. Liu, D. J. Myers, P. Zelenay, G. Wang, G. Wu, V. Ramani, and Y. Shao, *Nat. Catal.*, **3**, 1044–1054, 2020.

[86] J. Yang, B. Chen, X. Liu, W. Liu, Z. Li, J. Dong, W. Chen, W. Yan, T. Yao, X. Duan, Y. Wu, and Y. Li, *Angew. Chem., Int. Ed. Engl.*, **57**, 9495–9500, 2018.

[87] E. Luo, H. Zhang, X. Wang, L. Gao, L. Gong, T. Zhao, Z. Jin, J. Ge, Z. Jiang, C. Liu, and W. Xing, *Angew. Chem., Int. Ed. Engl.*, **58**, 12469–12475, 2019.

[88] Y. T. Kim, K. Ohshima, K. Higashimine, T. Uruga, M. Takata, H. Suematsu, and T. Mitani, *Angew. Chem., Int. Ed. Engl.*, **45**, 407–411, 2006.

[89] A. Von Weber, E. T. Baxter, S. Proch, M. D. Kane, M. Rosenfelder, H. S. White, and S. L. Anderson, *Phys. Chem. Chem. Phys.*, **17**, 17601–17610, 2015.

[90] L. Zhang, R. Si, H. Liu, N. Chen, Q. Wang, K. Adair, Z. Wang, J. Chen, Z. Song, and J. Li, *Nat. Commun.*, **10**, 1–11, 2019.

[91] E. Guan, L. Debefve, M. Vasiliu, S. Zhang, D. A. Dixon, and B. C. Gates, *ACS Catal.*, **9**, 9545–9553, 2019.

[92] H. Yan, H. Lin, H. Wu, W. Zhang, Z. Sun, H. Cheng, W. Liu, C. Wang, J. Li, X. Huang, T. Yao, J. Yang, S. Wei, and J. Lu, *Nat. Commun.*, **8**, 1070, 2017.

[93] Z. Lu, B. Wang, Y. Hu, W. Liu, Y. Zhao, R. Yang, Z. Li, J. Luo, B. Chi, Z. Jiang, M. Li, S. Mu, S. Liao, J. Zhang, and X. Sun, *Angew. Chem., Int. Ed. Engl.*, **58**, 2622–2626, 2019.

[94] J. Wang, Z. Huang, W. Liu, C. Chang, H. Tang, Z. Li, W. Chen, C. Jia, T. Yao, S. Wei, Y. Wu, and Y. Li, *J. Am. Chem. Soc.*, **139**, 17281–17284, 2017.

[95] M. Xiao, H. Zhang, Y. Chen, J. Zhu, L. Gao, Z. Jin, J. Ge, Z. Jiang, S. Chen, C. Liu, and W. Xing, *Nano Energy*, **46**, 396–403, 2018.

[96] W. Ye, S. Chen, Y. Lin, L. Yang, S. Chen, X. Zheng, Z. Qi, C. Wang, R. Long, M. Chen, J. Zhu, P. Gao, L. Song, J. Jiang, and Y. Xiong, *Chem*, **5**, 2865–2878, 2019.

[97] N. Zhang, T. Zhou, J. Ge, Y. Lin, Z. Du, C. a. Zhong, W. Wang, Q. Jiao, R. Yuan, Y. Tian, W. Chu, C. Wu, and Y. Xie, *Matter*, **3**, 509–521, 2020.

[98] J. C. Meier, C. Galeano, I. Katsounaros, J. Witte, H. J. Bongard, A. A. Topalov, C. Baldizzone, S. Mezzavilla, F. Schüth, and K. J. Mayrhofer, *Beilstein J. Nanotechnol.*, **5**, 44–67, 2014.

[99] S. G. Rinaldo, P. Urchaga, J. Hu, W. Lee, J. Stumper, C. Rice, and M. Eikerling, *Phys. Chem. Chem. Phys.*, **16**, 26876–26886, 2014.

[100] S. G. Rinaldo, W. Lee, J. Stumper, and M. Eikerling, *Phys. Rev. E, Stat. Nonlinear Soft Matter Phys.*, **86**, 041601, 2012.

[101] C. Chen, Y. Kang, Z. Huo, Z. Zhu, W. Huang, H. L. Xin, J. D. Snyder, D. Li, J. A. Herron, and M. Mavrikakis, *Science*, **343**, 1339–1343, 2014.

[102] M. Li, Z. Zhao, T. Cheng, A. Fortunelli, C.-Y. Chen, R. Yu, Q. Zhang, L. Gu, B. V. Merinov, and Z. Lin, *Science*, **354**, 1414–1419, 2016.

[103] D. Geng, Y. Chen, Y. Chen, Y. Li, R. Li, X. Sun, S. Ye, and S. Knights, *Energy Environ. Sci.*, **4**, 760–764, 2011.

[104] N. Cheng, M. N. Banis, J. Liu, A. Riese, S. Mu, R. Li, T.-K. Sham, and X. Sun, *Energy Environ. Sci.*, **8**, 1450–1455, 2015.

[105] N. Cheng, M. N. Banis, J. Liu, A. Riese, X. Li, R. Li, S. Ye, S. Knights, and X. Sun, *Adv. Mater.*, **27**, 277–281, 2015.

[106] Z. Song, B. Wang, N. Cheng, L. Yang, D. Banham, R. Li, S. Ye, and X. Sun, *J. Mater. Chem. A*, **5**, 9760–9767, 2017.

Index